ADVANCED ORGANIC CHEMISTRY

ADVANCED ORGANIC CHEMISTRY:
Reactions and Mechanisms

Second Edition

Bernard Miller
University of Massachusetts, Amherst

PEARSON
Prentice
Hall

Prentice Hall, Upper Saddle River, New Jersey 07458

Library of Congress Cataloging-in-Publication Data

Miller, Bernard, 1930–

    Advanced organic chemistry : reactions and mechanisms / Bernard Miller.—2nd ed.
      p. cm.
    Includes bibliographical references and index.
    1. Chemistry, Organic. I. Title.
QD251.2.M534 2003
547—dc21
                            2003045779

Project Manager: *Kristen Kaiser*
Editor-in-Chief: *John Challice*
Executive Marketing Manager: *Steve Sartori*
Production Editor: *Donna King*
Director of Creative Services: *Paul Belfanti*
Creative Director: *Carole Anson*
Cover Designer: *Bruce Kenselaar*
Art Editor: *Jessica Einsig*
Manufacturing Manager: *Trudy Pisciotti*
Manufacturing Buyer: *Lynda Castillo*

 © 2004 by Pearson Education, Inc.
Pearson Education, Inc.
Upper Saddle River, NJ 07458

Printed in the United States of America

10 9 8 7 6 5 4 3 2 1

ISBN 0-13-065588-0

Pearson Education Ltd., *London*
Pearson Education Australia Pty. Ltd., *Sydney*
Pearson Education Singapore, Pte. Ltd.
Pearson Education North Asia Ltd., *Hong Kong*
Pearson Education Canada, Inc., *Toronto*
Pearson Education de Mexico, S.A. de C.V.
Pearson Education Japan, *Tokyo*
Pearson Education Malaysia, Pte. Ltd.
Pearson Education Inc., *Upper Saddle River, New Jersey*

Contents

Preface to the Second Edition

Among the rewards I've reaped for having written, *Advanced Organic Chemistry: Reactions and Mechanisms*, are the conversations and correspondence I've had with the users (or perhaps just perusers) of this text.

I greatly appreciate the contributions from those of you who have taken the time to inform me of misprints in the first edition. (No errors of fact, I'm happy to say.) The publisher and I have tried to correct those errors in later printings.

I'm also grateful for your suggestions of additonal subjects to incorporate into the second edition. (I appreciate your faith that there would be a second edition.) Several of you were rightfully indignant that I had skipped so briefly over photochemistry, aside from pericylic photochemical reactions. The complaint is a valid one, and I have tried to make up for the earlier lapse by including a chapter on photochemistry in this edition.

A smaller number of readers offered a suggestion that I had thought of as well: the great and increasing importance of synthetic reagents incorporating phosphorus and sulfur atoms makes a discussion of the mechanisms by which those reagents act imperative. I have therefore added a chapter on those reactions to the text of this edition.

Of course, a principal purpose of a second edition is to bring the subject matter up-to-date. This edition includes material (and many problems) from the literature through the summer of 2002.

I can't really claim that our understanding of organic reaction mechanisms has been revolutionized in the past half-decade, but there have been some fascinating examples of applications of known processes. There *have* been notable advances in some areas, such as the dehydro Diels-Alder reaction and the Bergman and Meyers-Saito reactions, and in perhaps my favorite example (because of its elegant simplicity), Bakke's demonstration, contradicting a century of experience that pyridine can be easily nitrated in high yield and under mild conditions—provided you know how.

I am looking forward to hearing from you in regard to this latest version of my text. I can be reached at miller@chem.umass.edu.

Bernard Miller
Amherst, MA

Preface to the First Edition

While there aren't nearly as many advanced organic chemistry textbooks as introductory texts, several advanced texts are available. What makes this text different?

- It Avoids the Deadly "Review" Label

One common approach to teaching advanced organic chemistry is to use the course to review and expand on subjects covered in the introductory course—nucleophilic substitution and elimination reactions, additions to carbonyl compounds, aromatic substitution reactions, cyclohexane conformations, stereochemistry, and so on. There are good reasons for this approach. Experience has shown that many students, whether graduate or undergraduate, have only a shaky grasp of the material covered in their introductory courses and could greatly benefit from a thorough review.

However, experience has also shown that there are excellent reasons *not* to emphasize review of basic material in an advanced course. The principal reason is that students tend to hate that approach. They just do not want to go back over S_N1 and S_N2 reactions or photochlorination of hydrocarbons or additions of acids to alkenes, even if the subjects are covered at a more advanced level. And our best students—the students we most want to remain fascinated by organic chemistry—are least happy about reviewing introductory material.

- It Introduces "Advanced" Topics Early

My experience has convinced me that maintaining student interest in the advanced course requires the early introduction of "advanced" topics—topics that are not usually stressed in first-year courses. Fortunately, we can still discuss many of the most important aspects of the standard introductory course using this approach, but in a more intellectually stimulating manner than by a straightforward review.

- New and "Review" Topics Are Integrated

After a brief introductory chapter, this book proceeds with a discussion of the Woodward-Hoffmann rules and pericyclic reactions. While these are usually treated as advanced topics, they have the virtue of being relatively self-contained and do not require detailed knowledge of many other reactions. The topics allow review of Hückel's rule and of the concept of aromaticity, and the discussion leads naturally to sigmatropic shifts, which introduces the general subjects of carbocation and carbanion reactions without tagging them with the deadly "review" label.

- Heterocyclic Chemistry Is Covered in Depth

A review of the basic reactions covered in introductory courses is facilitated because this book includes major sections on heterocyclic chemistry.

At one time it would have been unthinkable for an introductory organic chemistry course *not* to include a significant discussion of heterocyclic chemistry. After all, a huge number—perhaps a majority—of organic molecules contain heterocyclic rings. However, in many current texts heterocyclic chemistry has been downgraded to a "special topic" and afforded only brief attention. The elimination of heterocyclic chemistry from the introductory curriculum is particularly unfortunate because increasing numbers of our students are specializing in biological aspects of chemistry, where a knowledge of the properties of heterocyclic molecules is critical.

Another good reason to include heterocyclic chemistry in an advanced organic chemistry course is that the study of heterocyclic chemistry involves the study of carbonyl addition and condensation reactions, aromatic substitution reactions, and the properties of amines and carboxyl groups. Thus, it provides the opportunity to go over many of the most important topics covered in the introductory course in a new, "nonreview" context. For that reason this book discusses methods of synthesis of heteroaromatic rings (although synthetic methods are not generally emphasized in this text).

- It Offers Flexibility in Course Organization

Some instructors may wish to discuss heterocyclic chemistry at the beginning of the course. They will find no difficulty in doing so using this text, since the chapters on heterocyclic chemistry do not, for the most part, require a knowledge of material in other chapters. (The sections on dipolar cycloaddition reactions and the Fischer indole synthesis are exceptions, but these topics can easily be postponed for later discussion.)

- "Theory" Is Introduced in Context

Some time ago I sent a letter to *Chemical and Engineering News* decrying the trend toward having chemistry courses begin with a large does of chemical theory. I contended that the significance of the theory would be more obvious, and the subject would be more interesting, if chemical phenomena were introduced first and the theory then introduced to explain the facts.

I was astonished at the volume of mail I received in response to my letter, almost all of it agreeing with my viewpoint. I was therefore encouraged to use that approach in writing this text. For instance, the stereospecificity of electrocyclic reactions is discussed first, and qualitative molecular orbital theory is then introduced to explain the facts, rather than the other way around.

- It Introduces the "Art" of Writing Mechanisms

The introductory chapter, which is often a vehicle for discussing large amounts of chemical theory, is relatively brief in this text. It covers only the "art" of proposing mechanisms for complex reactions. That art includes the use of resonance theory, which is introduced as a way to solve rather puzzling mechanism problems rather than simply as a review topic.

Of course, even the deliberately simple "mechanism" exercises in Chapter 1 require a knowledge of many of the basic reactions discussed in introductory courses. Discussion of these exercises serves as another useful but informal review of that material.

• In Conclusion ...

I've attempted to make this a teaching text rather than one that simply presents the material in an encyclopedic manner. I've used relatively informal language and, I hope, anticipated many of the problems students have with this material.

Over the past years, both undergraduate and graduate students have reported finding the discussions in this text useful and interesting. I hope your students will feel the same way.

I would like to thank the following reviewers for their input: William F. Bailey, University of Connecticut; Donald B. Denney, Rutgers University; John C. Gilbert, University of Texas-Austin; John M. McIntosh, University of Windsor; Jeffrey S. Moore, University of Illinois, Urbana; Donna J. Nelson, University of Oklahoma; Michael Ogliaruso, Virginia Polytechnic Institute; Robert S. Phillips, University of Georgia; and Dennis J. Sardella, Boston College.

I sincerely hope that you will let me know of your experiences and those of your students with this book. All comments and suggestions will be received with gratitude.

Bernard Miller
Amherst, MA

Introduction

1.1 MECHANISMS OF REACTIONS

On Understanding Organic Reactions

This is a book about the reactions of organic molecules. Millions of compounds containing carbon are currently known, and many others are being created or discovered every day. Since each compound can undergo many possible reactions, there are enormous numbers of possible reactions of organic molecules.

Understanding organic chemistry would therefore be an enormously difficult task, except that the varied reactions all result from a relatively small number of basic processes. These processes can be combined to result in overall reactions that may appear, at first glance, to be mysterious or even impossible. Fortunately, closer analysis will almost always show that, like a stage magician's tricks, even the most mysterious-appearing reactions are quite simple.

A major function of this book is to help you analyze the magician's secrets. Every chapter in this book is followed by problems that ask you to unravel the mystery behind sometimes quite strange-looking organic reactions. These problems are often in the form, "Propose reasonable mechanisms for the following reactions."

Before you can solve any of these problems, you have to be clear about three things: the meaning of the "mechanism" of a reaction, what "proposing" or "writing" a mechanism means, and how you can decide whether or not a mechanism is "reasonable."

Reaction Mechanisms

The mechanism of a reaction consists of everything that happens as the starting materials for the reaction are converted into the products. In principle, therefore, "writing" (or drawing) the mechanism means describing everything that happens in the course of the reaction. However, providing an exact description of a reaction on paper is an impossible goal. Instead, a proposal for the mechanism of a reaction should include certain critical types of information about the course of the reaction.

A proposal for the mechanism of a reaction should indicate whether the reaction proceeds in a single step or by a sequence of several consecutive reactions. If more than one step is required—and all the mechanism problems in this book require more than one step—*the proposal should include equations for every step in the overall mechanism.* (Identifying each step in a reaction sequence is, by itself, often considered an adequate description of the mechanism for that reaction.)

A product of an individual step in a reaction mechanism that is involved in succeeding steps is called an *intermediate* in the overall reaction. Thus, identifying intermediates in the formation of the products of a reaction is a crucial part of any proposal for a reaction mechanism.

In addition to identifying any reaction intermediates, a proposal for a reaction mechanism often includes a description of changes in the electronic structures of molecules as they pass from starting materials to products. The description is made by means of *curved arrows.* A curved arrow in a chemical equation means something quite different from the arrows representing the conversion of starting materials to products (or the pair of arrows representing the equilibrium between them). *A curved arrow shows the change in position of a pair of electrons.*[1] The arrow starts at an electron pair that takes part in the reaction. If the electron pair forms a bond between two atoms in the starting material and forms a new bond in the product, the arrow should start at the center of the line representing the original bond and should end up pointing to the location of the new bond being formed. If the electrons become a nonbonding (unshared) pair in the product, the arrow should point to the atom accepting the nonbonding pair.

In this book, unshared pairs of electrons are frequently shown as pairs of dots. However, molecular formulas need not specifically show the presence of unshared electron pairs. The presence of unshared electron pairs on every neutral—or negatively charged—oxygen, nitrogen, or halogen atom, for instance, is understood. (The unshared electron pairs on the chloride ion, for instance, are not shown in Eq. 1c.) A curved arrow can start at an atom bearing unshared pairs of electrons, even if the electrons are not specifically shown.

$$\text{Na}^{\oplus} \quad \overset{\ominus}{\text{HO:}} \curvearrowright \text{H}-\overset{O}{\overset{\|}{\text{CH}_2-\text{CH}}} \longrightarrow \text{H}_2\overset{..}{\text{O}:} + \overset{\ominus}{:\text{CH}_2}-\overset{O}{\overset{\|}{\text{CH}}} \qquad (1a)$$

$$:\overset{..}{\overset{..}{\text{Cl}}}-\text{H} \curvearrowleft \text{H}_2\text{C}=\text{C(CH}_3)_2 \longrightarrow :\overset{..}{\overset{..}{\text{Cl}}}:^{\ominus} + \text{H}_3\overset{\oplus}{\text{CC(CH}_3)_2} \qquad (1b)$$

$$\text{H}_2\overset{..}{\text{N}}-\text{CH}_2-\text{Cl} \longrightarrow \text{H}_2\overset{\oplus}{\text{N}}=\text{CH}_2 \ \ \text{Cl}^{\ominus} \qquad (1c)$$

Some examples of the use of curved arrows

Movements of single electrons in free radical reactions are often represented by single-barbed "fishhooks," as shown in Eq. 2 for the reaction of methane with a chlorine atom.

$$:\overset{..}{\overset{..}{\text{Cl}}}\cdot \ \ \text{H}-\text{CH}_3 \longrightarrow \text{Cl}-\text{H} + \cdot\text{CH}_3 \qquad (2)$$

Discussions of reaction mechanisms frequently include discussions of the nature of the *transition state* for each step in a reaction sequence—or, at least, for the slowest or "rate-limiting" step. A transition state is the point of highest energy in a reaction or in each step of a reaction involving more than one step. The nature of the transition state will determine whether the reaction is a difficult one, requiring a high activation enthalpy $(\Delta H^{\ddagger})$, or an easy one.

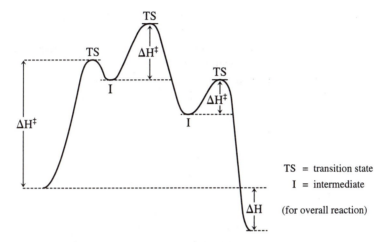

Energy diagram for a three-step reaction sequence

Unlike reaction intermediates, transition states can never be isolated. By definition, they exist only at the tops of "energy hills." There are no barriers to prevent them from immediately "rolling" downhill to form the reaction products or intermediates or to reform the starting materials.

Although a knowledge of (or at least an idea about) the structure of the transition state for an organic reaction is important in understanding its mechanism, you will usually not be expected to describe transition states as part of your proposed reaction mechanisms. One reason is that transition states always involve partially formed or partially broken bonds. Unfortunately, no really good way has been developed to represent partial bonds in molecules. The best that can be done is to represent partial bonds by dotted lines. However, consider the representation of the transition state for the reaction of sodium methoxide with methyl bromide (an S_N2 substitution reaction), shown in Eq. 3. The dotted lines indicate that the carbon–oxygen bond has been partially formed, and the carbon–bromine bond partially broken. Does that mean that both bonds are "half-formed" at the transition state? (And what would "half-formed" mean, anyway?) Or is one bond 75 or 95% formed and the other largely broken? And how would that be represented by the dotted-line symbolism?

$$CH_3O^{\ominus} \quad Na^{\oplus} \quad CH_3\!-\!Br \longrightarrow \left[CH_3O \cdots\cdots \overset{\overset{\displaystyle Na^{\oplus}}{}}{\underset{\displaystyle H}{\overset{\displaystyle H\ H}{C}}} \cdots\cdots Br \right]^{\ominus} \longrightarrow CH_3\!-\!O\!-\!CH_3 + NaBr \qquad (3)$$

the transition state structure

We will avoid these questions and consider a proposed mechanism complete if it identifies all reaction intermediates and all electron motions occurring in the course of a chemical change.

Microscopic Reversibility

The *principle of microscopic reversibility* states that the reverse of any reaction must proceed along exactly the same route as the forward reaction. This simply means that the lowest energy route in one direction in a reaction equilibrium must also be the lowest energy route in the other. Thus, every intermediate in the forward reaction must also exist in the mechanism for the reverse reaction. (It should be understood that the principle of microscopic reversibility applies if the conditions for both reactions are identical. Even apparently small changes in reaction conditions may result in a change in mechanism.)

Recognizing "Reasonable" Mechanisms

Assuming that a particular reaction proceeds in several steps, just how does one go about deciding what those steps are?

There is one basic principle to be followed in proposing a possible mechanism for a chemical reaction. *Each step in the reaction of any compound should be a well-known reaction of that type of compound under the conditions of the reaction*. (Like lawyers, chemists justify their proposals on the basis of precedent.)

As we've mentioned above, there are enormous numbers of possible chemical reactions. Many of these reactions will seem quite extraordinary, if not impossible, at first sight. However, it is important to avoid proposing extraordinary steps for the mechanisms of these reactions. If a proposed mechanism has novel, "interesting" steps that haven't been seen many times before, the odds are that the mechanism is incorrect.* Nearly all chemical changes that appear extraordinary are simply the result of combining well-known, routine reactions.

In previous courses, you studied many of the most important organic processes. A vast variety of organic reactions, many of which you have never seen before, can be explained simply as combinations of reactions you have already studied.

Proposing Reaction Mechanisms

Rejecting an unreasonable mechanism.   Let's work out a reasonable mechanism for a simple reaction—one you probably haven't seen before:

$$\underset{\text{OH}}{\overset{\text{OH}}{H_3C-\overset{|}{C}H-OC_2H_5}} + HCN \xrightarrow{\text{NaOH}} \underset{\text{OH}}{\overset{\text{OH}}{H_3C-\overset{|}{C}H-CN}} + C_2H_5OH \qquad (4a)$$

---

*Novel types of reactions are still being discovered. However, it takes convincing demonstrations that new reactions cannot be explained as combinations of known processes before chemists are willing to accept that they really are novel.

Since the overall result of the reaction in Eq. 4a is to replace an ethoxy group by a cyanide group, it may be tempting to propose an S_N2 mechanism (Eq. 4b):

$$\underset{\substack{\bigcirc \\ :CN}}{H_3C-\underset{\underset{}{\overset{OH}{|}}}{CH}-\ddot{O}C_2H_5} \overset{??}{\longrightarrow} H_3C-\underset{\overset{OH}{|}}{CH}-CN \ + \ \overset{\bigcirc \ ..}{:\ddot{O}C_2H_5} \qquad (4b)$$

However, an essential requirement for any S_N2 reaction is that the group being displaced by the nucleophile must be a *good leaving group*. Only a small group of anions, all the conjugate bases of strong acids, can act as leaving groups in typical S_N2 reactions (or, for that matter, in S_N1, E1, and E2 reactions). That group includes chloride, bromide, and iodide anions, as well as derivatives of sulfate anions, such as benzenesulfonate anions. Neutral molecules, such as water and alcohols, can act as good leaving groups in acid-catalyzed S_N1 and E1 reactions (e.g., Eq. 5). Ethers can act as leaving groups in S_N2 and E2 reactions under acidic or basic conditions (as in Eq. 6). Unless an anion or a molecule on this very short list acts as the leaving group, you can usually reject mechanisms involving nucleophilic substitution or elimination mechanisms.

$$(CH_3)_3C-OCH_3 \overset{\overset{\overset{\curvearrowright}{H-Cl}}{\longrightarrow}}{\longrightarrow} (CH_3)_3C-\underset{CH_3}{\overset{\oplus \ \overset{H}{\diagup}}{O}} \overset{Cl^{\bigcirc}}{\longrightarrow} (CH_3)_3C^{\oplus} \ Cl^{\bigcirc} \ + \ HOCH_3$$
$$\downarrow \qquad (5)$$
$$(CH_3)_3C-Cl \ (\text{or } CH_2{=}C(CH_3)_2 \ + \ HCl)$$

$$\underset{H_3CCHCH_3}{\overset{\overset{\oplus}{O}(CH_3)_2}{|}} \ Br^{\bigcirc} \overset{\Delta}{\longrightarrow} CH_3OCH_3 + \underset{Br}{\overset{Br}{|}}H_3CCHCH_3 \ (\text{or } H_2C{=}CHCH_3 + HBr) \qquad (6)$$

Alkoxide anions are the conjugate bases of alcohols, which are not strong acids. Thus, they usually cannot be leaving groups in S_N2 reactions (or in S_N1, E1, or E2 reactions).* Diethyl ether, for instance, does not undergo S_N2 reactions with sodium cyanide. Therefore, the displacement of an ethoxide anion by a cyanide anion, as shown in Eq. 4a, would have to be classified as a novel, and therefore highly suspect, process.

Proposing a reasonable mechanism. How can you select a reasonable mechanism for the reaction in Eq. 4a? *Don't* start out by trying to arrive at the products of the reaction. Instead, *focus on the starting materials, and select the reaction they are most likely to undergo under the reaction conditions.* Then, do the same with the products of the reaction you've proposed and, if

---

*Alkoxide anions *can* act as leaving groups in E2 reactions with such strong bases as alkyllithium reagents.

necessary, with the products of the second reaction and later reactions. It is surprising how often a sequence of the most likely steps will lead to the actual reaction products.*

When there are several reasonable alternatives to choose from at each step of a possible mechanism, you will have to choose one possibility and see where it leads. If it does not result in the product formation you are trying to explain, you'll have to go back and look at some of the other possibilities. (Incidentally, a problem that asks for the mechanism by which a particular product is formed does *not* suggest that it is the only, or even the major, product of the reaction.)

Let's apply this procedure to reaction 4a. Fortunately, there is only one reasonable initial reaction that can occur. The starting organic compound is an alcohol and, as shown in Eq. 7a, will be partially converted to its anion by reaction with cyanide ions or other bases:

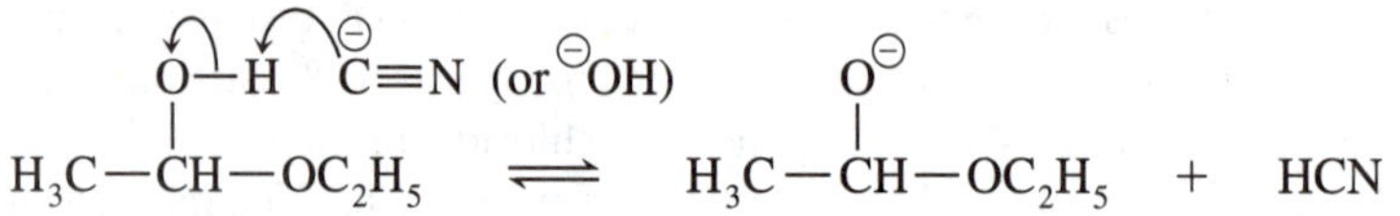

$$\text{O–H} \quad C{\equiv}N \text{ (or } {}^{\ominus}OH) \qquad O^{\ominus}$$
$$\text{H}_3\text{C–CH–OC}_2\text{H}_5 \quad \rightleftharpoons \quad \text{H}_3\text{C–CH–OC}_2\text{H}_5 \quad + \quad \text{HCN} \qquad (7a)$$

Ignore the fact that this reaction may not seem to be leading to the desired product, and consider what can happen to the new anion. It can, of course, pick up a proton to give back the starting alcohol, but obviously, that leads nowhere. The only other reaction the anion can undergo is to eject an alkoxide anion to form an aldehyde (Eq. 7b):

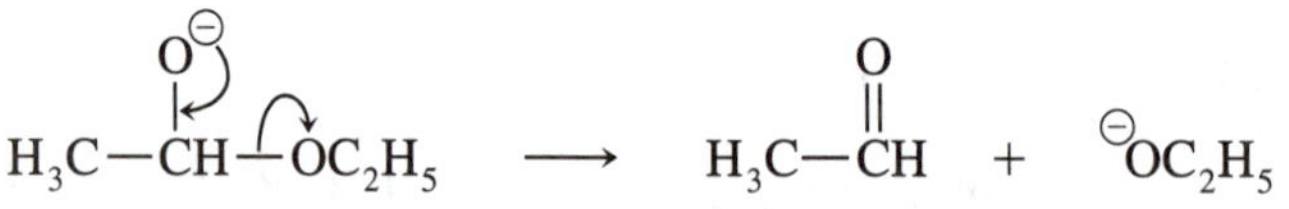

$$O^{\ominus} \qquad\qquad O$$
$$\text{H}_3\text{C–CH–OC}_2\text{H}_5 \quad \longrightarrow \quad \text{H}_3\text{C–CH} \quad + \quad {}^{\ominus}\text{OC}_2\text{H}_5 \qquad (7b)$$

Several possible reactions might take place at this point. Perhaps the most obvious, however, is the addition of a cyanide ion to the carbonyl group (Eq. 7c). Now, an acid–base reaction (Eq. 7d) would yield the specified product and regenerate the cyanide ion:

$$O \qquad\qquad O^{\ominus}$$
$$\text{H}_3\text{C–CH} \quad {}^{\ominus}\text{CN} \quad \longrightarrow \quad \text{H}_3\text{C–CH–CN} \qquad (7c)$$

$$\qquad H–C{\equiv}N \qquad\qquad OH$$
$$O^{\ominus} \qquad\qquad\qquad\qquad$$
$$\text{H}_3\text{C–CH–CN} \quad \longrightarrow \quad \text{H}_3\text{C–CH–CN} \quad + \quad {}^{\ominus}\text{CN} \qquad (7d)$$

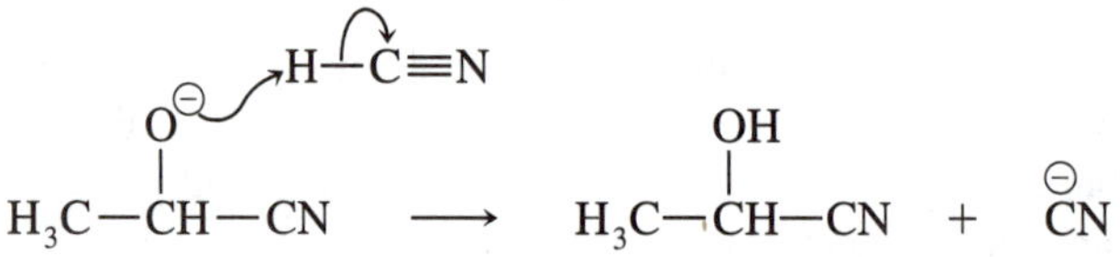

The reaction in Eq. 4a can be described as proceeding by a *retro*-hemiacetal formation (Eqs. 7a and 7b) followed by a cyanohydrin reaction. To experienced organic chemists, who are familiar with the mechanisms of both reactions, naming the sequence of reactions provides an excellent brief description of the mechanism of the overall reaction.

---

*Several reasonable mechanisms might account for a given reaction. Distinguishing among them to determine the "correct" mechanism often requires ingenious and prolonged experimental work.

Eqs. 7a to 7d show a reasonable mechanism for the reaction in Eq. 4a. *The mechanism is reasonable, because each step is a very common, routine reaction.*

Another example.   Let's try our procedure for proposing mechanisms on another base-catalyzed reaction.

$$
H_2C=CH-\underset{\underset{\displaystyle OH}{|}}{CH}-C\equiv N \quad \xrightarrow{\ominus OH} \quad N\equiv CCH_2CH_2\overset{\displaystyle O}{\overset{||}{C}}H \tag{8}
$$

We might briefly consider forming the product by an allylic shift of the cyano group, followed by a base-catalyzed tautomerism of the resulting enol:

$$
H_2C=CH-\underset{\underset{\displaystyle C\equiv N}{|}}{\overset{\overset{\displaystyle OH}{|}}{CH}} \quad \xrightarrow{??} \quad N\equiv CCH_2CH=\overset{\overset{\displaystyle OH}{|}}{CH} \tag{9a}
$$

$$
\downarrow \ominus OH
$$

$$
N\equiv CCH_2CH=\overset{\overset{\displaystyle O^{\ominus}}{|}}{CH} \quad \xrightarrow{H_2O} \quad N\equiv CCH_2CH_2\overset{\displaystyle O}{\overset{||}{C}}H
$$

You may not be certain whether a cyano group can move around that way. (It cannot, as will be discussed in Chapter 4.) However, the fact that it is difficult to suggest good precedents for that step justifies trying to find an alternative mechanism in which each step does have common precedents. In fact, Eq. 8 can be accounted for by a *retro*-cyanohydrin reaction, followed by conjugate addition of the cyanide ion to the conjugated carbonyl group.

$$
H_2C=CHCHC\equiv N \quad \xrightarrow{\ominus OH} \quad H_2C=CHCH-CN \quad \longrightarrow \quad H_2C=CHCH \tag{9b}
$$

$$
\downarrow
$$

$$
N\equiv CCH_2CH_2\overset{\displaystyle O}{\overset{||}{C}}H \quad \xleftarrow{H_2O} \quad N\equiv CCH_2\overset{\ominus}{C}H_2\overset{\displaystyle O}{\overset{||}{C}}H
$$

The mechanism in Eq. 9b is reasonable because each step is a common reaction, with many precedents. Of course, the fact that it is reasonable does not necessarily mean that it is correct—although in this case, it is difficult to propose any reasonable alternative.

More about leaving groups.   You may wonder why ethoxide anions can be ejected from anions of hemiacetals (Eq. 7b) and cyanide ions from anions of cyanohydrins (Eq. 9b) when neither ethoxide nor cyanide anions are conjugate bases of strong acids and therefore are not on the list of "good leaving groups" discussed previously.

The list of good leaving groups previously discussed applies to several specific reactions: S_N1 and E1 reactions, in which the departures of the leaving groups are not appreciably assisted by nucleophiles, and S_N2 and E2 reactions, in which the leaving groups are displaced (or eliminated) by external nucleophiles or bases. (The list also applies to intramolecular displacement reactions, such as that in Eq. 10, in which the nucleophiles are not bonded to the atoms bearing the leaving groups.)

$$\text{(structure)} \longrightarrow \text{(structure)} + \text{KBr} \qquad (10)$$

The situation is very different in a reaction such as that shown in Eq. 7b, in which the leaving group and a negatively charged atom are both bonded to the same atom, so that ejection of the leaving group results in formation of a new π bond. Reactions of this type occur much more easily than do substitutions by external nucleophiles or elimination by external bases.

Displacements of leaving groups by negatively charged atoms bonded to the same carbons as the leaving groups occur so readily, in fact, that it is easier to list the groups that typically *cannot* act as leaving groups in those reactions than it is to list the groups that can. *The only common groups that typically cannot act as leaving groups in these reactions are hydrocarbon anions and hydride anions*.* Aside from hydrocarbon and hydride anions, almost all common anions—cyanide ions, hydroxide ions, alkoxide anions, enolate anions—can act as leaving groups when they are displaced by negatively charged atoms bonded to the same atoms as in the leaving groups.

1.2 ELECTRON DELOCALIZATION AND RESONANCE

The Theory of Resonance

Hybrid structures. No discussion of reaction mechanisms would be complete without introducing the *theory of resonance*. Consider, for instance, the reaction shown in Eq. 11. The products of the reaction might seem to have little relationship to the starting materials. Furthermore, the reaction is a rapid one, while most primary halides and most ethers are quite unreactive under the same conditions.

$$\text{Cl}-\text{CH}_2-\text{CH}=\text{CH}-\text{OCH}_3 + \text{H}_2\text{O} \longrightarrow \overset{\overset{\displaystyle O}{\displaystyle \|}}{\text{H}_2\text{C}=\text{CH}-\text{CH}} + \text{CH}_3\text{OH} + \text{HCl} \qquad (11)$$

Fortunately, the theory of resonance provides a simple explanation for the formation of the reaction products as well as for the rapid rate of the reaction. Resonance theory originated as

---

*Even these very strong bases can act as leaving groups under exceptional conditions. (See Chapter 8, pp. 232 and 233.)

valence bond theory, a mathematical technique for approximating the quantum mechanical wave functions for molecules by combining the wave functions for several possible electronic structures.[2] A *qualitative theory* was derived from this idea to deal with some deficiencies in the standard method of depicting the structures of molecules and ions by means of individual structural (Lewis) formulas.

Consider a positively charged CH_2 group bonded to a double bond (an *allyl cation*), for instance. Although a standard Lewis formula, **1**, for the cation suggests that just one carbon bears a positive charge, the orbitals of the π bond are so close to the empty *p* orbital that the π electrons actually spread out over all three *p* orbitals. The electrons, and the bonds they form, are then said to be *delocalized*. The two π electrons are shared among three carbon atoms, rather than two. In an allyl anion, four electrons would be shared among three carbon atoms.

$$H_2\overset{\oplus}{C}-CH{=}CH_2 \qquad \equiv \qquad \text{(12)}$$

$$\mathbf{1} \qquad\qquad\qquad\qquad \mathbf{1}$$

Unfortunately, there doesn't seem to be a good way to show the structure of a molecule or ion of this type by means of a single structural formula. In the resonance method, this deficiency is overcome by using a combination of several Lewis formulas to represent the actual structure of a molecule or ion with delocalized bonds. The individual formulas are called *resonance forms* or *canonical forms*. The fact that two Lewis formulas are resonance forms of the same molecule or ion is commonly indicated by placing a double-headed arrow between them.

The resonance forms for a molecule or ion with delocalized bonds should include every "significant" structural formula that can be obtained simply by changing the positions of π electrons or unshared electrons. (Curved arrows are often helpful in converting one resonance form to another.)

$$\left[H_2\overset{\oplus}{C}{-}CH{=}CH_2 \quad \longleftrightarrow \quad H_2C{=}CH{-}\overset{\oplus}{C}H_2 \right]$$

$$\left[H_2\overset{\ominus}{\underset{..}{C}}{-}CH{=}CH_2 \quad \longleftrightarrow \quad H_2C{=}CH{-}\overset{\ominus}{\underset{..}{C}}H_2 \right] \qquad \text{(13)}$$

Resonance forms for the allyl cation and allyl anion

The actual structure is a "hybrid" of all the individual resonance forms, with the degree to which it actually resembles a particular resonance form dependent on the relative energies of the individual structures. The lower in energy a particular resonance form would be (if it actually existed as an individual structure), the more closely the actual structure will resemble that resonance form. (Since the two resonance forms for the allyl cation are indistinguishable, they contribute equally to the actual structure. The positive charge is shared equally by C1 and C3.)

It is important to stress that resonance forms are simply crutches used to compensate for our difficulties in drawing delocalized bonds. Resonance forms never have an individual existence. The π electrons in the allyl cation, for instance, do not flip between C1 and C3. There is always half a π bond between C1 and C2 and half a π bond between C2 and C3, and there is always half a positive charge on both C1 and C3.

The double-headed arrow is a rather unfortunate symbol for resonance, because it so closely resembles the double arrows used to signify the equilibrium between the starting materials and the products of a reaction. Don't be fooled by the double-headed arrow. *Because resonance forms have no real existence, there can never be an equilibrium between them.* Each resonance form is simply meant to provide some of the information necessary to depict the hybrid structure of a molecule with delocalized bonds.

In order to avoid drawing several resonance forms for a single molecule or ion, delocalized bonds are sometimes represented by dotted lines. This method probably has more disadvantages than virtues. Consider the dotted-line representation of cation **2**, for example. Can you easily see which four carbon atoms are positively charged? If you can, it is probably because you have mentally drawn the four resonance forms for the ion.

$$
\left[
\begin{array}{c}
CH_2 \\
\vdots \\
CH \\
\vdots \\
H_2C \!=\!\!= CH \!=\!\!= C \!=\!\!= CH \!=\!\!= CH_2
\end{array}
\right]^{\oplus}
\qquad (14)
$$

2

Significant and insignificant forms. Resonance forms that would be a great deal higher in energy than other forms (if they actually existed as individual structures) may be considered "insignificant," and their contributions to the hybrid structure may be ignored. Unfortunately, confusion often exists over which resonance forms should be considered "significant" and which "insignificant."

To a first approximation, we need consider only those resonance forms that have the maximum number of bonds that may be drawn for a particular structure. Ethene, for instance, can be considered to have only one significant resonance form. The diradical and dipolar resonance forms, which each have one less bond than the doubly bonded form, are so high in energy that they can be ignored. (The dipolar resonance forms of carbonyl groups are so useful in illustrating the properties of aldehydes and ketones that they are often treated as significant. The dipolar properties of carbonyls, however, can be explained entirely on the basis of inductive effects, without requiring a significant resonance contribution.*)

$$
\left[H_2\dot{C} - \dot{C}H_2 \quad \longleftrightarrow \quad H_2\overset{\oplus}{C} - \overset{\ominus}{\ddot{C}}H_2 \quad \longleftrightarrow \quad H_2\overset{\ominus}{\ddot{C}} - \overset{\oplus}{C}H_2 \right]
$$

Insignificant resonance
forms of ethene

$$
\left[R - \overset{\overset{\displaystyle O}{\|}}{C} - R \quad \longleftrightarrow \quad R - \overset{\overset{\displaystyle O^{\ominus}}{|}}{\underset{\oplus}{C}} - R \right]
\qquad (15)
$$

Nonpolar and dipolar resonance
forms of a ketone

*"Diradical" resonance forms of conjugated dienes and polyenes are often invoked to account for the greater stabilities of molecules with conjugated double bonds, compared to their isomers with isolated double bonds. However, it has been argued that these forms are insignificant, and that the stabilities of conjugated polyenes are due to the strengths of the $sp^2 - sp^2$ single bonds linking the double bonds.[3]

$$
\left[\dot{C}H_2 - \dot{C}H - CH = CH_2 \quad \longleftrightarrow \quad \dot{C}H_2 - CH = CH - \dot{C}H_2 \quad \longleftrightarrow \quad CH_2 = CH - \dot{C}H - \dot{C}H_2 \right]
$$

"Diradical" resonance forms of 1,3-butadiene

In contrast to forms lacking the maximum number of bonds, every resonance form that can be drawn without losing one or more bonds must be considered significant. The dipolar resonance forms for enol ethers and enamines, for instance, are certainly less important than the nonpolar forms. However, they have the same number of bonds and are, therefore, significant forms. This can be demonstrated from ^{13}C and ^{1}H NMR spectra. In those spectra, the peaks for the carbons bearing negative charges in the dipolar resonance forms and for the hydrogens on those carbons appear far upfield from typical vinyl carbon and hydrogen signals.

$$H_2C{=}CH{-}\overset{..}{\underset{..}{O}}CH_3 \quad \longleftrightarrow \quad H_2\overset{\ominus}{\overset{..}{C}}{-}CH{=}\overset{\oplus}{\underset{..}{O}}CH_3$$

an enol ether

(16)

$$H_2C{=}CH{-}\overset{..}{N}(CH_3)_2 \quad \longleftrightarrow \quad H_2\overset{\ominus}{\overset{..}{C}}{-}CH{=}\overset{\oplus}{N}(CH_3)_2$$

an enamine

Hyperconjugation resonance forms, involving the delocalization of σ bonds rather than π bonds, are not usually included among the significant resonance forms for neutral molecules. However, the appreciable stabilization of carbocations by alkyl groups is at present usually attributed to hyperconjugation rather than to inductive effects. Stabilization of free radicals by alkyl groups is similarly attributed to hyperconjugation.

$$\underset{H_2C{-}\overset{\oplus}{C}H{-}CH_2}{\overset{\overset{\displaystyle H}{\mid}\quad\quad\overset{\displaystyle CH_2CH_3}{\mid}}{}} \quad \longleftrightarrow \quad \underset{H_2C{=}CH{-}CH_2}{\overset{\overset{\displaystyle H\,\oplus}{}\quad\quad\overset{\displaystyle CH_2CH_3}{\mid}}{}} \quad \longleftrightarrow \quad \underset{H_2C{-}CH{=}CH_2}{\overset{\overset{\displaystyle H}{\mid}\quad\quad\overset{\displaystyle \oplus CH_2CH_3}{}}{}}$$

(17)

Some hyperconjugation resonance forms of the 2-pentyl cation

Although hyperconjugation is sometimes described as "no-bond resonance," hyperconjugation does *not* involve breaking bonds. As always, the actual structures are hybrids of all the resonance forms. Since resonance forms with σ bonds are more important than those with π bonds (because σ bonds have greater bond energies), the actual structures will closely resemble the forms with σ bonds, but with the electrons in those bonds drawn toward the empty, or half-empty, orbitals.

Using resonance structures.   Representing chemical structures as hybrids of several resonance forms may seem a rather crude device, but it has served remarkably well as a means for chemists to understand and predict the properties of molecules and ions with delocalized bonds. It is not infallible, and in recent years has been supplemented, and partially supplanted, by molecular orbital theory (discussed in Chapter 2). However, *qualitative* resonance theory remains an essential means by which chemists think about and understand many organic reactions.

There are two basic rules to follow in using resonance structures to predict possible products of organic reactions and to propose reasonable mechanisms:

1. Draw all significant resonance structures for every molecule or ion taking part in a reaction.
2. Imagine that each resonance form has an individual existence, with localized bonds and charges, and draw the reaction products that should arise from each form. Each such product is a possible product of the actual molecule or ion.

Let's apply these rules to the reaction in Eq. 11 (which is repeated below as Eq. 18a).

$$Cl-CH_2-CH=CH-OCH_3 \quad + \quad H_2O$$

$$\downarrow$$

$$\underset{\overset{\displaystyle O}{\overset{\|}{}}}{H_2C=CH-CH} \quad + \quad CH_3OH \quad + \quad HCl \tag{18a}$$

Only one reasonable reaction may be anticipated in the first step: dissociation of the carbon–chlorine bond to form a carbocation (Eq. 18b). Obeying Rule 1, we draw all the resonance forms (**3a** to **3c**) for the carbocation. Then, obeying Rule 2, we consider the possible reactions of the individual resonance forms.

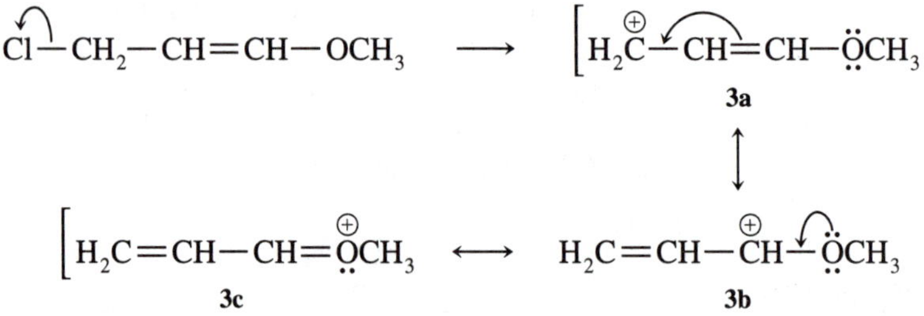

$$Cl-CH_2-CH=CH-OCH_3 \quad \longrightarrow \quad \left[H_2\overset{\oplus}{C}-CH=CH-\overset{..}{O}CH_3 \right]$$
$$\mathbf{3a}$$

$$\updownarrow \tag{18b}$$

$$\left[H_2C=CH-CH=\overset{\oplus}{\underset{..}{O}}CH_3 \right] \quad \longleftrightarrow \quad H_2C=CH-\overset{\oplus}{C}H-\overset{..}{O}CH_3$$
$$\mathbf{3c} \qquad\qquad\qquad\qquad \mathbf{3b}$$

In this case, we will show only the reactions of **3a** and **3b** (Eqs. 18c and 18d), because the reaction of **3c** would give the same products as that of **3a** and **3b**.

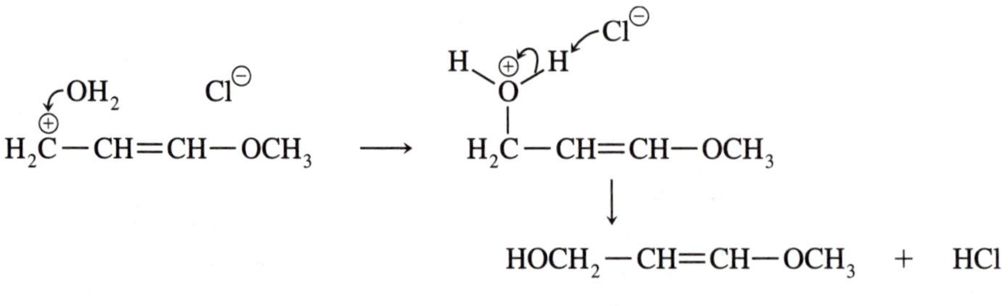

$$\underset{\oplus}{H_2\overset{OH_2}{C}}-CH=CH-OCH_3 \quad Cl^{\ominus} \quad \longrightarrow \quad \underset{\underset{|}{\overset{H\,\overset{\oplus}{O}\,H}{\quad}}}{H_2C}-CH=CH-OCH_3 \quad Cl^{\ominus} \tag{18c}$$

$$\downarrow$$

$$HOCH_2-CH=CH-OCH_3 \quad + \quad HCl$$

$$\underset{\oplus}{H_2C=CH-\overset{H_2O}{C}H}-OCH_3 \quad Cl^{\ominus} \quad \longrightarrow \quad \underset{\underset{|}{\overset{H\,\overset{\oplus}{O}\,H}{\quad}}}{H_2C=CH-CH}-OCH_3 \quad Cl^{\ominus} \tag{18d}$$

$$\downarrow$$

$$\underset{\overset{\displaystyle OH}{\overset{|}{}}}{H_2C=CH-CH}-OCH_3 \quad + \quad HCl$$

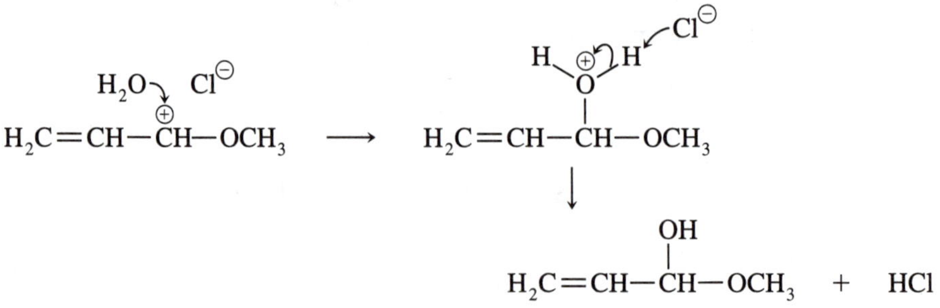

A quick inspection of the product of Reaction 18d shows that it is a hemiacetal; hemiacetals are rapidly converted to carbonyl compounds in acid solutions. (The solution has become acidic, because HCl was formed in the reaction steps shown in Eqs. 18c and 18d.)

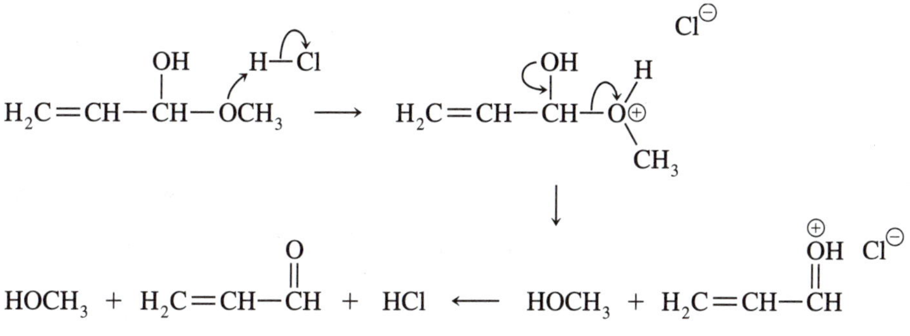

(18e)

As usual, what might (or might not, depending on your degree of chemical sophistication) seem a rather surprising reaction is simply the result of a combination of several very common reactions. The key to proposing the mechanism was to follow Rule 1 and to draw all the resonance forms before worrying about any succeeding steps.

Each product of a reaction can actually be obtained from any resonance form of the starting material, because the resonance forms are simply different representations of the same molecule or ion. This is shown in Eq. 18f, employing resonance form **3a**.

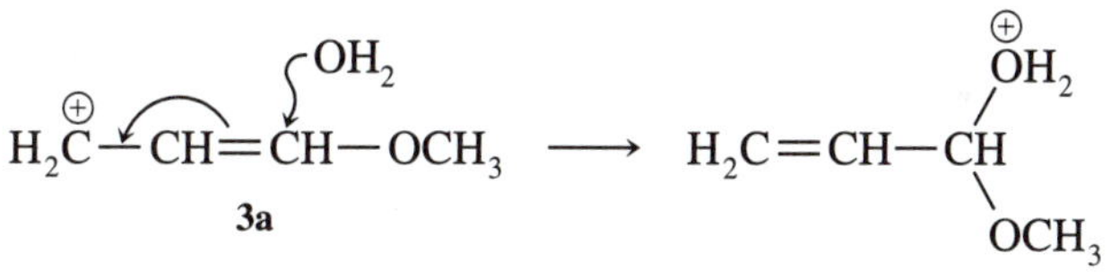

(18f)

Unless you have a good deal of experience in writing organic reaction mechanisms, however, it is best to begin by drawing the individual resonance forms.

Which products are actually formed?   Up to this point, we've ignored the fact that Eqs. 18c and 18d show the formation of different products. In this reaction, as it happens, both products are formed. This is not necessarily true for every reaction involving starting materials or intermediates with several resonance forms.

Rule 2 for the use of resonance theory shows how to predict possible products from reactions of a molecule or ion with delocalized bonds. However, it does not predict which of these products will actually be formed, or in what proportions. Unfortunately, there is no simple set of rules that will reliably predict which products, from among several reasonable possibilities, will actually be obtained from the reactions of a molecule or ion with several significant resonance structures.

Consider, for instance, the results of the protonation of a "dienolate" anion such as **4**, in solutions that remain basic throughout the reaction.

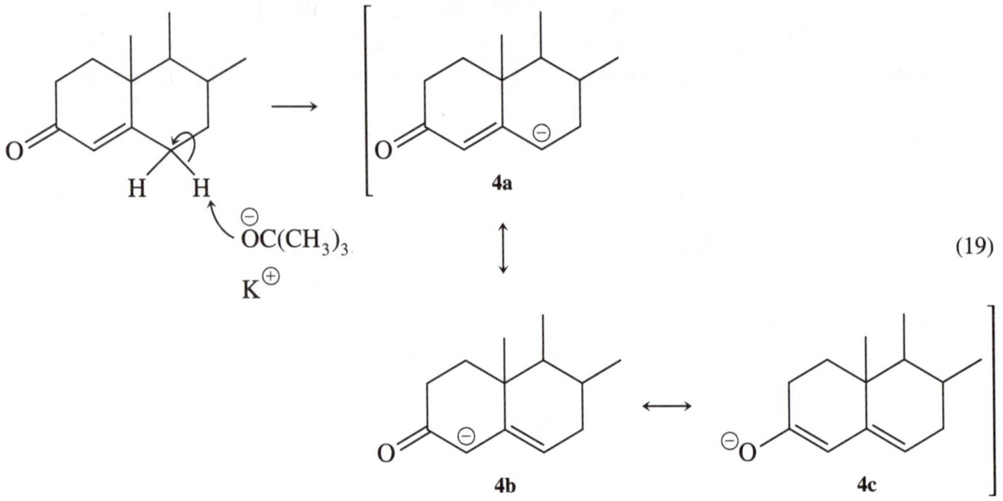

$$\text{(19)}$$

Careful studies have shown that the *kinetic* (fastest-formed) *product* of the reaction is the enol, **5**.[4] However, rapid equilibration of the enol with its anion results in conversion of the enol to ketone **6**, which, in turn, can be almost entirely converted to the more stable conjugated ketone, **7**, the *thermodynamic product*. Thus, the products that would result from the protonation of each individual resonance form are obtained, but the relative amounts of each depend on the reaction time and reaction conditions.

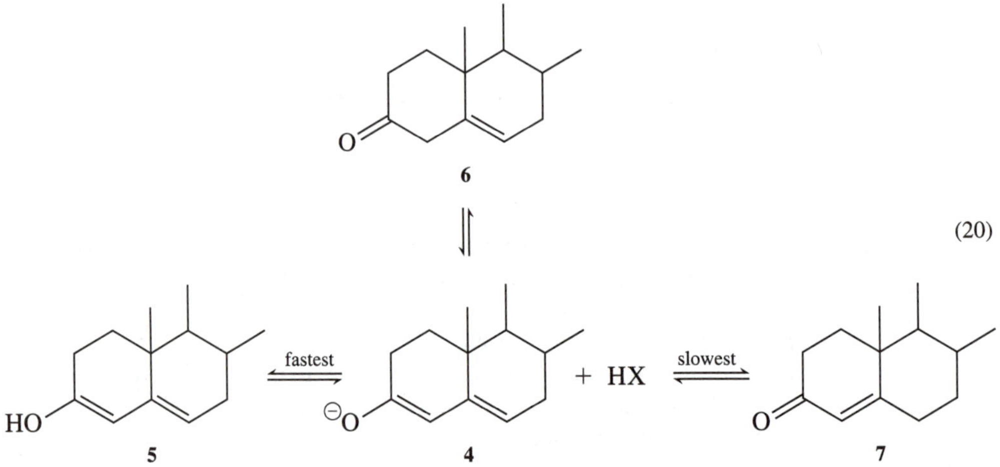

$$\text{(20)}$$

Not infrequently, the kinetic product of a reaction is the product that would appear to be formally derived from the principal resonance form of the starting material, as is the case in the protonation of **4**. However, that is far from a general rule. The reaction of anion **4** with methyl iodide (Eq. 21), for instance, yields the product from alkylation at the carbon α to the carbonyl group, rather than at the oxygen atom. (No alkylation occurs at the γ carbon.) Acylation of the anion, on the other hand, usually takes place on the oxygen atom rather than a carbon atom.

There are reasonable explanations for the different courses of these reactions.* However, there are no general rules that can reliably predict the relative yields of products from reactions of molecules or ions with delocalized bonds. Chemists attempting the synthesis of organic molecules, who must be able to predict the principal products of the reactions they are using, simply have to know which products were previously obtained from similar reactions.

$$+ \ CH_3I \ \longrightarrow \qquad \qquad \longrightarrow \quad \text{further reaction products} \qquad (21)$$

Resonance Stabilization

Resonance energies.   A molecule or ion with delocalized bonds is always more stable (that is, lower in energy) than its lowest energy resonance form. The difference in energy between the actual structure and the lowest energy form is called the *resonance stabilization energy* (or simply the resonance energy).

The resonance energy of a molecule or ion depends, in part, on the number of possible resonance forms. Other things being equal, any molecule (or ion) will have a larger resonance energy than a closely related molecule with a smaller number of resonance forms. The resonance energy of the allyl radical, for instance, is about 12 kcal/mol, while the resonance energy of the pentadienyl radical is about 19 kcal/mol.

$$\left[H_2C\!=\!CH\!-\!\dot{C}H_2 \ \longleftrightarrow \ H_2\dot{C}\!-\!CH\!=\!CH_2 \right]$$
$$\text{RSE} \approx 12 \text{ kcal/mol}$$

$$\left[H_2\dot{C}\!-\!CH\!=\!CH\!-\!CH\!=\!CH_2 \ \longleftrightarrow \ H_2C\!=\!CH\!-\!\dot{C}H\!-\!CH\!=\!CH_2 \right.$$
$$\updownarrow \qquad\qquad (22)$$
$$\left. H_2C\!=\!CH\!-\!CH\!=\!CH\!-\!\dot{C}H_2 \right]$$
$$\text{RSE} \approx 19 \text{ kcal/mol}$$

The value of the resonance stabilization energy also depends on the relative stabilities of the various resonance forms. The closer in energy the different resonance forms are, the greater the degree of delocalization of the electrons and the higher the resonance energy. Resonance stabilization of the α-keto radical **8**, for instance, appears to be appreciably less effective than that of the allyl radical, although it is difficult to assign a specific value to its resonance energy.[7,8]

---

*For instance, the different courses of alkylation, protonation, and acylation of enolate anions may be accounted for by the principle of hard and soft acids and bases[5] (see Chapter 13) or by the Hammond postulate.[6]

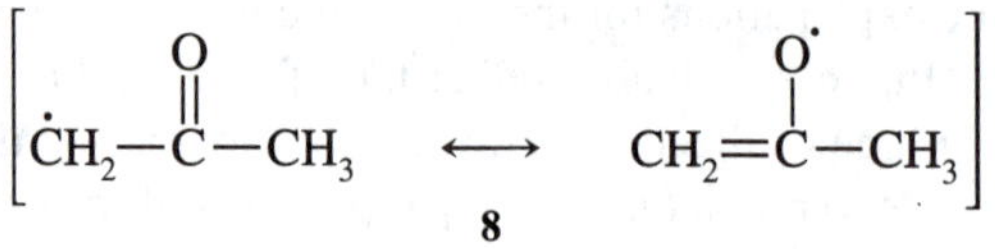

8

It is important not to confuse resonance with aromaticity. *Resonance* is a general term applied to the phenomenon of bond delocalization. *Aromaticity,* and *antiaromaticity,* are special properties of *rings* composed entirely of conjugated π bonds, empty orbitals, or unshared pairs of electrons. Unlike other delocalized systems that are always stabilized by resonance, the energies of aromatic and antiaromatic systems may be either higher or lower than those of their lowest energy resonance forms, depending on the number of electrons in the rings. (For a discussion of aromaticity, see Chapter 2.)

Nonplanar π systems.   In order for electrons to be delocalized over several *p* orbitals, the orbitals should be lined up parallel to each other. If one or more orbitals are twisted away from a parallel arrangement, the energy of the system will be increased. If two orbitals are at right angles to each other, delocalization of electrons between them cannot occur.

The possibility of nonparallel arrangements of orbitals rarely causes trouble in open-chain π systems. Since delocalization of electrons always lowers the energies of the molecules or ions, the orbitals will usually stay in the lowest-energy (parallel) conformations. However, in some cyclic molecules, it may be difficult or impossible for orbitals to maintain a parallel arrangement.

This fact can be demonstrated by the remarkable differences between the reactivities of triphenylmethyl bromide (**9**) and 9-bromotriptycene (**10**).[9]

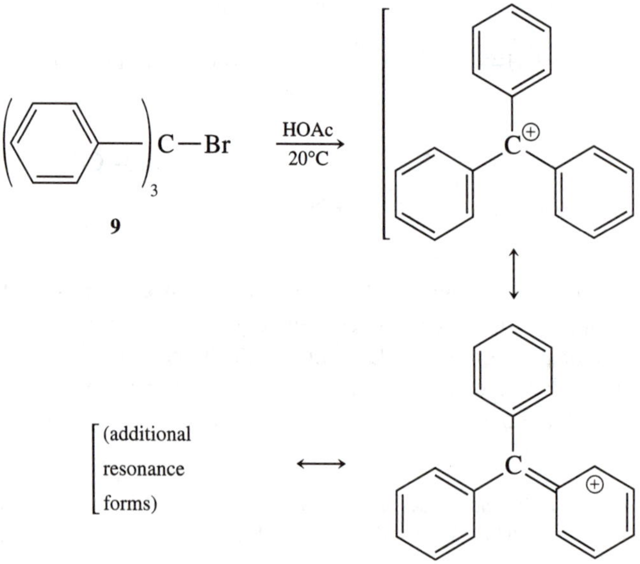

(23)

(24)

10   **11**

Bromide **9** dissociates easily to form the triphenylmethyl cation, because the positive charge in the cation can be distributed over no fewer than ten carbons. In contrast, **10** could not be forced to undergo ionic dissociation even when heated with strong Lewis acids.[8] The remarkable lack of reactivity of bromide **10** can be accounted for by the fact that the empty p orbital of its cation, **11**, would be essentially at right angles (orthogonal) to the π systems of the aromatic rings.* In effect, double bonds between the vacant p orbital of **11** and the adjacent aromatic ring would be so weak that resonance forms showing those double bonds would be insignificant.

Crowding by neighboring atoms can sometimes prevent molecules from adopting conformations in which p orbitals, or orbitals containing unshared electrons, are parallel to each other. This phenomenon is called the *steric inhibition of resonance*. For instance, 2,4,6-trinitroaniline (**12**) is more than 10^4 times less basic than its *N,N*-dimethyl derivative (**13**),[10] even though alkyl substitution usually has only a slight effect on the basicity of anilines.

(25)

12   **13**

The *ortho*-nitro groups in **13** prevent the molecule from adopting a conformation with the methyl groups in the plane of the aromatic ring. That would be the only conformation in which the unshared electrons on the amino group could interact effectively with the π systems of the ring and the nitro groups. Thus, more resonance energy is lost on protonation of **12** than of **13**, and **13** is far more basic.

---

*Another factor that would destabilize **11** is that the trivalent carbon could not maintain a planar structure. It is important that a carbocation assume a planar structure so that part of the energy originally required to break the bond to the leaving group is compensated for by moving electrons in the remaining three bonds from sp^3 orbitals to lower-energy sp^2 orbitals.

REFERENCES

[1] For early examples of the use of curved arrows to depict electron motions, see A. Lapworth, *J. Chem. Soc. 121*, 416 (1922); W.O. Kermack and R. Robinson, *J. Chem. Soc., 121*, 427 (1922).

[2] See G.W. Wheland, *Resonance in Organic Chemistry.* John Wiley & Sons, New York (1955).

[3] M.J.S. Dewar and G.J. Gleicher, *J. Am. Chem. Soc., 87*, 692 (1965).

[4] See H.O. House, *Modern Synthetic Reactions,* 2nd ed. W.A. Benjamin, Inc., Menlo Park, CA, pp. 502–506 (1972).

[5] R.G. Pearson, ed., *Hard and Soft Acids and Bases.* Dowden, Hutchinson, and Ross, Stroudsberg, PA (1973).

[6] G.S. Hammond, *J. Am. Chem. Soc., 77*, 334 (1955).

[7] D.F. McMillen and D.M. Golden, *Ann. Rev. Phys. Chem., 33*, 493 (1982).

[8] J.A. Kerr, *Chem. Rev., 66*, 465 (1966).

[9] P.D. Bartlett and E.S. Lewis, *J. Am. Chem. Soc., 72*, 1005 (1950).

[10] L.P. Hammett and M.A. Paul, *J. Am. Chem. Soc., 56*, 827 (1934).

PROBLEMS

1.1 Propose reasonable mechanisms for each of the following reactions. Use curved arrows to show any changes in the positions of electrons. Do not combine any two steps into one.

(a)

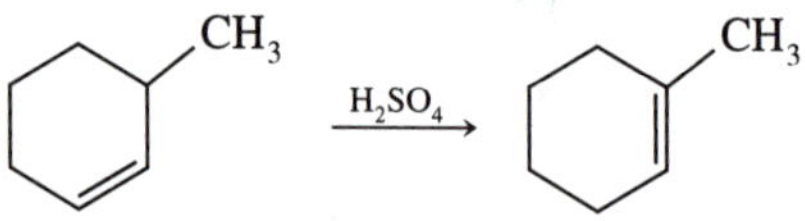

$$\text{(cyclohexene with } CH_3) \xrightarrow{H_2SO_4} \text{(cyclohexene with } CH_3)$$

(b) $CH_3CHCH{=}CH_2$ (Br substituent) $\xrightarrow[\Delta]{CH_3OH}$ $CH_3CH{=}CHCH_2OCH_3$

(c) (methylcyclopentadiene) $+$ HCl $\longrightarrow$ (Cl, CH_3 cyclopentene)

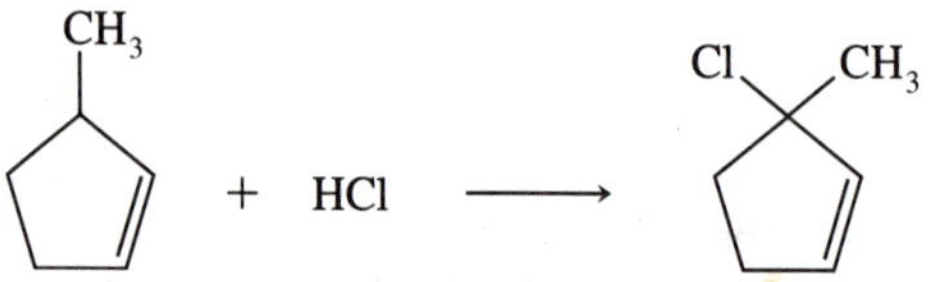

(d) $H_2C{=}CH{-}CH{=}CHCH_3 + H_2O \xrightarrow{H_2SO_4} H_3C{-}CH{=}CH{-}\underset{\underset{OH}{|}}{C}HCH_3$

(e) $CH_3CH{=}CH{-}OCH_3 + H_2O \xrightarrow{HCl} CH_3CH_2\overset{\overset{O}{\|}}{C}H + CH_3OH$

(f) 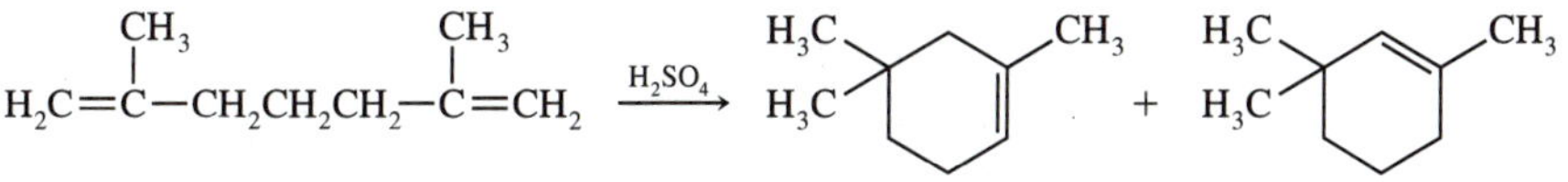 $H_2C=\overset{\underset{\displaystyle CH_3}{|}}{C}-CH_2CH_2CH_2-\overset{\underset{\displaystyle CH_3}{|}}{C}=CH_2$ $\xrightarrow{H_2SO_4}$

+

(g) $HC\equiv C-OCH_3 + H_2O$ $\xrightarrow{HCl}$ $H_3C\overset{\displaystyle O}{\overset{\displaystyle \|}{C}}OCH_3$

(h)

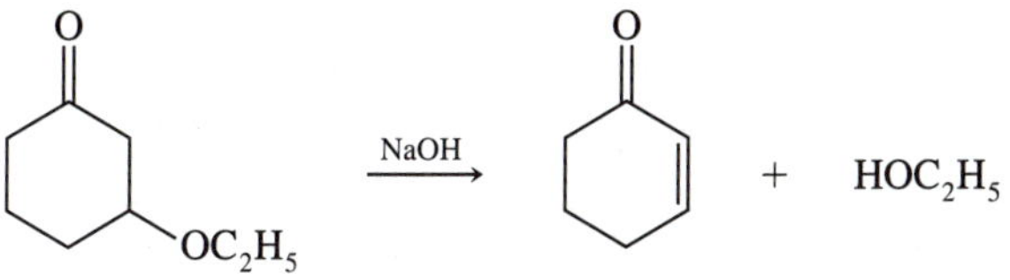

$\xrightarrow{NaOH}$ + HOC_2H_5

(i)

$\xrightarrow{NaOH}$

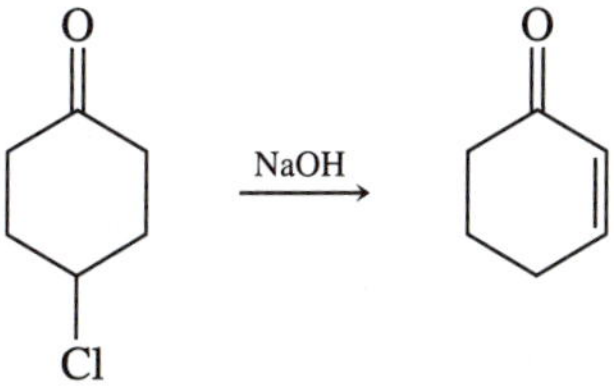

(j) Although the reactions in (h) and (i) occur readily, no elimination takes place when 4-ethoxy-cyclohexanone is heated in sodium hydroxide solution. Explain why this is so.

(k) $CH_3-\overset{\underset{\displaystyle CH_3}{|}}{\overset{\displaystyle OH}{\overset{|}{C}}}-CH_2\overset{\displaystyle O}{\overset{\|}{C}}CH_3$ $\xrightarrow{NaOH}$ $2\ CH_3\overset{\displaystyle O}{\overset{\|}{C}}CH_3$

(l)

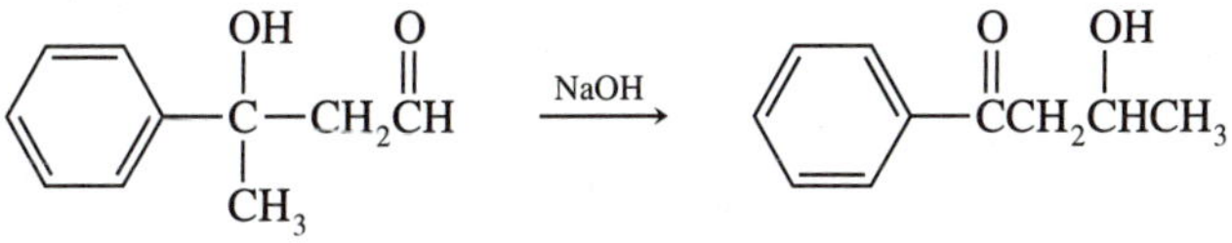

$-\overset{\underset{\displaystyle CH_3}{|}}{\overset{\displaystyle OH}{\overset{|}{C}}}-CH_2\overset{\displaystyle O}{\overset{\|}{C}}H$ $\xrightarrow{NaOH}$ $-\overset{\displaystyle O}{\overset{\|}{C}}CH_2\overset{\underset{}{\overset{\displaystyle OH}{\overset{|}{C}}}}HCH_3$

(m)

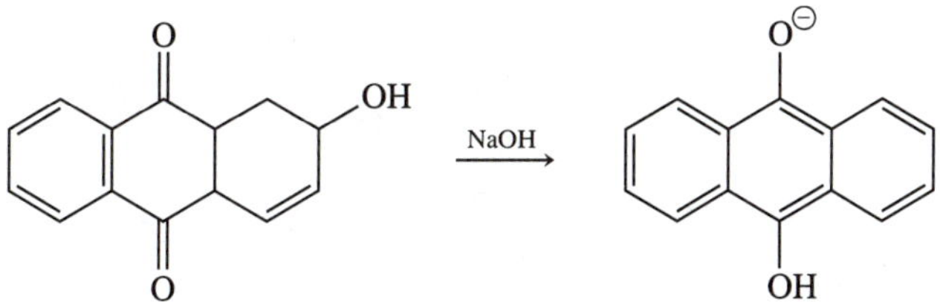

$\xrightarrow{NaOH}$

(n) $CH_3-\overset{\underset{\displaystyle C\equiv N}{|}}{\overset{\displaystyle OH}{\overset{|}{C}}}-CH=CH_2$ $\xrightarrow{NaOCH_3}$ $CH_3\overset{\displaystyle O}{\overset{\|}{C}}CH_2CH_2C\equiv N$

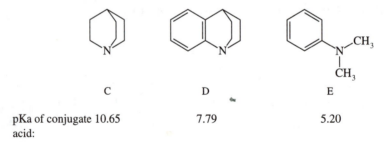

(o)

$$NaOCH_3$$

1.2 a. Phenol A has a pKa of 7.22, while its isomer, B, has a pKa of 8.25 [G.W. Wheland, R.M. Brownell, and E.C. Mayo, *J. Am. Chem. Soc., 70,* 2492 (1948)].

Explain why A is a much stronger acid than B.

b. Benzoquinuclidine (D) is a significantly weaker base than quinuclidine (C), but it is a significantly stronger base than *N,N*-dimethyaniline (E) [B.M. Wepter, *Rec. Trav. Chim., 71,* 1171 (1952)]. Account for these facts.

	C	D	E
pKa of conjugate acid:	10.65	7.79	5.20

Electrocyclic Reactions

2.1 INTRODUCTION

Reactions in which conjugated polyenes close to form rings or in which rings open to form polyenes simply by having electrons "chase each other's tails" around in a circle are called *electrocyclic reactions*. These are among the conceptually simplest organic reactions. (It doesn't matter in which direction the curved arrows depicting electron motions are drawn in electrocyclic reactions, as long as they all proceed in the same direction.)

$$\text{(1)}$$

$$\text{(2)}$$

Electrocyclic reactions belong to a broader class of reactions called *pericyclic reactions*. Several other types of pericyclic reactions will be discussed in later chapters. All pericyclic reactions proceed via transition states in which the electrons taking part in the reactions can be considered to form continuous rings.

The equilibrium between the open chain and ring forms* in an electrocyclic reaction will usually favor the ring, because closing the ring will convert a π bond, with a bond dissociation energy (BDE) of roughly 63 kcal/mol, to a σ bond, with a BDE averaging about 85 kcal/mol. Therefore, 1,3,5-hexatrienes will usually close to form 1,3-cyclohexadienes on heating, as shown in Eq. 1. However, a cyclobutene will usually open to form a 1,3-butadiene on heating, as

---

*The two isomers in an electrocyclic reaction are sometimes called *valence tautomers*. This is a rather unfortunate name, because tautomers (such as the keto and enol forms of carbonyl compounds) are isomers that are rapidly interconverted and usually differ in the location of a hydrogen atom. In contrast, the interconversion of two molecules by electrocyclic processes may, or may not, be rapid.

shown in Eq. 2, because relief of the strain in the cyclobutene ring more than compensates for converting a σ bond to a π bond.[1,2]

The differences in energy between the open-chain and closed-ring isomers in electrocyclic reactions are usually not great. Therefore, partial conversion of the more stable isomers in electrocyclic reactions to the less stable forms may easily occur as steps in longer mechanisms. For instance, conversion of *cis,cis*-2,4-hexadiene to *trans,trans*-2,4-hexadiene on heating (Eq. 3) can easily be accounted for as the result of two electrocyclic steps.

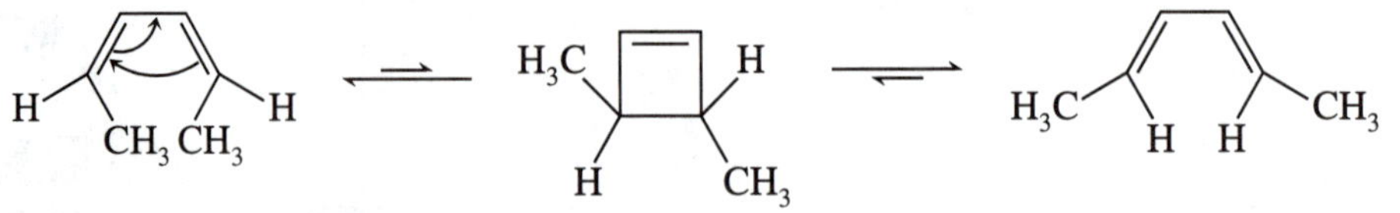

$$(3)$$

2.2 CONROTATORY AND DISROTATORY PROCESSES

When a polyene undergoes an electrocyclic ring closure to form a cycloalkene, the terminal carbons of the polyene chain must rotate approximately 90° to convert the *p* orbitals on those carbons into the sp³ orbitals forming the new σ bond. The substituents on those carbons must, therefore, be rotated into a plane that is approximately at right angles to the newly formed ring. Conversely, in the ring opening of a cycloalkene, the substituents on the atoms forming the bond being broken will rotate into the plane of the new double bonds.

In any electrocyclic reaction, these rotations might, in principle, proceed in two ways. For instance, the terminal carbons of a polyene, or the saturated carbons of the cyclized form, might rotate in the same direction (both moving in either a clockwise or counterclockwise direction), as in Eqs. 4 and 5. Reactions proceeding in this manner are called *conrotatory.*

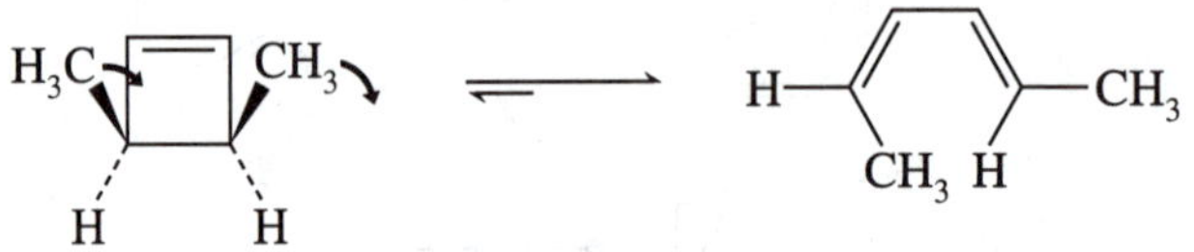

$$(4)$$

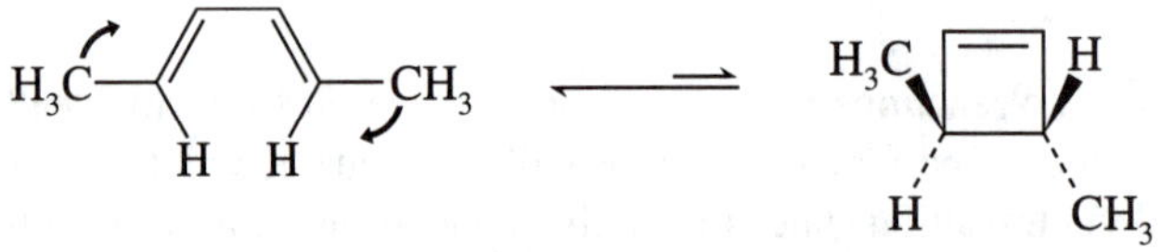

$$(5)$$

conrotatory reactions

Alternatively, the terminal carbons of a polyene chain or the saturated carbons of a cycloalkene may rotate so that one carbon follows a clockwise path and the other a counterclockwise path (Eqs. 6 and 7).

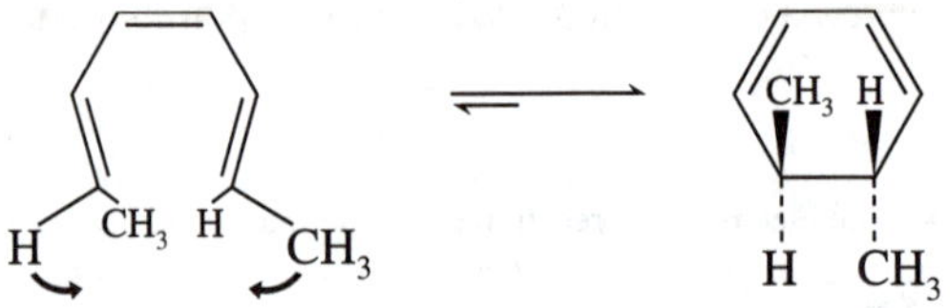

$$(6)$$

$$(7)$$

disrotatory reactions

Electrocyclic reactions of this type are called *disrotatory.* (The "heavy" arrows showing directions of group rotations in these equations should not be confused with the regular curved arrows indicating electron motions.)

The terms "conrotatory" and "disrotatory," as well as "electrocyclic" and "pericyclic," were coined by Robert B. Woodward and Roald Hoffmann of Harvard University in 1965.*[1] Woodward and Hoffmann pointed out that the *stereochemistries* of pericyclic reactions (that is, which stereoisomers may be formed in the reactions) are governed by selection rules, according to which the formation of some stereoisomers are "allowed" and others are "forbidden." (Despite the absolute sounds of these terms, *forbidden reactions* are simply expected to occur with much higher activation energies than *allowed reactions*.) These selection rules are now commonly known as the *Woodward–Hoffmann rules.*

The first Woodward–Hoffmann rule states that *thermal electrocyclic reactions involving* 4n *electrons (where* n *is an integer) are allowed if they proceed by conrotatory paths; thermal electrocyclic reactions involving* 4n + 2 *electrons are allowed if they proceed by disrotatory paths*. (Note the word "thermal," which indicates that the reactions under discussion do *not* require photoirradiation to proceed.)

Each electrocyclic ring closure of a 1,3-butadiene or ring opening of a cyclobutene involves four electrons: the electrons in the two π bonds of a butadiene or in one π bond and one σ bond of a cyclobutene. Therefore, these reactions proceed by conrotatory paths, as shown in Eqs. 3 to 5. Ring closures of 1,3,5-hexatrienes or ring openings of 1,3-cyclohexadienes proceed by disrotatory paths, as in Eqs. 6 and 7, because they involve six electrons in each case.[†]

Many studies, both before and after the formulation of the Woodward–Hoffmann rules, show that most thermal electrocyclic reactions do indeed follow the rules. For instance, the interconversion of stereoisomers **1a** and **1b**, which presumably proceeds by two consecutive electrocyclic reactions, as shown in Eq. 8, has been studied. After **1a** was heated at 124°C for 51 days, closure of the butadiene to the cyclobutene form and reopening of the cyclobutene had taken place an average of more than 2×10^6 times for each molecule. Yet, not the smallest trace of isomers **1c** or **1d**, which would arise by a forbidden disrotatory ring closure or opening, could be detected.[4]

---

*Robert Burns Woodward (1917–1979) was the preeminent organic chemist of his generation. He received the Nobel Prize in 1965 for his work on the synthesis of natural products, which included the first total syntheses of quinine, cholesterol, cortisone, strychnine, and chlorophyll. Had he lived, he would undoubtedly have shared a second Nobel Prize in 1981. Roald Hoffmann (b. 1937) shared the Nobel Prize in chemistry in 1981 with K. Fukui of Japan for their work on molecular orbital theory and pericyclic reactions.

[†]When writing the mechanism for an electrocyclic reaction, a curved arrow should start at each bond involved in the reaction. Thus, if the curved arrows are drawn properly, *thermal electrocyclic reactions requiring even numbers of curved arrows proceed by conrotatory paths; those involving odd numbers of curved arrows proceed by disrotatory paths.*

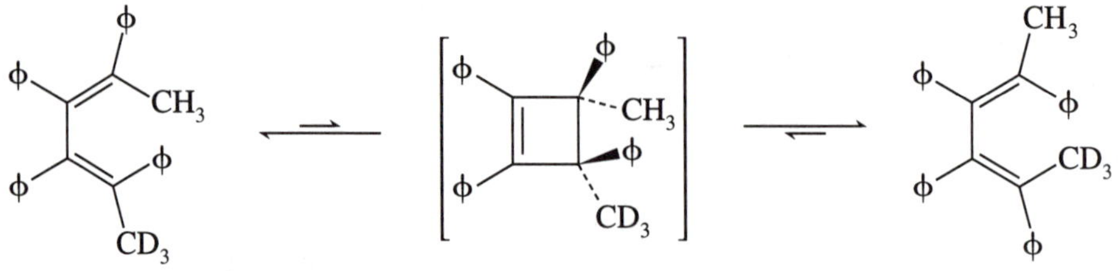

1a **1b**

(8)

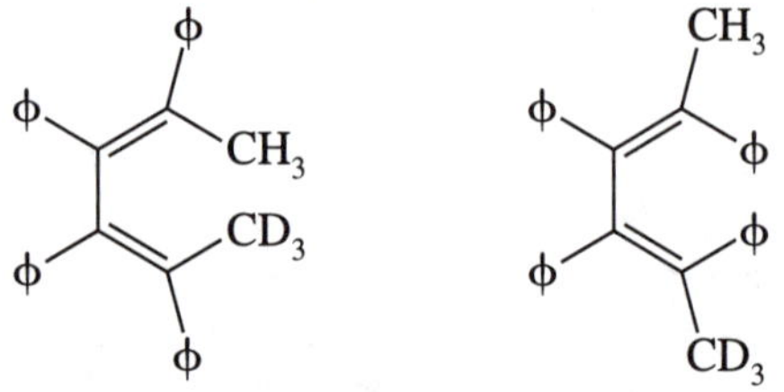

1c **1d**

The Woodward–Hoffmann rules should apply to thermal electrocyclic reactions involving any number of electrons. Indeed, within experimental limits, only the predicted conrotation products are formed on heating the three isomeric tetraenes shown in Eqs. 9 and 10.[5]

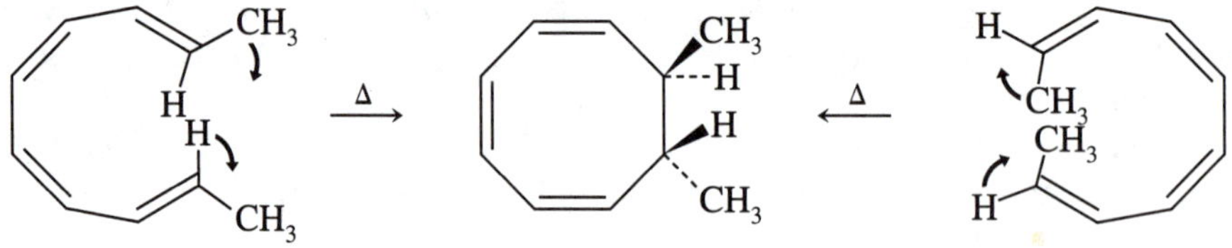

(9)

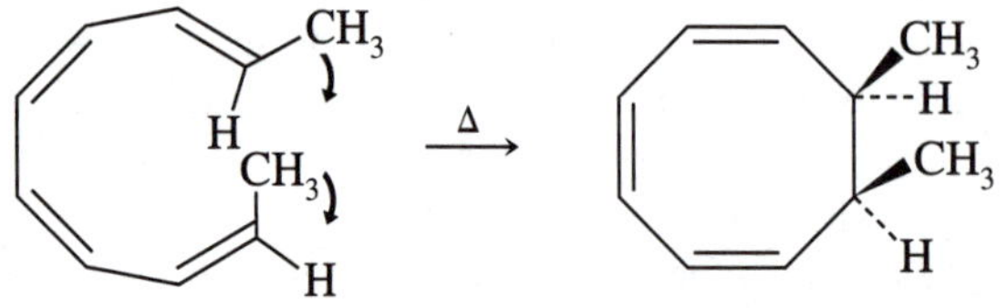

(10)

(There are two allowed modes of rotation in any electrocyclic reaction, because each carbon may rotate in a clockwise or counterclockwise direction. In thermal ring opening reactions of cyclobutenes, electron donating substituents or weakly electron accepting substituents tend to end up "outside" the butadiene chains, while strongly electron accepting substituents are more likely to be rotated toward the "inside" positions.[6] An interesting result of this fact is that different isomers of pentadienoic acid are obtained when cyclobutene-3–carboxylic acid is heated in acidic or basic solutions.)

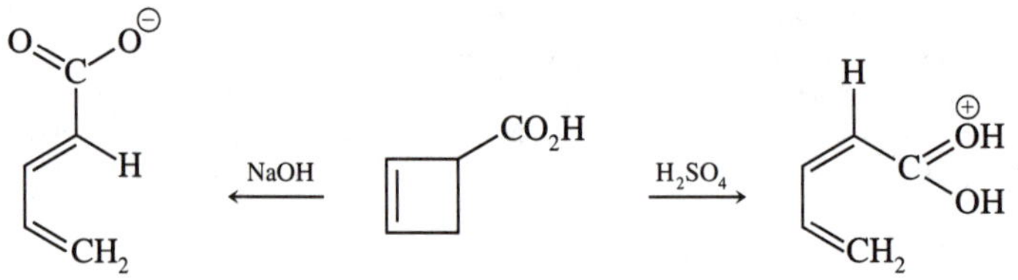

(11)

The Woodward–Hoffmann rules can account for the stabilities of some molecules that might otherwise be quite reactive. For instance, the bicyclic hydrocarbon **2** is stable at temperatures up to 260°C,[7] even though cyclobutenes normally open to butadienes on heating at steam-bath temperatures.

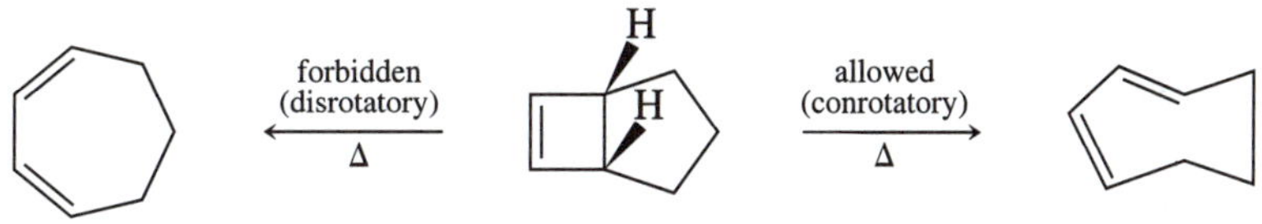

$$\text{forbidden (disrotatory)} \quad \Delta \qquad \qquad \text{allowed (conrotatory)} \quad \Delta \tag{12}$$

Z,Z-1,3-cycloheptadiene   (not obtained)
                                                                        E,Z-1,3-cycloheptadiene

                              2

Similarly, in the tricyclic cyclobutene derivatives shown below, each *cis*-fused isomer requires a temperature nearly 200°C greater than its *trans*-fused isomer to have the central ring open at a comparable rate.[8] In all of these reactions, the allowed conrotatory ring openings of the *cis*-fused isomers would require forming strained rings containing double bonds with *trans*-configurations. (In fact, only Z,Z-1,3-cycloheptadiene, formally the product of a forbidden electrocyclic reaction, is formed on thermolysis of **2**. Unfortunately, it cannot be determined whether this means that the Woodward–Hoffmann rules were violated, because the *E,Z* isomer would rapidly isomerize to the *Z,Z* isomer at the temperature necessary to open the cyclobutene ring of **2**.)

	E_a (kcal/mol)	Temperature required for $k = 10^{-4}$ sec^{-1}
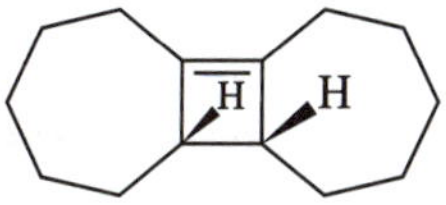	45	273°C
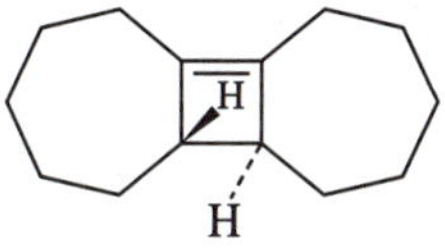	27	109°C
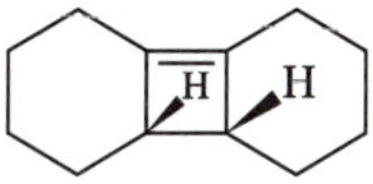	42	261°C
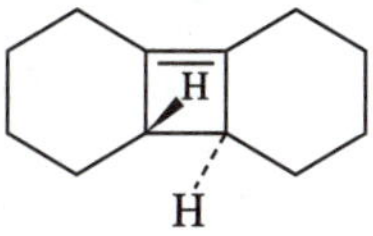	29	87°C

The Woodward–Hoffmann rules can explain the existence of some extraordinarily high-energy molecules. "Dewar-benzene" (**3**), for instance, is stable at room temperature,[9] although it could isomerize to benzene simply by changing some bond angles by a few degrees. Isomerization of Dewar-benzene to benzene has been estimated to be exothermic by about 79 kcal/mol,[10] since not only would an aromatic ring be formed, but the strain energy of two cyclobutene rings would be relieved. However, an allowed conrotatory opening of a cyclobutene ring in Dewar-benzene would yield an extraordinarily strained isomer of benzene with one *trans*-double bond. Thus, neither the allowed nor the forbidden processes occur rapidly at room temperature. However, at only slightly higher temperatures, Dewar-benzene does isomerize to benzene.

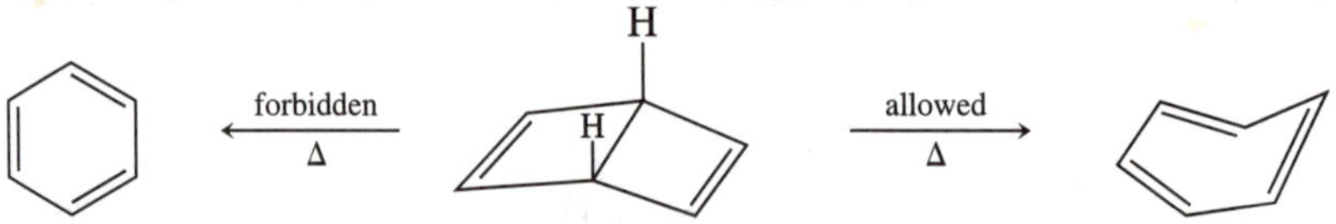

"Dewar-benzene"

3

$$(13)$$

Isomerization of the remarkable polycyclic alkene **4** to its isomer, bianthranyl, would not only release a large amount of strain energy but would result in the formation of *two* new aromatic rings. Nonetheless, this molecule exists, with a half-life of about 30 min at 80°C.[11] The stability of the molecule can be accounted for by the fact that only a conrotatory ring opening, which would give rise to an aromatic ring with a *trans* double bond, would be allowed.

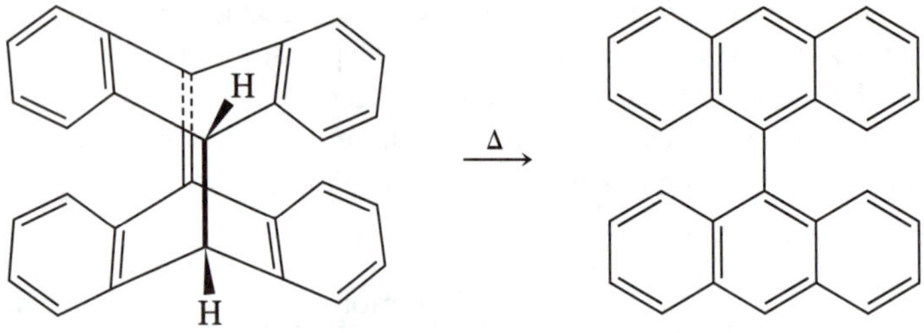

4

$$(14a)$$

(The ring opening in **4** can be regarded as simply an eight-electron process, as shown in Eq. 14b.)

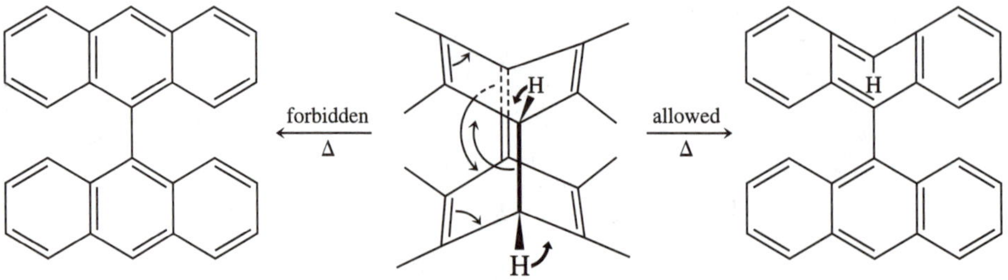

$$(14b)$$

It is not clear whether the isomerizations of Dewar-benzene and compound **4**, or other apparent examples of "forbidden" electrocyclic reactions, actually violate the Woodward–Hoffmann rules.* Up to now, we have assumed that both terminal carbons of a polyene (or sp³ carbons of a

---

*Theoretical calculations indicate that the lowest energy path for isomerization of Dewar-benzene to benzene is by an allowed conrotatory ring opening to form "*trans*-benzene" (Eq. 14c), which then undergoes a second isomerization step to form benzene.[12]

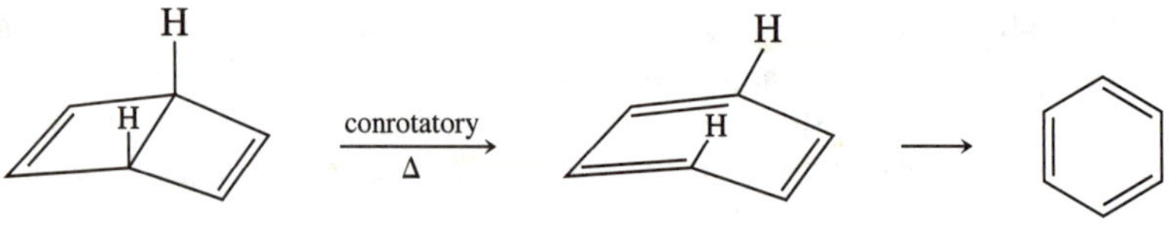

$$(14c)$$

cycloalkene) must rotate simultaneously. There is another possibility. Initially, just one end of a polyene, for instance, might rotate (or might rotate to a much greater extent than the other end), breaking a π bond to form a diradical, as shown in Eq. 15. Rotation of the other end of the chain could then result in formation of the ring isomer. Unlike concerted electrocyclic reactions, such two-step mechanisms would not be expected to be stereospecific but should give rise to mixtures of stereoisomers. (The Woodward–Hoffmann rules do not apply to nonconcerted, diradical reactions.)

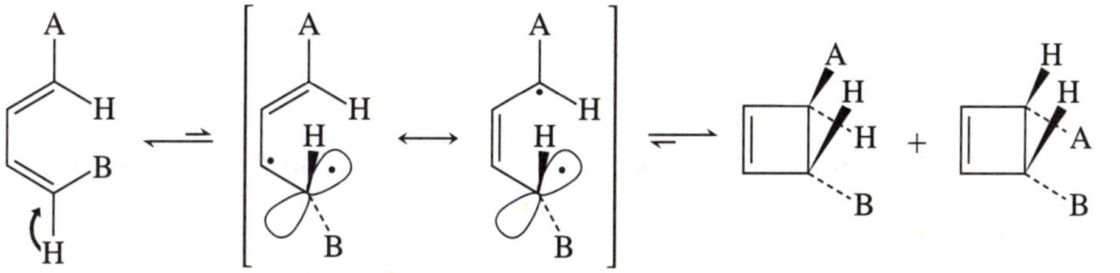

$$(15)$$

Activation energies for ring closures (or ring openings) by diradical mechanisms should normally be much higher than those for allowed electrocyclic reactions. However, diradical mechanisms may be preferable to reactions by forbidden paths.

2.3   EXPLANATIONS FOR THE WOODWARD–HOFFMANN RULES

Woodward and Hoffmann's original paper on electrocyclic reactions[3] was intended to explain the reasons for the existence and nature of selection rules in electrocyclic reactions, not just to describe the rules. Since then, several other ways to explain the rules have been proposed. These explanations are, for the most part, complementary to the original Woodward–Hoffmann explanation.

Symmetries of Molecular Orbitals

Before considering any of the explanations for the Woodward–Hoffmann rules, it is important to be familiar with qualitative procedures for determining the symmetries of molecular orbitals.
    Whenever two atoms with half-empty orbitals (two hydrogen atoms, for instance) approach each other, two results are possible. If the wave functions of the electrons are *in phase* (shown in the following diagram by shading both orbitals) the atoms can form a bond, resulting in the formation of a molecule that is lower in energy than the "zero energy level" of the two unbonded atoms. In effect, a new *molecular orbital* is formed by combining the two atomic orbitals of the individual atoms. Mathematically, this can be done by the linear combination of atomic orbitals (LCAO), so that if ϕ_1 and ϕ_2 are the wave functions for the atomic orbitals, the wave function for the molecular orbital, ψ, can be approximated by the equation $\psi = c_1\phi_1 + c_2\phi_2$. The coefficients, c_1 and c_2, are proportional to the contributions of the two atomic orbitals to the molecular orbital, and the squares of the coefficients are proportional to the electron density in the neighborhood of each atom.

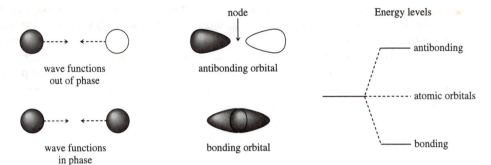

Formation of bonding and antibonding orbitals

If the wave functions of the two atoms are *out of phase*, there will be a net repulsion rather than a net attraction between the two atoms. (The molecular wave function would then be $\psi = c_1\phi_1 - c_2\phi_2$.) If the atoms are held at bonding distance, a new molecular orbital would still be formed, but this orbital would be an *antibonding orbital*, with an energy that exceeds that of the "zero level" to the same degree that the bonding orbital would be lower in energy. Whereas the electron density in a bonding orbital is highest at the midpoint between the two atomic nuclei (if the two atoms are identical), the electron density in an antibonding orbital is lowest at that point. The wave function of the antibonding orbital is said to have a *node* at that point.

It should be noted that the number of molecular orbitals will always be the same as the number of atomic orbitals from which they are formed. Mathematically, bonding molecular wave functions will form when the wave functions of the individual atoms have the same sign: either both + or both −. (Remember, these are purely mathematical signs. They have no relation to the negative charge on an electron or the positive charge on an atomic nucleus.)

Instead of discussing only the interactions of two half-filled orbitals, we can consider orbitals in the abstract, free of any electrons they may hold. Orbitals by themselves cannot form bonds, but we can imagine the interaction of any two orbitals as resulting in the formation of one low-energy bonding orbital and one high-energy antibonding orbital that can potentially form new bonds (or new "antibonds") if electrons are inserted into them.

In the preceding discussion, we pictured spherical *s* orbitals. If, instead, we consider the interaction of two *p* orbitals, we should remember that each lobe of a *p* orbital is opposite in phase to the other lobe. Thus, if two *p* orbitals approach each other in a "side-by-side" manner, whether bonding or antibonding orbitals are formed will depend on whether the lobes coming together are in phase or out of phase (that is, whether they have the same or opposite signs).

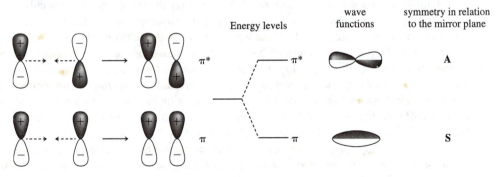

Formation of π and π^* orbitals

The bonding orbital formed from the side-by-side interaction of two p orbitals is called a π orbital, and the antibonding orbital is called a π^* orbital. The wave function for the π orbital will have its largest value at the center of the new orbital, while the wave function for the π^* orbital will have a node at the center of the orbital.

Suppose we try to combine atomic orbitals for *three* p orbitals side by side. Three atomic orbitals must give rise to *three* new molecular orbitals. It is easy to see how two of these orbitals are formed. If all three orbitals are in phase, as shown in the diagram below, a new, very low-energy molecular orbital (ψ_1) will be formed. If the center orbital is out of phase with *both* end orbitals, a high-energy molecular orbital (ψ_3) with two nodes—that is, one in which each orbital repels the orbital next to it—will be formed.

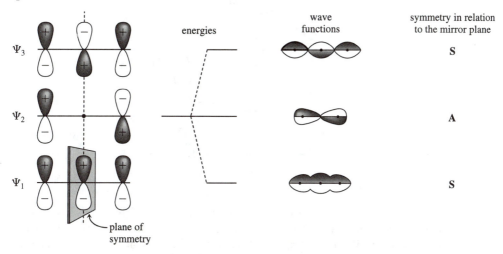

Combination of three atomic orbitals

But how about the third orbital? There are constraints as to how you can combine atomic orbitals. When p orbitals are lined up side by side, the array has a vertical plane of symmetry (a mirror plane) as a symmetry element. Therefore, the molecular orbitals formed by combining the atomic orbitals must also have a mirror plane as a symmetry element—that is, the central atomic orbital must have the same relationship to both end orbitals. However, molecular orbitals with the central orbital in phase and out of phase with the end orbitals were already constructed. The only other possibility is to eliminate the center atomic orbital completely; that is, to have the coefficient of the wave function of the molecular orbital drop to zero at that point, so that the new orbital has a node at that atom. Since the two remaining atomic orbitals are far from each other, the third molecular orbital (ψ_2) will essentially remain at the "zero energy level" of atomic orbitals.

Molecular orbitals from the linear combination of four atomic orbitals can be formed in a similar manner, as shown on page 30. (In the preceding diagram, and all future orbital diagrams, the p orbitals are shown connected by σ bonds, as they would be in polyenes.)

No matter how many atomic orbitals are combined, the lowest-energy orbital in an open chain of p orbitals (ψ_1) has no nodes, the second lowest (ψ_2) has one node, the next has two nodes, and so on. As a result, the lowest-energy orbital is *symmetric* around the mirror plane, and the next lowest is *antisymmetric*; that is, the mirror image of each side would replace each + lobe with a − lobe, and vice versa. The third-lowest orbital is again symmetric, and the next is antisymmetric, around the mirror plane, until the highest-energy orbital is reached.

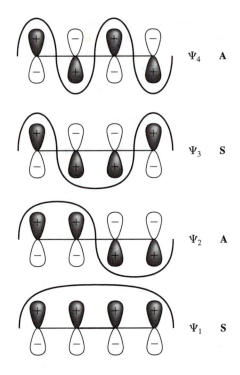

Ψ_4 A

Ψ_3 S

Ψ_2 A

Ψ_1 S

Of course, real molecules have electrons in some of their orbitals, so let's insert electrons into the orbitals to make neutral molecules. Following the standard "aufbau" approach, electrons will go into the lowest available orbitals, with, of course, a maximum of two electrons in each orbital. For the two-atom system, the two electrons will fill the π orbital, while the π^* orbital will remain empty. The π orbital will therefore be the *highest occupied molecular orbital* (HOMO) and the π^* orbital will be the *lowest unoccupied molecular orbital* (LUMO). In a four-orbital system, the four electrons necessary to make a neutral molecule will fill orbitals ψ_1 and ψ_2, so that ψ_2 would be the HOMO and ψ_3 the LUMO. In a six-orbital system, ψ_3 would be the HOMO and ψ_4 the LUMO.

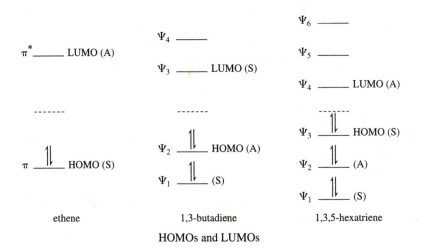

ethene   1,3-butadiene   1,3,5-hexatriene

HOMOs and LUMOs

Notice that in systems containing $4n + 2$ electrons the HOMO is symmetric, and in systems containing $4n$ electrons the HOMO is antisymmetric.

The Frontier Orbital Approach

It is common in chemistry for the chemical properties of atoms (at least of atoms in the first two rows of the periodic table) to be approximated by considering only the properties of the highest occupied orbitals—the valence orbitals. The lower orbitals can be pretty much ignored.

A similar approach can be employed with molecules. Ignore lower-energy orbitals, and consider only the properties of the HOMOs (the *frontier orbitals*).*

Suppose, for instance, that a linear system of six p orbitals (e.g., 1,3,5-hexatriene) is folded around so that the ends of the chain approach each other. In the neutral molecule, the HOMO will be ψ_3, which is symmetrical, so that the two ends of the system will be in phase. The top lobe at one end of the molecular orbital will attract the top lobe at the other end, and the bottom lobe at one end will attract the bottom lobe at the other. When a σ bond is formed between the two ends of the chain, the two top lobes (or two bottom lobes) will rotate so as to overlap end to end. This results in a disrotatory reaction, as Woodward and Hoffmann pointed out.

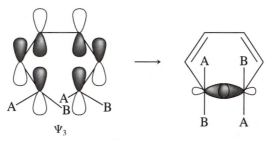

Ψ_3

Now, consider the effect of bringing the two ends of a neutral molecule with *four* molecular orbitals (e.g., butadiene) together.

In that case, the HOMO, ψ_2, is antisymmetric, so that the top lobe at one end of the orbital will *repel* the top lobe at the other. The top lobe at one end will attract the *bottom* lobe at the other. To form a new σ bond, the top lobe at one end will rotate to meet the bottom lobe on the other, resulting in a conrotatory process.

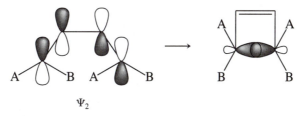

Ψ_2

The frontier orbital approach is very useful. It is simple, easy to visualize, and applicable to a wide variety of processes—not just to electrocyclic reactions but to many addition reactions and rearrangements as well. However, it is not really satisfying theoretically. Entire molecules are involved in electrocyclic reactions, and all of their molecular orbitals change, not just the

---

*The frontier orbital approach was pioneered by Kenichi Fukui[12] of Kyoto Imperial University, who shared the Nobel Prize in 1981 with R. Hoffmann.

frontier orbitals. Thus, even though frontier orbital approaches give the right answers, it would be nice to have a theoretically more exact approach.

Correlation Diagrams

The correlation diagram approach to electrocyclic reactions was proposed by Longuet-Higgins and Abrahamson[13] a few years after the appearance of the original Woodward–Hoffmann paper. A *correlation diagram* is a chart that follows the molecular orbitals of the starting materials in a reaction and shows how they are converted to the molecular orbitals of the products.

For instance, let's again consider the electrocyclic transformation of 1,3,5-hexatriene into 1,3-cyclohexadiene, this time following what happens to *all* of the molecular orbitals. We already know about the orbitals of the triene. What are the relevant orbitals of 1,3-cyclohexadiene? Of course, the molecule contains a conjugated diene system, with the familiar four orbitals, ψ_1 to ψ_4. In addition, a new σ bond will be formed during formation of the ring from the triene. Orbitals for the σ bond will have to be included in our picture. These orbitals are the bonding orbital, σ, and the antibonding orbital σ^*. The bonding orbital of a σ bond will be lower in energy than any of the orbitals for π bonds, while the σ^* orbital will be correspondingly higher in energy than any of the other orbitals. (The orbitals of the starting hexatriene and the product cyclohexadiene are shown on page 33.)

Each orbital in the triene is either symmetric or antisymmetric around the mirror plane, and so is each orbital of the product. Furthermore, molecular symmetry around the mirror plane is maintained throughout the reaction as the end carbons of the triene rotate to form a new σ bond. As the reaction proceeds, the coefficient for each atom in each wave function changes, and each orbital of the triene changes into an orbital of the cyclohexadiene. *Each orbital of the starting material must be converted to an orbital with the same symmetry.* (This principle is called the *conservation of orbital symmetry.*[14])

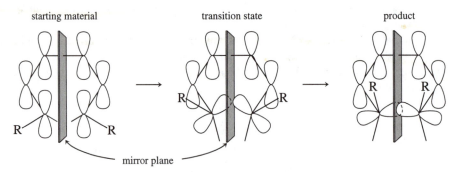

starting material   transition state   product

mirror plane

Symmetry around the mirror plane is maintained throughout the reaction

The lowest-energy orbital of the triene, ψ_1, is symmetric around the mirror plane and can thus be converted to the lowest-energy orbital of the cyclohexadiene, σ, which is also symmetric around the mirror plane. (Do not worry about the fact that ψ_1 contains six atomic orbitals while σ has only two. It's perfectly legitimate for the coefficients at some atoms to drop to zero or, going the other way from σ to ψ_1, for coefficients to increase from zero to some finite value.)

The second-lowest energy orbital of the triene, ψ_2, however, cannot be converted to the second-lowest orbital of the product, because those two orbitals have different symmetries. Instead, the second-lowest orbital of the starting material is converted into the third-lowest orbital of the product, while the third orbital of the triene is converted to the second orbital of the cyclohexadiene.

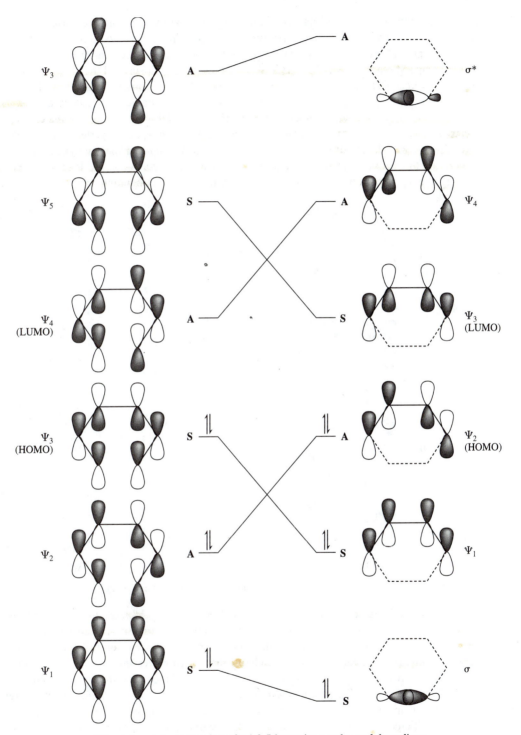

Disrotatory interconversion of a 1,3,5-hexatriene and a cyclohexadiene

The remaining three orbitals of the starting material can similarly be correlated with the three highest orbitals of the product. (However, we do not really have to be concerned with those orbitals at present. Because they contain no electrons, they will not affect the energy of the system.) The principal point to be noted is that in this disrotatory ring closure, all the occupied orbitals of the starting material are converted nicely to the lowest-energy orbitals of the product.

Now, let's try a similar disrotatory ring closure of 1,3-butadiene to cyclobutene. ψ_1 can be converted to σ, but ψ_2 cannot be converted to π, the second-lowest orbital of the cyclobutene. Instead, to conserve symmetry around the mirror plane, ψ_2 must be converted to π^*, while ψ_3 is converted to π. The result is that the filled orbitals of the starting material cannot be directly converted to bonding orbitals of the product. Instead, this process would lead to a very high-energy excited state of the product, with two electrons in an antibonding orbital.

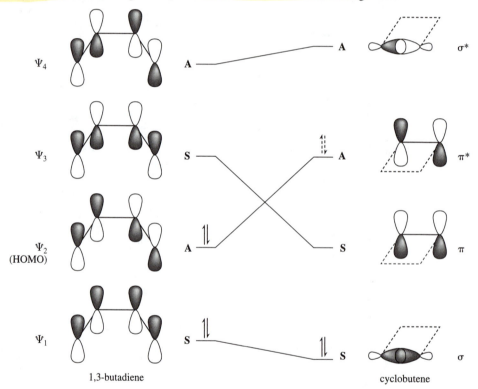

Disrotatory conversion of 1,3-butadiene to cyclobutene

Similarly, in the reverse reaction—the ring opening of a cyclobutene to a 1,3-butadiene—the electrons in the π orbital would have to be raised to a ψ_3 orbital, forming an excited state of the product. In fact, this does not appear to happen. Somewhere along the reaction path, electrons will shift from the original π orbital to the ψ_2 orbital being formed. However, this process would still require a very high activation energy—usually much higher than is required for symmetry-allowed processes.

Up to now, when discussing the symmetries of molecular orbitals, their symmetry in relation to a mirror plane was always discussed. However, molecular orbitals of polyenes have a second symmetry element. All of the orbitals are either symmetric or antisymmetric around a twofold axis of rotation. (A twofold axis of rotation can be regarded as a pin in the plane of a polyene or cycloalkene, passing through the middle of the polyene chain or through the sp^3—sp^3 bond of a cycloalkene and the bond on the other side of the ring.) If each molecular orbital were spun 180°

around the twofold axis, it would yield either an identical orbital or an orbital with all signs the opposite of what they were originally.

As we've seen, the symmetries of orbitals around the mirror plane are maintained during disrotatory ring closures (or ring openings). In contrast, as can be seen in the next figure, the mirror plane is *not* maintained as an element of symmetry during a conrotatory ring closure. On the other hand, the twofold axis of rotation *is* maintained as a symmetry element during a conrotatory ring closure but not during a disrotatory ring closure.

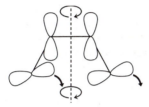

Transition state for a conrotatory reaction. Symmetry around the mirror plane is not maintained, but symmetry around the axis of rotation is maintained.

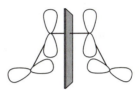

Transition state for a disrotatory reaction. Symmetry around the mirror plane is maintained, but symmetry around the axis of rotation is not maintained.

Now, let's take another look at the correlation diagram for the conrotatory ring closure of 1,3-butadiene to cyclobutene. This time, we'll show the symmetries of the orbitals in relation to since the axis of rotation.

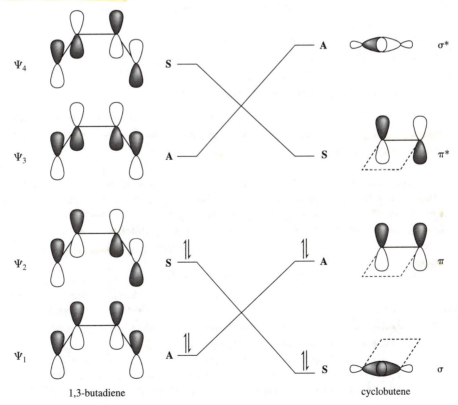

Conrotatory interconversion of 1,3-butadiene and cyclobutene

Notice that ψ_1 is antisymmetric in relation to its axis of rotation, since if it were rotated 180° around the axis, the shaded lobes would be on the bottom instead of on top. The same is true of π. However, ψ_2 and π^* are symmetric, because in each case a 180° rotation would bring shaded lobes back to the top left and bottom right of the orbital. Similarly, σ is symmetric and σ^* is antisymmetric in relation to the axis of rotation.

Therefore, the two filled orbitals of 1,3-butadienes can be directly converted to the two lower-energy (bonding) orbitals of cyclobutene, so that the conrotatory ring closure (or ring opening) is a symmetry-allowed process.

Although the fates of all molecular orbitals are depicted in correlation diagrams, the frontier orbitals are of critical importance. If the electrons in the HOMO must according to the diagram, be raised to an antibonding orbital in the product (leaving an unfilled orbital of lower energy), the reaction will be classified as forbidden.[15]

Correlation diagrams can provide elegant demonstrations of the basis for selection rules in pericyclic reactions. However, like frontier orbital theory, the correlation diagram method requires abstracting certain elements of a reacting molecule and assuming that these elements alone control its reactions. Both methods, for instance, ignore the σ bond framework that holds the π bonds of polyenes together. Furthermore, most molecules involved in electrocyclic reactions (e.g., 1,3-pentadiene rather than 1,3-butadiene) are not actually symmetric around either a mirror plane or an axis of rotation. Correlation diagram analysis must assume that the effects of substituents may be disregarded when the "fundamental symmetry" of the system is discussed. This assumption is probably reasonable for the replacement of a hydrogen atom by an alkyl group, but it might not be equally reasonable if the substituent were a nitro group or an amino group. Furthermore, there are many reactions (some to be discussed in later chapters) that follow Woodward–Hoffmann selection rules but in which neither starting materials nor products have significant elements of symmetry, so that correlation diagrams cannot be constructed. For these reactions, the frontier orbital approach is still of value.

Aromatic Rings and Aromatic Transition States

You have probably noticed that in distinguishing between $4n + 2$ and $4n$ electron systems, the Woodward–Hoffmann rule for electrocyclic reactions follows the same form as *Hückel's rule* for aromatic and antiaromatic molecules. It is easy to see that the stabilities of aromatic rings can be predicted by considering the frontier orbitals of open-chain polyenes. Bringing the ends of a polyene chain together results in an attractive interaction if the chain holds $4n + 2\pi$ electrons. This attraction results in the cyclic system's being lower in energy than the open-chain analog. In contrast, in chains of $4n$ π electrons, the HOMO is antisymmetric, and the ends of the frontier orbital would repel each other if brought together. Thus, the ring of $4n$ electrons would be antiaromatic—that is, it would be even higher in energy than the open-chain polyene.

Instead of using the Woodward–Hoffmann rules to account for the existence of aromatic and antiaromatic rings, we can turn the argument around and account for the Woodward–Hoffmann rules in terms of aromaticity and antiaromaticity.[16] Disrotatory electrocyclic processes of $4n + 2$ electrons, in which the lobes of orbitals on the same sides of polyene chains attract each other, can be said to have aromatic *Hückel transition states*.

What about conrotatory processes? Suppose a chain of conjugated double bonds containing $4n$ electrons—a chain of eight double bonds, for instance—were given a single twist, so that one end of the chain was turned upside down. Now the ends of the HOMO would be *in* phase, and the ends of the chain would attract each other if they were brought together. The energy of the system would be lowered compared to the open chain. If a hydrogen atom were removed from each end carbon and the carbons were joined, a new aromatic ring system with a single twist would be formed. In mathematical topology, a ring with a single twist is called a *Möbius strip.**

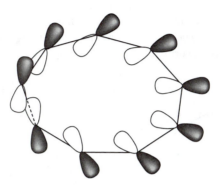

A Möbius ring of *p*-orbitals

A Möbius ring of *p* orbitals containing 4*n* electrons should theoretically be aromatic. However, no such compound has ever been prepared. The reduced overlap between the *p* orbitals caused by the twist in the chain, the steric strain caused by abnormal bond angles, and the possibility of electrocyclic reactions occurring along parts of the polyene chain make it unlikely that one will ever be prepared.

The essential point about Möbius strips, from our point of view, is that if one were formed from a polyene chain, the top lobe on one end of the chain would interact with the bottom lobe on the other. Because this is what happens in a conrotatory reaction, we can describe conrotatory reactions as having *Möbius transition states.*

Thus, one way to phrase the Woodward–Hoffmann rules for electrocyclic reactions is that the *reactions are allowed if they proceed by aromatic transition states and are forbidden if they proceed by antiaromatic transition states. For systems of 4n + 2 electrons, Hückel transition states are aromatic; for systems of 4n electrons, Möbius transition states are aromatic.*[16]

Of course, in theory, a chain of *p* orbitals might have more than one twist. A second twist would bring the original top lobes back on top again. In general, a ring with an even number of twists (that is, an even number of phase inversions of its orbitals) would be a Hückel ring; a ring with an odd number of twists (phase inversions) would be a Möbius ring. We are unlikely ever to prepare polyenes with more than one twist in the ring, but as we'll see later, the concept of counting phase inversions can be useful in analyzing some types of pericyclic reactions.

---

*Möbius strips have fascinating characteristics. Because they have only one surface, a line drawn along the outside of the strip the line will continue on the inside. If the strip is cut lengthwise into two, a single ring twice as long as the original ring will be formed.

2.4   ELECTROCYCLIC REACTIONS WITH ODD NUMBERS OF ATOMS

Theoretical Predictions

The Woodward–Hoffmann rules can be applied to electrocyclic ring openings and closing of rings with odd numbers of atoms (which may be cations, anions, or radicals) as well as the even-numbered systems we've just examined.

To begin with, let's look at the correlation diagram for a disrotatory ring closure of a three-atom chain of p orbitals. The resulting three-membered ring will have a new σ bond, of course, as well as a nonbonded (n) orbital.

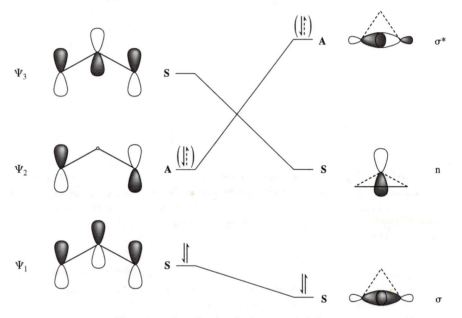

Disrotatory ring closure of a three atom chain

Whether this reaction is allowed depends on the number of electrons in the system. If the starting material is a cation, with only two electrons, the reaction is allowed, because the ψ_1 and σ orbitals have the same symmetry. However, if the chain is an anion, the diagram indicates that an excited state, with two electrons in a high-energy orbital, would be produced. The disrotatory ring closure of the anion is therefore forbidden. If the starting material is a radical, with three electrons, the diagram again indicates that the product would be formed in an excited state. However, only one electron would be located in a high-energy orbital. Therefore, although the cyclization of the radical is forbidden, the energy barrier to the reaction should be lower than that for the anion.

In contrast, a correlation diagram employing symmetries in relation to the twofold axis of rotation would show that a conrotatory process is allowed for the anion but is forbidden for the cation.

Of course, the same conclusions could be obtained by the frontier orbital approach, although the relative energy barriers to radical and anion cyclizations would not be as easily estimated. Since Hückel's rule applies only to even numbers of electrons, we cannot directly classify the electrocyclic reactions of free radicals as proceeding by aromatic or antiaromatic transition states. However, the aromatic transition-state approach can still be applied if it is assumed that free radicals will follow the same rules as anions.

Reactions of Cations and Anions

The simplest possible electrocyclic reactions would be the cyclization of an allyl cation to form a cyclopropyl cation or the ring opening of a cyclopropyl cation to form an allyl cation. Since these reactions would be two-electron processes, they should proceed by disrotatory paths.

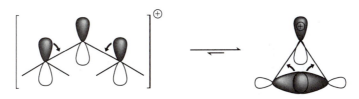

Cyclopropyl cations are very unstable, as a result of ring strain and the fact that the empty orbital of a cyclopropyl cation would have appreciable *s* orbital character. Therefore, there are no known examples of allylic cations cyclizing to form cyclopropyl cations. Indeed, there is little evidence that cyclopropyl cations are ever actually formed in any reaction. Instead, reactions that would be expected to yield cyclopropyl cations proceed with simultaneous ring opening to form allylic cations.

Although cyclopropyl cations may not actually be formed as discrete intermediates, the ring openings of cyclopropyl derivatives to allylic cations do usually proceed in disrotatory fashion. The structures of cations **6a** to **6c**, formed by reaction of **5a** to **5c** with antimony penta-fluoride at −100°C, have been determined by NMR spectroscopy.[17] As predicted, each ring opening proceeds in a disrotatory manner. (This is shown on page 39.) At higher temperatures, cation **6a** isomerizes to **6b**, which in turn, as the temperature is raised, isomerizes to the least-crowded cation, **6c**.

(Molecular orbital calculations show that of the two possible modes of disrotatory ring openings of cyclopropyl derivatives, the preferred path will be that in which the substituents *trans* to the leaving group rotate apart.[1] This results in the bonding lobes of the cyclopropyl bond being positioned *trans* to the leaving group, where they can participate in an "S_N2-like" displacement of the leaving group. A good deal of experimental evidence, including the reactions of **5a**, **5b**, and **5c**, are in agreement with this analysis.[18])

In contrast to cyclopropyl cations, cyclopropyl anions appear to have finite lifetimes.* However, they *can* open to form allylic anions. The different stereoisomers of allylic anions are easily interconverted, so that it is usually not possible to determine from the reaction products whether the ring openings proceeded by conrotatory or disrotatory paths. However, opening of the cyclopropyl ring in anion **7**, which can proceed by an allowed conrotatory path, is about 10^4 times as fast as the ring openings of anions **8** and **9**, which must proceed by forbidden disrotatory paths. It has been estimated that the activation energy for a disrotatory ring opening is at least 5.9 kcal/mol greater than for a conrotatory ring opening.[19]

---

*Many studies of cyclopropyl anions deal with their lithium "salts," which have only limited anionic character. However, there does not appear to be any reason to believe that conclusions drawn from the lithium compounds will not apply to more ionic derivatives.

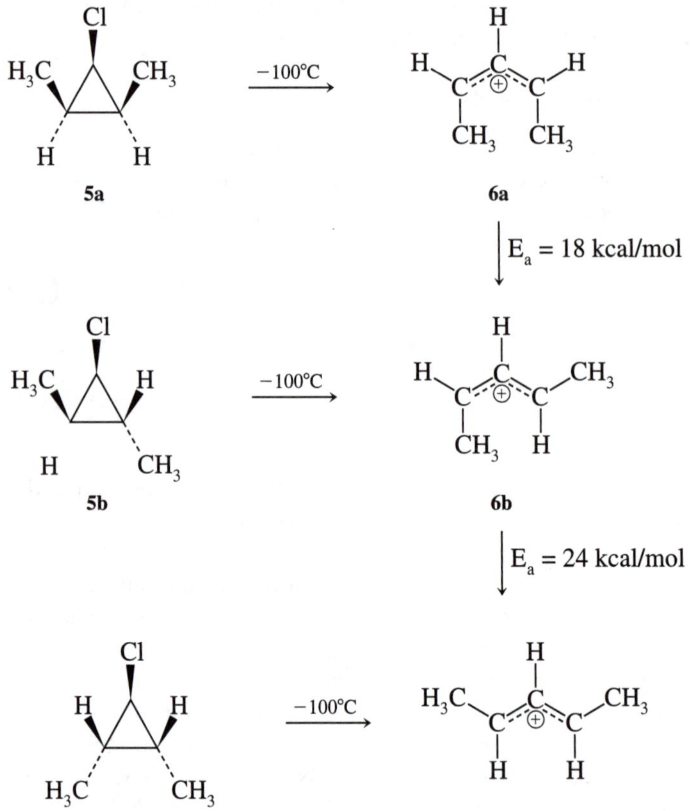

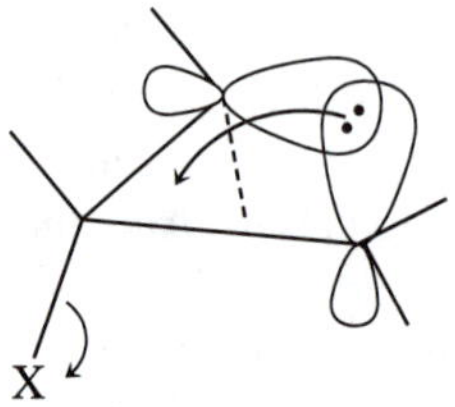

5a →(−100°C)→ 6a

$E_a = 18$ kcal/mol

5b →(−100°C)→ 6b

$E_a = 24$ kcal/mol

5c →(−100°C)→ 6c

Preferred transition state for ring opening of a cyclopropyl halide

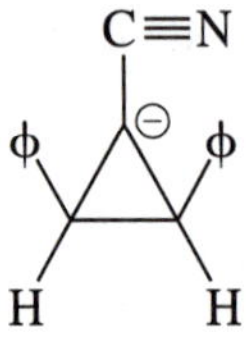

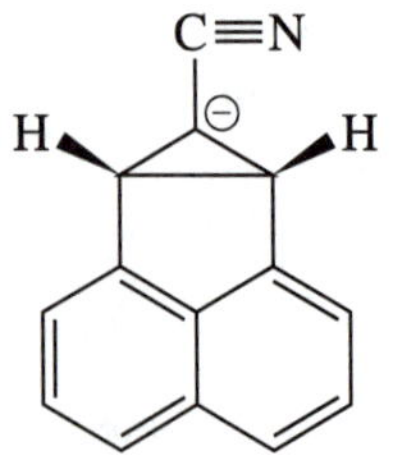

8 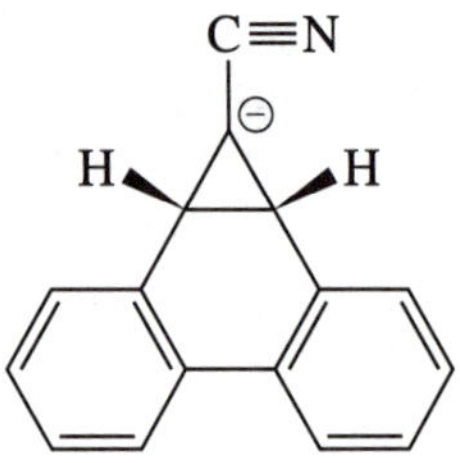   9

Several studies have examined electrocyclic reactions of pentadienyl cations and anions. Pentadienols **10a** and **10b** undergo stereospecific cyclizations in acid solutions to yield cations resulting from conrotatory ring closures,[20] as would be expected of four-electron systems.

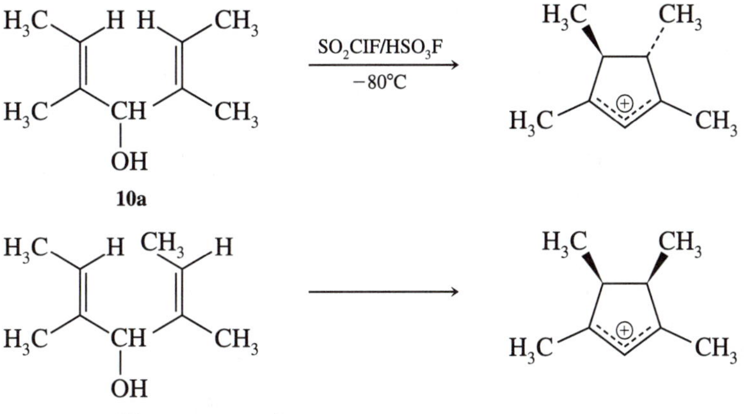

(16)

In contrast to pentadienyl cations, the pentadienyl anion **11** cyclizes in disrotatory fashion to yield a *cis*-fused bicyclic anion.[21]

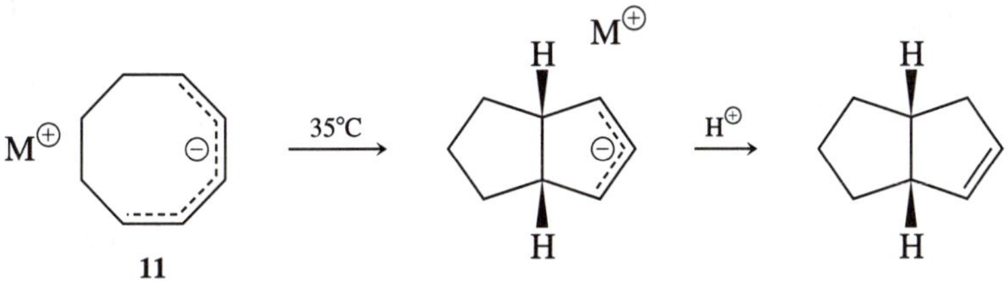

(17)

Formation and Cyclization of Dipolar Molecules

Aziridines (three-membered ring amines) are isoelectronic with cyclopropyl anions. The three-membered rings will open on heating to form dipolar molecules. (The dipoles are members of a class generically known as *1,3-dipoles*, although their principal resonance forms normally have their charges on adjoining atoms, rather than on atoms 1 and 3.)

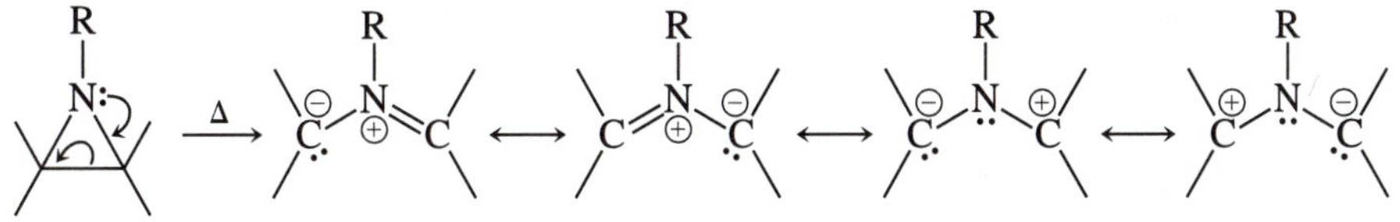

(18)

Studies by Rolf Huisgen at the University of München demonstrated that the 1,3-dipoles from ring-opening reactions of aziridines can undergo addition reactions with molecules such as dimethyl acetylenedicarboxylate. The structures of the products demonstrated that the ring openings proceeded by allowed conrotatory paths.[22]

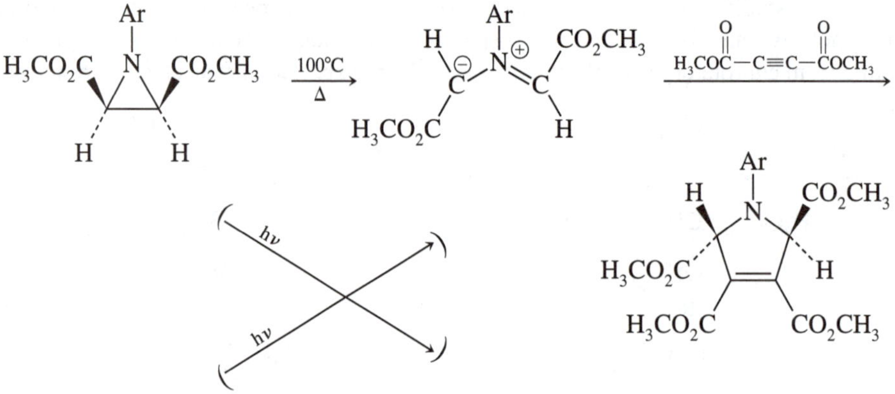

(19)

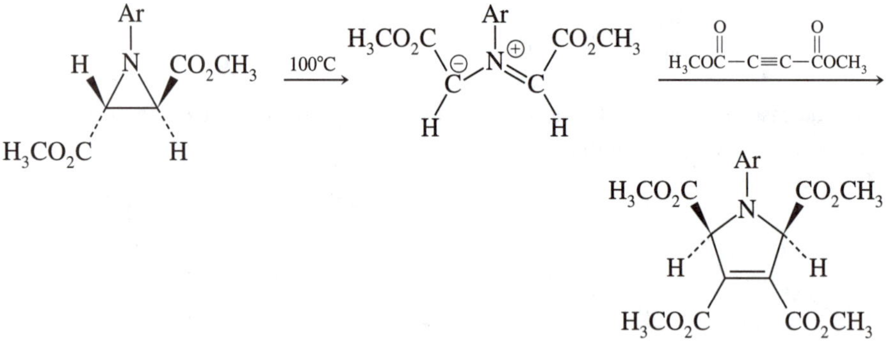

(20)

Cleavage of the carbon–carbon bonds of oxirane (three-membered ether) rings requires higher temperatures than the similar reactions of aziridines. Experimental studies indicate that disrotatory ring openings of oxiranes require activation energies about 6 kcal/mol higher than conrotatory ring openings,[23] and in some cases, such as the reaction of oxirane **12**, the ring openings were shown to proceed stereospecifically by conrotatory paths.[24]

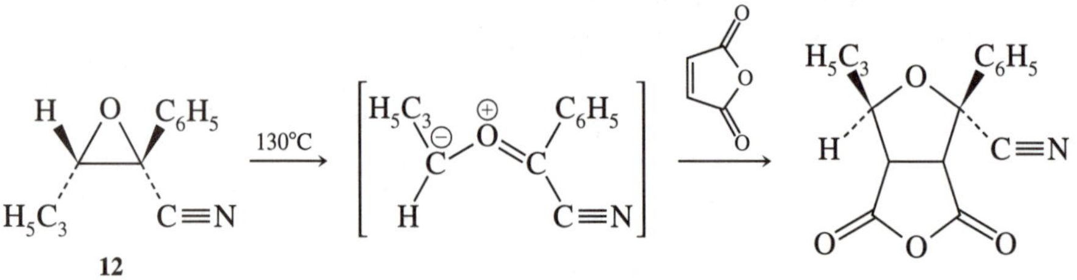

(21)

However, in other reactions, mixtures of stereoisomers are obtained, possibly because the intermediate 1,3-dipoles cannot maintain their geometries, or because diradical processes compete with the concerted electrocyclic paths.[23]

2.5 PHOTOCHEMICAL CYCLIZATIONS

Photochemical Reactions

Conjugated polyenes can undergo cyclization reactions on irradiation with ultraviolet light as well as on heating.

Photochemical cyclizations can be valuable synthetic tools—particularly in the synthesis of four-membered rings. As we've seen, cyclobutenes usually cannot be prepared by thermal ring closures of butadienes, because the strained cyclobutene rings are usually less stable than the dienes. However, when conjugated dienes are irradiated with UV light at common wavelengths (above 220 nm), much more of the light is absorbed by the polyenes than by the cyclobutenes. The rate at which a polyene is converted to a cyclobutene on photoirradiation is therefore usually greater than the rate at which the cyclobutene ring reopens—at least until most of the polyene has reacted. Useful yields of cyclobutenes may thus be formed by photochemical cyclizations of butadienes, particularly when the two double bonds are fixed *cis* to each other, as in diene **13**.[25]* If, however, cyclobutenes are irradiated by light in the far UV region (below 200 nm), where unconjugated double bonds absorb light, they can be largely reconverted to butadienes.[†]

$$\xrightarrow{h\nu} \tag{22}$$

13

In order to apply the Woodward–Hoffmann rules to photochemical cyclizations, we have to identify the HOMOs involved in the reaction. When a molecule absorbs light, an electron is normally raised from the highest occupied orbital to the next-highest energy orbital. A molecule

---

*Molecules with double bonds in a *cis* relationship around the single (σ) bonds connecting the double bonds are said to be in *s-cis* forms. Most open-chain dienes exist primarily in *s-trans* conformations.

s-*cis*   s-*trans*

[†]Ultraviolet light is very energetic. Absorption of a "mole" of photons of light at 273 nm, for example, will raise the energy of the absorbing compound by 105 kcal/mol. Thus, very high-energy products can be obtained from photochemical reactions. Irradiation of 1,3-butadiene, for instance, can result in the formation of bicyclobutane as well as cyclobutene, and irradiation of benzene gives traces of not only Dewar-benzene but also the even higher-energy products prismane and benzvalene.

bicyclobutane

prismane   benzvalene

that has undergone this process is said to be in an *excited state*, in contrast to the unexcited molecule, which is in the *ground state*. The molecule will only absorb light with a wavelength that corresponds to the energy necessary to raise the molecule from the ground state to the excited state. It is assumed here that the photoexcitation state of a π bond system will yield the π^* state. (However, as described in Chapter 10, use of very high energy, low wavelength, irradiation may produce another type of excited state.)

The process of raising an electron to a higher-energy orbital does not change the spin of the electron, so that the excited state, like the ground state, is a *singlet*, with all electron spins paired. However, once an electron is removed from a doubly occupied orbital, where it is constrained to have a spin paired to that of the other electron in the orbital, it is theoretically free to change its spin to be parallel to that of the other unpaired electron. If it did, the molecule would be in a *triplet* state. For most polyenes (in the absence of heavy atoms), *intersystem crossing*—the conversion of singlet states to triplets, or vice versa—is usually slower than other reactions of their excited states. Therefore, most photochemical cyclization reactions appear to take place from singlet states.

Since photochemical excitation of a molecule normally raises an electron from the HOMO to the LUMO, the LUMO of the ground state becomes the HOMO of the excited state of the molecule. And, because each molecular orbital of a polyene is opposite in symmetry to the orbital below it, the HOMO of the excited state of a polyene must be opposite in symmetry to the HOMO of the ground state. Thus, the frontier orbital approach leads to a very simple rule for photochemical electrocyclic reactions: *the stereochemistry of photochemical cyclizations should be opposite to that of thermal cyclizations.* Photochemical cyclizations of systems of $4n$ electrons should be *disrotatory*, and photochemical cyclizations of systems of $4n + 2$ reacting electrons should be *conrotatory*.

The correlation diagram approach leads to the same conclusion as the frontier orbital approach. Photochemical conversion of butadiene to cyclobutene by a conrotatory process would require raising an electron to the very high-energy σ^* orbital of cyclobutene (or at least to a high-energy transition state). The molecule formed, according to the diagram, would not be in the lowest-energy excited state but would be in an even higher-energy state. In contrast, the disrotatory reaction would yield the lowest-energy excited state of cyclobutene.

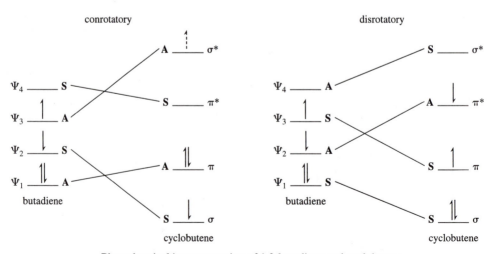

Photochemical interconversion of 1,3-butadiene and cyclobutene

Stereochemistry of Photochemical Electrocyclic Reactions

Photochemical electrocyclic reactions usually proceed according to the predictions of the Woodward–Hoffmann rules.

Precalciferol (an isomer of calciferol, vitamin D_2), for instance, undergoes an electrocyclic ring closure on UV irradiation to form the predicted product of conrotatory cyclization.[26] In contrast, on heating, precalciferol yields only products of disrotatory ring closure.

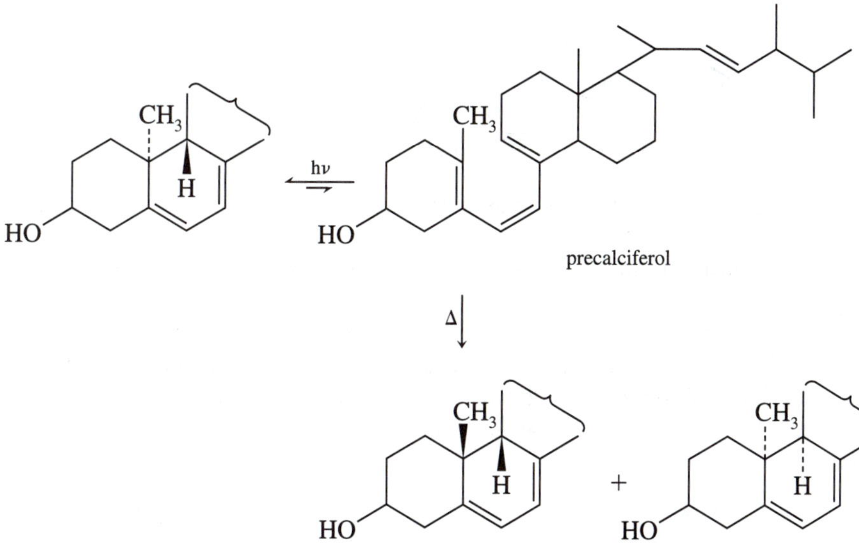

precalciferol (23)

A similar distinction between photochemical and thermal reactions is shown by all *cis*-cyclodecapentaene.* On photoirradiation, all *cis*-cyclodecapentaene (**13**) yields *trans*-9,10-dihydronaphthalene, resulting from a conrotatory ring closure, while on heating the *cis*-isomer is obtained.[27]

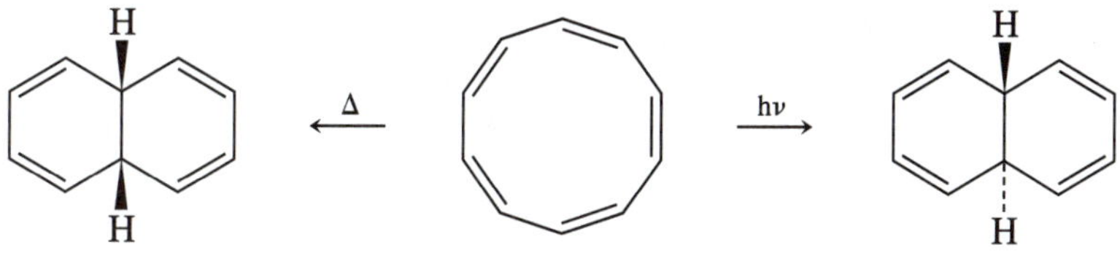

(24)

13

---

*All *cis*-cyclodecapentaene, despite having 10π electrons in a closed, conjugated ring, does not have aromatic properties, because the ring is twisted into a tub-like structure in order to maintain normal bond angles. The double bonds in the ring can hardly all be considered to be conjugated, because several π bonds in the molecule are essentially at right angles (*orthogonal*) to others.

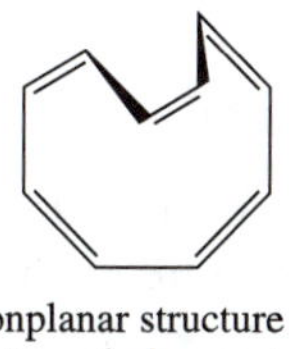

nonplanar structure of
all *cis*-cyclodecapentaene

A remarkable type of photocyclization takes place when *cis*-stilbenes are photoirradiated in the presence of air. Under those conditions, phenanthrene derivatives are formed in high yields. It was long assumed that the initial photochemical step was cyclization of the stilbene to form a dihydrophenanthrene, which is then oxidized to a phenanthrene, as shown in Eq. 25. (The huge amount of energy that can be imparted on photoirradiation is dramatically illustrated by this reaction, since the first step simultaneously converts *two* aromatic rings to nonaromatic rings.)

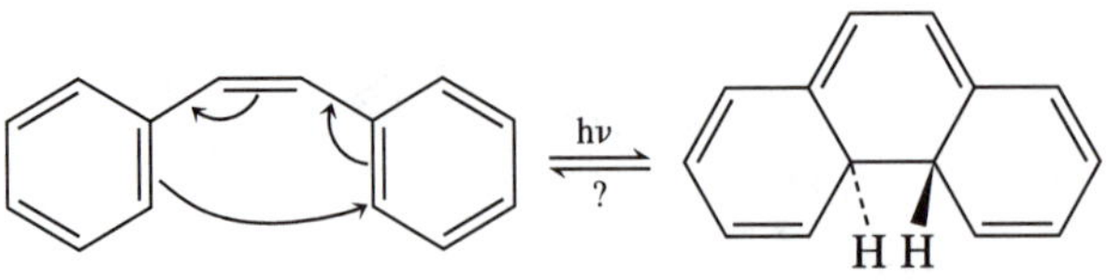

$$\text{(25)}$$

Conversion of *cis*-stilbenes to dihydrophenanthrenes may be considered to be 6π, 10π, or 14π electron cyclizations. No matter how the electrons are counted, this should result in conrotatory ring closure to form *trans*-dihydrophenanthrenes. However, it was not easy to test this prediction, because the dihydrophenanthrene intermediates, if not rapidly oxidized to phenanthrenes, were quickly reconverted to stilbenes. The problem of determining the geometries of the intermediates was solved in an ingenious manner by irradiation of diethylstilbestrol (**14**). The intermediate dienol formed from this reaction tautomerized to the diketone form, which was stable, so that its geometry could be determined. The predicted *trans* isomer was indeed formed.[28]

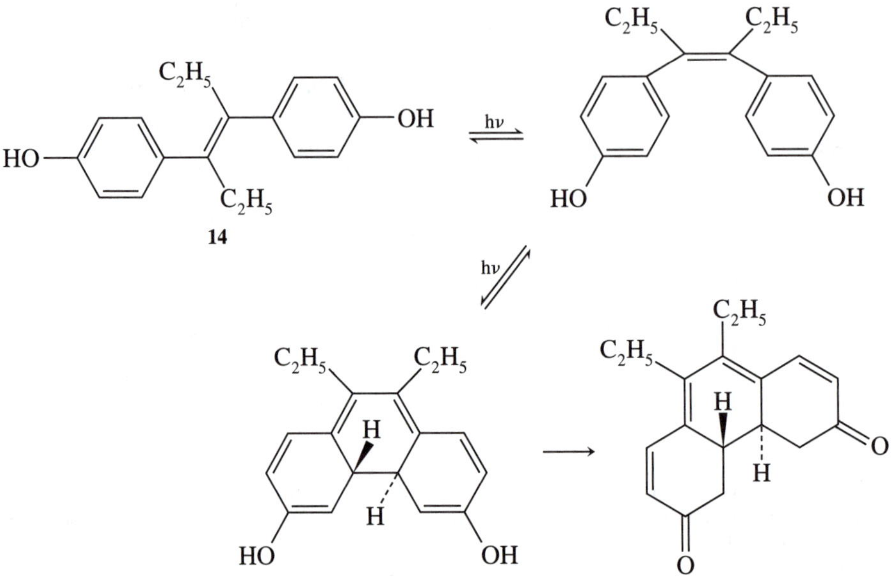

$$\text{(26)}$$

Nonstereospecific Ring Openings

1,3-Butadienes cyclize on photoirradiation with UV light with wavelengths above 220 nm to form the predicted disrotatory ring closure products (Eqs. 27 and 28).[29]

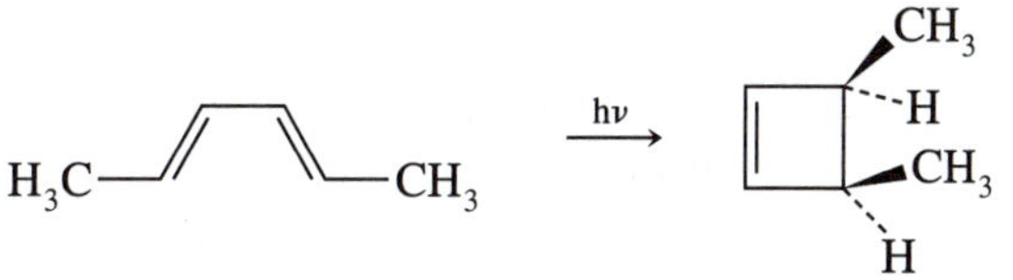

(27)

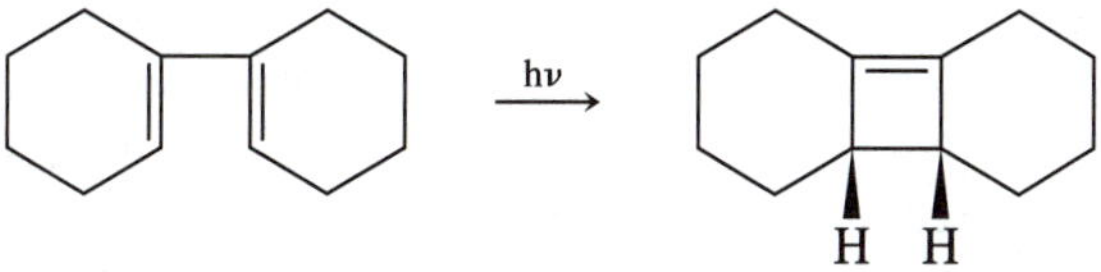

(28)

However, despite the predictions of the Woodward–Hoffmann rules, irradiation of cyclobutene derivatives by far UV light (wavelengths below 200 nm) gave rise to mixtures of stereoisomeric products. The products of the "forbidden" conrotatory ring openings were actually the major isomers, as in Eqs. 29 and 30. Slightly different ratios of stereoisomers were obtained at different wavelengths.[30]

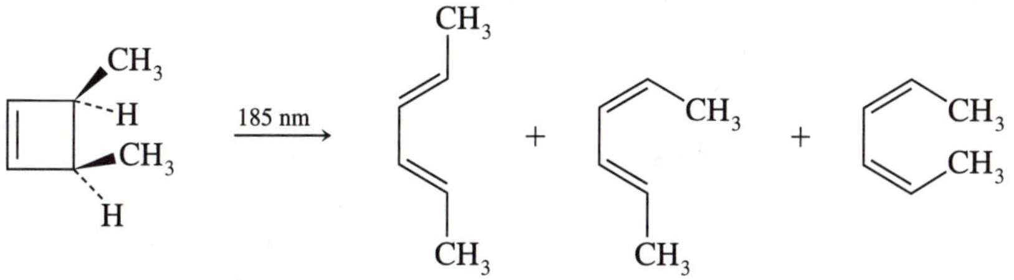

(29)

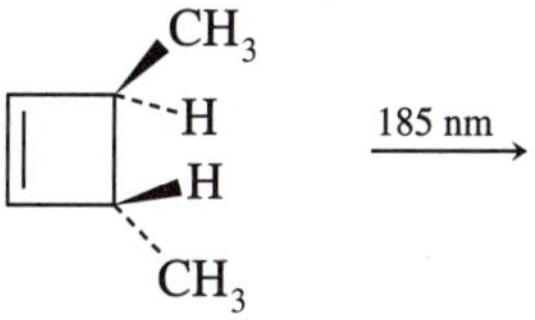

(30)

There are several possible reasons why photochemical ring openings, employing such high-energy radiation, might not follow predictions from orbital symmetry conversion rules. Theoretical calculations indicate that at least two different mechanisms are involved, depending on the wavelength of light employed.[31] At the lowest wavelengths, it appears the reaction proceeds by direct excitation of the double bond to the *second* (higher-energy) π^* state, which has the opposite symmetry from the lowest-energy excited state. Thus, the reaction would be symmetry allowed. At somewhat longer wavelengths, the photochemical energy may be converted into rapid bond rotations and vibrations. Thus, instead of an electronically excited state, the cyclobutene rings would be converted to high-energy ground states. The "hot" molecules would then undergo thermally allowed conrotatory ring opening.

It thus appears that the Woodward–Hoffmann rules form useful guides about what can be expected of the geometry of photochemical electrocyclic reactions. However, the possible complexities in photochemical reactions are such that exceptions to the Woodward–Hoffmann rules may occur.

REFERENCES

[1] For exceptions to this rule, see L.N. Misra, A. Chandra, and R.S. Thakur, *Tetrahedron Lett., 30,* 1437 (1989); M.M. Rahman, B.A. Secor, K.M. Morgan, P.R. Shater, and D.M. Lemal, *J. Am. Chem. Soc., 112,* 5986 (1990).

[2] K. Somekawa, Y. Okumura, K. Uchida, and T. Shimo, *J. Hetero. Chem., 25,* 731 (1988).

[3] R.B. Woodward and R. Hoffmann, *J. Am. Chem. Soc., 87,* 395 (1965).

[4] G.A. Doorakian and H.H. Freedman, *J. Am. Chem. Soc., 90,* 5310, 6896 (1968).

[5] R. Huisgen, A. Dahmen, and H. Huber, *J. Am. Chem. Soc., 89,* 7130 (1967). E.N. Marvell and J. Seubert, *J. Am. Chem. Soc., 89,* 3377 (1967).

[6] A.B. Buda, Y. Wang, and K.N. Houk, *J. Org. Chem., 54,* 2264 (1989); (b) R. Hayes, S. Ingham, S.T. Saengchantrara, and T.W. Walker, *Tetrahedron Lett., 32,* 2953 (1991); (c) F. Binns, R. Hayes, S. Ingham, S.T. Saengchantara, R.W. Turner, and T.W. Wallace, *Tetrahedron Lett., 33,* 883 (1992); (d) S. Niwayama and K.N. Houk, *Tetrahedron Lett., 33,* 883 (1992); (e) E. Piers and K.A. Ellis, *ibid., 34,* 1875 (1993).

[7] R. Bronton, H.M. Frey, D.C. Montaque and I.D.R. Stevens, *Trans. Faraday Soc., 62,* 659 (1966).

[8] R. Criegee and H.G. Reinhardt, *Chem. Ber., 101,* 102 (1968).

[9] E.E. van Tamelen, *Acc. Chem. Res., 5,* 186 (1972).

[10] I. Frank, S. Grimme, and S.D. Peyerimhoff, *J. Am. Chem. Soc., 116,* 5949 (1994); (b) activation energies for disrotatory ring openings of cyclobutenes appear to be at least 15 kcal/mol higher than for the allowed conrotatory processes: J.I. Braumann and W.C. Archie, *J. Am. Chem. Soc., 94,* 4262 (1972). For additional references, see J. Breulet and H.F. Schaefer III, *J. Am. Chem. Soc., 106,* 1221 (1984).

[11] N.M. Weinschenker and F.D. Greene, *J. Am. Chem. Soc., 90,* 506 (1968).

[12] R.P. Johnson and K.J. Daoust, *J. Am. Chem. Soc., 118,* 7381 (1996).

[13] H.C. Longuet-Higgins and E.W. Abrahamson, *J. Am. Chem. Soc., 87,* 2045 (1965).

[14] R.B. Woodward and R. Hoffmann, *The Conservation of Orbital Symmetry.* Academic Press, New York (1971).

[15] For further discussion of the construction of correlation diagrams, see Ref 14 (Woodward & Hoffmann pp. 10–36), or T.L. Gilchrist and R.C. Storr, *Organic Reactions and Orbital Symmetry.* Cambridge University Press, London (1972).

[16] See M.J.S. Dewar, *Tetrahedron Suppl., 8,* 75 (1966); H.E. Zimmerman, *Acc. Chem. Res., 4,* 272 (1971).

[17] P.V.R. Schleyer, T.M. Su, M. Saunders, and J.C. Rosenfeld, *J. Am. Chem. Soc., 91,* 5174 (1969).

[18] G.H. Whithham and M. Wright, *Chem. Comm., 294* (1967); C.H. De Puy, *Acc. Chem. Res., 1,* 33 (1968).

[19] M. Newcomb and W.T. Ford, *J. Am. Chem. Soc., 96,* 2968 (1974).

[20] N.W.K. Chiu and T.S. Sorensen, *Can. J. Chem., 51,* 2776 (1973).

[21] L.H. Slaugh, *J. Org. Chem., 32,* 108 (1967); R.B. Bates and D.A. McCombs, *Tetrahedron Lett.,* 977 (1969).

[22] R. Huisgen, W. Scheer, and H. Huber, *J. Am. Chem. Soc., 89,* 1753 (1967).

[23] R. Huisgen, *Angew. Chem. Int. Ed. Engl., 16,* 572 (1977).

[24] M.S. Medimagh and J. Chuche, *Tetrahedron Lett.,* 793 (1977).

[25] D.H. Aue and R.N. Reynolds, *J. Am. Chem. Soc., 95,* 2027 (1973).

[26] E. Havinga and J.L.M.A. Schlattman, *Tetrahedron, 16,* 146 (1961).

[27] S. Masamune and R.T. Seidner, *Chem. Comm.*, 542 (1969); S. Masamune, *Acc. Chem. Res., 5,* 272 (1972).

[28] T.D. Doyle, N. Filipescu, W.R. Benson, and D. Banes, *J. Am. Chem. Soc., 92,* 6371 (1970).

[29] W.G. Dauben, R.G. Cargill, R.M. Coates, and J. Saltiel, *J. Am. Chem. Soc., 88,* 2742 (1966).

[30] K.B. Clark and W.J. Leigh, *J. Am. Chem. Soc., 109,* 6086 (1987); W.G. Dauben and J.E. Haubrich, *J. Org. Chem., 53,* 600 (1988); W.J. Leigh, K. Zheng, and K.B. Clark, *Can. J. Chem., 68* (1990); *J. Org. Chem., 56,* 1574 (1991).

[31] W.J. Leigh, and B.H.O. Cook, *J. Org. Chem., 64,* 5256 (1999).

PROBLEMS

2.1 Explain why each of the following reactions is, or is not, allowed to occur as a concerted process, if the starting material is heated and if the starting material is subjected to photoirradiation.

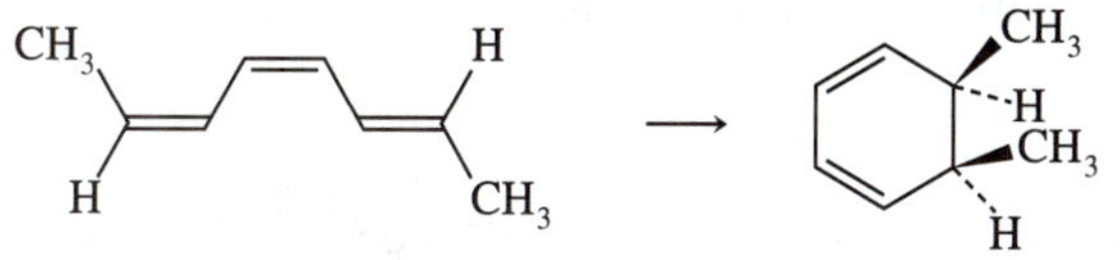

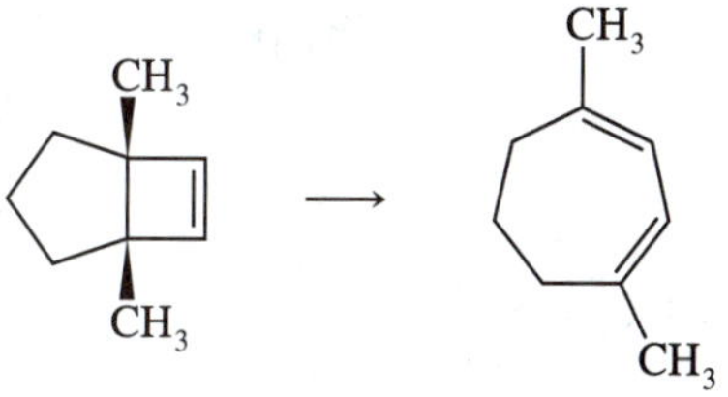

2.2 Explain why the equilibrium point in the following reaction is different if the starting material is heated or if the starting material is subjected to photoirradiation. Predict the major component of the reaction mixture in each case.

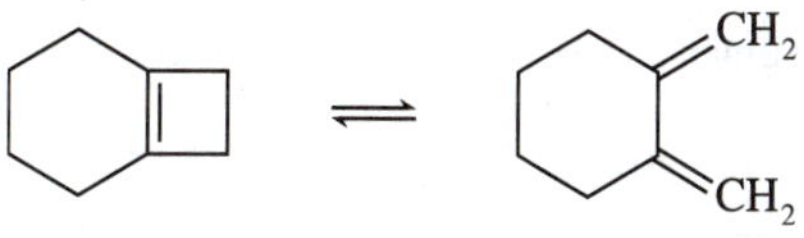

2.3 Draw correlation diagrams for conrotatory and disrotatory paths for reactions (a) and (b). Label the HOMOs and LUMOs of the starting materials. Predict the geometry of the product of each reaction.

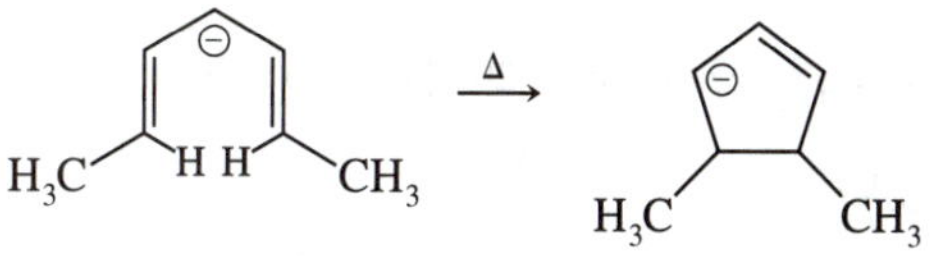

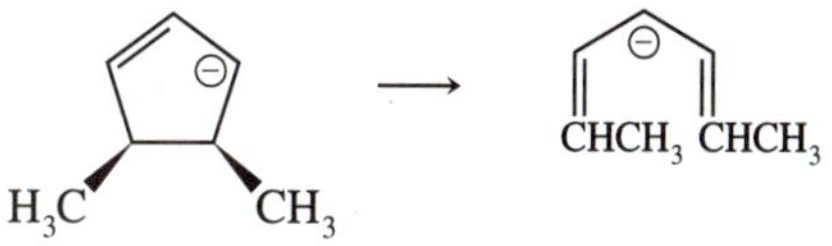

2.4 Write reasonable mechanisms for the following reactions, and indicate the expected geometries of the product.

(a)

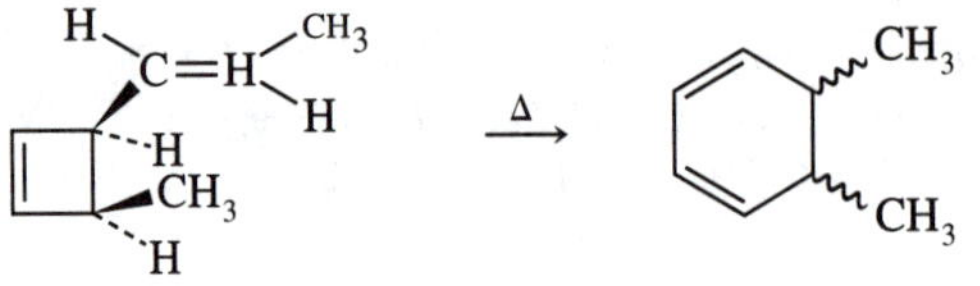

(b)

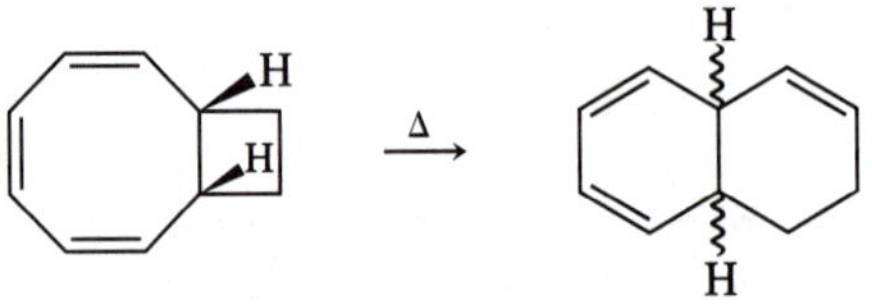

(S.W. Staley and T.J. Henry, *J. Am. Chem. Soc.*, *93*, 1294 [1971].)

(c)

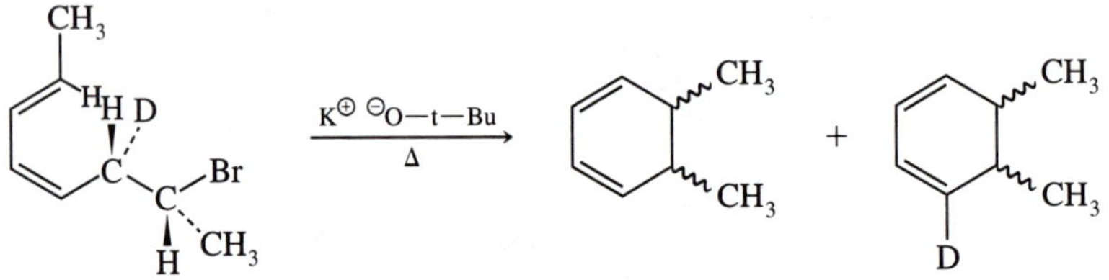

(d)

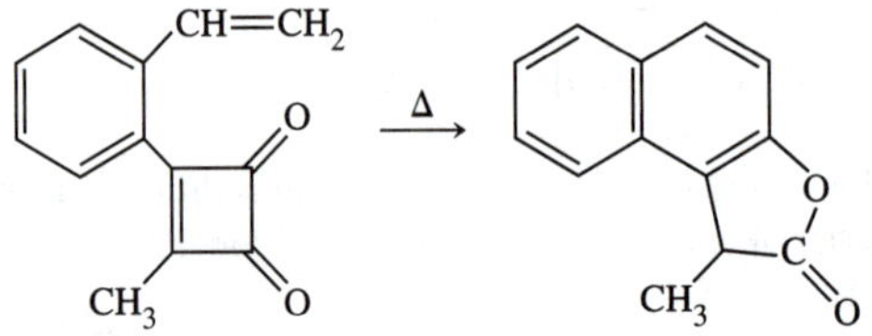

(M.J. Heilman and H.W. Moore, *Tetrahedron Lett.*, *39*, 3643 [1998].)

(e)

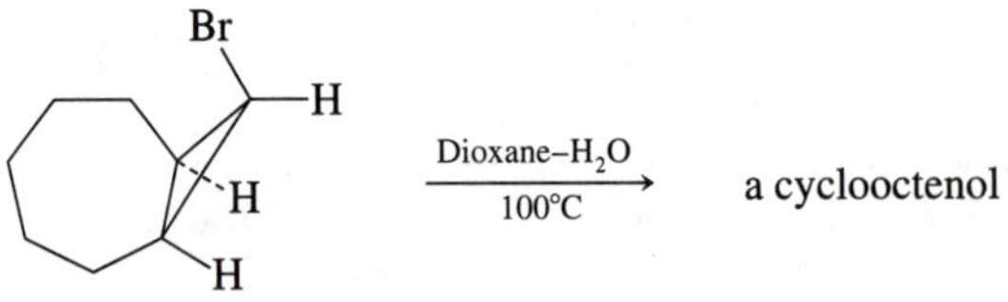

(G.H. Whitham and M. Wright, *Chem. Comm.*, 294 [1967].)

(f)

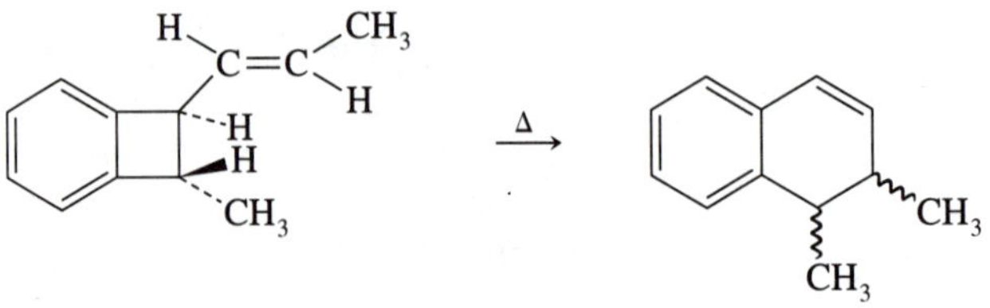

(Adapted from T. Hamaura et al., *Org. Lett.*, *4*, 1675 [2002].)

2.5 Write reasonable mechanisms for the following reactions, employing electrocyclic steps in each mechanism.

(a)

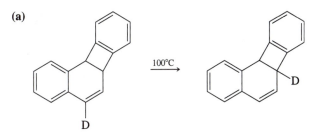

(K.R. Motion, I.R. Robertson, J.T. Sharp, and M.D. Walkinshaw, *Perkin Trans., 1,* 1709 [1992].)

(b)

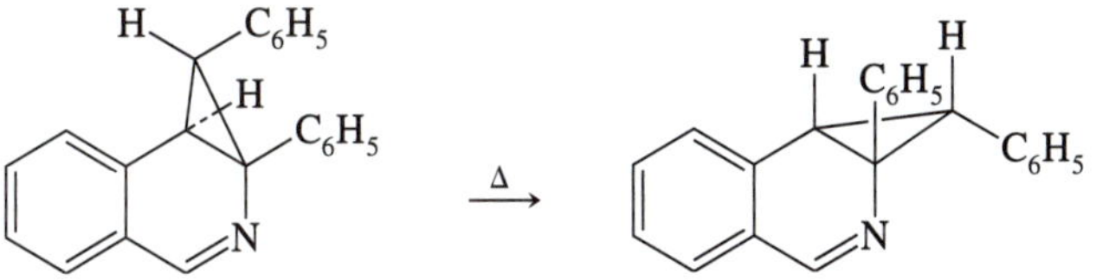

(W. Grime, J. Lex, and T. Schmidt, *Angew. Chem. Int. Ed. Engl., 26,* 1268 [1987].)

(c)

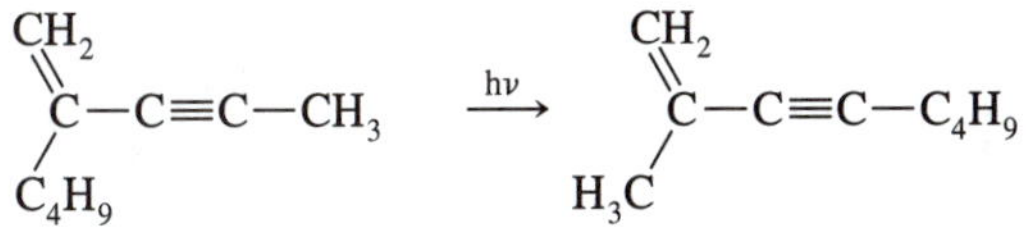

(M. Zhan, J.D. Rica, M.M. Kirchoff, K.M. Phillips, L.M. Cuff, and R.P. Johnson, *J. Am. Chem. Soc., 115,* 1216 [1993].)

Cycloaddition and Cycloreversion Reactions

3.1 INTRODUCTION

Cycloadditions are reactions in which at least two new bonds are formed simultaneously, so as to convert two or more open-chain molecules or portions of molecules into rings. The reverse of cycloadditions are retrocycloadditions, or *cycloreversions*. Cycloadditions and cycloreversions proceeding via transition states in which the electrons involved in the reactions form continuous rings are part of the general class of pericyclic reactions.

Cycloadditions and cycloreversions can be classified according to the number of electrons in each reacting molecule that change locations during the reactions. This classification scheme is illustrated by schematic reactions **a** through **d** below, in which not only the number but the types of electrons are identified in each case.

(a)

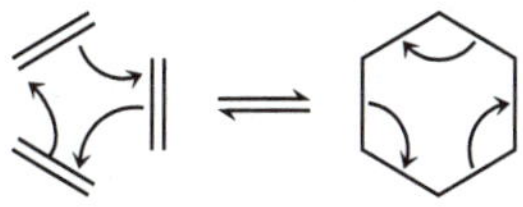

$[_\pi 2 + _\pi 2 + _\pi 2]$ $[_\sigma 2 + _\sigma 2 + _\sigma 2]$

(b)

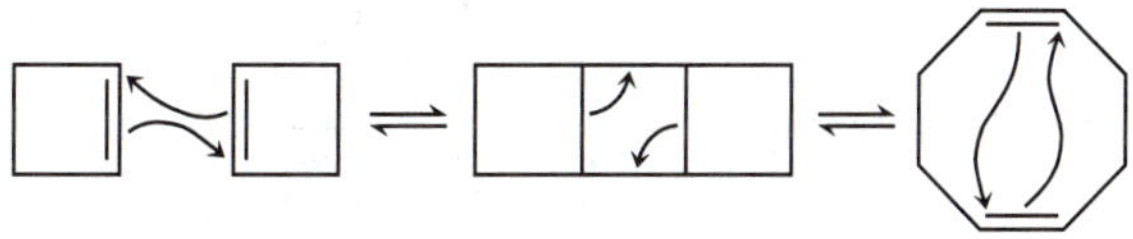

$[_\pi 2 + _\pi 2]$ $[_\sigma 2 + _\sigma 2]$ $[_\pi 2 + _\pi 2]$

(c)

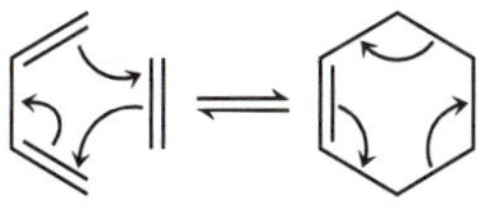

$[_\pi 4 + _\pi 2]$ $[_\sigma 2 + _\pi 2 + _\sigma 2]$

(d)

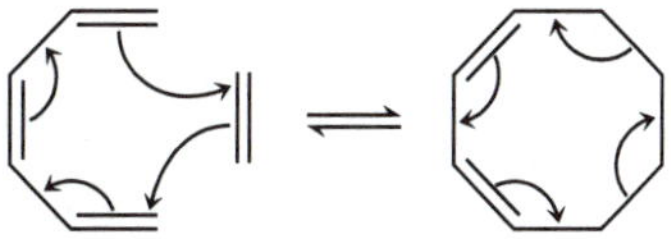

$[_\pi 6 + _\pi 2]$ $[_\sigma 2 + _\pi 4 + _\sigma 2]$

In a more general sense, we can regard *all* reactions with transition states involving closed cyclic arrays of electrons as cycloaddition and cycloreversion reactions, even if new rings are not formed during the reactions. Thus, the following reactions (**e** and **f**) may be considered to be cycloaddition or cycloreversion processes.

(e) 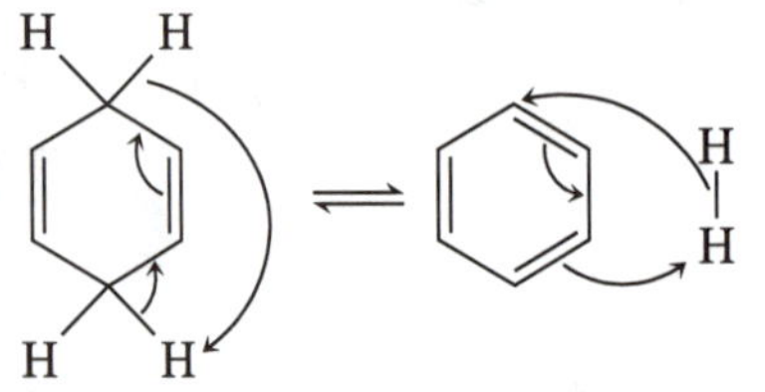 $[_\sigma 2 + _\pi 2 + _\sigma 2]$ $[_\pi 4 + _\sigma 2]$ (f) 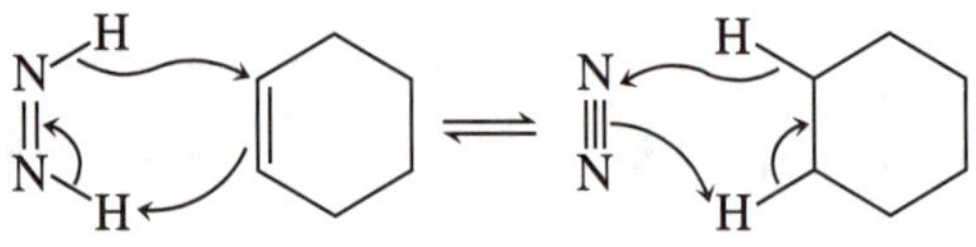 $[_\sigma 2 + _\pi 2 + _\sigma 2]$ $[_\sigma 2 + _\pi 2 + _\sigma 2]$

This expanded view of cycloaddition reactions is very broad. In it, all pericyclic reactions can be defined as concerted cycloaddition or cycloreversion reactions, as shown here for the electrocyclic reactions **g** and **h**.

(g) 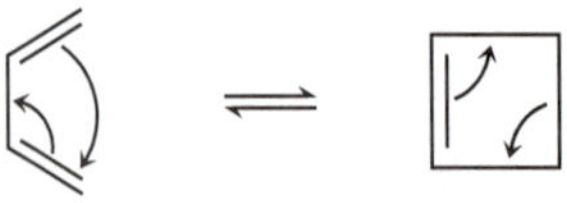 $[_\pi 2 + _\pi 2]$ $[_\sigma 2 + _\pi 2]$ (h) 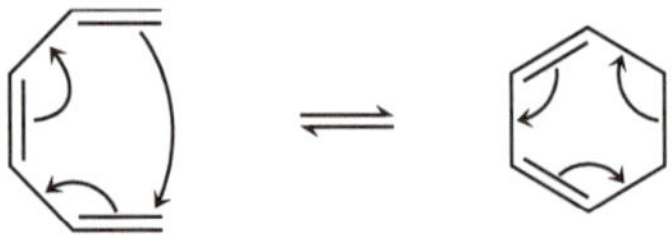 $[_\pi 2 + _\pi 2 + _\pi 2]$ $[_\sigma 2 + _\pi 2 + _\pi 2]$

Although it is often conceptually useful to regard all pericyclic reactions as cycloadditions, the term *cycloaddition* is usually used in the more restricted sense, to refer only to ring-forming reactions between separate molecules or between isolated π units in a single molecule.

3.2 SUPRAFACIAL AND ANTARAFACIAL ADDITION

There are two possible ways to form bonds to the two atoms of a π bond or to the two terminal atoms of a set of conjugated π bonds. The two new bonds may be formed either from lobes on the same side of the π bond system or from lobes on opposite sides. Woodward and Hoffmann designated addition to lobes on the same side of a π system as *suprafacial* addition, and called addition to lobes on opposite sides of a π system *antarafacial* addition.[1]

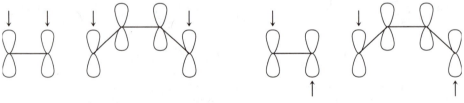

Suprafacial sites   Antarafacial sites

These modes of addition are identified by the symbols *s* and *a*, respectively. Thus, cycloaddition of two π bonds, each reacting suprafacially, would be classified as a $[_\pi 2_s + _\pi 2_s]$ reaction. (The *s* and *a* symbols are written as subscripts, so reactions involving suprafacial reactions of two electrons should not be confused with electrons in a $2s$ orbital.) Cycloaddition of a four-electron "unit" reacting antarafacially with a two-electron unit reacting suprafacially would be classified as a $[4_a + 2_s]$ reaction. If the four-electron unit were reacting suprafacially and the two-electron unit antarafacially, the reaction would be classified as a $[4_s + 2_a]$ reaction. (A "unit" may be a pair of unshared electrons as well as a bond or a set of conjugated π bonds.)

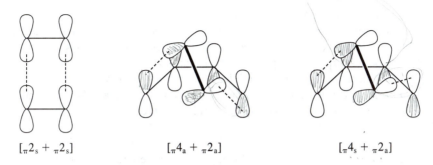

$$[_\pi 2_s + _\pi 2_s] \qquad\qquad [_\pi 4_a + _\pi 2_a] \qquad\qquad [_\pi 4_s + _\pi 2_a]$$

When σ bonds take part in cycloaddition reactions, the two "inside" lobes of the bonding orbitals (those forming the bond) are considered to be on the same side of the bond. Similarly, the two "outside" lobes are both on the same side of the bond. The latter statement may seem surprising, but picture the rotation of the atomic orbitals forming a σ bond, with the orbitals end to end, into a π bond arrangement, with the orbitals side by side. It is easy to see that the two inside lobes of the σ bond end up on the same side of the π arrangement, as do the two outside lobes.

3.3 SELECTION RULES FOR CYCLOADDITION AND CYCLOREVERSION REACTIONS

Woodward–Hoffmann Rules

Woodward–Hoffmann selection rules governing the stereochemistry of cycloaddition and cycloreversion reactions are easy to state, although perhaps sometimes difficult to visualize in action:

Thermal cycloadditions or cycloreversions involving $4n + 2$ electrons are allowed if they proceed with an *even* number (including zero) of antarafacial interactions of reacting units, while those involving $4n$ electrons are allowed if they proceed with an *odd* number of antarafacial interactions.

According to this rule, both $[4_s + 2_s]$ and $[4_a + 2_a]$ reactions are allowed, because both types of reactions would involve six electrons and an even number (zero and two, respectively) of antarafacial interactions. Thermal reactions of order $[2_s + 2_s]$ or $[4_a + 2_s]$ are forbidden, but $[2_a + 2_s]$ or $[4_a + 4_s]$ reactions are allowed.

It is worth stressing that these rules simply state which types of reactions are allowed or forbidden by orbital symmetry considerations. They do *not* imply that all allowed reactions actually take place. As we shall see, reactions involving antarafacial interactions of bonds, in particular, are often sterically difficult and, therefore, are rare.

Derivations of the Rules

Like the rules for electrocyclic reactions, the Woodward–Hoffmann rules for cycloaddition reactions can be derived from frontier orbital, orbital correlation diagram, or "aromatic transition state" approaches.

In analyzing frontier orbital effects on cycloaddition reactions, we have to consider the interactions of molecular orbitals from two (or more) molecules rather than interactions of two ends of a single orbital. Like the interactions of any two filled orbitals, the interaction of two HOMOs would result in net repulsion rather than formation of new bonds. However, following standard chemical principles, a filled orbital may react with an empty one; that is, the HOMO of one reacting unit may interact with the LUMO of another to form bonds. (As we shall see, if only the symmetries of the orbitals are considered, it does not matter which HOMO and which LUMO are chosen.)

To begin, consider thermal cycloadditions in which both reacting units are interacting suprafacially. For a $[_\pi2_s + _\pi2_s]$ cycloaddition of two ethene molecules, the HOMO of one will be π and the LUMO of the other will be π^*. (For a $[_\sigma2_s + _\sigma2_s]$ cycloreversion, the orbitals would be σ and σ^*, respectively.) Because the HOMO is symmetric, and the LUMO is antisymmetric, the two orbitals cannot combine to form two new bonds. *The reaction is forbidden.*

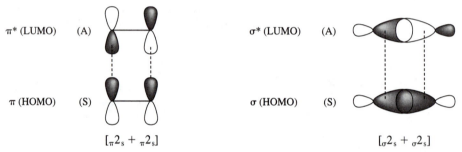

$$[_\pi2_s + _\pi2_s] \qquad [_\sigma2_s + _\sigma2_s]$$

Frontier orbital interactions in thermal $[2_s + 2_s]$ reactions

Similarly, the interaction of two butadiene molecules in a $[_\pi4_s + _\pi4_s]$ manner is forbidden, because each HOMO (ψ_2) is antisymmetric, and each LUMO (ψ_3) is symmetric.

In contrast, the suprafacial reaction of butadiene with ethene (a $[4_s + 2_s]$ reaction) is allowed, because the HOMOs and LUMOs (ψ_2 and π^*, or π and ψ_3) have the same symmetries. The conclusion is the same regardless of which molecule is considered to be the electron donor and which the acceptor.

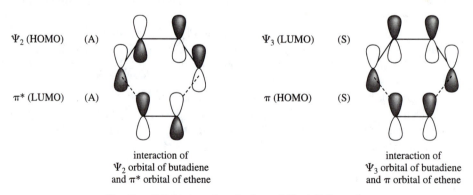

interaction of
Ψ_2 orbital of butadiene
and π^* orbital of ethene

interaction of
Ψ_3 orbital of butadiene
and π orbital of ethene

Frontier orbital interactions in thermal $[4_s + 2_s]$ reactions

Two orbitals of opposite symmetry *would* interact properly if one component reacted in an antarafacial manner, in which the upper lobe at one end of the orbital and the lower lobe at the other end interact with the second orbital. Therefore, a $[_{\pi}2_s + _{\pi}2_a]$ reaction would be allowed. In contrast, $[4_s + 2_a]$ and $[4_a + 2_s]$ reactions would be forbidden, because the molecular orbitals would have the wrong symmetries.

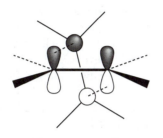   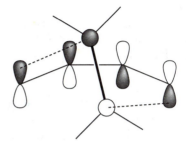

Allowed antarafacial
interaction of a π^* orbital
with a π orbital (π^* shown
in top view)

Forbidden antarafacial
interaction of a π^* orbital
with Ψ_2 of butadiene (π^*
shown in top view)

The Woodward–Hoffmann rules also apply to cycloadditions of systems with odd numbers of atomic orbitals. Suprafacial addition of an allyl anion to ethene would be allowed, because the π^* and ψ_2 orbitals (as well as the π and ψ_3 orbitals) have the same symmetries. Purely in terms of orbital symmetry, it doesn't matter whether or not the HOMO of the anion and the LUMO of the π bond are chosen, or vice versa. However, in this case we will show the allyl anion acting as the electron donor and the π bond as the electron acceptor, because ψ_2 of the anion should be higher in energy than π, and π^* should be lower in energy than ψ_3. (This is simply another way of stating that carbanions tend to be much better electron donors—but much poorer electron acceptors—than neutral alkenes.)

Unlike the allyl anion, the allyl cation is forbidden from reacting suprafacially with a π bond (or a σ bond for that matter), because the HOMO of one component would have opposite symmetry from the LUMO of the other.

Ψ_2 (HOMO)   (A)
of an allyl anion

π^* (LUMO)   (A)
of ethene

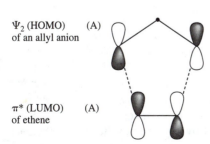

Ψ_1 (HOMO)   (S)
of an allyl cation

π^* (LUMO)   (A)
of ethene

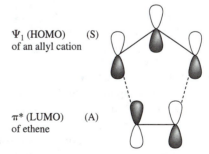

Allowed suprafacial interaction
of an allyl anion with a π bond

Forbidden suprafacial interaction
of an allyl cation with a π bond

Of course, the conclusion that suprafacial reactions of alkenes with allyl anions is allowed and with allyl cations is forbidden refers only to pericyclic reactions, in which both ends of the reacting molecules interact. The Woodward–Hoffmann rules do not apply if reaction takes place at only one end of each orbital. It is well known that carbocations react much more readily with alkenes in such reactions than do carbanions.

The Woodward–Hoffmann rules also follow directly from the "aromatic transition state" approach. The presence of any antarafacial interaction in a cyclic array of electrons is the equivalent of a "twist" in the array. Thus, a transition state with an odd number of antarafacial interactions can be considered to be a Möbius transition state and will be aromatic if the reaction involves $4n$ electrons. Cycloadditions requiring an even number (usually zero) of antarafacial interactions will proceed through aromatic transition states if $4n + 2$ electrons are involved.

The Woodward–Hoffmann rules for cycloadditions, like those for electrocyclic reactions, can also be obtained from the consideration of orbital correlation diagrams. However, we will not take the time to discuss the application of orbital correlation diagrams to cycloaddition reactions.[2]

3.4   EXAMPLES OF THERMAL CYCLOADDITION REACTIONS

[2 + 2] Reactions

Diradical reactions.   In principle, the mechanism of a cycloaddition reaction should be easily established from its stereochemistry, because a suprafacial reaction of a π bond would result in retention of geometry about that bond, while an antarafacial reaction of a π bond would result in inversion of geometry about that bond, as shown in Eq. 1 for a $[_\pi 2_s + _\pi 2_a]$ reaction.

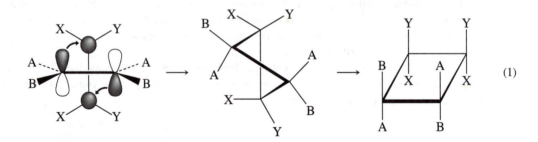

(1)

One difficulty in investigating cycloadditions of alkenes to form cyclobutanes is that they are rather difficult reactions that rarely proceed in high yields.[3] (This fact by itself is perhaps good evidence that the geometrically reasonable $[2_s + 2_s]$ cycloaddition reaction is forbidden.) Orbital overlap in the transition state for a theoretically allowed $[2_a + 2_s]$ reaction would be extremely poor, and the reaction would initially give rise to a badly twisted cyclobutane, as shown in Eq. 1. Therefore, $[2_a + 2_s]$ reactions are rare, if they exist at all.

While [2 + 2] cycloaddition reactions of hydrocarbons to form cyclobutanes are rare, cycloreversions of cyclobutanes into alkenes at high temperatures are common reactions that have been extensively investigated.[4] Their enthalpies and entropies of activation have been found to coincide with those expected if one carbon–carbon bond initially breaks to form a diradical (as in Eq. 2) rather than with those expected from concerted cleavage of both bonds.*

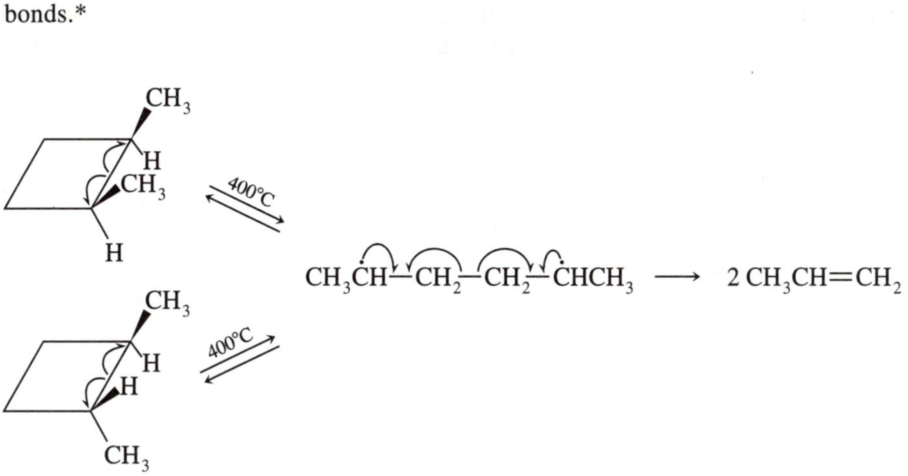

$$CH_3\overset{\frown}{C}H\overset{\frown}{\smile}CH_2\overset{\frown}{-}CH_2\overset{\frown}{\smile}\overset{\cdot}{C}HCH_3 \longrightarrow 2\,CH_3CH{=}CH_2 \qquad (2)$$

There *are* a few types of alkenes that will easily add to other alkenes to form cyclobutane derivatives. One type consists of molecules with extremely strained or twisted, and therefore highly reactive, double bonds. One example of such a molecule is benzyne (**2**). The addition of benzyne to either *cis*- or *trans*-cyclooctene was found to give a mixture of stereoisomers, although the product resulting from retention of the geometry of the cyclooctene was the principal product from each reaction (Eq. 3a).[6]

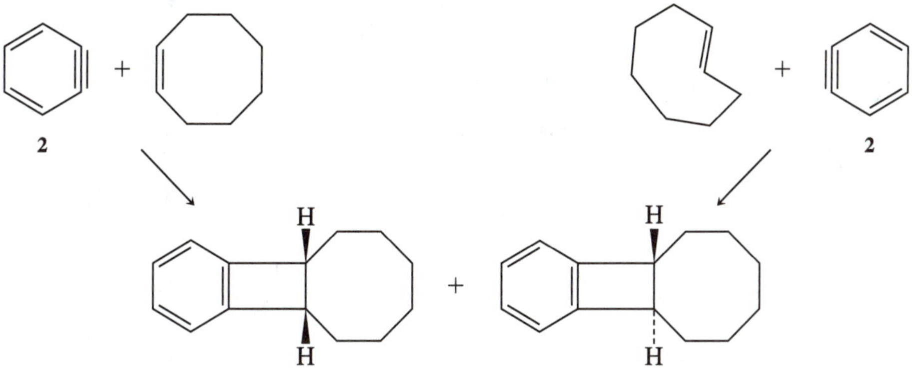

principal product
from *cis*-cyclooctene

principal product
from *trans*-cyclooctene

(3a)

---

*Geometrical isomers of disubstituted cyclobutanes undergo interconversion under the conditions for pyrolysis, as would be expected if tetramethylene diradicals that can reform cyclobutanes are formed as intermediates in the pyrolysis reactions. Tetramethylene diradicals, which can be detected by femtosecond spectroscopy,[5] can also be formed from other reactions. Their relative rates of cleavage to alkenes and recyclization to cyclobutanes were found to be the same as those of the intermediates in cyclobutane pyrolysis.

Although mixtures of stereoisomers might conceivably have resulted from simultaneous antarafacial additions and (principally) suprafacial additions, a diradical mechanism (Eq. 3b), in which ring closure of the diradical occurred even more rapidly than rotation around a single bond for the majority of reacting molecules, seemed more likely.[6]

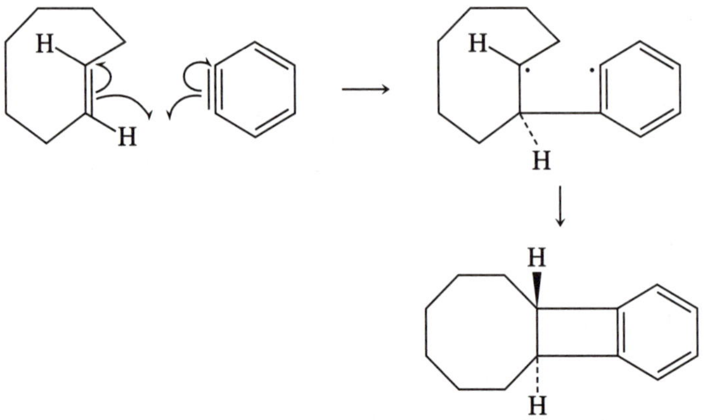

$$(3b)$$

Similarly, dimerization of the strained cyclooctadiene **3** yielded mixtures of stereoisomers (Eq. 4), suggesting that the reaction proceeded via diradical intermediates.[7]

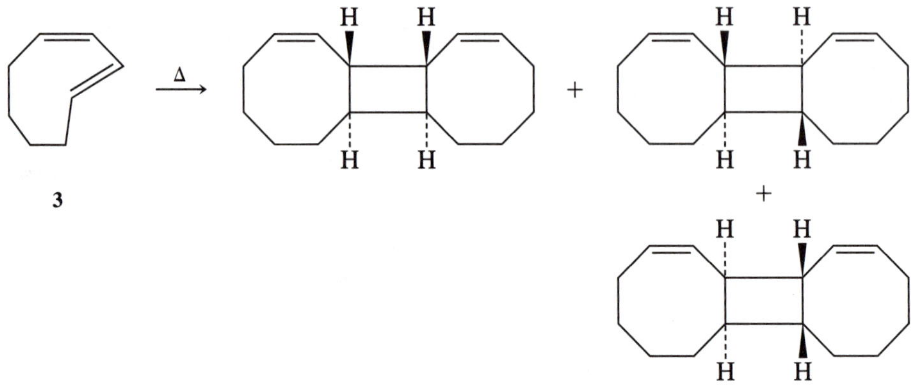

$$(4)$$

Although they are not strained or twisted molecules, tetrafluoroethenes and other polyfluoroalkenes with at least two fluorine atoms on one vinyl carbon can react with other alkenes to form four-membered rings.

Studies by Paul D. Bartlett and his group at Harvard University showed that cycloaddition reactions of polyfluoroalkenes yield mixtures of stereoisomers, as shown in Eq. 5.[8]

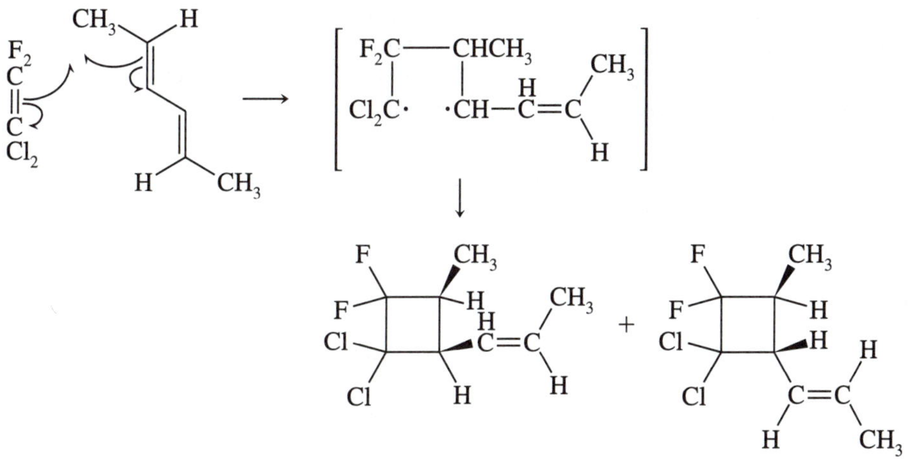

(5)

Reaction of tetrafluoroethene with either *cis*- or *trans*-1,2-dideuteroethene (Eq. 6) yielded the same 50:50 mixture of stereoisomers, providing strong evidence that these reactions proceeded via diradical intermediates rather than by simultaneous suprafacial and antarafacial additions. It would be a most unlikely coincidence for the antarafacial and suprafacial additions to proceed at precisely the same rates. In contrast, rotation around a single bond in a diradical intermediate, thereby giving a 1:1 mixture of stereoisomers, should be extremely rapid in this reaction.

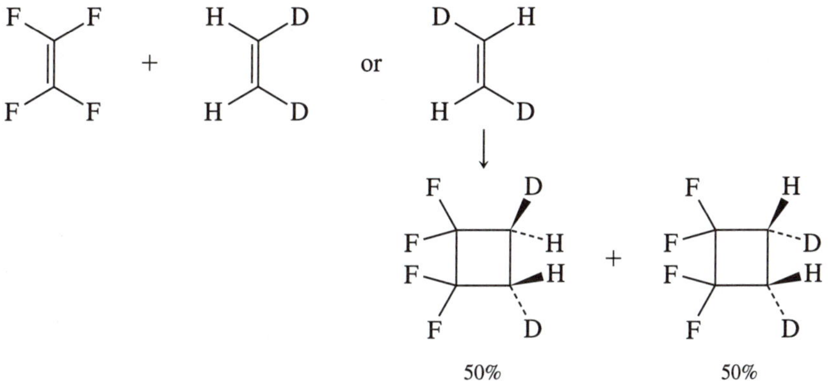

(6)

50%   50%

Cycloaddition reactions occurring by diradical mechanisms are rare, because, for most alkenes, formation of the intermediate diradicals would be endothermic by about 40 kcal/mol (the difference in energy between the two π bonds broken and the one σ bond formed). Therefore, at least one of the two reacting double bonds must be very weak, to reduce the endothermicity of diradical formation. The π bonds in polyfluoroalkenes were calculated to be about 17 kcal/mol weaker than other unstrained double bonds.[9] This is apparently the result, in part, of strong repulsions between the unshared electrons on the fluorine atoms and the electrons of the π bonds. In addition, the bond energies are lowered by resonance stabilization of the radical fragments formed when the π bonds of the polyfluoroalkenes are broken.

Cycloadditions with ketenes. Ketenes—molecules in which double bonds and carbonyl functions are located on the same carbon atoms—add to double bonds to yield cyclobutanones.[10] Intramolecular additions of ketenes to alkenes resulted in the formation of some interesting molecules, such as ketone **4** (Eq. 7).[11]

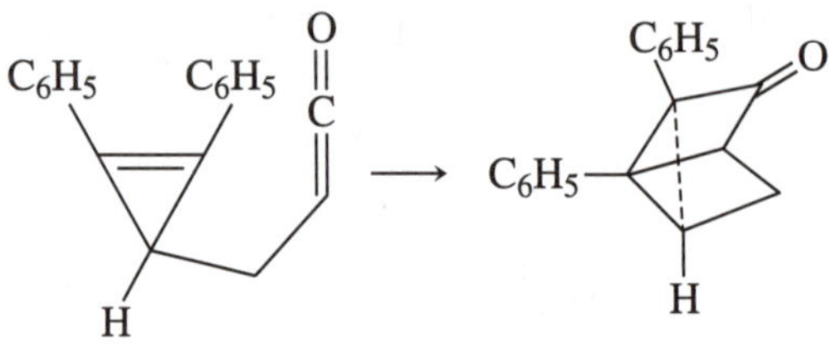

(7)

4

Cycloaddition reactions of ketenes with alkenes are usually stereospecific, with the geometries of the alkenes retained in the products, as shown in Eqs. 8 and 9.[10,12] Woodward and Hoffmann, therefore, suggested that the reactions might proceed by concerted $[_\pi2_a + _\pi2_s]$ mechanisms, that could be facilitated by attractions between the carbonyl groups of the ketenes and the π bonds of the alkenes.[2]

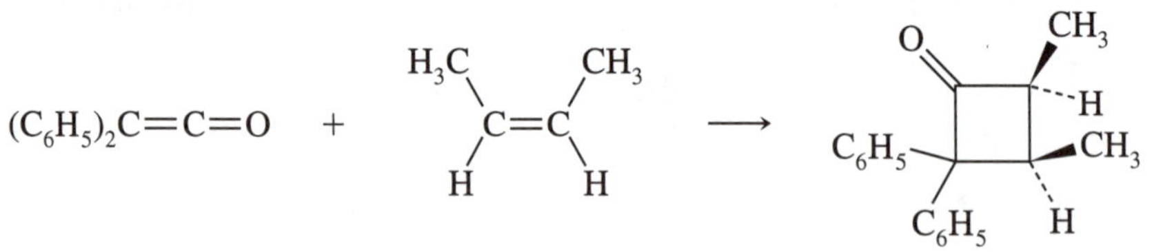

(8)

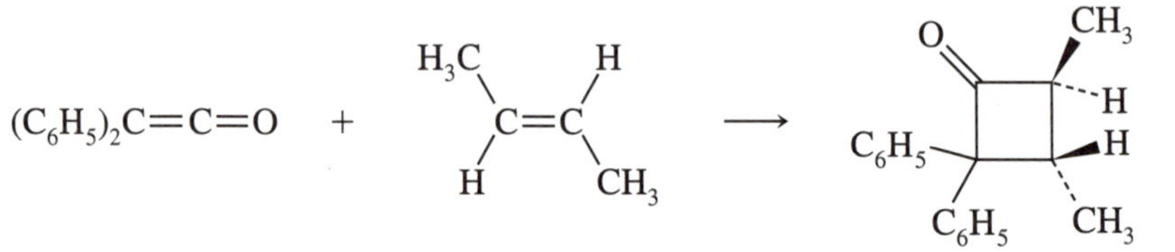

(9)

Unfortunately, the usual experimental test for an antarafacial addition—inversion of the geometry of one component in the reaction—cannot conclusively establish the stereochemistry of cycloaddition reactions of ketenes, because carbonyl groups are neither *cis* nor *trans* to substituents on adjacent atoms. However, theoretical calculations indicate that these reactions do *not* proceed by $[_\pi2_a + _\pi2_s]$ mechanisms. Apparently, the reactions can best be described as proceeding via transition states closely resembling diradicals, with the bonds between the carbonyl carbons and the alkenes almost fully formed, while the bonds to the α carbons of the ketenes have barely begun to form. However, the transition states do exhibit significant interactions between the carbonyl carbons and *both* doubly bonded carbons of the alkenes, which results in the cyclobutanones retaining the configurations of the starting alkenes.[13]

The products from additions of ketenes to double bonds normally have the structures that would result from formation of the most stable intermediate diradicals, as illustrated in Eq. 10. (Other [2 + 2] cycloadditions, such as the reaction shown in Eq. 5, similarly proceed by way of the most stable intermediate diradicals.)

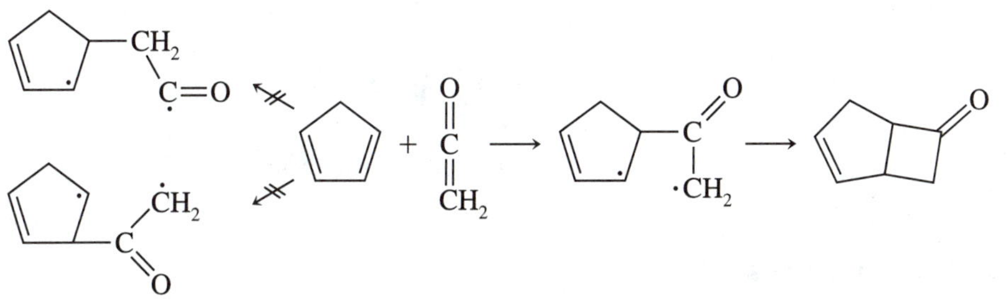

(10)

Polar mechanisms.   Alkenes substituted with powerful electron-withdrawing groups can react with electron-rich alkenes to form cyclobutanes (e.g., Eqs. 11a and 11b).[14,15]

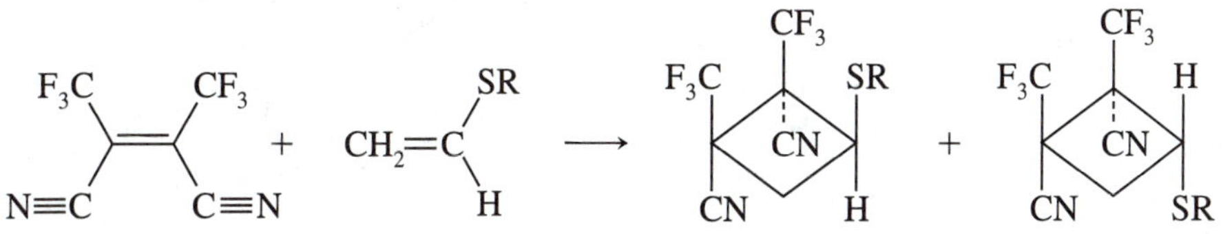

(11a)

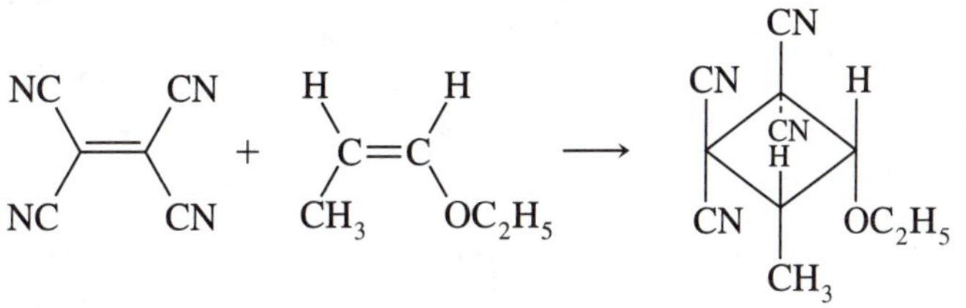

(11b)

Several facts suggest that these reactions are not concerted cycloadditions, but proceed via formation of zwitterionic intermediates. For one thing, the rates of these reactions increase markedly with increasing solvent polarity. Furthermore, the reactions are not completely stereospecific, although the principal reaction products frequently retain the configurations of the starting alkenes. The extent to which mixtures of stereoisomers are formed increases with increasing polarity of the solvents. Finally, when the reactions are carried out in hydroxylic solvents, open-chain products are frequently formed in addition to the cyclobutanes. These observations strongly suggest the formation of dipolar intermediates (as in Eq. 12), which may react with the solvents rather than close to form cyclobutanes.[16]

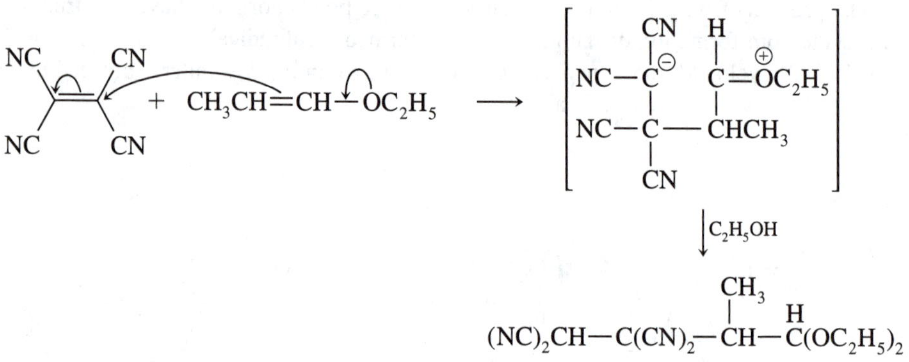

$$(NC)_2CH—C(CN)_2—\overset{\overset{CH_3}{|}}{CH}—\overset{\overset{}{}}{\underset{}{C}}(OC_2H_5)_2$$

(12)

[4 + 2] Cycloadditions

The Diels–Alder reaction. In marked contrast to the general failure of alkenes to undergo thermal $[_\pi2_s + _\pi2_s]$ cycloadditions, thermal $[_\pi4_s + _\pi2_s]$ cycloaddition reactions occur readily. They proceed smoothly, not only as reactions of carbon–carbon double bonds and triple bonds but also as reactions of carbonyl and imine groups, and even when neither end of a double bond is a carbon atom, as shown below. These reactions, known generally as *Diels–Alder reactions,*[*] are among the most useful reactions in synthetic organic chemistry.[17]

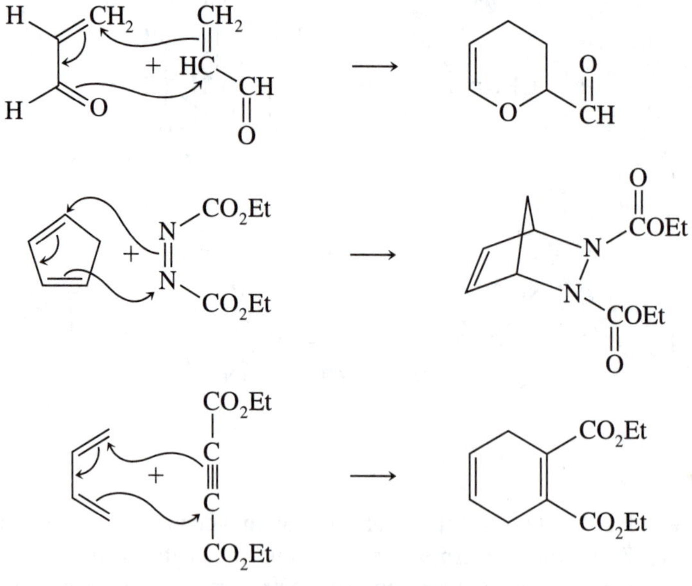

Some Diels-Alder reactions

Even some aromatic rings, such as anthracene and furan (but not usually benzene or naphthalene),[18] can readily undergo Diels–Alder reactions. However, cycloreversion processes

---

[*]Kurt Alder (1902–1958) and Otto Paul Herman Diels (1876–1954) of the University of Kiel, Germany, were awarded the Nobel Prize in 1950 for their development of the Diels–Alder reaction.

(*retro*-Diels–Alder reactions) are particularly likely to occur at reasonable temperatures when the dienes form parts of aromatic rings, as shown in Eqs. 13 and 14.

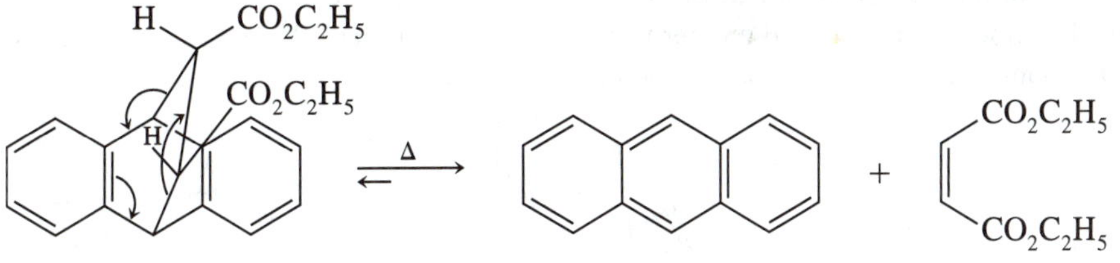

(13)

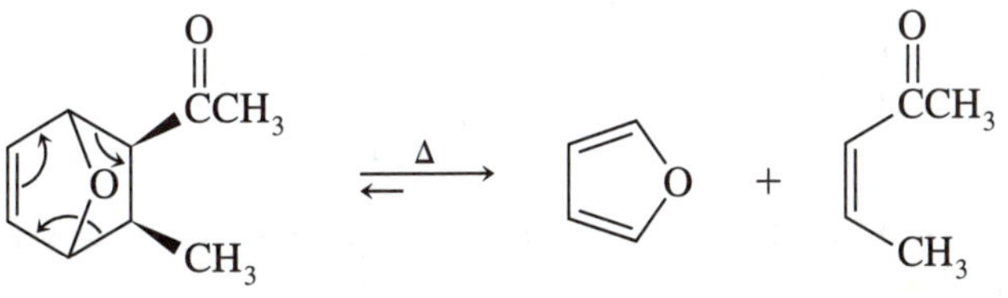

(14)

Thousands of Diels–Alder reactions have been studied.[15] In almost every instance, the reactions were found to be completely stereospecific, resulting in retention of the geometries of the conjugated diene and of the two-electron component, which is called the *dienophile*.

An interesting example of the distinction between 2 + 2 cycloadditions, which typically do not proceed by concerted mechanisms, and 4 + 2 cycloadditions, which do, is found in the reactions of butadienes with polyfluoroalkenes. These reactions can yield six-membered rings (Diels–Alder products) as well as the predominant four-membered rings (Eq. 15). The Diels–Alder reactions proceed nearly stereospecifically,* in contrast to the nonstereospecific formation of cyclobutanes.

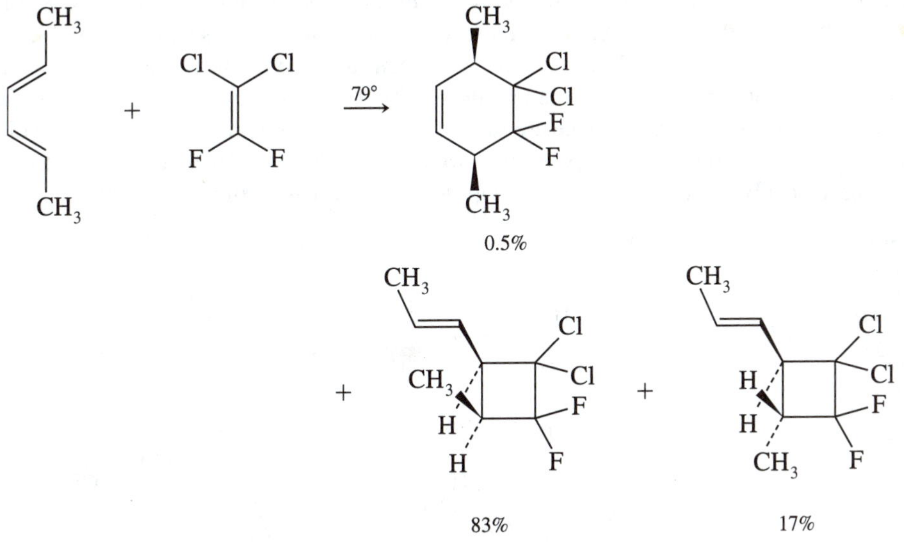

(15)

_____

*Approximately 1% of the cyclohexenes formed from the reaction shown in Eq. 15 had the methyl groups in a *trans*-relationship, suggesting that they were formed by diradical mechanisms.

The Alder "*Endo* Rule". Many Diels–Alder reactions of cyclic dienes can yield two stereoisomeric products (see Eq. 16). In one of the products (the *endo* isomer), carbonyl groups or other unsaturated substituents on the dienophile are *cis* to the double bond of the newly formed cyclohexene ring. In the other product (the *exo* isomer), the unsaturated substituents on the dienophile are *trans* to the double bond.

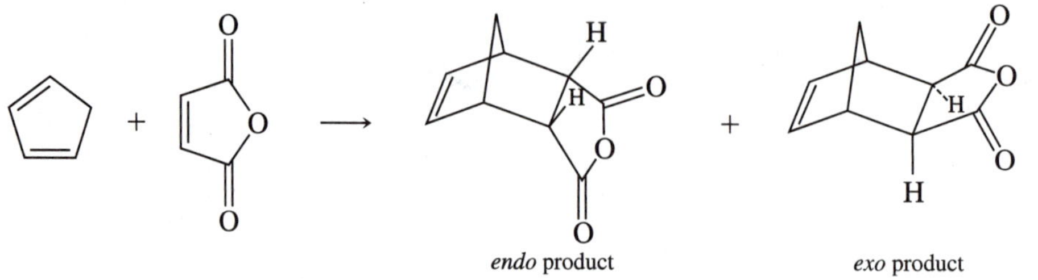

endo product   exo product   (16)

Endo and *exo* isomers can also be formed in Diels–Alder reactions of acyclic dienes, as is shown in Eq. 17.

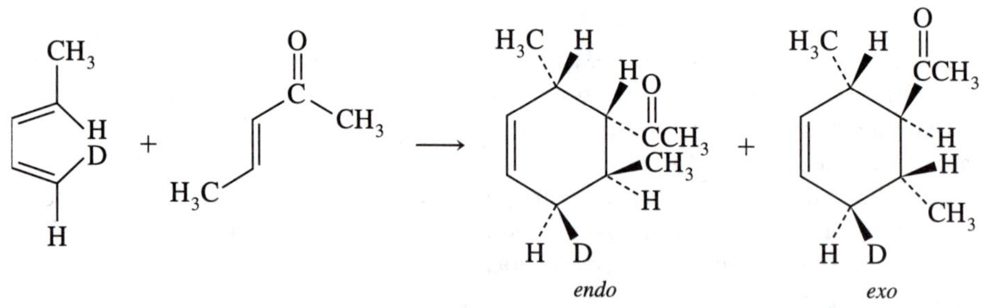

endo   exo   (17)

In the *endo* isomer, the two substituents that were *trans* to the bond linking the two double bonds of the diene (the "outside" substituents) end up *cis* to the major substituents on the dienophile. In the *exo* isomer, the former "outside" substituents of the diene are *trans* to the major substituents on the dienophile. (To confirm that these are indeed the results of *exo* and *endo* additions, imagine that the two "inside" substituents on the diene are fused together, as shown in Eq. 18, so that the diene is part of an imaginary ring. After drawing the *exo* and *endo* Diels–Alder products from the "ring" structures, mentally separate the "fused" groups to see the structures of the products.)

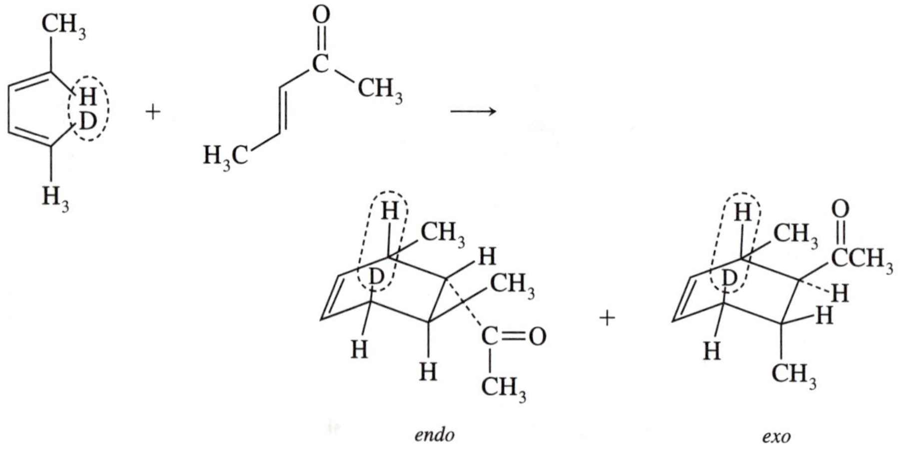

endo   exo   (18)

In the 1930s, Alder and Stein formulated the *endo rule*, which, in effect, states that *endo* products from Diels–Alder reactions are usually obtained in higher yields than *exo* products.[19] The *endo* rule may at first sight seem surprising, because *endo* isomers tend, for steric reasons, to be thermodynamically less stable than *exo* isomers. This can be demonstrated by the fact that *endo* isomers are frequently converted to *exo* isomers on heating, as shown in Eqs. 19 and 20.

$$\begin{array}{ccccc} & & H & & CO_2CH_3 \\ & \rightleftharpoons & \left[\begin{array}{c} + \quad C \\ \quad \parallel \\ CH_2 \end{array} \right] & \rightleftharpoons & \end{array} \tag{19}$$

$$\rightleftharpoons \left[2 \right] \rightleftharpoons \tag{20}$$

(Thermal *endo–exo* isomerizations of products of Diels–Alder reactions usually appear to start by *retro* Diels–Alder reactions.[20] As the regenerated dienes and dienophiles continue to undergo Diels–Alder condensations, the proportions of the more stable *exo* isomers increase.)

Woodward and Hoffmann pointed out that the *endo* rule may be rationalized by frontier orbital theory.[21] Consider the transition state for formation of the *endo* product from the Diels–Alder reaction of a conjugated diene with a conjugated dienophile. Let us assume that the HOMO (ψ_2) of the diene interacts with the LUMO (ψ_3) of the dienophile. The atomic orbitals at the end atoms of the diene (atoms 1 and 4) will be in phase with the orbitals of one double bond (atoms 1 and 2) of the dienophile, so that new bonds may form. In addition, the atomic orbitals at atoms 3 and 4 of the dienophile are in phase with those of atoms 2 and 3 of the diene. Thus, there is a secondary orbital interaction that lowers the energy of the transition state for *endo* cycloaddition compared to the transition state for *exo* addition, in which the secondary orbitals—those not forming new bonds—are far apart. (Similar secondary orbital effects would exist if the HOMO of the dienophile and the LUMO of the diene were considered to interact.)

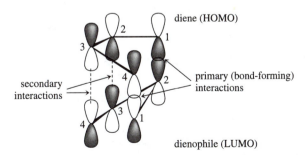

HOMO–LUMO interactions in a transition state for formation of an *endo* Diels-Alder adduct

Once the transition state is passed and the *endo* adduct is formed, the changes in bond angles that take place on conversion of π bonds to σ bonds reduce the strengths of the secondary orbital interactions (which would then be repulsive, in any case). Thus, secondary orbital interactions may affect the rates of Diels–Alder reactions but do not increase the thermodynamic stabilities of *endo* adducts in comparison to *exo* adducts.*

Reactivity in Diels–Alder reactions.   The rates of Diels–Alder reactions can be affected by both steric and electronic effects. A diene must have its double bonds on the same side of the central single bond—that is, it must be in an *s-cis* conformation—in order to take part in a Diels–Alder reaction. Dienes in which one or both substituents at C1 and C4 are *cis* to the other double bond react very slowly or do not undergo Diels–Alder reactions at all, since formation of the *s-cis* conformations is even more difficult than in the absence of such substituents. In contrast, the presence of a substituent at C2 of a diene usually increases its reactivity in Diels–Alder reactions, since the energy difference in favor of the *s-trans* conformation is reduced by such substitutents.

s-trans (low-energy conformation)	*s-cis* (high-energy conformation)	*s-trans* and *s-cis* conformations similar in energy

The most reactive dienes are those in which the diene unit is *forced* to maintain an *s-cis* conformation, such as those in which both double bonds are contained in a ring structure. Cyclopentadiene, for instance, will undergo Diels–Alder dimerization simply on prolonged standing at room temperature.

In general, however, Diels–Alder reactions between two hydrocarbon molecules are slow and proceed in poor yields. The reaction of butadiene with ethene, for instance, requires heating at 200°C for 17 h at 90 atm pressure to yield "up to" 18% of vinylcyclohexene.[23] Similarly, butadiene can undergo a Diels–Alder dimerization reaction at high temperatures, but the reaction is slow and the yields are poor, even if free radical inhibitors are present to minimize the rate of free radical polymerization of the diene.[17]

Usually, in order for Diels–Alder reactions to take place in high yields and at reasonable rates, the dienophile must be substituted with powerful electron-withdrawing groups, such as carbonyl or carboxyl groups. This fact may be accounted for by the general principle that reactions occur most readily when one reacting unit is a good electron donor and the other a good electron acceptor.

Frontier molecular orbital theory provides another way of looking at that principle. The interaction of any two orbitals will result in formation of two new orbitals in the transition state—one

---

*A closely related approach (similar to that employed below to rationalize the *regioselectivities* of Diels–Alder reactions) is to assume that just one bond has been appreciably formed at the transition state, which would thus resemble a pair of allylic radicals. Interactions of the LUMOs of those radicals during *endo* adduct formation would stabilize the transition states.[22]

lower in energy than either of the original orbitals and one higher in energy. The extent of the energy differences between the original orbitals and the new orbitals will depend on the relative energies of the two original orbitals. If the LUMO of one molecule is very much higher in energy than the HOMO of the other, the newly formed HOMO will not be much lower in energy than the original HOMO. If the original orbitals are closer in energy, the differences in energy between the original orbitals and the transition state orbitals increase. The reaction should therefore be faster, because the electrons in the original HOMO will go into transition state orbitals that are significantly lower in energy.

Placing electron-donating groups on one component in a Diels–Alder reaction would raise the energy of its HOMO, while placing electron-withdrawing groups on the other component would lower the energy of its LUMO. Diels–Alder reactions would therefore be expected to proceed most rapidly when one component in the reaction bears strongly electron-donating substituents and the other strongly electron-withdrawing substituents.

Diels–Alder reactions also proceed rapidly if the dienophile bears electron-donating substituents and the diene bears electron-withdrawing substituents. (These are called *inverse electron-demand* reactions.[24]) However, the synthesis of dienes bearing electron-donating groups and of dienophiles bearing electron-withdrawing groups is usually easier than the reverse, so that most Diels–Alder reactions involve the former type of substitution pattern.

Catalysis by strong Lewis acids greatly increases the rates of many Diels–Alder reactions.[24] Formation of a complex between the Lewis acid and one of the components (usually the dienophile, as shown in Eq. 21) lowers the energy of its LUMO and thereby decreases the activation energy. Lewis acid catalysis also frequently increases the *endo–exo* ratios of Diels–Alder reactions as well as their *regioselectivities*—that is, the extent to which some positional isomers are favored over others in the products.[25]

(21)

Regioselectivity. Recent theoretical work has shown that the *regiochemistry* of Diels–Alder reactions—the type and ratio of positional isomers among the products—can usually be predicted on the basis of frontier orbital calculations.[26] The general principle is that the principal isomer formed will result from connecting the atoms with the largest coefficients in the frontier orbitals of the two components.

An older approach, which in most cases is equivalent to the approach based on the coefficients of wave functions, is useful in understanding the regiochemistry of Diels–Alder reactions. That approach is based on the rule that *the major product from a Diels–Alder reaction will arise from the transition state that resembles the most stable of the possible diradical intermediates that might be formed in the reaction.*

To see how this rule works let's consider the reaction between 1,3-pentadiene and acrolein. Reaction of acrolein with one end of the diene system could form four possible diradicals: **A**, **B**, **C**, or **D** (shown in Eq. 22).

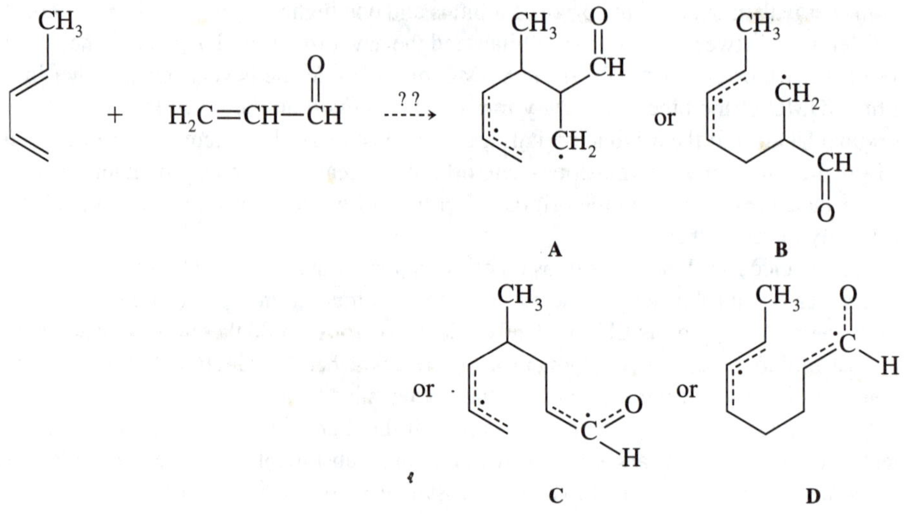

A   B   (22)

C   D

Of these four structures, **A** and **B** are clearly less stable than **C** and **D**, since in **A** and **B**, the radical fragment formed from acrolein has no resonance stabilization. Diradical **C** is less stable than **D**, since in **C** the radical fragment formed from the pentadiene can be written as a hybrid of one secondary radical resonance form and one primary radical form, while in **D** the radical fragment from the pentadiene can be written as a hybrid of two secondary radical resonance forms.

The Diels–Alder product that would be formed from diradical **D** is indeed the principal product formed from the reaction of 1,3-pentadiene and acrolein (Eq. 23). Similarly, the principal regioisomers obtained from most Diels–Alder reactions can be predicted on the basis of the "diradical-like transition state rule," even though on steric grounds, these isomers frequently appear less likely to form as in Eqs. 23 and 24.

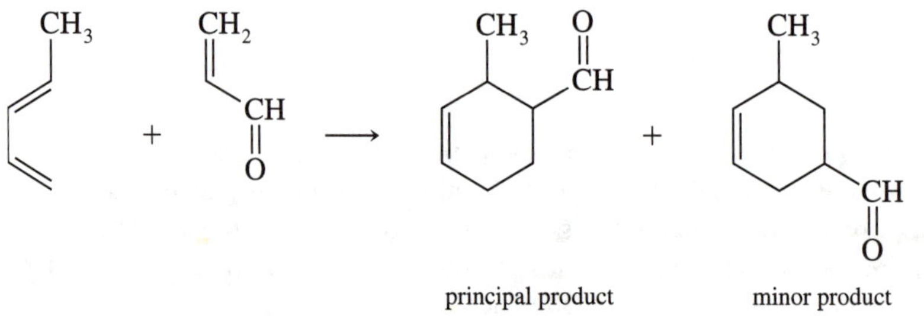

principal product   minor product

(23)

It cannot be stressed too strongly that use of the diradical-like transition state model does *not* imply that the Diels–Alder reaction proceeds via free-radical intermediates. The model simply recognizes that even in concerted reactions, in which several bonds are made or broken at the same time, some bonds are likely to be less completely formed than others in the transition state. Thus, the transition state for Diels–Alder reactions can benefit from the factors that would stabilize a diradical. Reactions that proceed by pericyclic mechanisms but have transition states in which some bonds are more fully formed than others have been described as "concerted but *nonsynchronous.*"

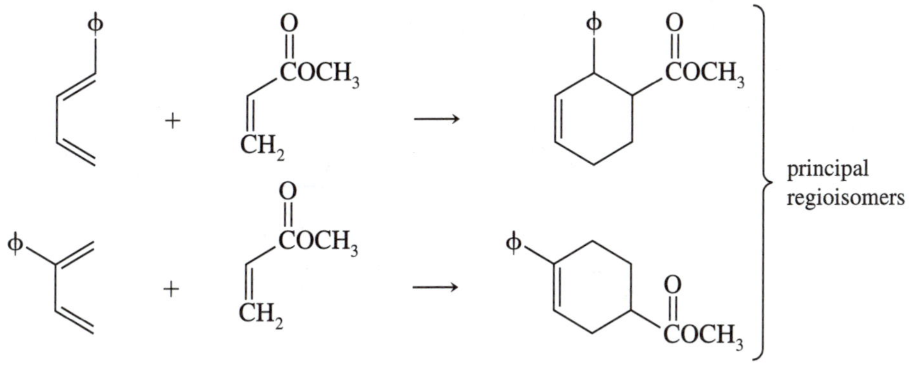

(24)

principal
regioisomers

(25)

Cycloadditions of Enynes.   It has been known for over a century that conjugated enynes can react with acetylenic compounds to form aromatic rings, as shown in Eq. 26.[27a]

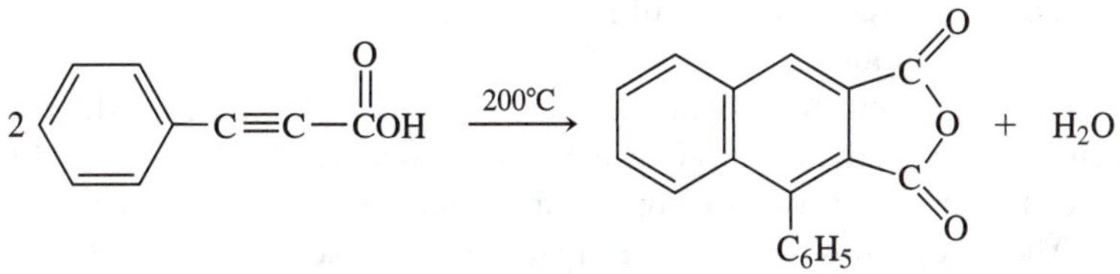

(26)

Early examples were usually carried out in strongly acidic solutions, in which reactions involving carbocations might occur.[27b] However, several examples (Eqs. 27 and 28)[28,29] are now known that do appear to proceed under purely thermal conditions.

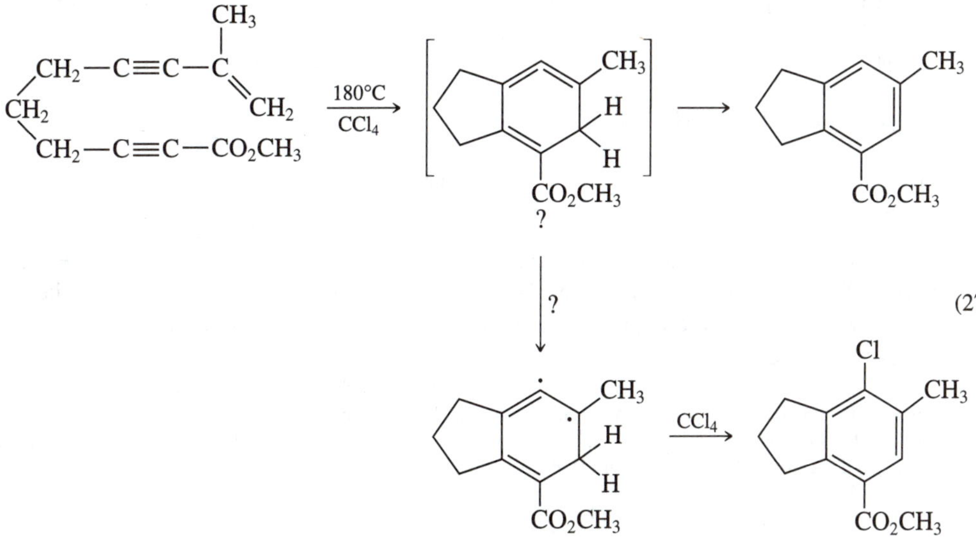

(27)

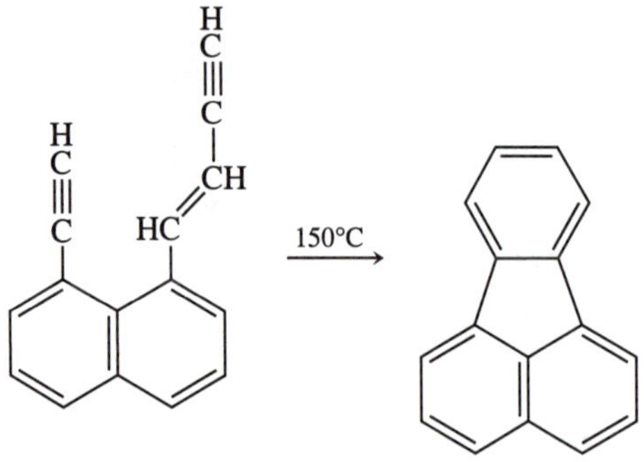

(28)

Not surprisingly, it has been suggested that the initial steps in these reactions, which have been labeled "dehydro Diels–Alder reactions,"[30] may be concerted cycloaddition reactions.* However, the aromatic rings formed from reactions of enynes with acetylenes lack stereochemical markers that might be used to test whether they proceed by concerted mechanisms. While reactions of enynes with olefinic dienophiles might, in principle, provide such evidence, there appear to be no clear-cut examples of reactions of simple enynes with olefinic molecules under purely thermal conditions.†

While simple enynes do not appear to undergo thermal condensations with olefinic dienophiles, conjugated *di*enynes (1,5-diene-3-ynes) were shown in the 1940s to yield cycloaddition products in which two dienophile molecules had combined with each molecule of dienyne.[34] These reactions were reinvestigated five decades later and shown to proceed with complete stereospecificity. The geometries of dienophiles with *cis* configurations were retained in the products (as shown in Eq. 29) as were those of dienophiles with *trans* configurations.[35]

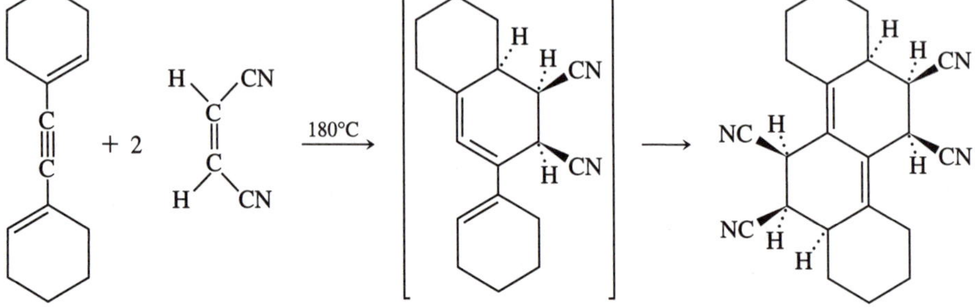

(29)

Thus, it appears that "dehydro Diels–Alder reactions" of dienynes do involve concerted cycloaddition processes. However, cycloaddition reactions of enynes with olefinic dienophiles have never been observed to proceed under purely thermal conditions. Reactions of enynes with

---

*1,2,4-Cyclohexatrienes ("isoaromatic molecules"), which might be formed from concerted cycloadditions of enynes with acetylenic dienophiles, have been identified as short-lived intermediates in other reactions.[31]

†One apparent reaction of that type[32] was shown to result from addition of HCl to the triple bond of the enyne, forming a diene that underwent a standard Diels–Alder reaction.[33]

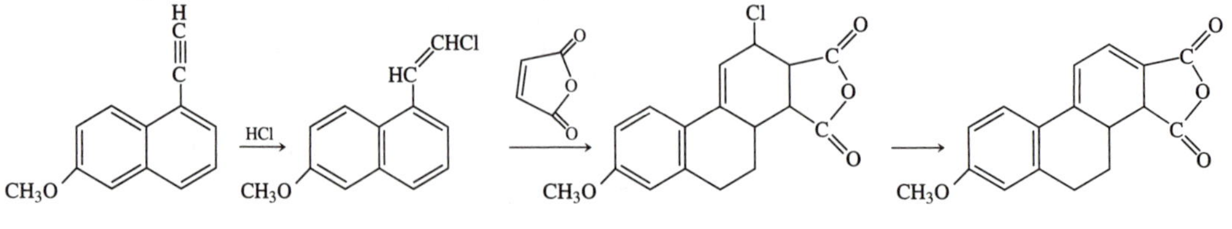

acetylenic dieophilles, as well as reactions of dienynes with olefinic dienophiles, proceed in reasonable yields. These results suggest that the equilibria for the initial cycloaddition steps are unfavorable. The reactions will proceed, however, if the initially formed allenic structures are "trapped" by a second cycloaddition reaction (as shown in Eq. 29) or by [1,5] hydrogen migrations to form aromatic products from reactions with acetylenic dienophiles.*

[2 + 2 + 2] Cycloadditions

$[_\pi 2_s + _\pi 2_s + _\pi 2_s]$ **reactions.** Although thermal $[2_s + 2_s + 2_s]$ cycloadditions are theoretically allowed, the simultaneous combination of three molecules would suffer from a large negative entropy effect. This would be particularly unfavorable at the high temperatures necessary for many cycloaddition reactions. Thus, there appear to be no examples of concerted thermal cycloadditions of three molecules.

Several examples of thermal cycloaddition reactions of unconjugated dienes, such as the reaction of norbornadiene with maleic anhydride or acrylonitrile (Eq. 30) are known.[36] The thermal cycloreversion reaction shown in Eq. 31 proceeds stereospecifically to form the all-*cis* product. This appears to be a concerted $[_\sigma 2_s + _\sigma 2_s + _\sigma 2_s]$ reaction.[37]

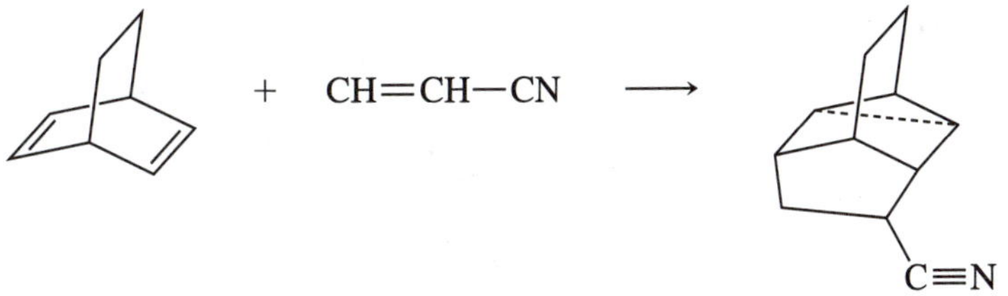

$$+ \quad CH\!=\!CH\!-\!CN \quad \longrightarrow \qquad\qquad (30)$$

$$\xrightarrow{\ 125°C\ } \qquad\qquad (31)$$

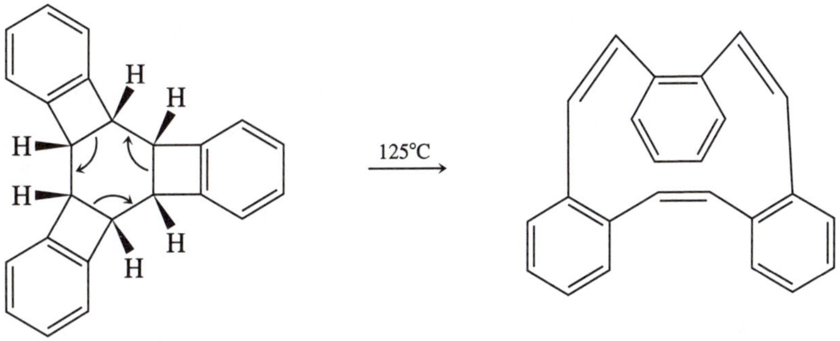

---

*It might have been anticipated that [1,5] hydrogen shifts converting "isoaromatic" to aromatic molecules would be very rapid processes. In fact, they appear to be relatively slow.[31b] This may result from the fact that in these fascinating molecules, the allowed [1,5] shifts are simultaneously forbidden [1,3] shifts.

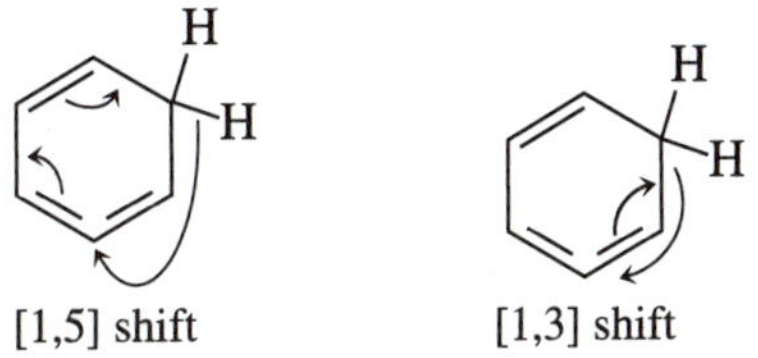

[1,5] shift   [1,3] shift

However, molecular orbital calculations suggest that the conversion of 1,6,11-dodecatriyne into an aromatic compound (Eq. 32) proceeds by initial formation of a diradical.[38] In this case, steric strain may prevent a concerted $[_{\pi}2 + {_\pi}2 + {_\pi}2]$ reaction.

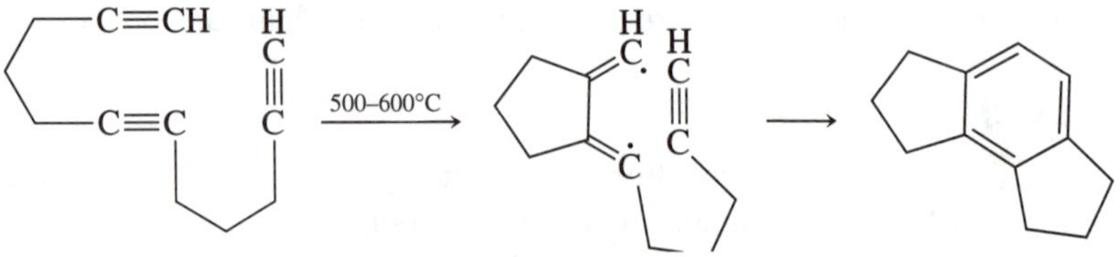

$$(32)$$

The Alder "Ene Reaction".   Alkenes can form addition products, resulting from the transfer of hydrogen atoms, on reaction with strong dienophiles (Eqs. 33a and 33b). However, much higher temperatures are required than are needed for reactions of dienophiles with conjugated dienes.

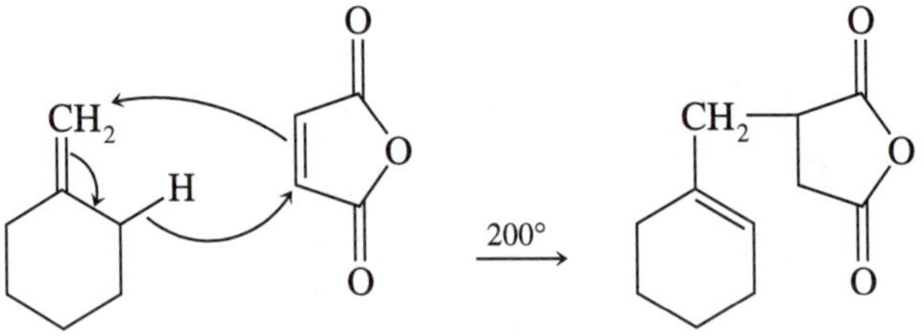

$$(33a)$$

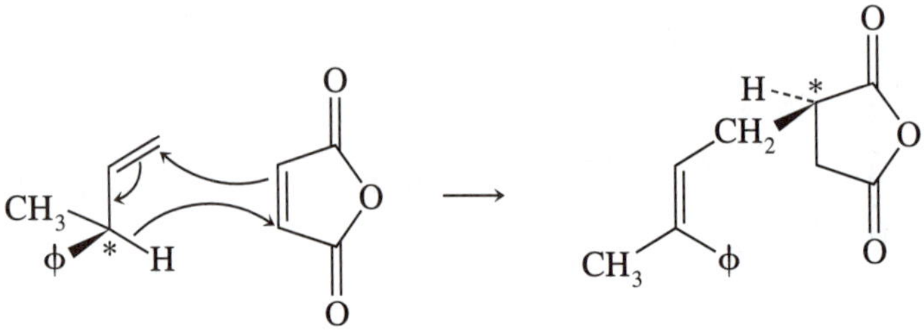

$$(33b)$$

By analogy with the Diels–Alder reaction, which is called the "diene reaction" or the "diene synthesis" in German-speaking countries, this reaction is called the "Alder ene reaction," or simply the "ene reaction."[39]

Ene reactions invariably result in migrations of the double bonds of the alkenes, providing evidence that they proceed via concerted $[_{\pi}2_s + {_\sigma}2_s + {_\pi}2_s]$ mechanisms, because free-radical mechanisms would be expected to yield mixtures of products. A radical mechanism should also yield achiral products from the reaction of a chiral alkene. However, as shown in Eq. 28b, chiral products can be obtained from such reactions.[39]

Carbonyl groups, particularly in aldehydes, can act as "enophiles" in ene reactions.[40] In these reactions, allylic hydrogens of alkenes are transferred to carbonyl oxygens, and the carbon atoms of the carbonyls react with the double bonds, converting the carbonyl groups to alcohols.

Ene reactions with carbonyl groups can be catalyzed by dimethylaluminum chloride or ethylaluminum dichloride, as shown in Eqs. 34a and 34b.[41]

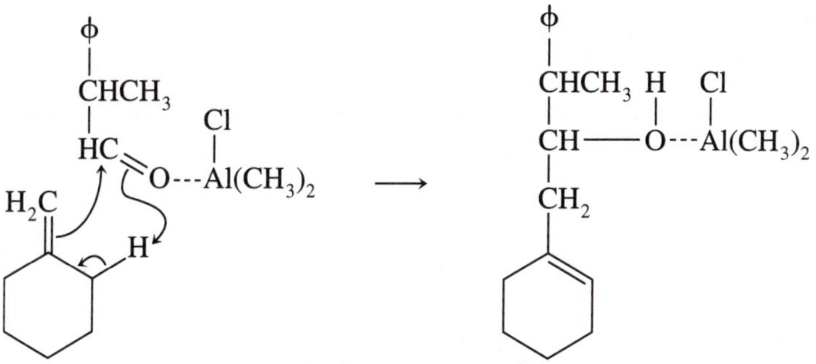

(34a)

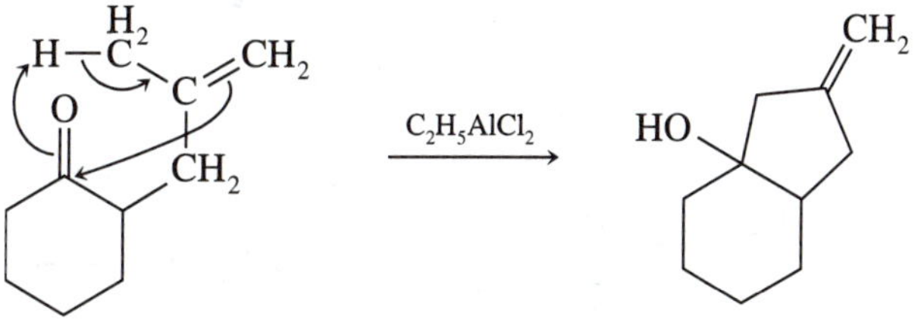

(34b)

These unusual Lewis acids will react with protic acids to form methane and ethane (e.g., Eq. 34c). This protects the reaction products from undergoing decomposition by the protic acids.

$$(CH_3)_2AlCl \;+\; HCl \;\longrightarrow\; CH_4 \;+\; CH_3AlCl_2 \qquad (34c)$$

Other Thermal Cycloaddition Reactions

The Diels–Alder reaction and other $[4_s + 2_s]$ reactions constitute by far the most important group of cycloaddition and cycloreversion processes. Cycloadditions involving larger numbers of electrons are rare and are of comparatively little synthetic importance, but they can be quite fascinating. Unfortunately, it can rarely be convincingly demonstrated that these reactions proceed by concerted mechanisms, and their stereochemistries are often difficult to establish.

Suprafacial eight-electron $[4_s + 4_s]$ or $[6_s + 2_s]$ cycloadditions are forbidden by the Woodward–Hoffmann rules and are rare. Two examples are the dimerizations of 1,3-diphenylisoindenone (Eq. 35a)[42] and of o-xylylene (Eq. 35b).[43] The latter reaction was shown to proceed by a diradical mechanism.

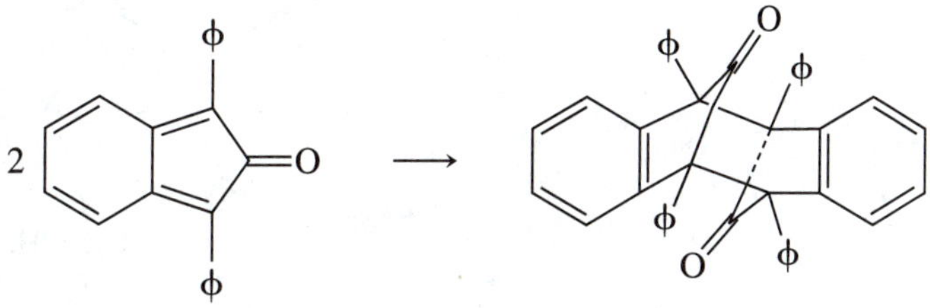

(35a)

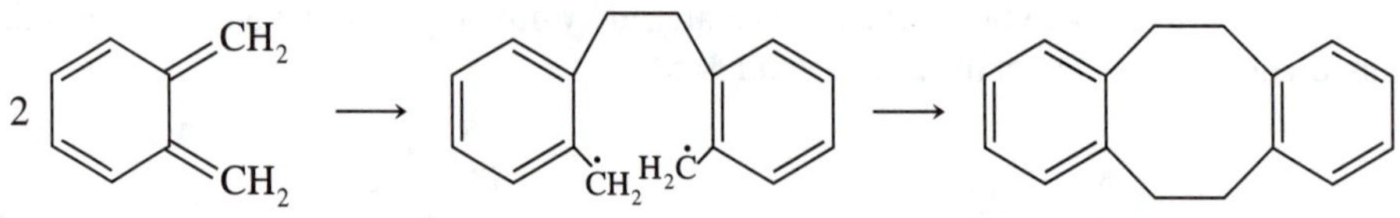

(35b)

The thermally allowed $[_\pi 8 + _\pi 2]$ addition in Eq. 36 proceeds readily.[44] Other $[_\pi 8 + _\pi 2]$ reactions are presumably the initial steps in the reactions shown in Eqs. 37[45] and 38.[46]

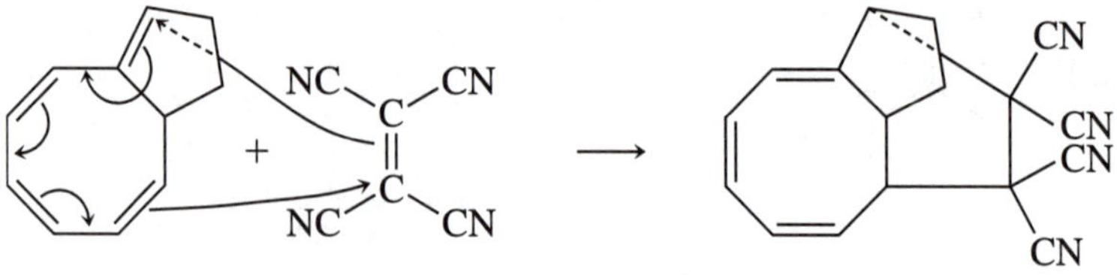

(36)

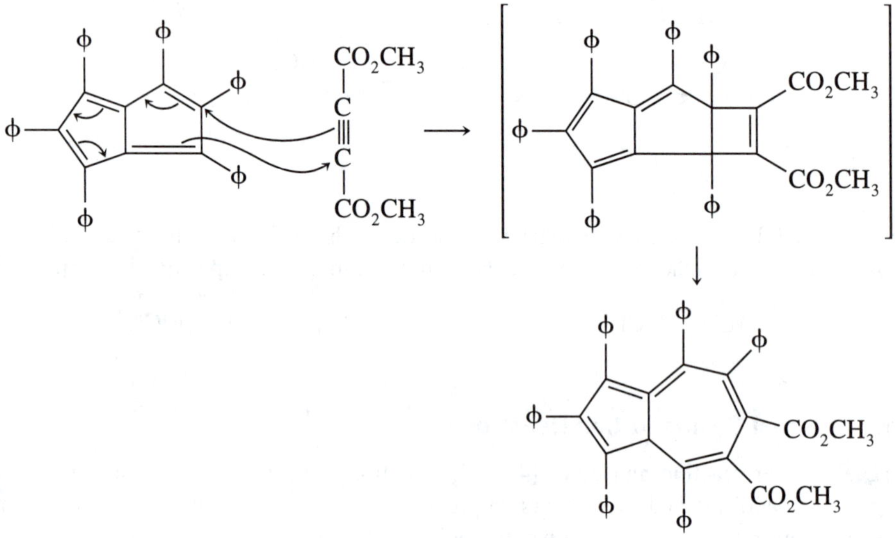

(37)

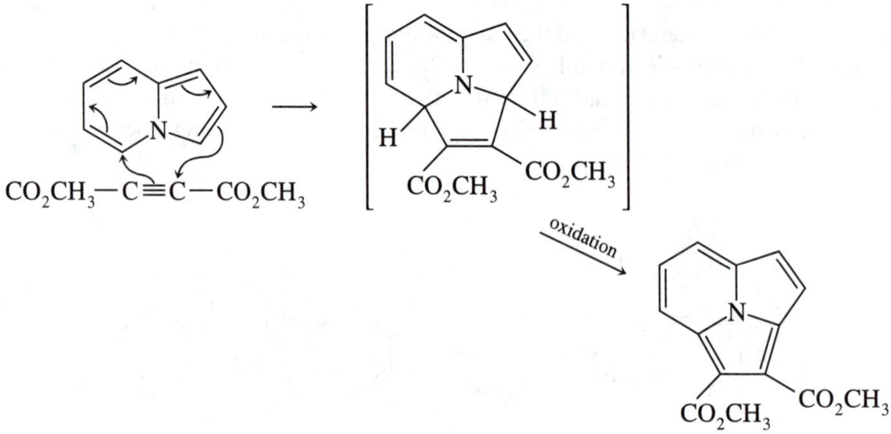

(38)

After the Woodward–Hoffmann rules for cycloaddition were published, examples of [6 + 4] reactions were sought. Several were found, including those shown in Eqs. 39, 40, and 41.[47–49]

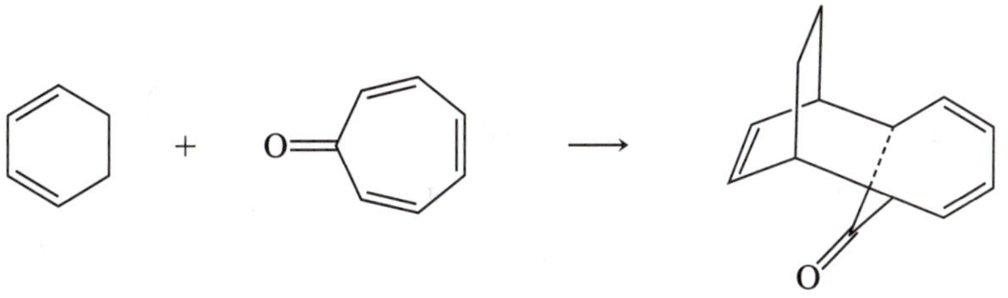

(39)

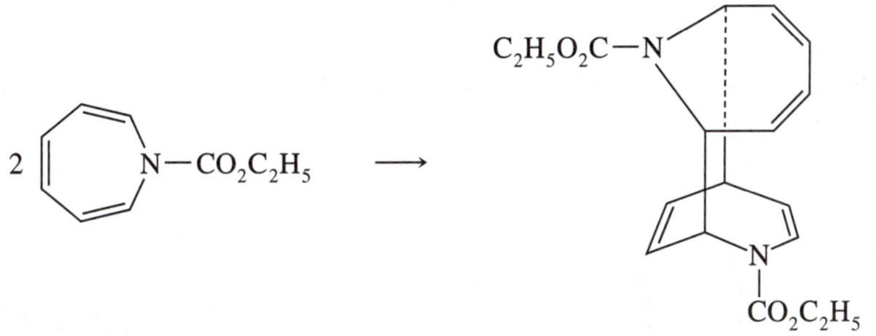

(40)

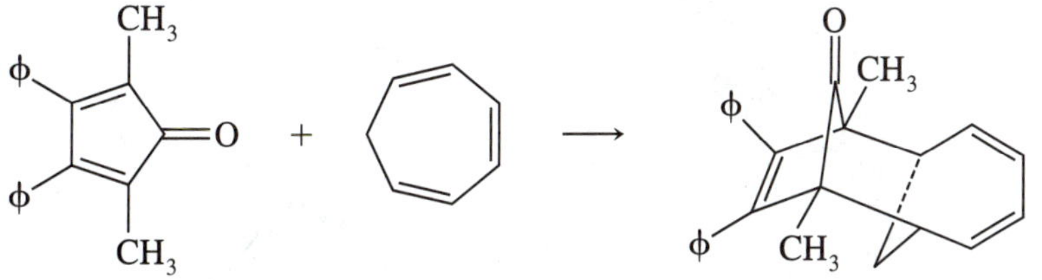

(41)

Unfortunately, only suprafacial addition is possible in these reactions. Therefore, though these reactions are consistent with the Woodward–Hoffmann rules, they do not rule out the possibility of diradical mechanisms. The [6 + 4] cycloadditions appear to be favored over possible Diels–Alder reactions. The preference for [6 + 4] cycloaddition was accounted for on the basis that [6 + 4] reactions connect the atoms having the largest coefficients in the LUMOs and HOMOs of the starting materials. It might also be explained by the reasoning that the transition state with the largest number of delocalized electrons will be lower in energy than other aromatic transition states.

Cycloadditions involving still larger numbers of electrons are quite rare. A $[_\pi 12 + _\pi 2]$ reaction (Eq. 42), as predicted, proceeds suprafacially.[50]

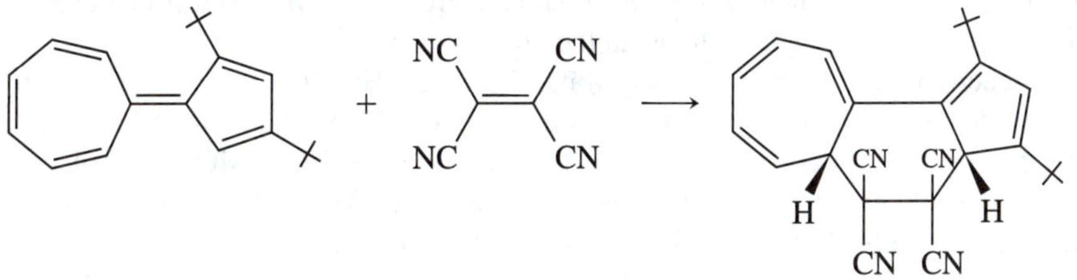

(42)

In contrast, Woodward and Hoffmann[51] quoted work indicating that the $[_\pi 14 + _\pi 2]$ reaction shown in Eq. 43 proceeds antarafacially. However, the work has not yet been published.

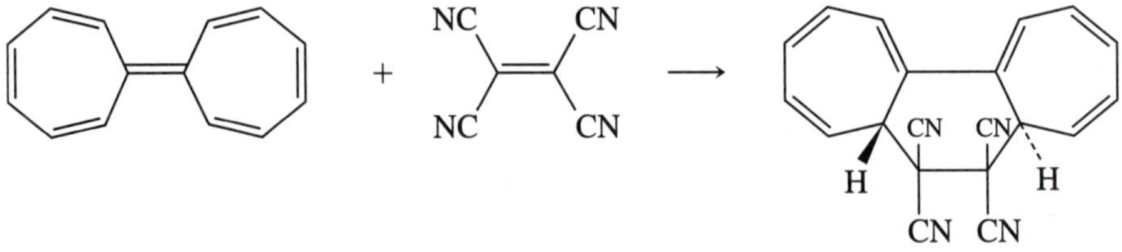

(43)

Finally, a $[_\pi 18 + _\pi 2]$ cycloaddition was reported (Eq. 44), but the geometry of the product has not yet been established.

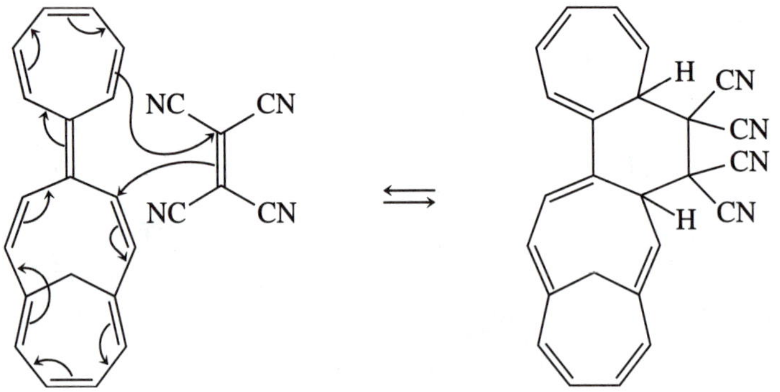

(44)

3.5   PHOTOCHEMICAL CYCLOADDITIONS

[2 + 2] Reactions

Since molecules are unlikely to have long lifetimes in excited states, the probability of two molecules encountering one another when both are in excited states is small. Photochemical cycloaddition reactions, when they occur, should therefore normally result from the reaction of one molecule in an excited state with one in the ground state.

According to frontier molecular orbital theory, the HOMO (π^*) of an alkene in an excited state would have the correct symmetry for a suprafacial [2 + 2] addition to an alkene molecule in its ground state. In contrast, in [4 + 2] cycloadditions, the HOMO of a molecule in an excited state would have the opposite symmetry from the LUMO of a molecule in the ground state. Only antarafacial [4 + 2] cycloadditions would be allowed.

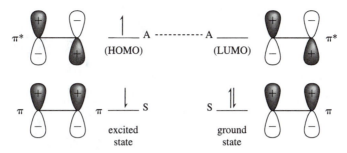

A suprafacial photochemical [2 + 2] cycloaddition is allowed.

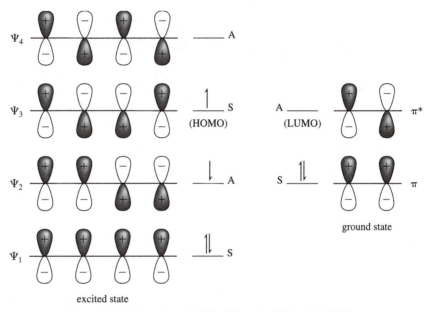

A suprafacial photochemical [4 + 2] cycloaddition is forbidden.

Reactions from Singlet and Triplet States

The photochemical excitation of a π bond will initially raise it to an excited *singlet* state. However, alkenes in excited singlet states are likely to drop back to ground states before they add to another alkene. As a result, singlet bimolecular photoaddition reactions of alkenes tend to be inefficient processes that proceed with low quantum yields. (That is, many quanta of light must be absorbed—and many molecules temporarily raised to excited states—for each molecule that undergoes a photoaddition reaction.)

Alkenes *can* undergo photochemical cycloaddition reactions efficiently if the double bonds are held in close proximity. Some fascinating molecules, such as hydrocarbons **5**[52] and **6**,[53] can be formed in this way.

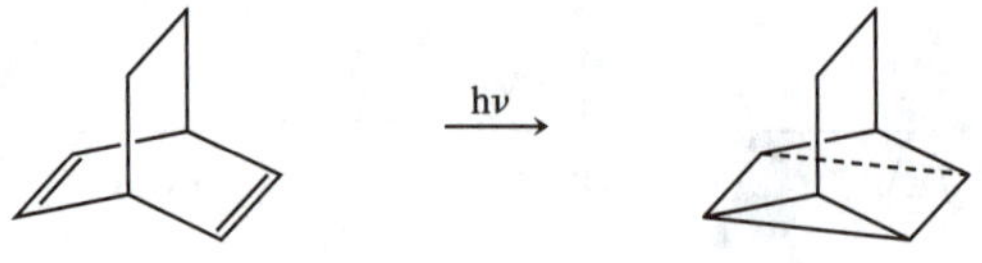

(45)

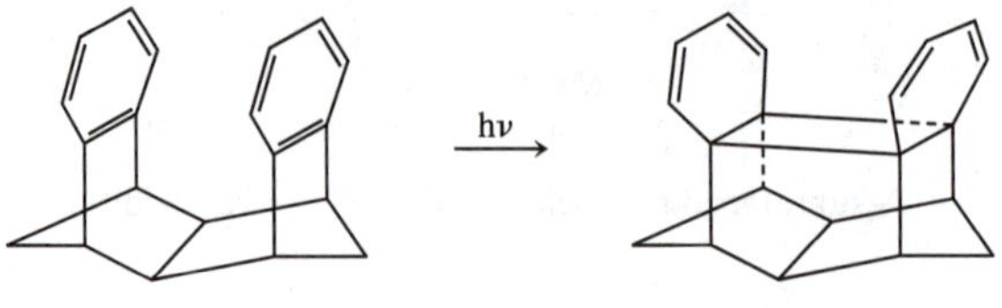

5

(46)

6

Singlet-state cycloaddition reactions of open-chain alkenes, such as the dimerizations of *cis*- and *trans*-2-butene (Eqs. 47 and 48), proceed by stereospecific $[_\pi 2_s + {}_\pi 2_s]$ paths. (These reactions are only stereospecific at low conversions—that is, after very short reaction times—because alkenes can undergo *cis–trans* isomerizations on photoirradiation.[54])

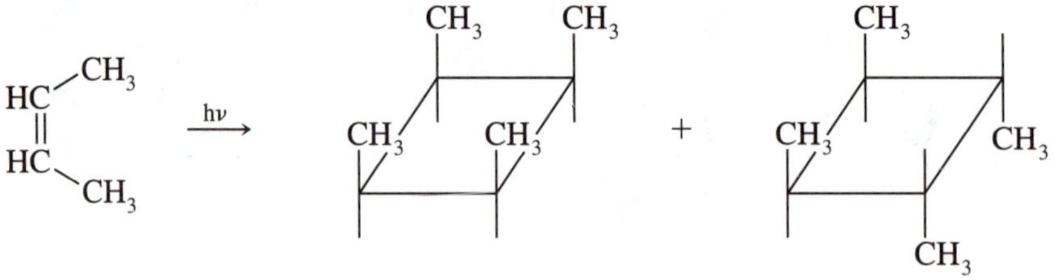

(47)

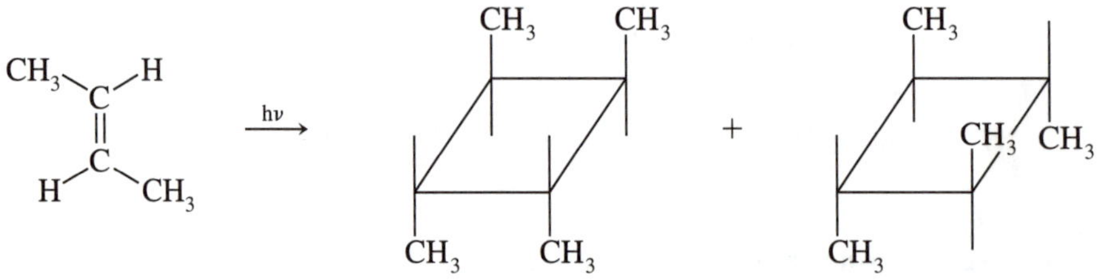

(48)

Some photochemical [2 + 2] cycloreversions, such as those in Eqs. 49 and 50, have also been shown to proceed suprafacially.[55]

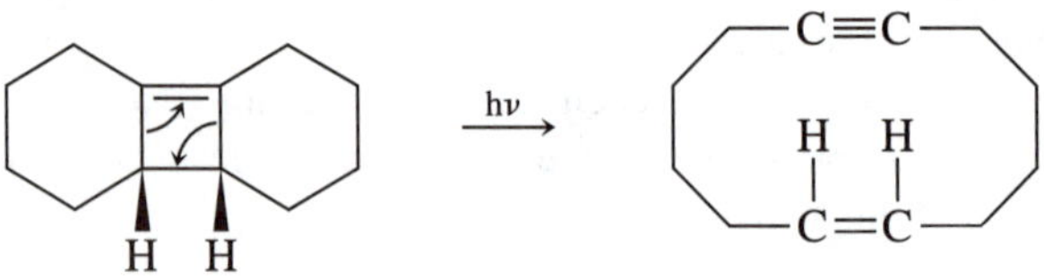

(49)

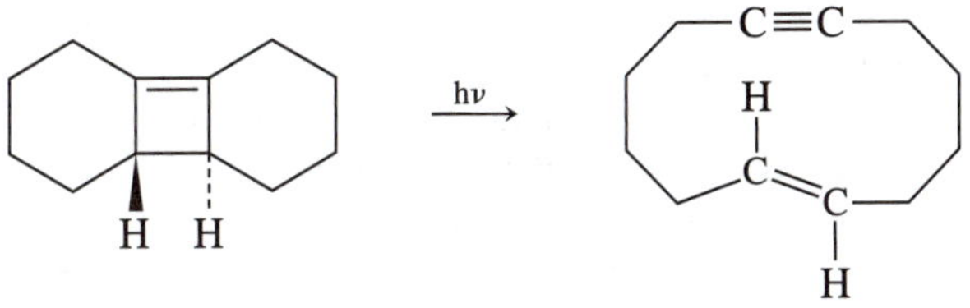

$$\text{(50)}$$

As noted earlier, intermolecular [2 + 2] reactions of alkenes proceeding by singlet paths are usually inefficient processes. However, triplet excited states of alkenes are usually longer lived than singlet excited states. They are therefore more likely to undergo cycloaddition reactions with other alkene molecules. Since conversion of the excited singlet state of an alkene to a triplet state (intersystem crossing) is relatively slow, photocycloaddition reactions are often assisted by the addition of *photosensitizers*. Photosensitizers (such as benzophenone) are molecules that are easily excited to triplet states and can then transfer their energy to other molecules, converting them to triplets, while themselves dropping back to ground states. If the resulting triplets then react with other alkenes, they will form triplet diradicals, as shown below.

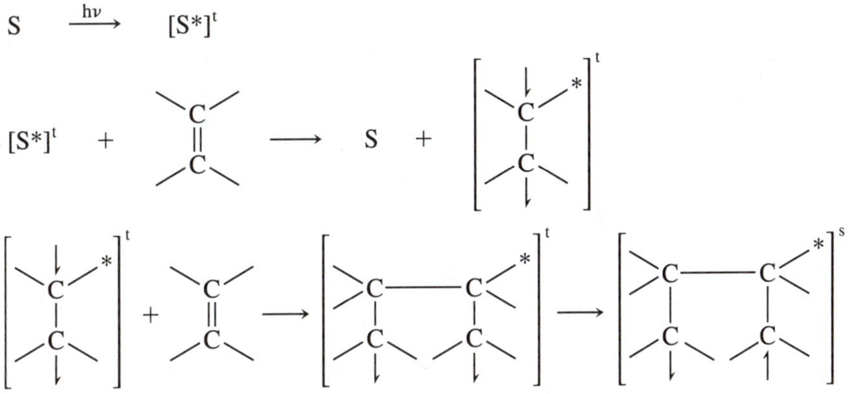

S = sensitizer   t = triplet state   s = singlet state

Stages in photosensitized $2\pi + 2\pi$ cycloadditions

Triplet diradicals cannot directly cyclize to form cyclobutane rings, because the spins of the odd electrons are unpaired, but they can undergo intersystem crossing to singlet states by contact with other molecules and can then cyclize. Cyclobutane formation via triplet states is typically not a stereospecific reaction (e.g., Eq. 51).[56]

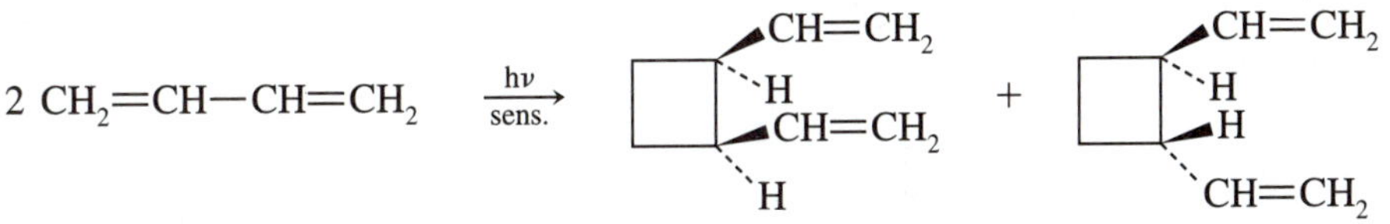

$$\text{(51)}$$

The excited states of α, β-unsaturated carbonyl compounds are more easily converted to triplets than are those of alkenes or conjugated dienes. The yields of cycloaddition products from the photoirradiation of unsaturated carbonyl compounds are therefore frequently higher than those from the photoirradiation of alkenes.[57] These reactions may result in the formation of dimers of unsaturated ketones or other carbonyl compounds (Eq. 52)[58] or in the additions of the carbon–carbon double bonds of the carbonyl compounds to other double bonds, as shown in Eqs. 53[59] and 54.[60]

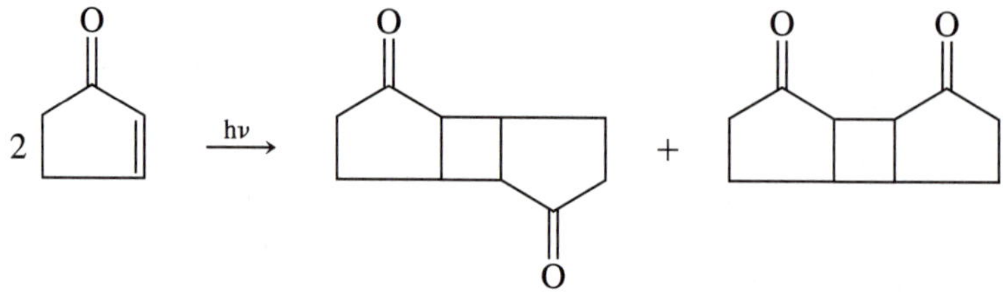

(52)

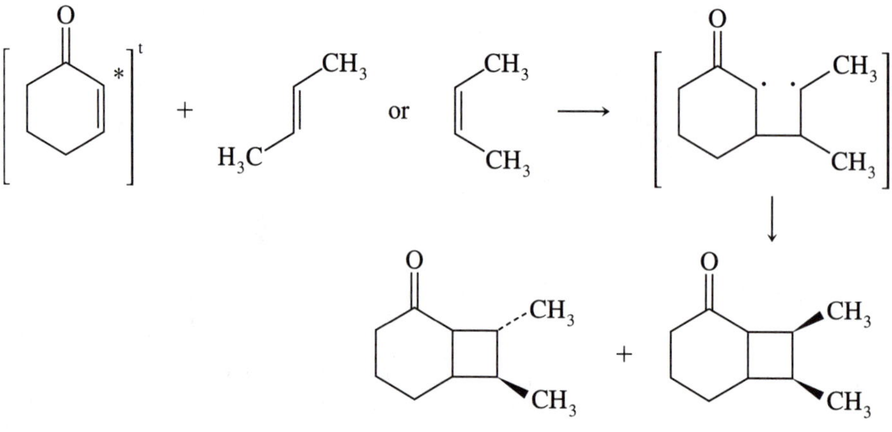

(53)

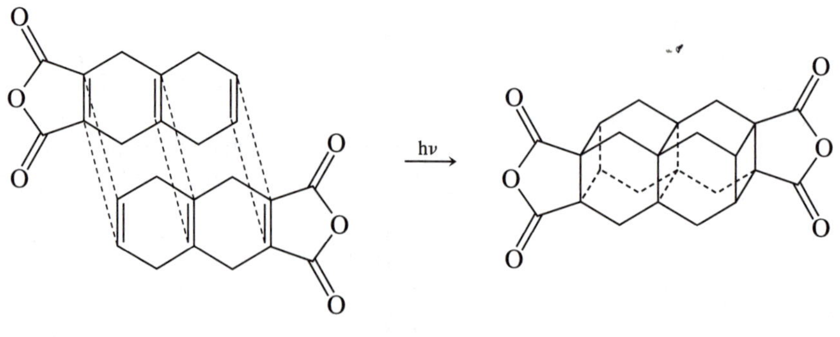

(54)

REFERENCES

[1] R.B. Woodward and R. Hoffmann, *J. Am. Chem. Soc., 87,* 2511 (1965).

[2] R.B. Woodward and R. Hoffmann, *The Conservation of Orbital Symmetry.* Academic Press, New York (1971); T.L. Gilchrist and R.C. Storr, *Organic Reactions and Orbital Symmetry,* 2nd ed. Cambridge University Press, New York (1971).

[3] J.E. Baldwin, in *Comprehensive Organic Synthesis,* vol. 5, L.A. Paquette, ed. Pergamon Press, New York (1991), pp. 63–84.

[4] P.B. Dervan and D.A. Dougherty, in *Diradicals,* W.T. Borden, ed. John Wiley, New York (1982); K. Hassenrueck, H.D. Martin, and R. Walsh, *Ber., 121,* 369 (1988).

[5] S. Pedersen, J.L. Herek, and A.H. Zewail, *Science, 266,* 1359 (1994).

[6] P.G. Gassman and H.P. Benecke, *Tetrahedron Lett.,* 1089 (1969); J. Leitich, *Tetrahedron Lett., 21,* 3025 (1971).

[7] A. Padwa, W. Koehn, J. Masaracchia, C.L. Osborn, and D.J. Trecker, *J. Am. Chem. Soc., 93,* 3633 (1971).

[8] L.K. Montgomery, K. Schueller, and P.D. Bartlett, *J. Am. Chem. Soc., 86,* 622 (1964); P.D. Bartlett, *Science, 159,* 833 (1968); P.D. Bartlett et al., *J. Am. Chem. Soc., 94,* 2899 (1972).

[9] S.J. Getty and W.T. Borden, *J. Am. Chem. Soc., 113,* 4334 (1991).

[10] L. Ghosez and M.J. O'Donnell, in *Pericyclic Reactions,* vol. II, A.P. Marchand and R.E. Lehr, eds. Academic Press, New York (1977), pp. 85–109. B.B. Snider, *Chem. Rev., 88,* 793 (1988); S. Xu, H. Xia, and H.W. Moore, *J. Org. Chem., 56,* 6094 (1991).

[11] S. Masamune and K. Fukumoto, *Tetrahedron Lett.,* 4647 (1965).

[12] R. Huisgen and L.A. Feiler, *Ber., 102,* 3391 (1969).

[13] X. Wang and K.N. Houk, *J. Am. Chem. Soc., 112,* 1754 (1990); F. Bernardi, A. Bottini, M.A. Robb, and A. Venturini, *ibid.,* 2106.

[14] S. Proskow, H.E. Simmons, and T.L. Cairns, *J. Am. Chem. Soc., 85,* 2341 (1963).

[15] R. Huisgen and G. Steiner, *J. Am. Chem. Soc., 95,* 5054 (1973).

[16] R. Huisgen, *Acc. Chem. Res., 10,* 117 (1977); R. Huisgen and G. Mloston, *Tetrahedron Lett., 35,* 4971 (1994).

[17] For reviews, see A.S. Onischenko, *Diene Synthesis.* Daniel Davey & Co., New York (1964); W. Carruthers, *Cycloaddition Reactions in Organic Synthesis.* Pergamon Press, New York (1990). See also chapters by W. Oppolzer, S.M. Weinreb, D.L. Boger, W.A. Roush, and R.W. Sweger and A.W. Czarnik, pp. 315–592 in *Comprehensive Organic Synthesis,* vol. 5, L.A. Paquette, ed. Pergamon Press, New York (1991).

[18] P.D. Bartlett and J.J.-B. Mallett, *J. Am. Chem. Soc., 98,* 143 (1976).

[19] K. Alder and G. Stein, *Angew. Chem., 50,* 510 (1937).

[20] A. Wasserman, *Diels-Alder Reactions.* Elsevier Publishing Co., New York (1965); J.A. Berson and W.A. Mueller, *J. Am. Chem. Soc., 83,* 4947 (1961).

[21] R. Hoffmann and R.B. Woodward, *J. Am. Chem. Soc., 87,* 4388 (1965).

[22] P. Caramella, P. Quadrelli, and L. Toma, *J. Am. Chem. Soc., 104,* 1130 (2002).

[23] L.M. Joshel and L.W. Butz, *J. Am. Chem. Soc., 63,* 3350 (1971).

[24] D.L. Boger and M.J. Kochanny, *J. Org. Chem., 59,* 4950 (1994).

[25] H.B. Kagan and O. Riant, *Chem. Revs., 92,* 1007 (1992).

[26] (a) K.N. Houk, *Acc. Chem. Res., 8,* 361 (1975); I.-M. Tegmo-Larsson, M.D. Rozeboom, N.G. Rondan, and K.N. Houk, *Tetrahedron Lett., 22,* 2047 (1981).

[27] A. Michael and J.E. Bucher, *Ber., 28,* 2511 (1895).

[28] R.F. Danheiser, A.E. Gould, R.F. de la Pradilla, and A.L. Helgason, *J. Org. Chem., 59,* 5514 (1994).

[29] J.J. González, A. Francesch, D.J. Cárdenas, and A.M. Eschavarren, *J. Org. Chem., 63,* 2854 (1998).

[30] H.W. Whitlock, Jr., E.M. Wu, and B.J. Whitlock, *J. Org. Chem., 34,* 1857 (1969).

[31] B. Miller and X. Shi, *J. Am. Chem. Soc., 109,* 578 (1987); M. Christl, M. Braun, and G. Müller, *Angew. Chem., Int. Ed. Engl., 31,* 473 (1992).

[32] E. Dané, O. Höss, A.W. Bendseil, and J. Schmitt, *Liebigs Ann., 532,* 39 (1937).

[33] B. Miller and D. Ionescu, *Tetrahedron Lett., 35,* 6615 (1994).

[34] L.W. Butz, A.M. Gaddis, E.W.J. Butz, and R.E. Davis, *J. Org. Chem., 15,* 379 (1940); L.W. Butz, A.M. Gaddis, E.W.J. Butz, *J. Am. Chem. Soc., 69,* 924 (1947).

[35] D. Ionescu, J.V. Silverton, L.C. Dickinson, and B. Miller, *Tetrahedron Lett., 37,* 1559 (1196).

[36] H.K. Hall, Jr., *J. Org. Chem., 25,* 42 (1960).

[37] D.L. Mohler, K.P.C. Vollhardt, and S. Wolff, *Angew. Chem. Intl. Ed. Engl., 29,* 1151 (1990).

[38] M.G. Kociolek and R.P. Johnson, *Tetrahedron Lett., 40,* 4141 (1999).

[39] B.B. Snider, in *Comprehensive Organic Synthesis,* vol. 5, L.A. Paguette, ed. Pergamon Press, New York (1991), pp. 1–25.

[40] Z. Song and P. Beak, *J. Am. Chem. Soc., 112,* 8126 (1990).

[41] C.P. Cataya-Marin, A.C. Jackson, and B.B. Snider, *J. Org. Chem., 49,* 2443 (1984); A.C. Jackson, B.E. Goldman, and B.B. Snider, *ibid., 49,* 3988 (1984).

[42] J.M. Holland and D.W. Jones, *Chem. Comm.,* 587 (1969).

[43] L.A. Errede, *J. Am. Chem. Soc., 83,* 949 (1961).

[44] P.H. Ferber, G.E. Gream, P.K. Kirkbride, and E.R.T. Tiekink, *Aust. J. Chem., 43,* 463 (1990).

[45] E. Le Goff, *J. Am. Chem. Soc., 84,* 3975 (1962).

[46] A. Galbraith, T. Small, R.A. Barnes, and V. Boekelheide, *J. Am. Chem. Soc., 83,* 453 (1961).

[47] R.C. Cookson, B.V. Drake, J. Hudec, and A. Morrison, *Chem. Comm.,* 15 (1966). See also T. Machiguchi and S. Yamabe, *Tetrahedron Lett., 41,* 4169 (1990); M. Wollenweber, H. Fritz, G. Rihs, and H. Prinzbach, *Ber., 124,* 2465 (1991).

[48] T. Mukai, T. Tezuka, and Y. Akasaki, *J. Am. Chem. Soc., 88,* 5025 (1966).

[49] K.N. Houk and R.B. Woodward, *J. Am. Chem. Soc., 92,* 4143 (1970).

[50] H. Prinzbach and H. Knoefel, *Angew. Chem., Intl. Ed. Engl., 8,* 881 (1969).

[51] Reported in R.B. Woodward and R. Hoffmann. *The Conservation of Orbital Symmetry.* Academic Press, New York (1971), p. 85.

[52] W.G. Dauben and R.L. Cargill, *Tetrahedron, 15,* 197 (1961).

[53] G. Sedelmeir et al., *Tetrahedron Lett., 27,* 1277 (1986).

[54] H. Yamazaki and R.J. Cvetanovic, *J. Am. Chem. Soc., 91,* 520 (1969).

[55] J. Saltiel and L.-S. Ng Lim, *J. Am. Chem. Soc., 91,* 5404 (1969).

[56] G.S. Hammond, N.J. Turro, and A. Fischer, *J. Am. Chem. Soc., 83,* 4674 (1961); G.S. Hammond, N.J. Turro, and R.S.H. Liu, *J. Org. Chem., 28,* 3297 (1963).

[57] See D.I. Schuster, G. Lem, and N.A. Kaprindis, *Chem. Rev., 93,* 3 (1993).

[58] P.E. Eaton, *J. Am. Chem. Soc., 84,* 2344 (1962).

[59] E.J. Corey, J.D. Bass, R. Le Mahieu, and R.B. Mitra, *J. Am. Chem. Soc., 86,* 5570 (1964).

[60] V.T. Hoffmann and H. Musso, *Angew. Chem. Intl. Ed. Engl., 26,* 1006 (1987).

PROBLEMS

3.1 Explain why orbital symmetry conservation rules allow or forbid each of the following reactions to occur as a concerted process.

(a)

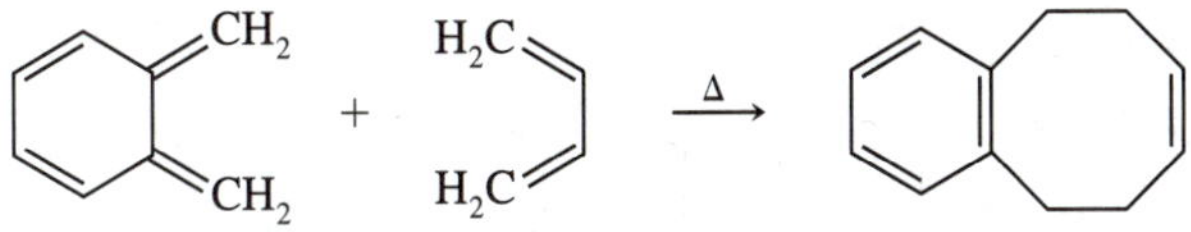

(b)

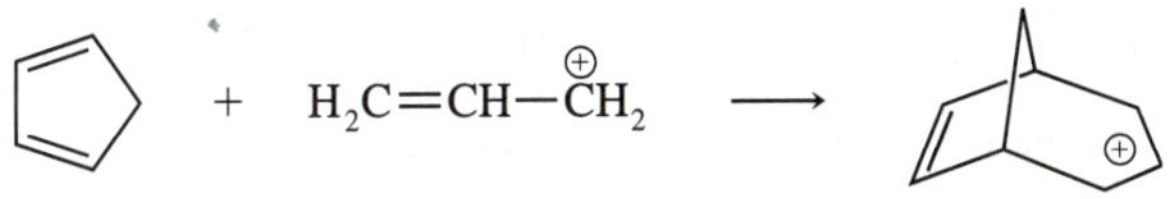

(c)

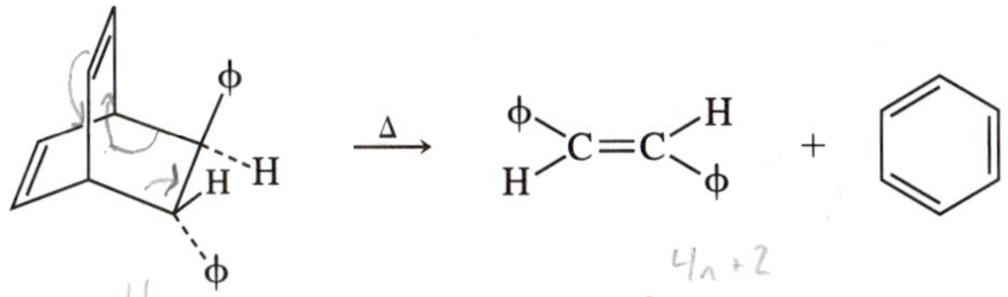

(d)

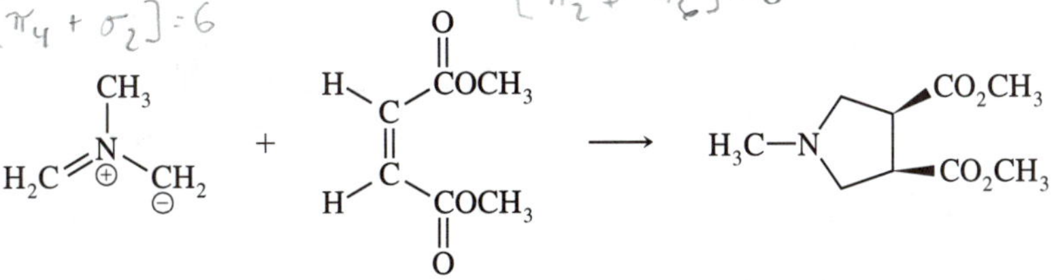

(e) 2

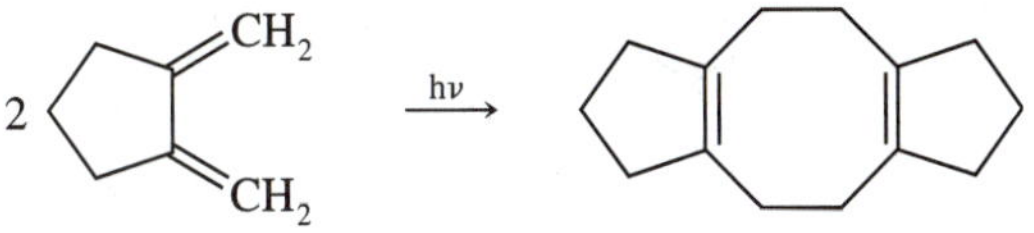

(f)

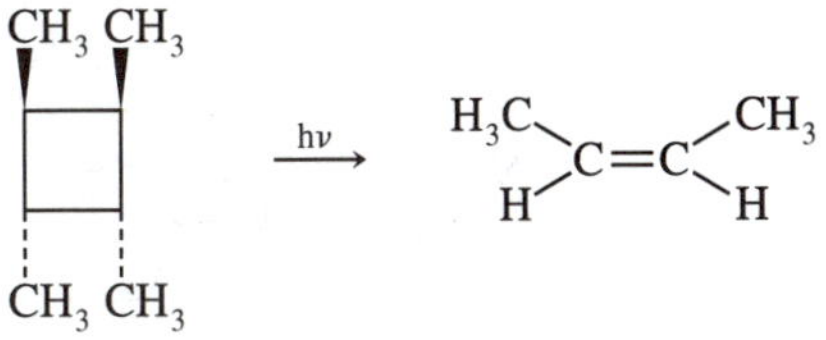

3.2 Write reasonable mechanisms for the following reactions, using curved arrows to show movements of electrons. Do not combine steps.

(a)

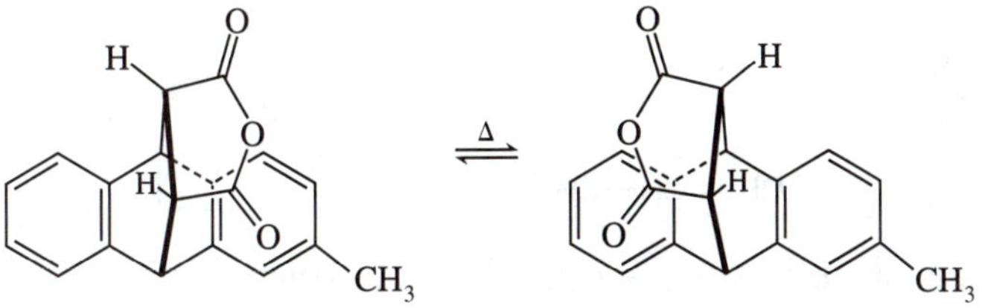

(b)

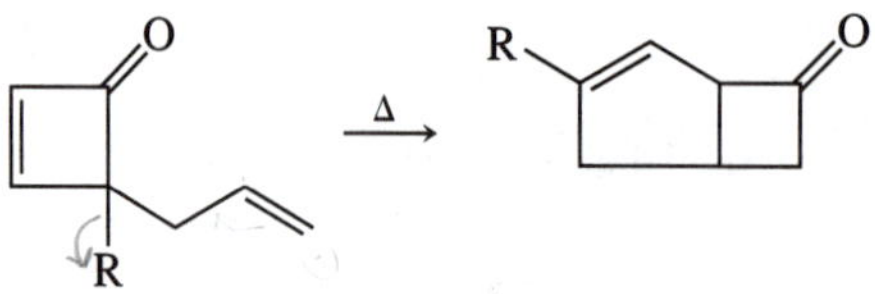

(See ref. 9c)

(c)

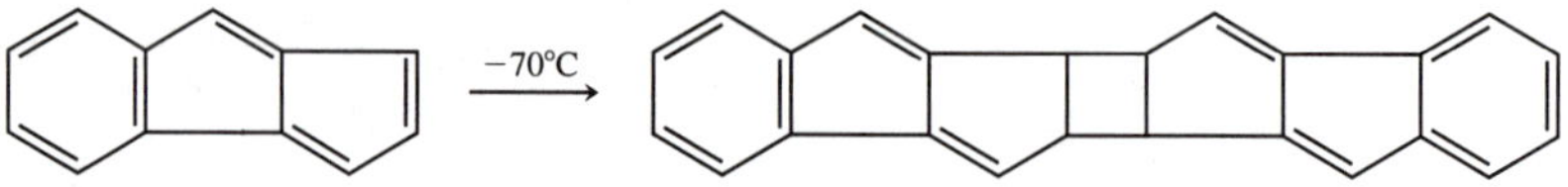

(R.F. Brown et al., *Tetrahedron Lett.*, *35*, 4405 [1994])

(d) Account for the ease with which reaction 2c takes place.

(e)

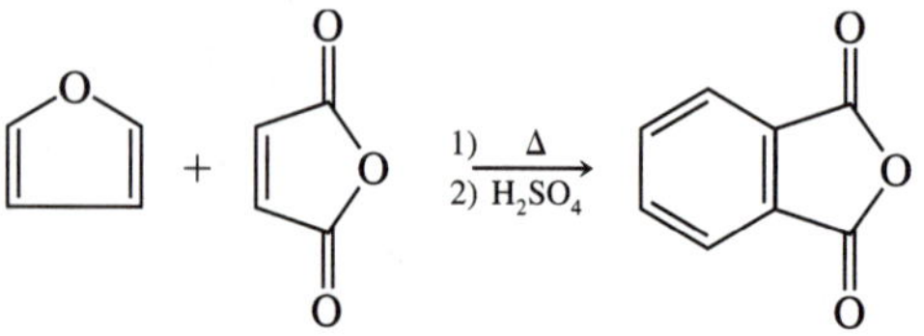

(f)

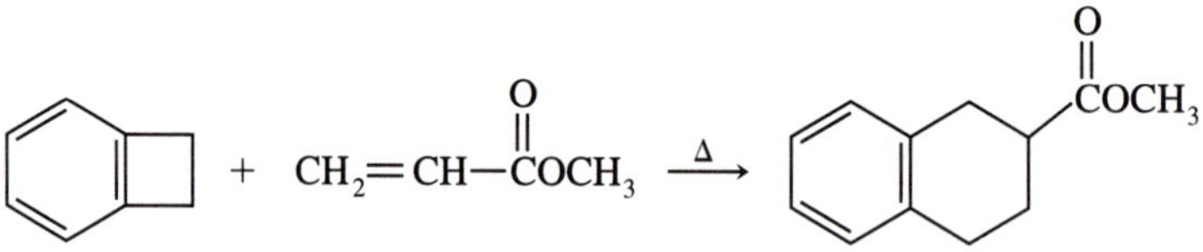

(g)

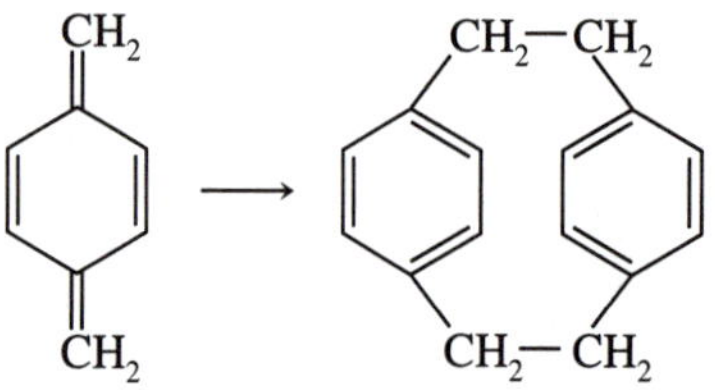

(h)

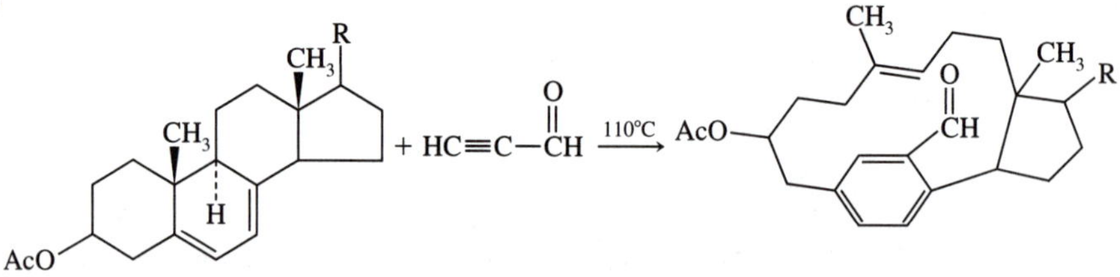

(D. Schomburg, M. Thielmann, and E. Winterfeldt, *Tetrahedron Lett.*, *26*, 1705 [1985])

(i)

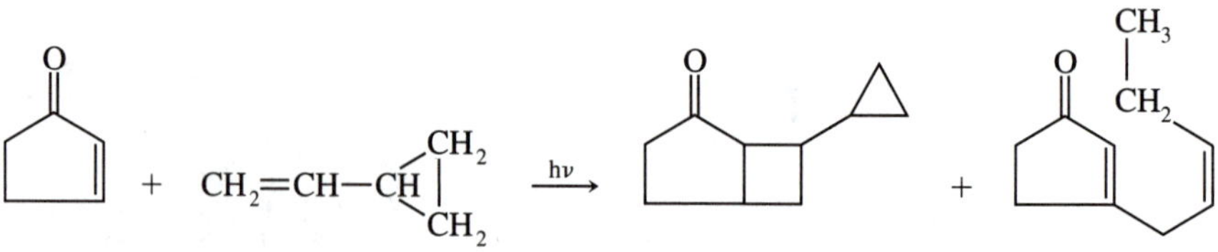

(A. Rudolf and A. C. Weedon, *Can. J. Chem.*, *68*, 1590 [1990])

(j)

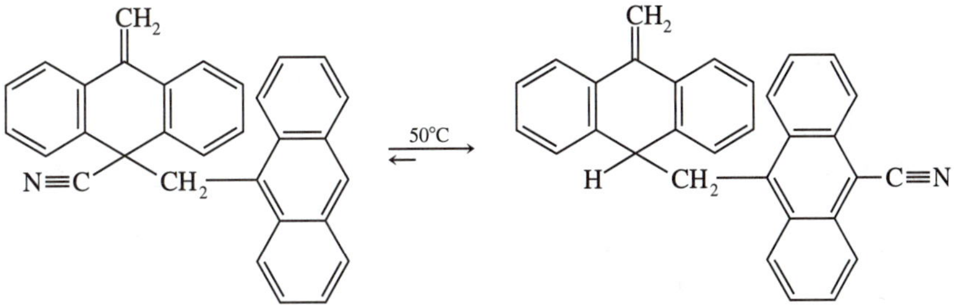

(H. D. Becker and K. Anderson, *Tetrahedron, 42,* 1555 [1986])

3.3 Draw structures for the principal products of the following reactions, or write "no reaction." Show the configurations of the reaction products if they can be determined from the structures of the starting materials.

(a)

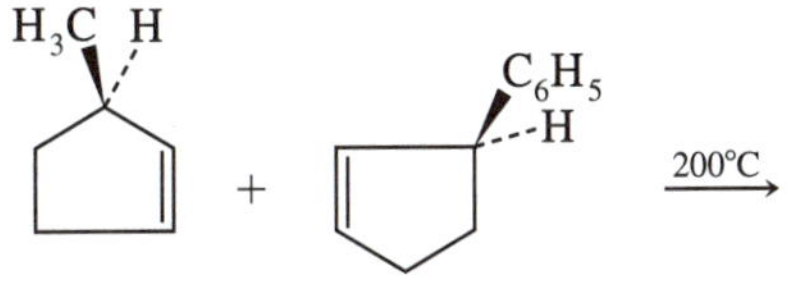

(b)

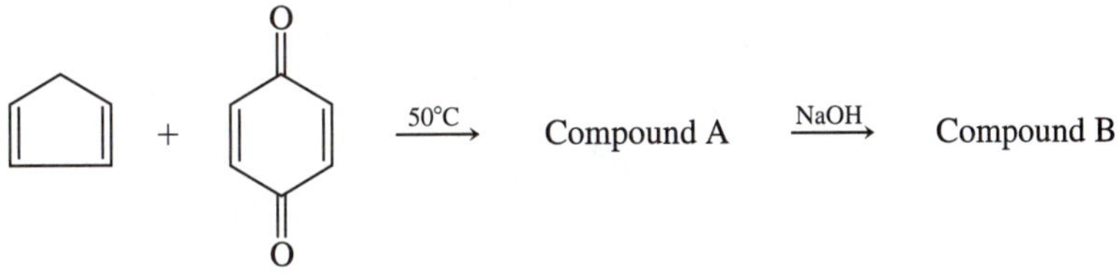

(c)

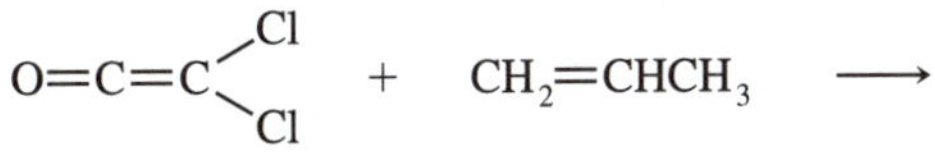

(d)

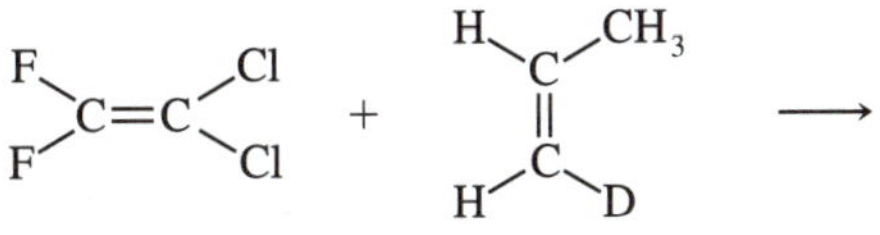

(e)

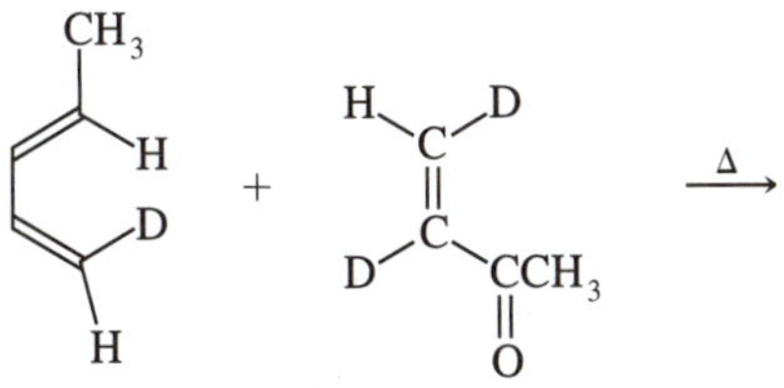

(f)

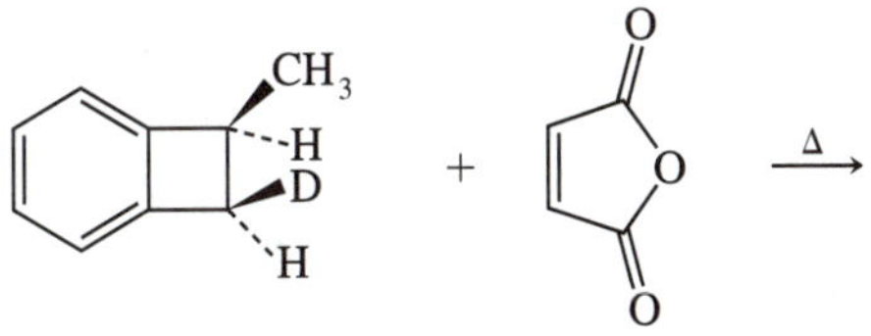

Sigmatropic Reactions

4.1 THEORY OF SIGMATROPIC SHIFTS

Pericyclic Rearrangements

While many common rearrangements require the formation of carbocations or other reactive intermediates, some rearrangement processes proceed directly by concerted pericyclic mechanisms. One example is the *Cope rearrangement* (Eq. 1), which results in the interconversion of two 1,5-hexadiene derivatives on heating.[1]

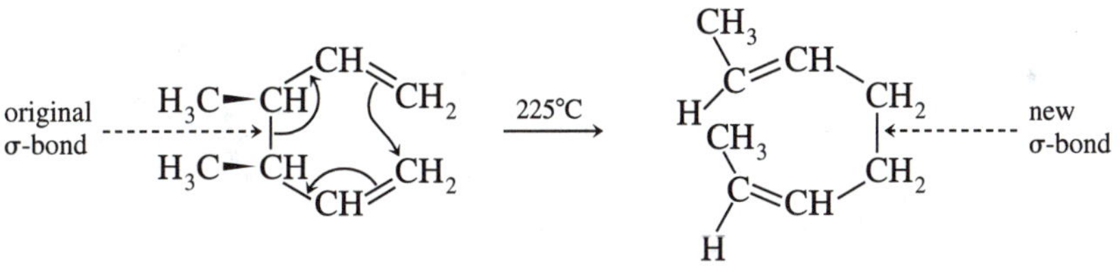

$$\text{(1)}$$

Migrations of hydrogen atoms in conjugated dienes can also proceed by concerted pericyclic mechanisms, as shown in Eq. 2.

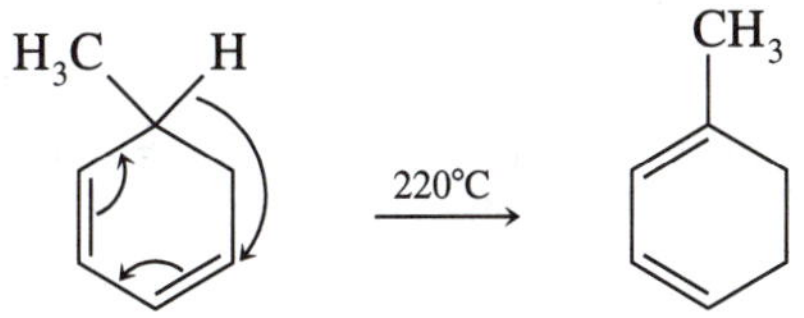

$$\text{(2)}$$

Since each of these reactions results in the change in position of one σ bond (as well as of several π bonds), Woodward and Hoffmann coined the term "sigmatropic shifts" to describe them.[2,3] In a sigmatropic shift, a σ bond in an allylic position* migrates to the further end of the

---

*The migrating σ bond may instead be linked to an atom bonded to an atom with an empty orbital or to one bonded to an atom bearing unshared electrons (see Chapters 6 through 8).

adjoining π bond or to the end of a set of conjugated π bonds. (In a Cope rearrangement, a "doubly allylic bond" migrates to the ends of both π bonds.)

Woodward and Hoffmann classified different types of sigmatropic shifts as being rearrangements of different "orders." To identify the order of a particular sigmatropic shift, the two atoms forming the bond being broken are both numbered atom 1. Then, the atoms in each direction from the bond being broken, up to and including the atoms that form the new σ bond in the product, are numbered consecutively as atoms 2, 3, and so forth. The numbers assigned to the atoms forming the new bond, separated by commas, are placed within brackets to designate the reaction order.

According to this rule, Cope rearrangements are sigmatropic shifts of order [3,3].

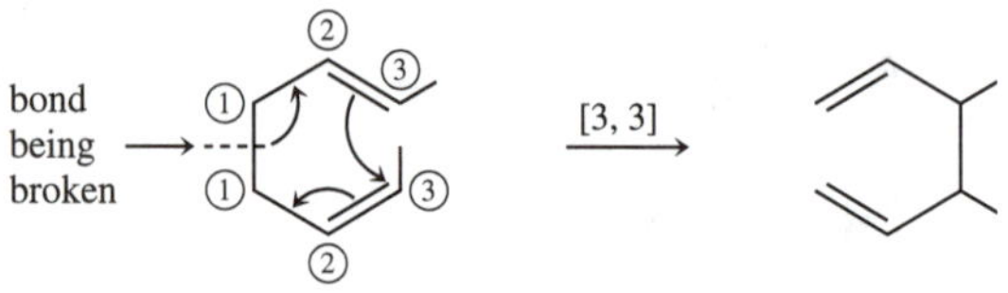

The migration of a hydrogen shown in Eq. 2 is a sigmatropic shift of order [1,5]. Note that this is *not* because the hydrogen migrates from carbon 1 to carbon 5, but because the hydrogen (one of the two atoms bearing the number 1) will form part of the σ bond being made as well as forming part of the bond being broken.

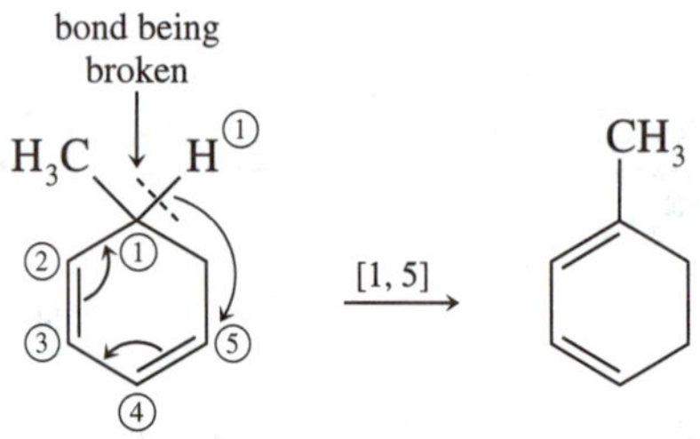

$$(2)$$

All of the atoms taking part in the reaction (that is, atoms forming part of a bond being made or broken), and only those atoms, must be counted when determining the order of a sigmatropic shift. Thus, the rearrangement of the cyclohexadiene shown in Eq. 2 cannot be called a [1,3] shift rather than a [1,5] shift, since the CH_2 group linking atoms 1 and 5 does not take part in the reaction.

Suprafacial and Antarafacial Shifts

In theory, every sigmatropic shift might result in retention or inversion of the geometry of the migrating group. (That is, the new bond might be formed using the original bonding lobe of the migrating atom or using its back lobe, as shown on page 91.)

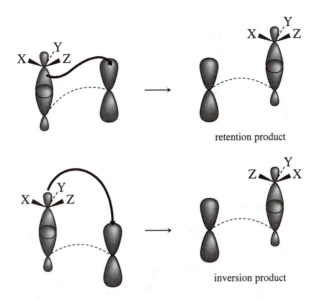

retention product

inversion product

There is another aspect to the stereochemistry of sigmatropic shifts. In principle, the migrating group might remain on the same "face" of the π system as that from which it started, or it might end up on the opposite face. Woodward and Hoffmann defined a sigmatropic reaction in which the migrating group remains on the original face of the π system as a *suprafacial* rearrangement and one in which the migrating group ends up on the opposite face as an *antarafacial* rearrangement.

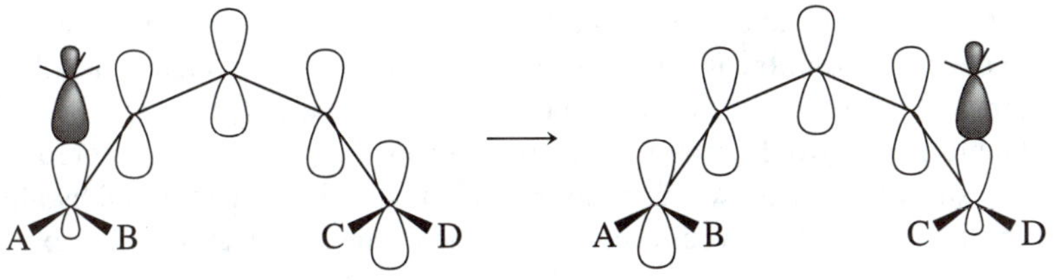

A suprafacial [1, 5] shift

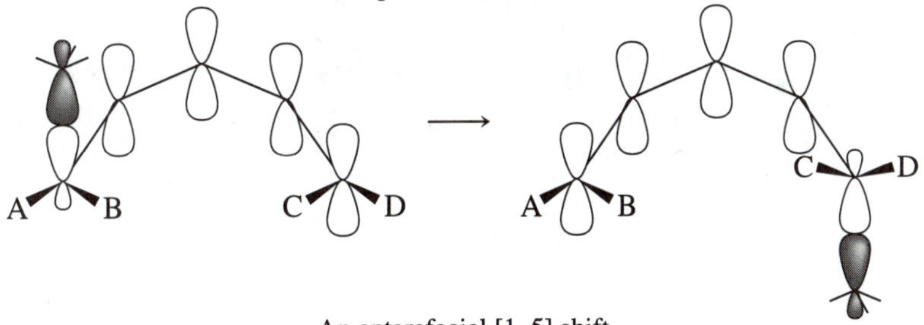

An antarafacial [1, 5] shift

An antarafacial rearrangement is actually an inversion process, but rather than causing the inversion of the geometry of a single chiral atom, it results in moving the migrating group to the opposite face of a ring or double-bond system. This can be most easily seen in the reactions of cyclic systems, such as the (very hypothetical) [1,5] antarafacial migration of a methyl group, shown below. If this reaction actually took place—*it does not*—it would convert a *cis* derivative of cyclohexadiene into a *trans* form.

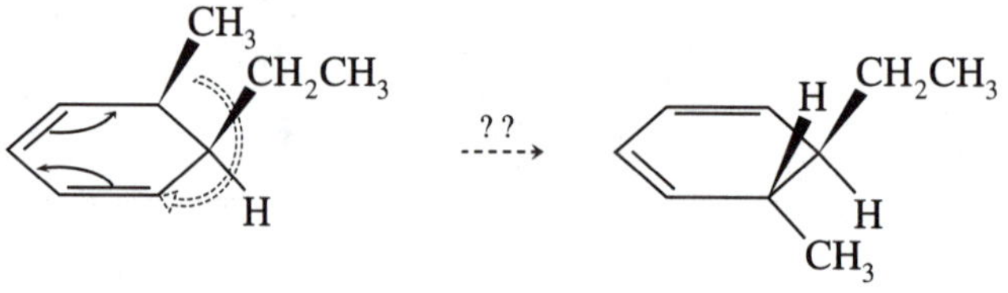

Inversion of geometry resulting from a hypothetical antarafacial rearrangement

Application of Symmetry Conservation Rules

The geometries of sigmatropic rearrangements, like those of other pericyclic reactions, can be predicted from electronic theory.

The aromatic transition state approach.   Perhaps the simplest way to predict the stereochemistry of a sigmatropic rearrangement is by counting the number of electrons involved in the reaction and applying the *aromatic transition state* approach. (The electrons involved in a sigmatropic shift include the electrons in the σ bond and in all reacting π bonds, as well as any unshared electrons that become bonding electrons as a result of the reaction.)

The rules are essentially the same as for cycloaddition reactions. Sigmatropic shifts involving $4n$ electrons proceed via an odd number of antarafacial interactions. That will result in an odd number of inversions, either of the geometry of the migrating group or of the geometry of a ring or double bond. Sigmatropic shifts involving $4n + 2$ electrons result in an even number of inversions. For practical purposes, that even number will almost always be zero.

A [1,5] shift of a hydrogen atom, for instance, is a six-electron process (four π electrons and two electrons from the σ bond). The reaction with zero inversions is allowed. The hydrogen atom would be expected to remain on the face of the π system from which it started, and both double bonds would retain their *cis* geometries. (Three other stereoisomers, resulting from rearrangements with two inversions, could theoretically be formed by allowed processes. However, their formation would require strained transition states and is extremely unlikely.)

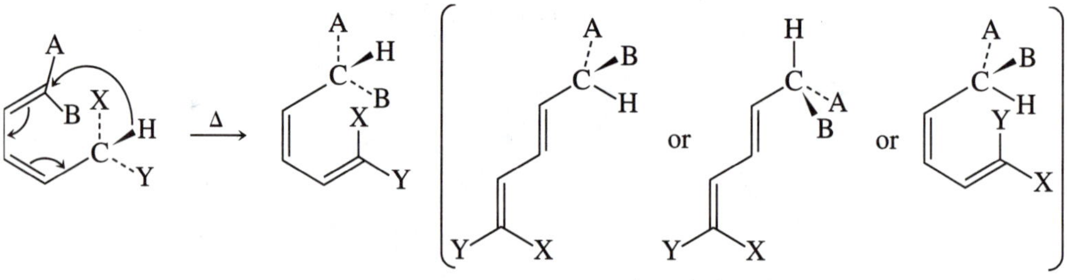

theoretical possibilities only

A technique that is sometimes useful in visualizing the stereochemistry of sigmatropic rearrangements is to regard them as cycloaddition reactions of the individual bonds. This is essentially equivalent to counting electrons. A [1,5] shift, for instance, can be regarded as a $[_\pi 2 + _\pi 2 + _\sigma 2]$ cycloaddition. The expected $[_\pi 2_s + _\pi 2_s + _\sigma 2_s]$ process would yield the most likely product, while reactions involving two antarafacial interactions would, in theory, yield the other possible stereoisomers.

The frontier orbital approach.   The stereochemistries of sigmatropic shifts can also be predicted by frontier orbital methods. One way to do this would be to regard all the reacting bonds—π bonds and σ bonds together—as comprising one big orbital system, and then to consider the symmetry of its HOMO. A more common approach, suggested by Woodward and Hoffmann, is to treat the reaction as though the bond being broken partially ruptures to form a transition state consisting of two associated free radicals. The phase relationships between the HOMOs of the two "radicals" then determine whether the reaction is suprafacial or antarafacial.

In a Cope rearrangement, for example, partial cleavage of the central bond would give rise to a transition state resembling two allylic radicals (see the following diagram). The HOMOs of the two radicals would have the same symmetry and would recombine in a suprafacial manner.

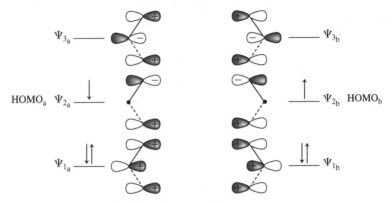

"Diradical-like" transition state for a cope rearrangement

In a [1,5] hydrogen shift, the transition state would resemble a hydrogen atom associated with a pentadienyl radical. The hydrogen orbital has a single lobe, which must be in phase with the lobe of the carbon atom to which it was initially bonded. The HOMO (ψ_3) of a pentadienyl radical is symmetric. Therefore, in a suprafacial migration, the hydrogen orbital would also be in phase with the atomic orbital at the other end of the chain. A suprafacial [1,5] shift would be allowed.

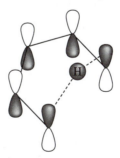

HOMOs in the transition state for a [1, 5] hydrogen shift

The situation would be different in a suprafacial [1,3] hydrogen shift. In that case, since the HOMO (ψ_2) of the allyl orbital is antisymmetric, the hydrogen atom could not be simultaneously in phase with the atomic orbitals at both C1 and C3 of the allyl radical. A suprafacial migration would therefore be forbidden. In theory, an antarafacial reaction, in which the hydrogen atom moved to the opposite side of the π system, would be allowed, but it is geometrically improbable.

An additional possibility would exist if an alkyl group, rather than a hydrogen atom, were the potential migrator. Suprafacial migration with retention of the configuration of the migrating group would again be forbidden. An antarafacial migration with retention, while theoretically

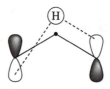

A *theoretically* allowed antarafacial [1, 3] hydrogen migration

allowed, would be very unlikely to occur. A suprafacial migration forming a new bond to the "back" lobe of the migrating orbital, as in the diagram below, would also be theoretically allowed. This process would result in inversion of the configuration of the migrating group. It would obviously require grossly distorted bonds in the transition state and would be an unlikely process, although perhaps not quite so unlikely as a migration to the opposite face of the π system.*

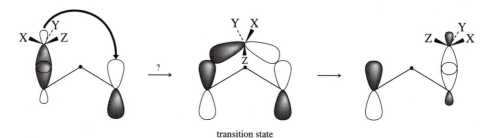

transition state

Suprafacial [1, 3] migration with inversion is allowed

4.2   EXPERIMENTAL OBSERVATIONS

[1,3] Sigmatropic Shifts

The conclusion that suprafacial [1,3] shifts of hydrogen atoms are forbidden is consistent with the fact that double bonds of alkenes do not normally undergo migration at temperatures below about 500°C. Even hydrocarbon **1** is sufficiently stable that it can be prepared by pyrolysis of an ester at 350°C[4] (Eq. 3), despite the fact that its isomerization to toluene by a [1,3] hydrogen shift would be highly exothermic.

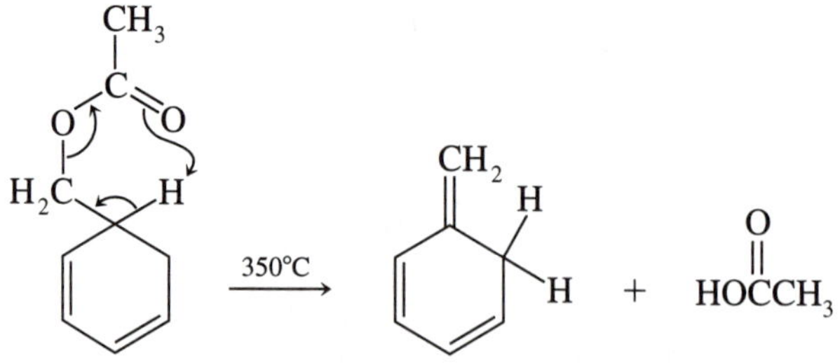

(3)

1

---

*Thermolyses of allenes *can* result in intramolecular [1,3] hydrogen shifts. It was suggested that both π bonds may participate in these reactions, resulting in rather peculiar six-electron processes.[5]

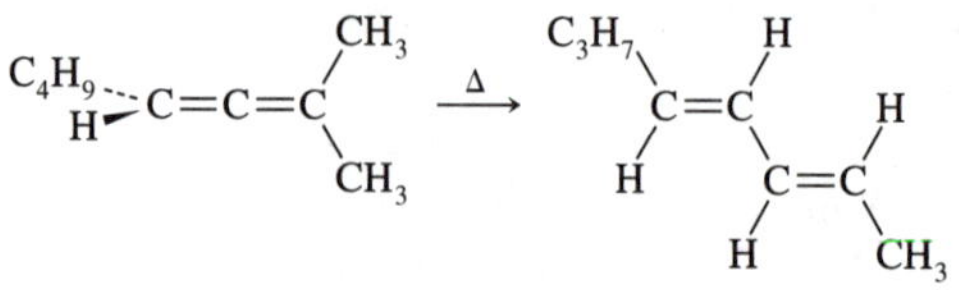

The fact that suprafacial [1,3] hydrogen shifts are forbidden accounts for the fact that enols do not isomerize to their more stable carbonyl tautomers unless the reactions are catalyzed by acids or bases (Eq. 4).

$$
\underset{CH_3-C=CH_2}{\overset{OH}{|}} \quad \underset{\xleftarrow{\hspace{1cm}}}{\overset{\overset{\oplus}{H} \text{ or } \overset{\ominus}{OH}}{\xrightarrow{\hspace{1cm}}}} \quad \underset{CH_3-C-CH_3}{\overset{O}{\|}} \tag{4}
$$

Not surprisingly, there do not seem to be any examples of thermal [1,3] alkyl shifts in open-chain alkenes. However, [1,3] shifts of alkyl groups *can* occur if the reactions involve expansions of strained three- or four-membered rings, as in the rearrangements of **2a** and **3**.

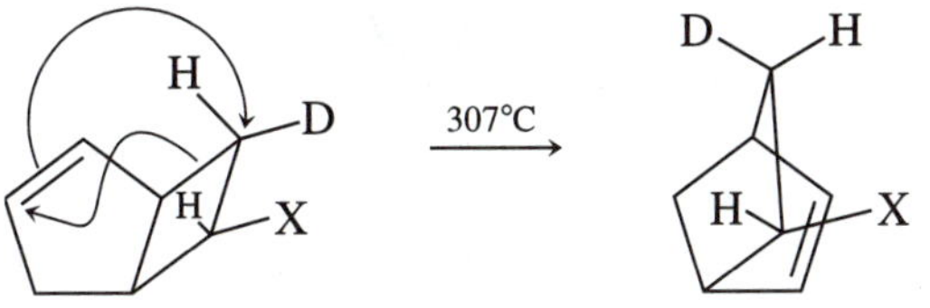

$$\tag{5}$$

2a, X = OAc   2b, X = D

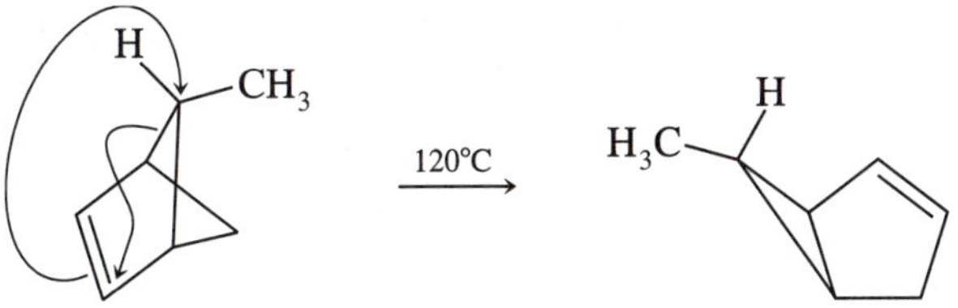

$$\tag{6}$$

3

As predicted by orbital symmetry conservation rules, these reactions proceed almost entirely with inversion of the configurations of the migrating groups.[6,7]*

The [1,3] migration of a deuterated benzyl group from nitrogen to carbon in diamine **4** similarly results entirely in the inversion of the geometry of the chiral benzyl group. This reaction takes place at remarkably low temperatures, possibly because the rearrangement eliminates "antiaromatic" character in the dihydropyrazine ring of **4**.

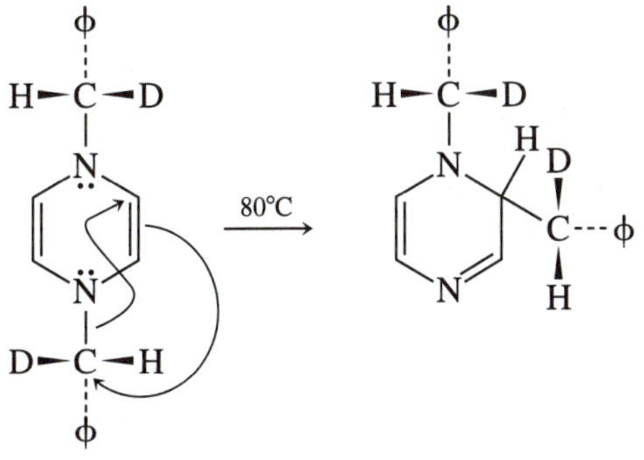

$$\tag{7}$$

4

---

*The degree of inversion in rearrangement of 3 and its analogs, however, does not vary with temperature. This has been interpreted as meaning that inversion and retention proceed via a single intermediate, presumably a diradical.[7]

In the transition states for the rearrangements of **2a** and **3**, the normal bond angles would have to be greatly distorted to bring the "back lobes" of the migrating σ bonds into bonding distance with the π bonds. This seems to require nearly enough energy to completely split the σ bonds.* Indeed, small changes in the structures of the starting materials appear to be sufficient to make diradical mechanisms competitive with the concerted mechanisms. Rearrangement of hydrocarbon **2b**, for instance, while proceeding principally with inversion, does demonstrate significant amounts of rearrangement with retention. It is not clear whether rearrangement with retention results from a diradical rearrangement process, or whether "the orbital rules are somewhat permissive for the [1,3] sigmatropic migrations of carbon."[9]

Effects of Polar Substituents

While migrating groups in nonpolar molecules usually undergo inversions in the course of thermal [1,3] rearrangements, the same does not seem to be true of molecules containing strongly electron-withdrawing or electron-donating substituents. The thermal rearrangement of compound **5**, for instance, proceeds with about 95% retention of configuration,[10] and migration of the chiral 1-phenylethyl group in the alkoxide ion **6** proceeds with about 65% retention.[11]

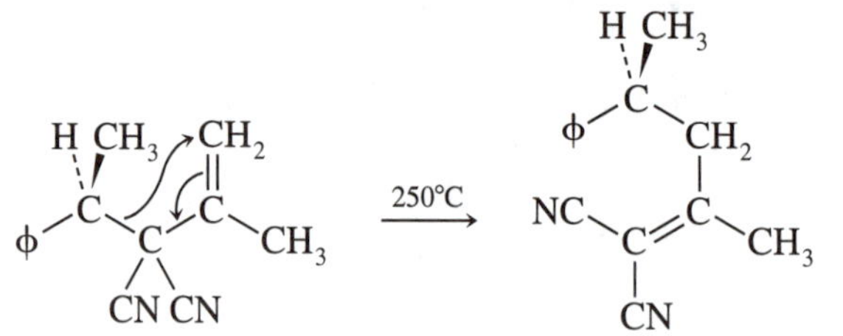

(8)

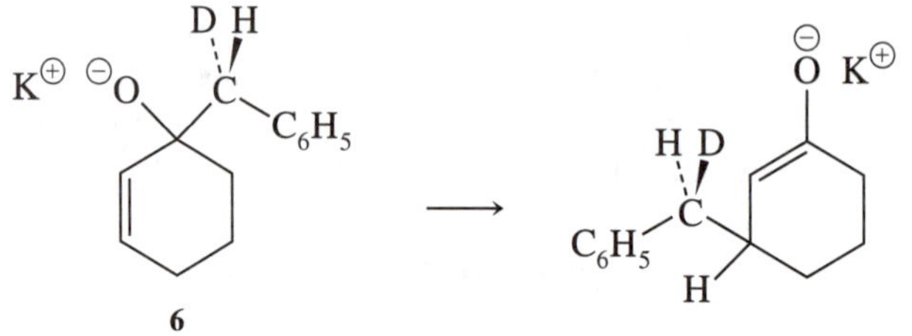

5

(9)

6

If the transition states for rearrangement of **5** and **6** are considered to resemble complexes of free radicals, it can be seen that the two "radical units" in each transition state will differ markedly in polarity.

It has been noted that forbidden reactions should be much "less forbidden" in highly polar systems than in hydrocarbons or other nonpolar molecules.[12] The presence of strongly electron-withdrawing groups on one "radical unit" in the transition state would lower the energy levels of its molecular orbitals. Lowering the energy level of the LUMO of one "radical" would bring its energy level closer to that of the HOMO of the other "radical". There should therefore be appreciable interaction between the HOMO and LUMO of the two "radicals", which would be in phase in the formally forbidden reaction. (This is illustrated below for a reaction in which the allylic unit is substituted with electron-withdrawing groups.)

---

*The activation energies for the rearrangements of **2a** and **3** are only 2 to 3 kcal/mol lower than the energies calculated to be necessary to cleave the migrating bonds to form diradicals. They fall within the margin of error of those calculations.[7,8]

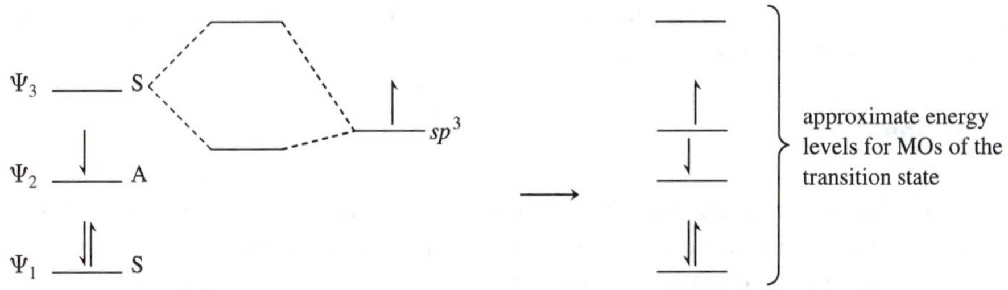

Formation of the transition state for a suprafacial [1, 3] migration in a polar molecule

Similarly, the presence of strongly electron-donating groups, such as a negatively charged oxygen, on one radical would raise the energy of its HOMO closer to that of the LUMO of the other radical unit and would facilitate their interaction in the transition state.

Despite these interactions, suprafacial [1,3] thermal shifts would still require that electrons in the frontier orbitals be raised to relatively high levels in the transition states. Those transition states, however, would be much lower in energy than if nonpolar compounds were undergoing similar rearrangements. They might well be lower in energy than the distorted transition states required for [1,3] migrations with inversion of the migrating groups.

Photochemical [1,3] Shifts

According to the general principle that concerted photochemical and thermal processes should proceed with opposite geometries, photochemical [1,3] shifts, if concerted, should yield products resulting from suprafacial migrations.[3,13] [1,3] Shifts in alkenes are common in photochemical processes.[14,15] However, many of these reactions proceed via excited triplet states. The triplet states apparently form diradicals or radical pairs, which then recombine to form the rearrangement products.[14] Orbital symmetry conservation rules are irrelevant for these reactions.

[1,3] Shifts resulting from excited singlet states of alkenes can also proceed via recombination of free radical intermediates.[15] However, in some cases rearrangements from singlet states do appear to proceed by concerted mechanisms. The rearrangements of **7a** and **7b** for instance,

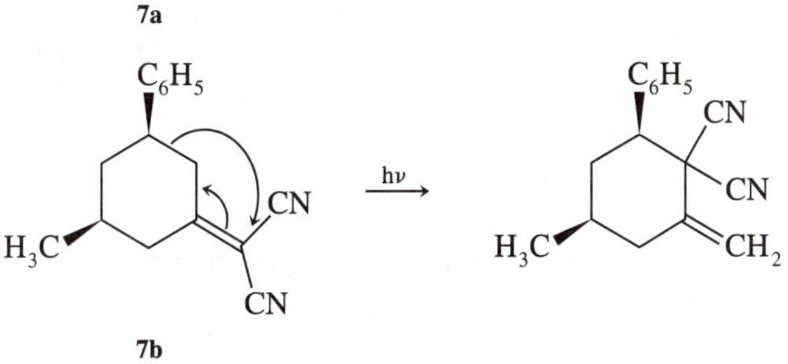

(10)

have been shown to proceed stereospecifically, with retention of the configurations of the migrating benzylic carbons.[16]

[1,5] Shifts

Hydrogen migrations. [1,5] Shifts of hydrogen atoms, usually at temperatures of 200°C and above, are common reactions. In small cyclic systems, such as those shown in Eqs. 11 and 12,[17] the hydrogen migrations must almost certainly be suprafacial.

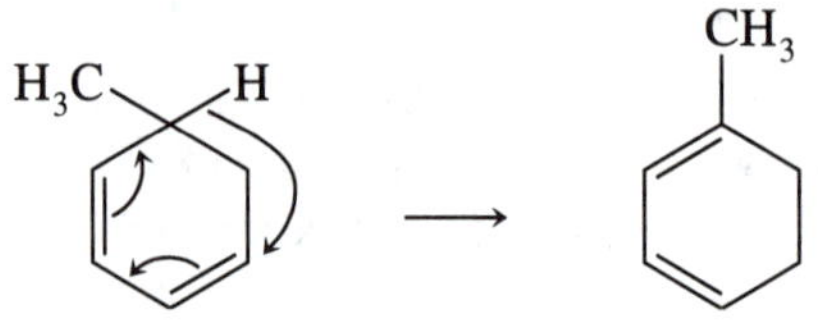

$$(11)$$

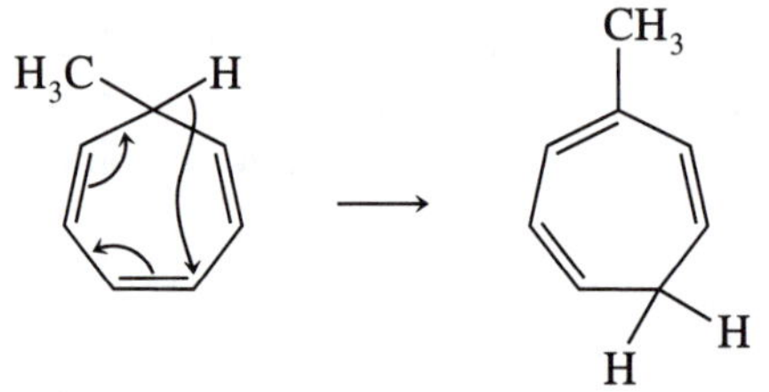

$$(12)$$

Thermal [1,5] hydrogen shifts in open-chain systems also appear to proceed via purely suprafacial paths. That was demonstrated by the fact that at 250°C, the chiral diene **8** yielded specifically the two stereoisomers shown in Eqs. 13a and b (arising from different conformations of **8**) and none of the products that would result from antarafacial hydrogen migrations.[18]

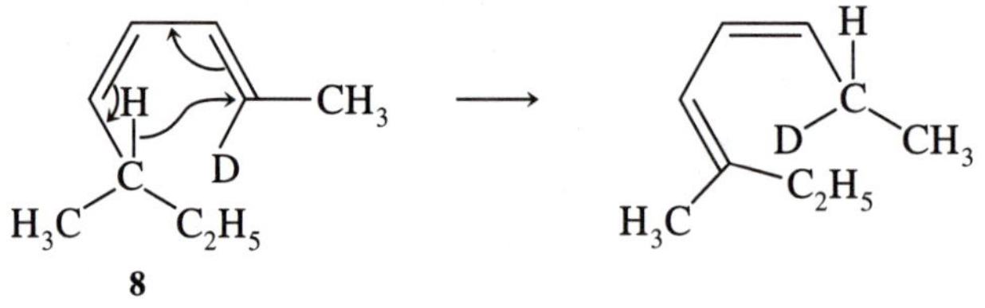

$$(13a)$$

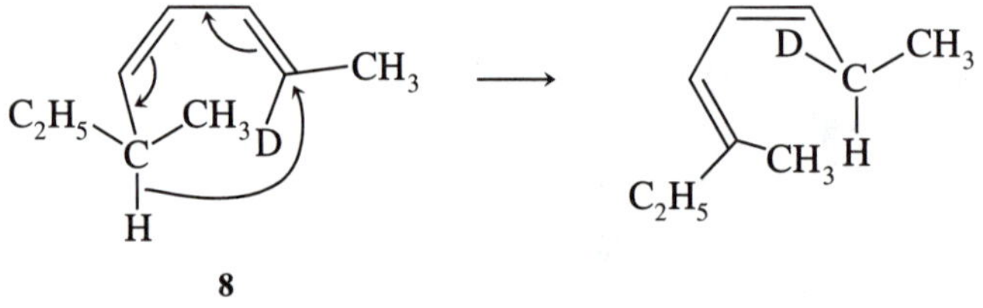

$$(13b)$$

When indene derivatives are heated to 100°C, they undergo rearrangements that appear, at first, to be [1,3] hydrogen migrations (Eq. 14).

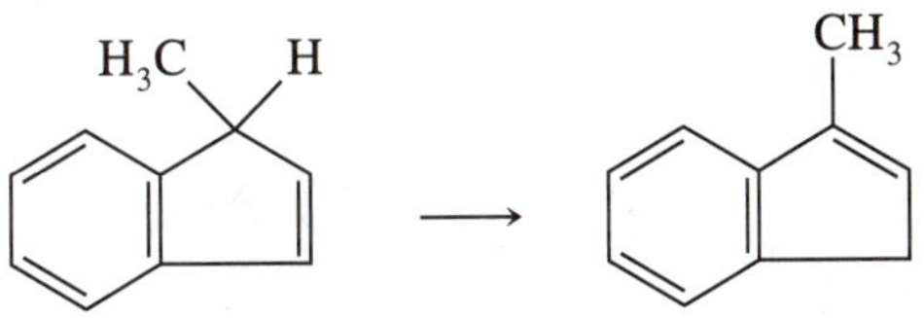

(14)

However, experimental studies have shown that these reactions proceed entirely by [1,5] shifts, as in Eq. 15.[19]

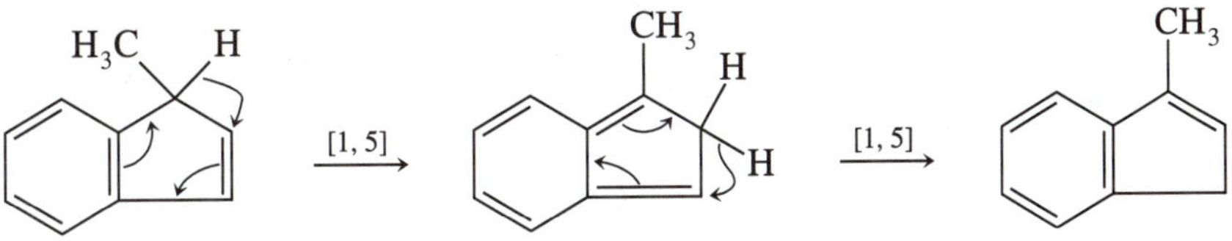

(15)

In these rearrangements, a mechanism requiring disruption of an aromatic ring is preferred over a mechanism that leaves the aromatic ring intact throughout the reaction. This is dramatic evidence of the barriers to rearrangements proceeding by "forbidden" mechanisms.

Photochemical [1,5] shifts of hydrogen atoms would be allowed only if they proceeded by antarafacial paths. Such reactions are therefore quite rare. However, several examples are known in which thermal [1,5] shifts of hydrogen atoms are difficult but in which [1,5] shifts proceed readily on photoirradiation.[20,21] In these cases, the geometries of the molecules (e.g., diene **9**)[20] actually favor antarafacial [1,5] hydrogen shifts.

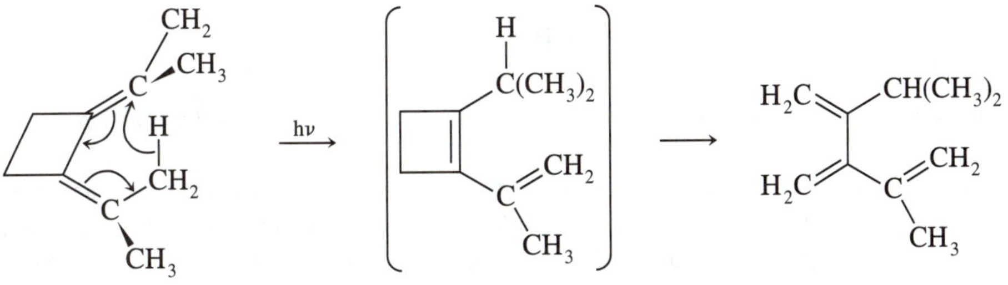

(16)

[1,5] migrations of alkyl, aryl, and acyl groups.   In contrast to hydrogens, *alkyl and aryl groups have never been observed to undergo thermal [1,5] shifts in acyclic systems.* However, such migrations are common in reactions of cyclopentadiene derivatives. Rearrangements of the isomeric spiro compounds **10a** and **10b** proceed, as predicted, with retention of the configurations of the migrating groups. (The products actually obtained from these reactions result from [1,5] hydrogen shifts in the initial products. The hydrogen shifts are very rapid at the temperatures required for alkyl migrations.[22])

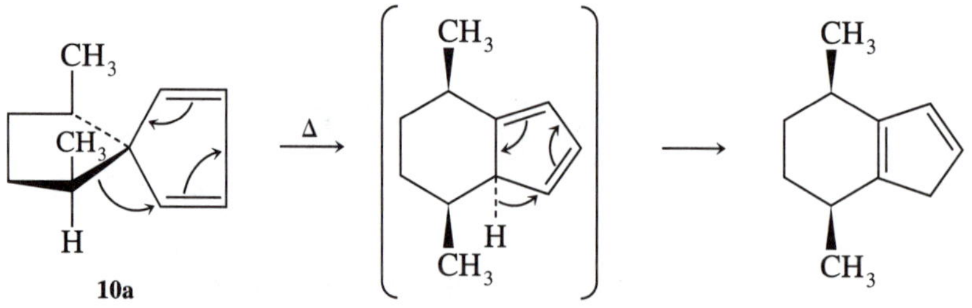

(17a)

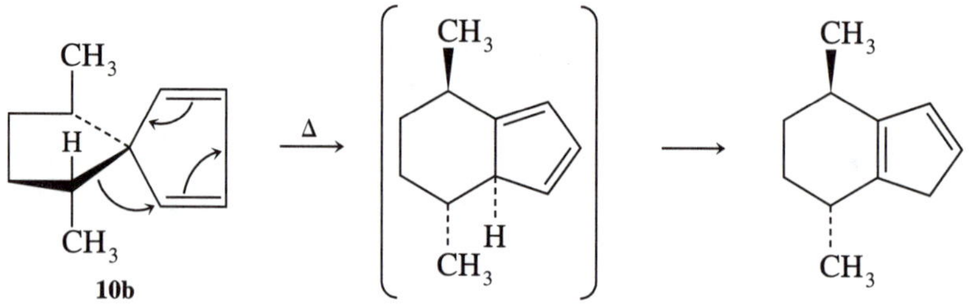

(17b)

Migratory aptitudes in [1,5] shifts. Several groups have studied the relative abilities of different migrating groups to undergo [1,5] shifts in rearrangements of cyclopentadienes and indenes.[23-25] These studies show that carbonyl and carboxyl groups are the very best migrators (as in Eq. 18), with hydrogen atoms second best.*

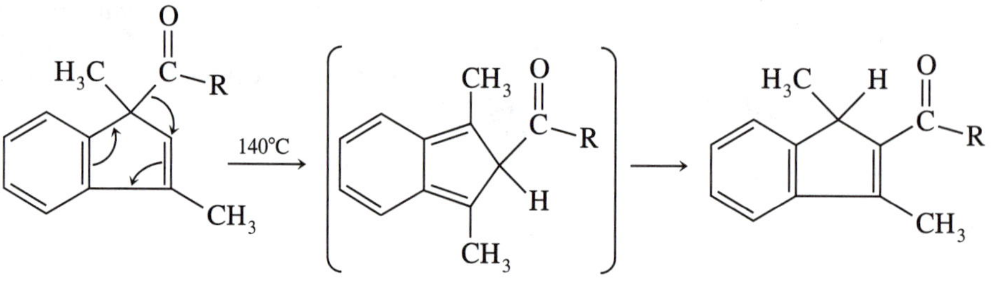

(18)

Aromatic rings and vinyl groups are also good migrators, but not as good as carbonyl groups or hydrogens.

Alkyl groups are by far the poorest migrators in thermal [1,5] shifts. They apparently do not undergo migration at all in indenes, while temperatures above 250°C (more than 150°C higher than for migrations of carbonyl groups) are necessary for their migrations to take place in reactions of cyclopentadienes.†

---

*However, hydrogen atoms appear to be better migrators than carboxyl groups in rearrangements of 1,3-cyclo-hexadienes.[26]

†It was suggested that unsaturated groups are more effective migrators than alkyl groups because there are secondary orbital interactions between the HOMOs of the cyclopentadienyl and indenyl systems and the LUMOs of the π bond systems of the migrating groups.[23] This hypothesis is supported by the fact that [1,5] shifts of aromatic rings are most rapid when the migrating rings bear electron-withdrawing substituents.[24,25] Electron-withdrawing substituents would lower the energies of the LUMOs of the migrating groups, thus bringing them into closer correlation with the HOMOs of the indenyl radicals. Of course, carbonyl and carboxyl groups have particularly low-lying LUMOs and would thus be expected to be effective migrators.

While [1,5] migrations of alkyl groups do not take place in indenes, they proceed readily in the very exothermic rearrangements of isoindenes, such as **13**, since these reactions result in the formation of new aromatic rings.[28]

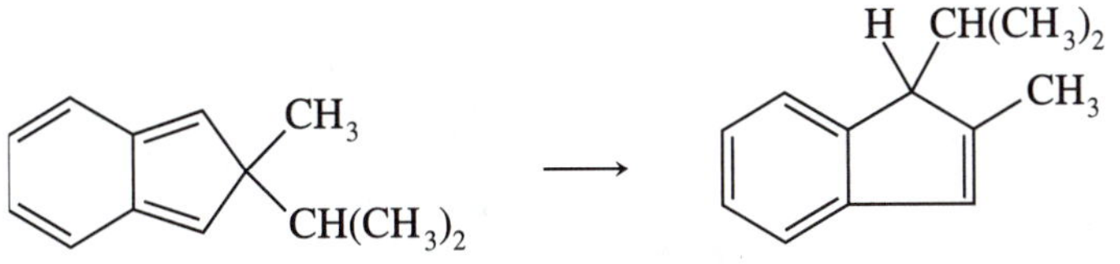

(19)

13

In these rearrangements, the relative "migratory aptitudes" of the migrating groups decrease in the order benzyl > cyclopropylmethyl > isopropyl > ethyl > methyl. The best migrators among alkyl (and substituted alkyl) groups are those that would form the most stable free radicals and that are therefore linked to the isoindenyl framework by the weakest bonds.

Rearrangements of 1,3-cyclohexadienes.   Hydrogen atoms can undergo [1,5] migrations in 1,3-cyclohexadienes, as can carbonyl groups (Eq. 20).

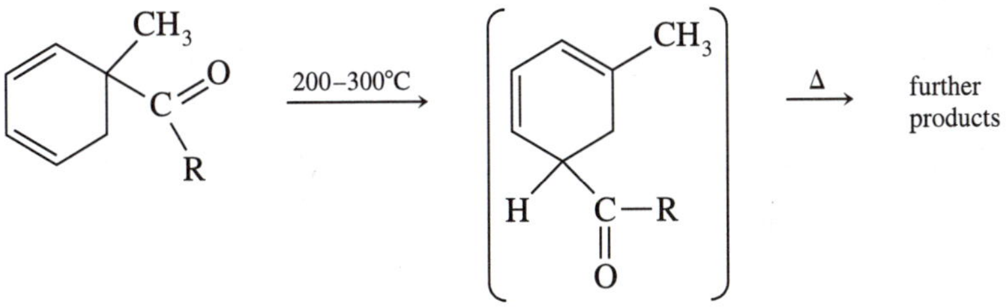

(20)

However, appreciably higher temperatures are required than are needed for the corresponding rearrangements of cyclopentadienes.[26]

Surprisingly, alkyl groups, which are very poor migrating groups in cyclopentadienes, have been reported to undergo [1,5] migrations in cyclohexadienes at temperatures not much higher than those required for carbonyl migrations.[29] This anomaly was resolved when it was demonstrated that the apparent migrations of alkyl groups in 1,3-cyclohexadienes are not simple [1,5] shifts. For instance, it was found that rearrangement of the cyclohexadiene derivative **14**, bearing a methyl group labeled with radioactive carbon, yielded a rearrangement product in which the radioactive carbon was located in the ring rather than in the methyl group.[30]

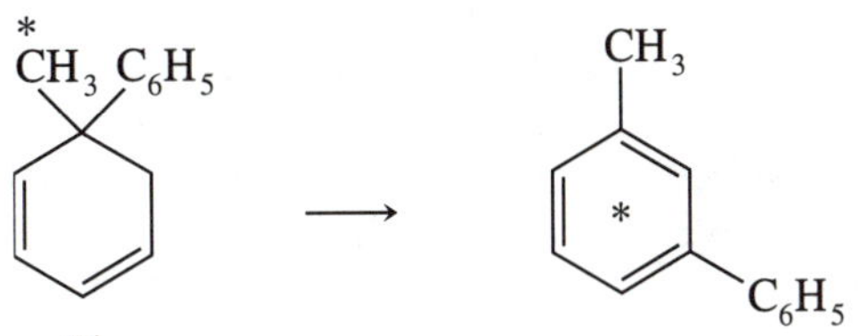

(21)

14

It has been proposed that this reaction proceeds by initial electrocyclic ring opening, followed by a [1,7] hydrogen shift and reclosure of the ring, as shown in Eq. 22. (The final loss of hydrogen gas presumably occurs via a free-radical chain reaction.)

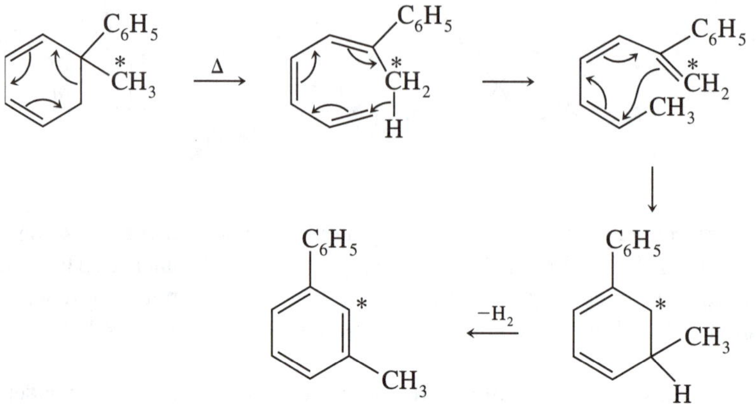

(22)

Migrations in blocked aromatic systems.   [1,5] Migrations can take place at relatively low temperatures in *blocked aromatic molecules*, such as methylenecyclohexadienes and cyclohexadienones. (Isoindenes can also be classified as blocked aromatic compounds.)

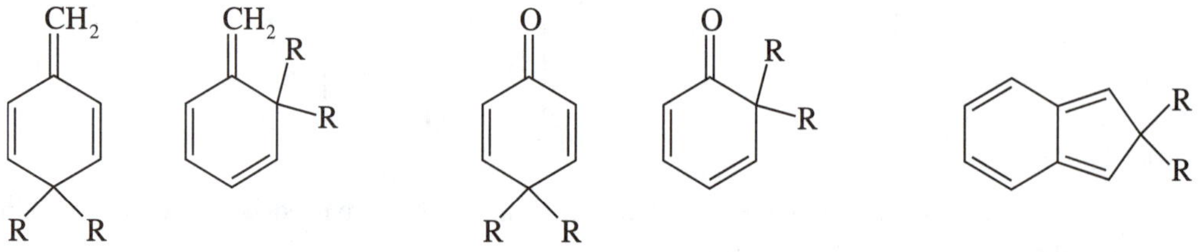

methylenecyclohexadienes   cyclohexadienones   an isoindene
("semibenzenes")

In these molecules, the presence of "blocking" substituents prevents the isomerization of nonaromatic rings to aromatic structures. The rearrangements of these compounds do not always result in the immediate formation of aromatic rings (although the final products of the reaction sequences are indeed usually aromatic). However, the ease with which they undergo rearrangements suggests that the blocked aromatic rings attain an appreciable degree of aromatic stability in the transition states for the rearrangements.

For instance, [1,5] migrations of benzyl[31] or cyclopropylmethyl[32] groups take place when *ortho*-cyclohexadienones such as **15** and **16** are heated. This converts the starting compounds to isomers that can tautomerize (or can lose isobutene molecules) to form phenols.

An exceptionally rapid [1,5] benzyl group migration takes place when compound **17** is treated with base at 0°C. In this case, the rearrangement step is so rapid that the presumed intermediate, **18**, cannot be detected. The remarkable rate of this reaction is presumably due to the fact that a new aromatic ring is formed in the reaction.[33]

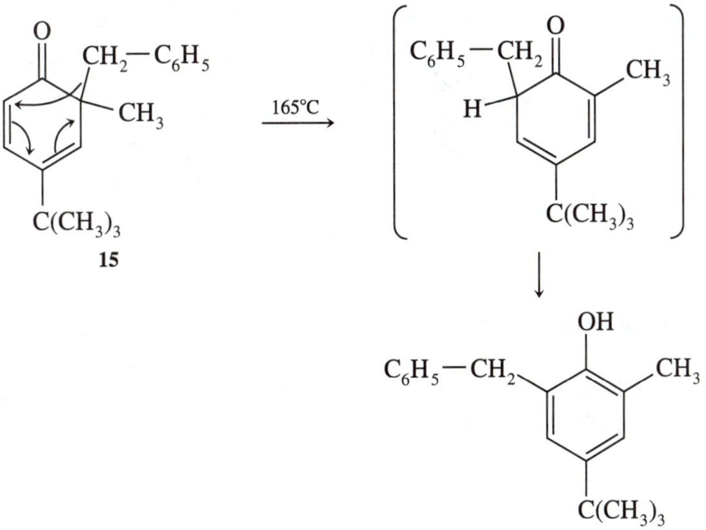

$$(23)$$

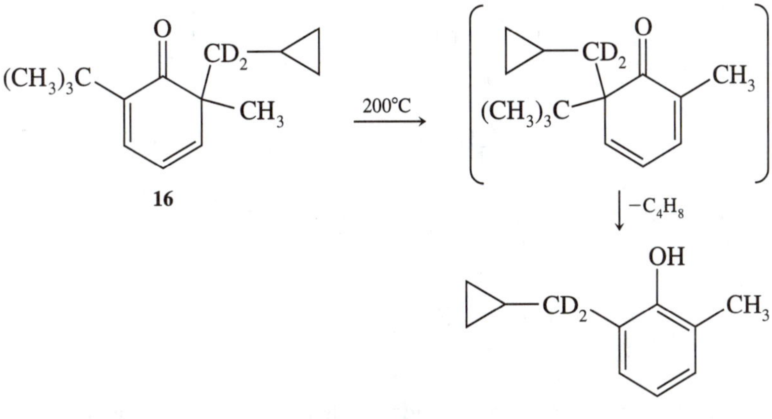

$$(24)$$

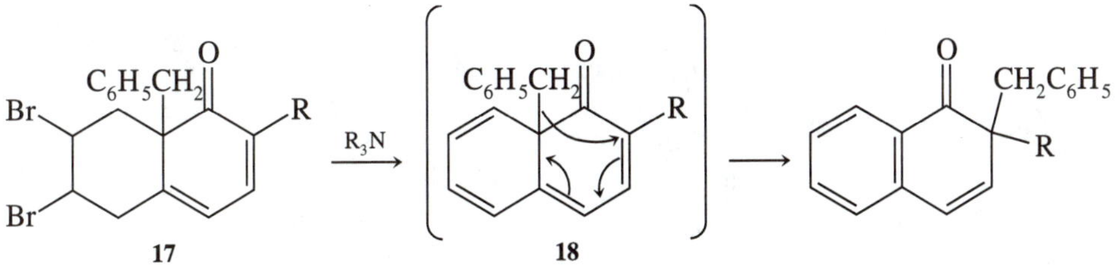

$$(25)$$

[1,5] Migrations of saturated alkyl groups in cyclohexadienes are very rare. However, even gentle heating of **19** results in [1,5] alkyl shifts to form **20**.[34] As with **17**, the exceptional rapidity of the rearrangement can be attributed to the fact that a new benzene ring is formed during the rearrangement step.

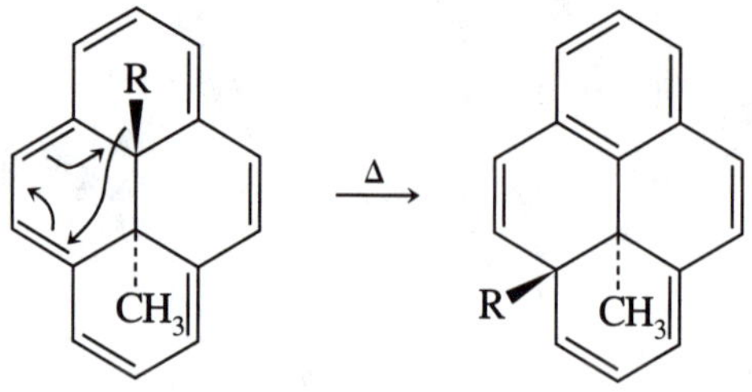

$$(26)$$

19 (R = CH_3, C_2H_5, C_3H_7)   **20**

It is important to note that [1,5] migrations of groups other than hydrogen have been observed only in cyclopentadienes or cyclohexadienes, where the dienes are fixed in *s–cis* structures.

The "semibenzene rearrangements" of compounds such as **21**[35a] and **22**[35b] originally appeared to contradict this statement.

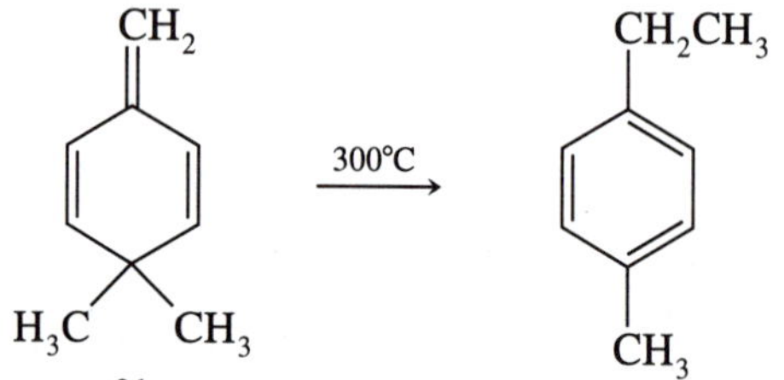

$$(27a)$$

21

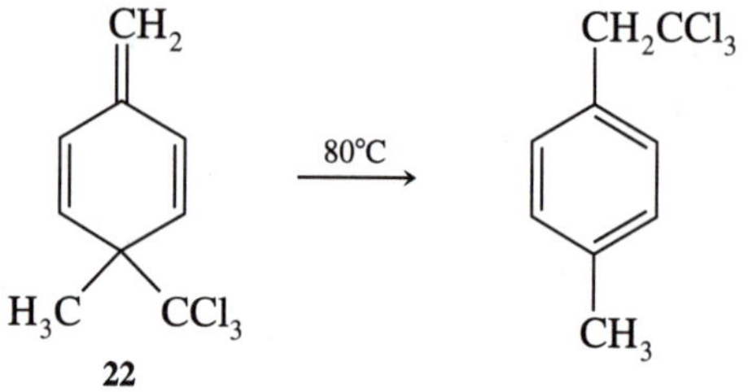

$$(27b)$$

22

However, there is now convincing evidence that these reactions actually proceed via radical chain mechanisms, as in Eq. 28.[36]

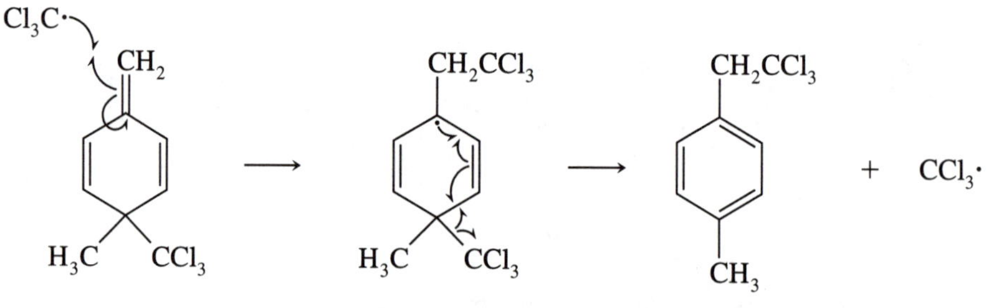

$$(28)$$

For instance, at 0°C semibenzene **23** rearranges to yield both **24** and **25**, demonstrating that a 2-butenyl free radical is an intermediate in the reaction.[36b]

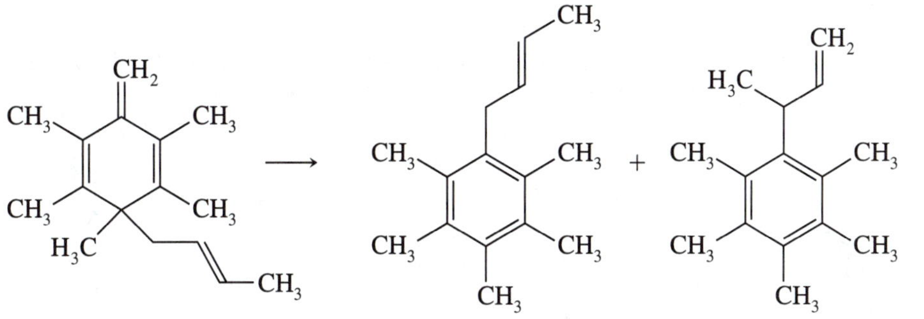

$$(29)$$

23   **24**   **25**

"Walk Rearrangements" of Norcaradienes.   When 1,3,5-cycloheptatrienes (*tropilidenes*) are heated or photoirradiated, the saturated carbon atoms (C7) appear to "walk" around the ring. It is believed that these reactions proceed by initial electrocylic formation of bicyclo[4,1,0]heptadienes (*norcaradienes*), followed by 1,5-shifts of one of the cyclopropyl bonds (e.g., bond *a*), as shown in Eq. 30.[37]

$$X = CN_1CO_2CH_3 \qquad\qquad\qquad\qquad\qquad\qquad\qquad\qquad (30)$$

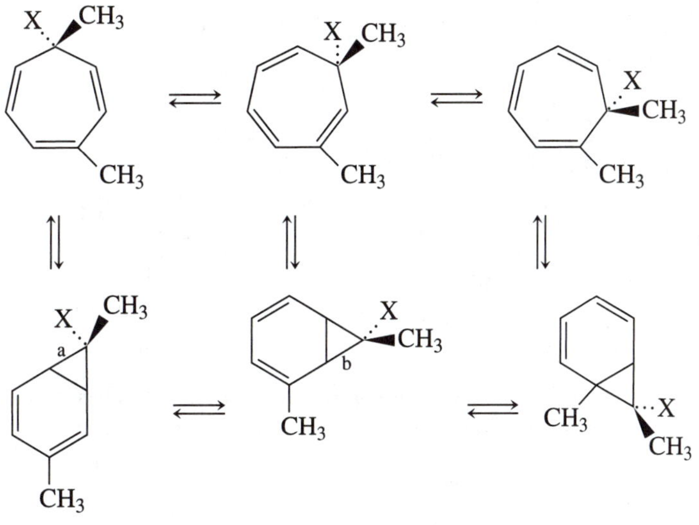

Of course, a concerted [1,5] migration should proceed with retention of the geometry of the migrating atom. However, both thermal and photochemical "walk" rearrangements proceed with *inversion* of the geometries at C7.[37] (Notice that bond *b* in Eq. 30 has been formed at the opposite side of C7 from bond *a*.) Thus, despite the almost complete stereospecificity of the rearrangements, it does not appear that the thermal migrations, at least, are concerted processes. One theoretical study indicates that "there is no significant bonding in the transition state between C7 and C2 or C6," and that "the 1,5-shifts of norcaradienes are diradical processes, but do not involve any diradical minima on the potential energy surface."[38]

[1,7] Shifts

A thermal [1,7] shift of a hydrogen atom is believed to take place in the biological conversion of previtamin D_3 to vitamin D_3.[39]

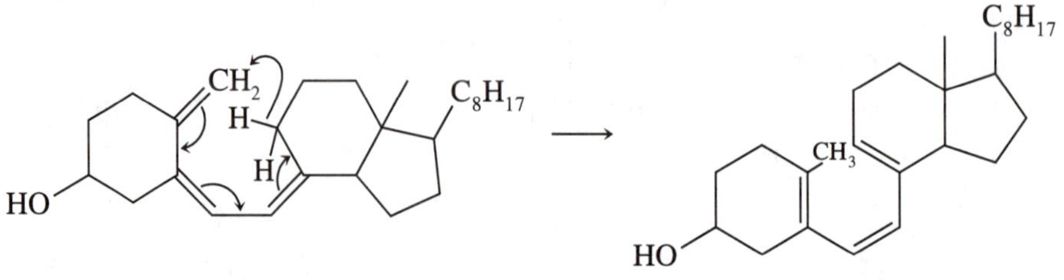

$$(31)$$

previtamin D_3 vitamin D_3

While the stereochemistry of the migration cannot be determined in that system, it has been shown that [1,7] hydrogen shifts in the rearrangement of **26** proceed entirely by antarafacial paths,[40] as predicted by orbital symmetry conservation rules. (Suprafacial [1,5] shifts, of course, would also be allowed, but the [1,7] shifts are more rapid.)

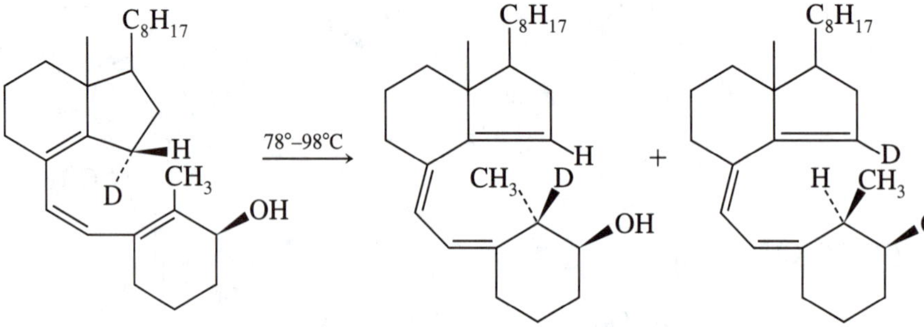

$$(32)$$

Thermal [1,7] shifts of carbon atoms are extremely rare but do occur in the interconversions (walk rearrangements) of the bicyclic nonatriene derivatives **27a–27d**.[41a]

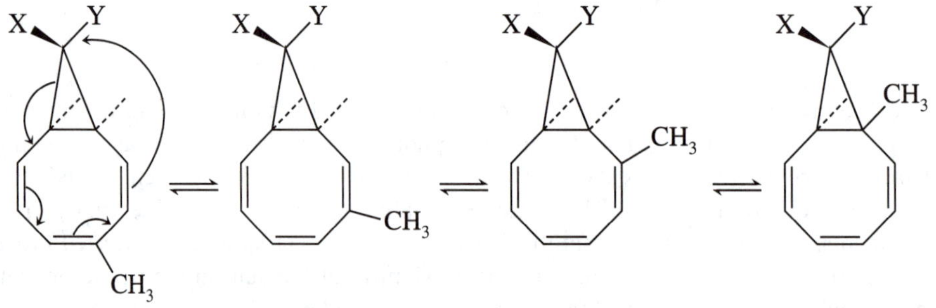

$$(33)$$

27a　　　　**27b**　　　　**27c**　　　　**27d**

$(X = CN, Y = CH_3 \text{ or } X = CH_3, Y = CN)$

These reactions occur without any detectable interchange of *endo* and *exo* substituents. They thus proceed with *inversion* of the configuration of the chiral center in each migration, as required by orbital symmetry conservation rules.

Photochemical [1,7] shifts should proceed suprafacially; therefore, they are geometrically feasible in cyclic systems. Photochemical [1,7] shifts of both hydrogen atoms and alkyl groups have been observed in cycloheptatriene derivatives (Eq. 34).[41]

$$(34)$$

[3,3] Sigmatropic Shifts

Cope rearrangements.   [3,3] shifts in 1,5-hexadienes (Cope rearrangements) are very common reactions.[40] The Cope rearrangement of 1,5-hexadiene itself requires temperatures in the range of 250–300°C to proceed at a reasonable rate. However, substituted 1,5-hexadienes often rearrange at significantly lower temperatures, particularly when the substituents are aromatic rings or other unsaturated groups.[42,43]

Substituents at C1 (or C6) and C3 (or C4) of 1,5-hexadienes would be expected to increase the rates of Cope rearrangements, since the presence of substituents at those positions would weaken the bonds being broken in the reactions. However, phenyl substituents at C2 and C5 appear to be at least as effective as phenyl substituents at C1 and C3 in lowering the activation energies for Cope rearrangements (Eqs. 35 and 36).[43,44]

$$(35)$$

$$(36)$$

These facts, as well as the results of some theoretical calculations,[45] have been considered to support a nonconcerted mechanism, in which the initial step is the formation of a cyclohexane-1,4-diyl diradical (or *biradicaloid**), as shown on page 108. At present, however, the bulk of the experimental and theoretical evidence seems to support a concerted mechanism for the Cope rearrangement of 1,5-hexadiene itself.[46] However, it is possible for substitution or other modifications to tilt the mechanism of a Cope rearrangement to a diradical path.[47] The rearrangement of the cyclic

---

*A *biradicaloid* has been defined as "a singlet biradical in which through-bond coupling between the radical centers is so strong that the system behaves in many respects as a closed shell species."[45]

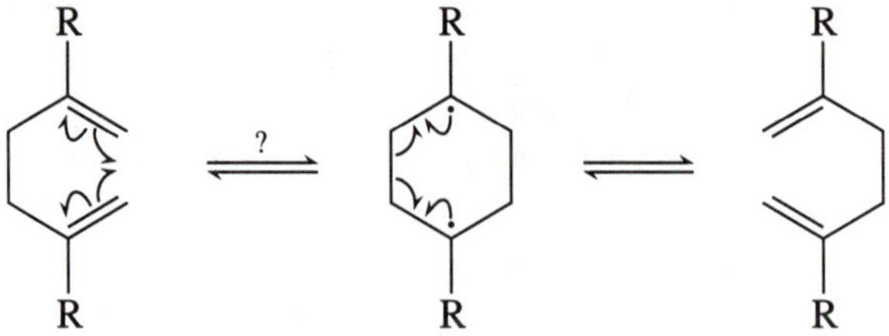

A cyclohexane-1,4-diyl diradical

diallene **28a** for instance, has been shown to proceed by a nonconcerted (presumably diradical) mechanism.[48] The choice of reaction paths is so delicately balanced, however, that the stereo-isomeric diallene **28b** rearranges by both diradical and concerted mechanisms.[36b]

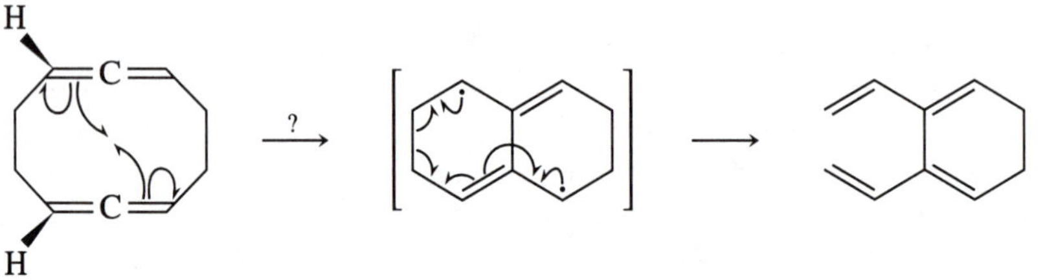

28a

(37)

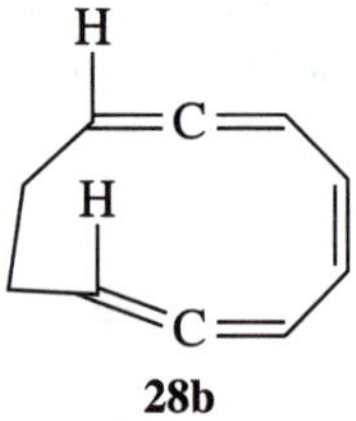

28b

Cope rearrangements occur at exceptionally low temperatures when the single bond is part of a small strained ring and the two double bonds are *cis* to each other, as in Eqs. 38 and 39.[49]

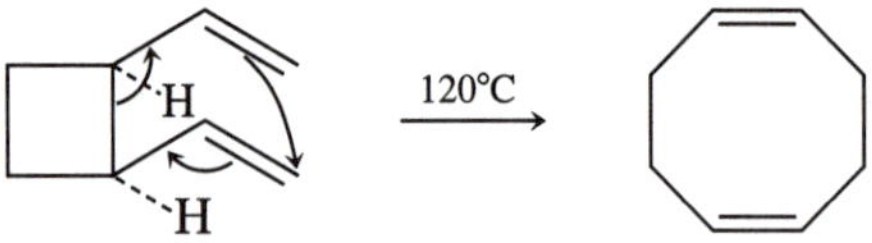

(38)

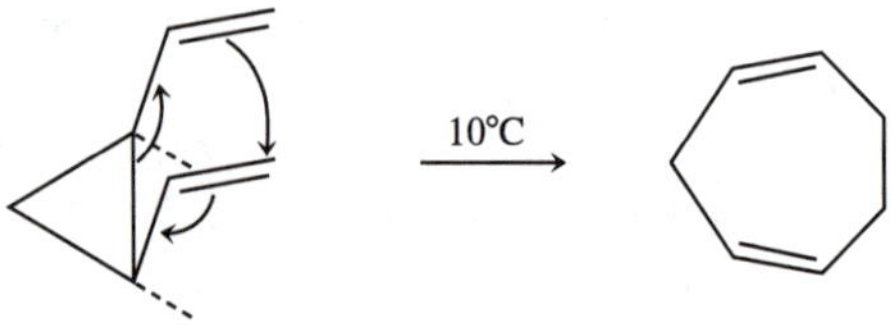

(39)

The fastest known Cope rearrangements are those of derivatives of *semibullvalene*.[50,51] The degenerate rearrangement of semibullvalene itself proceeds with an activation enthalpy of approximately 5.0 kcal/mol,[50,51] and the rearrangements of some of its derivatives have even lower enthalpies of activation.

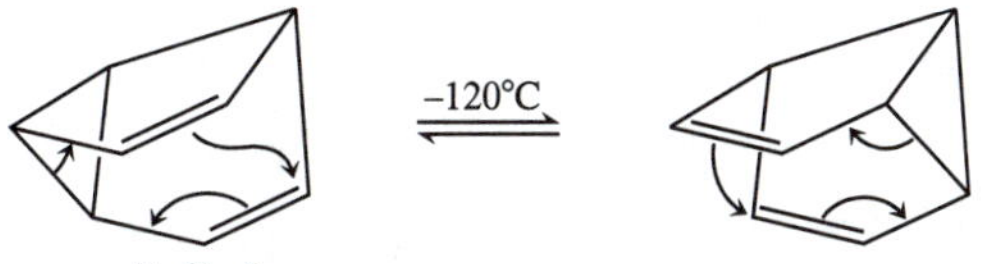

semibullvalene

(40)

It has been suggested that suitable substitution might lower the activation energy all the way to zero,[52] in which case, the two forms would not have independent existences. Instead, they would be resonance forms of a structure with delocalized σ and π electrons. So far, no such delocalized structure has been detected.

Geometries of Cope rearrangements.   In a Cope rearrangement of an open-chain diene, the transition state might resemble a chair form or a boat form of a six-membered ring. Elegant studies by Doering and Roth at Yale University showed that 99.7% of the product from rearrangement of *meso*-3,4-dimethyl-1,4-hexadiene (**29**) had the *cis,trans* geometry.

(41a)

The *cis,trans* product would result from a chair-like transition state (**29a**),

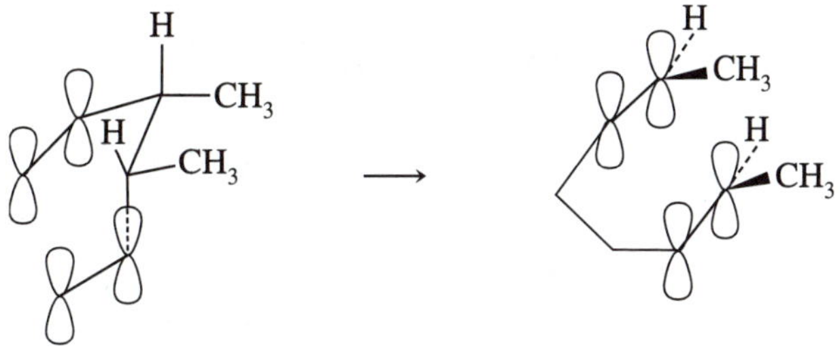

29a

(41b)

while a boatlike transition state (**29b**) would yield principally the *trans,trans* isomer of the product.[53]

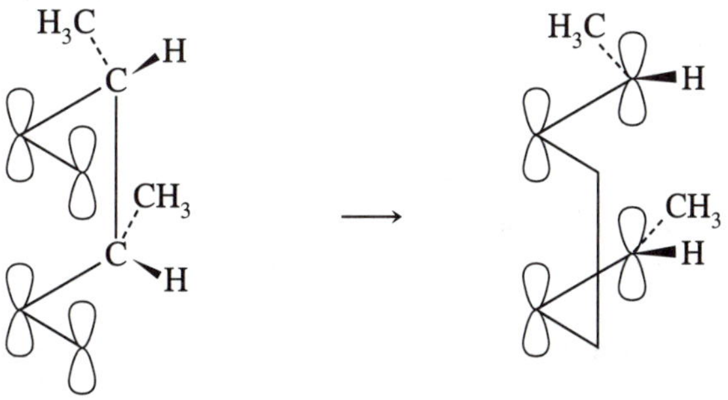

29b

(41c)

Claisen rearrangements.   Claisen rearrangements are thermal [3,3] shifts in allyl vinyl ethers such as **30**, or in allyl phenyl ethers.[1,54]

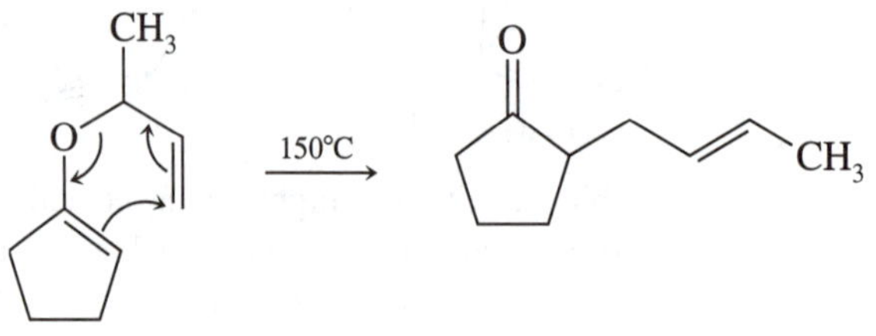

30

(42)

Aza-Claisen[56] and thio-Claisen rearrangements (e.g., Eq. 43),[57] in which the oxygen atoms of allyl vinyl ethers are replaced by nitrogen and sulfur atoms, respectively, are also known.

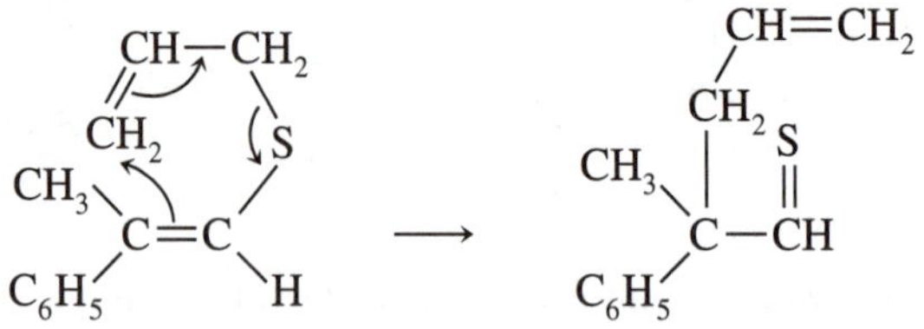

(43)

The rearrangements of allyl phenyl ethers are the most common types of Claisen rearrangements. The conversion of an allyl phenyl ether to an *ortho*-allylphenol is frequently identified simply as a "Claisen rearrangement" but is more precisely called an "*ortho*-Claisen rearrangement" (Eq. 44).*

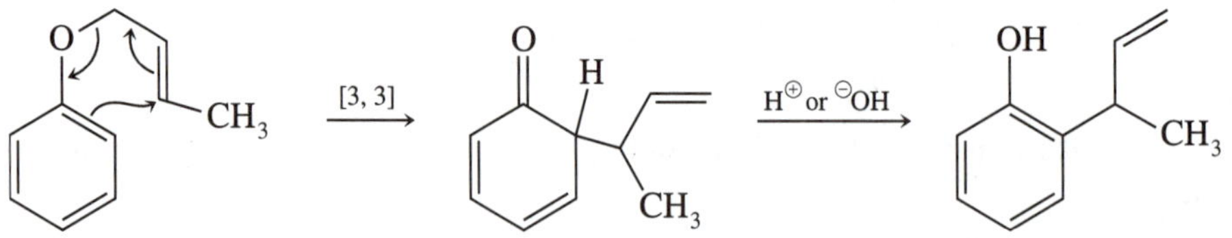

(44)

An *ortho*-Claisen rearrangement

When both *ortho* positions on the aromatic ring are already substituted (and even, to a small degree, when one or both are not substituted), the migrating allylic group will shift to the *para* position, resulting in a *para*-Claisen rearrangement (Eq. 45).

---

*The term *Claisen rearrangement* is commonly used to refer both to the [3,3] shift steps in rearrangements of allyl phenyl ethers and to the overall processes, including tautomerism steps, resulting in the formation of the phenolic products.

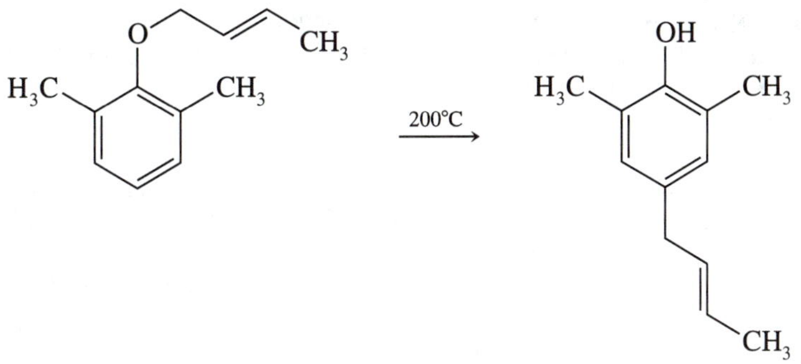

A *para*-Claisen rearrangement

(45)

While the [3,3] shifts in *ortho*-Claisen rearrangements result in inversions of the allylic struc-tures of the migrating groups, the migrating groups retain their original structures in *para*-Claisen rearrangements. This would be the expected result if *para*-Claisen rearrangements proceeded by [1,5] shifts, but the direct [1,5] shift mechanism seems unlikely, in view of the distance between the oxygen atoms and the *para* positions. In fact, *para*-Claisen rearrangements have been demon-strated to proceed by two successive [3,3] shifts, as in Eq. 46. The allylic group first migrates to the *ortho* position and then undergoes a second [3,3] shift (a Cope migration step) to the *para* position.

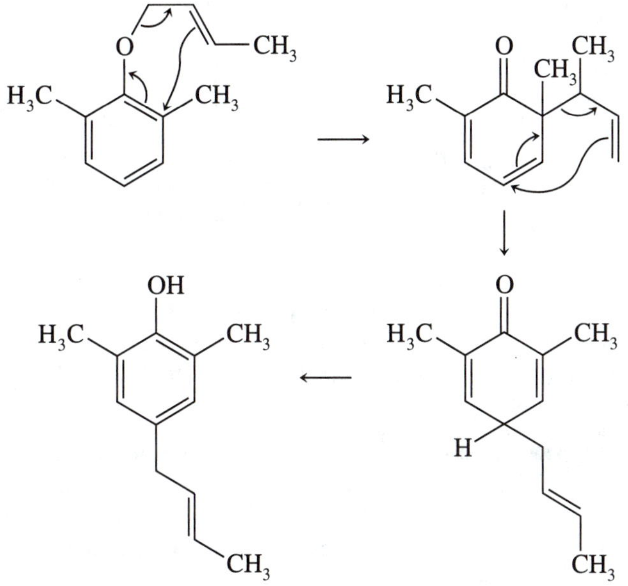

(46)

The result of the two successive inversions is to regenerate the original structure of the allylic group. Finally, the *para*-cyclohexadienone formed in the second migration step tautomerizes to form a phenol—a process presumably catalyzed by bases or acids. (The phenolic products them-selves are acidic enough to catalyze the tautomerism steps.)

Several lines of evidence demonstrate that *para*-Claisen rearrangements proceed via two successive [3,3] shifts rather than a single [1,5] shift:

1. The *ortho*-cyclohexadienone, **32**, formed as the initial intermediate in rearrangement of ether **31**, has been trapped as its Diels–Alder adduct with maleic anhydride.[57]

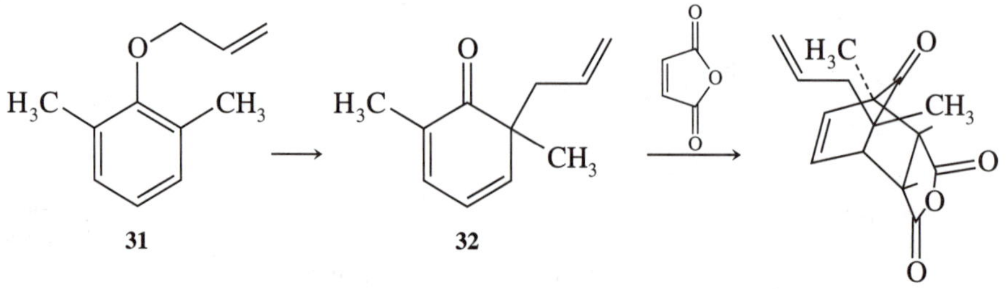

(47)

2. It has been shown that allylic substituents in *ortho*-cyclohexadienones such as **33** migrate to the *para* positions of the rings at temperatures much lower than those required for *para*-Claisen rearrangements.[58] (Rather surprisingly, these Cope migrations of allylic groups are usually faster than their *retro*-Claisen migrations to reform allyl aryl ethers, in spite of the fact that aryl ethers are appreciably lower in energy than are *para*-cyclohexadienones.)

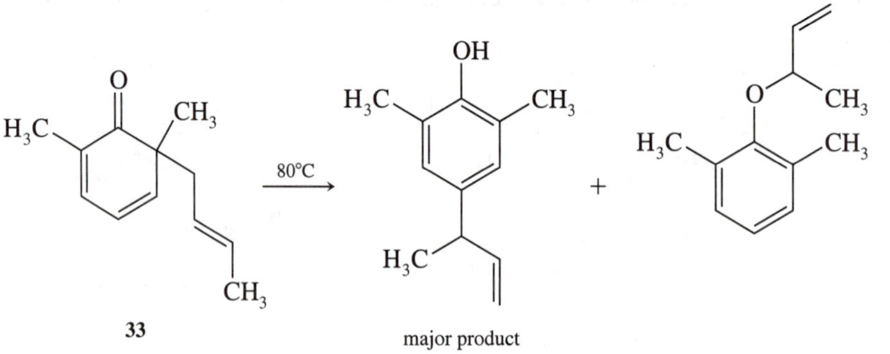

major product

(48)

3. When substituents at *ortho* positions of allyl phenyl ethers are themselves allylic groups, they may undergo Cope migrations in place of the original migrating groups, as shown in Eq. 49.[59]

If tautomerism of the *ortho*-cyclohexadienone formed in the first step of a Claisen rearrangement is possible, tautomerism will normally be much faster than a [3,3] shift of an allyl group. As a result, the principal products obtained from heating allyl phenyl ethers with unsubstituted *ortho* positions are those from *ortho*-Claisen rearrangements.

Abnormal Claisen rearrangements.   Thermal rearrangements of allyl vinyl ethers and allyl aryl ethers sometimes give rise to products with rearranged (rather than simply inverted) structures of the allylic groups. These reactions, known as *abnormal Claisen rearrangements*, are particularly likely to occur if the reaction temperatures are higher than those usually employed for Claisen rearrangements.

Abnormal Claisen rearrangements, such as that shown in Eq. 50, have been shown to proceed by initial [3,3] shifts to yield "normal" Claisen rearrangement products, which then undergo further rearrangements (Eq. 51).[60]

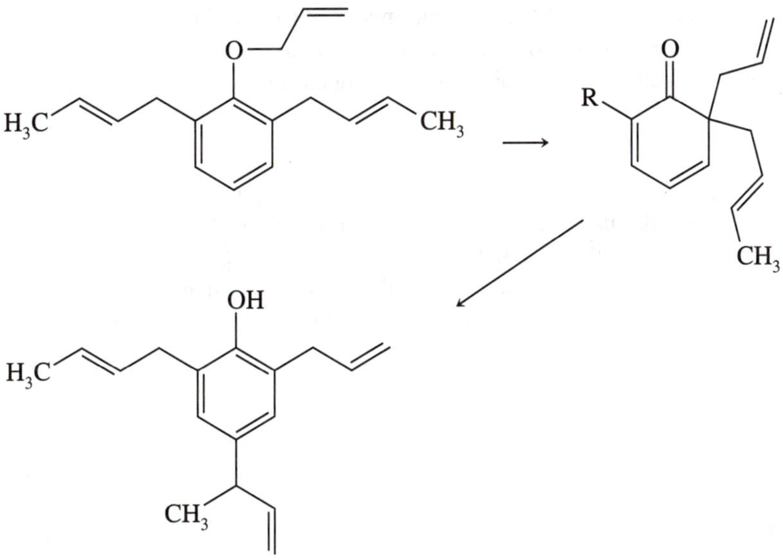

(after tautomerism)

(49)

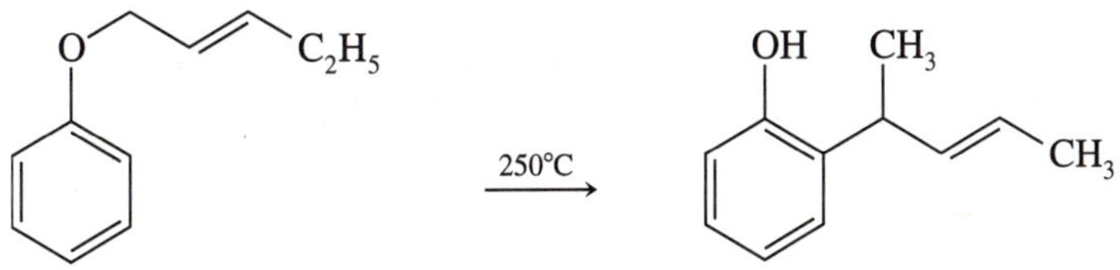

250°C

An abnormal Claisen rearrangement

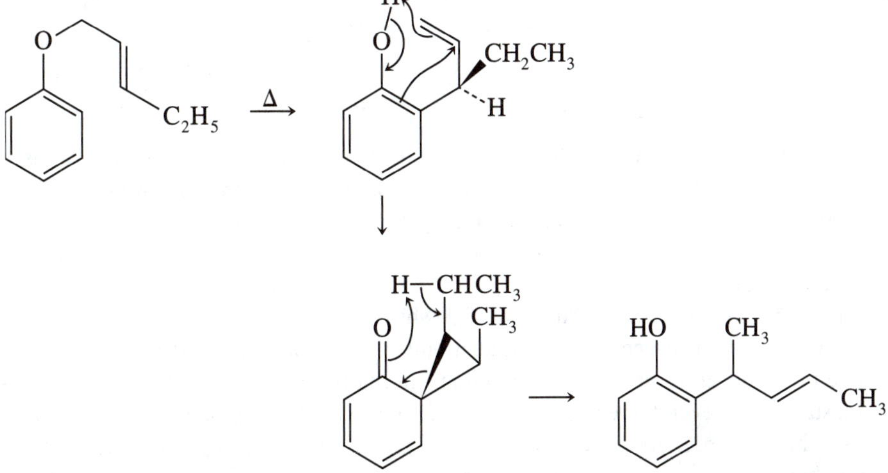

(50)

Δ

(51)

The final two steps in the abnormal Claisen rearrangement can be classified as "*homo*-[1,5] hydrogen shifts." (The term *homo* is a shortened form of *homologous*, meaning that the reaction involves one more carbon atom than a typical [1,5] hydrogen shift.)

[5,5] Shifts

Thermal [5,5] shifts are much less common than [3,3] shifts. However, this appears to result from the fact that molecules capable of exhibiting such reactions are relatively rare, rather than because of any intrinsic difficulty with the reactions.

On heating at 186°C, phenyl pentadienyl ethers yield principally 4-pentadienylphenols, along with smaller yields of products from *ortho*-Claisen rearrangements (Eq. 52).[61]

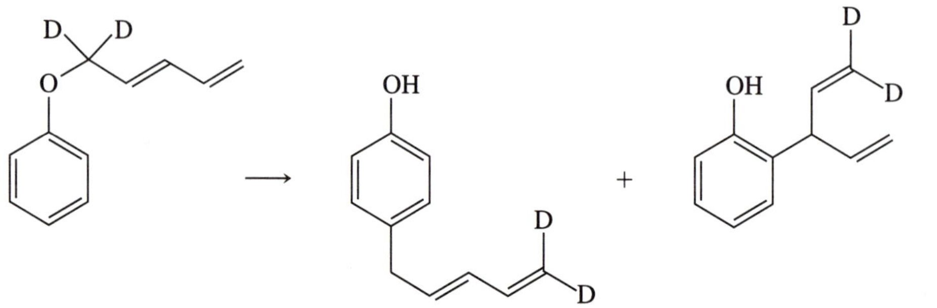

(52)

Deuterium labeling experiments demonstrated that the principal products result from direct [5,5] shifts (Eq. 53) rather than from two consecutive [3,3] shifts.[62]

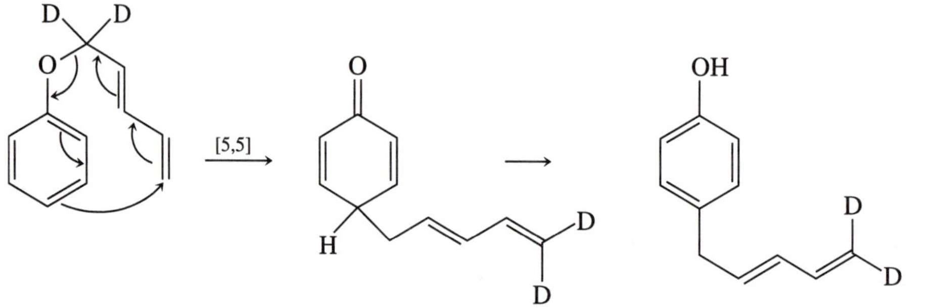

(53)

If the geometry of the molecule is favorable, [5,5] shifts can occur remarkably easily. The rearrangement of **34a**, for instance, is quite rapid at room temperature. In contrast, **34b** does not undergo any rearrangement, even at elevated temperatures. Apparently, interference between the two methyl groups in **34b** prevents the molecule from attaining the proper conformation for rearrangement to occur.[63]

It might be considered surprising that [5,5] shifts in these reactions are preferred over [3,3] shifts, because reactions proceeding via transition states resembling six-membered rings should have more favorable entropies of activation than similar reactions proceeding via 10-membered transition states. However, there are other examples of competing sigmatropic shifts in which the reactions of highest order are preferred, and it has been suggested that this preference represents a general phenomenon.[64] (Instances of [1,7] hydrogen shifts being favored over [1,5] hydrogen

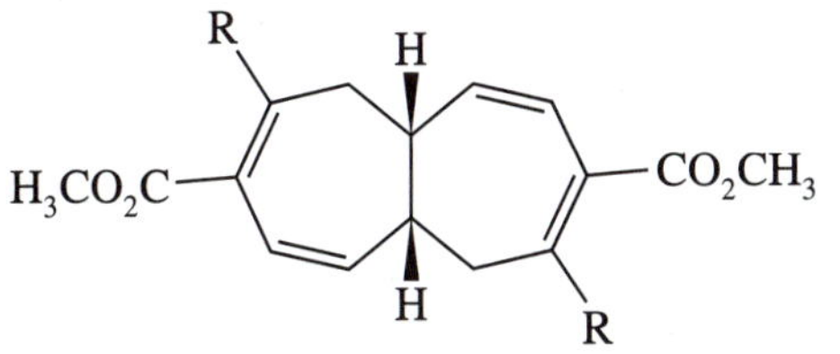

34a, R = H or D
b, R = CH₃

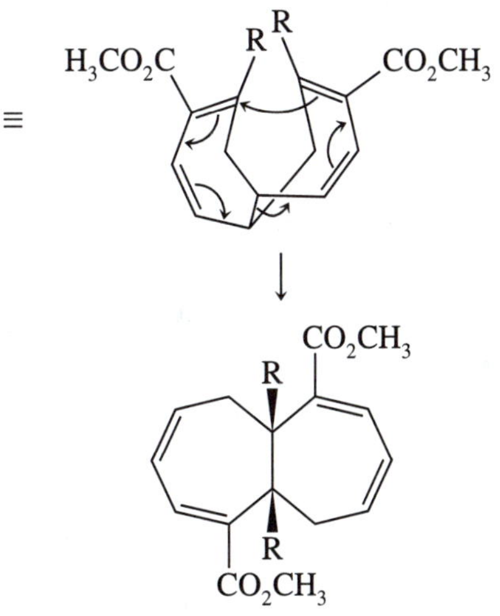

(54)

shifts were pointed out on page 106.) Possibly, transition states for pericyclic reactions involving large numbers of participating electrons have higher "resonance energies" than transition states for similar reactions involving smaller numbers of delocalized electrons.

4.3   SIGMATROPIC REARRANGEMENTS OF CHARGED SYSTEMS

Acid Catalysis

It was long believed that the rates of Claisen and Cope rearrangements were relatively insensitive to acid and base catalysts.[55] However, it is now known that the addition of Lewis acid catalysts, such as boron trichloride, lowers the temperatures necessary for Claisen rearrangements of allyl phenyl ethers from about 200°C to below normal room temperatures (Eq. 55).[65]

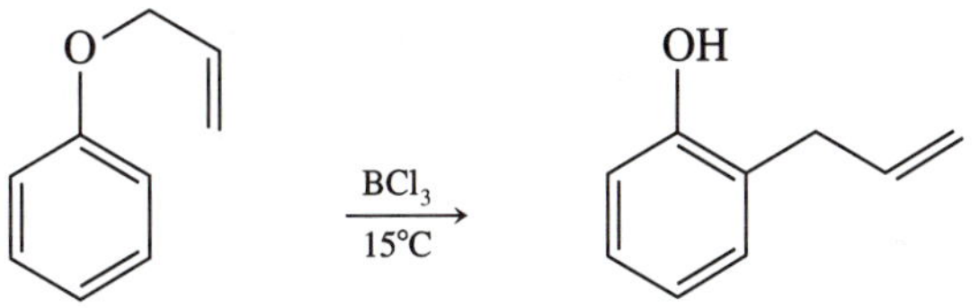

(55)

Cope rearrangements of cyclohexadienones, such as that shown in Eq. 56, take place rapidly at room temperature when catalyzed by 0.01 M hydrochloric acid solutions.[58] [1,5] Shifts of benzyl groups in cyclohexadienones (Eq. 57), which require temperatures above 150°C in the absence of catalysis, proceed at room temperature when catalyzed by solutions of sulfuric acid in acetic acid.[66]

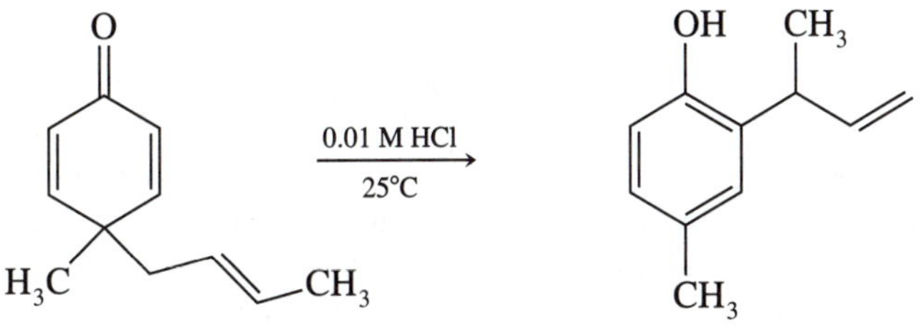

(56)

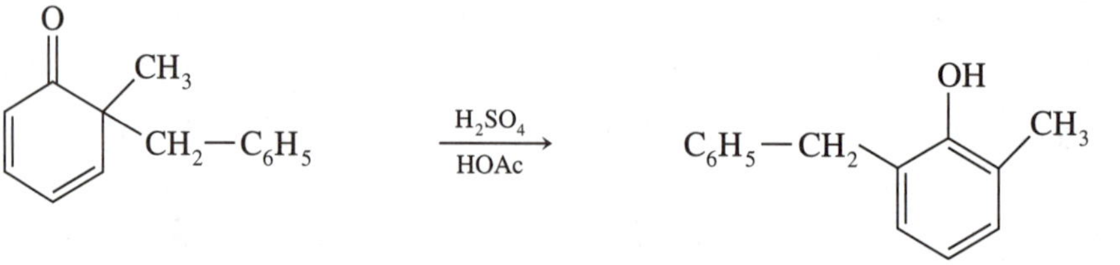

(57)

(Some sigmatropic shifts are accelerated by positive charges even in the absence of acid catalysis. Solvolysis of mesylate **35**, for instance, yields **37**, presumably by way of an "aza-Cope" rearrangement of **36**. However, the rearrangement is so rapid at room temperature that the
formation of **36** cannot be detected.)[67]

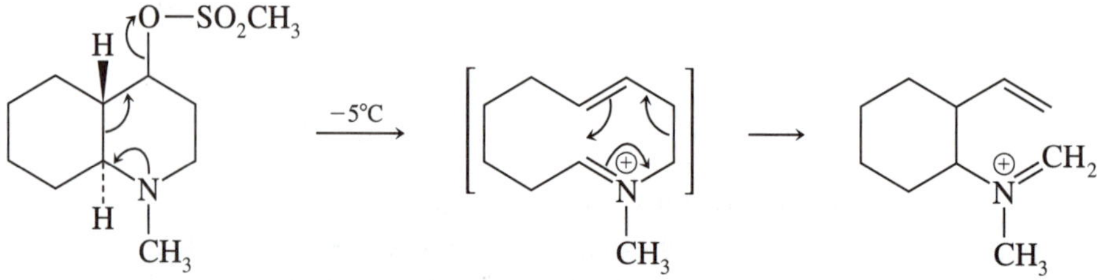

(58)

Catalysis by Bases

The rates of sigmatropic shifts can be accelerated by negative as well as positive charges. The
"oxy-Cope rearrangements"[1] of the potassium salts of 3-hydroxy-1,5-hexadienes, such as **38** and
39, have been found to proceed as much as 10^{12} times as rapidly as the rearrangements of the
parent alcohols. The temperatures needed for the reactions decreased by as much as 250°C.[68,69]

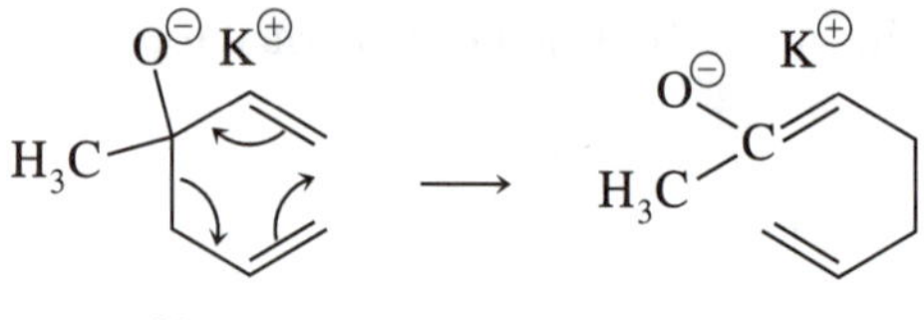

(59)

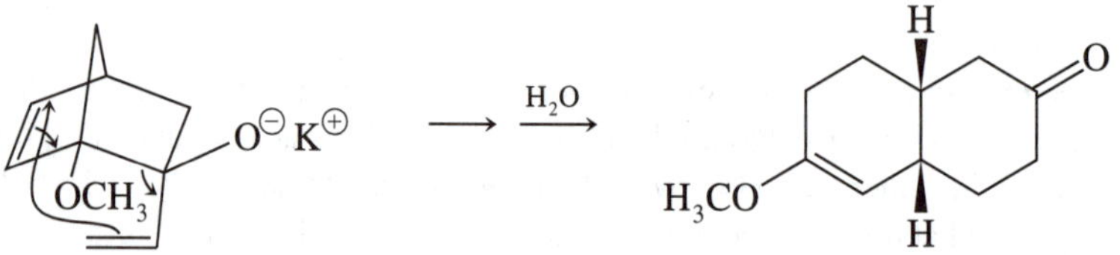

(60)

Converting alcohols to their anions should significantly reduce the bond strengths of adjoining diallylic bonds. A study of secondary isotope effects indicates that bond breaking proceeds to a
greater extent, and bond making to a lesser extent, in the transition states for "anionic oxy-Cope rearrangements" than for Cope rearrangements of neutral molecules.[68b] (As shown on page 96, [1,3]
migrations are also accelerated by negative charges. The synthetic usefulness of the anionic oxy-
Cope rearrangement is often reduced because [1,3] shifts may accompany the [3,3] shifts.[70])

[5,5] shifts can also proceed very rapidly in negatively charged compounds, as is demonstrated by the fact that the rearrangement of anion **40** is complete in one hour at room temperature. (It has been proven that the rearrangement does not proceed by a sequence of consecutive [3,3] shifts.[71])

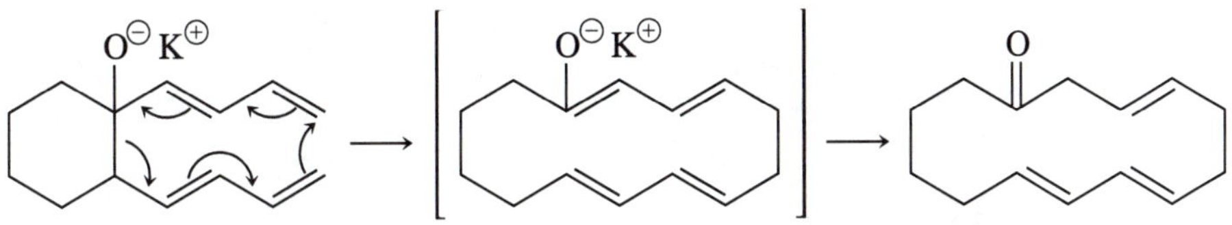

(61a)

Other types of rearrangements, such as the reaction shown in Eq. 61b,[71b] which can be regarded either as a [1,5] shift involving two σ bonds or as a retro-ene reaction, are similarly accelerated by conversion of the starting materials to anions.

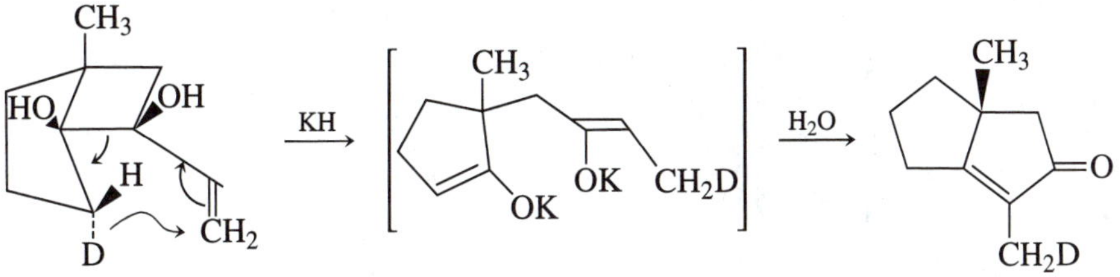

(61b)

Benzidine Rearrangements

The fact that Claisen and Cope rearrangements can be catalyzed by acids is a relatively recent discovery. In contrast, some types of sigmatropic shifts were originally reported to occur solely as acid-catalyzed processes. The best known of these is the complex of rearrangements collectively designated *benzidine rearrangements*.[72]

Rearrangement products. Hydrazobenzene (1,2-diphenylhydrazine) reacts with mineral acids to yield salts of "benzidine" (4,4'-diaminobiphenyl) and "diphenyline" (2,4'-diaminobiphenyl) in approximately 70:30 ratios. Under some conditions, small amounts of *ortho*-benzidine are also formed.

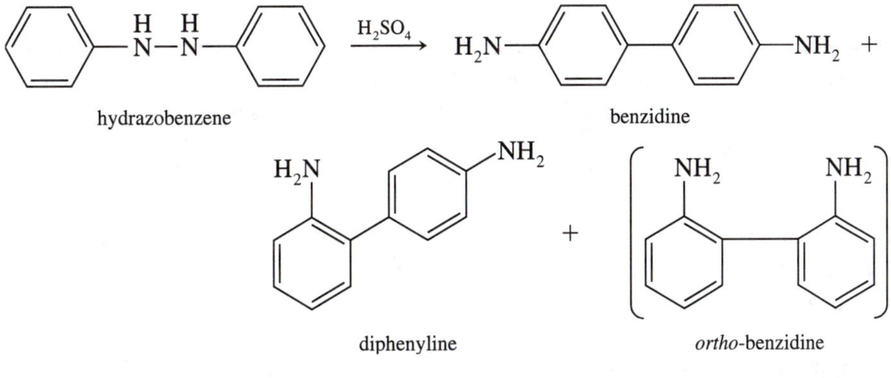

hydrazobenzene

benzidine

(62)

diphenyline

ortho-benzidine

Ortho-benzidines are major products from rearrangements of hydrazonaphthalenes (Eq. 63).

(63)

2,2'-hydrazonaphthalene

When substituents on at least one *para* position of a hydrazobenzene prevent the formation of benzidines, derivatives of diphenylamine ("*ortho*- and *para*-semidines") may be formed (Eq. 64).[72]

(64)

an *ortho*-semidine   a *para*-semidine

Intramolecularity and kinetics of benzidine rearrangements. Rearrangements of mixtures of two hydrazobenzenes, both rearranging at similar rates, do not result in the formation of benzidines or semidines combining aromatic rings from different starting molecules. (Studies employing radioactively labeled substrates would have detected very small amounts of products of intermolecular reactions.) Furthermore, the rearrangements of many hydrazobenzenes with two differently substituted aromatic rings have been studied; no "crossover products" combining rings from different molecules have ever been detected. Thus, it appears to be well established that *ortho*- and *para*-benzidines[72] and *ortho*- and *para*-semidines,[73] at least, are formed entirely by intramolecular processes.

The rearrangement rates of hydrazobenzene and of other comparatively slow-reacting hydrazines follow third-order kinetics—first order in the concentration of the amines and second order in the concentration of acids. Presumably, charge repulsion resulting from protonation of both nitrogen atoms, as in Eq. 65, weakens the nitrogen–nitrogen bond and facilitates rearrangement.

The rearrangement rates of very reactive diarylhydrazines, such as hydrazonaphthalenes, however are first order in acid (Eq. 66).

With diarylhydrazines of intermediate reactivity, such as *ortho*-hydrazotoluene, the reaction rates appear to be principally first order in acid if the reactions are carried out in dilute acid solutions, and principally second order in acid if more concentrated acid solutions are used.

The most reactive diarylhydrazines, such as 2,2'-hydrazonaphthalene, can undergo benzidine rearrangements in the absence of acid catalysts at approximately 100°C. Even these "thermal" rearrangements, however, proceed much more rapidly in alcohols than in hydrocarbon or ether solvents, suggesting that hydrogen bonding can catalyze the rearrangements.[72]

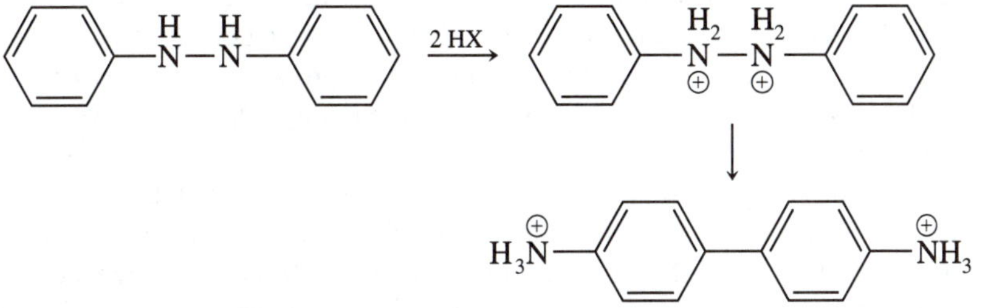

(65)

$$\frac{-d[\text{HB}]}{dt} = k[\text{HB}][\text{HX}]^2 \quad (\text{HB} = \text{hydrazobenzene})$$

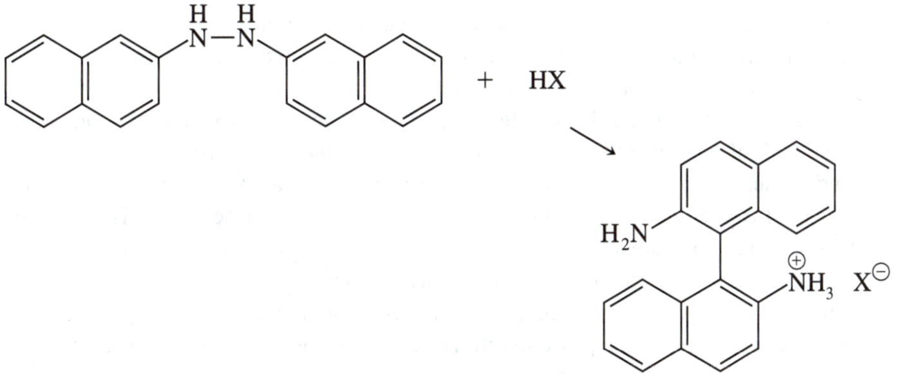

(66)

$$\frac{-d[\text{HN}]}{dt} = k[\text{HN}][\text{HX}] \quad (\text{HN} = \text{hydrazonaphthalene})$$

Mechanisms of benzidine rearrangements. Few reactions have puzzled organic chemists for a longer period than benzidine rearrangements. How can so many different types of products, many resulting from combining atoms that were widely separated in the starting materials, be formed in intramolecular processes?

M.J.S. Dewar suggested that the reactions proceed by way of π complexes of the two rings. (Hypothetical complexes from a diprotonated hydrazobenzene are shown in Eq. 67.)

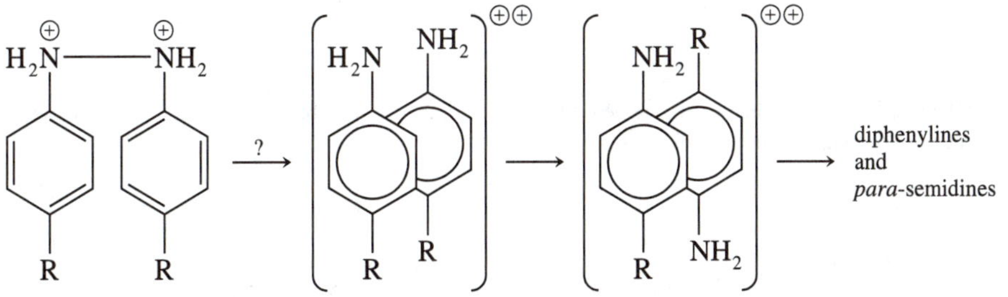

(67)

hypothetical π-complex intermediates

Rotation of the rings within the complex could, Dewar suggested, bring any position of one ring into proximity to each atom of the other ring so that the formation of any of the observed products would be geometrically feasible.[74]

Banthorpe, Hughes, and Ingold proposed that *ortho-* and *para-*benzidines, at least, were formed via "polar transition states."[75] The name is not very informative, because all transition states for the reactions of cations and dications must be polar. Presumably, it was intended to stress the absence of intermediates—that is, that the rearrangements proceed by what we would now call *concerted mechanisms*. It was assumed that products such as *para-*semidines, the formation of which via concerted reactions was geometrically improbable, were produced by other, unspecified, mechanisms.

Studies of heavy atom isotope effects have provided important evidence in regard to the mechanisms of the various members of the benzidine rearrangement complex. Substitution of ^{15}N for ^{14}N was found to slightly decrease the rates of all rearrangements of hydrazobenzenes. This demonstrated that rupture of the nitrogen–nitrogen bond takes place during the rate-determining step of each rearrangement.[76-79]

In addition, the reactions of hydrazobenzene were compared to those of hydrazobenzene labeled with ^{14}C at both *para* positions. The occurrence of a detectable isotope effect demonstrated that bond formation at the *para* positions was taking place during the rate-determining step. The combined isotope-effect studies provided strong evidence that benzidine is formed from hydrazobenzene by a concerted mechanism.[76]

In contrast, there was no detectable ^{14}C isotope effect for formation of diphenyline. It is true that the lack of a measurable isotope effect cannot *conclusively* prove that bond formation does not occur during the rate-determining step. However, the (at most) very small isotope effect for formation of diphenyline compared with that for formation of benzidine is most consistent with a nonconcerted mechanism for diphenyline formation, in which rupture of the nitrogen–nitrogen bond precedes formation of the carbon–carbon bond.[76]

Similar studies demonstrated that the *ortho-*benzidine rearrangement of 2,2'-hydrazonaphthalene[77] and the *para-*semidine rearrangement of 4-methoxyhydrazobenzene (Eq. 68)[78] proceed by concerted mechanisms.*

(68)

---

*However, the *p*-semidine rearrangement of 4,4'-dichlorohydrazobenzene, which requires reductive displacement of a chlorine atom, does not exhibit a ^{14}C isotope effect.[79]

On the other hand, the *ortho*-semidine rearrangement of 4,4'-dichlorohydrazobenzene does *not* show a ^{14}C isotope effect and is presumably not concerted.[79]

These results appear to be a triumph for the Woodward–Hoffmann rules. The *para*-benzidine, *ortho*-benzidine, and *para*-semidine rearrangements—allowed [5,5], [3,3], and [1,5] shifts, respectively—proceed by concerted paths, while the diphenyline and *ortho*-semidine rearrangements—forbidden [3,5] and [1,3] shifts—apparently do not.

After more than 100 years of study of benzidine rearrangements, many questions remain unanswered. The apparently concerted nature of the *para*-semidine rearrangement, in particular, is surprising, since it would seem to require a gross distortion of the aromatic ring to bring the nitrogen atom and the *para* carbon into close proximity. The mechanisms for formation of *ortho*-semidines and diphenylines are also still obscure. Are π complexes involved in these reactions, or are other types of intermediates still waiting to be identified?

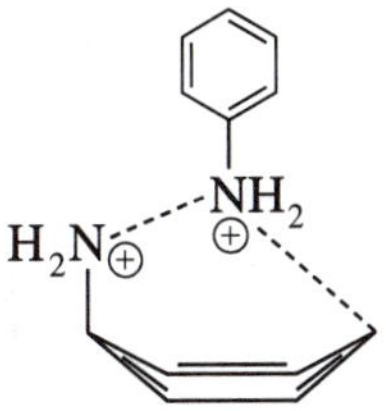

A possible distorted transition state
for a *para*-semidine rearrangement

Quinamine Rearrangements

Halogenation of 2,4,6-trisubstituted phenols yields halocyclohexadienone derivatives, such as **41**. Reactions of these compounds with aromatic amines yield *para*-quinamines, such as **42**.[79]

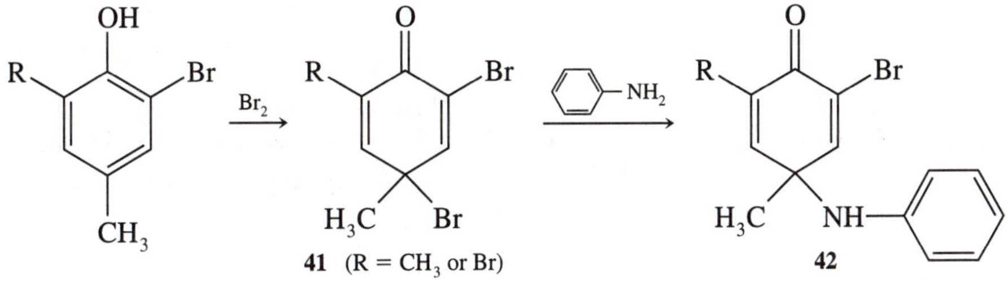

$$(69)$$

Para-quinamines have been found to rearrange rapidly in even very dilute solutions of acids. When *para* positions of the aniline rings are unsubstituted, the principal products are diaryl ethers, such as **43**.[80,81] These remarkable rearrangements have been shown to proceed by intramolecular mechanisms.[81]

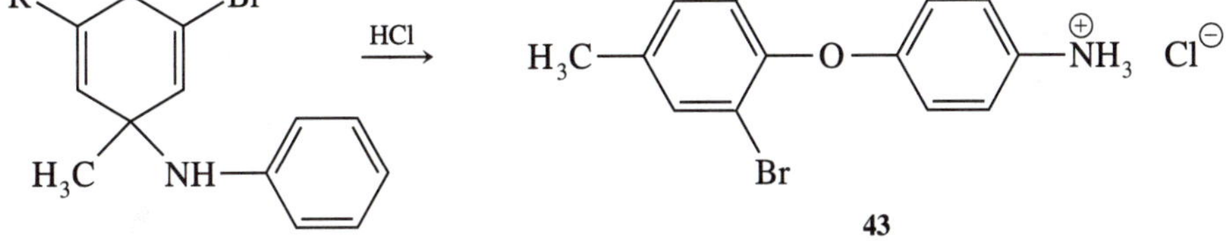

$$(70)$$

A second type of rearrangement of *para*-quinamines results in the reductive displacement of halogen atoms to form biphenyl derivatives, such as **45**. (Biphenyl derivatives are the *principal* products from *para*-quinamines formed from *para*-methylaniline.[82]) Carbazole derivatives, such as **46**, may also be formed.[83] *ortho*-Cyclohexadienones such as **44** are presumably intermediates in the formation of both biphenyls and carbazoles.

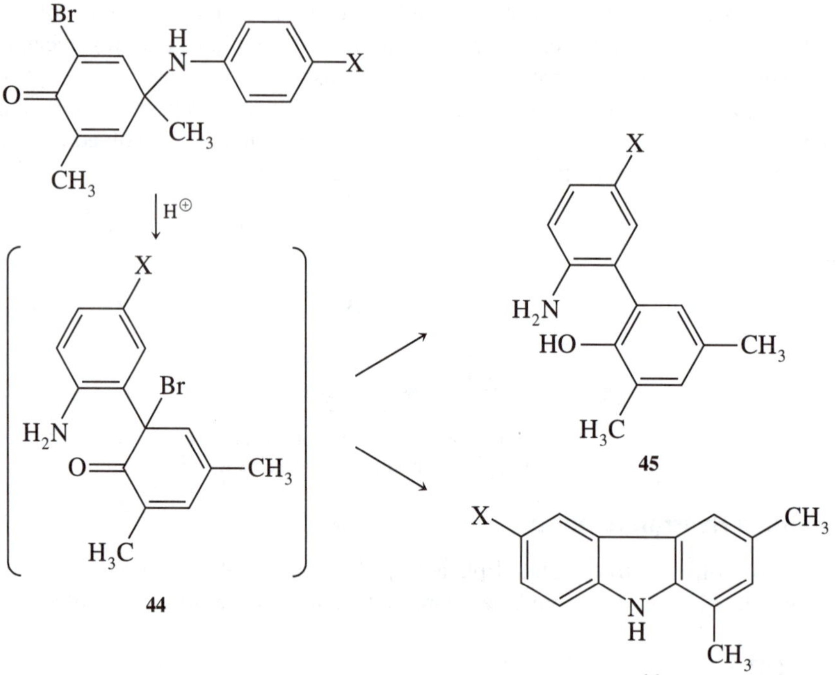

(71)

The effects of heavy atom isotopes on rates of rearrangements of *para*-quinamines have been measured. These studies[83] have demonstrated that the rearrangement steps leading to both diphenyl ethers and carbazoles (and, presumably, to diphenyl derivatives) are concerted processes, proceeding by [5,5] and [3,3] shifts, respectively.*

Para-quinamines in which the aniline rings bear electron-withdrawing substituents can undergo a third type of rearrangement, leading to the formation of diphenylamines. In these reactions, remarkably, the carbonyl oxygen is replaced by halogen from the acid catalyst.[84] As yet, little is known about the mechanisms of these reactions.

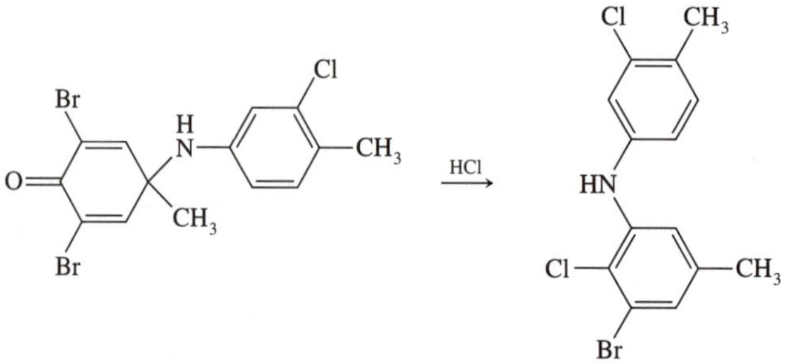

REFERENCES

[1] R.K. Hill, in *Comprehensive Organic Synthesis*, vol. 5, L.A. Paquette, Ed. Pergamon Press, New York, 827–873 (1991).

[2] R.B. Woodward and R. Hoffmann, *J. Am. Chem. Soc., 87*, 2511 (1965).

[3] R.B. Woodward and R. Hoffmann, *The Conservation of Orbital Symmetry.* Academic Press, New York (1970).

[4] W.J. Bailey and R.A. Baylouny, *J. Org. Chem., 27*, 3476 (1962).

[5] D.J. Pasto and J.E. Brophy, *J. Org. Chem., 56*, 4554 (1991).

[6] J.A. Berson and G.L. Nelson, *J. Am. Chem. Soc., 89*, 5303 (1967); J.A. Berson, *Acc. Chem. Res., 1,* 152 (1968); W.R. Roth and A. Friedrich, *Tetrahedron Lett.,* 2607 (1969).

[7] J.W. Lown, M.H. Akhtar, and R.S. McDaniel, *J. Org. Chem., 39,* 1998 (1974).

[8] B.K. Carpenter, *J. Org. Chem., 57*, 4645 (1992).

[9] J.A. Berson, *Acc. Chem. Res., 5,* 406 (1972); J.A. Baldwin and K.D. Belfield, *J. Am. Chem. Soc., 110,* 296 (1988); F.-G. Klaerner, R. Drewes, and D. Hasselmann, *ibid.,* 297 (1988); J.A. Bender, P.A. Leber, R.R. Lirio, and R.S. Smith, *J. Org. Chem., 65,* 5396 (2000).

[10] R.C. Cookson and J.E. Kemp, *Chem. Comm.,* 385 (1971).

[11] M.T. Zoeckler and B.K. Carpenter, *J. Am. Chem. Soc., 103,* 7661 (1981).

[12] N.D. Epiotis, *J. Am. Chem. Soc., 95,* 1206 (1973).

[13] F. Bernardi, M. Olivucci, M.A. Robb, and G. Tonachini, *J. Am. Chem. Soc., 114,* 5805 (1992).

[14] D.I. Schuster, in *Rearrangements in Ground and Excited States,* vol. 3, P. de Mayo, Ed. Academic Press, New York (1980), 221–226.

[15] H.E. Zimmerman and J.M. Cassel, *J. Org. Chem., 54,* 3800 (1989); V. Singh and M. Procini, *Chem. Comm.,* 134 (1993).

[16] R.C. Cookson, J. Hudec, and M. Sharma, *Chem. Comm.,* 107 (1971); M. Sharma, *J. Am. Chem. Soc., 97,* 1153 (1975).

[17] E.N. Marvell, G. Caple, B. Schatz, and W. Pippin, *Tetrahedron, 29,* 378 (1973); L. Skattebol, *J. Org. Chem., 31,* 2789 (1966).

[18] W.R. Roth, J. Koenig, and K. Stein, *Ber., 103,* 426 (1970).

[19] W.R. Roth, *Tetrahedron Lett.,* 1009 (1964).

[20] E.F. Kiefer and C.H. Tanna, *J. Am. Chem. Soc., 91,* 4478 (1969).

[21] W.G. Dauben, C.D. Poulter, and C. Suter, *J. Am. Chem. Soc., 92,* 7408 (1970).

[22] M.A.M. Boersma, J.W. de Haan, H. Kloosterziel, and L.J.M. van de Ven, *Chem. Comm.,* 1168 (1970).

[23] D.J. Fields, D.W. Jones, and G. Kneen, *Chem. Comm.,* 873 (1976).

[24] L.L. Miller, R. Greisinger, and R.F. Boyer, *J. Am. Chem. Soc., 91,* 1578 (1969).

[25] C. Manning, M.R. McClary, and J.J. McCullough, *J. Org. Chem., 46,* 919 (1981).

[26] P. Schiess and P. Fuenfschilling, *Tetrahedron Lett.,* 5195 (1972).

[27] J.W. de Haan and H. Kloosterziel, *Rec. Trav. Chim., 84,* 1594 (1965); *ibid., 87,* 298 (1968).

[28] W.R. Dolbier, Jr., K.E. Anapalle, L.C. McCullagh, K. Matsui, J.M. Riemann, and D. Robison, *J. Org. Chem., 44,* 2845 (1979).

[29] C.W. Spangler and D.L. Boles, *J. Org. Chem., 37,* 1020 (1972).

[30] P. Schiess and R. Dinkel, *Tetrahedron Lett.,* 2503 (1975); see also B.C. Baumann and A.S. Dreiding, *Helv. Chim. Acta, 57,* 1872 (1974).

[31] B. Miller, *Acc. Chem. Res., 8,* 245 (1975).

[32] B. Miller and K.-H. Lai, *Tetrahedron Lett.,* 517 (1972).

[33] B. Miller and J. Baghdadchi, *J. Org. Chem., 52,* 3390 (1987).

[34] V. Boekelheide and T.A. Hylton, *J. Am. Chem. Soc., 92,* 3669 (1970).

[35] (a) H. Hart and J.D. DeVrieze, *Tetrahedron Lett.,* 4259 (1968); (b) K.V. Auwers and G. Keil, *Ber., 36,* 1861, 3902 (1903).

[36] (a) M.S. Newman and R.M. Layton, *J. Org. Chem., 33,* 2338 (1968); (b) B. Miller and K.-H. Lai, *J. Am. Chem. Soc., 94,* 3472 (1972).

[37] (a) For reviews, see J.A. Berson, *Acc. Chem. Res., 1,* 152 (1968); F.G. Klärner, *Topics in Stereochemistry, 15* (1984).

[38] A. Kless, M. Nendel, S. Wilsey, and K.N. Houk, *J. Am. Chem. Soc., 121,* 4524 (1999).

[39] For references, see W.H. Okamura, H.Y. Elnagar, M. Ruther, and S. Dobreff, *J. Org. Chem., 58,* 600 (1993); M. Okabe, R.-C. Sun, and S. Wolff, *Tetrahedron Lett., 35,* 2865 (1994).

[40] C.A. Hoeger, A.D. Johnston, and W.H. Okamura, *J. Am. Chem. Soc., 109,* 4690 (1987).

[41] (a) F.G. Klärner, *Angew. Chem. Intl. Ed. Engl., 11,* 832 (1972); (b) L.B. Jones and V.K. Jones, *J. Am. Chem. Soc., 90,* 1540 (1968) and references therein.

[42] M.J.S. Dewar and L.E. Wade, Jr., *J. Am. Chem. Soc., 99,* 4417 (1977).

[43] W.R. Roth et al., *J. Am. Chem. Soc., 112,* 1722 (1990).

[44] W. von E. Doering et al., *J. Am. Chem. Soc., 116,* 4289 (1994).

[45] For references, see M.J.S. Dewar and C. Jie, *Acc. Chem. Res., 25,* 537 (1992).

[46] D.A. Hrovat, K. Morokuma, and W.T. Borden, *J. Am. Chem. Soc., 116,* 1072 (1994).

[47] A. Padwa and T.J. Blacklock, *J. Am. Chem. Soc., 102,* 2797 (1980).

[48] W.R. Roth, T. Schaffers, and M. Heiber, *Ber., 125,* 739 (1992).

[49] E. Vogel, *Liebigs Ann., 615,* 1 (1958); *644,* 172 (1961); G.S. Hammond and C.D. DeBoer, *J. Am. Chem. Soc., 86,* 899 (1964); J.M. Bown, B.T. Golding, and J.J. Stofko, Jr., *Chem. Comm.,* 319 (1973).

[50] A.K. Cheng, F.A.L. Anet, J. Mioduski, and J. Meinwald, *J. Am. Chem. Soc., 96,* 2887 (1974).

[51] D. Moskau, et al., *Ber., 122,* 955 (1989).

[52] For references, see M.J.S. Dewar and C. Jie, *Tetrahedron, 44,* 1351 (1988).

[53] W. von E. Doering and W.R. Roth, *Tetrahedron Lett., 18,* 67 (1962); see also M.J. Goldstein and M.Z. Benzon, *ibid., 94,* 7149 (1972).

[54] S.J. Rhoads, in *Molecular Rearrangements,* vol. 1, P. de Mayo, Ed. John Wiley & Sons, New York, 656 (1963).

[55] See J.A. Chamizo and M.F. Lappert, *J. Org. Chem., 54,* 4684 (1989).

[56] S. Desert and P. Metzner, *Tetrahedron, 48,* 10, 327 (1992).

[57] H. Conroy and R.A. Firestone, *J. Am. Chem. Soc., 78,* 2290 (1956).

[58] B. Miller, *J. Am. Chem. Soc., 67,* 5115 (1965).

[59] (a) D.Y. Curtin and H.W. Johnson, Jr., *J. Am. Chem. Soc., 78,* 2611 (1956); (b) H.-J. Hansen, in *Mechanisms of Molecular Migrations,* vol. 3, B.S. Thyagarajan, Ed. Wiley-Interscience, New York, 177–200 (1971).

[60] K. Seki, M. Tooya, T. Sato, M. Ueno, and T. Uyehara, *Tetrahedron Lett., 39,* 8673 (1998).

[61] G. Frater and H. Schmid, *Helv. Chim. Acta, 51,* 190 (1968).

[62] D. Alker, W.D. Ollis, and H. Shahriari-Zavareh, *J. Chem. Soc., Perkin Trans., 1,* 1637 (1990).

[63] K. Hafner, H.J. Linder, W. Luo, P.K. Meinhardt, and T. Zink, *Pure Appl. Chem., 65,* 17 (1993).

[64] J.C. Gilbert, K.R. Smith, G.W. Klumpp, and M. Schabel, *Tetrahedron Lett.,* 125 (1972).

[65] P. Fahrni, A. Habich, and H. Schmid, *Helv. Chim. Acta, 43,* 448 (1960).

[66] B. Miller, *J. Am. Chem. Soc., 92,* 6252 (1970).

[67] J.A. Marshall and J.H. Babler, *J. Org. Chem., 34,* 4186 (1969).

[68] D.A. Evans and A.M. Golub, *J. Am. Chem. Soc., 97,* 4765 (1975); S.M. Sclhulze, N. Santella, J.J. Grabowski, and J.K. Lee, *J. Org. Chem., 66,* 7247 (2001).

[69] J.J. Gajewski and K.R. Gee, *J. Am. Chem. Soc., 113,* 967 (1991).

[70] R.W. Thies and E.P. Seitz, *Chem. Comm.,* 846 (1976).

[71] (a) P.A. Wender, R.J. Ternansky, and S.M. Sieburth, *Tetrahedron Lett., 26,* 4319 (1985); (b) M.E. Jung and B. Davidov, *Org. Lett., 3,* 3025 (2001).

[72] H.J. Shine, *Aromatic rearrangements.* Elsevier Publishers, New York, 124–179 (1967); R.A. Cox and E. Buncel, in *The Chemistry of the Hydrazo, Azo and Azoxy Groups,* S. Patai, Ed. John Wiley & Sons, New York, 775–806 (1975).

[73] A. Heesing and U. Shine, *Ber., 105,* 3838 (1972); *110,* 3319 (1977).

[74] M.J.S. Dewar, in *Molecular Rearrangements,* Part 1, Paul de Mayo, Ed. Interscience Publishing, New York, 295–299, 333–343 (1963).

[75] D.V. Banthorpe, E.V. Hughes, and C.K. Ingold, *J. Chem. Soc.,* 2864 (1964).

[76] H.J. Shine, H. Zmuda, K.H. Park, H. Kwart, A.G. Horgan, and M. Brechbiel, *J. Am. Chem. Soc., 104,* 2501 (1982).

[77] H.J. Shine, E. Gruszecka, W. Subotkowski, M. Brownawell, and J. San Filippo, Jr., *J. Am. Chem. Soc., 107,* 3218 (1985).

[78] H.J. Shine, H. Zmuda, H. Kwart, A.G. Horgan, and M. Brechbiel, *J. Am. Chem. Soc., 104,* 5181 (1982).

[79] E.S. Rhee and H.J. Shine, *J. Am. Chem. Soc., 108,* 1000 (1986).

[80] K. Fries and G. Oehmke, *Liebigs Ann., 462,* 1 (1928).

[81] B. Miller, *J. Am. Chem. Soc., 86,* 1127 (1964).

[82] K. Fries and A. Kuester, *Liebigs Ann., 470,* 20 (1929).

[83] B. Boduszek and H.J. Shine, *J. Am. Chem. Soc., 110,* 3247 (1988).

[84] K. Fries, R. Boeker, and F. Wallbaum, *Liebigs Ann., 509,* 73 (1934).

PROBLEMS

4.1 Demonstrate by means of the frontier orbital approach that suprafacial [3,5] shifts are forbidden under thermal conditions and allowed under photochemical conditions, and that the opposite is true for antarafacial [3,5] shifts.

4.2 Indicate the order of each of the sigmatropic shifts shown in the equations below. (b) Indicate which of the reactions are theoretically allowed as concerted processes, and which are theoretically forbidden.

(b)

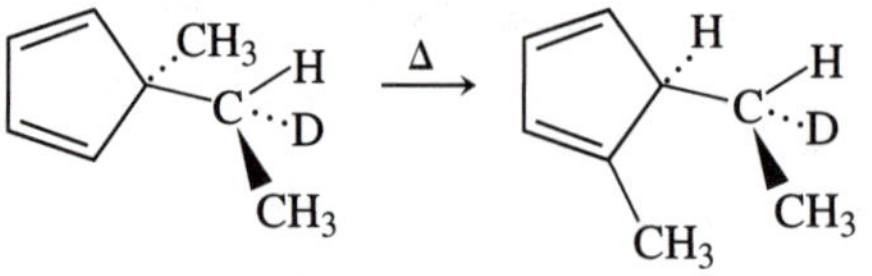

(c)

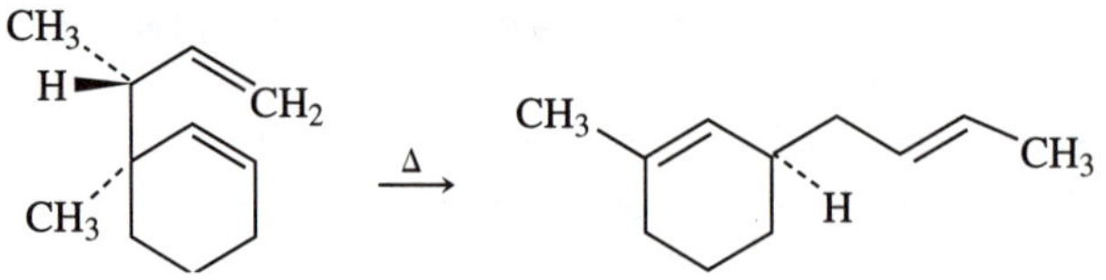

(d)

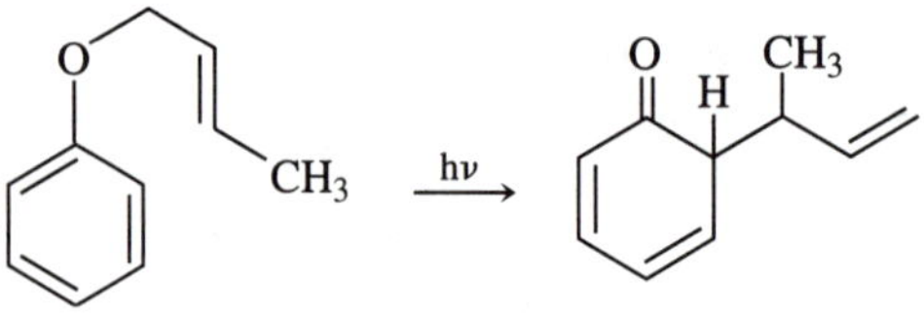

(e) $H_2C{=}CH{-}O{-}CH_2{-}CH{=}CH{-}CH{=}CH_2$ $\xrightarrow{\Delta}$ $\overset{\overset{\displaystyle O}{\|}}{HC}{-}CH_2CH_2{-}CH{=}CH{-}CH{=}CH_2$

4.3 Draw structures for the expected products from each of the following reactions:

(a)

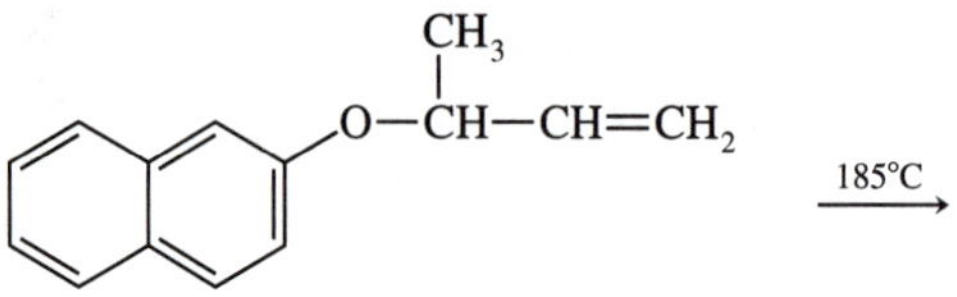

$\xrightarrow{185°C}$

(b)

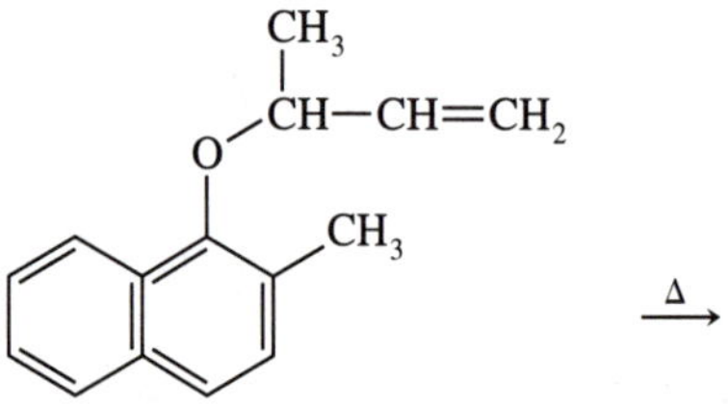

$\xrightarrow{\Delta}$

(c)

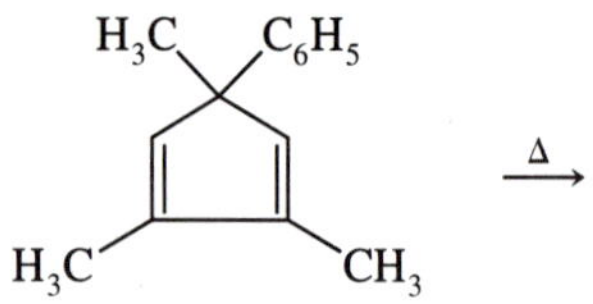

$\xrightarrow{\Delta}$

(d)

$\underset{270°C}{\rightleftharpoons}$

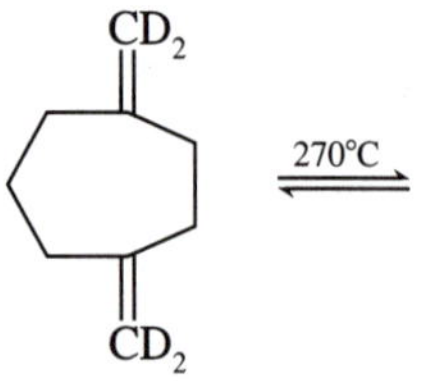

(W. von E. Doering and C.A. Troise, *J. Am. Chem. Soc.*, *107*, 5739 [1985])

(e)

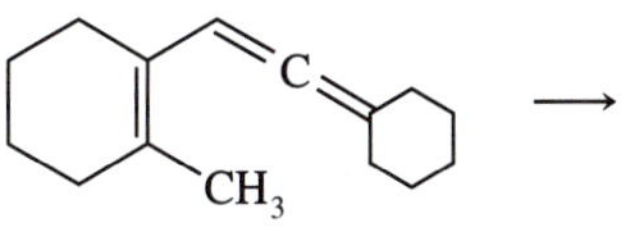

(S.A. Barrack and W.H. Okamura, *J. Org. Chem.*, *51*, 3201 [1986])

(f)

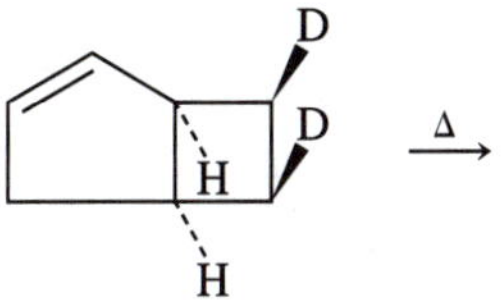

4.4 Write reasonable mechanisms for the following reactions, using curved arrows to show movements of electrons.

(a)

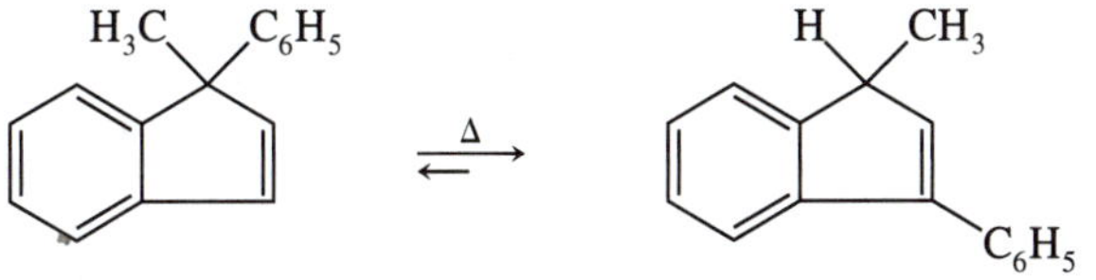

(b)

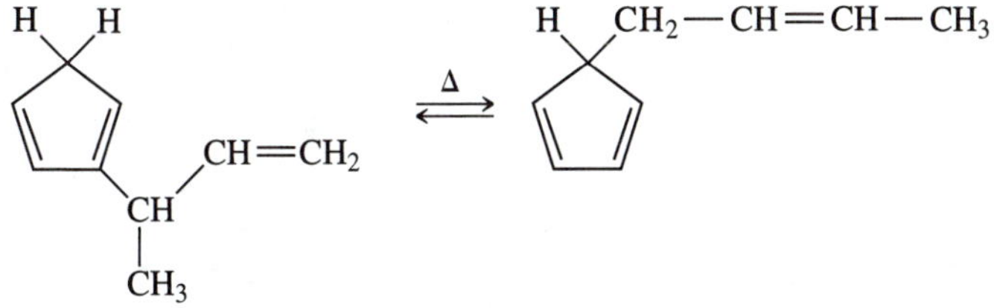

(c)

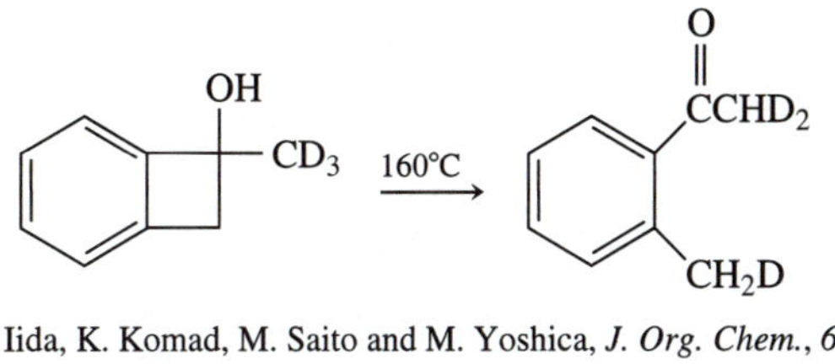

(K. Iida, K. Komad, M. Saito and M. Yoshica, *J. Org. Chem.*, *64*, 7407 [1999])

(d)

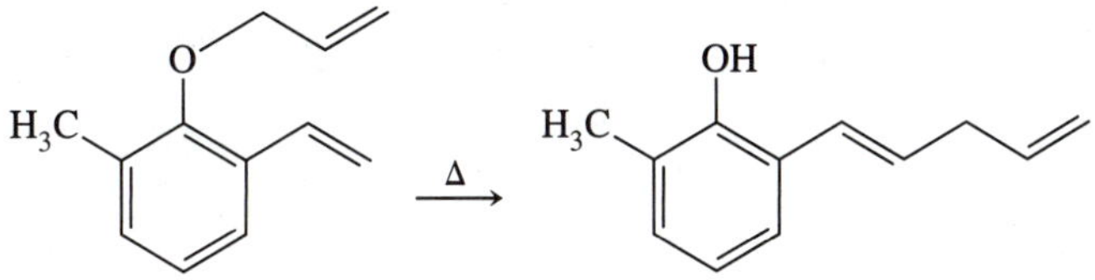

(e)

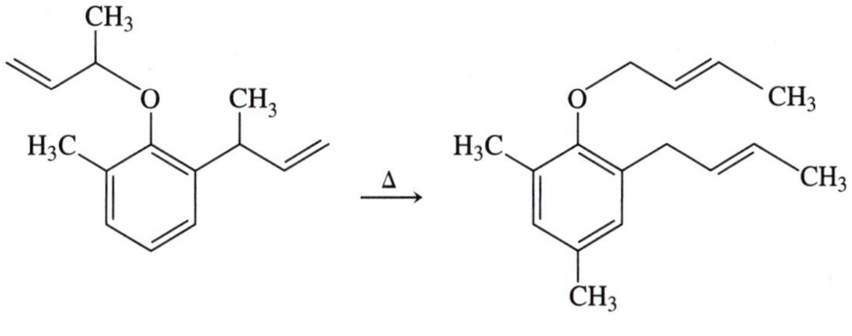

(f)

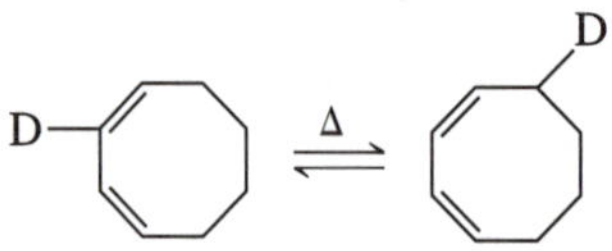

(J.E. Baldwin, D.A. Leher, and T.W. Lee, *J. Org. Chem., 66,* 5269 [2001])

(g)

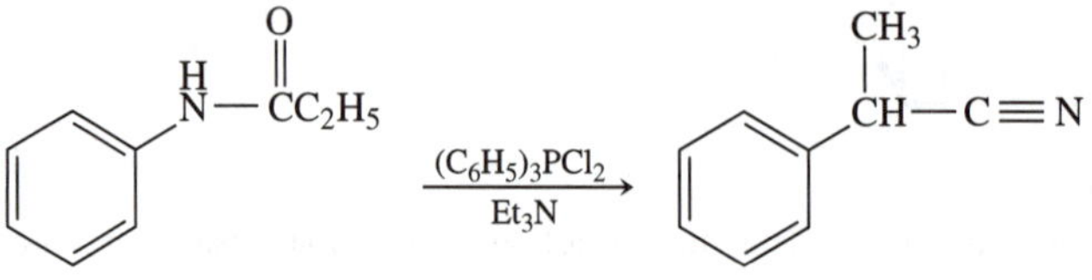

$$\xrightarrow{\text{(C}_6\text{H}_5)_3\text{PCl}_2}{\text{Et}_3\text{N}}$$

(h) $(CH_3)_2N - CH_2CH = CHCH_3 \quad + \quad$

$\longrightarrow$

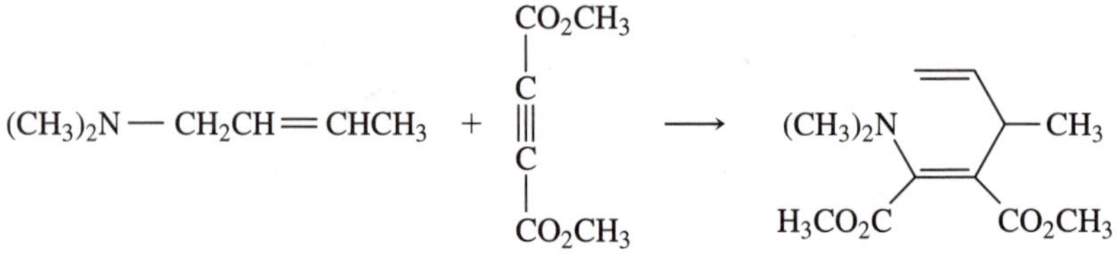

(E. Vedejs and M. Gingras, *J. Am. Chem. Soc., 116,* 5 [1994])

(i)

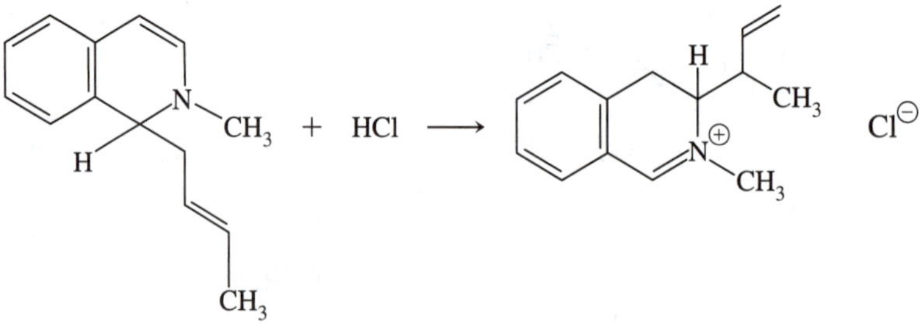

$+$ HCl $\longrightarrow$

Cl$^{\ominus}$

(J. Knabe and H.-D. Hoeltje, *Tetrahedron Lett.,* 2107 [1969])

(j)

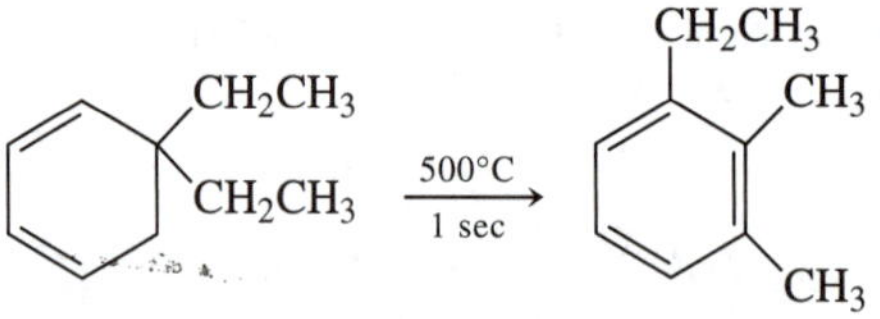

$$\xrightarrow[\text{1 sec}]{500°C}$$

(P. Schiess and R. Dinkel, *Tetrahedron Lett.,* 2503 [1975])

Linear Free-Energy Relationships

5.1 THE HAMMETT EQUATION

The Substituent Constant, σ

Chapters 1 through 4 deal principally with reactions carried out under thermal conditions. The remaining chapters deal in large part with polar reactions. Those reactions are strongly affected by substituents on organic molecules. This is therefore a good time to examine the effects of different substituents on the equilibrium constants and rates of reactions of organic molecules and ions.

In 1937 Louis P. Hammett at Columbia University suggested that the effects of *meta* and *para* substituents on the ionization constants of benzoic acids could be general predictors of the electronic influences of substituents in a variety of reactions.[1]

There are several good reasons to choose the ionizations of *meta*- and *para*-substituted benzoic acids as reference reactions. A large number of substituted benzoic acids are readily available, and their ionization constants are easily determined. In addition, substituents on aromatic rings are held at fixed distances from the points of reaction, which might not be the case with more flexible aliphatic molecules.

Hammett specifically excluded *ortho* substituents from his studies, since they might influence reactions by steric inhibition of access to the reaction center, or by steric inhibition of resonance, or by hydrogen bonding effects, or even by direct participation in the reactions.

Hammett defined a substituent constant, σ (sigma), as the logarithm of the ratio of the ionization constant of a substituted benzoic acid to that of benzoic acid itself in water solution at 25°C (Eq. A1):

$$\sigma = \log \frac{K}{K_H} \tag{A1}$$

where K is the acidity constant for a substituted benzoic acid and K_H the acidity constant for benzoic acid. Equation A1 is equivalent to

$$\sigma = \log K - \log K_H \qquad (A2)$$

and

$$\sigma = pK_{aH} - pK_a \qquad (A3)$$

Since acidity constants for *meta*- and *para*-substituted benzoic acids differ from each other, each substituent has *two* sigma constants: σ_{meta} and σ_{para}.

By 1991 values of σ constants had been determined for more than 530 substituent groups.[2] Table 5.1 lists values of σ_{meta} and σ_{para} for a number of substituents. Even a casual examination of these values indicates that they represent a measure of the electron-donating and electron-attracting powers of the substituents. Strongly electron-attracting groups, such as the $N{\equiv}N^{\oplus}$, NO_2, and CF_3 groups, have large positive σ values—that is, they markedly increase the ionization constants of benzoic acids (by stabilizing negative charges in the benzoate anions). Strongly electron-donating groups, such as NH_2 and OH groups, have large negative values: they decrease the ionization constants of benzoic acids, since the negative charges they contribute to the rings would repel the negative charges in benzoate anions. Hydrogen, of course, has a σ value of 0 (the logarithm of 1).

TABLE 5.1 SOME HAMMETT SUBSTITUENTS CONSTANTS

Substituent	σ para	σ meta	σ^{+*}	σ^{-*}
$N{\equiv}N^{\oplus}$	1.91	1.76	—	3.43
$(CH_3)_3N^{\oplus}$	0.82	0.88	0.41	0.77
NO_2	0.78	0.71	0.79	1.27
$C{\equiv}N$	0.66	0.56	0.66	1.00
CF_3	0.54	0.43	0.61	0.65
CO_2H	0.45	0.37	0.42	0.77
$CH{=}O$	0.42	0.35	0.73	1.03
Cl	0.23	0.39	0.15	0.25
Br	0.23	0.37	0.11	0.19
$C{\equiv}CH$	0.23	0.21	0.18	0.53
I	0.18	0.35	0.14	0.27
CH_2Cl	0.12	0.11	−0.01	—
F	0.06	0.34	−0.07	−0.03
OCH_2F	0.02	0.20	—	—
$CH{=}CHCH_3$ (*trans*)	0.09	0.02	—	—
CH_3	−0.17	−0.07	−0.31	−0.17
OH	−0.37	0.12	−0.92	−0.37
NH_2	−0.66	−0.16	−1.30	−0.15
$S^{\ominus}$	−1.21	−0.36	—	—
$O^{\ominus}$	(−0.81)†	−0.47	—	—

*applicable to *para* substituents

†doubtful value

Source: H.H. Jaffe, *Chem. Rev.*, 53, 191 (1953).

It is particularly interesting to compare the effects of the same substituents in *meta* and *para* positions. For some substituents, such as the NH_2 group, the absolute value of σ_{para} is much larger than that of σ_{meta}. That is because the electron-donating ability of the amino group is largely due to resonance effects. An amino group in the *para* position can directly distribute electrons into the carboxyl group of benzoic acid, thus making it more difficult for the carboxyl group to become a negatively charged carboxylate anion. (Another way to phrase the same idea is to say that conversion of the carboxyl group to its anion would reduce the resonance interaction of the *para*-amino and carboxyl groups.) In contrast, there is no direct interaction between a *meta*-amino group and the carboxyl group of benzoic acid.

$$(1)$$

Resonance structures for *p*-aminobenzoic acid

Most substituents do have somewhat larger effects in *para* positions than in *meta* positions, even when there is apparently no direct resonance interaction between the substituent and the carboxyl group. Inspecting polarized resonance forms of the benzene rings will show that almost all substituents tend to induce positive or negative charges in *para* (and *ortho*) positions of aromatic rings and have smaller effects on *meta* positions.

Polarized resonance structures for aromatic rings

For a few substituents, such as the OH and OCH_3 groups, σ_{meta} and σ_{para} have opposite signs. That is because electronic effects of substituents are due to at least two different factors: inductive (electronegativity) effects, and resonance effects. In alkoxy and hydroxy groups, the two types of electronic effects work in opposite directions. The very electronegative oxygen atom has a large electron-attracting inductive effect, which shows up in the values of σ_{meta} for the hydroxy and alkoxy groups. However, the oxygen atom also has a strong electron-donating resonance effect, which overpowers the inductive effect when the substituent is in the *para* position. Thus, the σ_{para} constants for alkoxy and hydroxy groups have large negative values.

Halogen atoms also have conflicting inductive and resonance effects. However, halogen atoms are poorer electron donors than oxygen atoms, because double bonds to halogens (aside from fluorine) are weak. Thus, σ constants for halogens always have positive signs, but the values are larger for halogens in *meta* positions than in *para* positions.

The σ values for substituents in polysubstituted aromatic molecules are approximately additive.* Thus, the ionization constant for 3-methyl-5-nitrobenzoic acid, for instance, can be estimated

*However, steric inhibition of resonance may, in some cases, result in poor additivity of σ values for substituents on adjacent carbons.

from knowledge of the ionization constant of benzoic acid and of the values of σ_{meta} for nitro and methyl groups.

The Reaction Constant, ρ

Hammett diagrams. σ Constants have been shown to be appropriate measures of the electronic effects of substituents in many reactions of aromatic molecules.[3–6] This can be demonstrated by *Hammett diagrams*. In a Hammett diagram, the *abscissa* (the *x*-axis) is a scale of σ values for substituents on aromatic rings. The *ordinate* (the *y*-axis) is a scale of the logarithms of the ratios of equilibrium (or rate) constants for reactions of substituted aromatic compounds to those of similar compounds in which the "substituent" is a hydrogen atom.

For example, Figure 5.1 shows a Hammett plot for the ionization constants of phenylacetic acids.[7] The points fit a straight line, which can be reasonably represented* by the *Hammett equation* (Eq. B1 or B2).

$$\log \frac{K}{K_H} = \rho\sigma \tag{B1}$$

$$\log K - \log K_H = \rho\sigma \tag{B2}$$

As we shall see, the Hammett equation has a wide variety of uses. For instance, if the equilibrium constants for several examples of a particular reaction of aromatic molecules are known, so that the proportionality constant, ρ (rho), can be determined, the Hammett equation can be used to estimate the equilibrium constants with different ring substituents with known σ constants.[3]

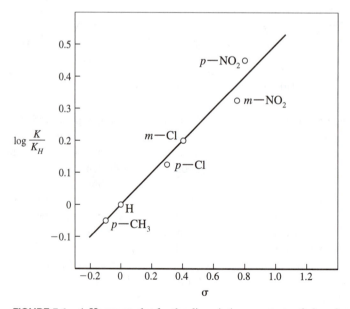

FIGURE 5.1   A Hammett plot for the dissociation constants of phenylacetic acids.

---

*Even aside from some inevitable scattering of points, the fit of empirical data to the Hammett equation is rarely exact, in part because the Hammett equation includes no term for an *x*- or *y*-intercept, on the assumption that the line will pass through the intercept. In practice, this assumption usually fits the facts quite well, but not perfectly.

The significance of ρ values.   The slope of the line in a Hammett diagram, ρ, is called *the reaction constant*. The numerical values and signs of ρ differ from reaction to reaction and show how each reaction is affected by the nature of substituents on the aromatic rings. The value of ρ for the ionization of phenylacetic acids in water at 25°C, for instance, is only about half as large as the value of ρ for the ionization of benzoic acids (which, from the definition of σ, would be exactly 1.00). That is a reasonable result, since the substituents in phenylacetic acids are further from the carboxyl group. The value of ρ for ionizations of 3-phenylpropanoic acids is smaller still (see Table 5.2).[7]

TABLE 5.2   SOME HAMMETT REACTION CONSTANTS (ρ)

Reaction	Solvent	$T(°C)$	ρ	Ref.
A. For ionization constants				
$ArCH_2\overset{O}{\overset{\|}{C}}OH \rightleftharpoons ArCH_2\overset{O}{\overset{\|}{C}}O^- + H^+$	H_2O	25	0.562	a
$ArCH_2CH_2\overset{O}{\overset{\|}{C}}OH \rightleftharpoons ArCH_2CH_2\overset{O}{\overset{\|}{C}}O^- + H^+$	H_2O	25	0.237	a
B. For rate constants				
$Ar\overset{O}{\overset{\|}{C}}OC_2H_5 + NaOH \longrightarrow Ar\overset{O}{\overset{\|}{C}}ONa + C_2H_5OH$	85% C_2H_5OH	25	2.562	a
$Ar\overset{O}{\overset{\|}{C}}OC_2H_5 + H_2O \overset{H^+}{\longrightarrow} Ar\overset{O}{\overset{\|}{C}}OH + C_2H_5OH$	60% $CH_3\overset{O}{\overset{\|}{C}}CH_3$	100	0.106	b
	60% C_2H_5OH	100	0.144	b
$Ar_2CHCHCl_2 + NaOH \longrightarrow Ar_2C{=}CHCl$	92.5% C_2H_5OH	20	2.456	b
$Ar\overset{O}{\overset{\|}{C}}OH + CH_3OH \overset{H^+}{\longrightarrow} Ar\overset{O}{\overset{\|}{C}}OCH_3 + H_2O$	CH_3OH	25	−0.577	a
$Ar_2CH_2OTs + H_2O \longrightarrow Ar_2CH_2OH + HOTs$	92.5% C_2H_5OH	25	−2.2	b
$ArH + Cl_2 \longrightarrow ArCl + HCl$	$CH_3\overset{O}{\overset{\|}{C}}OH$	25	−8.06[c]	d

[a]P.R. Wells, *Chem. Rev.*, 63, 171 (1963)
[b]H.H. Jaffe, *Chem. Rev.*, 53, 191 (1953)
[c]Values of σ[+] used.
[d]H.C. Brown and Y. Okamoto, *J. Am. Chem. Soc.*, 80, 4979 (1958)

A reaction with a positive ρ value is assisted by electron-attracting substituents and adversely affected by electron-donating substituents. The reverse is true for reactions with negative ρ values. Hammett plots for the ionization of carboxylic acids, for instance, all have positive values of ρ even though the numerical values differ from acid to acid. In contrast, plots of the basicities of carboxylate anions would have negative ρ values.

The Hammett equation as a linear free-energy relationship.   Free-energy changes (ΔG) resulting from chemical reactions are proportional to the logarithms of the equilibrium constants of the reactions.

$$\Delta G = -RT \ln K = -2.303 \, RT \log K \tag{C1}$$

or

$$\log K = \frac{\Delta G}{-2.303 \, RT} \tag{C2}$$

Inserting Eq. C2 into Eq. B2 yields

$$\rho\sigma = \frac{\Delta G}{(-2.303 \, RT)} - \frac{\Delta G_H}{(-2.303 \, RT)} \tag{C3}$$

For two reactions run at the same temperature,

$$\rho\sigma = A\Delta G_H - A\Delta G \tag{C4}$$

where $A \,(= 1/{-2.303} \, RT)$ is a constant. Since Eq. C4 is a first-order linear equation, the Hammett equation describes linear relationships between the electronic effects of substituents on some chemical reactions and free-energy changes in those reactions.

Just as free-energy changes of reactions are proportional to logarithms of equilibrium constants, free energies of activation ($\Delta G^{\ddagger}$) are proportional to logarithms of rate constants (Eq. D).

$$\Delta G^{\ddagger} = -2.303 \, RT \log \frac{h}{\text{CBT}} k \quad \begin{array}{l} \text{(Where h = Planck's constant and B} \\ \text{= Boltzmann's constant.)} \end{array} \tag{D}$$

Thus, the version of the Hammett equation shown in Eq. E,

$$\log \frac{k}{k_H} = \rho\sigma \tag{E}$$

where k is the rate constant for the reaction of a *meta-* or *para-*substituted benzene derivative, can be applied to rates of reactions rather than equilibrium constants.

Of course, free-energy changes may result from changes of enthalpies of entropies, or both.

$$\Delta G = \Delta H - T\Delta S \quad \text{and} \quad \Delta G^{\ddagger} = \Delta H^{\ddagger} - T\Delta S^{\ddagger} \tag{F}$$

It was initially believed that the electronic effects of substituents principally affected enthalpy changes and that the Hammett equation gave linear plots because entropy changes (or changes in entropies of activation) caused by substituents were either small or paralleled enthalpy changes. It has since been demonstrated, however, that the actual facts are more complicated. In some cases, substituents exert their influence principally through entropy effects, and in others both entropies and enthalpies are affected in ways that seem difficult to predict.[6]

Reaction mechanisms and the Hammett equation. The fact that the Hammett equation can be applied to reaction rates means that it can offer important information about the mechanisms of chemical reactions. The signs and magnitudes of the reaction constants, ρ, can be of particular significance.

For instance, consider the hydrolysis of benzoyl chlorides in water (Eq. 1).

$$\text{(2)}$$

Several possible mechanisms can be proposed for this type of reaction. Among the possibilities are (1) that water molecules might add to the carbonyl groups to form tetrahedral intermediates in slow, rate-limiting steps, followed by the rapid elimination of chloride ions (or hydrogen chloride) (Eq. 3a)

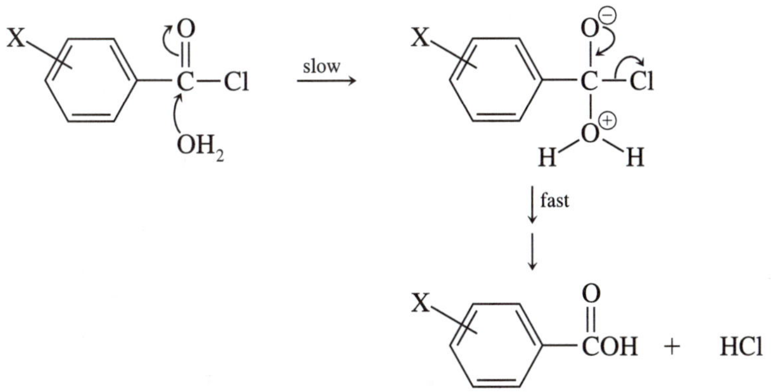

(3a)

and (2) that carbon–chlorine bonds might dissociate in slow, rate-limiting steps, followed by the rapid reaction of water with the resulting carbocations (Eq. 3b).

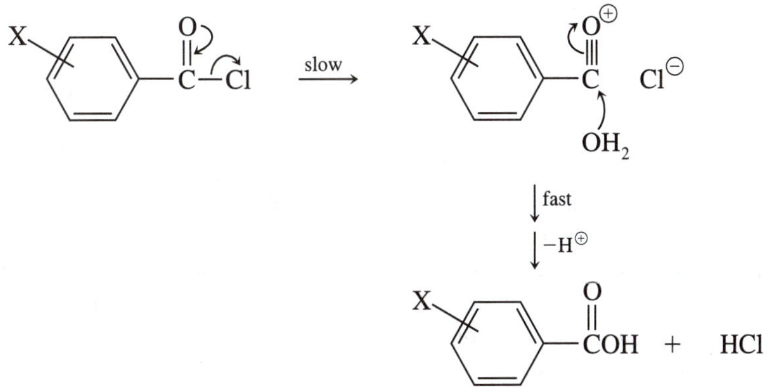

(3b)

The dissociation of benzoyl halides to form carbocations would be strongly assisted by electron-donating substituents. On the other hand, the addition of water to the carboxyl groups should be fastest in benzoyl chlorides with electron-withdrawing substituents. Thus, the fact that a Hammett plot for the rates of aqueous hydrolysis of benzoyl chlorides has a positive value of ρ is consistent with mechanism (1) and eliminates mechanism (2).

σ^+ and σ^- Constants

As we have seen, σ constants for substituents on aromatic rings are a measure of their "electronic effects" on reaction rates and equilibria. However, the electronic effect of a substituent is a combination of inductive and resonance effects. The fact that so many reactions of aromatic molecules are correlated by the Hammett equation suggests that the relative importance of inductive and resonance effects is similar in all those reactions. However, in some reactions strongly electron-donating groups in *para* positions accelerate the reaction rates far more than would be predicted from their σ values. Examples of such reactions include S_N1 reactions of *tert*-cumyl chlorides (Eq. 4) and electrophilic substitution reactions of aromatic rings (Eq. 5). In those reactions direct conjugation may exist between the *para* substituents and empty (or partially empty) orbitals in the transition states. Modified *para*-substituent constants, σ^+, whose values were derived

$$X-\text{\large\textcircled{}}-\underset{\underset{CH_3}{|}}{\overset{\overset{CH_3}{|}}{C}}-Cl \xrightarrow{H_2O} X-\text{\large\textcircled{}}-\underset{\underset{CH_3}{|}}{\overset{\overset{CH_3}{}}{C}}^{\oplus}\text{—}CH_3 \quad Cl^{\ominus} \tag{4}$$

$$\downarrow H_2O$$

$$X-\text{\large\textcircled{}}-\underset{\underset{CH_3}{|}}{\overset{\overset{CH_3}{|}}{C}}-OH \quad + \quad HCl$$

$$\text{\large\textcircled{}} + Br-Br \xrightarrow{HOAc} \text{\large\textcircled{}} \overset{H \; Br}{\underset{X}{}} \quad Br^{\ominus} \longrightarrow \text{\large\textcircled{}}\overset{Br}{\underset{X}{}} + HBr \tag{5}$$

from the substitution reactions of *tert*-cumyl chloride in 90 percent aqueous acetone (Eq. 4),[8] are often useful for reactions of this type.

If the use of σ^+ values in a Hammett plot for the rates of a reaction gives a better fit to a straight line than the use of σ values, it provides strong evidence that a direct resonance interaction exists between a partially empty orbital and *para* substituents in the transition states for the reaction. In fact, most reactions that correlate well with σ^+ proceed by way of carbocation intermediates.

There are other reactions in which *negative* charges can be directly stabilized by resonance with *para* substituents. For these reactions, a modified substituent constant, σ^-, defined as log K/K_H for the acid-base reactions of *para*-substituted phenols or anilinium salts in water (Eqs. 6 and 7), often provides better plots than standard σ values.[3]

$$X-\text{\large\textcircled{}}-OH + NaOH \rightleftharpoons X-\text{\large\textcircled{}}-O^{\ominus} \quad Na^{\oplus} + H_2O \tag{6}$$

$$X-\text{\large\textcircled{}}-\overset{\oplus}{N}H_3 \underset{Cl^{\ominus}}{} + NaOH \rightleftharpoons X-\text{\large\textcircled{}}-NH_2 + H_2O + NaCl \tag{7}$$

Curved Hammett Plots

For a number of reactions, Hammett plots yield curved lines even though points for both *meta* and *para* substituents fit onto the same line. Frequently, a curved Hammett plot indicates that the reaction

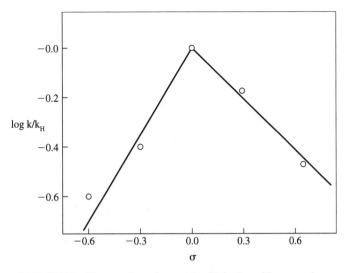

FIGURE 5.2 The reaction of aromatic aldehydes with ammonia.

has undergone a change in mechanism as the electronic effects of the substituents changed. The observed curve can be interpreted as resulting from the intersection of two straight-line Hammett plots.

The deviation of Hammett plots from linearity may be so small that it is difficult to be sure the curvatures are real. However, in some cases there may be quite sharp breaks in the plots. For instance, Figure 5.2 shows a Hammett plot for the rates of formation of imines from aromatic aldehydes and ammonia.[9] While the sequence of reaction steps (shown in Eq. 8) is the same throughout the range of substituents, addition of ammonia to the aldehydes is rate-limiting when the substituents are electron-donating, and dehydration of the addition product is rate-limiting when the substituents are electron-attracting.[9]

$$(8)$$

Other Applications

Acrylic acids. The Hammett equation provides an excellent correlation between σ_{para} values for substituents at C3 of *trans*-substituted acrylic acids (**1**) and the pK_a's of the acids. Since substituents at C3 of acrylic acids are closer to the carboxyl groups than substituents in benzoic acids, a comparatively large ρ value, $+2.23$, was observed.[10]

$$X \diagdown \quad \diagup H$$
$$C = C$$
$$H \diagup \quad \diagdown CO_2H$$

1

Biological systems. There has been a great deal of interest in using the Hammett equation and other linear free-energy relationships to interpret and predict biological effects of organic molecules. In particular, correlations between free-energy relationships and the effectiveness of synthetic drugs have been eagerly sought.

It should be no surprise that the rates of reactions between a wide variety of enzymes and *meta*- and *para*-substituted aromatic molecules, as well as the overall effects of those molecules on living organisms, are closely correlated by the Hammett equation.[11,12] This is simply an illustration that chemical reactions in biological systems are governed by the same mechanistic principles as in nonbiological systems.

5.2   SEPARATION OF POLAR, RESONANCE, AND STERIC EFFECTS

Aliphatic Systems with Fixed Geometries

Several groups have attempted to use variations on the Hammett theme to separate the inductive and resonance effects combined in Hammett σ constants. Roberts and Moreland measured the pK_as of 4-substituted bicyclo[2.2.2]octane-1-carboxylic acids (**2**).

2   **3**

It was assumed that the distance of the substituents from the carboxyl groups and the rigidity of these systems would eliminate problems with steric effects, and that the absence of double bonds would eliminate resonance effects. Therefore, values of σ′ in Eq. G were considered to represent solely the "inductive effects"* of the substituents.[13]

$$\log \frac{K}{K_H} = \rho' \sigma' \tag{G}$$

Values of σ′ are listed in Table 5.3. (To obtain these values, ρ′ was set equal to 1.464, the value of ρ for ionization of benzoic acids in the 50 percent aqueous ethanol solvent used to measure pK_a's of **2**.) Values of σ_{meta} are included in the table for comparison.

The values of σ′ are not very different from those of σ_{meta} for the bromine atom and the carbethoxy and cyano groups. However, the hydroxy group is a much more powerful

---

*Roberts and Moreland[13] defined electrostatic influences operating "through bonds" as *inductive effects* and defined electrostatic influences operating "through space" as *field effects*. The term *inductive effects* as used here includes both types of electrostatic influences.

TABLE 5.3   SUBSTITUENT CONSTANTS FOR ALIPHATIC ACIDS
WITH RIGID GEOMETRIES

Substituent	(σ_{meta})	$\sigma'*$	$\sigma''^{\dagger}$
H	(0.000)	0.000	0.000
OH	(0.120)	0.283	0.227
COC$_2$H$_5$	(0.398)	0.297	0.244
Cl	(0.390)	—	0.335
Br	(0.370)	0.454	—
CN	(0.560)	0.579	0.440

*for reactions of acids with structure **2**.[13]
†for reactions of acids with structure **3**.[14]

electron-withdrawing group in the aliphatic system, in which its inductive effect is not partially offset by its resonance effect, than when it is bonded to an aromatic ring.

Trans-4-substituted cyclohexanecarboxylic acids (**3**) exist largely in chair conformations, with both large groups in equatorial positions. Thus, although they are not quite as rigid as **2**, their pK_a's, as well as the rates of saponification of their esters, give good correlations with Eq. H.[14]

$$\log \frac{K}{K_H} = \rho''\sigma'' \quad \text{or} \quad \log \frac{k}{k_H} = \rho''\sigma'' \tag{H}$$

Values of σ'' for reactions of **3** (listed in the last column in Table 5.3) parallel values of σ' for reactions of **2** reasonably well. However, σ'' values are somewhat smaller, since substituents at C4 in chair conformations of cyclohexanes are farther from C1 than are substituents in the boat conformation required for cyclohexane rings in **2**.

The Taft Equation

R.W. Taft analyzed the effects of substituents on the rates of reactions of a series of esters, $X\text{CO}_2R$. If R is held constant, the rates of reactions of the esters would depend on the inductive, resonance, and steric effects of group X. Furthermore, if a saturated carbon in X were bonded to the carboxyl group, there could be no direct resonance interaction between X and the carboxyl group. Thus, applying an earlier suggestion by Ingold, Taft suggested that rates of *basic* hydrolysis of the esters could be defined by Eq. I, in which $\sigma*$ and E_s are constants reflecting the inductive effects and steric effects, respectively, of X.

$$\log\left[\frac{k}{k_H}\right]_{\text{in base}} = \rho*\sigma* + E_s \tag{I}$$

A value of 2.48 was assigned to $\rho*$ to bring values of $\sigma*$ onto a scale comparable to those of Hammett σ values.

To obtain values of E_s, Taft took advantage of the fact that the rates of *acid* hydrolysis of aromatic esters are not strongly influenced by the polar effects of substituents. That can be demonstrated by the very small values of ρ for those reactions (see Table 5.2). Thus, if changes in X strongly affect the rates of acid-catalyzed hydrolyses of ester, the changes are presumably

TABLE 5.4 SOME TAFT STERIC AND POLAR CONSTANTS

Substituent (X)	E_s	σ^*
H	1.240	0.490
CH_3	0.000	0.000
C_2H_5	−0.007	−0.100
$CH_3CH_2CH_2$	−0.360	−0.115
$(CH_3)_2CH$	−0.470	−0.190
$(CH_3)_3C$	−1.540	−0.300
FCH_2	−0.240	1.100
$BrCH_2$	−0.270	1.000
$ClCH_2$	−0.240	1.050
ICH_2	−0.370	0.600
Cl_2CH	−1.540	1.940
F_3C	−1.160	—
Cl_3C	−2.060	2.650

Source: R.W. Taft, Jr., in *Steric Effects in Organic Chemistry*, M.S. Newman, ed., John Wiley, New York (1956) and J. Shorter, in *Advances in Linear Free Energy Relationships*, N.B. Chapman and J. Shorter, eds., Plenum Press, New York, (1972).

due largely to steric effects. Therefore, E_s can be defined by Eq. J.

$$\log\left[\frac{k}{k_H}\right]_{\text{in acid}} = E_s \qquad (J)$$

Clearly, some very broad assumptions are made in defining E_s as a steric constant. However, an examination of E_s values listed in Table 5.4 shows that they do appear to reflect the sizes of the various groups and have proved useful for that purpose.[16] Similarly, values of σ^* have proved to be useful measures of inductive effects.*

REFERENCES

[1] L.P. Hammett, *J. Am. Chem. Soc.*, *59*, 96 (1937).

[2] C. Hansch, A. Leo, and R.W. Taft, *Chem. Rev., 91*, 165 (1991).

[3] H.H. Jaffe, *Chem. Rev.*, *53*. 191 (1953).

[4] L.P. Hammett, *Physical Organic Chemistry*, 2nd ed. McGraw-Hill, New York (1971).

[5] O. Exner, in *Advances in Linear Free Energy Relationships*, N.B. Chapman and J. Shorter, Eds., Plenum Press, New York, p. 1 (1972).

[6] K.F. Johnson, *The Hammett Equation*. Cambridge University Press, New York (1973).

[7] J.F.J. Dippy and J.E. Page, *J. Chem. Soc.*, 357 (1938); H.H. Jaffe, *J. Chem. Phys.*, *21*, 415 (1953).

[8] H.C. Brown and Y. Okamoto, *J. Am. Chem. Soc.*, *80*, 4979 (1958).

[9] Y. Ogata, A. Kawasaki, and N. Okumura, *J. Org. Chem.*, *29*, 1985 (1964).

*More complete lists of σ^* and E_s values include substituents that may have resonance interactions with carboxyl groups. For such substituents, σ^* and E_s are not accurate predictors of inductive and steric effects.

[10] M. Charton and H. Meislich, *J. Am. Chem. Soc.*, *80*, 5940 (1958); M. Charton, *Prog. Phys. Org. Chem.*, *13*, 119 (1981).

[11] J.G. Topliss, Ed., *Quantitative Structure—Activity Relationships of Drugs.* Academic Press, New York (1983).

[12] J.F. Kirsch, A. Cammarat and K.S. Rogers, in *Advances in Linear Free Energy Relationships.* N.B. Chapman and J. Shorter, Eds., Plenum Press, New York, 369–444 (1972).

[13] J.D. Roberts and W.T. Moreland, Jr., *J. Am. Chem. Soc.*, *75*, 2167 (1953).

[14] S. Siegel and J.M. Komarmy, *J. Am. Chem. Soc.*, *82*, 2547 (1960).

[15] R.W. Taft, Jr., in *Steric Effects in Organic Chemistry*, M.S. Newman, ed. John Wiley, New York (1956).

[16] See J. Shorter, in *Advances in Linear Free Energy Relationships*, N.B. Chapman and J. Shorter, Eds., Plenum Press, New York, 71–118 (1972).

PROBLEMS

5.1 Calculate the relative rates of saponification of ethyl benzoate and ethyl-3,4-dichlorobenzoate by sodium hydroxide in 85% ethanol at 25°C.

5.2 Indicate whether Hammett plots for rates of each of the following reactions should have positive or negative slopes. Decide whether the best fit to straight lines would be obtained by using values of σ, σ^+, or σ^-. Explain your reasoning in each case.

(a) 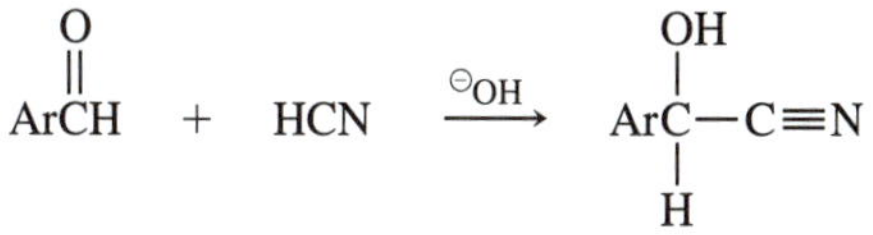
$$\text{ArCH} \ (=\!\!O) \ + \ \text{HCN} \xrightarrow{\ominus \text{OH}} \text{ArC(OH)(H)} - \text{C}\equiv\text{N}$$

(b)
$$\text{Ar}\overset{\oplus}{\text{N}}(\text{CH}_3)_3 \ + \ \ominus\text{OCH}_3 \xrightarrow{\Delta} \text{ArN}(\text{CH}_3)_2 \ + \ \text{CH}_3\text{OCH}_3$$

(c)
$$\text{Ar}_3\text{COH} \ + \ \text{HCl} \longrightarrow \text{Ar}_3\text{CCl} \ + \ \text{H}_2\text{O}$$

5.3 A Hammett plot for rates of the reactions of *meta*- and *para*-substituted benzoic acids with diphenyldiazomethane in methanol solution had a ρ value of 0.844. [A. Buckley et al., *J. Chem. Soc. (B)*, 631 (1968).]

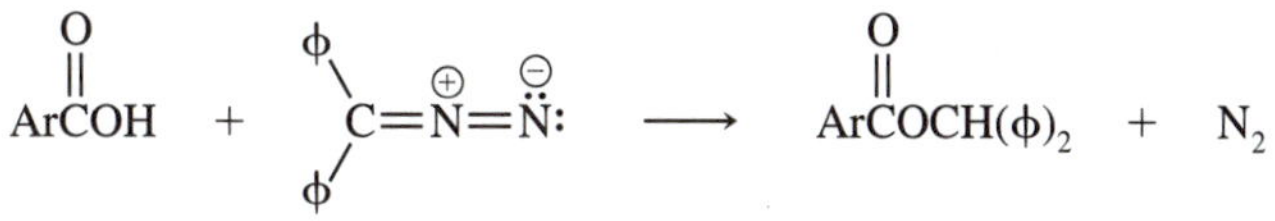

$$\text{ArCOH} \ (=\!\!O) \ + \ (\phi)(\phi)\text{C}=\overset{\oplus}{\text{N}}=\overset{\ominus}{\text{N}}: \longrightarrow \text{ArCOCH}(\phi)_2 \ (=\!\!O) \ + \ \text{N}_2$$

When the same reaction was carried out in toluene solution, the value of ρ was 2.20 [C.K. Hancock and E. Foldvary, *J. Org. Chem.*, *30*, 1180 (1965)]. Write mechanisms for the reaction that are consistent with the signs of ρ, and explain the dependence of the values of ρ on the nature of the solvents.

5.4 Suggest a reasonable mechanism for the hydrolysis of a carboxylic ester in dilute acid. Explain how that mechanism might account for the small value of ρ for the reaction.

Migrations to Electron-Deficient Centers

6.1 MIGRATIONS TO ELECTRON-DEFICIENT CARBONS

Wagner-Meerwein Rearrangements

1,2-Shifts of migrating groups to empty orbitals in *carbocations,** or toward partially empty orbitals in developing carbocations, are the most common rearrangements of organic molecules. (The movements of electrons in molecular rearrangements are sometimes represented by "looped" curved arrows indicating that a substituent moves along with the electron pair. The fact that a rearrangement is taking place may also be indicated by placing a loop on the reaction arrow, as shown in Eq. 1.)

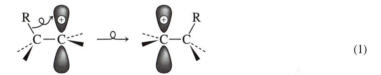

$$ \tag{1} $$

Rearrangement of a carbocation

Migrations of hydrogen atoms or of alkyl or aryl groups in carbocations are often called *Wagner–Meerwein rearrangements*, in honor of the early terpene chemists who first observed and studied the reactions.[2]

Early investigators considered the possibility that Wagner–Meerwein rearrangements proceed by cleavage of the bonds joining the migrating groups to the remainder of the molecules, followed by formation of new bonds at different positions. However, it was found that chiral

---

*Ions containing trivalent carbon atoms were originally called *carbonium ions*. However, in 1972 it was pointed out that the suffix *-onium* is generally used for ions such as ammonium ions, in which all the orbitals of the central atom are filled. It was suggested that use of "carbonium ions" be similarly restricted, and the term *carbenium ions* was invented to refer to trivalent carbon-centered ions.[1] In order to avoid confusion between carbonium and carbenium ions, textbook authors commonly use the more general term *carbocations* (analogous to *carbanions*) to refer to any ion containing a positively charged carbon.

migrating groups normally retain their configurations during Wagner–Meerwein rearrangements, as illustrated in Eq. 2.[3]

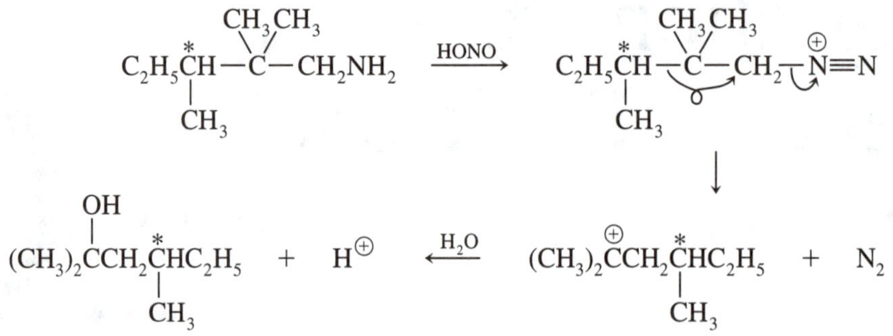

$$
\underset{\overset{|}{CH_3}}{\overset{CH_3\,CH_3}{C_2H_5\overset{*}{C}H-\overset{|}{\underset{}{C}}-CH_2NH_2}} \xrightarrow{\text{HONO}} \underset{\overset{|}{CH_3}}{\overset{CH_3\,CH_3}{C_2H_5\overset{*}{C}H-\overset{|}{C}-CH_2\overset{\oplus}{N}\!\equiv\!N}}
$$

$$
(2)
$$

$$
\underset{\substack{|\\CH_3\\ \text{98\% retention of}\\ \text{configuration}}}{\overset{OH}{(CH_3)_2\overset{|}{C}CH_2\overset{*}{C}HC_2H_5}} + H^{\oplus} \xleftarrow{\;H_2O\;} \underset{\substack{|\\CH_3}}{(CH_3)_2\overset{\oplus}{C}CH_2\overset{*}{C}HC_2H_5} + N_2
$$

Furthermore, in Wagner–Meerwein migrations of substituents on carbocyclic rings, the substituents invariably end up on the face of the ring from which they started, as demonstrated in Eq. 3.[4]

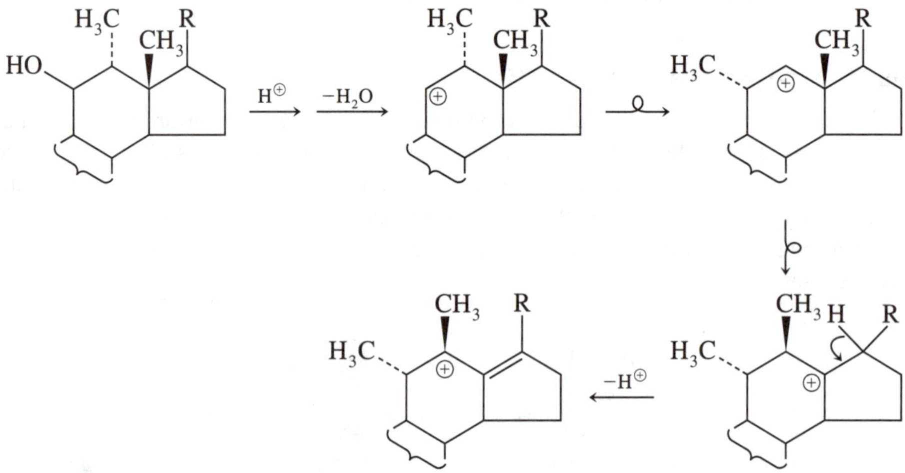

$$
(3)
$$

These facts demonstrate that Wagner–Meerwein rearrangements are intramolecular reactions.* They have transition states that resemble cyclic arrays of three atomic orbitals, each array containing the two electrons from the migrating bond. (The transition states for Wagner–Meerwein rearrangements, in fact, have been described as "corner-protonated cyclopropanes," as shown on page 145. That description requires that there be at least one proton on the migrating carbon.)

Since transition states for Wagner–Meerwein rearrangements consist of cyclic arrays of orbitals, the Woodward–Hoffmann rules may be applied to the reactions. Wagner–Meerwein

---

*However, without experimental evidence, it is often difficult to distinguish between a Wagner–Meerwein shift of a hydrogen and a deprotonation–reprotonation sequence:

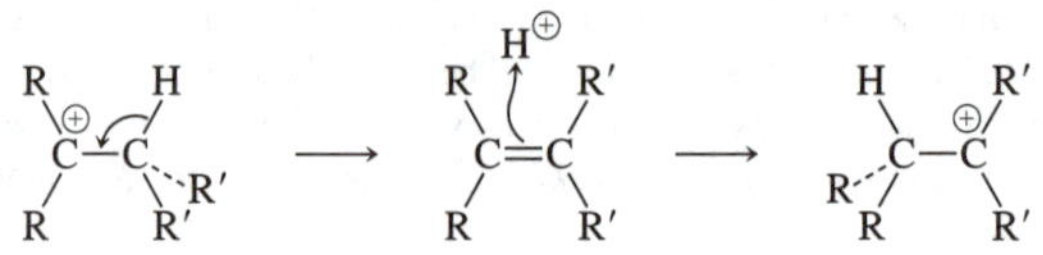

$$\left[\begin{array}{c} R' \\ R\diagdown\overset{|}{\underset{C}{C}}\diagup H \\ \diagup C \!-\! C \diagdown \end{array}\right]^{\oplus} \equiv \left[\begin{array}{c} R' \\ R\diagdown\overset{|}{\underset{C}{C}}\diagup H \\ \diagup C \!-\! C \diagdown \end{array}\right]^{\oplus} \tag{4}$$

The transition state for a Wagner-Meerwein rearrangement
(a corner-protonated cyclopropane)

rearrangements can be classified as sigmatropic shifts of order [1,2]. As two-electron processes, Wagner–Meerwein rearrangements should proceed in a suprafacial manner, with retention of the configurations of the migrating groups. As we have seen, this is indeed the case. (Of course, any other geometry for a concerted [1,2] migration would be extremely unlikely.)

Rearrangements proceed most easily when the carbocations are converted to more stable forms; for example, when a secondary cation rearranges to form a tertiary cation, as in Eq. 5, or when a simple carbocation is converted to a resonance-stabilized cation, as in Eq. 6.

$$(CH_3)_2\overset{CH_3}{\underset{|}{C}}-\overset{\oplus}{C}HCH_3 \quad\xrightarrow{\;\;}\quad (CH_3)_2\overset{\oplus}{C}-\overset{CH_3}{\underset{|}{C}}HCH_3 \tag{5}$$

$$\text{Ph}-CH_2-\overset{\oplus}{C}HCH_3 \quad\xrightarrow{\;\;}\quad \text{Ph}-\overset{\oplus}{C}H-CH_2CH_3 \tag{6}$$

However, it is also very common for secondary cations to rearrange to form other secondary cations and for tertiary cations to rearrange to form other tertiary cations (Eqs. 7 and 8).

$$\text{(7)}$$

$$\overset{H}{\underset{|}{(C_2H_5)_2}}\overset{\oplus}{C}-\overset{H}{\underset{|}{C(CH_3)_2}} \quad\rightleftharpoons\quad (C_2H_5)_2\overset{\oplus}{C}-\overset{H}{\underset{|}{C(CH_3)_2}} \tag{8}$$

Wagner–Meerwein rearrangements, like all reactions, must form final products that are thermodynamically more stable than the starting materials. Some processes proceeding via Wagner–Meerwein rearrangements, however, do appear to require "uphill" steps, in which the carbocations rearrange to less stable isomers. For example, when cyclohexanol is dissolved in a solution of antimony pentachloride in fluorosulfonic acid at −60°C it is instantly converted to the 1-methylcyclopentyl cation. No trace of the cyclohexyl cation, which is presumably the first cation formed, can be detected.[5] The most obvious path for this rearrangement would be by initial conversion of the cyclohexyl cation to a higher-energy primary cation, as shown in Eq. 9.

$$\text{(9)}$$

In trying to predict possible products from Wagner–Meerwein rearrangements, therefore, it is necessary to consider the possibility that products may arise via "uphill" rearrangement steps, including steps that form intermediates that to a first approximation can be described as primary carbocations.*

Degenerate Rearrangements

In many rearrangements, the products have the same structures as the starting materials. The existence of these *degenerate* rearrangements can be detected only by sophisticated techniques. The 2-propyl cation, for instance, appears to be stable in very strong acid solutions and does not undergo any obvious rearrangement. However, NMR studies by Martin Saunders and his coworkers at Yale University have shown that the primary and secondary hydrogens in the 2-propyl cation are rapidly interchanged at temperatures as low as $-100°C$. This reaction can be made visible by deuterium labeling (Eq. 10).[6]

$$D_3C-\overset{\oplus}{C}H-CH_3 \;\rightleftharpoons\; D_2CH-\overset{\oplus}{C}D-CH_3 \;+\; D_2CH-\overset{\oplus}{C}H-CH_2D$$

$$+ \quad H_2CD-\overset{\oplus}{C}D-CH_2D \tag{10}$$

Remarkably, even the carbon atoms of the 2-propyl cation constantly change locations at low temperatures. This was proved by labeling the central carbon with ^{13}C. The NMR spectra of the cation show that the isotopically labeled carbon rapidly interchanges positions between C1 and C2 (Eq. 11).[6]

$$H_3C-^{13}\overset{\oplus}{C}H-CH_3 \;\rightleftharpoons\; H_3^{13}C-\overset{\oplus}{C}H-CH_3 \tag{11}$$

The rearrangement reactions of the 2-propyl cation can be accounted for by the mechanism shown in Eq. 12, in which the initial step is a hydrogen migration to convert the 2-propyl cation to a 1-propyl cation.

$$H_3C-^{13}\overset{\oplus}{C}H-CH_3 \;\rightleftharpoons\; H_3C-^{13}CH_2-\overset{\oplus}{C}H_2 \;\rightleftharpoons\; ^{13}\overset{\oplus}{C}H_2-CH_2-CH_3$$

$$\updownarrow$$

$$^{13}CH_3-\overset{\oplus}{C}H-CH_3 \tag{12}$$

---

*There are, in fact, serious doubts about whether primary carbocations are actually formed as discrete intermediates. It has been proposed instead that their formation and rearrangement occur simultaneously (see page 147).

A Closer Look at "Primary Carbocations"

Even while a portion of the deuterium and ^{13}C-labeled 2-propyl cation remains unrearranged, the deuterium atoms of the portion that *has* rearranged become completely scrambled among the three carbons of the cation.[6,7] This observation would be difficult to explain by the mechanism of Eq. 12. It has therefore been proposed that methyl groups and hydrogens migrate simultaneously to form corner-protonated cyclopropanes. The rapid scrambling of hydrogens and deuteriums among the carbons might then be accounted for by migration of hydrogens from corner to corner around the "cyclopropane" ring.[6] (Eq. 13).

$$\tag{13}$$

Of course, if hydrogens can migrate in corner-protonated cyclopropanes, the protonated cyclopropanes must have more than infinitesimal lifetimes—that is, they must be intermediates rather than just transition states. If that is the case, each alkyl group migration must pass through several transition states. Molecular orbital calculations[8] and experimental evidence[9] are consistent with this hypothesis.

6.2   THE NATURE OF THE MIGRATING GROUPS

Migratory Aptitudes

There have been frequent attempts to measure the *migratory aptitudes* of various groups. This was done by carrying out rearrangements in which several competing groups could migrate and determining the relative yields of products resulting from migrations of the different groups.

Unfortunately, the relative yields of products from migrations of different groups are not necessarily related to the "intrinsic" abilities of the groups to migrate. For one thing, the percentages of the different products may be determined principally by the effectiveness of the *nonmigrating* groups in stabilizing the product cations. Many reports of apparent migratory aptitudes are suspect, because they do not take into consideration the stabilities of the product cations.

Geometric factors can also be important in determining the course of Wagner–Meerwein rearrangements. It is frequently observed, for instance, that a methyl group at the ring juncture of two fused carbocyclic rings will migrate in preference to the migration of a primary or secondary alkyl group forming part of one of the rings (Eq. 14).[10]

However, this does not necessarily demonstrate that methyl groups are intrinsically better migrators than other alkyl groups. Instead, the critical consideration appears to be that the methyl

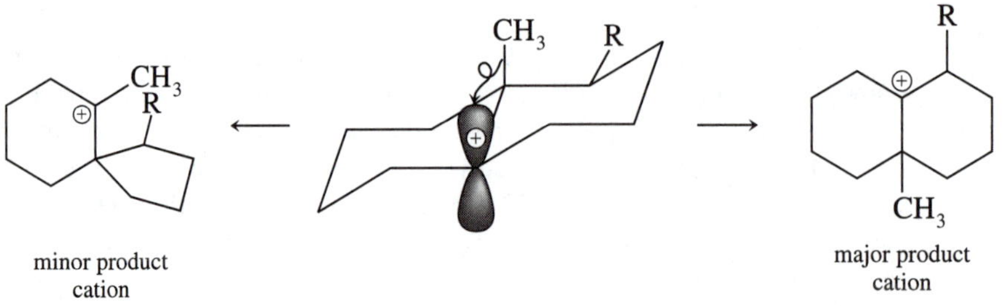

$$(14)$$

minor product
cation

major product
cation

groups are located in axial positions of six-membered rings. Since the bonds between the methyl groups and the rings are nearly coplanar with the empty orbitals of the cations, the methyls are almost ideally positioned to migrate. In contrast, bonds to substituents in equatorial positions overlap poorly with the empty orbitals. Equatorial substituents are therefore unlikely to migrate unless the carbocation ring inverts to form the other, less stable, chair conformation, thus placing the formerly equatorial substituents into axial positions.

Steric factors may also significantly affect the rearrangements of open-chain cations. Non-migrating substituents are forced into "eclipsed" positions in the transition states for Wagner–Meerwein rearrangements, as illustrated in the following diagram. Thus, there may be an intrinsic advantage for bulky groups to migrate rather than to remain stationary.

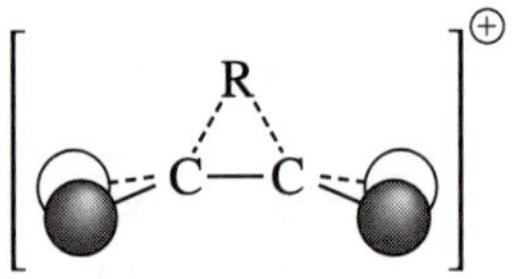

"Eclipsing" of stationary groups in Wagner-Meerwein migrations

Interpretations of studies of migratory aptitudes must therefore be made with appreciable caution.

Migration Tendencies

The "intrinsic" abilities of different groups to migrate are called their *migration tendencies*.[11] In some cases, migration tendencies can be determined by comparing the rates of migration of different groups in reactions with identical stationary groups. In NMR studies of the "phenanthrenium" ions **1** (Eq. 15) and **2** (Eq. 16), for instance, ethyl groups were found to migrate 13 to 55 times as rapidly as methyl groups.[12]

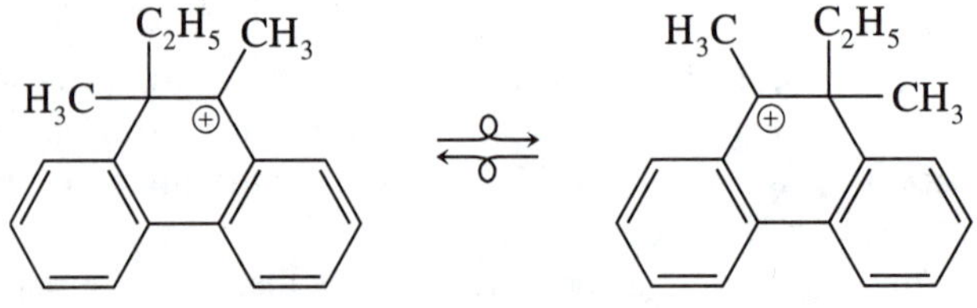

$$(15)$$

1   **1**

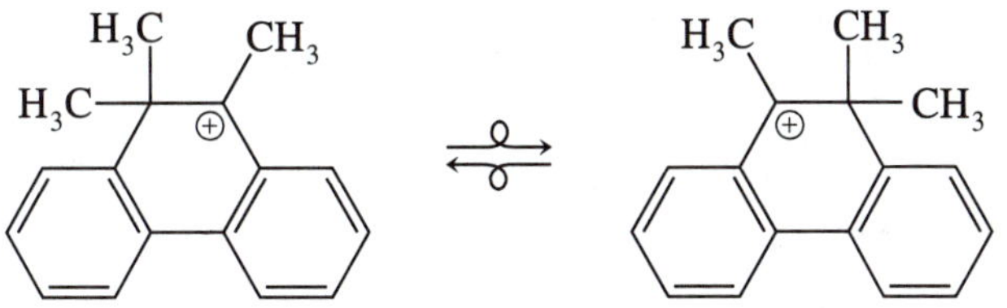

(16)

Kinetic studies of degenerate rearrangements such as those in **1** and **2** are applicable to only a limited group of reactions. However, in other reactions, it is possible to correct observed "migratory aptitudes" for the electronic and steric effects of stationary groups. Using this approach,[11] the relative migration tendencies of methyl, ethyl, and *tert*-butyl groups in pinacol rearrangements (see pages 156 to 157) were found to increase in the ratios of 1 to 17 to 4000, respectively.

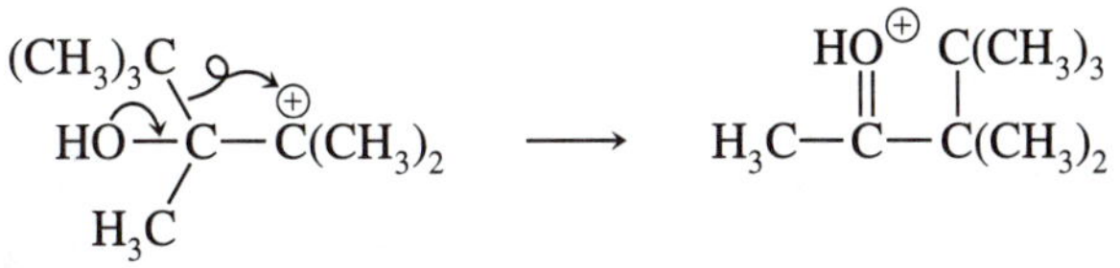

(17)

Preferential migration of a *tert*-butyl group in a pinacol rearrangement

The general trend observed in these studies is that migration tendencies of alkyl groups in carbocation rearrangements increase with increasing substitution at the migrating carbon atoms. This is the result expected if the migrating groups resemble carbocations in the transition states for the migrations. The positive charge of the carbocation is apparently shared by the migrating group as well as by the carbons at the *migration origin* and the *migration terminus* in the transition state for a Wagner–Meerwein rearrangement.

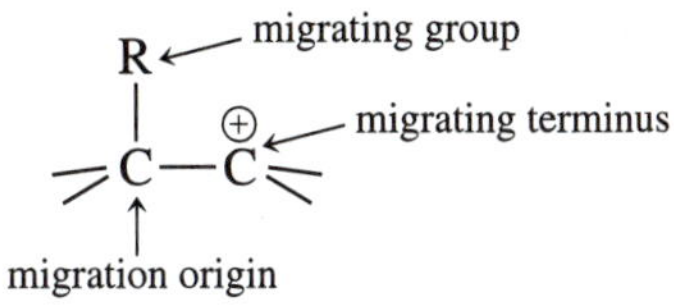

The transition states have sometimes been described as hybrids of three resonance forms, including a resonance form bearing a positive charge on the migrating group:

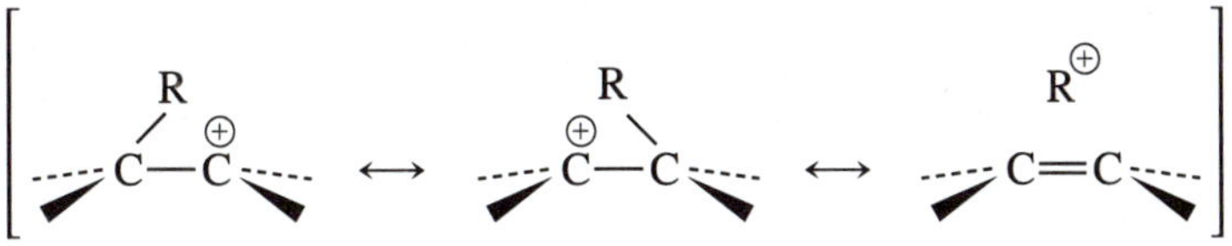

Resonance description of the transition state for a Wagner-Meerwein rearrangement

If the migrating groups in Wagner–Meerwein rearrangements are positively charged, then the groups best able to bear such charges should have the highest migration tendencies, as has indeed been observed.

Shifts of hydrogens in carbocations are often described as "hydride ion migrations," and alkyl shifts are sometimes described as "alkyl anion" migrations, because the bonding electrons move along with the migrating groups. While these terms may be reasonable descriptions of the stochiometries of the reactions, they tend to give the unjustified impression that migrating groups bear negative charges during carbocation rearrangements. As we have seen, quite the opposite is true.

Hydrogens usually migrate more rapidly than alkyl groups in Wagner–Meerwein rearrangements. However, migration of a hydrogen will normally result in the formation of a more substituted carbocation than will migration of an alkyl group from the same carbon.

$$(CH_3)_2\overset{\oplus}{C}-CH_2C_2H_5 \xleftarrow{\quad} (CH_3)_2\overset{\oplus}{C}-\overset{\overset{\displaystyle H}{|}}{C}HC_2H_5 \xrightarrow{\quad} CH_3\overset{\oplus}{C}H-\overset{\overset{\displaystyle CH_3}{|}}{C}HC_2H_5 \quad (18)$$

Hydrogen migration yields   Methyl migration yields
a tertiary carbocation   a secondary carbocation

NMR studies indicate that, when migrations of hydrogens and of methyl groups result in the formation of similarly substituted cations, as in the degenerate reactions shown in Eqs. 19 and 20, the hydrogen atom is only a slightly better migrator than the methyl group[13] and is probably a less effective migrator than most larger alkyl groups.

$$(CH_3)_2\overset{\overset{\displaystyle H}{|}}{C}-\overset{\oplus}{C}(CH_3)_2 \quad \rightleftharpoons \quad (CH_3)_2\overset{\oplus}{C}-\overset{\overset{\displaystyle H}{|}}{C}(CH_3)_2 \qquad (19)$$

$$(CH_3)_2\overset{\overset{\displaystyle CH_3}{|}}{C}-\overset{\oplus}{C}(CH_3)_2 \quad \rightleftharpoons \quad (CH_3)_2\overset{\oplus}{C}-\overset{\overset{\displaystyle CH_3}{|}}{C}(CH_3)_2 \qquad (20)$$

Aryl and vinyl groups are much better migrators than alkyl groups. Migrations of these groups will be discussed in Chapter 7.

6.3   REARRANGEMENTS OF CARBOCATIONS

Competition with Other Reactions

The major products from many reactions in which carbocations (or developing carbocations) are intermediates, such as S_N1 and E1 reactions, are usually formed without carbon skeleton rearrangements. Under typical conditions for these reactions, carbocations and developing carbocations react with nucleophiles much more rapidly than they undergo rearrangements (Eq. 21).

$$(CH_3)_2CH\overset{\oplus}{C}HCH_3 \xrightarrow{X^\ominus} (CH_3)_2CH\overset{\overset{\displaystyle X}{|}}{C}HCH_3 + (CH_3)_2C{=}CHCH_3 \qquad (21)$$

usual products

In general, rearrangements of carbocations are likely to be faster than their reactions with nucleophiles only when some features of the carbocations make rearrangements particularly easy, or when the reactions with nucleophiles are particularly slow.

Rearrangements Under Minimally Nucleophilic Conditions

Usually, if nucleophiles react exceptionally slowly with carbocations, it is because the reaction mixtures contain strong Lewis acids that complex with the nucleophiles and minimize their reactivities. Reactions carried out under Friedel–Crafts conditions are classic examples. In Friedel–Crafts reactions, Lewis acids (usually metal halides) tie up halide ions and other "normal" nucleophiles. This extends the lifetimes of carbocations in the reaction mixtures sufficiently so that the very slow reactions of the cations with aromatic rings can take place. However, the relatively long lifetimes of the carbocations also allow time for Wagner–Meerwein rearrangements to precede the reactions of the carbocations with aromatic molecules.

(22)

A Friedel-Crafts alkylation with rearrangement

Carbocations can also have long lifetimes in *superacids*, which are polar solvents with very low nucleophilicities.[14]* Superacids can be roughly defined as solutions that are stronger acids than 100% sulfuric acid.† The list of superacids includes several individual acids, such as fluorosulfonic acid (FSO_3H) and trifluoromethanesulfonic acid (F_3CSO_3H). However, many of the most useful superacids are mixtures of protic acids with powerful Lewis acids. The Lewis acids complex with conjugate bases of the protic acids, lowering their basicities (and nucleophilicities) and thereby increasing the acidities of the solutions.

Common superacid mixtures include solutions of boron trifluoride in HF, arsenic pentafluoride in fluorosulfonic acid, and sulfuryl fluorochloride (SO_2ClF) in liquid sulfur dioxide. *Magic acid*, which consists of a solution of antimony pentafluoride in fluorosulfonic acid, is among the strongest known acids. It has an H_0 value below -20.

---

*Superacids were developed by George A. Olah and his coworkers at the University of Southern California, as well as by other groups. Professor Olah (born in 1926 in Budapest, Hungary) was awarded the Nobel Prize in 1994 for his studies on carbocation rearrangements in superacids.

†Technically, superacids can be described as acids with H_0 (Hammett acidity function) values below -12. The H_0 scale, which is similar to a scale of pH values extended to apply to very strongly acidic solutions, is discussed in detail in other texts.[15]

Carbocations can have extremely long lifetimes in superacid solutions because the solutions do not contain good nucleophiles to react with the cations. Even when carbocations *do* react with nucleophiles or bases in superacids, the resulting substitution and elimination products will usually react rapidly with the powerful acids to regenerate carbocations. As a result, carbocations in superacid solutions not only have ample opportunity to rearrange, but are likely to continue rearranging until the thermodynamically most stable cations have formed.

All of the chloropentane isomers, for instance, are converted to the *tert*-pentyl cation—the most stable saturated C_5 carbocation—when they are dissolved in a solution of antimony pentafluoride in SO_2ClF, as shown in Eq. 23.[16] Obviously, some of these rearrangements require many consecutive migration steps.

$$CH_3(CH_2)_3CH_2Cl \quad \searrow \qquad \qquad \overset{Cl}{\underset{|}{(CH_3)_2CCH_2CH_3}}$$

$$(CH_3)_2\overset{\oplus}{C}CH_2CH_3 \qquad \qquad (23)$$

$$\underset{CH_3CH_2\overset{|}{C}HCH_2CH_3}{\overset{Cl}{\nearrow}} \qquad \qquad \nwarrow (CH_3)_2CHCH_2CH_2Cl$$

Rearrangements in superacids

Many carbocations are so stable in superacid solutions that their NMR spectra can be recorded. In several cases, including the degenerate rearrangements discussed on pages 148 to 150, it has been possible not only to detect otherwise "invisible" rearrangements but also to measure their rates and to compare them with the rates of other reactions. Thus, the development of superacids has made possible major advances in the study of carbocation reactions.

Rearrangements of Alkanes

Even alkanes, which are noted for their lack of reactivity in ionic reactions, appear to rearrange rapidly in superacids or under Friedel–Crafts conditions. In 1946, however, Pines and Wackher demonstrated that very pure alkanes do not undergo aluminum halide-catalyzed rearrangements unless a minute amount of a more reactive organic compound, such as an alkene or an oxygenated molecule, is present. Unlike alkanes, these compounds, called *promoters* or *cocatalysts,* can easily be converted to carbocations by Lewis acids.[17]

Prior to Pines and Wackher's findings, it had been shown that carbocations could abstract hydride ions from alkane molecules to form new carbocations.[18] Pines and Wackher therefore proposed a cationic chain mechanism for the isomerization of alkanes in the presence of Lewis acids and promoters. The rearrangement of *n*-butane by that mechanism is shown on page 153.

In the first step, the acid catalyst converts the promoter to a carbocation, which then abstracts a hydride ion from a molecule of the alkane, converting it to a new carbocation. This carbocation undergoes one or more Wagner–Meerwein rearrangement steps to form a more stable carbocation and then propagates the chain by abstracting a hydride ion from another molecule of the alkane.

(Later, it was found that small amounts of hydrogen gas are generated from alkanes in the presence of *exceptionally* strong protic acids, such as magic acid or mixtures of aluminum chloride and HCl, although not with aluminum chloride by itself.[19] It was therefore suggested that

$$\text{initiation steps} \begin{cases} \text{promoter} \xrightarrow{\text{acid}} R^{\oplus} \\ R^{\oplus} + CH_3CH_2CH_2CH_3 \longrightarrow RH + CH_3\overset{\oplus}{C}HCH_2CH_3 \end{cases}$$

$$\text{propagation steps} \begin{cases} CH_3\overset{\oplus}{C}HCH_2CH_3 \xrightarrow{\quad} (CH_3)_3C^{\oplus} \\ (CH_3)_3C^{\oplus} + CH_3CH_2CH_2CH_3 \longrightarrow (CH_3)_3CH + CH_3\overset{\oplus}{C}HCH_2CH_3 \end{cases}$$

Cationic chain mechanism for the isomerization of n-butane

under these unusual conditions protons can directly abstract hydride ions from alkanes (Eq. 24), thereby initiating cationic chain reactions even in the absence of promoters.)

$$R{-}H + H^{\oplus} \longrightarrow R^{\oplus} + H_2 \tag{24}$$

Under the very strongly acidic (nonnucleophilic) conditions needed for rearrangements of alkanes, it is common for alkanes to undergo multiple rearrangements to form the thermodynamically most stable products, or mixtures of products in amounts proportional to their relative thermodynamic stabilities.[20]

The Lewis-acid-catalyzed rearrangements of alkanes are important industrial processes. They occur in the catalytic reformulation of petroleum, in which low-octane, straight-chain hydrocarbons are isomerized to higher-octane, branched-chain molecules (Eq. 25).

$$CH_3(CH_2)_6CH_3 \xrightarrow[\Delta]{\text{Lewis acids}} (CH_3)_3CCH_2CH(CH_3)_2 + \text{ other isomers} \tag{25}$$

$$\qquad\quad n\text{-octane} \qquad\qquad\qquad\qquad\quad \text{``isooctane''}$$

Cationic chain reactions also play important roles in the catalytic "cracking" of high-molecular-weight components of petroleum to the more valuable low-molecular-weight fractions that form gasoline.

$$\tag{26}$$

A "fragmentation" (cracking) reaction of a carbocation

Structures Favoring Carbocation Rearrangements

As noted earlier, in the presence of common nucleophiles, such as water or other hydroxylic solvents, most carbocations form substitution and elimination products more rapidly than they undergo rearrangements. However, there are several types of organic molecules in which carbocation rearrangements are particularly likely to occur, even in the presence of good nucleophiles.

The neopentyl system. 2,2-Dimethylpropyl halides and 2,2-dimethylpropyl sulfonates (commonly called neopentyl halides and sulfonates), if they react with ionic reagents at all, typically yield only products of Wagner–Meerwein rearrangements, as shown in Eq. 27.

neopentyl p-bromobenzenesulfonate

$$(CH_3)_2C{=}CHCH_3 \quad + \quad (CH_3)_2\overset{OAc}{\underset{|}{C}}CH_2CH_3$$

Neopentyl derivatives do not undergo Wagner–Meerwein rearrangements because the re-arrangement processes are particularly favorable, but rather *because other reactions are particu-larly unfavorable*.

Neopentyl halides and sulfonates cannot, of course, undergo E2 reactions, since there are no protons on the β-carbons. Furthermore, it has been estimated that neopentyl halides undergo S_N2 reactions only about 10^{-5} times as rapidly as other primary halides.[21] This is presumably the result of strong steric interference between the *tert*-butyl groups and the entering necleophiles in transition states for S_N2 reactions of neopentyl derivatives. Other compounds with quaternary carbons bonded to the carbons bearing the leaving groups similarly undergo S_N2 reactions very slowly, if at all.

Transition state for an S_N2
reaction of neopentyl bromide

The bulky *tert*-butyl groups similarly inhibit nucleophilic attack on developing primary cations in S_N1 reactions of neopentyl derivatives. Thus, while prolonged heating of neopentyl halides and tosylates in good ionizing solvents will yield substitution and elimination products, the prod-ucts invariably have structures resulting from Wagner–Meerwein rearrangements. Rearranged products are formed even in the presence of strong bases or other good nucleophiles.

Relief of ring strain. Wagner–Meerwein rearrangements are particularly likely to occur if they result in a three- or four-membered ring expanding to form a less strained structure.

Some of the best-known examples are found in the reactions of terpenes. For example, α-pinene (a major constituent of turpentine) rearranges to form bornyl chloride on reaction with hydrogen chloride (Eq. 28).[2] The rearrangement is very rapid because there is such a large driv-ing force for expansion of the four-membered ring to a less strained five-membered ring. How-ever, it is worth noting that even in this example, which is exceptionally favorable for rearrangement, addition of HCl to the double bond—that is, reaction of the carbocation with a

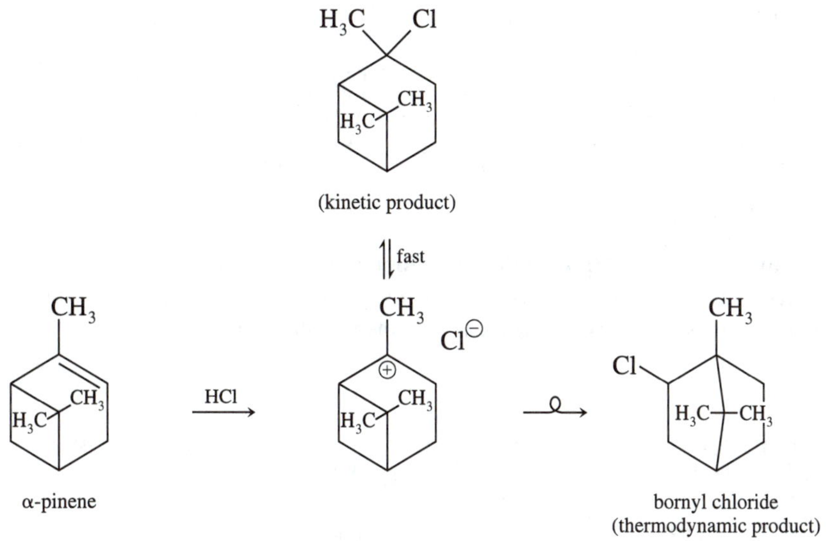

(28)

α-pinene

bornyl chloride
(thermodynamic product)

(kinetic product)

nucleophile—occurs more rapidly than rearrangement. (It is easy to overlook the initial formation of the addition product, since it also rearranges rapidly to form bornyl chloride.[2])

The expansion of a four-membered ring is observed in a more complex reaction sequence during reaction of the sesquiterpene derivative caryophyllene oxide with acids (Eq. 29).[22]

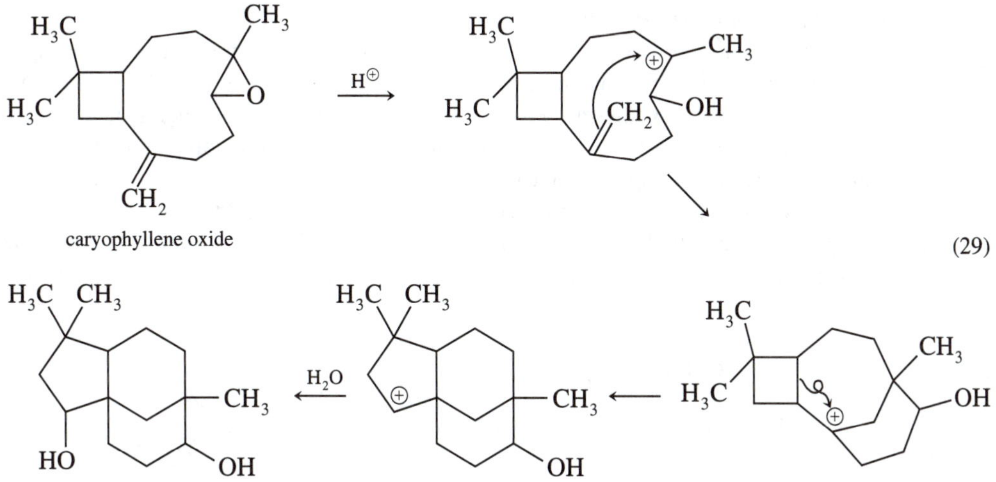

caryophyllene oxide

(29)

While contractions of four-membered rings to three-membered rings would normally be endothermic, there are almost no absolute rules in organic chemistry. In the very strained polycyclic system shown in Eq. 30, relief of angle strain is achieved by *contraction* of a cyclobutane ring.[23]

Rearrangements involving expansions of three-membered rings will be discussed in Chapter 7.

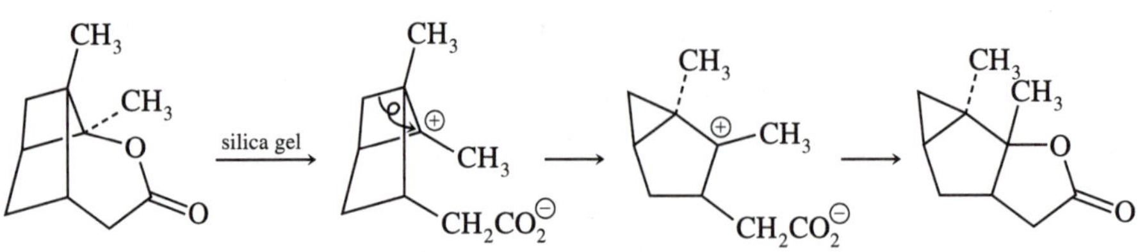

(30)

Pinacol and semipinacolic rearrangements.   Tetramethylethylene glycol (pinacol) rearranges in acid to form *tert*-butyl methyl ketone (pinacolone). The pinacol–pinacolone transformation is the oldest-known rearrangement of carbocations.

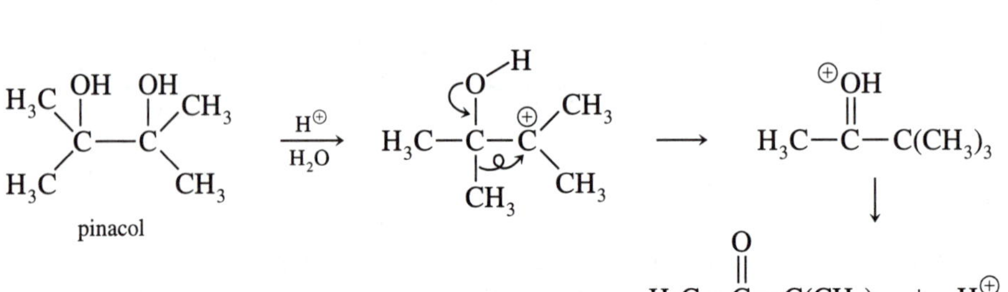

(31)

Other 1,2-glycols similarly rearrange to carbonyl compounds in acid, in reactions that are generically called *pinacol rearrangements*.

Formation of a new bond (the π bond of the protonated carbonyl group) during the rearrangement step provides the driving force for the pinacol rearrangement. Formation of an additional bond is a highly exothermic process, so that pinacol rearrangements occur even under conditions in which most Wagner–Meerwein rearrangements cannot compete with direct substitution and elimination reactions of the carbocations.

Reactions proceeding via β-hydroxy cations, but in which the cations are not formed from 1,2-glycols, are called *semipinacolic rearrangements*. β-halo alcohols (as in Eq. 32) and 1,2-epoxides (as in Eq. 33) are common starting materials for semipinacolic rearrangements.[25]

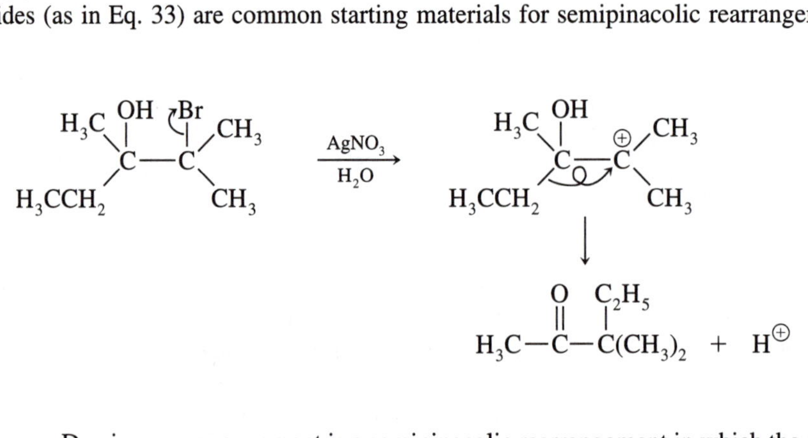

(32)

The Tiffeneau–Demjanov rearrangement is a semipinacolic rearrangement in which the cation is formed by diazotization of a β-amino alcohol.[26]

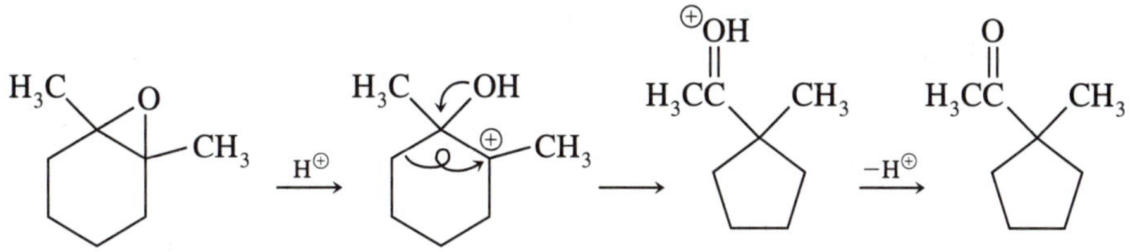

(33)

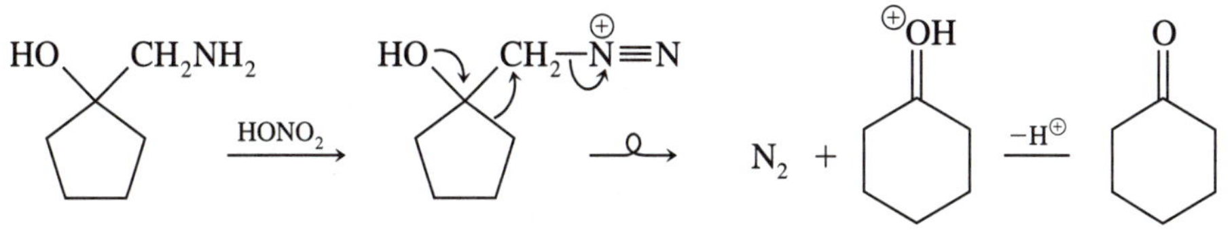

(34)

A Tiffeneau-Demjanov rearrangement

Rearrangements of α-hydroxyaldehydes and α-hydroxyketones.   In acid-catalyzed rearrangements of α-hydroxyaldehydes or ketones (sometimes called *ketol–ketol* or *acyloin* rearrangements), the migrating groups migrate to carbon atoms of protonated carbonyl groups, rather than to "real" carbocation centers.

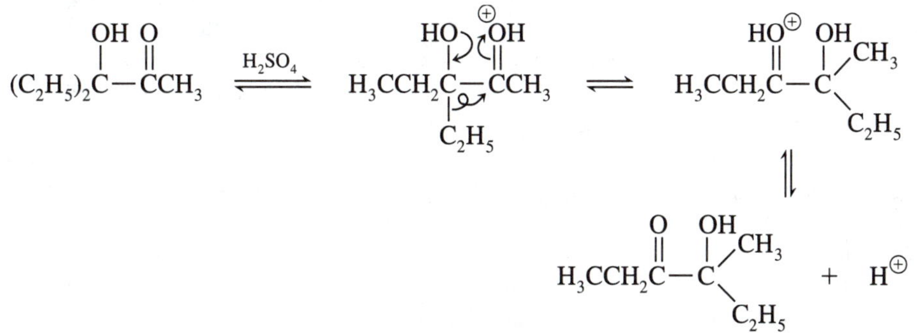

(35)

These reactions are closely related to pinacol and semipinacolic rearrangements. In both types of rearrangements, the migration steps are assisted by conversion of hydroxy groups to carbonyl groups.

One difference between acyloin and pinacolic rearrangements is that competing elimination and substitution reactions of the alcohol functions are unlikely to occur in acyloin rearrangements, because the formation of carbocations α to carbonyl groups is quite difficult.* A second difference is that pinacol rearrangements are usually highly exothermic, because the bond energy of a carbonyl double bond is normally significantly greater than the combined bond energies of two carbon–oxygen single bonds. In contrast, except when aldehydes are converted to ketones, the starting materials and products in acyloin rearrangements are often similar in energy, so that acyloin rearrangements are readily reversible.

---

*The α-keto and α-cyano carbocations actually gain some stabilization from resonance:[27]

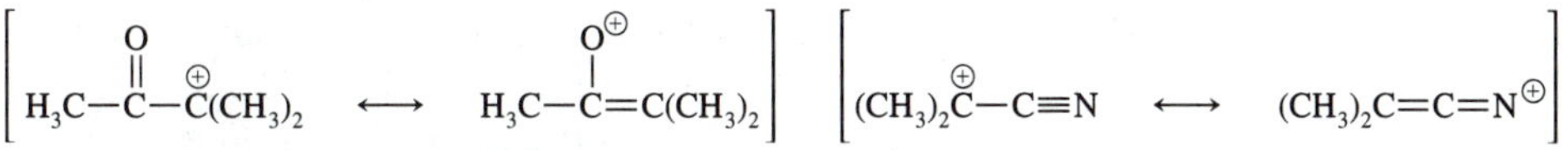

However, the resonance effects of carbonyl and cyano groups in carbocations are far outweighed by their destabilizing inductive efforts.

The most important difference between acyloin rearrangements and most rearrangements of carbocations is that acyloin rearrangements can take place under basic as well as acidic conditions (Eq. 36).[28]

$$\text{H}_3\text{C}-\overset{\overset{\text{O}}{\|}}{\text{C}}-\overset{\overset{\text{OH}}{|}}{\text{C}}(\text{C}_2\text{H}_5)_2 \ + \ ^{\ominus}\text{OH} \ \rightleftharpoons \ \text{H}_3\text{C}-\overset{\overset{\text{O}}{\|}}{\text{C}}-\overset{\overset{\text{O}^{\ominus}}{|}}{\underset{\overset{|}{\text{C}_2\text{H}_5}}{\text{C}}}-\text{C}_2\text{H}_5$$

(36)

$$\underset{\text{H}_3\text{CCH}_2}{\overset{\text{H}_3\text{C}}{\diagdown}}\overset{\overset{\text{OH}}{|}}{\underset{/}{\text{C}}}-\overset{\overset{\text{O}}{\|}}{\text{C}}\text{C}_2\text{H}_5 \ + \ ^{\ominus}\text{OH} \ \underset{\overset{\text{H}_2\text{O}}{\rightleftharpoons}}{} \ \underset{\text{H}_3\text{CCH}_2}{\overset{\text{H}_3\text{C}}{\diagdown}}\overset{\overset{\text{O}^{\ominus}}{|}}{\underset{/}{\text{C}}}-\overset{\overset{\text{O}}{\|}}{\text{C}}\text{C}_2\text{H}_5$$

Base-catalyzed acyloin rearrangements are still migrations to electron-deficient sites, but a good deal of the impetus for the migrations comes from the "push" provided by the negatively charged oxygens.

Acyloin rearrangements can take place under very mild conditions, since they do not require the formation of carbocation intermediates but only the protonation of carbonyl groups or the removal of protons from hydroxy groups. They often occur unexpectedly, and a synthetic chemist planning the synthesis of an α-hydroxyaldehyde or ketone has to be careful to avoid conditions that might give rise to acyloin rearrangements.

Many other aldehydes and ketones can undergo rearrangements in acid to form isomeric carbonyl compounds, as shown in Eqs. 37 and 38. (The initial migration steps in those reactions are the reverse of the migration steps in pinacol rearrangements.) However, rearrangements of simple carbonyl compounds usually require much more strongly acidic conditions, and often much higher temperatures, than are needed for acyloin rearrangements,[29] because the migration steps are not assisted by unshared electron pairs on hydroxy groups.

(37)

$$\underset{\Delta}{\overset{\text{O}}{\underset{(\text{CH}_3)_3\text{CCH}}{\|}}} \ \xrightarrow[\Delta]{\text{H}_2\text{SO}_4} \ \overset{\text{O}}{\underset{(\text{CH}_3)_2\text{CHCCH}_3}{\|}}$$

(38)

The benzilic acid rearrangement.   If diphenyl-1,2-dione (benzil) is heated in strong basic solutions, it is quantitatively converted to salts of benzilic acid (Eq. 39). The corresponding reactions employing alkoxide and amide anions as the bases yield esters and amides, respectively, of benzilic acid. Other diaryl diketones will also undergo benzilic acid rearrangements in excellent yields. (Similar rearrangements can occur with aliphatic α-diketones, as shown in Eq. 40, but yields are often low because of competing aldol condensation reactions.[28])

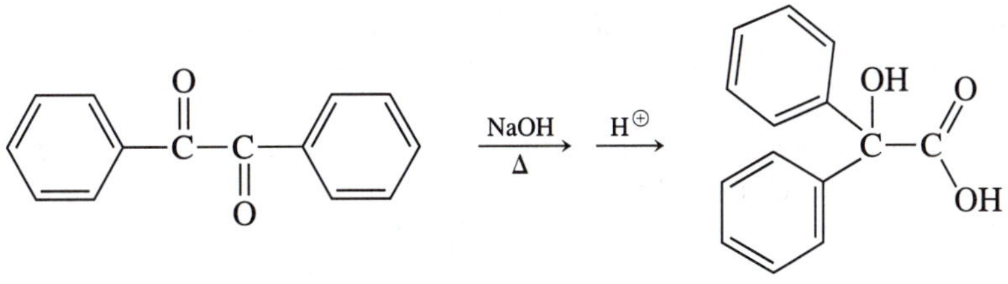

benzil   benzilic acid

(39)

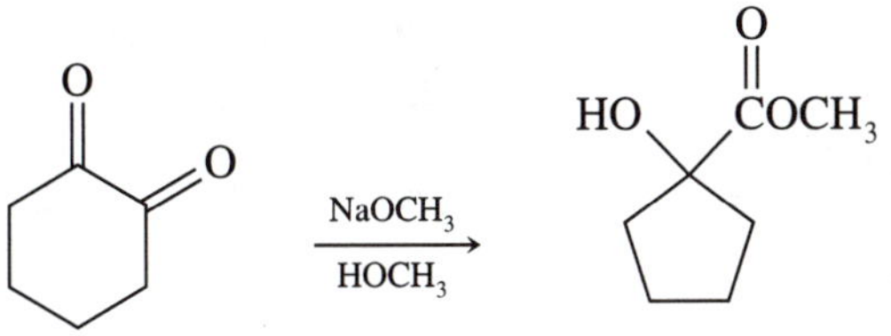

(40)

The benzilic acid rearrangement is initiated by addition of the base to a carbonyl group (Eq. 41). The succeeding rearrangement step is closely related to the base-catalyzed acyloin rearrangement.

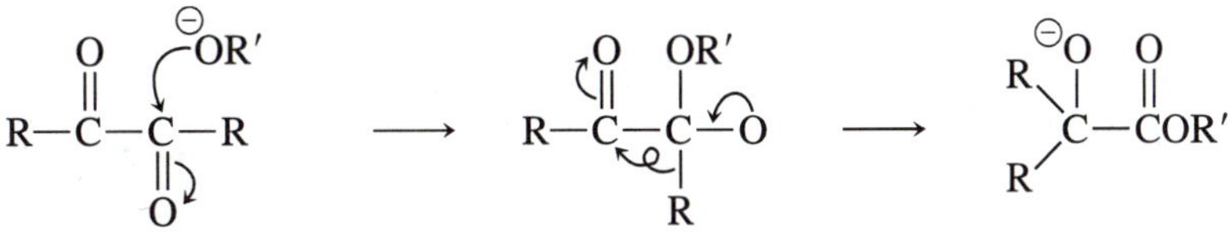

(41)

Studies employing α-diketones labeled with radioactive carbon have shown that aryl groups invariably migrate more rapidly than alkyl groups in benzilic acid rearrangements. When two different aryl groups might migrate, the principal product usually arises from migration of the aromatic ring bearing the most strongly electron-withdrawing substituents (Eq. 42). That is

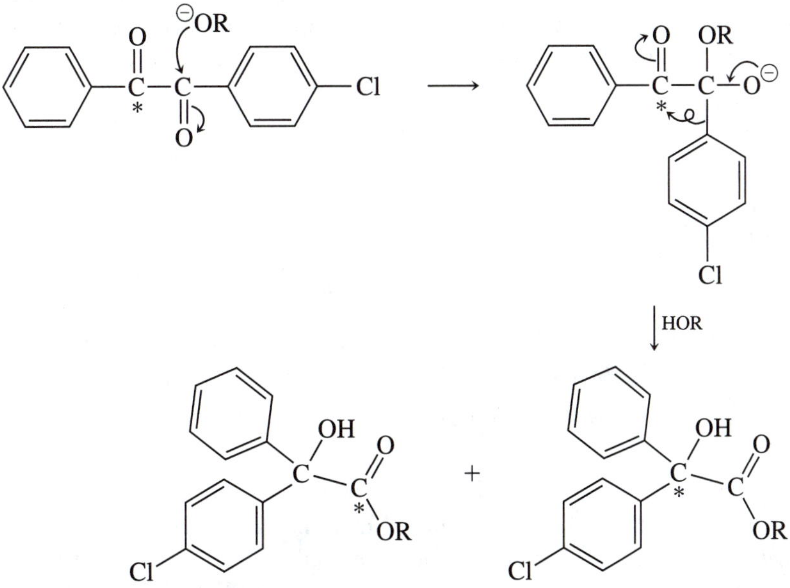

(42)

minor product   major product

presumably because the base will add predominantly to the carbonyl group bonded to the ring with the more strongly electronegative substituents.[29]

Dienone–phenol rearrangements. *Ortho*- and *para*-cyclohexadienones (cyclohexa-2,4-dienones and cyclohexa-2,5-dienones) undergo acid-catalyzed rearrangements to yield phenols or esters of phenols, as in Eqs. 43 and 44.[30]

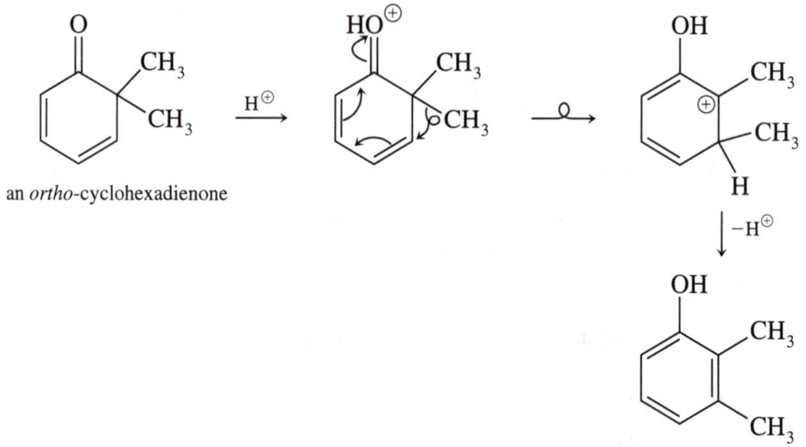

an *ortho*-cyclohexadienone

(43)

While most rearrangements of ketones require very strong acid catalysts, dienone–phenol rearrangements will take place in only moderately strong acids, such as solutions of sulfuric acid in acetic acid or acetic anhydride. Unlike most acid-catalyzed rearrangements of ketones, dienone–phenol rearrangements are highly exothermic reactions, since nonaromatic molecules are converted to aromatic isomers.

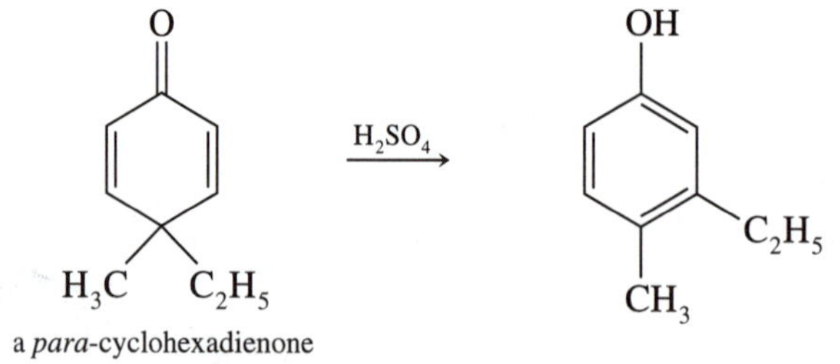

a *para*-cyclohexadienone

(44)

Many examples of dienone–phenol rearrangements occur in derivatives of steroids. The rearrangement products from these reactions often appear to arise from 1,3-shifts of migrating cycloalkyl rings, as in Eq. 45.[31,32]

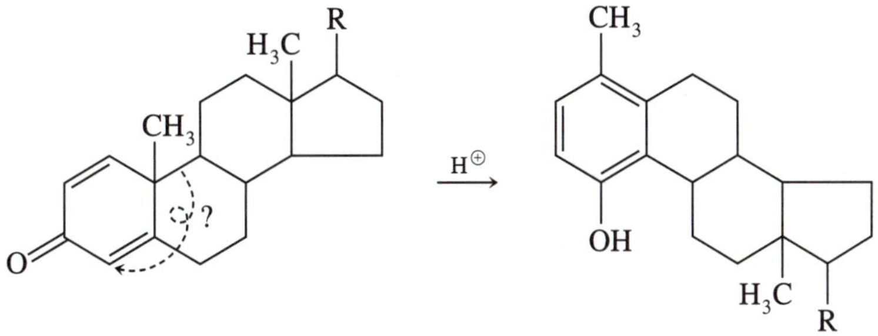

$$(45)$$

Several studies have shown that apparent 1,3-migrations in dienone–phenol rearrangements actually result from two successive 1,2-migrations. For example, it was found that rearrangement of the radioactive dienone **3** resulted in the radioactive carbon being equally distributed between the methylene groups *ortho* and *meta* to the hydroxy group of the product phenol (**5**).[33] The equal distribution of the radioactivity suggests that the rearrangement proceeds by way of the spirocyclic carbocation **4**, in which the two methylene groups are equivalent.

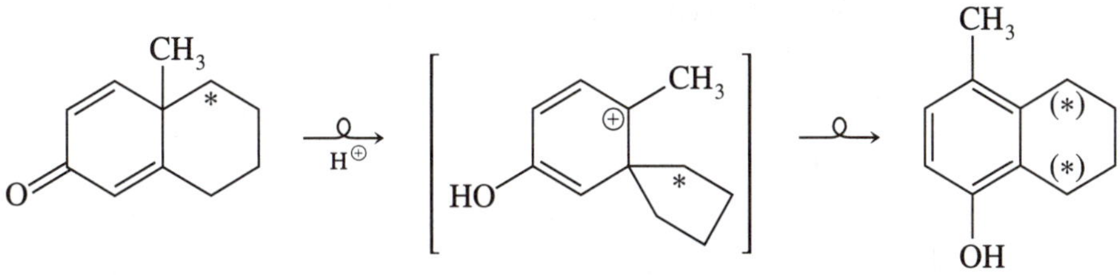

$$(46)$$

3   **4**   **5**

It has also been shown that rearrangement of dienone **6** in acid yields phenol **7**, again demonstrating that the apparent 1,3-migration proceeds by a sequence of two 1,2-shifts.[34] As expected, the more substituted alkyl group migrates in the second step.

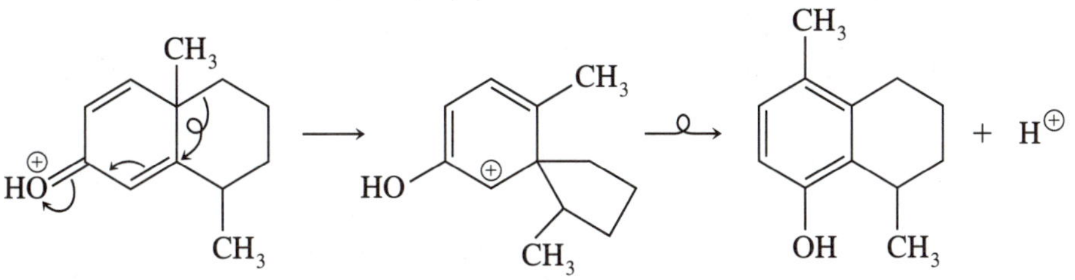

$$(47)$$

6   **7**

Allyl and Benzyl Group Migrations in Dienone–Phenol Rearrangements

Acid-catalyzed migrations of benzyl groups in *ortho*-cyclohexadienones can proceed by [1,5] shifts as well as [1,2] shifts (Eq. 48).[35]*

---

*Migration of the benzyl group to the *para* position of the phenol ring was also observed. It was suggested that this apparent [1,3] shift proceeded by a sequence of two [1,2] shifts.[35]

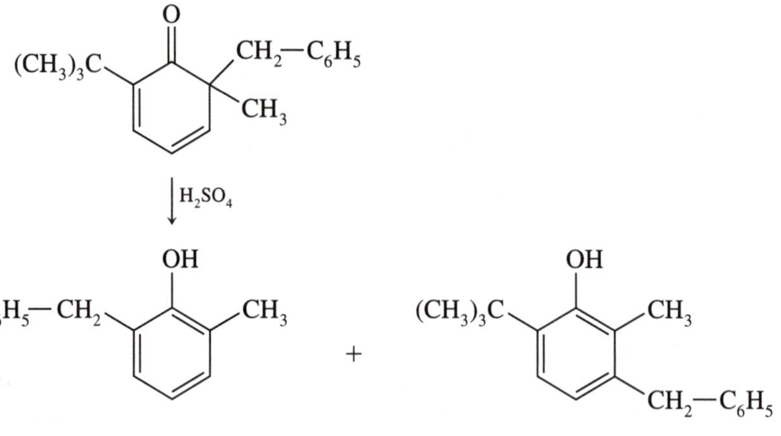

(48)

Migrations of allyl groups in both *ortho*- and *para*-cyclohexadienones usually proceed via [3,3] shifts, although [1,2] shifts can also take place, as shown in Eq. 49.[36]

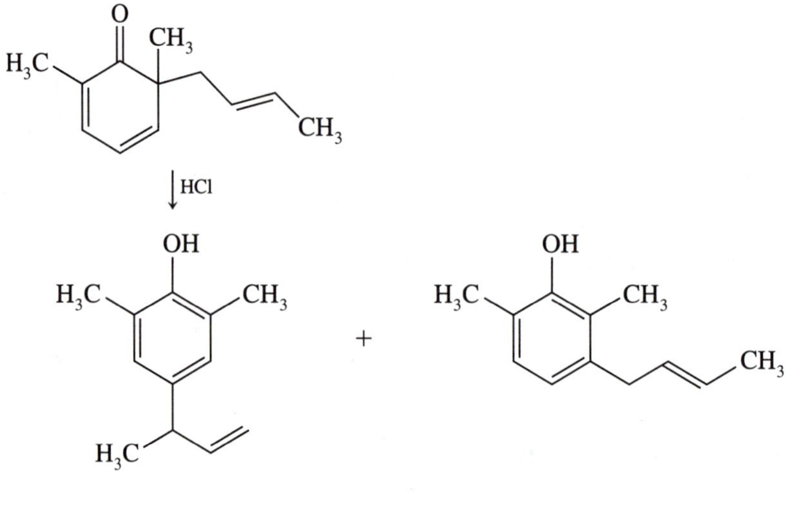

(49)

In general, the use of Lewis acids or other strongly electrophilic agents, such as trifluoroacetic anhydride, is more likely to result in [1,2] shifts. Thus, protic acids tend to catalyze the usual thermal reactions of cyclohexadienones, while stronger electrophiles, which would place a greater degree of positive charge on the rings, yield products expected from reactions of carbocations.

Dienol–benzene rearrangements.   Cyclohexadienols are very acid-sensitive compounds that easily rearrange to form aromatic rings, as shown in Eqs. 50 and 51. In fused-ring systems, in which the intermediates formed from the first rearrangement steps might not be able to lose protons to form aromatic rings, multiple migrations, leading to apparent [1,3] shifts, are possible (Eq. 52).[37]

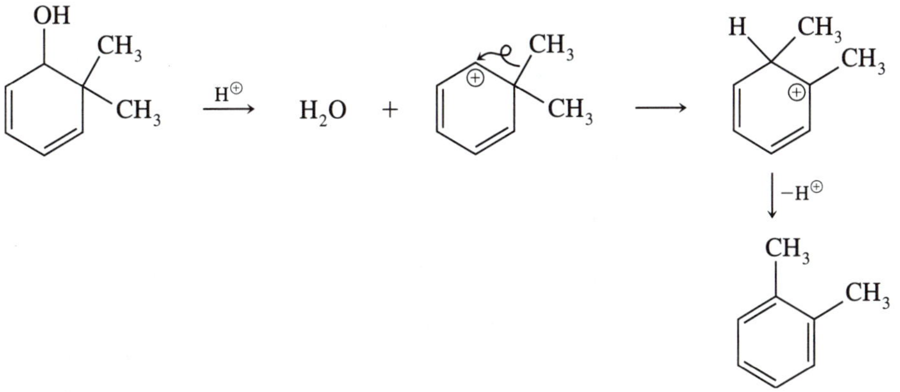

(50)

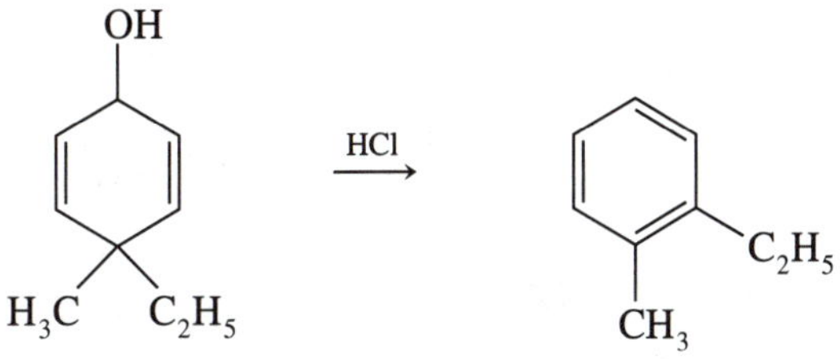

(51)

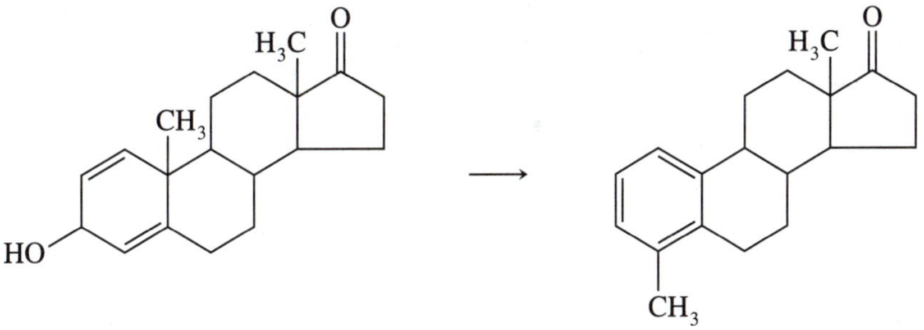

(52)

Backbone rearrangements. Migrations of hydrogens and methyl groups at ring junctures of molecules containing several fused cyclohexane rings are known as *backbone rearrangements*. These reactions are of particular interest, because they occur during the biosynthesis of steroids and terpenes, as well as in organic chemistry laboratories.

Among the most spectacular examples is the rearrangement of friedelanol (a derivative of a triterpene obtained from cork) to oleanene (Eq. 53).[38] This rearrangement requires seven individual migration steps—three hydrogen migrations and four methyl migrations. The driving force for this reaction appears to be a decrease in steric repulsions between axial substituents.

Early workers suggested that backbone rearrangements might proceed in a concerted manner, with several groups migrating simultaneously. The available evidence indicates, however, that the migrations occur sequentially rather than in a concerted fashion.[39] (However, the rearrangement is depicted in Eq. 53 as if all the rearrangements occur in a single step, in order to avoid drawing eight different intermediate carbocations.)

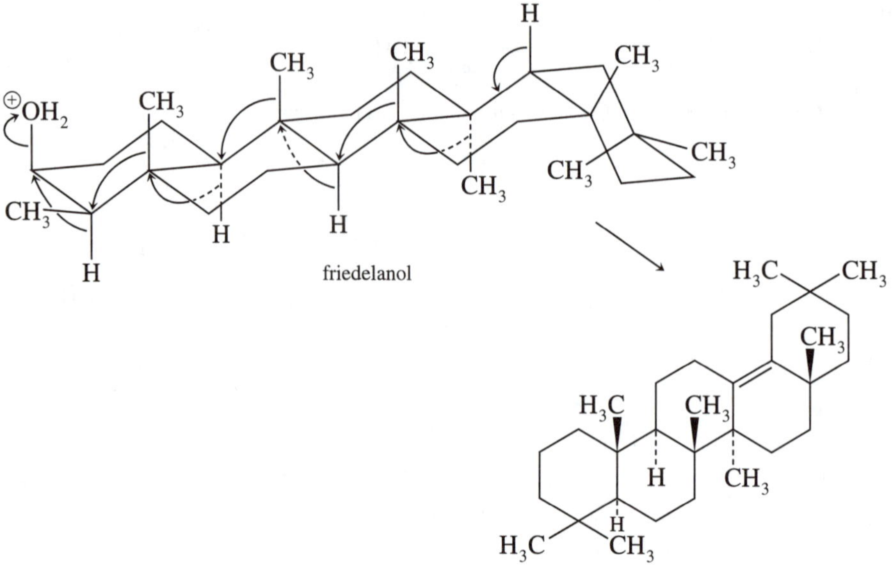

friedelanol

(53)

6.4 LONG-DISTANCE MIGRATIONS

Migrations in Open-Chain Cations

There is no convincing evidence that alkyl groups in open-chain carbocations can migrate in any manner other than by 1,2-shifts to adjacent carbons. Of course, alkyl groups in carbocations can, and often do, end up at locations distant from the migration origins. We know, however, that those rearrangements always proceed by sequences of 1,2-shifts.

While alkyl groups undergo only 1,2-migrations, there are some reactions of open-chain carbocations, such as the Friedel-Crafts reactions shown in Eq. 54a,[40a] that appear to proceed by direct 1,5-migrations of hydrogen or deuterium atoms. 1,6-Migrations are much rarer, but one was reported in the reaction shown in Eq. 54b.[40b]*

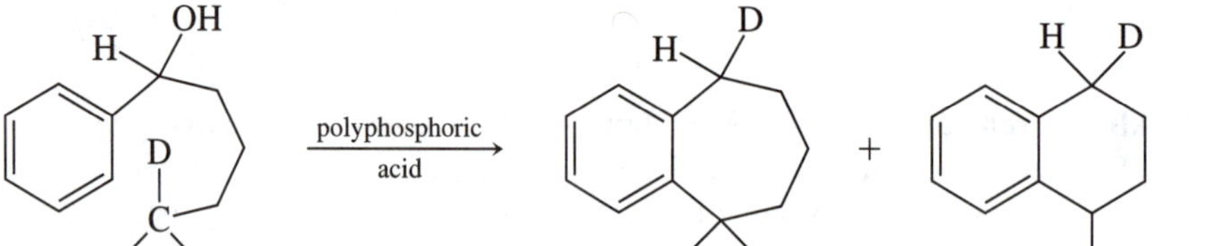

polyphosphoric acid

(54a)

---

*It was suggested that reaction 54b proceeds via the intermediate oxonium ion shown below, so that the geometry of the rearrangement resembles a 1,3 hydride migration more than a 1,6-shift. [40b]

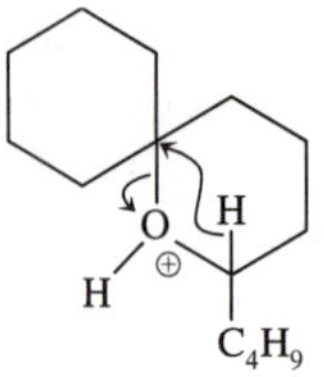

$$\text{(54b)}$$

In 1973, Saunders and Stofko confirmed that hydrogen atoms can migrate between distant sites in open-chain carbocations. They reported that at temperatures between -90 and $-55°C$, the NMR spectra of ions **8b**, **8c**, and **8d** showed hydrogens being interchanged between two tertiary positions (Table 6.1). The rearrangements could not have proceeded by sequences of 1,2-shifts, because rearrangements converting tertiary carbocations to secondary carbocations at such low temperatures are too slow to account for the reactions.[41]

TABLE 6.1   INTRAMOLECULAR HYDROGEN MIGRATIONS IN CARBOCATIONS

		Ea (kcal/mol)
$(CH_3)_2\overset{H}{C}-\overset{\oplus}{C}(CH_3)_2$ **8a**	$\rightleftharpoons$ $(CH_3)_2\overset{\oplus}{C}-\overset{H}{C}(CH_3)_2$	3.1 ± 0.1^a
$(CH_3)_2\overset{H}{C}\underset{CH_2}{\diagdown}\overset{\oplus}{C}(CH_3)_2$ **8b**	$\rightleftharpoons$ $(CH_3)_2\overset{\oplus}{C}\underset{CH_2}{\diagdown}\overset{H}{C}(CH_3)_2$	8.5 ± 0.1^{34}
$(CH_3)_2\overset{H}{C}\underset{CH_2-CH_2}{\diagdown}\overset{\oplus}{C}(CH_3)_2$ **8c**	$\rightleftharpoons$ $(CH_3)_2\overset{\oplus}{C}\underset{CH_2-CH_2}{\diagdown}\overset{H}{C}(CH_3)_2$	$12-13^{34}$
$(CH_3)_2\overset{H}{C}\underset{\underset{CH_2}{CH_2}\quad CH_2}{}\overset{\oplus}{C}(CH_3)_2$ **8d**	$\rightleftharpoons$ $(CH_3)_2\overset{\oplus}{C}\underset{\underset{CH_2}{CH_2}\quad CH_2}{}\overset{H}{C}(CH_3)_2$	$6-7^{34}$

^{a} Value given is for $\Delta G^{\ddagger}$.[10]

As indicated in Table 6.1, 1,5 hydrogen shifts, in particular, can be very rapid reactions with low activation energies. However, they are still much slower than 1,2 hydrogen shifts between carbocations of equal stability. 1,5-Shifts of hydrogens in open-chain cations are therefore only likely to occur when uphill rearrangements of carbocations to form less stable isomers are the only alternatives.

Direct migrations of hydrogens to distant positions in carbocations are actually intramolecular analogs of hydride-ion transfer reactions (see page 152). The geometrically favorable 1,5-migrations (as well as, of course, 1,2-migrations) can take place under milder conditions than are necessary for the corresponding intermolecular processes.

It should be noted that 1,3-, 1,4-, and 1,5-hydrogen shifts in carbocations are not [1,3], [1,4], and [1,5] sigmatropic shifts. In electronic terms, 1,3-shifts in carbocations may be classified as *homo* [1,2] shifts—that is, as homologs of [1,2] shifts. Despite the extra methylene groups between the migration origins and the migration termini, the reaction is allowed by the Woodward–Hoffmann rules. Similarly, 1,4-shifts may be regarded as *bishomo*[1,2] shifts, and [1,5] shifts as *trishomo* [1,2] shifts.

Transannular Migrations

Migrations of hydrogens between distant sites are of greatest importance in reactions of medium-sized rings; those containing between 7 and 11 carbons.

In the 1950s, Arthur C. Cope's group at MIT reported that the principal product from re-action of *cis*-cyclooctene with performic acid, followed by base hydrolysis, was *cis*-1,4-cyclo-octanediol. 3-Hydroxy- and 4-hydroxyclooctene were also obtained, along with the expected *trans*-1,2-cyclooctanediol. It was assumed that *cis*-cyclooctene oxide was first formed and then rearranged under the mildly acidic conditions to form the observed products (Eq. 55).[42]

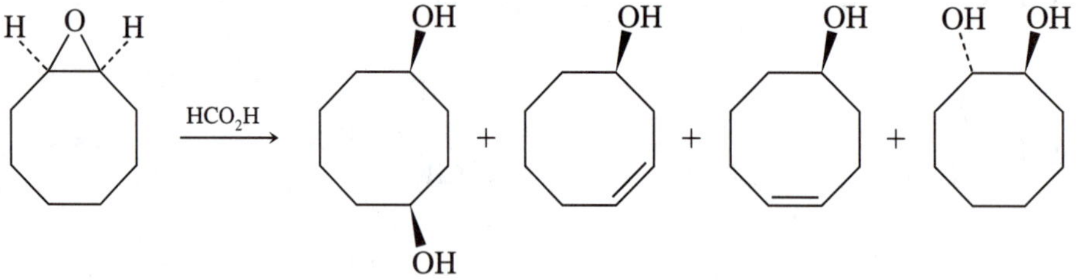

$$\tag{55}$$

The unexpected products in Eq. 55 might have resulted from 1,3- or 1,5-migrations of hy-drogens. (Sequences of 1,2-shifts were considered to be unlikely, because no 1,3-cyclooctanediol was obtained.) To distinguish between these possibilities, the distribution of deuterium in the products from the reaction of performic acid with cyclooctene-5,6-d_2 was determined. The re-sults indicated that 39% of the rearranged products resulted from 1,3-hydrogen shifts and 61% from 1,5-hydrogen shifts (see Eq. 56).

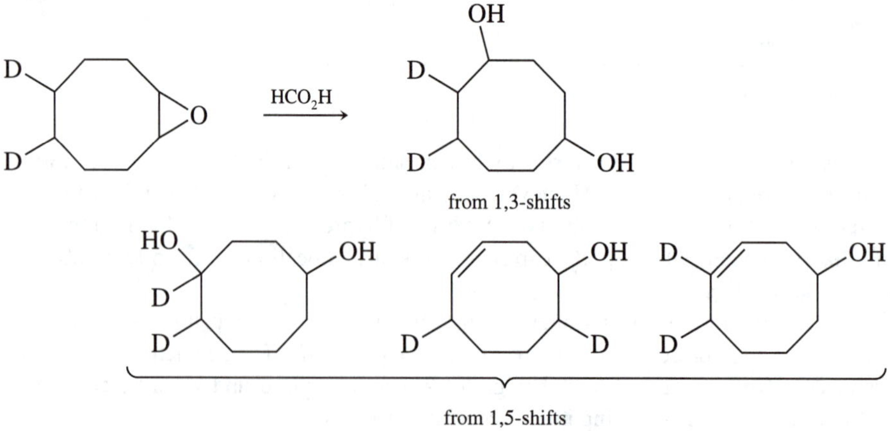

from 1,3-shifts

$$\tag{56}$$

from 1,5-shifts

Similar transannular rearrangements occur with epoxides of other medium-sized rings, with the percentages of rearrangement dependent on the sizes of the rings. V. Prelog and his coworkers at the Technische Hochshule in Zurich showed that the reactions of performic acid with cyclononenes yielded only 1,5-diols (Eq. 57) and with cyclodecenes only 1,6-diols (Eq. 58), the products (in both cases) of 1,5-hydrogen shifts.*

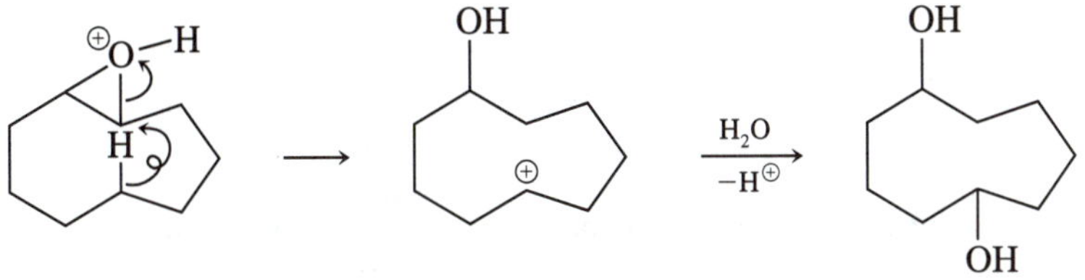

(57)

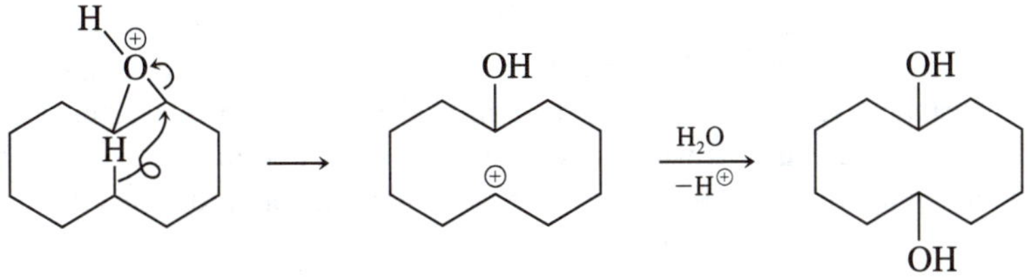

(58)

Reaction of the 11-carbon cycloalkene, cycloundecene, with performic acid similarly yielded only products resulting from hydrogen migrations, although their precise structures were not established. In contrast, reaction of cycloheptene oxide with hydrochloric acid yielded only 2.4% of 1,4-cycloheptanediol (resulting from a 1,4- or 1,5-shift) along with the "normal" 1,2-diol.[43]

Unlike epoxides of medium rings, cyclohexene oxide and cyclododecene (a 12-carbon ring) oxide yielded essentially no rearrangement products on reaction with acids.[43]

Epoxide rings normally open to form diols in stereospecific reactions, but the geometries of the reactions depend on whether or not they involve transannular hydride migrations. In forming a "normal" 1,2-diol, the nucleophile displaces the bond being broken from the rear, while *in a transannular rearrangement process the nucleophile enters from the same face of the ring as the bond being broken.*

Cis-cyclooctene oxide, for instance, yields *cis*-cyclooctane-1,4-diol, while *trans*-cyclooctene oxide yields only *trans*-cyclooctane-1,4-diol (Eqs. 59 and 60).[42]

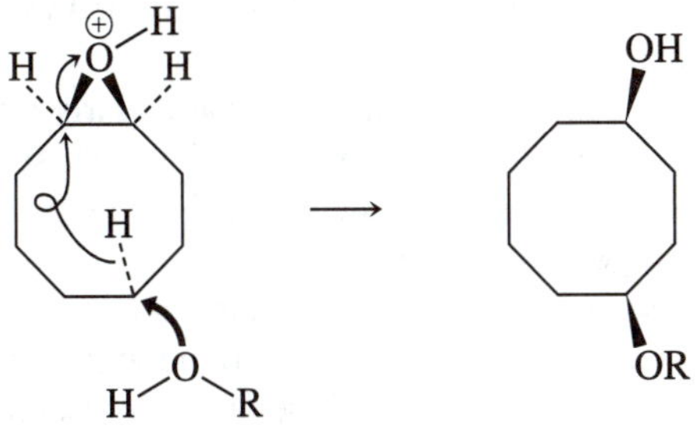

(59)

---

*1,5-shifts in cyclonanes are equivalent to 1,6-shifts and in cyclodecanes are equivalent to 1,7-shifts. Migrations in ring systems are usually defined as proceeding via the smaller number of intervening atoms.

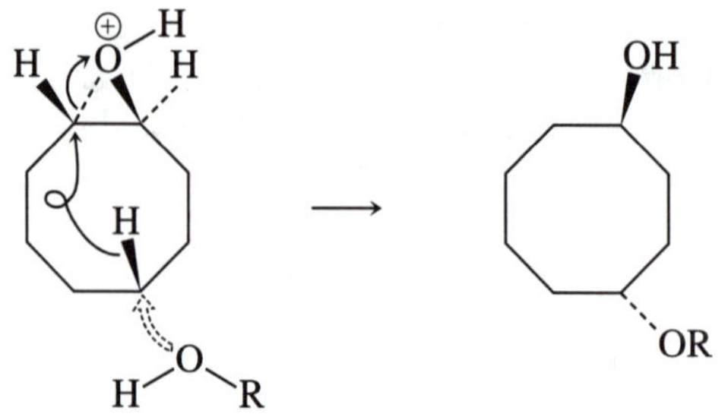

$$(60)$$

The net "retentions" of geometry in these reactions presumably result from double-inversion processes. The carbon–oxygen bonds being broken are displaced by hydrogens *trans* to the epoxide rings, while nucleophiles, in turn, displace the hydrogens with inversion.

Transannular rearrangements also take place during electrophilic addition reactions of medium-sized cycloalkenes. The reaction of *cis*-cyclooctene with deuterated trifluoroacetic acid, for instance, yields (after saponification of the initially formed ester) an alcohol resulting from a 1,5-hydrogen shift as well as the alcohol resulting from the "normal" 1,2-addition. The deuterium and hydroxy functions are *cis* to each other in both products (Eq. 61).[44]

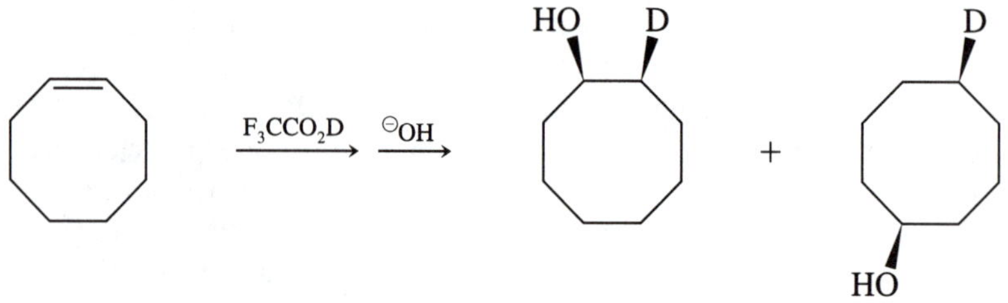

$$(61)$$

Conformations and Reactivities of Medium Rings

The enthalpies of formation of medium-ring cycloalkanes are higher than the expected values calculated by adding the enthalpies for the comparable numbers of CH_2 groups in open-chain alkanes. The differences between the measured and expected values may be defined as the *strain energies* of the cycloalkanes.

As shown in Table 6.2, the differences are particularly large for rings containing 8 to 11 carbons, as well as for three- and four-membered rings. These rings are therefore all *strained rings*. It is more than a coincidence that the strained medium-sized rings are precisely those in which transannular hydrogen shifts are prevalent.

Ring strain in three- and four-membered rings is largely due to distortion of bond angles from the normal 109.5° of a tetrahedral carbon. Bond angles in cyclopentanes are not significantly distorted, but the geometry of the cyclopentane ring requires that most hydrogen atoms be nearly eclipsed by hydrogens on adjacent carbons. Cyclopentane therefore exhibits a moderate strain energy. In normal chair conformations of cyclohexane, in contrast, all the carbon–hydrogen bonds are staggered, and cyclohexane is essentially strain free.

TABLE 6.2   STRAIN ENERGIES OF CYCLOALKANES

Ring size	Strain energy (kcal/mol)
C_3	27.5
C_4	26.3
C_5	6.2
C_6	0.1
C_7	6.2
C_8	9.7
C_9	12.6
C_{10}	12.4
C_{11}	11.3
C_{12}	4.1
C_{14}	1.9

Data taken from E.L. Eliel and S.H. Wilen, *Stereochemistry of Organic Compounds.*
John Wiley. New York (1994), p. 677.

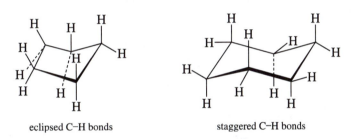

eclipsed C–H bonds   staggered C–H bonds

In principle, bond angles in medium-sized rings need not be distorted from the normal angles of sp^3 bonds. The relatively large strain energies of those rings, like that of cyclopentane, must, therefore, be due to steric repulsions between nonbonded atoms. The strain in medium-sized rings, however, cannot be accounted for entirely on the basis of eclipsing of hydrogens on adjacent carbons.

The origin of much of the strain energies of medium-sized rings can be deduced by examining the principal conformations and repulsions between nonbonded atoms in those rings. Using a set of molecular models would be extremely valuable. However, significant information can be obtained by examining structures on the printed page. Representations of the most stable conformations of cyclooctane and cyclodecane,[46] for example, are shown in the following diagrams.

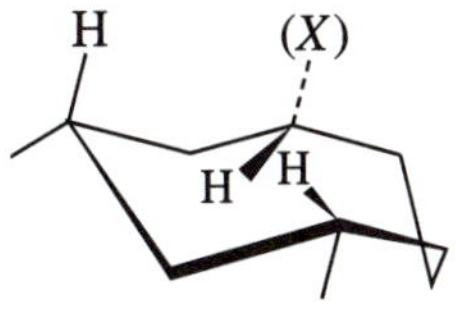

cyclooctane 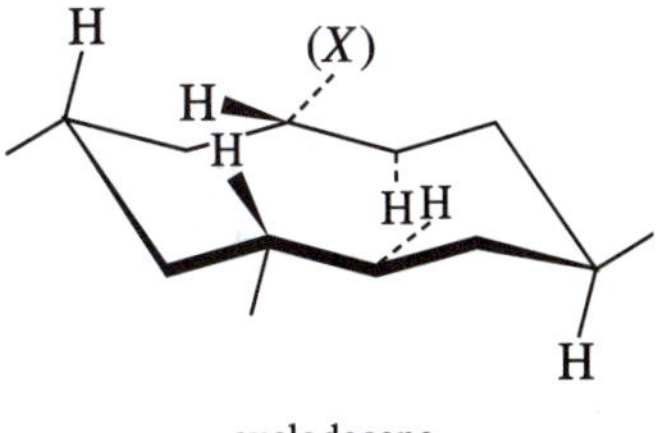   cyclodecane

In cyclooctane, hydrogens on two carbons, which can arbitrarily be labeled C1 and C5, are directed over the face of the ring and come into close proximity. This *transannular strain* destabilizes the cyclooctane ring. Furthermore, one of these hydrogens is beautifully positioned to take part in transannular migrations (displacing group "X"). In the conformation of cyclodecane shown on page 169, one pair of hydrogens above and one pair below the face of the ring closely approach each other. (In space-filling models, they actually appear to bump into each other.) Again, they are well positioned to take part in 1,5-transannular shifts. Cyclononane and cycloundecane similarly exhibit transannular strain and have hydrogens properly positioned to take part in 1,5-transannular shifts.

As is shown in the illustration, other hydrogens in cyclooctane and cyclodecane are well positioned to take part in 1,3-shifts, displacing group "X" in each case.

Conversion of tetravalent carbon atoms in medium-sized rings to trivalent cationic carbons—for instance, by loss of group "X" from cyclooctane or cyclodecane—would reduce the strain in the rings. As a result, carbocations of medium-sized rings are formed relatively easily. Studies by R. Heck and V. Prelog have shown that tosylates of medium-ring alcohols undergo solvolysis far more rapidly than tosylates of cyclohexanol or of large-ring alcohols (Table 6.3).[47]

TABLE 6.3 RELATIVE RATES OF SOLVOLYSIS OF CYCLOALKYL TOSYLATES IN ACETIC ACID AT 50°C

Number of carbons in ring	Relative reaction rate
6	1
7	31
8	285
9	266
10	539
11	67
12	3.1

There is a clear correlation between the degree of rate acceleration in solvolyses of cycloalkyl tosylates and the extent to which the corresponding cycloalkyl epoxides undergo transannular rearrangements.

"Hydrogen-Bridged" Cations

Spectroscopic studies have shown that at −120°C, medium-ring carbocations can exist in *hydrogen-bridged* forms, such as **9** and **10**.

A hydrogen atom in each structure is simultaneously linked to carbons on both sides of the ring via three-atom, two-electron bonds.[48,49] In the bicyclic cations **11** and **12**, the hydrogen-bridged structures are stable even at room temperature.[50,51]

Although these ions can be directly observed only in superacid solutions, similar hydrogen-bridged structures are now considered to be intermediates in transannular hydrogen migrations in ordinary solvents.[44] The facts that nucleophiles always enter *syn* to the leaving groups in the

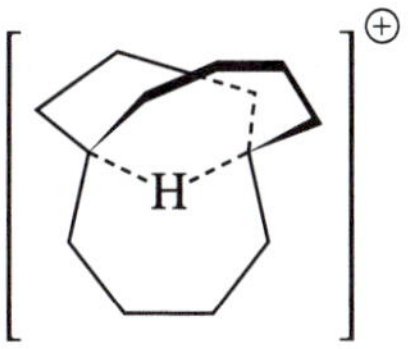

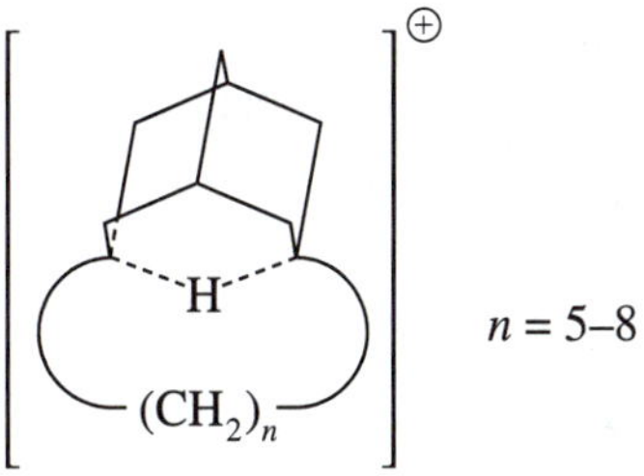

11   **12** $n = 5–8$

formation of diols from transannular rearrangements of epoxides, and that transannular re-arrangements in reactions of acids with medium-ring cycloalkenes result in products of *cis* addi-tion (as in Eq. 62), presumably result from nucleophilic attack on such bridged intermediates.

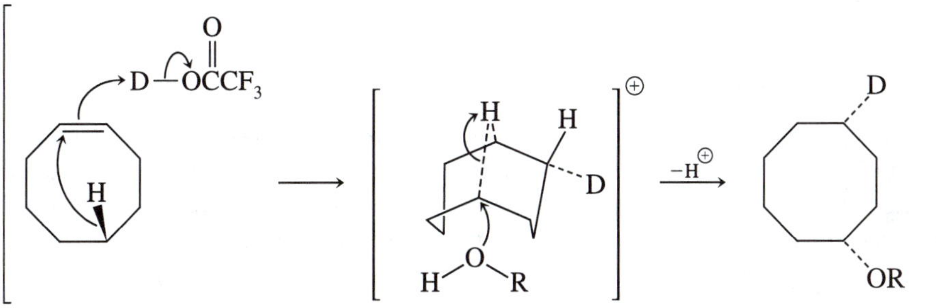

(62)

6.5   MIGRATIONS TO NITROGEN AND OXYGEN

Alkyl and aryl groups can migrate to positively charged nitrogen or oxygen atoms as well as to trivalent carbons.

Migrations to Electron-Deficient Nitrogens

The reaction of phosphorus pentachloride with *N*-chloro or *N*-bromo triphenylmethylamine re-sults in the migration of a phenyl group to nitrogen to yield, after hydrolysis, benzophenone and aniline (Eq. 63). These reactions are known as *Stieglitz rearrangements*.[52] A similar rearrange-ment occurs in the reaction of triphenylmethylhydroxylamine with acids.[53]

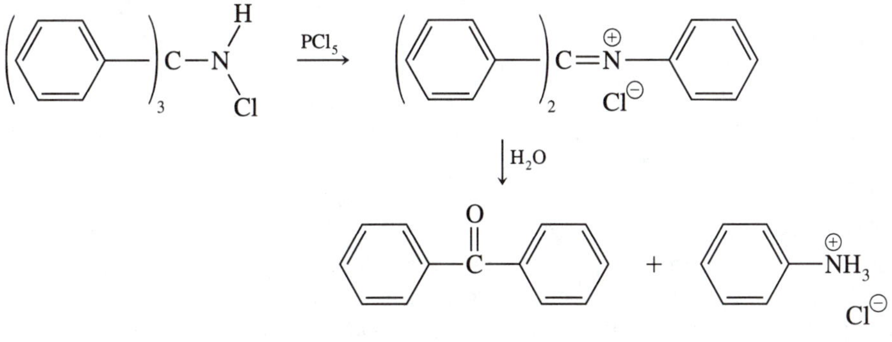

(63)

It was originally reported that alkyl groups did not migrate in acid-catalyzed reactions of *N*-haloamines or *N*-hydroxylamines. However, it has been found that alkyl migrations *can* take place in the solvolysis of bicyclic chloramines such as **13**.[54] The initial step in the rearrangement has been shown to be the formation of the nitrenium ion **14** (Eq. 64).*

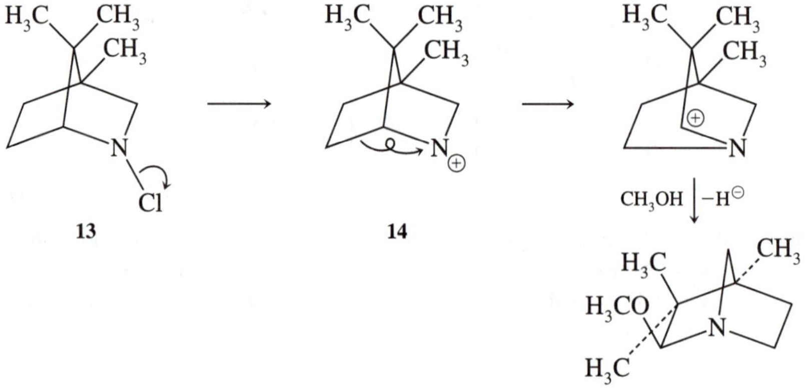

(64)

The Beckmann Rearrangement

In 1886, Ernst Beckmann reported that the reaction of benzophenone oxime with phosphorus pentachloride in ether resulted in the migration of a phenyl group from carbon to nitrogen to yield the imino chloride **16** (possibly formed via the nitrilum salt **15**). Aqueous hydrolysis of **16** converted it to *N*-phenylbenzamide, presumably by tautomerism of **17**.[55]

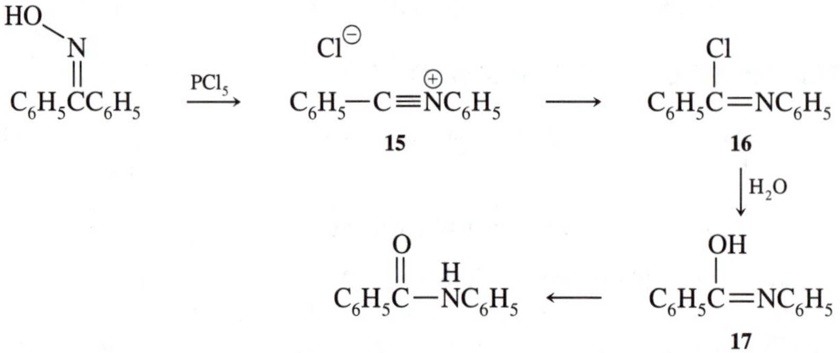

(65)

Beckmann later showed that protic acids are also effective in converting oximes to amides, as shown in Eq. 66.[56]

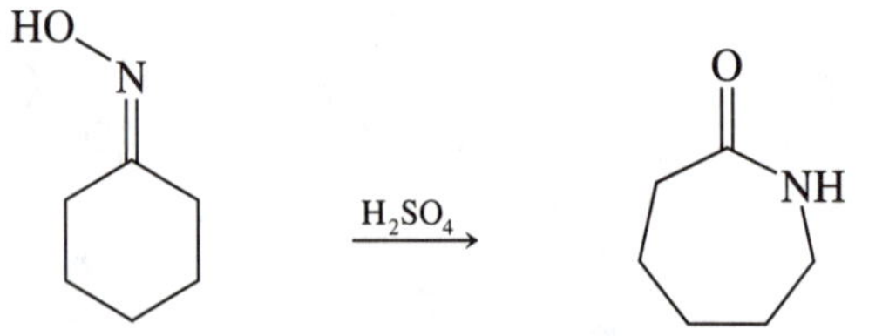

(66)

The geometries of the migrating groups are usually retained in Beckmann rearrangements[57,58] (see Eq. 67), as they are in Wagner–Meerwein rearrangements.

Oximes of unsymmetrical ketones can exist as geometrical isomers with *syn* and *anti* configurations of the carbon–nitrogen double bonds. The *syn* and *anti* isomers normally yield different amides

---

*Ion **14** is actually likely to have a bridged "nonclassical" structure (see Chapter 7).

when subjected to Beckmann rearrangement conditions. Early workers speculated that the group *syn* to the hydroxy migrated to replace the hydroxy group on the nitrogen atom. However, it was later determined that it is the group *anti* to the hydroxy group that migrates, as shown in Eqs. 67 to 69.[59]

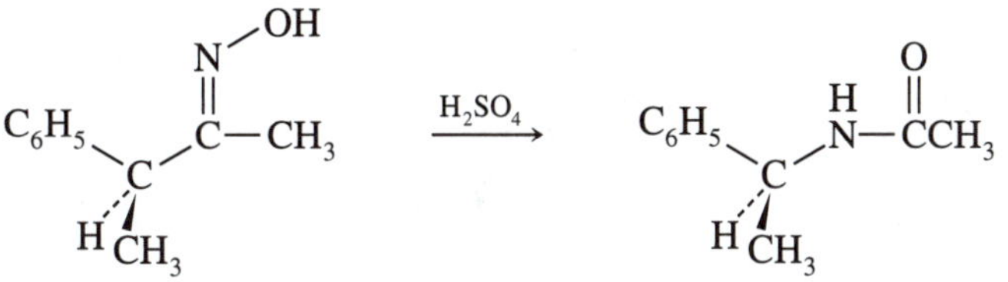

(67)

(68)

(69)

This fact has been so firmly established so that the products obtained from Beckmann rearrangements of some oximes have themselves been used to prove the geometries of the oximes. (Complications can arise, however, because oximes may undergo *syn–anti* isomerism in acid.)

The stereospecificity of the Beckmann rearrangement demonstrates that the migration step must occur simultaneously with the breaking of the N–O bond, as shown in Eq. 70.

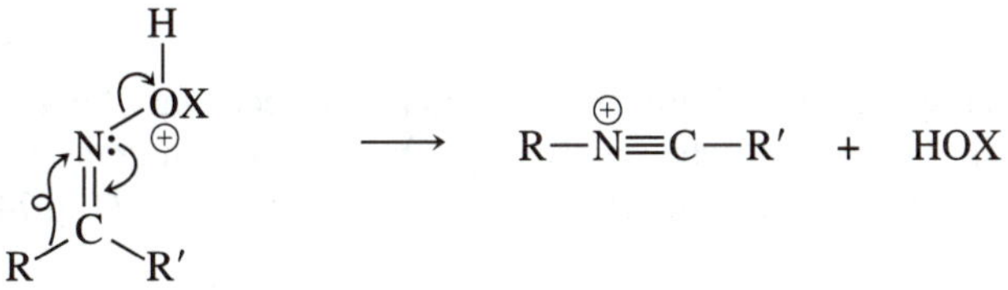

$$R-\overset{\oplus}{N}\equiv C-R' \quad + \quad HOX$$

(70)

Reactions of acids with aldoximes commonly result in dehydration of the aldoximes to form nitriles rather than in Beckmann rearrangements (Eq. 71).

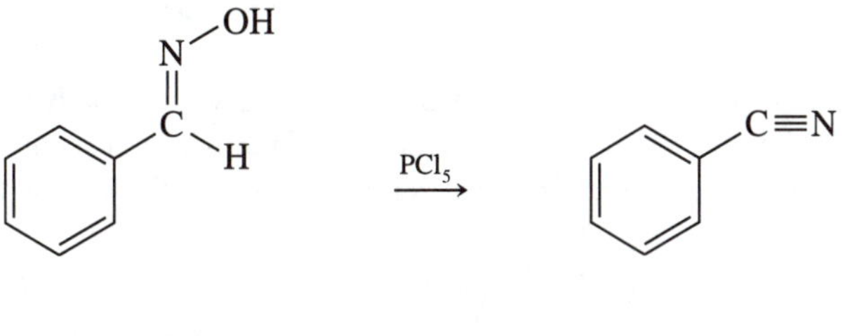

(71)

Related fragmentation processes, rather illogically called "second-order Beckmann rearrangements," can take place with ketoximes if a potential migrating group can form a relatively stable carbocation, as in Eq. 72.[59]

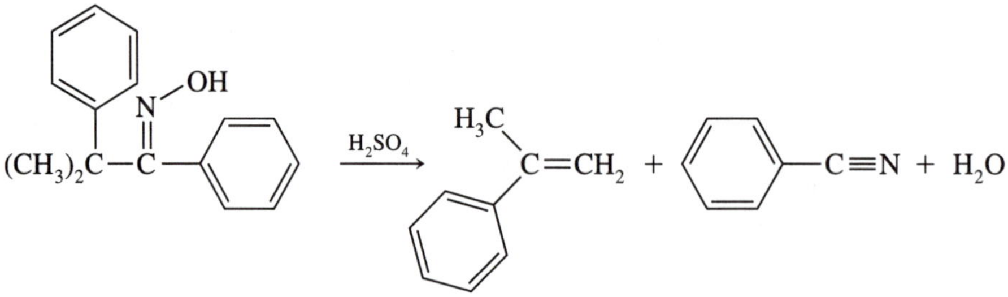

$$(72)$$

Rearrangements of Peroxides

Alkyl and aryl groups can migrate from carbon to oxygen when hydrocarbons with peroxy or hydroperoxy groups (particularly at tertiary positions) are treated with acids. This rearrangement is used commercially in the production of acetone and phenol from cumene hydroperoxide (Eq. 73).

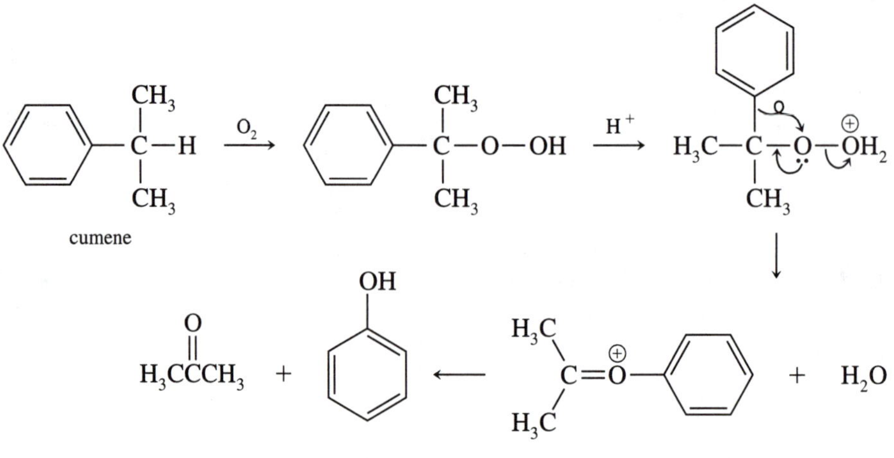

$$(73)$$

Aryl groups migrate so readily that even primary or secondary hydroperoxides can undergo rearrangements. In contrast, when only alkyl groups might migrate, primary and secondary hydroperoxides typically yield unrearranged carbonyl derivatives (Eq. 74).[60]

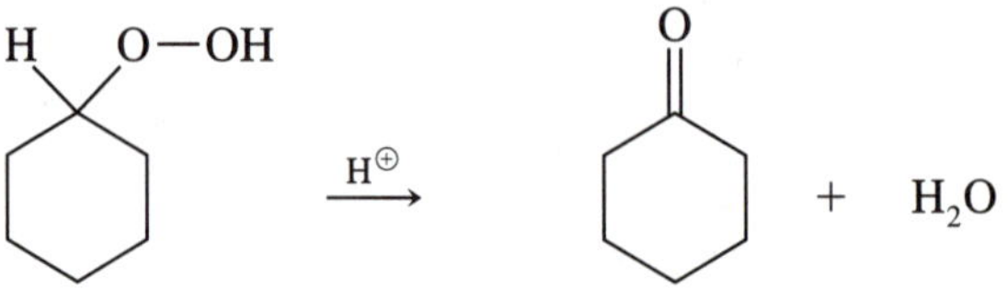

$$(74)$$

In view of the high electronegativity of oxygen, it is very unlikely that real "oxenium ions" are intermediates in the acid-catalyzed rearrangements of peroxides. It seems probable that, as in Beckmann rearrangements, the migrations occur simultaneously with the departure of the leaving groups.

The Baeyer–Villiger Oxidation

In the *Baeyer–Villiger reaction*, ketones are converted to esters (and cyclic ketones are converted to lactones) by reaction with peracids (Eqs. 75 and 76).[61a]*

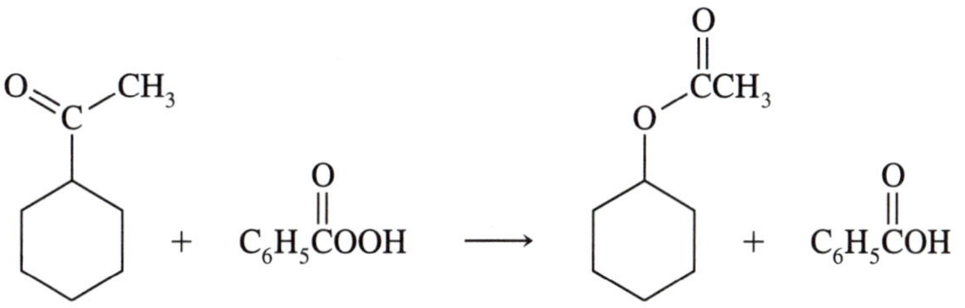

$$(75)$$

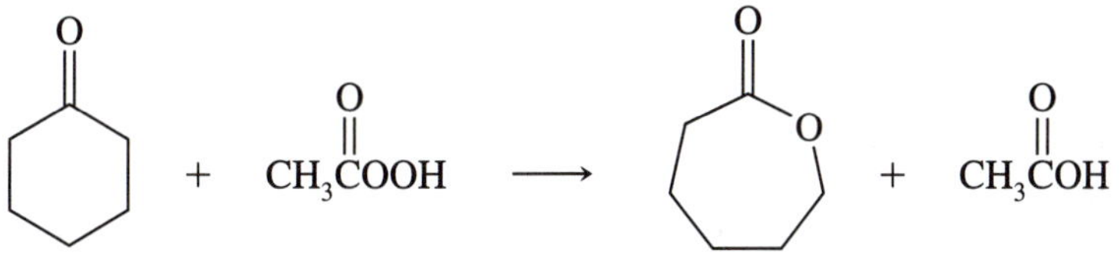

$$(76)$$

The Baeyer–Villiger reaction begins with the acid-catalyzed addition of a peracid to a carbonyl group to form a peroxy ester (such as **18**).[62]

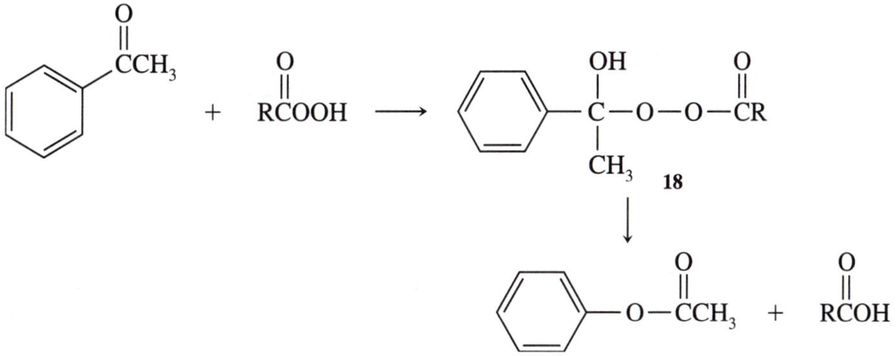

$$(77)$$

Baeyer and Villiger originally suggested that the rearrangement then proceeded via the formation of a cyclic peroxide intermediate.[63] However, in 1953, W. von E. Doering and E. Dorfman studied the reaction of peracetic acid with benzophenone labeled with ^{18}O. They found that all

---

*Ketones that cannot easily be converted to enolate anions may also undergo Baeyer–Villiger reactions on treatment with hypohalite solutions. [61b]

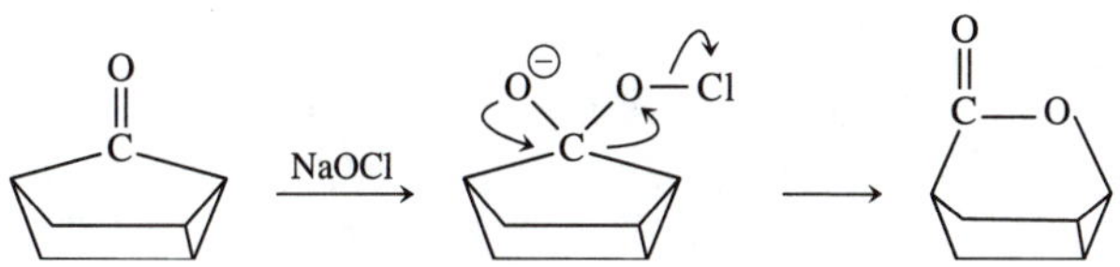

the ^{18}O was retained as the carbonyl oxygen of the product ester.[64] This result eliminated the possible formation of a symmetrical intermediate and supported a mechanism similar to that proposed for rearrangements of other peroxides (Eq. 78).

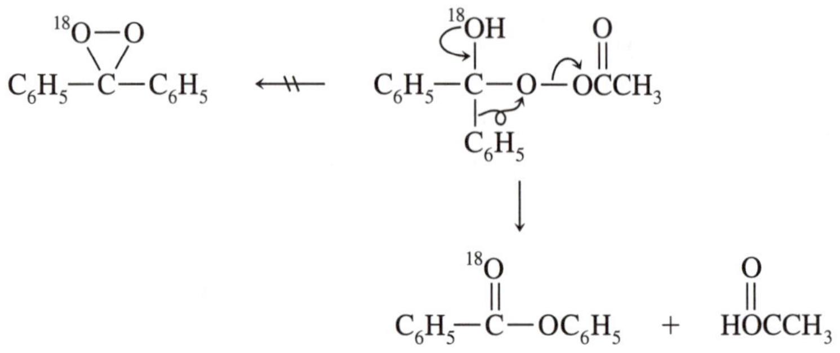

$$
\begin{array}{ccc}
{}^{18}O\!-\!O & & {}^{18}OH \quad\quad O \\
\diagdown\ \diagup & & \\
C_6H_5\!-\!C\!-\!C_6H_5 & \xleftarrow{\quad\quad} & C_6H_5\!-\!C\!-\!O\!-\!OCCH_3 \\
& & C_6H_5 \\
\end{array}
$$

$$
\begin{array}{cc}
{}^{18}O & O \\
\parallel & \parallel \\
C_6H_5\!-\!C\!-\!OC_6H_5 & + \quad HOCCH_3 \\
\end{array}
$$

(78)

As in Wagner–Meerwein rearrangements, the alkyl groups most likely to migrate in Baeyer–Villiger reactions are those that can most readily bear positive charges (Table 6.4).[65]

TABLE 6.4 RELATIVE MIGRATORY APTITUDES IN THE BAEYER-VILLIGER REARRANGEMENT

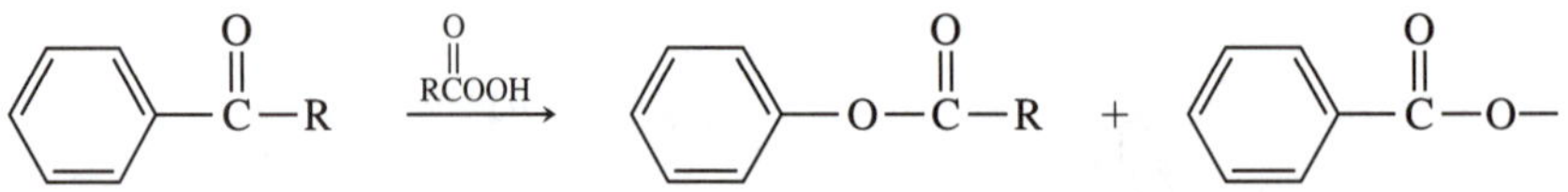

CH_3-	very small
C_2H_5-	7×10^{-2}
$CH_3CH_2CH_2-$	7×10^{-2}
$(CH_3)_2CH-$	1.9
$(CH_3)_3C-$	39
$C_6H_5CH_2-$	1.3

REFERENCES

[1] G.A. Olah, *J. Am. Chem. Soc., 94*, 808 (1972).

[2] G. Wagner and W. Brickner, *Ber., 32*, 2302 (1899). See also H. Meerwein and K. van Emster, *Ber., 55*, 2500 (1922).

[3] W. Kirmse, W. Gruber, and J. Knist, *Ber., 106*, 1376 (1973).

[4] N.L. Wendler, in *Molecular Rearrangements*, vol. 2, Paul de Mayo, Ed. Interscience, New York (1964), p. 1025.

[5] G.A. Olah, J.M. Bollinger, C.A. Cupas, and J. Lukas, *J. Am. Chem. Soc., 89*, 2692 (1967).

[6] M. Saunders, P. Vogel, W.L. Hagen, and J. Rosenfeld, *Acc. Chem. Res., 6*, 53, (1973). See also M. Saunders and H.A. Jimenez-Vazques, *Chem. Rev., 91*, 375 (1991).

[7]   C. Dean and D. Whittaker, *J. Chem. Soc., Perkin Trans, 2*, 1541 (1991).

[8]   J. Planelles, J. Sanchez-Marin, F. Tomas, and A. Corma, *Perkin Trans, 2*, 333 (1985).

[9]   D.P. Dymerski, R.M. Prinstein, P.F. Bente III, and F.W. McLafferty, *J. Am. Chem. Soc., 98*, 6834 (1976).

[10] For example, see P.D. Bladon, H.B. Henbest, and G.W. Wood., *J. Chem. Soc.*, 2737 (1952).

[11] M. Stiles and R.P. Mayer, *J. Am. Chem. Soc., 81*, 1497 (1959).

[12] D.V. Korchagina et al., *J. Org. Chem. USSR, 12*, 378 (1976).

[13] M. Saunders and M.R. Kates, *J. Am. Chem. Soc., 100*, 7082 (1978).

[14] G.A. Olah, G.K. Surya Prakash, and J. Sommer, *Superacids*. John Wiley, New York (1985).

[15] C.H. Rochester, *Acidity Functions*. Academic Press, New York (1970).

[16] P. Vogel, *Carbocation Chemistry*. Elsevier, New York (1985), p. 338.

[17] H. Pines and R.C. Wackher, *J. Am. Chem. Soc., 68*, 595, 599 (1946).

[18] P.D. Bartlett, F.E. Condon, and A. Schneider, *J. Am. Chem. Soc., 66*, 1531 (1944).

[19] H.S. Bloch, H. Pines, and L. Schmerling, *J. Am. Chem. Soc., 68*, 153 (1946).

[20] R.A. Fort and P. von R. Schleyer, *Chem. Rev., 64*, 277 (1964). See also M.A. McKervey, *Tetrahedron, 36*, 971 (1980).

[21] A. Streitwieser, Jr., *Solvolytic Displacement Reactions*. McGraw-Hill, New York (1962), p. 13.

[22] A. Aebi, D.H.R. Barton, and A.S. Lindsey, *J. Chem. Soc.*, 3124 (1953).

[23] G. Buchi and I.M. Goldman, *J. Am. Chem. Soc., 79*, 4741 (1957).

[24] R. Fittig, *Liebigs Ann., 110*, 17 (1859); *114*, 54 (1860).

[25] R.E. Parker and N.S. Isaacs, *Chem. Rev., 59*, 737 (1959).

[26] P.A.S. Smith and D.R. Baer, *Org. Reactions, 11*, 157 (1960).

[27] P. Gassmann and T.T. Tidwell, *Acc. Chem. Res., 16*, 279 (1983). See also X. Creary, *Chem. Revs., 91*, 1625 (1991).

[28] For example, see M.-C. Carre, B. Jamart-Gregoire, P. Geoffroy, and P. Caubere, *Tetrahedron, 44*, 127 (1988); D.H.G. Crout, E.R. Lee, and D.P.J. Pearson, *Perkin Trans, 2*, 53 (1991); D.H.G. Crout, C.R. McIntyre, and N.W. Alcock, *Perkin Trans, 2*, 381 (1991).

[29] A. Fry, in *Mechanisms of Molecular Migrations*, vol. 4, B.S. Thyagarajan, Ed. Interscience, New York, 1971.

[30] S. Selman and J.F. Eastham, *Quart. Rev. Chem. Soc., 14*, 22 (1960).

[31] B. Miller, in *Mechanisms of Molecular Migrations*, vol. 1, B.S. Thyagarajan, Ed. Interscience, New York (1968).

[32] N.L. Wendler, in *Molecular Rearrangements*, vol. 2, P. de Mayo, Ed. Interscience, New York (1963), p. 1033.

[33] R. Futaki, *Tetrahedron Lett.*, 3059 (1964).

[34] S.M. Bloom, *J. Am. Chem. Soc., 81*, 4728 (1959).

[35] B. Miller, *J. Am. Chem. Soc., 92*, 6252 (1970).

[36] B. Miller, *Acc. Chem. Res., 8*, 245 (1975).

[37] M.J. Gentles, J.B. Moss, H.L. Herzog, and E.B. Hershberg, *J. Am. Chem. Soc., 80*, 3702 (1958).

[38] E.J. Corey and J.J. Ursprung, *J. Am. Chem. Soc., 78*, 5041 (1956). See also H. Dutler, O. Jeger, and L. Ruzicka, *Helv. Chim. Acta, 38*, 1268 (1955).

[39] H.W. Whitlock, Jr. and A.H. Olson, *J. Am. Chem. Soc., 92*, 5383 (1970).

[40] (a) Q. Branca and D. Arigoni, *Chimia, 23*, 189 (1969); (b) J.F. Gil, D.J. Ramon, and M. Yus, *Tetrahedron, 50*, 7307 (1994).

[41]   M. Saunders and J.J. Stofko, Jr., *J. Am. Chem. Soc., 95*, 252 (1973).

[42]   A.C. Cope, M.M. Martin, and M.A. McKervey, *Quart. Revs. Chem. Soc., 20*, 119 (1966).

[43]   V. Prelog and J.G. Traynham, in *Molecular Rearrangements*, vol. 1, P. de Mayo, Ed., Interscience Publishers, New York (1963), p. 593.

[44]   J.E. Nordlander, K.D. Kotian, D.E. Raff II, F.G. Njoroge, and J.J. Winemiller, *J. Am. Chem. Soc., 106*, 1427 (1984).

[45]   E.L. Eliel and S.H. Wilen, *Stereochemistry of Organic Compounds*. John Wiley, New York (1994), p. 677.

[46]   E.L. Eliel and S.H. Wilen, *Stereochemistry of Organic Compounds*. John Wiley, New York (1994), pp. 762–769.

[47]   R. Heck and V. Prelog, *Helv. Chim. Acta, 38*, 1541 (1955).

[48]   R.P. Kirchen, K. Ranganayukulu, A. Rauk, B.P. Singh, and T.S. Sorensen, *J. Am. Chem. Soc., 103*, 588 (1981).

[49]   P. Buzek, P. von R. Schleyer, H. Vancik, and D.E. Sunko, *Chem. Comm.*, 1538 (1991). See also the references in T.S. Sorensen and S.M. Whitworth, *J. Am. Chem. Soc., 112*, 8135 (1990).

[50]   J.E. McMurry, T. Lectka, and C.N. Hodge, *J. Am. Chem. Soc., 111*, 8867 (1989).

[51]   T.S. Sorensen and S.M. Whitworth, *J. Am. Chem. Soc., 112*, 8135 (1990).

[52]   J. Stieglitz and P.N. Leech, *Ber., 46*, 2147 (1913).

[53]   J. Stieglitz and I. Vosburg, *Ber., 46*, 2151 (1913).

[54]   P.G. Gassman, *Acc. Chem. Res., 3*, 26 (1970).

[55]   E. Beckmann, *Ber., 20*, 1507 (1887).

[56]   E. Beckmann, *Ber., 19*, 988 (1886).

[57]   A. Campbell and J. Kenyon, *J. Chem. Soc.*, 25 (1946).

[58]   R.K. Hill and O. Chortyk, *J. Am. Chem. Soc., 84*, 1064 (1962).

[59]   L.G. Donaruma and W.Z. Heldt, *Org. Reactions, 11*, 1 (1960).

[60]   P.A.S. Smith, in Ref. 25, pp. 568–577.

[61]   (a) C.H. Hassall, *Org. Reactions, 9*, 73 (1957). See also H. Seto and H. Yoshioka, *Chem. Letters*, 1797 (1990) and P. Hamley, A.B. Homes, D.R. Marshall, and J.W.M. MacKinnon, *Perkin Trans. 1*, 1793 (1991); (b) M.G. Rosenberg, J. Haslinger, and U.H. Briker, *J. Org. Chem., 67*, 450 (2002).

[62]   S.L. Friess and A.H. Soloway, *J. Am. Chem. Soc., 73*, 3968 (1951).

[63]   A. Baeyer and V. Villiger, *Ber., 32*, 3625 (1889).

[64]   W. von E. Doering and E. Dorfman, *J. Am. Chem. Soc., 75*, 5595 (1953).

[65]   M.F. Hawthorne, W.D. Emmons, and K.S. McCallum, *J. Am. Chem. Soc., 80*, 6393 (1958).

PROBLEMS

6.1   Draw structures for the principal organic products of the following reactions.

(a) $(CH_3)_2CH-\underset{\underset{\displaystyle CH_3}{|}}{\overset{\overset{\displaystyle CH_3}{|}}{C}}-CH_2Cl \xrightarrow[\Delta]{ZnCl_2}$

(b)

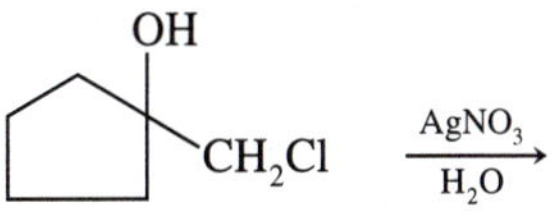

$$\xrightarrow[\text{H}_2\text{O}]{\text{AgNO}_3}$$

(c)

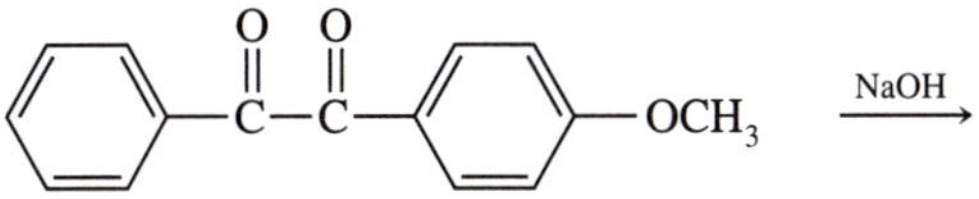

$$\xrightarrow{\text{NaOH}}$$

(d)

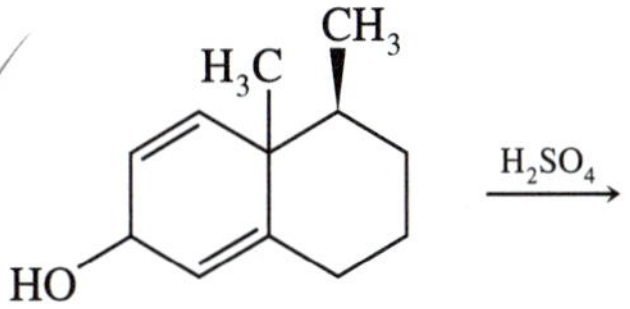

$$\xrightarrow{\text{H}_2\text{SO}_4}$$

(e)

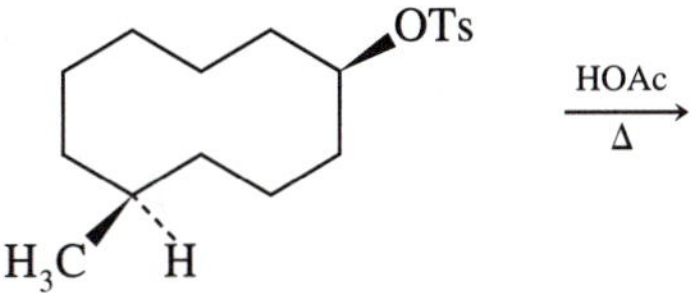

$$\xrightarrow[\Delta]{\text{HOAc}}$$

(f) 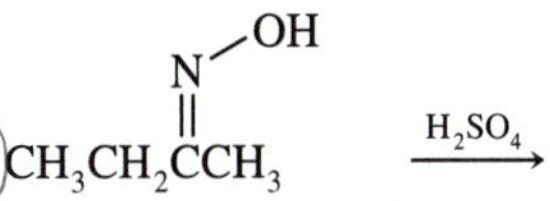 $\text{CH}_3\text{CH}_2\overset{\displaystyle \|}{\underset{}{\text{C}}}\text{CH}_3$ with N–OH

$$\xrightarrow{\text{H}_2\text{SO}_4}$$

(g)

$$\xrightarrow{\text{H}_3\text{CCOOH}}$$

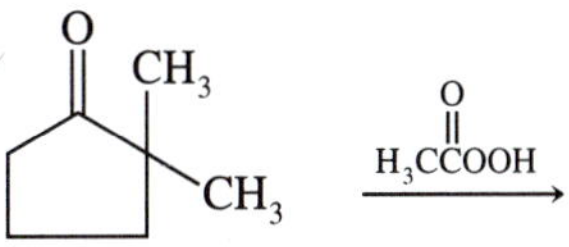

6.2 Explain why the deuterium and oxygen atoms are *cis* to each other in the product of the following reaction.

$$\xrightarrow[\text{(2) }^{\ominus}\text{OH}]{\text{(1) F}_3\text{CCOD}}$$

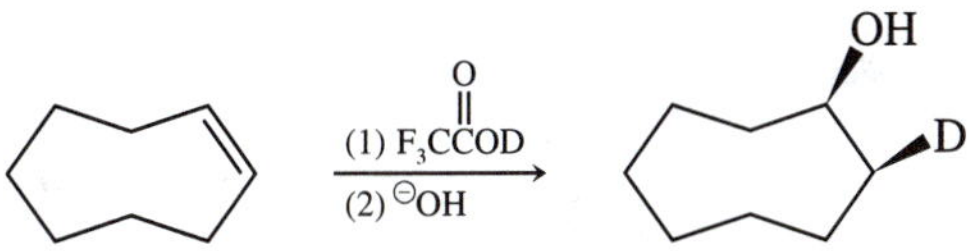

6.3 Write reasonable mechanisms for the following reactions.

(a) $\text{H}_3\text{C(CH}_2)_3\text{CH}_2\text{Cl} \xrightarrow{\text{SbF}_5} (\text{CH}_3)_2\overset{\text{F}}{\underset{}{\text{C}}}\text{CH}_2\text{CH}_3$

(b)

$$\xrightarrow{\text{EtAlCl}_2}$$

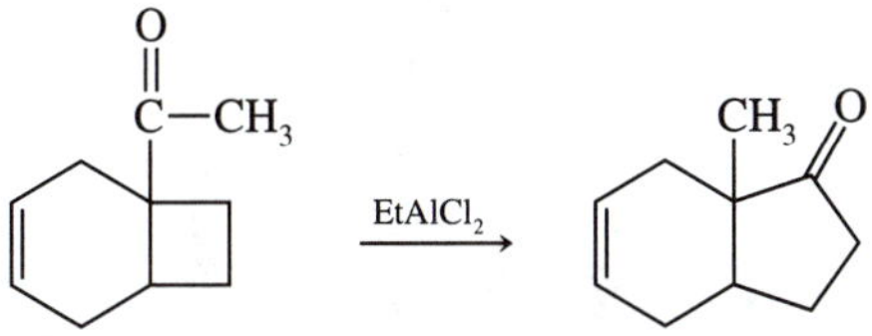

(T. Fujiwara et al., *Tetrahedron Lett.*, *33*, 2583 [1992])

(c)

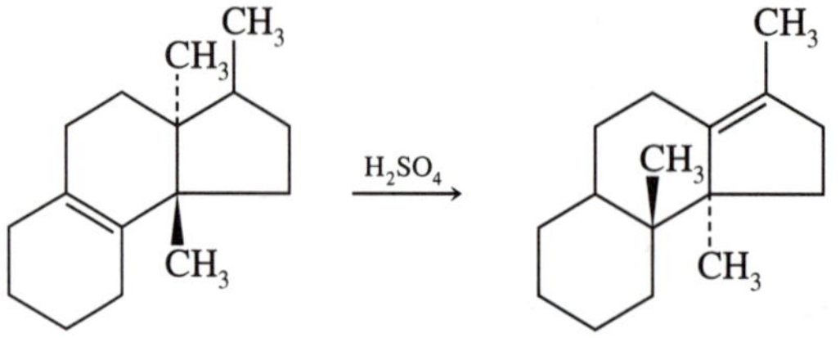

$$\xrightarrow{\text{H}_2\text{SO}_4}$$

(d)

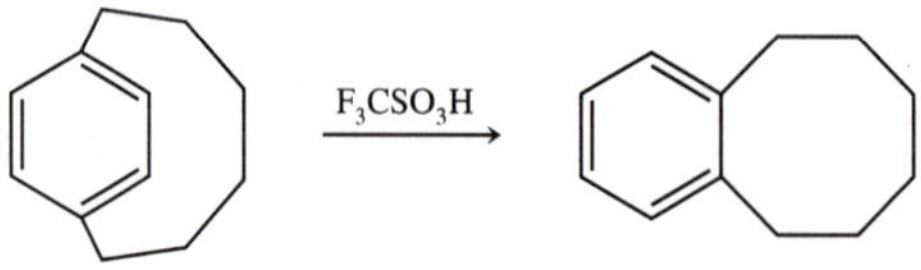

$$\xrightarrow{\text{F}_3\text{CSO}_3\text{H}}$$

(e)

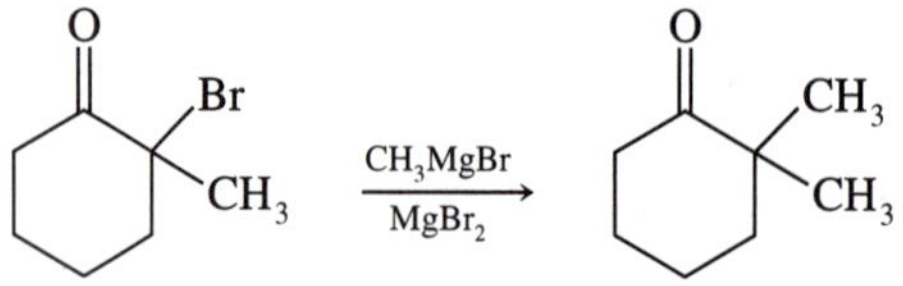

$$\xrightarrow[\text{MgBr}_2]{\text{CH}_3\text{MgBr}}$$

(f)

$$\xrightarrow[\text{H}^{\oplus}]{\text{H}_2\text{O}_2}$$

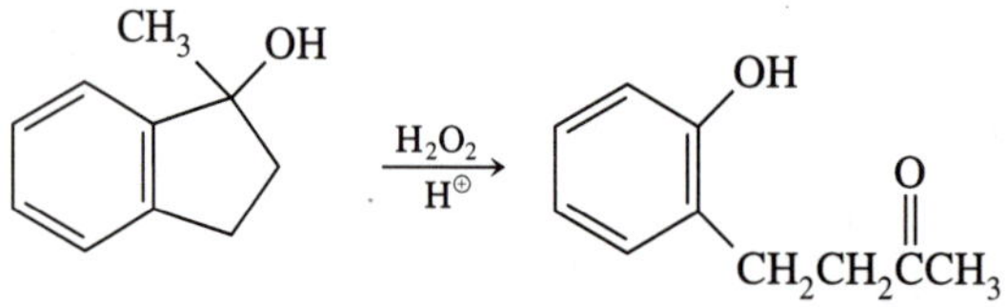

(H.-J. Hamman & J. Liebscher, *J. Org. Chem.*, *65*, 1873 [2000])

(g)

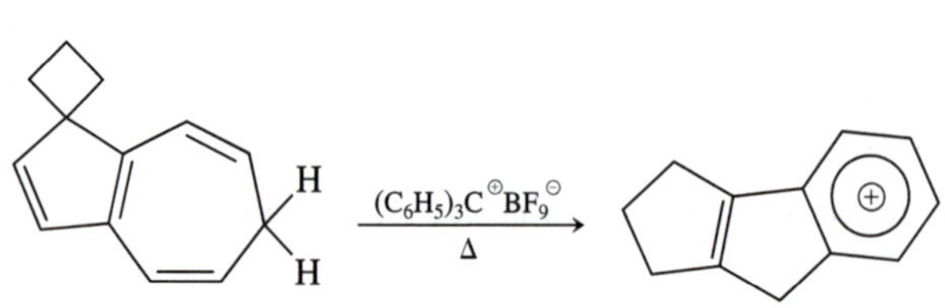

$$\xrightarrow[\Delta]{(\text{C}_6\text{H}_5)_3\text{C}^{\oplus}\text{BF}_9^{\ominus}}$$

(K. Oda, A. Sakamoto, R. Miyatake, and S. Kuroda, *Tetrahedron Lett.*, *39*, 6195 [1998])

(h)

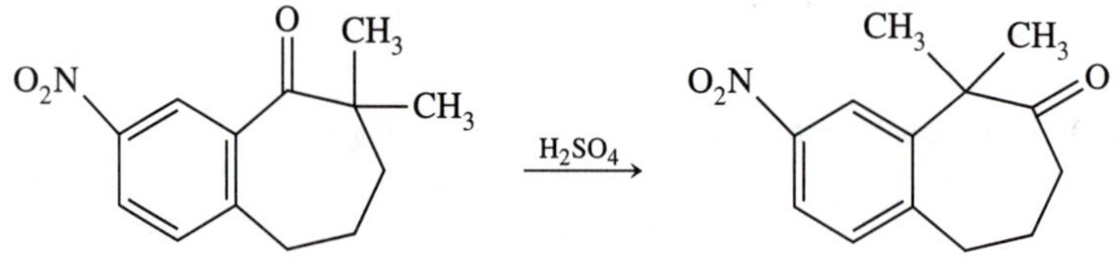

$$\xrightarrow{\text{H}_2\text{SO}_4}$$

Suggest a mechanism that does *not* involve methyl migrations.

(J.C. Garcia, J.D. Enas, F.R. Fronczen and H.F. van Brocklin, *J. Org. Chem.*, *59*, 8299 [1994])

(i)

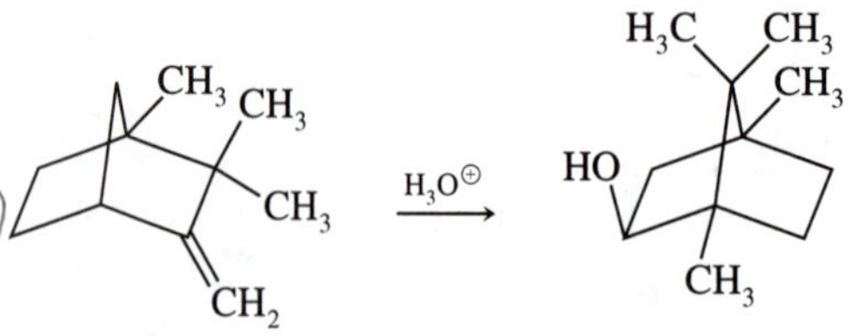

$$\xrightarrow{\text{H}_3\text{O}^{\ominus}}$$

(j)

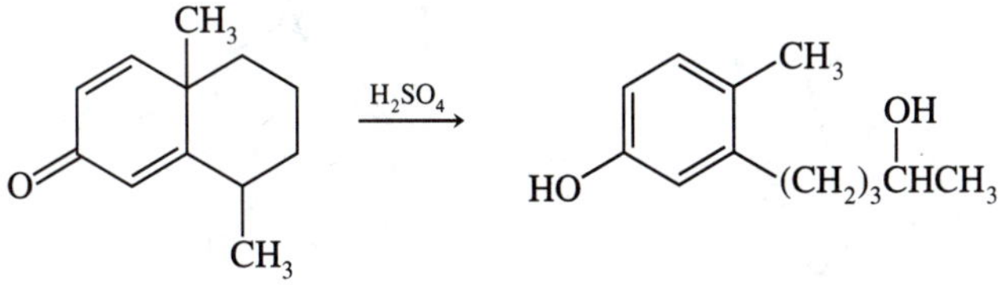

(P.J. Kropp, *J. Am. Chem. Soc.*, *85*, 3280 [1963])

(k)

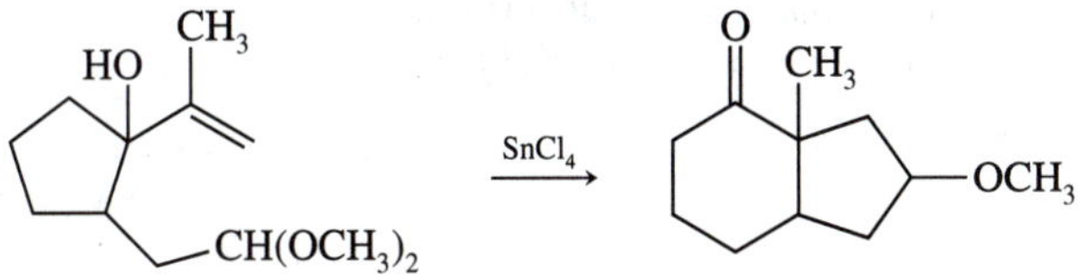

(G.C. Hirst, P.N. Howard, and L.E. Overman, *J. Am. Chem. Soc.*, *111*, 1514 [1989])

(l)

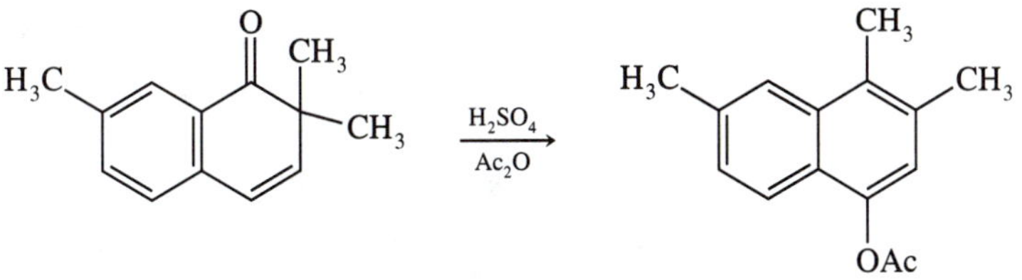

(E.N. Marvell and E. Magoon, *J. Am. Chem. Soc.*, *76*, 5118 [1954])

(m)

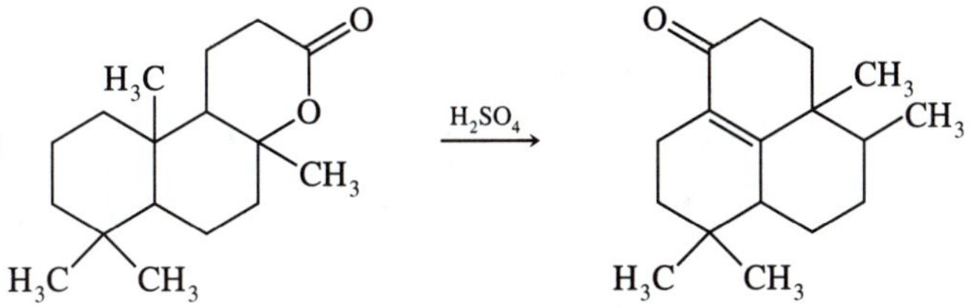

(G. Büchi, W.S. Saar, and A. Eschenmoser, *Experientia*, *12*, 136 [1956])

(n)

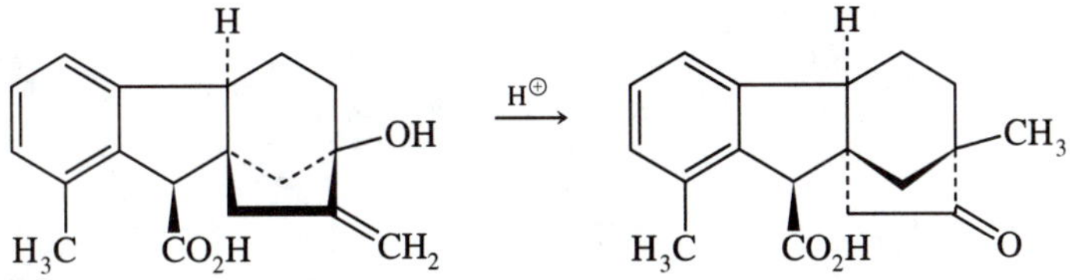

(A.J. Birch et al., *Chem. & Ind. (London)*, 401 [1960])

(o)

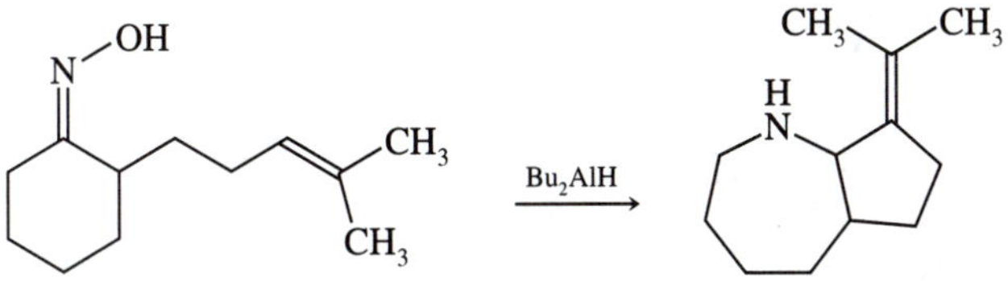

(Adapted from D. Schinzer and Y. Bo, *Angew. Chem. Intl. Ed. Engl.*, *30*, 687 [1991])

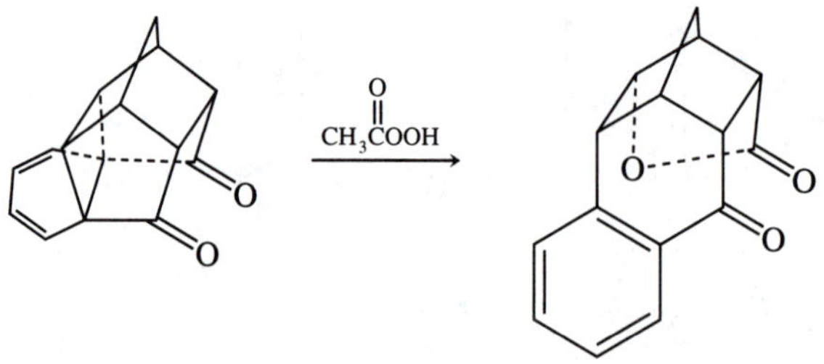

(G. Mento and V. Singh, *Tetrahedron Lett.*, 4591 [1978]
B. Pandey and P.V. Dalvi, *J. Org. Chem.*, *54*, 2968 [1989])

Neighboring Group Effects and "Nonclassical" Cations

7.1 SUBSTITUTION WITH RETENTION OF CONFIGURATION

S_N2 reactions at chiral carbons yield products with inverted configurations.* Even in S_N1 reactions, which normally yield at least partially racemized products, it is usually easiest for the nucleophile to attack from the direction opposite that by which the leaving group is departing. Thus, the percentage of S_N1 products formed with inversion of configuration should always be at least as great as the percentage formed with retention.[†]

While most nucleophilic substitution reactions, whether they proceed by S_N1 or S_N2 mechanisms, thus result predominantly in inversion of configuration, that is not necessarily the case if the β-carbon—the carbon bonded to the carbon bearing the leaving group—has a substituent with unshared electron pairs. In such cases, nucleophilic substitution reactions frequently proceed entirely, or almost entirely, with *retention* of configuration! For example, the

---

*Nucleophilic substitution reactions proceeding with retention may be viewed as pericyclic reactions. As suprafacial four-electron processes, they would be forbidden by the Woodward–Hoffmann rules.

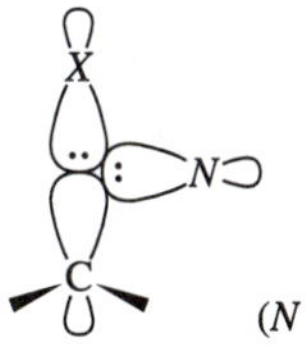

(N = nucleophile, X = leaving group)

A forbidden "cyclic" transition state for nucleophilic
substitution with retention

In contrast, electrophilic substitution reactions, which can be regarded as pericyclic reactions of two-electron systems, frequently proceed with retention of configuration, in spite of possible unfavorable steric effects.

[†]The term *racemization* refers here to processes resulting in both inversion and retention at the chiral center at which substitution is taking place. Of course, the products of a racemization process may be chiral if the starting materials contain more than one chiral center.

reaction of *trans*-2-iodocyclohexanol with hydrochloric acid yields *trans*-1-chloro-2-iodocyclo-hexane (Eq. 1). None of the *cis*-isomer is obtained.[1]

$$
\begin{array}{ccc}
\text{I} & & \text{I} \\
\text{(cyclohexane ring) ··OH} & \xrightarrow{\text{HCl}} & \text{(cyclohexane ring) ··Cl}
\end{array}
\tag{1}
$$

Saul Winstein and H.J. Lucas at UCLA demonstrated that nucleophilic substitutions in acyclic molecules can also proceed with retention of configuration if the β-carbons bear halogen atoms. Reaction of the chiral bromohydrin **1** with fuming hydrobromic acid, for instance, yield-ed only *meso*-2,3-dibromobutane, the product of substitution with retention.[2]

$$
\begin{array}{ccc}
\underset{\text{H}_3\text{C}}{\overset{\text{Br}}{\underset{\text{H}}{\text{C}}}}-\underset{\text{OH}}{\overset{\text{CH}_3}{\underset{}{\text{C}}}}\text{H} & \xrightarrow{\text{HBr}} & \underset{\text{H}_3\text{C}}{\overset{\text{Br}}{\underset{\text{H}}{\text{C}}}}-\underset{\text{Br}}{\overset{\text{CH}_3}{\underset{}{\text{C}}}}\text{H}
\end{array}
\tag{2}
$$

$$
\textbf{1} \qquad\qquad\qquad \textit{meso}\text{-2,3-dibromobutane}
$$

Reaction of fuming hydrobromic acid with bromohydrin **2** (a diasteromer of **1**)* yielded none of the *meso* dibromide. Instead, **2** was converted into *racemic* 2,3-dibromobutane (that is, into an equimolar mixture of 2**R**,3**R**- and 2**S**,3**S**-dibromobutane).[2]

$$
\underset{\text{H}_3\text{C}}{\overset{\text{Br}}{\underset{\text{H}}{\text{C}}}}-\underset{\text{OH}}{\overset{\text{H}}{\underset{}{\text{C}}}}\text{CH}_3 \xrightarrow{\text{HBr}} \underset{\text{H}_3\text{C}}{\overset{\text{Br}}{\underset{\text{H}}{\text{C}}}}-\underset{\text{Br}}{\overset{\text{H}}{\underset{}{\text{C}}}}\text{CH}_3 \; + \; \underset{\text{Br}}{\overset{\text{H}_3\text{C}}{\underset{}{\text{C}}}}-\underset{\text{CH}_3}{\overset{\text{Br}}{\underset{}{\text{C}}}}\text{H}
\tag{3}
$$

$$
\textbf{2}
$$

That is indeed a remarkable result. Not only did the attack at C2 (the carbon bearing the hydroxy group) apparently proceed with exactly 50% inversion and 50% retention, but C3 must also have been completely racemized during the reaction.

7.2   CYCLIC HALONIUM IONS

To explain their results, Winstein and Lucas proposed that the bromine atoms in **1** and **2** act as nucleophiles, displacing leaving groups from the neighboring carbons to form "cyclic bromo-nium ions," as shown in Eqs. 4 and 5.

$$
\underset{\text{H}_3\text{C}}{\overset{:\ddot{\text{B}}\text{r}:}{\underset{\text{H}}{\text{C}}}}-\underset{\overset{\oplus}{\text{OH}_2}}{\overset{\text{CH}_3}{\underset{}{\text{C}}}}\text{H} \longrightarrow \underset{\text{H}_3\text{C}}{\overset{\overset{\oplus}{\text{Br}}}{\underset{}{\text{C}}}}\text{—}\underset{\text{H}}{\overset{}{\underset{}{\text{C}}}}\text{CH}_3 \underset{\underset{\ominus}{\text{Br}}\text{ or }\underset{\ominus}{\text{Br}}}{} \longrightarrow \begin{array}{c}\textbf{R,S}\text{-2,3-dibromobutane} \\ (\textit{meso}\text{ structure})\end{array}
\tag{4}
$$

$$
\textbf{1}
$$

---

*In the original papers, bromohydrin **1** was described as having the *threo* configuration, and bromohydrin **2** the *erythro* configuration, by analogy with the structures of the four-carbon sugars threose and erythrose.

$$\text{2} \longrightarrow \text{cyclic bromonium ion} \longrightarrow \textbf{R,R and S,S-2,3-dibromobutane} \tag{5}$$

(The existence of cyclic bromonium ions had previously been postulated to explain the stereospecific *anti* addition of bromine to alkenes.[3]) Each cyclic bromonium ion then undergoes nucleophilic attack by bromide ions to form the observed dibromides.

The cyclic bromonium ions from **1** and **2** are each formed stereospecifically with inversion, since they are formed by intramolecular processes resembling S_N2 reactions. The reactions of the bromonium ions with nucleophiles (bromide ions) also result in inversion of the geometries of the carbons being attacked. In reactions with achiral reagents, each end of either cyclic bromonium ion is indistinguishable from the other end, so exactly 50% of each ring-opening reaction takes place at each end. The result is that the bromonium ion from **1** yields solely *meso*-2,3-dibromobutane, while the bromonium ion from **2** yields exactly equal amounts of 2**R**,3**R**- and 2**S**,3**S**-2,3-dibromobutane.

It is important to understand that there are no steps in the mechanisms shown in Eqs. 4 and 5 in which displacements proceed with retention of configuration. The overall retention of configuration in each case occurs as the result of two successive displacements with inversion.

Iodine atoms are better nucleophiles than bromine atoms, so cyclic iodonium ions are formed even more easily than cyclic bromonium ions. (Reaction 1, for example, proceeds via a cyclic iodonium ion.) In contrast, chlorine atoms are poor nucleophiles. The formation of cyclic chloronium ions is therefore relatively difficult, though not impossible. For example, 2S,3R-3-chloro-2-butanol (**3**) does not react with fuming hydrochloric or hydrobromic acid. However, thionyl chloride, a more powerful reagent, converts **3** into *meso*-2,3-dichlorobutane (Eq. 6).

$$\text{3} \xrightarrow{\text{SOCl}_2} \text{cyclic chloronium ion} \longrightarrow \textit{meso}\text{-2,3-dichlorobutane} \tag{6}$$

Similarly, 2**R**,3**R**-3-chloro-2-butanol is converted to the racemic dichloride. These reactions undoubtedly proceed by the stereospecific formation and ring opening of cyclic chloronium ions.[4]

Although cyclic halonium ions are only short-lived intermediates in reactions 1 through 6, they have much longer lifetimes in less nucleophilic solvents, such as superacids. Olah and his group have demonstrated this and, have recorded the NMR spectra of cyclic bromonium, chloronium, and iodonium ions in superacid solutions.[5]

Nucleophilic substitution with retention of configuration may also occur if other atoms that may act as nucleophiles, such as oxygen atoms in alkoxy groups, are located on carbons linked to carbons bearing leaving groups. This is illustrated in Eq. 7.[6]

$$\xrightarrow[\text{AgOAc}]{\text{HOAc}} \qquad + \qquad \tag{7}$$

7.3   SULFUR AND NITROGEN MUSTARDS

Anchimeric Assistance

Winstein described the participation of a nucleophilic atom in displacement of a leaving group on an adjacent carbon as a "neighboring group effect."[7] As we've seen, one common result of a neighboring group effect is substitution with retention of configuration. Another common result is a greatly increased rate of reaction. Such accelerated rates are most dramatically evident in reactions of *sulfur and nitrogen mustards*.

During World War I, "mustard gas," which is actually a liquid aerosol rather than a gas, caused thousands of deaths and cases of blindness. The active component of mustard gas is *bis*-2,2'-(chloroethyl)sulfide. The toxic effects of this compound are due to the fact that it reacts remarkably rapidly with nucleophiles, including nucleophiles in the human body. Other 2-haloethyl sulfides, generically known as sulfur mustards, have similar properties.

$$Cl-CH_2CH_2-S-CH_2CH_2-Cl \qquad\qquad R-S-CH_2CH_2-X$$

$$\text{mustard gas} \qquad\qquad\qquad \text{a sulfur mustard}$$

In water, the hydrolysis of mustard gas, as well as that of other sulfur mustards, proceeds with first-order kinetics. The rate, which is unusually rapid, is not affected by the addition of hydroxide ions or of most other nucleophiles, even when the nucleophiles replace the leaving groups in the reacting molecule. That is very surprising behavior for a primary halide.

To explain the unusual behavior of sulfur mustards, it was suggested that the initial reaction is displacement of the leaving group by the neighboring sulfur atom, forming a three-membered cyclic sulfonium ion (Eq. 8).

$$\text{(8)}$$

$$RSCH_2CH_2OH + HCl \quad \longleftarrow \quad RSCH_2CH_2\overset{\oplus}{O}H_2 \quad Cl^{\ominus}$$

Like cyclic halonium ions, the cyclic sulfonium ion reacts extremely rapidly with even quite weak nucleophiles. Thus, the initial displacement by the neighboring sulfur atom—a first-order process—is the rate-determining step.[8]

Sulfur atoms are good nucleophiles, and their intramolecular displacements of chloride ions, or of other good leaving groups, usually take place more rapidly than displacements by external nucleophiles. (In general, intramolecular substitution reactions have more favorable entropies of activation than intermolecular reactions.[9]) The acceleration of rates of nucleophilic displacement reactions due to neighboring group participation has been described as due to *anchimeric assistance* (from the Greek *anchi*, "neighbor").[10]

Like other displacement reactions involving neighboring groups, the substitution reactions of sulfur mustards proceed with overall retention of configuration, as in Eq. 9a.[11]

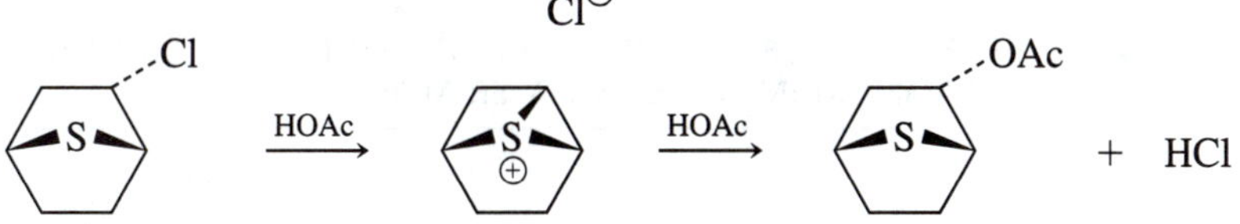

$$\tag{9a}$$

If the nucleophile in the second step of the reaction of a sulfur mustard does not attack the same carbon atom as that which originally held the leaving group, the reaction can result in the migration of the sulfur atom, as shown in Eq. 9b.[12]

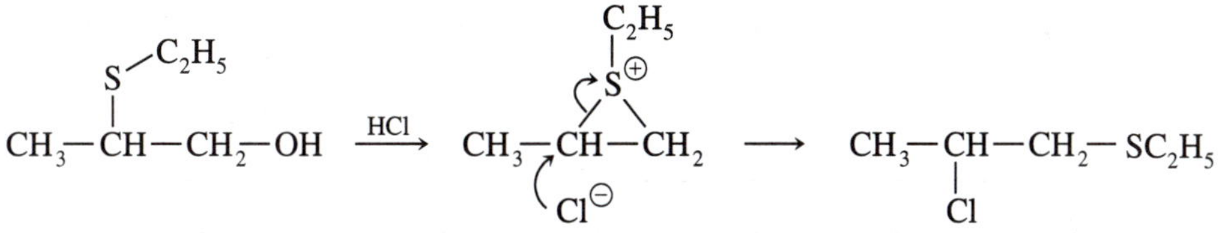

$$\tag{9b}$$

Like their sulfur analogs, primary alkyl halides with dialkylamino groups on β-carbons (nitrogen mustards) undergo rapid intramolecular reactions to form charged, three-membered rings, as shown in Eq. 10. However, in contrast to cyclic bromonium or sulfonium ions, the intermediate cyclic ammonium ions are relative stable and may not react very rapidly with weak nucleophiles, such as water.*

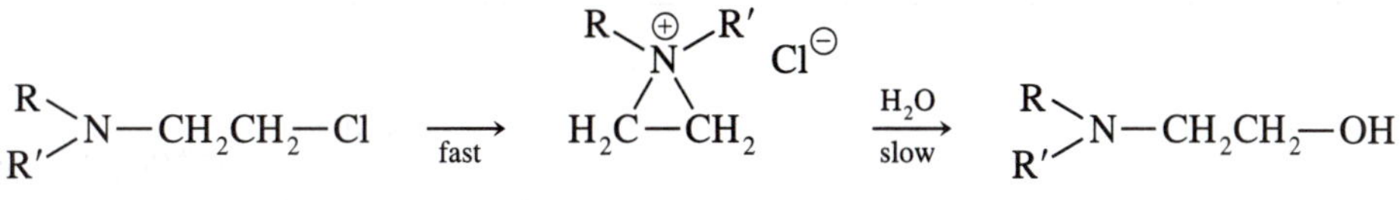

$$\tag{10}$$

Like sulfur mustards, nitrogen mustards often undergo rapid rearrangement reactions.[14]

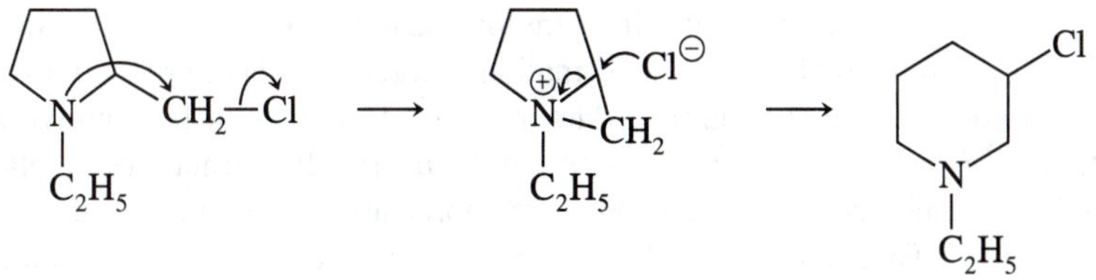

$$\tag{11}$$

---

*Thus, in the absence of powerful nucleophiles, the reactions of nitrogen mustards may exhibit complex kinetics. In contrast, when high concentrations of good nucleophiles are present, they can react rapidly with the cyclic ammonium salts, leaving the intramolecular displacements as the rate-limiting steps.[13] An interesting paradox: nitrogen mustards can undergo nucleophilic displacement reactions in which the addition of good nucleophiles makes it more likely that first-order kinetics will be observed.

Effects of Ring Size

It is interesting to compare the effects of ring size on the rates of formation of cyclic sulfonium and of cyclic ammonium salts (Tables 7.1 and 7.2).

TABLE 7.1   EFFECTS OF RING SIZE ON RATES OF FORMATION OF CYCLIC AMMONIUM SALTS (IN WATER AT 25°C)[15]

$$Br-(CH_2)_n-NH_2 \longrightarrow (CH_2)_n \overset{\oplus}{N}H_2$$

n	Number of atoms in ring	Rate relative to formation of 3-membered ring
2	3	1
3	4	0.014
4	5	833
5	6	14
6	7	0.027

TABLE 7.2   EFFECTS OF RING SIZE ON RATES OF FORMATION OF CYCLIC SULFONIUM SALTS (IN 20% AQUEOUS DIOXANE AT 100°C)[16,17]

$$Cl-(CH_2)_n-S-C_6H_5 \longrightarrow (CH_2)_n \overset{\oplus}{S}-C_6H_5$$

n	Number of atoms in ring	Rate relative to formation of 3-membered ring
2	3	1.0000
3	4	0.0053
4	5	0.2000
5	6	0.0170*
1-chlorohexane		0.0096

*Extrapolated from rate in 50% aqueous acetone at 80°C.

Three-membered ammonium rings are formed more rapidly than four- or seven-membered rings. This is presumably because the formation of three-membered rings freezes out the motions of fewer atoms than the formation of larger rings and results in less unfavorable entropies of activation.[9,10] However, the enthalpies of activation for the formation of relatively strain-free five- and six-membered ammonium rings are appreciably lower than for the formation of three-membered rings. Thus, the five- and six-membered rings are formed much faster than the three-membered ammonium salts.

In contrast, three-membered sulfonium rings are actually formed faster than sulfonium rings of any other size! This appears to be due, at least in part, to the fact that normal bond angles of divalent sulfur atoms are much smaller than bond angles of trivalent nitrogen atoms, as is illustrated on page 189. Therefore, less distortion of bond angles is required to form three-membered rings containing sulfur atoms than is needed to form three-membered rings containing trivalent nitrogen atoms.

$$H \overset{O}{\underset{100°}{\diagdown}} H \qquad H \overset{\overset{\displaystyle H}{|}}{\underset{106°}{\overset{N}{\diagdown}}} H \qquad H \overset{S}{\underset{92°}{\diagdown}} H$$

$$H_3C \overset{O}{\underset{111°}{\diagdown}} CH_3 \qquad H_3C \overset{\overset{\displaystyle CH_3}{|}}{\underset{109°}{\overset{N}{\diagdown}}} CH_3 \qquad H_3C \overset{S}{\underset{100°}{\diagdown}} CH_3$$

Bond angles in molecules containing oxygen, nitrogen, and sulfur atoms[18]

7.4   *TRANS/CIS* RATE RATIOS

We've seen that the rates of nucleophilic substitution reactions can be greatly increased by neighboring sulfur and nitrogen atoms. That is not necessarily true for neighboring halogen atoms.

The effects of anchimeric assistance by halogen atoms can be demonstrated by the data in Table 7.3, which compare the solvolysis rates of *cis*- and *trans*-2-halocyclohexyl brosylates.

TABLE 7.3   RELATIVE RATES OF REACTIONS OF CYCLOHEXYL BROSYLATES (IN ACETIC ACID AT 75°C)[19]

	Relative rates of reaction	k_{trans}/k_{cis}
$X = H$	1	
$X = trans$-Cl	4.8×10^{-4}	
cis-Cl	1.3×10^{-4}	4
$X = trans$-Br	0.10	
cis-Br	1.2×10^{-4}	800
$X = trans$-I	$1.2 \times 10^{3*}$	
cis-I	$4.3 \times 10^{-4†}$	3×10^{6}

*At 24°C.

†Calculated from data from Andrew Streitwieser, Jr., *Solvolytic Displacement Reactions.* McGraw Hill, New York (1962), p. 121.

Although *trans*-2-iodocyclohexyl brosylate* *does* undergo S_N1 reactions much more rapidly than cyclohexyl brosylate, *trans*-2-bromocyclohexyl brosylate undergoes solvolysis in acetic acid at only one-tenth the rate of cyclohexyl brosylate, and *trans*-2-chlorocyclohexyl brosylate at only about one-thousandth the rate of cyclohexyl brosylate.[19] However, electron-withdrawing substituents, such as halogen atoms, markedly slow the rates of S_N1 (and of S_N2) reactions. The effects of anchimeric assistance by halogen atoms are still appreciable, even though they only partially offset the inductive effects of the halogen atoms.

---

*The term *brosylate* (abbreviated as OBs) is a shorthand way of referring to the *p*-bromobenzenesulfonate group. Similar terms are used to refer to the *p*-toluenesulfonate and *p*-nitrobenzenesulfonate groups.

ethyl brosylate
(EtOBs)

methyl tosylate
(MeOTs)

n-propyl nosylate
(*n*-PrONs)

In each pair of halides, the *trans* isomer, in which anchimeric assistance by the halogen atom is possible, reacts more rapidly than the *cis* isomer, in which anchimeric assistance is not possible. The difference in rates between the two isomers is small for 2-chlorocyclohexyl brosylate, indicating that the chlorine atom is not an effective neighboring group. However, the rate difference is much larger for a neighboring bromine and is truly impressive for a neighboring iodine atom.*

Whether in open-chain or cyclic molecules, neighboring group effects will be most evident when the leaving group and the neighboring group can easily be placed in a *trans*, coplanar arrangement. For instance, the steroid derivative **4a**, in which both bromine atoms are in axial positions, reacts rapidly when heated in acetic acid. In contrast, its isomer **4b** (which is formed from **4a** on heating, presumably via an intermediate cyclic bromonium salt, as shown in Eq. 12) is quite unreactive under the same conditions.[21]

(12)

In summary, neighboring group effects can result in several changes from the normal course of nucleophilic substitution reactions:

1. Displacements involving participation by neighboring groups proceed with overall retention of configurations.
2. Effective neighboring groups, such as sulfur, nitrogen, and iodine atoms, can greatly increase the rates of displacement reactions.
3. Cyclic compounds in which neighboring groups are *trans* to leaving groups often react much more rapidly than do their stereoisomers in which the neighboring groups are *cis* to the leaving groups.
4. Displacements involving neighboring groups often result in migrations of the neighboring groups.

7.5 NEIGHBORING ACETOXY GROUPS

When *trans*-2-acetoxycyclohexyl tosylate (**5**) is heated in acetic acid containing potassium acetate *trans*-1,2-diacetoxycyclohexane (**7**), the product of substitution with retention is formed. This demonstrates that the acetoxy group can participate as a neighboring group.

---

*In assigning the differences in rates between *cis* and *trans* isomers to anchimeric assistance, it is assumed that the inductive effects of *cis* and *trans* halogen atoms will be comparable.

However, when the reaction is carried out in acetic acid containing a small amount of water, pure *cis*-2-acetoxycyclohexanol is obtained.[7] *cis*-2-Acetoxycyclohexanol is also obtained if **5** is heated in acetic acid containing ethanol and the reaction then worked up by the addition of water.[22]

To explain these facts, Winstein proposed that the displacement of tosylate by the neighboring acetoxy group yields the cyclic five-membered acetoxonium ion **6** rather than a three-membered ring. He suggested that acetate anions attack cyclohexyl ring carbons of **6** to yield the double-inversion product **7**. However, water adds to the "carboxyl" carbon of **6** to form **8**, which then opens to form *cis*-2-acetoxycyclohexanol. Ethanol similarly adds to **6** to form **9**, which can be isolated on very careful workup.[23] Addition of water converts **9** to **8** and then to *cis*-2-acetoxycyclohexanol.

(13)

7.6   CYCLIC PHENONIUM IONS

Participation by Aromatic Rings

The demonstration that displacements of leaving groups can be assisted by neighboring groups with unshared electron pairs led to attempts to determine whether neighboring carbon atoms, though lacking unshared electrons, can participate in displacement reactions.

Donald Cram* at UCLA showed that 2**S**,3**R**-3-phenyl-2-butyl tosylate (**10**) yielded largely racemic 3-phenyl-2-butyl acetate on heating in acetic acid solution. In contrast, the 2**R**,3**R**-diastereomer **11** was converted to optically active 3-phenyl-2-butyl acetate.

These results can be explained by assuming the formation of cyclic phenonium ions as intermediates. The phenonium ion obtained from **10** would have a plane of symmetry and would yield equal amounts of two enantiomeric acetates. In contrast, the phenonium ion obtained from **11** would not have a plane of symmetry. Attack by solvent at either end of the three-membered ring would yield an acetate with the same geometry as the starting tosylate.[24,25]

---

*Donald James Cram (born in 1919) shared the 1987 Nobel Prize with Charles J. Pedersen of the DuPont Corporation and Jean-Marie Lehn of the University of Strasbourg, France, for their work on crown ethers and host–guest chemistry.

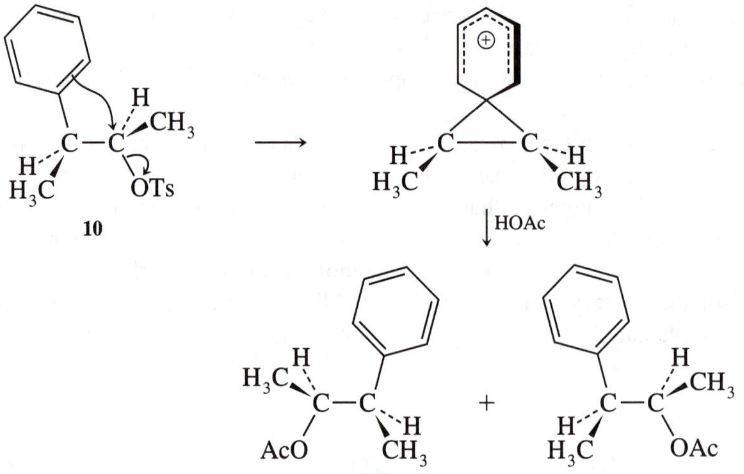

(14a)

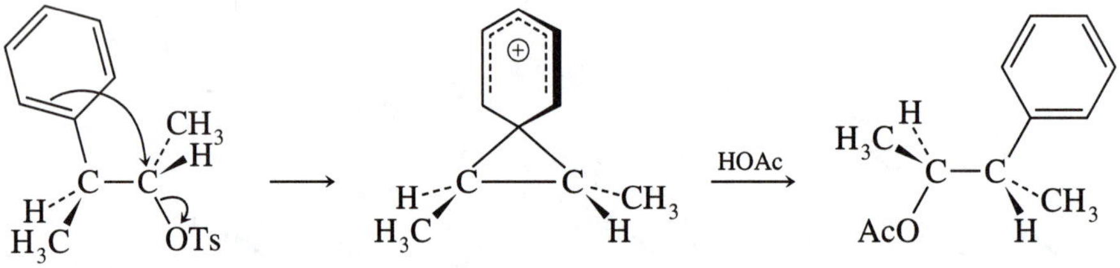

(14b)

Cram also found that if the reactions were stopped before they were complete and unreacted **10** and **11** were recovered, **10** was recovered in a completely racemic form (Eq. 14c), while tosylate **11** had retained its optical activity.

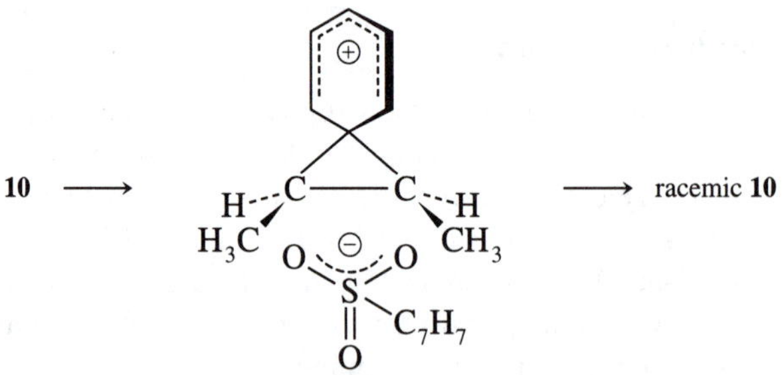

These results could be accounted for if the phenonium ion and the tosylate ion in each case stayed together as an *intimate ion pair*. The tosylate ions would be in a position to attack the three-membered rings before solvent molecules could intervene. (Winstein later showed that "internal return" by tosylate ions was about four times as fast as attack by acetic acid.[26])

    Cyclic, three-membered phenonium ions can now be prepared, and their spectra recorded, in superacid solutions. Their spectra are consistent with those expected of cyclopropane derivatives.[27]

Reactions resulting in the formation of three-membered phenonium ions can be classified as *intramolecular Friedel–Crafts reactions*. However, in most Friedel–Crafts reactions, nucleophiles can abstract protons from the phenonium ions to reform aromatic rings, as illustrated in Eq. 15.

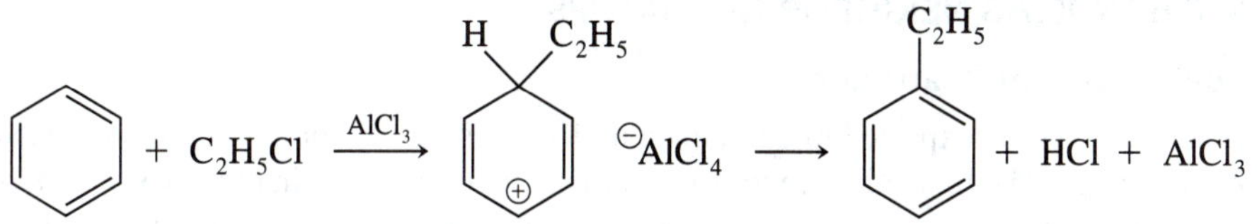

$$(15)$$

A cyclic phenonium ion cannot undergo loss of a proton from the six-membered ring. Instead, the three-membered ring is attacked by a nucleophile, with the aromatic ring acting as a leaving group.

Migrations of Aryl Groups

Except when steric factors interfere, phenyl groups and other aryl groups are usually much better migrators in rearrangements of carbocations than are alkyl groups. The migrations of aryl groups usually yield rearrangement products with inverted configurations at both the migration origins and the migration termini, as illustrated in Eq. 16.[25]

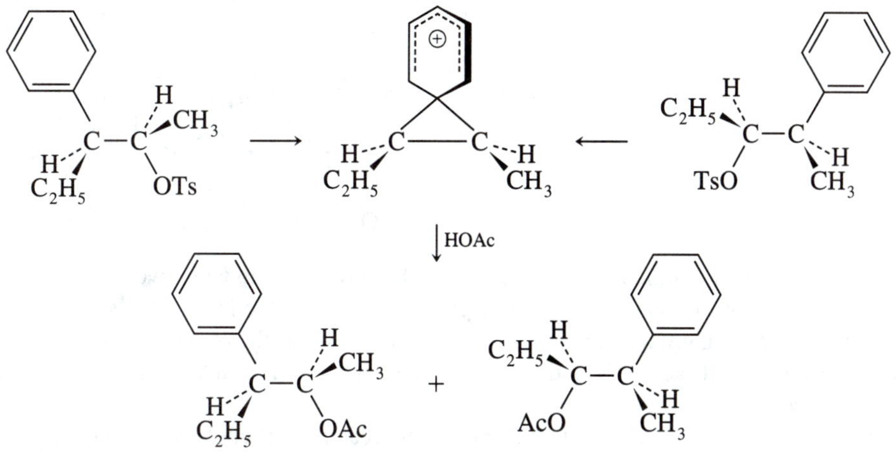

$$(16)$$

This demonstrates that migrations of aryl groups proceed by way of intermediate "arylonium" ions. In contrast, migrations of alkyl groups in open-chain molecules usually result in appreciable racemization at both migration termini and migration origins.

Studies of competitive migrations of phenyl groups and substituted phenyl groups in pinacol rearrangements (Eq. 17) show that electron-donating substituents increase the migratory aptitudes of the substituted rings, and electron-withdrawing substituents decrease their migratory aptitudes.[29] Hammett plots show excellent correlations between relative rates of migration and σ^+ constants, as would be expected if the rates of migrations of the phenyl groups depend on the rates at which they undergo electrophilic attack.

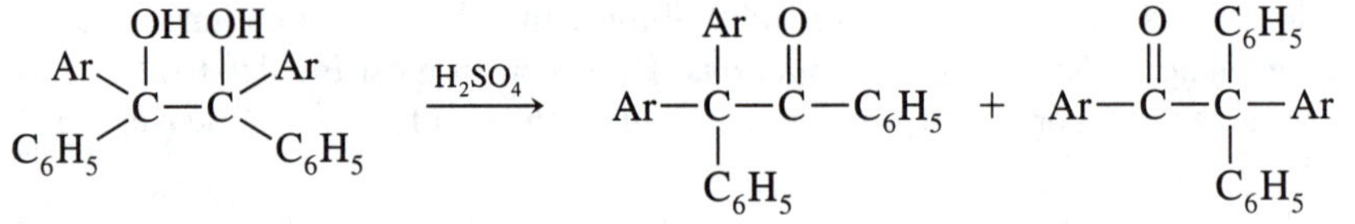

$$\underset{\underset{C_6H_5}{|}}{\overset{\overset{OH}{|}}{Ar}}\underset{}{\overset{}{C}}-\underset{\underset{C_6H_5}{}}{\overset{\overset{OH}{|}}{C}}\overset{}{Ar} \xrightarrow{H_2SO_4} \underset{\underset{C_6H_5}{|}}{\overset{\overset{Ar}{|}}{Ar}-\overset{}{C}}-\overset{\overset{O}{||}}{C}-C_6H_5 \;+\; \underset{\underset{C_6H_5}{|}}{\overset{\overset{O}{||}}{Ar}-\overset{}{C}}-\overset{\overset{C_6H_5}{|}}{C}-Ar \qquad (17)$$

7.7   DOUBLE BONDS AS NEIGHBORING GROUPS

The *Iso*-Cholesterol Rearrangement

Double bonds would be expected to participate in displacement reactions even more effectively than aromatic rings. The first evidence for such participation was found in reactions of cholesterol and its derivatives. Cholesteryl tosylate (or cholesteryl chloride or bromide) reacts with buffered methanol solutions to form a rearranged methyl ether containing a three-membered ring (Eq. 18).

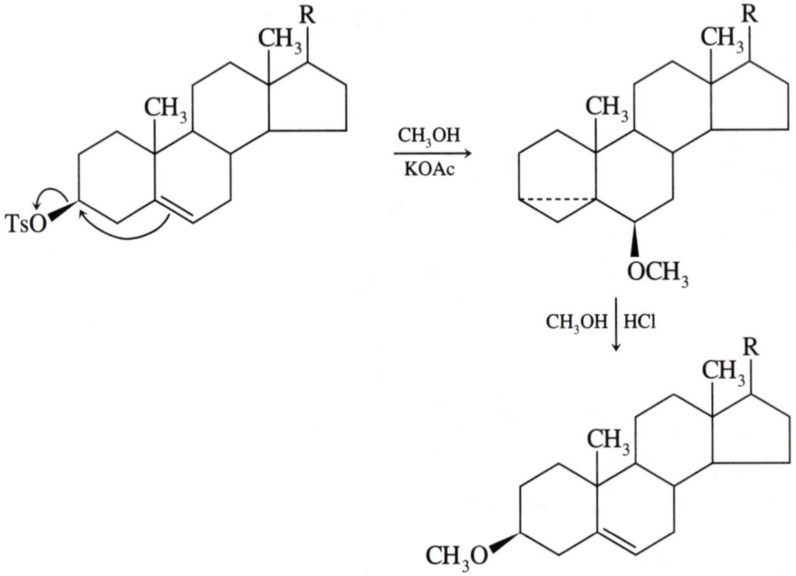

$$(18)$$

The cyclopropyl ring is formed by displacement of the leaving group with inversion of the geometry of the chiral carbon, while the methoxy group enters from the face of the starting material from which the leaving group departed. The process is known as the *iso-cholesterol* (or *i-cholesterol*) *rearrangement*. If acid is added, the *iso*-cholesteryl ether is converted back into a cholesteryl derivative.[30]

Winstein and Adams showed that displacement reactions of cholesteryl tosylate are about 100 times as rapid as reactions of cyclohexyl tosylate. They suggested that the double bond participates in displacement of the leaving group to form an *allylcarbinyl* (or *homoallylic*) cation, in which the charge is distributed between the two ends of the cation (Eq. 19).[31]

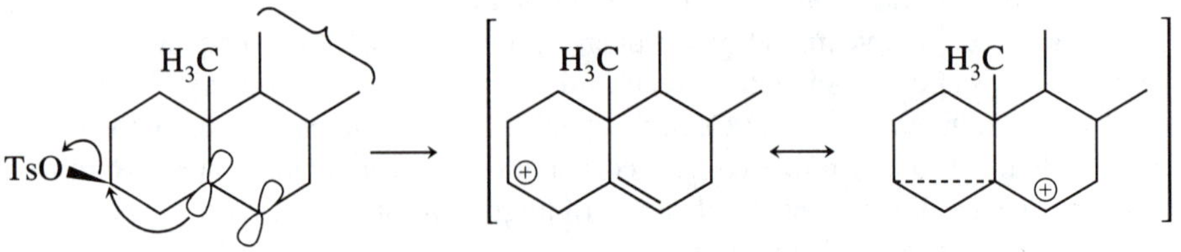

$$(19)$$

Formation of an allylcarbinyl cation

In contrast to the rapid reaction of cholesterol tosylate, no obvious rate enhancement was observed in displacement reactions of 4-chloro-1-butene, in which the charge in an intermediate allylcarbinyl cation would be distributed between two primary carbons.[32] Tosylate **12**, on the other hand, was found to react about 1200 times as rapidly as ethyl tosylate. The reaction yielded a cyclopropane derivative as well as the simple displacement product.[33]

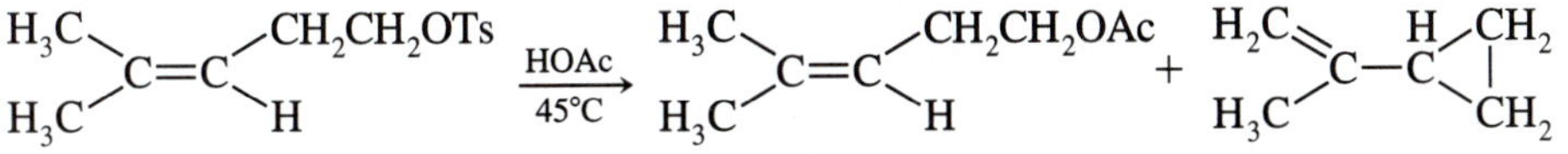

$$(20)$$

7-Norbornenyl Cations

The largest rate enhancement caused by a neighboring double bond is observed in the solvolysis of *anti*-7-norbornenyl tosylate (**13**),* which reacts 10^7 times as rapidly as its *syn* isomer, **14**,[34] and 10^{11} times as rapidly as 7-norbornyl tosylate (**15**).[35†] Tosylate **13** reacts stereospecifically, with retention of configuration.

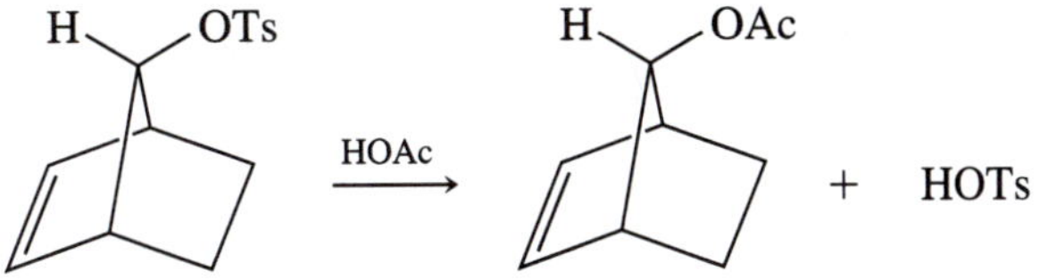

$$(21)$$

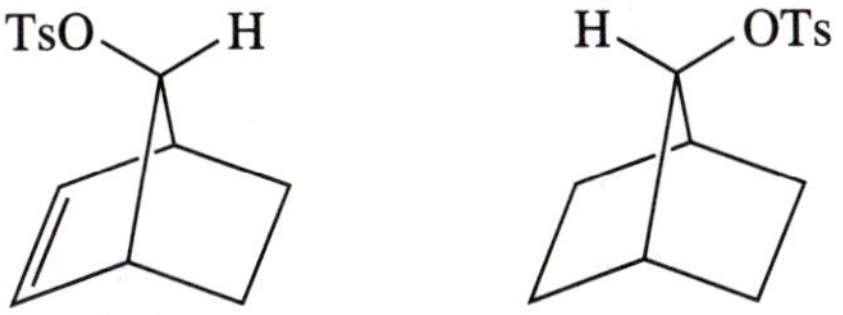

---

*The trivial names *norbornane* and *norbornene* are derived from the name of the terpene derivative *bornane*. They refer to molecules with the same bicyclic ring structure as bornane but lacking three methyl groups found in bornane.

In the IUPAC system, Compounds **13** to **15** are named as derivatives of bicyclo[2.2.1]heptane. In names of bicyclic molecules, the numbers in brackets specify the number of atoms in each "bridge" linking the "bridgehead" atoms, which are common to two or more rings. The number of atoms in the largest bridge is specified first, then the number of atoms in the next largest bridge, and finally in the smallest bridge. In identifying the atoms, a bridgehead atom is labeled atom 1, and numbering is continued along the largest bridge to the other bridgehead, then along the next largest bridge, and so on. Two examples are illustrated here:

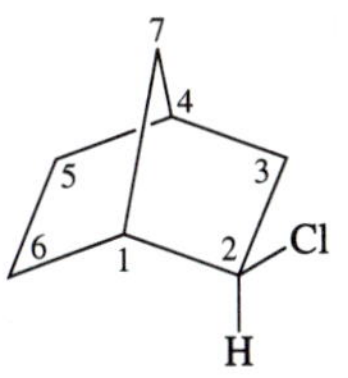

exo-2-chlorobicyclo[2.2.1]heptane 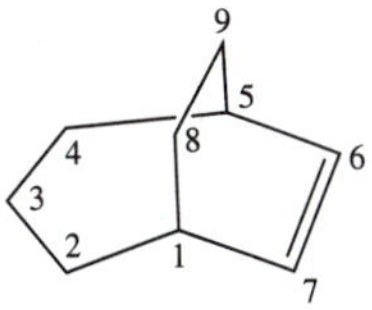   Δ-6-bicyclo[3.2.2]nonene

†7-Norbornyl sulfonates and halides are less reactive than other secondary alkyl halides and sulfonates, presumably because the rigid ring structure compresses the C–C–C bond angle at C7. This inhibits the normal expansion of the bond angle that occurs when a tetrahedral carbon is converted to a planar carbocation. The increased reactivity of **14** compared to **15** has been attributed to participation of C6 in the ionization process, but it seems more likely that it results from the fact that the double bond in **14** spreads apart the "sides" of the ring, thus increasing the bond angle at C7 compared to **15**.

Winstein and his coworkers proposed that the "nonclassical" ion **16a** was formed as an intermediate in reactions of *anti*-7-norbornenyl derivatives.

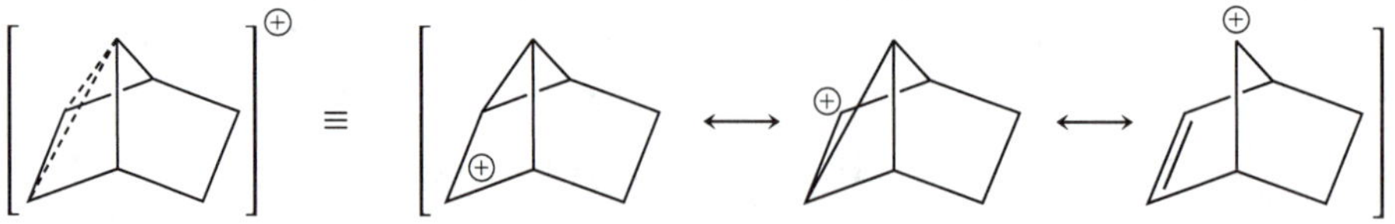

(22)

16a

That a carbocation having some "cyclopropane-like" characteristics is formed is supported by the fact that reactions of **13** in basic media can yield products containing cyclopropane rings, as shown in Eq. 23.[36]

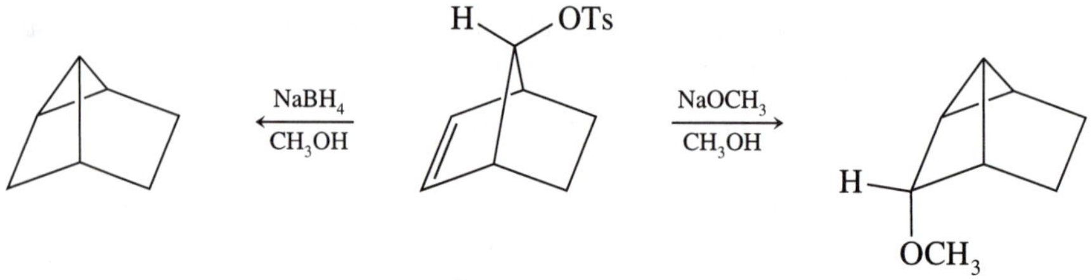

(23)

Methoxide ions attack the carbocation exclusively from the *endo* side of the ring, although substitution reactions of norbornyl derivatives are usually much faster from the *exo* side than from the *endo* side (see page 203).

Several factors may contribute to the remarkably rapid reaction rates of *anti*-7-norbornenyl derivatives. For one thing, the developing cation at C7 of the 7-norbornenyl system can interact with the center of the double bond, which is the region of greatest electron density. In a typical homoallylic cation, in contrast, the cationic carbon can interact with only one end of the double bond. In addition, in norbornenyl ring systems, the five-membered rings are distorted far from planarity, so that C7 is relatively close to the double bond. In comparison, 4-bromocyclopentene (**17**), in which the ring is nearly planar, appears to exhibit no special rate acceleration on heating in water.[37]

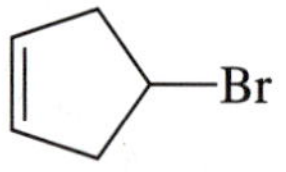

17

(However, the chiral tosylate **18** does react with retention of configuration in buffered formic acid solution.[38])

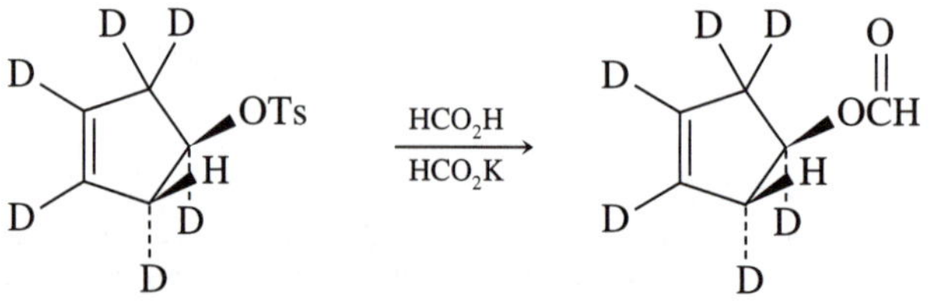

(24)

18

Finally, it should be noted that the cyclic array of three *p* orbitals containing two electrons in the 7-norbornenyl cation can be regarded as a "bishomoaromatic" ring system. If the 7-norbornenyl cation does indeed have some degree of aromatic stability, the transition states in reactions of *anti*-7-norbornenyl tosylates should share part of that stabilization energy.

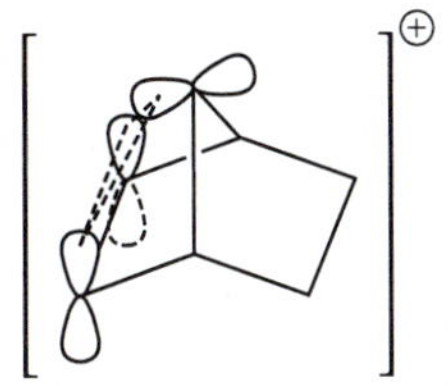

A cyclic array of electrons in the 7-norbornenyl cation

The suggestion that the 7-norbornenyl cation has a highly delocalized, "nonclassical" structure was not universally accepted. H.C. Brown suggested that it might actually be a rapidly equilibrating mixture of cations **16b** and **16c**.[39]

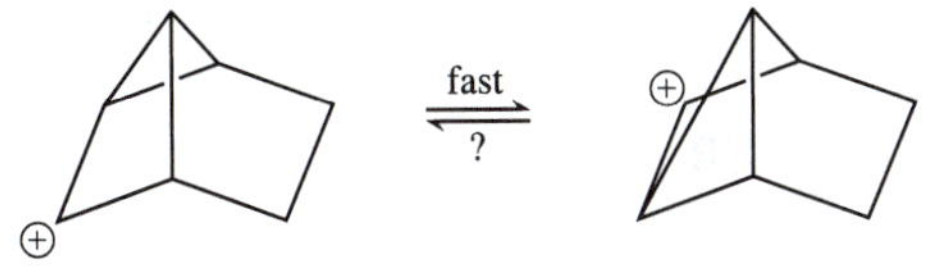

16b   **16c**

Paul G. Gassman and his coworkers tested this possibility by introducing methyl groups onto the double bond of an *anti*-7-norbornenyl derivative. If a positive charge were developed at only one end of the double bond at any given time, an alkyl group at that carbon should assist in the formation of the 7-norbornenyl cation. However, a second alkyl group on the other carbon of the double bond should have little effect. It was found that a second methyl group on the double bond had almost as large an effect on the solvolysis rate as the first methyl group.[40]

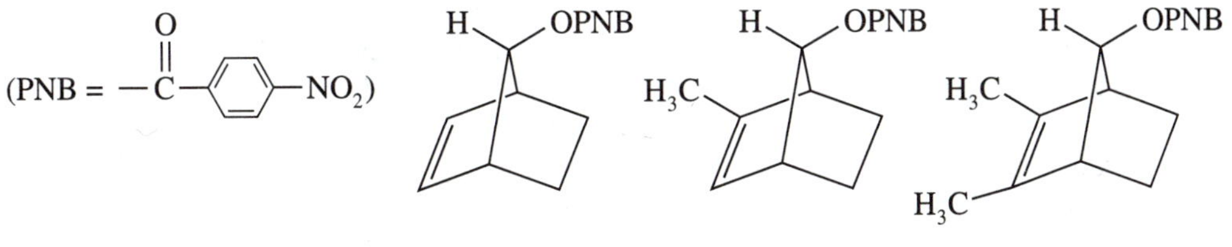

$$(PNB = -\overset{\overset{\displaystyle O}{\displaystyle \|}}{C}-\!\!\left\langle\!\!\bigcirc\!\!\right\rangle\!\!-NO_2)$$

Relative rates of reaction: 1.0   13.3   148
(in dioxane/water at 140°C)

This result supported a structure for the transition state in which both ends of the double bond share the positive charge at all times.

The solvolysis reactions of the *anti*-7-norbornenyl derivatives **19** yielded products that retained the original *anti* configurations if the phenyl groups at C7 were unsubstituted or were substituted with electron-withdrawing groups.

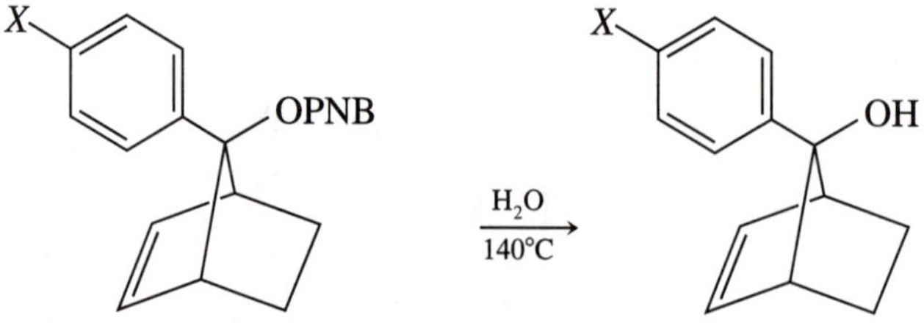

$$(25)$$

19 (X = H, p-CF$_3$, 3,5-di-CF$_3$)

However, mixtures of *anti*- and *syn*-isomers were formed when the substituents were *p*-dimethylamino or *p*-methoxy groups (Eq. 26).

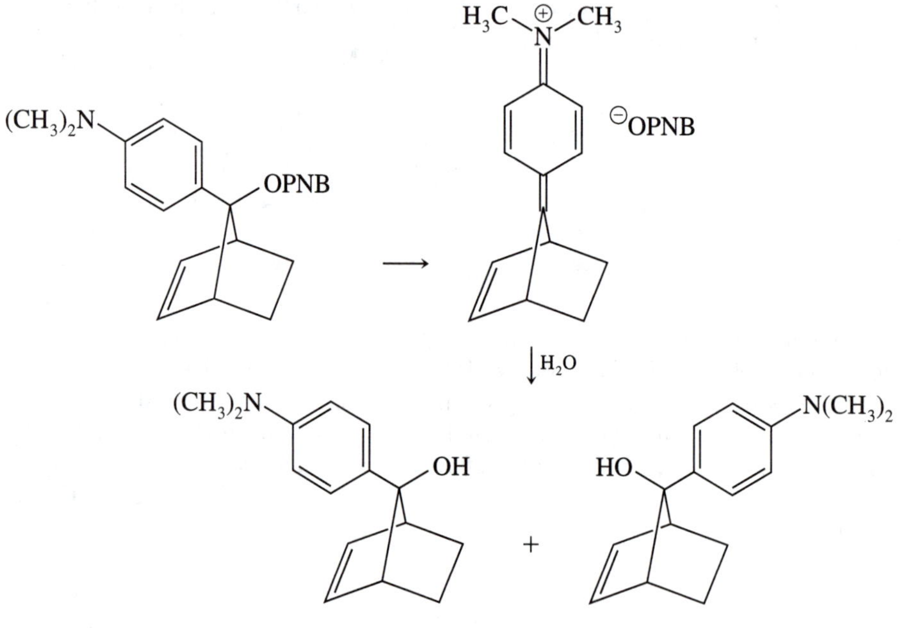

$$(26)$$

This suggested that a strongly electron-donating substituent can stabilize the carbocation sufficiently so that participation by the double bond is no longer a factor.[41]

Aromatic rings fused to *anti*-7-norbornenyl rings can also participate in displacement of leaving groups at C7 (Eq. 27), although, not surprisingly, the degree of anchimeric assistance is much smaller than for participation by a double bond.[42]

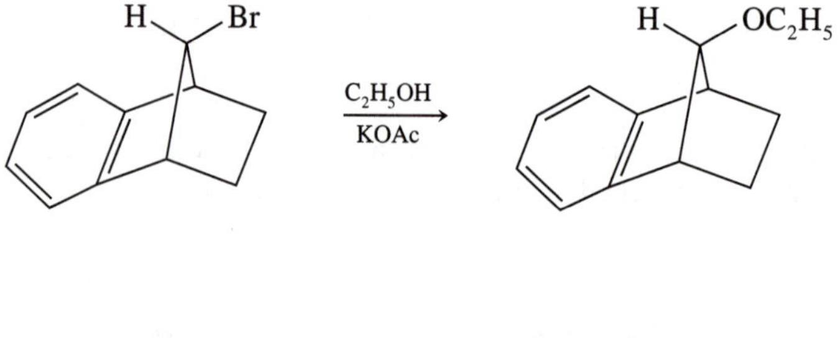

$$(27)$$

Remarkably, cycloheptatrienyl *cations* can also act as neighboring groups in the 7-norbornyl system (Eq. 28).[43]

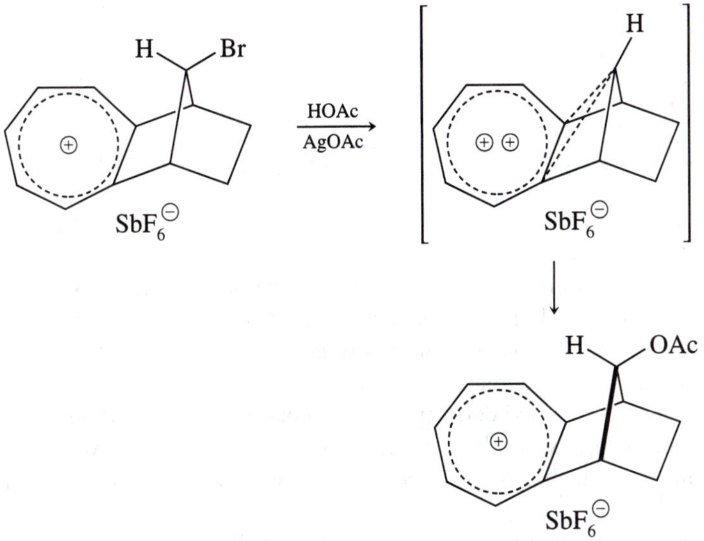

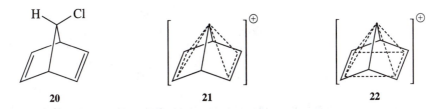

(28)

7-Chloronorbornadiene (**20**) reacts about 750 times as rapidly as *anti*-7-chloronorbornene in acetic acid solution.[44] It was originally suggested that both double bonds might assist in stabilizing the carbocation, as in structure **21**, or even that they might interact with each other as well, as in structure **22**.

20   21   22

However, NMR studies indicate that the orbital at C7 interacts strongly with only one double bond at a time.[45] It requires an activation energy of at least 19.5 kcal/mol to "flip" the C7 bridge from interaction with one double bond to interaction with the other.[46–48] The increased reactivity of the 7-norbornadienyl system, therefore, is presumably due primarily to the effect of the second double bond in expanding the bond angle at C7.

7.8   CYCLOPROPANE RINGS AS NEIGHBORING GROUPS

The cyclopropyl group appears to provide greater anchimeric assistance to the formation of carbocations than any other hydrocarbon group. Cyclopropylmethyl bromide, for instance, reacts about 40 times as rapidly as allyl bromide in aqueous ethanol solution,[32] while cyclopropylmethyl tosylate undergoes solvolysis in acetic acid nearly 90 times as rapidly as benzyl tosylate and 123,000 times as rapidly as ethyl tosylate![49]

Reactions that appear to proceed via the formation of cyclopropylmethyl cations usually yield mixtures of cyclopropylmethyl, cyclobutyl, and allylcarbinyl derivatives[32,50] (Eq. 29).

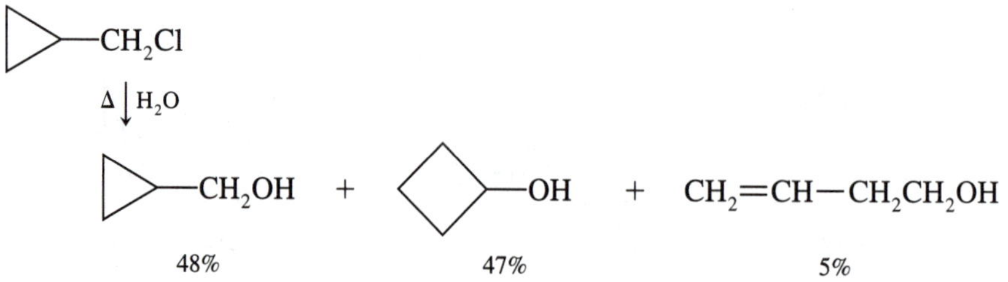

$$\text{(29)}$$

48%   47%   5%

Products with allylcarbinyl structures are likely to be minor products when the rearrangements are carried out in nucleophilic solvents. However, they are usually the most stable products, and they become the major products if the reactions are run under conditions in which the kinetic products can undergo further rearrangements.

Reactions of cyclobutyl derivatives under conditions likely to form carbocations yield the same products, in approximately the same ratios, as those formed from cyclopropylmethyl derivatives. This suggests that the same carbocations are intermediates in reactions of both types of molecules.[32,50]

John D. Roberts and his coworkers examined the products obtained from cyclopropylmethyl cations formed by diazotization of cyclopropylmethylamine labeled with ^{14}C at C1.[51] They found that the radioactive carbon was unequally distributed among the three methylene groups in the cyclobutanol and between the carbinol carbon and the ring carbons in the cyclopropylmethanol obtained from the reaction, as shown in Eq. 30.

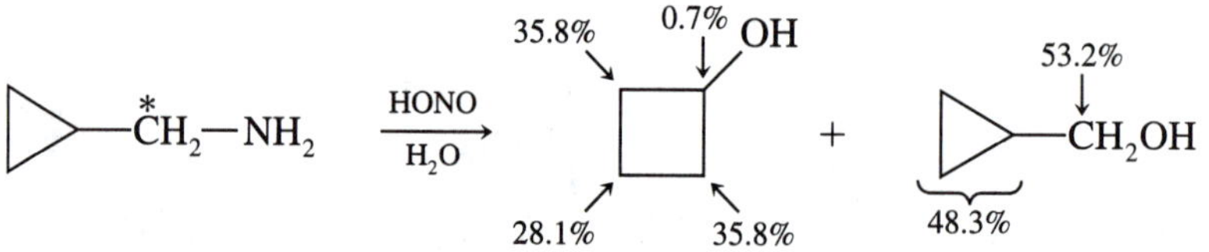

$$\text{(30)}$$

35.8%   0.7% OH   53.2%

28.1%   35.8%   48.3%

It was later shown that solvolysis reactions of deuterium-labeled cyclopropylmethyl tosylate and cyclobutyl tosylate similarly resulted in unequal distributions of the isotopic label among the methylene groups of the products.[52]

Roberts and his group had initially suggested[32] that dissociation of a cyclopropylmethyl derivative results in the formation of a "tricyclobutonium" ion, in which the cationic center interacts with the bond linking the two methylene groups of the cyclopropane ring.

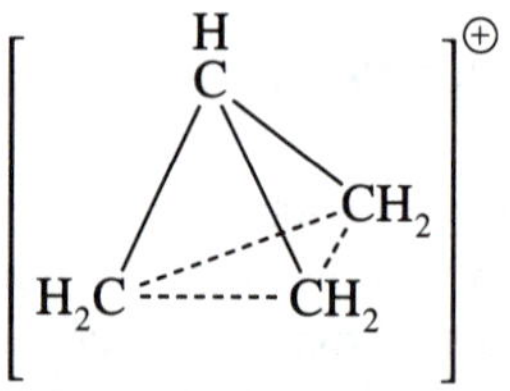

A tricyclobutonium ion

However, if tricyclobutonium cations had been formed, any radioactive label should have been equally distributed among the three methylene groups of the cations and of their reaction products. Since radioactivity was not equally distributed among the methylene groups, Roberts proposed that ionization of a cyclopropylmethyl (or cyclobutyl) derivative yields a "bicyclobutonium" ion, in which the cationic center interacts with one neighboring bond of the cyclopropyl ring. Isomerization of one bicyclobutonium ion to another could distribute isotopic labels among the methylene groups.

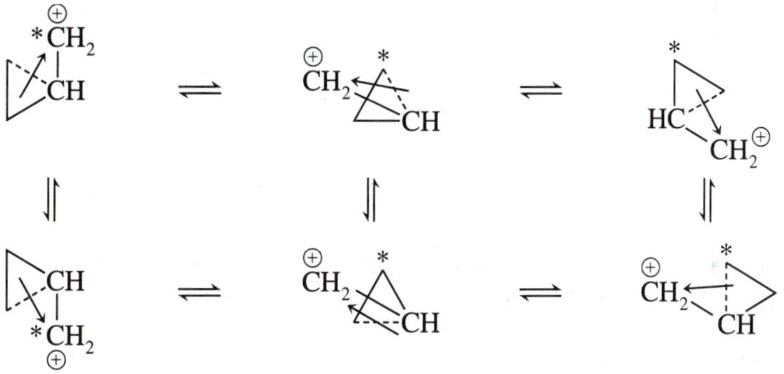

Equilibration of bicyclobutonium ions (arrows in structures indicate interactions of bonds with empty orbitals)

The NMR spectra of several cations formed from cyclopropylmethyl derivatives in superacid solutions, however, are inconsistent with bicyclobutonium structures. Instead, the ions appear to have structures in which the planes of the cationic carbons bisect the cyclopropane rings.[53,54] The stability of these "bisected" structures may be attributed to the fact that the cationic center can simultaneously interact with *two* "bent bonds" of the cyclopropyl ring.*

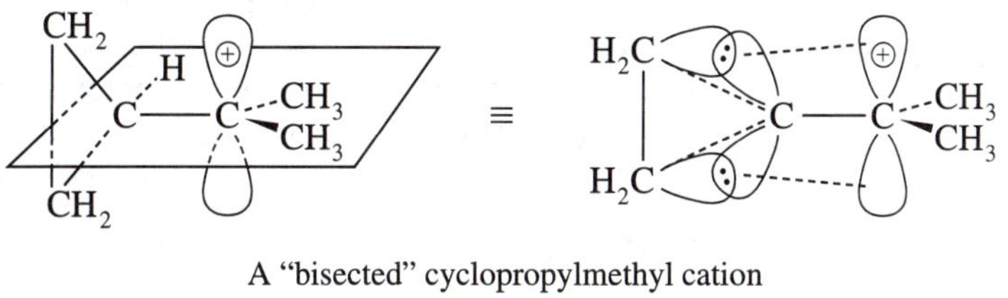

A "bisected" cyclopropylmethyl cation

The effects of substituents on the solvolysis rates of cyclopropylmethyl derivatives also suggest that the transition states for the reactions are lowest in energy when the developing cationic center can interact with two sides of the neighboring cyclopropyl ring, rather than one — that is, when the transition states resemble the bisected cations. For instance, molecules **23** to **25** react more rapidly in acetic acid than does the parent cyclopropylmethyl derivative, and they all

*In contrast to a substituted cyclopropylmethyl cation, such as that shown in Eq. 7-30b, the bisected and bicyclobutonium structures of the parent cyclopropylmethyl cation appear to be equal in energy and are rapidly interconverted in superacid solutions.[55]

react at approximately the same rate. This demonstrates that the positive charges in the transition states are equally shared by two carbons of the cyclopropyl ring.[56]

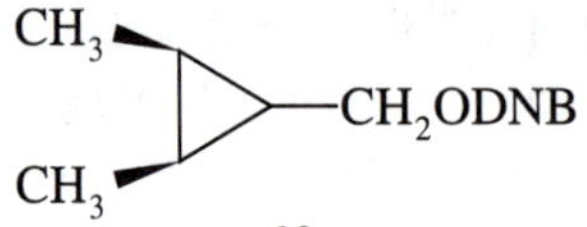

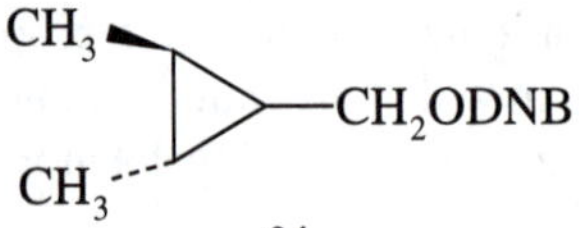

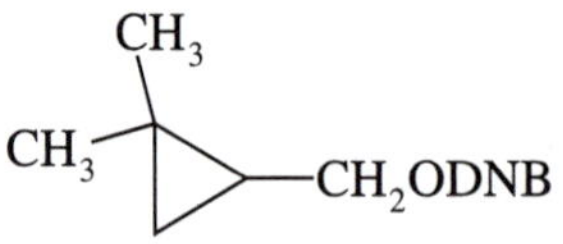

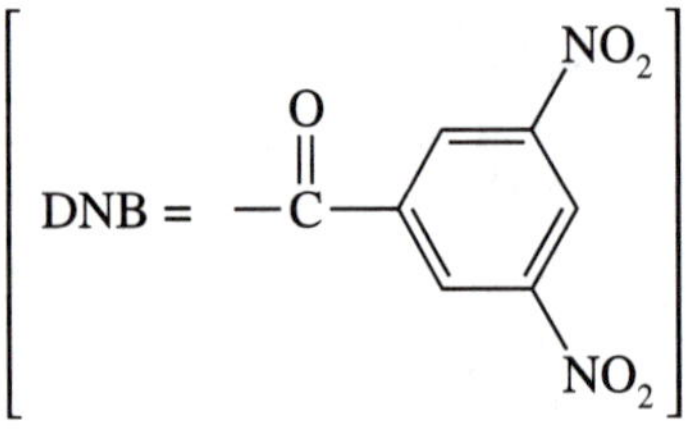

(H.C. Brown and his group demonstrated that a cyclopropyl substituent in the *para* position greatly increases the solvolysis rate of a tertiary benzylic chloride.

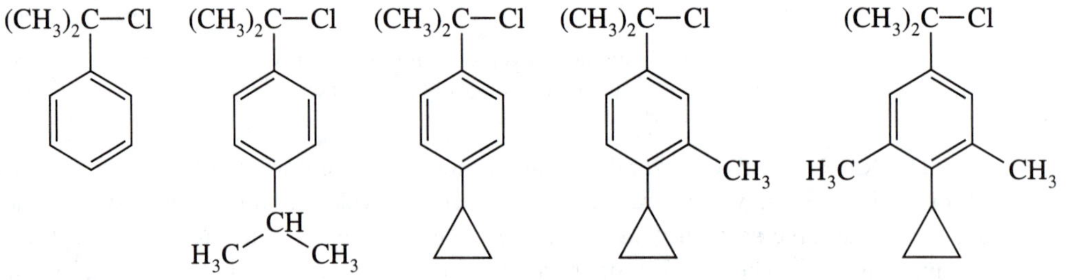

Relative solvolysis rates (in aqueous acetone)

Placing a single methyl substituent at a *meta* position, in addition to the cyclopropyl group at the *para* position, results in a slight further increase in the rate of reaction. However, a second *meta*-methyl substituent markedly decreases the rate, presumably because it forces the transition state to adopt a conformation similar to that of ion **26** rather than one similar to the bisected ion **27**.[57])

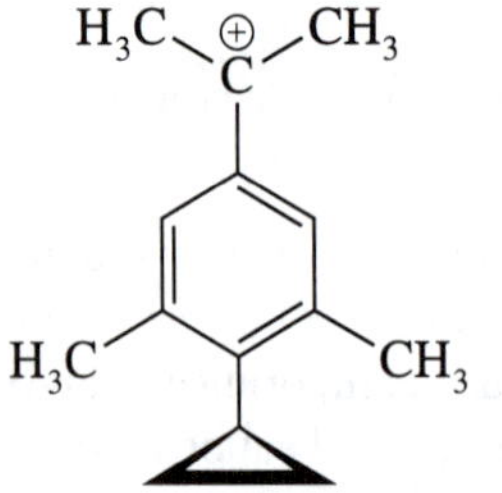

The fact that positive charges in transition states for the formation of cyclopropylmethyl cations can be shared among *three* carbon atoms is presumably responsible for the great reactivity of cyclopropylmethyl derivatives.

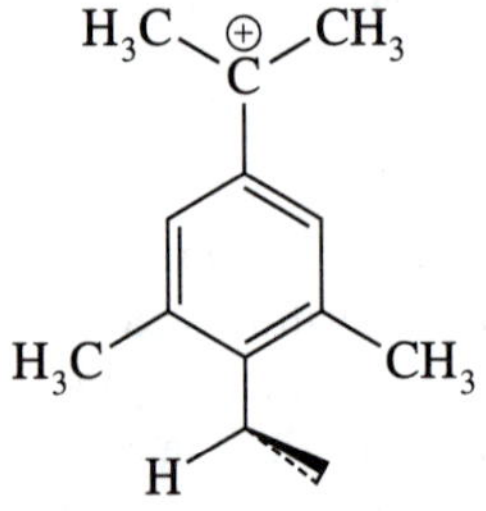

7.9 NEIGHBORING ALKYL GROUPS: 2-NORBORNYL CATIONS

Anchimeric Assistance by Alkyl Groups:
The "Nonclassical Ion" Hypothesis

In contrast to phenyl, vinyl, and cyclopropyl groups, alkyl groups rarely provide anchimeric assistance to the formation of carbocations. Even in the solvolyses of neopentyl derivatives, in which the migrations of alkyl groups appear to be more or less concerted with the loss of leaving groups, there does not appear to be significant evidence for anchimeric acceleration of the reaction rates.

Studies of the 2-norbornyl system, however, have demonstrated that alkyl groups can, under special circumstances, act as effective neighboring groups.

In 1949, Winstein and Trifan reported that solvolysis of *exo*-2-norbornyl brosylate (**28**) in acetic acid proceeds 350 times as rapidly as solvolysis of its *endo* isomer **29**.

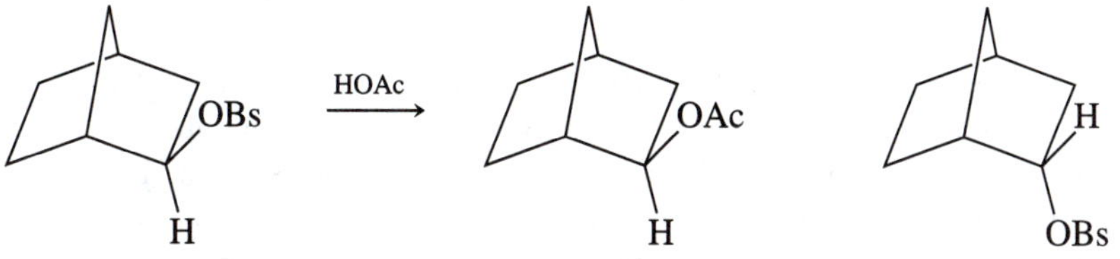

$$\text{(31)}$$

28   **29**

Brosylate **28** yields exclusively *exo*-2-norbornyl acetate. Furthermore, reaction of chiral **28** in acetic acid (or aqueous acetone) yields a completely racemic product. In contrast, the products from reaction of **29** in aqueous acetone, although significantly racemized, retain some optical activity.[58]

These observations are precisely analogous to the results of other substitution reactions exhibiting participation by neighboring groups:

1. One stereoisomer reacts far faster than the other stereoisomer.
2. The more reactive stereoisomer yields substitution products that have completely retained the original configuration at the site of substitution.
3. A compound with several chiral centers, but in which the participation of neighboring groups can form an intermediate with a plane of symmetry, yields completely racemic products.

Winstein and Trifan therefore suggested that the C1–C6 bond acts as a neighboring group in reactions of **28**, resulting in the formation of the "nonclassical" cation **30** (Eq. 32).*

---

*A similar structure was previously suggested as an intermediate in the camphene hydrochloride–isobornyl chloride rearrangement shown here:[59]

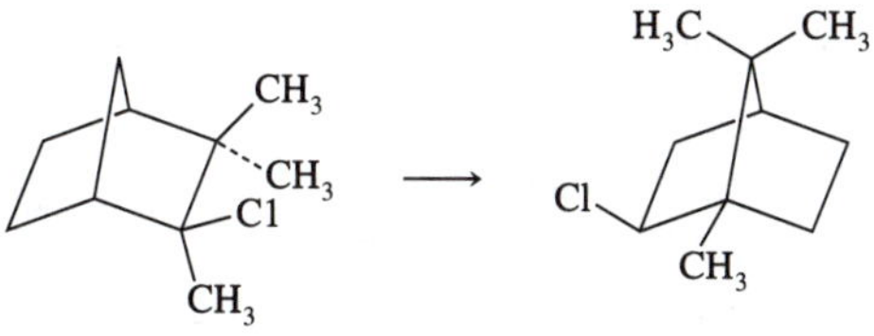

The term *nonclassical* ion has been applied to a great many types of cations with structures that seemed surprising to the chemists writing about them. Since there is no agreement as to what constitutes a classical ion, there is also no agreement as to which ions are nonclassical. It has been suggested that only structures such as 2-norbornyl cations, in which some carbons appear to have pentavalent ("carbonium ion") character, should be described as nonclassical ions.

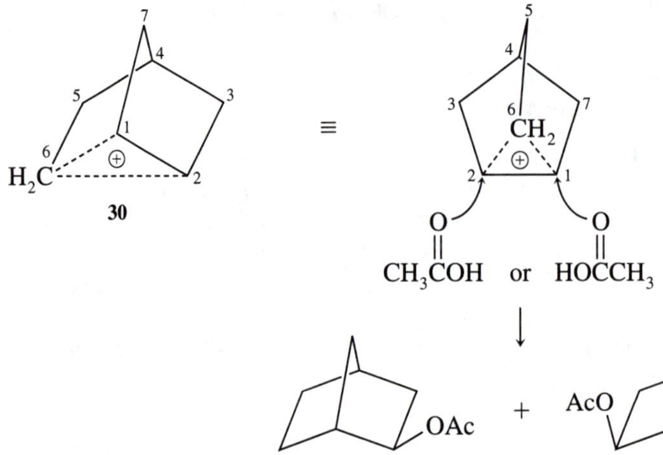

$$(32)$$

The formation of cation **30**, which has a plane of symmetry through C4, C5, and C6, would explain why reaction products from **28** are formed as racemic mixtures.

J.D. Roberts and his co-workers studied the rearrangements of 2-norbornyl derivatives labeled with radioactive carbon at C2. As expected, the radioactivity in the products was equally distributed between C1 and C2. However, radioactivity was also found at other positions—a result attributed to migrations of hydrogen atoms from C6 to C1 and C2 in the intermediate carbocation (Eq. 33).

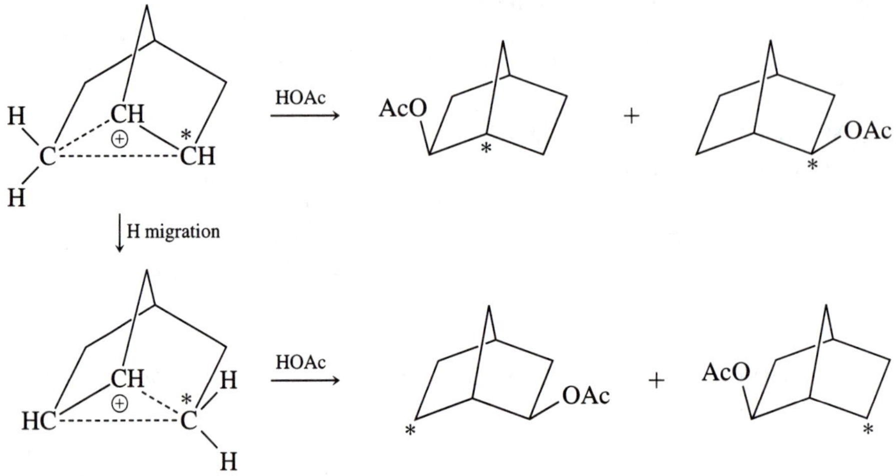

$$(33)$$

Roberts' results ruled out any hydrogen migrations from C3 to C2 (as in Eq. 34), which might have accounted for the racemization of the 2-norbornyl cation.[60,61]

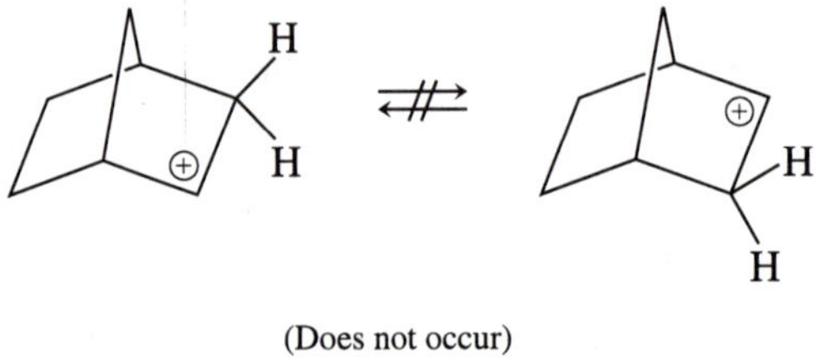

$$(34)$$

(Does not occur)

Why should an alkyl group provide anchimeric assistance to the formation of a 2-norbornyl cation, while the formation of other carbocations is not assisted by neighboring alkyl groups? Several factors might be responsible. As Winstein pointed out, the C1–C6 bonds in *exo*-2-norbornyl derivatives are *trans* to, and almost exactly coplanar with, the bonds between C2 and the leaving groups. Thus, the geometry of the *exo*-2-norbornyl system is exceptionally well suited to participation of the C1–C6 bond in the formation of a carbocation at C2. In addition, the norbornyl ring system, in which a single carbon links the opposite sides of a cyclohexyl ring, is quite strained. As a result, the C1–C6 bond, like bonds in cyclopropyl rings, is "bent" and weak, and its electrons are relatively available to participate in the displacement of a leaving group. Therefore, neighboring group participation by a carbon–carbon σ bond might reasonably be observed in the norbornyl system but not in open-chain systems.

The Challenge to the "Nonclassical Ion" Hypothesis

The arguments that the C1–C6 bond participates in forming the 2-norbornyl cation, and that the cation has a nonclassical structure, were generally accepted for over a decade, until they were reexamined by H.C. Brown in the mid-1970s.[62]

Winstein and his group had demonstrated that solvolysis rates of *endo*-2-norbornyl derivatives were similar to those of cyclohexyl or open-chain analogs. This was taken as indicating that reaction rates of *endo*-2-norbornyl derivatives were "normal" and that reactions of their *exo*-isomers were therefore "abnormally" fast. Brown, however, argued that since 2-norbornyl derivatives contain five-membered rings, their reactions should be compared to those of cyclopentyl halides and tosylates. (It had long been known that cyclopentyl derivatives are converted to carbocations more rapidly than their cyclohexyl or open-chain analogs.*)

Exo-2-norbornyl chloride, in fact, undergoes solvolysis only about five times as rapidly as cyclopentyl chloride, while the *endo* chloride is only about one-hundredth as reactive as cyclopentyl chloride.[63] From this, Brown inferred that the *endo* isomer is indeed unusually unreactive, rather than that the *exo* isomer is unusually reactive.

Brown also demonstrated that the solvolysis of **31** is 284 times as fast as that of its *endo* isomer and that substitution reactions of **31** proceed with complete retention of configuration.

(35)

31

---

*For a discussion of the relative reactivities of five- and six-membered rings, see H.C. Brown, J.H. Brewster, and H. Schechter, *J. Am. Chem. Soc., 76,* 467 (1954).

Yet the *p*-anisyl group in **31** stabilizes the cation so effectively that participation by the neighboring C1–C6 bond should be a negligible factor.[64] Furthermore, a wide variety of free-radical, anionic, and "molecular" reagents (such as diborane) attack the norbornyl system from the *exo* side, even though it has never been suggested that those reactions proceed by way of nonclassical intermediates. The exclusive formation of *exo* isomers from *exo*-2-norbornyl derivatives thus seemed like weak evidence for the formation of nonclassical norbornyl cations.[62]

Finally, Brown argued that the racemization of *exo* 2-norbornyl derivatives could be accounted for by the rapid equilibration of two "classical" 2-norbornyl cations (Eq. 36) rather than by formation of a single symmetrical cation.[61]

$$\text{(36)}$$

Brown's challenge to the existence of the nonclassical 2-norbornyl cation set off a frenzy of activity aimed at determining whether substitution reactions of *exo*-2-norbornyl derivatives are accelerated by anchimeric assistance, and whether the 2-norbornyl cation has a symmetrical "nonclassical" structure. (The two questions are related but not identical. It is not inconceivable, for instance, that two equilibrating ions with relatively "classical" structures might be formed with anchimeric assistance by the C1–C6 bond or that a classical cation might be formed without anchimeric assistance and might then rearrange to a nonclassical structure.)

At present, the evidence seems to strongly favor the Winstein view on both counts. For instance, as shown in Table 7.4, placing electron-donating groups at C6 increases the *exo–endo*

TABLE 7.4 EFFECTS OF SUBSTITUENTS ON *EXO/ENDO* RATE RATIOS IN REACTIONS OF 2-NORBORNYL TOSYLATES

	R	k_{exo}/k_{endo}
exo (X = OTs, Y = H)		
endo (X = H, Y = OTs)		
	CH_3	181[a]
	CH_2Br	16[a]
	CO_2CH_3	3.7[a]
	F	0.48[a]
	CN	0.37[a]
	H	100[b]
	CO_2CH_3	1.2[b]
exo (X = OTs, Y = H)	CN	1.1[b]
endo (X = H, Y = OTs)		

[a]In dioxane-water at 25°C. (Ref. 65).
[b]In ethanol-water at (Ref. 66).

rate ratios for reactions of 2-norbornyl tosylates, while placing electron-withdrawing groups at C6 decreases the *exo–endo* rate ratios so markedly that *exo*-2-norbornyl tosylates can become less reactive than their *endo*-isomers![65] Similarly, the presence of electron-withdrawing substituents at C1 effectively eliminates the *exo–endo* rate differences.[66]

These results demonstrate that the positive charge in the developing 2-norbornyl cation must be shared among C1, C2, and C6.

Thermochemical studies indicate that 2-norbornyl cations in solution are ca. 7.7 kcal/mole more stable than would be predicted if they had classical structures,[67] and that the 2-norbornyl cation (a secondary cation) is more stable in the gas phase than the *tert*-butyl cation.[68]

Finally, a variety of ^{1}H and ^{13}C NMR in super acids,[69] as well as the results of theoretical calculations,[70] strongly suggest that the 2-norbornyl cation has a symmetrical "nonclassical" structure.

REFERENCES

[1] S. Winstein, E. Grunwald, R.E. Buckles, and C. Hanson, *J. Am. Chem. Soc., 63*, 2541 (1963).

[2] S. Winstein and H.J. Lucas, *J. Am. Chem. Soc., 61*, 1576, 2845 (1939).

[3] I. Roberts and G.E. Kimball, *J. Am. Chem. Soc., 59*, 947 (1937).

[4] H.J. Lucas and C.W. Gould, Jr., *J. Am. Chem. Soc., 63*, 2541 (1941).

[5] G.A. Olah, G.K.S. Prakash, and J. Sommer, *Superacids*. John Wiley & Sons, New York (1985), pp. 200–201.

[6] S. Winstein and R.B. Henderson, *J. Am. Chem. Soc., 65*, 2196 (1943).

[7] S. Winstein and R.E. Buckles, *J. Am. Chem. Soc., 64*, 2780 (1942).

[8] P.D. Bartlett and C.G. Swain, *J. Am. Chem. Soc., 71*, 1406 (1949).

[9] M.I. Page, *Chem. Soc. Rev., 2*, 295 (1973).

[10] S. Winstein, C.R. Lindegren, H. Marshall, and L.L. Ingraham, *J. Am. Chem. Soc., 75*, 147 (1953).

[11] I. Tabushi, Y. Tomaru, Z. Yoshida, and T. Sugimoto, *J. Am. Chem. Soc., 97*, 2886 (1975).

[12] R.C. Fuson, C.C. Price, and D.M. Burness, *J. Org. Chem., 11*, 475 (1946).

[13] P.D. Bartlett, S.D. Ross, and C.G. Swain, *J. Am. Chem. Soc., 69*, 2971 (1947).

[14] R.C. Fuson and C.L. Zirkle, *J. Am. Chem. Soc., 70*, 2760 (1948).

[15] H. Freundlich and H. Kroepelin, *Z. Phys. Chem., 122*, 39 (1926).

[16] H. Boehme and K. Sell, *Ber., 81*, 123 (1948).

[17] G.M. Bennett, F. Heathcoat, and A.N. Mosses, *J. Chem. Soc.*, 2567 (1929).

[18] From H.A. Bent, in *The Chemistry of Organic Sulfur Compounds*, N. Kharasch and C.A. Meyers, Eds. Pergamon Press, Oxford (1966).

[19] S. Winstein, E. Grunwald, and L.L. Ingraham, *J. Am. Chem. Soc., 70*, 821 (1948).

[20] Andrew Streitwieser, Jr., *Solvolytic Displacement Reactions*. McGraw-Hill, New York (1962), p. 121.

[21] G.H. Alt and D.H.R. Barton, *J. Chem. Soc.*, 4284 (1954).

[22] S. Winstein, H.V. Hess, and R.G. Buckles, *J. Am. Chem. Soc., 64*, 2796 (1942).

[23] S. Winstein and R.G. Buckles, *J. Am. Chem. Soc., 65*, 613 (1943).

[24] D.J. Cram, *J. Am. Chem. Soc., 71*, 3863 (1949).

[25] D.J. Cram, *J. Am. Chem. Soc., 74*, 2129 (1952).

[26] S. Winstein and K. Schreiber, *J. Am. Chem. Soc., 74*, 2165 (1952).

[27] G.A. Olah, G.K.S. Prakash, and J. Sommer, *Superacids*. John Wiley & Sons, New York (1985), p. 101.

[28] D.J. Cram, *J. Am. Chem. Soc., 71,* 3875 (1949).

[29] W.E. Bachmann and F.H. Moser, *J. Am. Chem. Soc., 54,* 1124 (1932). See also W.E. Bachmann and H.R. Sternberger, *J. Am. Chem. Soc., 55,* 3821 (1933); *56,* 170 (1934); W.E. Bachmann and J.W. Ferguson, *56,* 2081 (1934).

[30] L.F. Fieser and M. Fieser, *Steroids*. Reinhold Publ. Co., New York (1959), pp. 314–316.

[31] S. Winstein and R. Adams, *J. Am. Chem. Soc., 70,* 838 (1948).

[32] J.D. Roberts and R.H. Mazur, *J. Am. Chem. Soc., 73,* 2509 (1951).

[33] J.B. Rogan, *J. Org. Chem., 27,* 3910 (1962).

[34] S. Winstein and E.T. Stafford, *J. Am. Chem. Soc., 79,* 505 (1957).

[35] S. Winstein, M. Shatavsky, C. Norton, and R.B. Woodward, *J. Am. Chem. Soc., 77,* 4183 (1955). See also S. Winstein and M. Shatavsky, *J. Am. Chem. Soc., 78,* 592 (1956).

[36] M. Brookhart, A. Diaz, and S. Winstein, *J. Am. Chem. Soc., 88,* 3135 (1966).

[37] P.D. Bartlett and M.R. Rice, *J. Org. Chem., 28,* 3351 (1963).

[38] J.B. Lambert, R. Finzel, and C.A. Belec, *J. Am. Chem. Soc., 102,* 3281 (1980).

[39] H.C. Brown, *J. Am. Chem. Soc., 85,* 2324 (1963).

[40] P.G. Gassman and D.S. Patton, *J. Am. Chem. Soc., 91,* 2160 (1969).

[41] P.G. Gassman and A.F. Fentiman, Jr., *J. Am. Chem. Soc., 92,* 2549 (1970).

[42] J.W. Wilt and P.J. Chenier, *J. Org. Chem., 35,* 1571 (1970).

[43] J.W. Wilt, C. George, and M. Peeran, *J. Org. Chem., 52,* 3739 (1987).

[44] S. Winstein and C. Ordronneau, *J. Am. Chem. Soc., 82,* 2084 (1960).

[45] P.R. Story and M. Saunders, *J. Am. Chem. Soc., 84,* 4876 (1962). P.R. Story et al., *J. Am. Chem. Soc., 85,* 3630 (1963).

[46] G.A. Olah et al., *J. Am. Chem. Soc., 95,* 8698 (1973).

[47] R.K. Lustgarden, M. Brookhart, and S. Winstein, *J. Am. Chem. Soc., 94,* 2347 (1972).

[48] D.D. Roberts, *J. Org. Chem., 30,* 23 (1965); *56,* 5661 (1991).

[49] G.A. Olah, V.P. Reddy, and G.K. Surya Prakash, *Chem. Rev., 92,* 69 (1992).

[50] R.H. Mazur, W.N. White, D.A. Semenow, C.C. Lee, M.S. Silver, and J.D. Roberts, *J. Am. Chem. Soc., 81,* 4390 (1959).

[51] Z. Majerski, S. Borčić, and D.E. Sunko, *Chem. Comm.,* 1636 (1970).

[52] C.U. Pittman, Jr. and G.A. Olah, *J. Am. Chem. Soc., 98,* 2998 (1965). See also G.A. Olah, D.P. Kelly, C.L. Jevell, R.D. Porter, *J. Am. Chem. Soc., 92,* 2544 (1970).

[53] For example, see N.C. Deno, H.G. Richey, Jr., J.S. Liu, J.D. Hodge, J.J. Houser, M.J. Wisotsky, *J. Am. Chem. Soc., 84,* 2016 (1962), and L.R. Schmitz and T.S. Sorensen, *J. Am. Chem. Soc., 104,* 2600 (1982).

[54] J.S. Staral, I. Yavari, J.D. Roberts, G.K.S. Prakash, D.J. Donovan, G.A. Olah, *J. Am. Chem. Soc., 100,* 8016 (1978). See also M. Saunders and H.U. Siehl, *J. Am. Chem. Soc., 102,* 6868 (1980).

[55] P. von R. Schleyer and G. van Dine, *J. Am. Chem. Soc., 88,* 2321 (1966).

[56] H.C. Brown and J.D. Cleveland, *J. Am. Chem. Soc., 88,* 2051 (1966).

[57] S. Winstein and D. Trifan, *J. Am. Chem. Soc., 71,* 2953 (1949); *74,* 1147, 1154 (1952).

[58] T.P. Nevell, E. de Salas, and C.L. Wilson, *J. Chem. Soc.,* 1188 (1939).

[59] J.D. Roberts and C.C. Lee, *J. Am. Chem. Soc., 73,* 5009 (1951).

[60] J.D. Roberts, C.C. Lee, and W.H. Saunders, Jr., *J. Am. Chem. Soc., 76,* 4501 (1954).

[61] For example, H.C. Brown, *Tetrahedron, 32,* 179 (1976).

[62] H.C. Brown and F.J. Chloupek, *J. Am. Chem. Soc., 85,* 2322 (1963).

[63] H.C. Brown and K. Takeuchi, *J. Am. Chem. Soc., 90,* 2691 (1968).

[64] W. Fischer, C.A. Grob, R. Hanreich, G.V. Sprecher, and A. Waldner, *Helv. Chim. Acta, 64,* 2298 (1981).

[65] D. Lenoir, *Ber., 108,* 2055 (1975).

[66] E.M. Arnett, N. Pienta, and C. Petro, *J. Am. Chem. Soc., 102,* 398 (1980).

[67] J.J. Solomon and F.H. Field, *J. Am. Chem. Soc., 98,* 1567 (1976).

[68] For example, M. Saunders and M.R. Kates, *J. Am. Chem. Soc., 102,* 6867 (1980). A. Olah, G.K.S. Prakash, M. Arvanaghi, F.A.L. Anet, *J. Am. Chem. Soc., 104,* 7105 (1982). C.S. Yannoni, V. Macho, and P.C. Myhre, *J. Am. Chem. Soc., 104,* 7380 (1982).

[69] For example, J.D. Goddard, Y. Osamura, and H.F. Schaefer, III, *J. Am. Chem. Soc., 104,* 3285 (1982). K. Raghavachari, R.C. Haddon, P.V.R. Schleyer, and H.F. Schaefer III, *J. Am. Chem. Soc., 105,* 5915 (1983).

PROBLEMS

7.1 Write reasonable structures for all products expected to be obtained in significant yields from the reactions below. Show the geometries of the products if they can be predicted from the structures of the starting materials.

(a)

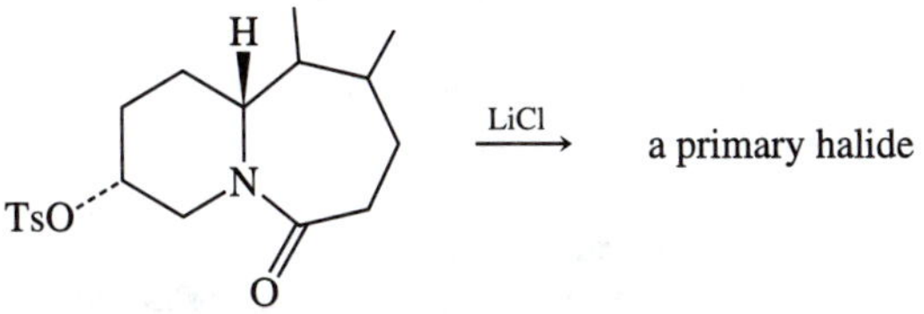

$\xrightarrow{\text{LiCl}}$ a primary halide

(T.G. Back, J.H.L.Chav, and J.W. Morzycki, *Tetrahedron Lett.*, **32**, 6517 [1991])

(b)

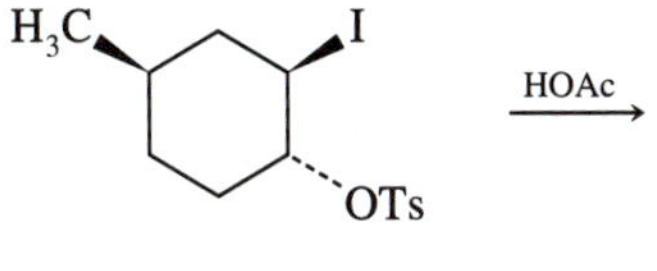

$\xrightarrow{\text{HOAc}}$

(c)

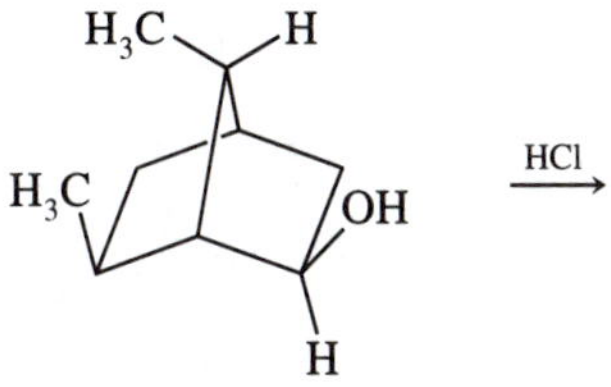

$\xrightarrow{\text{HCl}}$

(d)

$\xrightarrow{\text{HOAc}}$

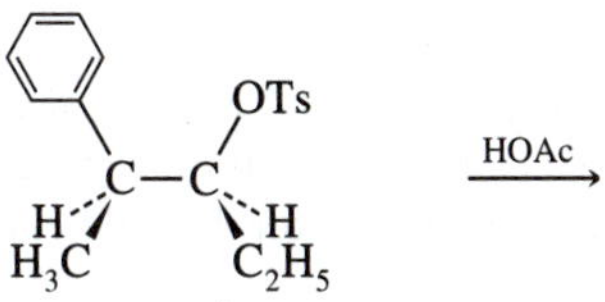

7.2 Write reasonable mechanisms for the following reactions.

(a)

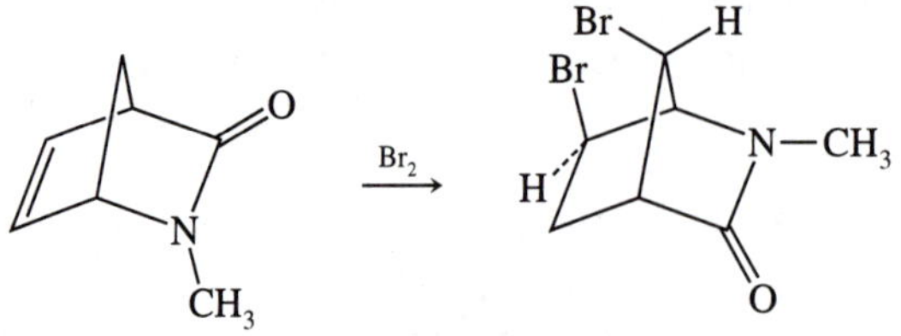

(C.F. Palmer et al., *PerkinsTrans. 1*, 1021, [1992])

(b)

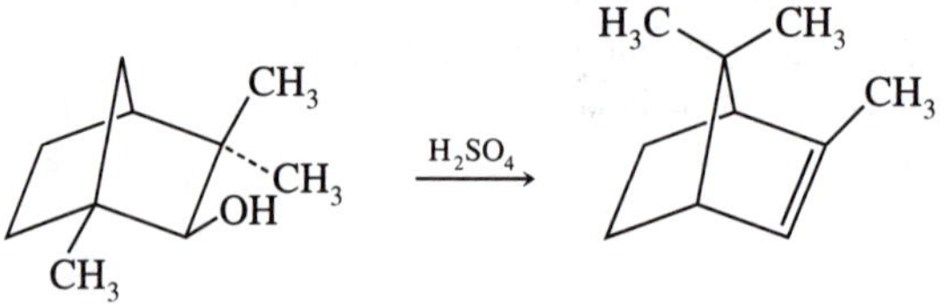

(c)

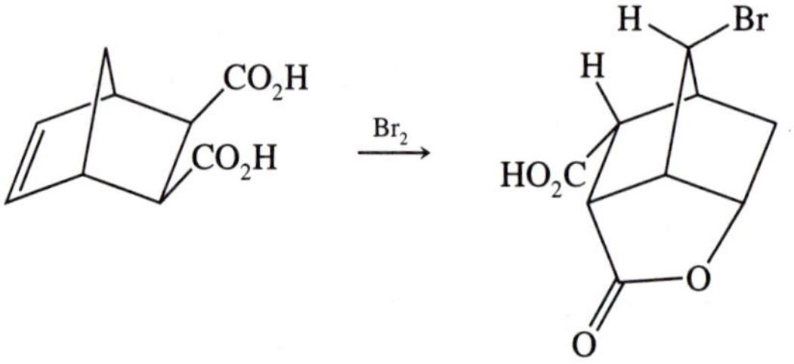

It can be argued that Reaction (c) provides evidence for the formation of a classical 2-norbornyl cation. What is the difficulty with assuming solely nonclassical ions as intermediates in this reaction?

(d) $ClCH_2\overset{S}{\underset{\|}{C}}OR$ $\xrightarrow{H_2O}$ $HSCH_2\overset{O}{\underset{\|}{C}}OR$

(S. Creary and M.E. Merhsheikh-Mohammadi, *J. Org. Chem.*, **51**, 7, [1986])

(e)

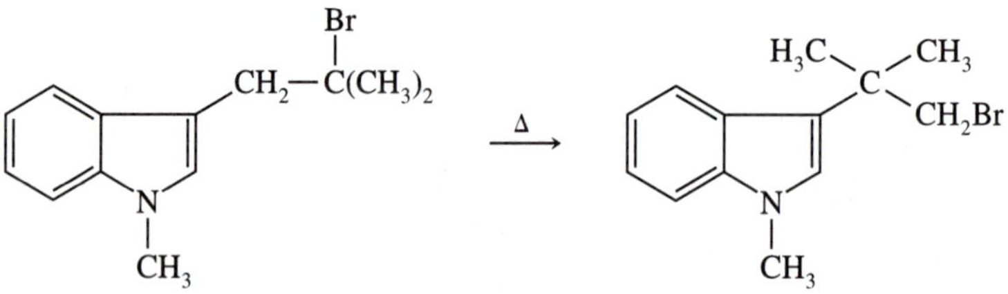

(Adapted from J.P. Sanchez and R.F. Parcell, *J. Hetero. Chem.*, **27**, 1601, [1990])

7.3 Reaction of cation **6** with sodium acetate in acetic acid yields products resulting from the attack of acetate anion (or acetic acid) at a saturated carbon. In contrast, water and ethanol add to the carboxyl carbon of **6** (page 191). Suggest an explanation for the differences between these reactions.

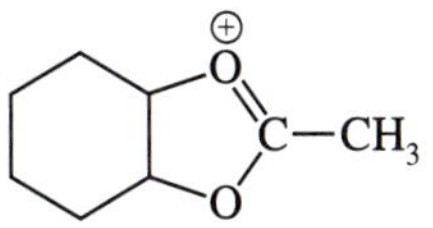

6

7.4 A Hammett plot of log k_{subst}/k_H versus σ^+ for the reaction of **19** in dioxane/water yields a curved line. (a) Explain this fact. (b) Should the slope of the line (ρ) be positive, negative, or change signs with changing substituents on the aromatic ring? (c) Would the absolute value of the slope of the line be greater when the substituents are electron-donating or when they are electron-withdrawing?

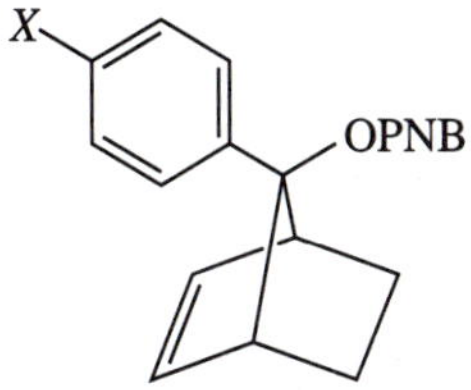

19

Rearrangements of Carbanions and Free Radicals

8.1 CARBANION REARRANGEMENTS

Sigmatropic Shifts in Hydrocarbon Anions

Since suprafacial [1,2] migrations in carbanions are forbidden by orbital symmetry conservation rules, rearrangements of carbanions are far less common than rearrangements of carbocations. In fact, 1,2-shifts of alkyl groups or hydrogen atoms in "carbanions"—that is, in alkali or alkali earth derivatives of hydrocarbons—simply do not occur. Suprafacial [1,4] migrations in allylic carbanions *would* be allowed by the Woodward-Hoffman rules. However, there does not seem to be clear evidence that such reactions actually occur.[1]

NMR studies have revealed several examples of [1,6] hydrogen migrations in pentadienyl anions,* including the rearrangements shown in Eq. 1.[2] No exchange of deuterium takes place between anion **1a** and its deuterated analog **1b** in a mixture of the two, demonstrating that the hydrogen migrations are intramolecular processes.[2]

[1,6] Hydrogen shifts in open-chain pentadienyl anions occur readily at temperatures as low as 35°C. In contrast, the cyclic anion **2**, in which any hydrogen migrations would have to proceed by suprafacial paths, undergoes no thermal rearrangements, even when heated to 150°C! This suggests that thermal rearrangements of pentadienyl anions proceed by antarafacial paths, as predicted by orbital symmetry conservation rules.[2] (Anion **2** does undergo a *photochemical* [1,6] hydrogen migration, as shown in Eq. 2.)

---

*Pentadienyl anions can exist in "W," "S," and "U" conformations. "U" conformations appear to be the highest in energy. However, the conformations are readily interconvertible at the temperatures needed for [1,6] hydrogen shifts.[1,3] The rearrangements of **1a** and **1b** are therefore shown as proceeding via U conformations of the anions.

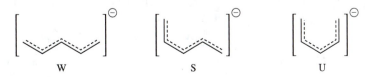

2

(2)

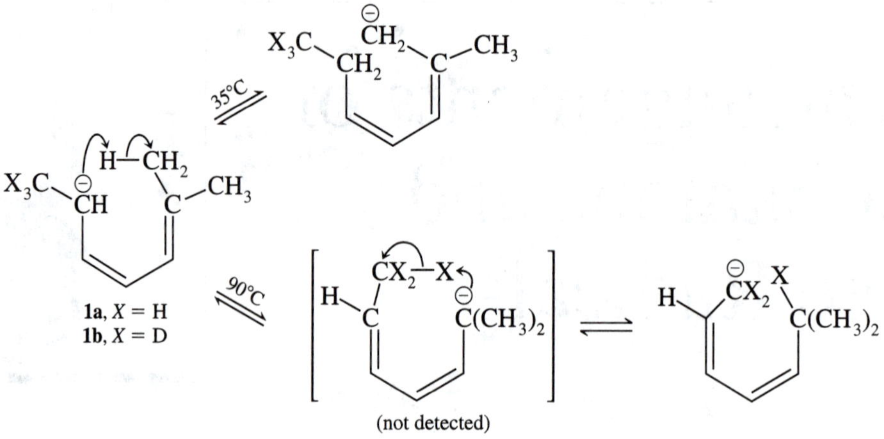

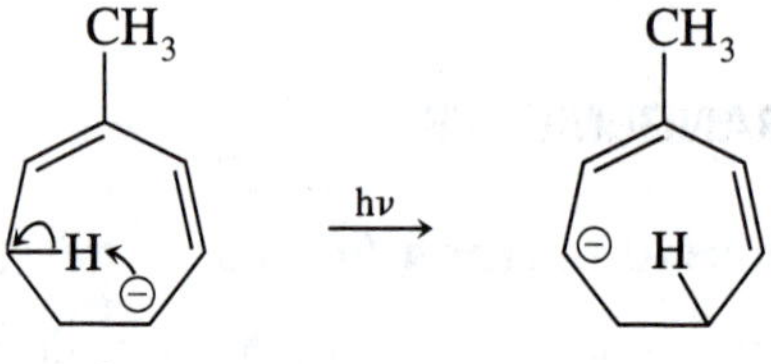

A possible [1,8] alkyl group migration occurs in the rearrangement of carbanion **3**, although the possibility of a two-step elimination-addition mechanism has not been eliminated.[4]

3

(3)

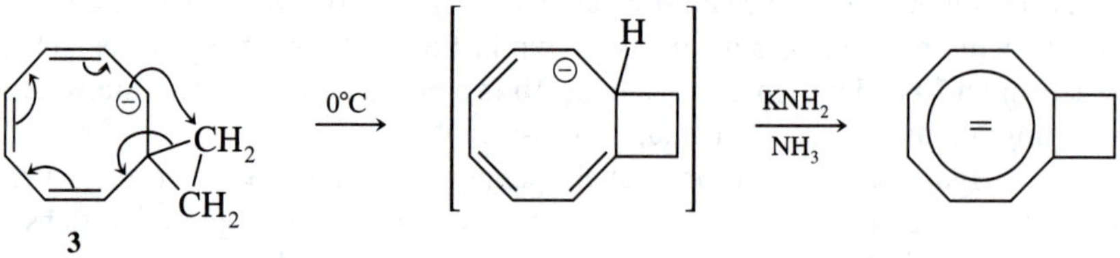

Migrations of allylic groups in carbanions proceed principally by allowed [2,3] shifts, as shown in Eqs. 4 and 5.[5,6]

major product   minor product

(4)

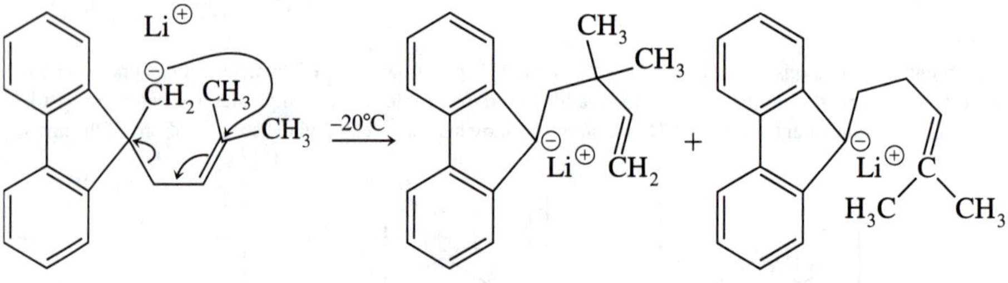

$$(C_6H_5)_2\overset{\ominus}{C}\cdots \overset{CH_2^{\ominus}}{\underset{\overset{*}{C}H_2}{}} \quad Li^{\oplus}$$

$$(* = {}^{14}C)$$

$$\Big\downarrow -50°C \tag{5}$$

$$\underset{\text{major product}}{Li^{\oplus} \quad (C_6H_5)_2\overset{\ominus}{C}CH_2\overset{*}{C}H_2{-}CH{=}CH_2} \quad + \quad \underset{\text{minor product}}{Li^{\oplus} \quad (C_6H_5)_2\overset{\ominus}{C}CH_2CH_2CH{=}\overset{*}{C}H_2}$$

However, as the reaction temperatures are raised, increasing amounts of 1,2-migration products are formed. These products might arise by free-radical dissociation–recombination mechanisms, as shown in Eq. 6, or by ejection of allyl anions, which then add to the resulting alkenes.

$$(C_6H_5)_2C\cdots \underset{CH_2{-}CH}{\overset{CH_2^{\ominus}}{}} \overset{CH_2}{\underset{}{\parallel}}$$

$$\Big\downarrow ? \tag{6}$$

$$[(C_6H_5)_2\overset{\bullet}{C}{-}\overset{\ominus}{\overset{\bullet\bullet}{C}}H_2 \quad \longleftrightarrow \quad (C_6H_5)_2\overset{\ominus}{\overset{\bullet\bullet}{C}}{-}\overset{\bullet}{C}H_2]$$

$$+ \quad\quad\quad\quad\quad \longrightarrow \quad \text{products}$$

$$[\overset{\bullet}{C}H_2{-}CH{=}CH_2 \quad \longleftrightarrow \quad CH_2{=}CH{-}\overset{\bullet}{C}H_2]$$

Addition–Elimination Mechanisms

Migrations of vinyl groups.   In contrast to alkyl groups, vinyl groups and aromatic rings do undergo 1,2-migrations in carbanions. However, there is no evidence that these rearrangements proceed by concerted sigmatropic shifts.

In 1960 John D. Roberts and his coworkers at MIT found that C1 and C2 of 3-butenyl (homoallylic) Grignard reagents slowly interchange positions. They explained this as due to initial addition of the "carbanionoid" carbon to the double bond (Eq. 7).[7]* It was already known that the resulting cyclopropylmethyl Grignard reagent would be rapidly converted back to a homoallylic Grignard.

---

*Grignard and organolithium reagents are often shown as covalent molecules in this chapter, while organopotassium and organocesium reagents are usually shown as ionic compounds. A more precise presentation would show all these reagents with covalent carbon–metal bonds, with the degree of negative charge on the carbon atoms increasing as the electronegativities of the metal atoms decrease.

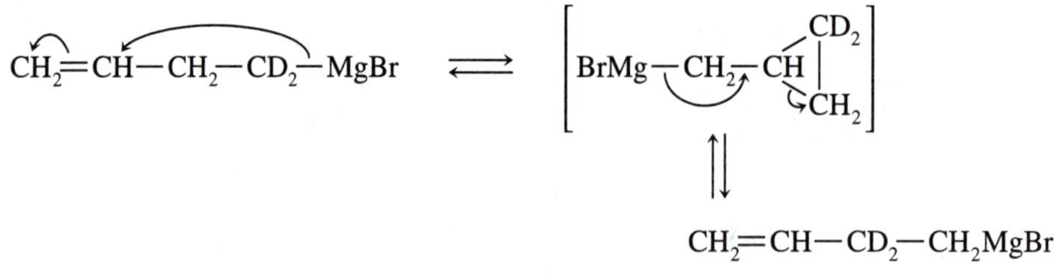

$$(7)$$

The rearrangements of homoallylic Grignard reagents are slow at room temperature. Similar rearrangements of lithium reagents can be quite rapid, as shown by the examples in Eqs. 8 and 9.[8,9]*

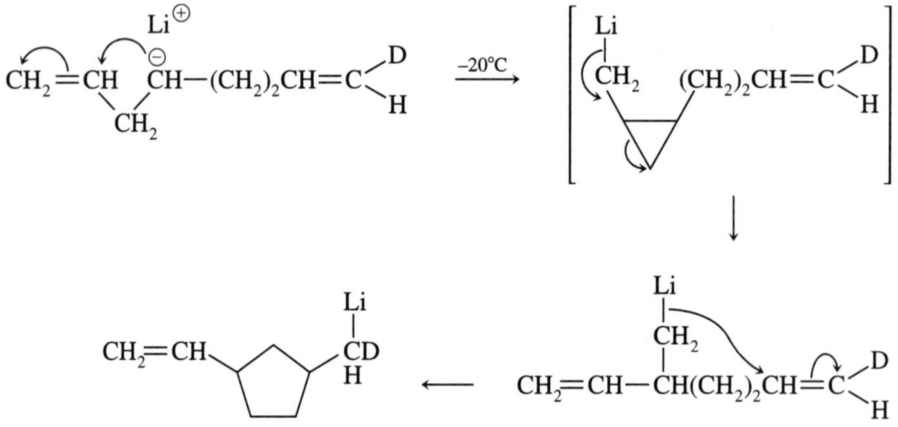

$$(8)$$

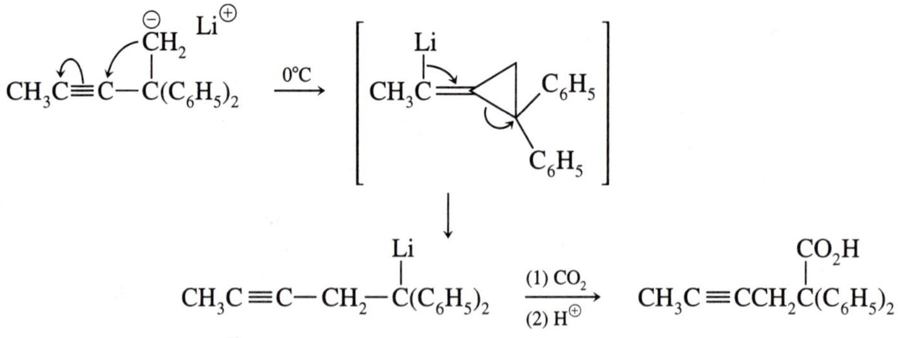

$$(9)$$

---

*It should be noted that Grignard and lithium reagents normally do *not* add to isolated carbon–carbon double bonds. Lithium reagents (but not Grignard reagents) *will* add to conjugated double bonds, even in intermolecular reactions.[10]

Aryl group migrations.   Unlike vinyl groups, phenyl groups or other aromatic rings have not been observed to undergo 1,2-migrations in Grignard reagents. However, phenyl groups do undergo slow 1,2-migrations in lithium reagents, and rapid 1,2-migrations in organosodium, organopotassium, or organocesium reagents, provided that at least one aryl group remains at the migration origin to stabilize the resulting anion (Eq. 10).[11]

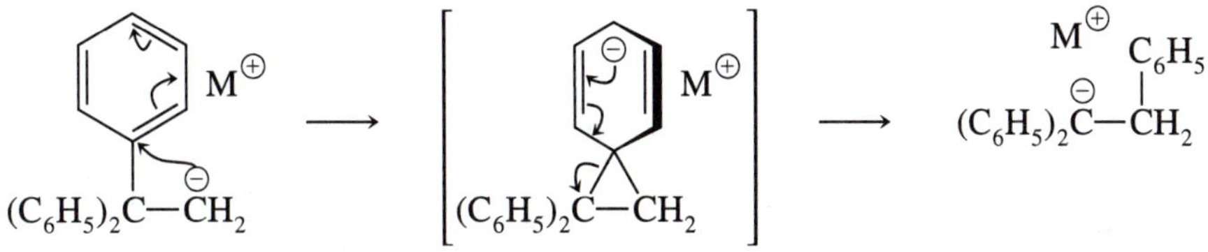

$$(10)$$

Like migrations of vinyl groups, these rearrangements proceed by addition–elimination mechanisms involving three-membered ring intermediates rather than by [1,2] sigmatropic shifts. This conclusion is supported by the fact that a *para*-biphenyl group, which yields an intermediate cyclohexadienyl anion stabilized by resonance with the second aromatic ring, migrates almost exclusively in preference to the migration of a *meta*-biphenyl group (Eq. 11).

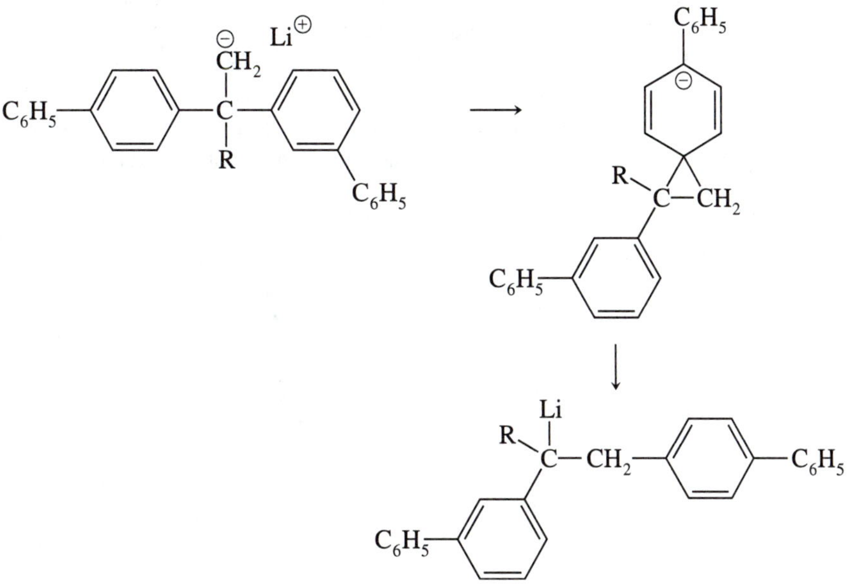

$$(11)$$

In addition, it was found that the cesium reagent **4** reacts with carbon dioxide to yield a cyclopropyl derivative as the principal product.[11]

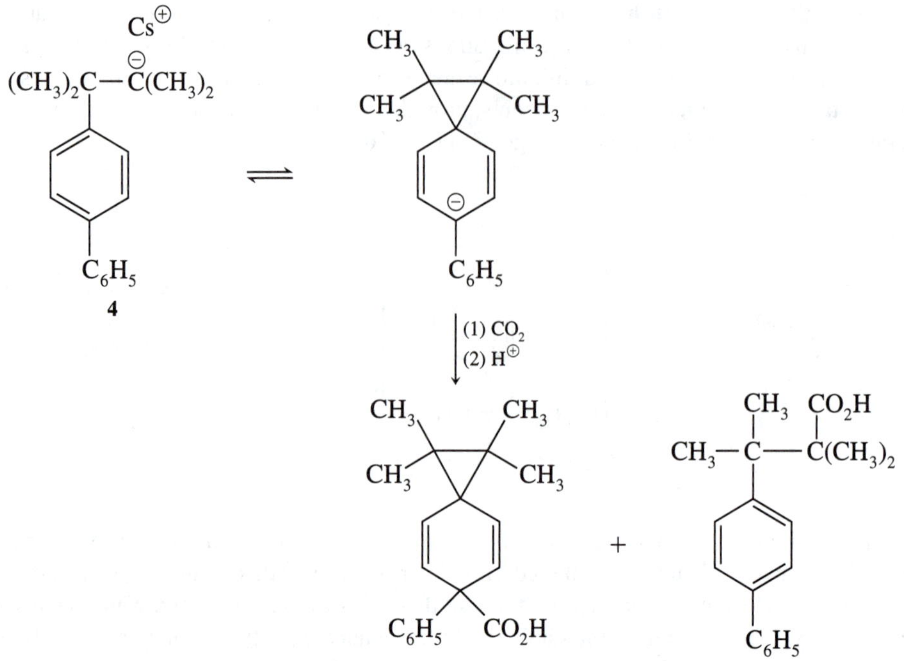

$$(12)$$

There was initially some doubt whether migrations of aromatic rings were really re-arrangements of carbanions or whether the migrations were actually taking place in free radicals formed during the formation of the organometallic reagents. However, the relative migratory aptitudes of different aromatic rings are consistent with migrations in carbanions rather than in free radicals.[12,13] For instance, phenyl groups migrate more rapidly than *para*-methylphenyl groups (Eq. 13).[12] The *para*-methyl group would stabilize an intermediate from free-radical attack at an aromatic ring, but it would destabilize an intermediate from carbanion attack.

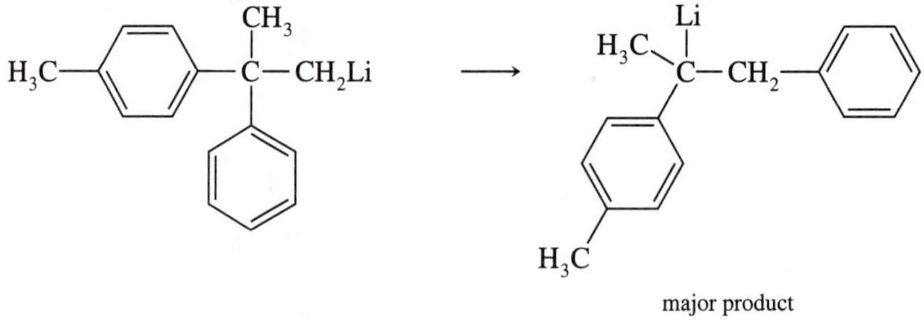

$$(13)$$

major product

The rearrangement of carbanion **5** resulted in migration of the benzyl group rather than of a phenyl group. However, radioactivity was incorporated into the product when the rearrangement was carried out in the presence of radioactive benzyllithium. It was therefore suggested that the reaction proceeded by an elimination–addition mechanism, as shown in Eq. 14.[11]

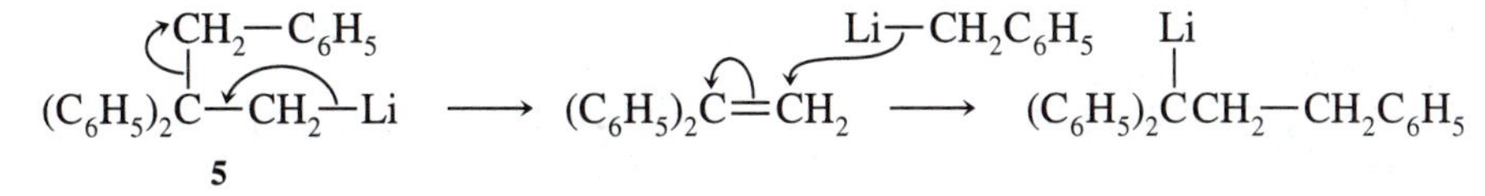

$$(C_6H_5)_2\overset{\overset{\displaystyle CH_2-C_6H_5}{|}}{C}-CH_2-Li \longrightarrow (C_6H_5)_2\overset{\displaystyle Li-CH_2C_6H_5}{C=CH_2} \longrightarrow (C_6H_5)_2\overset{\overset{\displaystyle Li}{|}}{C}CH_2-CH_2C_6H_5 \qquad (14)$$

5

A radical dissociation–recombination mechanism (see page 215) does not seem to have been excluded, however.

Rearrangements via homoenolate anions. Since carbanionic centers in homoallylic anions can add to double bonds to form three-membered rings, one might expect similar reactions to occur much more rapidly in *homoenolate* anions (anions β to carbonyl groups), in which the carbanionic sites could add to the carbonyl groups to form cyclopropoxide anions.

The first evidence for the formation of homoenolate anions (or, rather, for the formation of cyclopropoxide anions by abstraction of protons from carbons β to carbonyl groups)* was the observations that optically active camphenilone (**6**) undergoes racemization on heating with potassium *tert*-butoxide at 185°C, and that a single hydrogen from a carbon β to the carbonyl group can be exchanged with deuterium at a rate equal to that of racemization. If the temperature is raised to 250°C, all eight β protons, as well as the bridgehead α proton, can be exchanged with deuterium.[14]

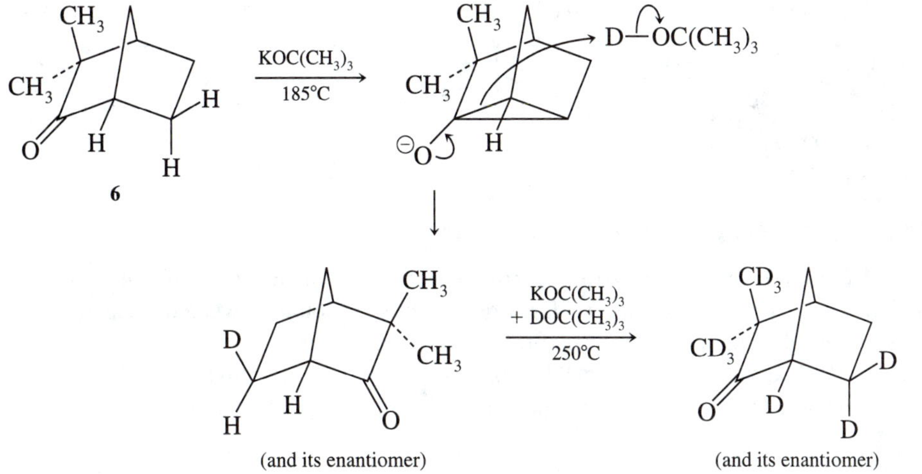

$$(15)$$

(and its enantiomer)   (and its enantiomer)

Other ketones are converted to less symmetrical cyclopropoxide anions by abstraction of β protons[15] (or, in one example, by loss of nitrogen from a diimide anion[16]). In these cases, shown respectively in Eqs. 16a and 16b, protonation of the anions may result in carbon skeleton rearrangements.

---

*There does not appear to be conclusive evidence as to whether homoenolate anions are actually formed as discrete intermediates or if removal of protons from carbons β to carbonyl groups leads directly to cyclopropoxide anions.

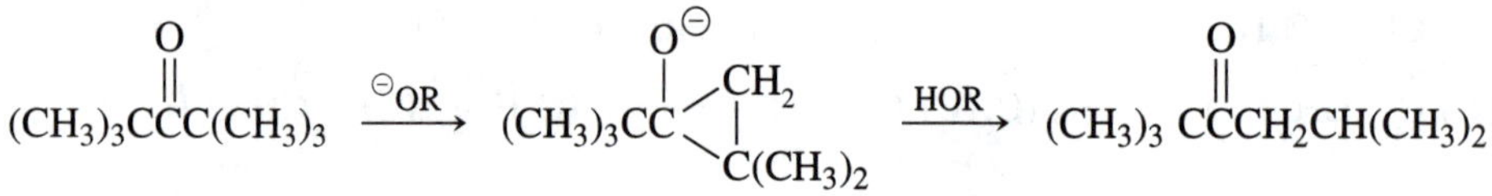

$$(CH_3)_3CCC(CH_3)_3 \xrightarrow{\ominus OR} (CH_3)_3CC \overset{CH_2}{\underset{C(CH_3)_2}{\diagdown}} \xrightarrow{HOR} (CH_3)_3 CCCH_2CH(CH_3)_2 \qquad (16a)$$

$$(CH_3)_2C-CH_2CCH_3 \xrightarrow{-N_2} (CH_3)_2C-CCH_3 \xrightarrow{HOR} (CH_3)_3CCCH_3 \qquad (16b)$$

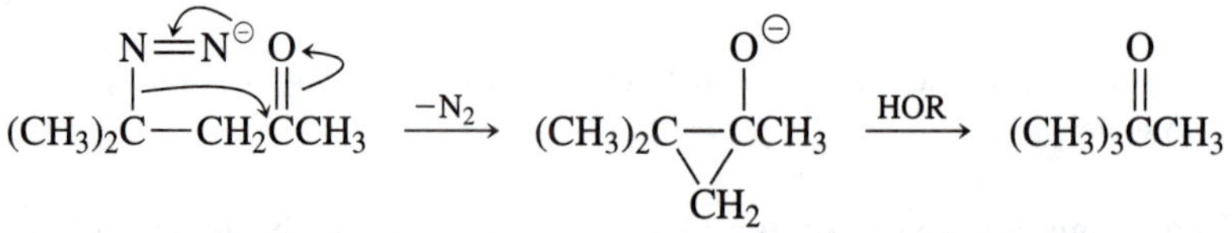

In a few rare instances, as in Eqs. 17a and 17b, even protons at γ carbons can be abstracted by bases.[17]

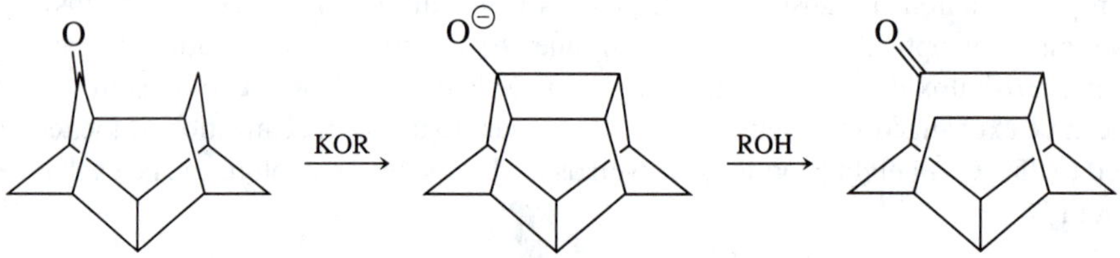

$$\xrightarrow{KOR} \qquad \xrightarrow{ROH} \qquad (17a)$$

$$\xrightarrow{KOR} \qquad \xrightarrow{H-OR} \longrightarrow \qquad (17b)$$

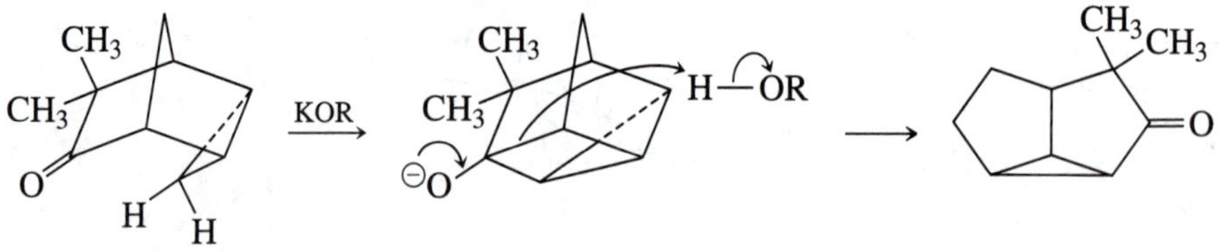

Rearrangements of α-Hetero Carbanions

1,2- and 1,4-Shifts in Wittig rearrangements. Although 1,2-shifts of alkyl groups do not occur in hydrocarbon anions, such rearrangements are common when the negatively charged carbons are substituted with oxygen, nitrogen, or sulfur atoms.

In 1942, Georg Wittig reported that the reaction of benzyl methyl ether with organolithium reagents resulted in migration of the methyl group from oxygen to carbon (Eq. 18).[18]

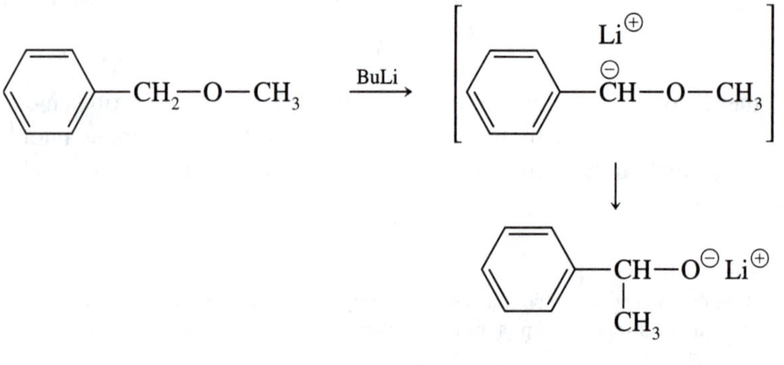

$$(18)$$

Similar rearrangements have been observed in a wide variety of α-oxycarbanions. However, yields of rearrangement products are often low if the ethers can undergo elimination reactions, as in Eq. 19.[19]*

$$CH_3CH_2CH_2-O-CH_2CH_2CH_3$$

$$\downarrow RLi$$

$$CH_3CH_2CH_2OLi \quad + \quad CH_2{=}CHCH_3 \quad + \quad CH_3CH_2CH-OLi \qquad (19)$$
$$\underset{\displaystyle CH_2CH_2CH_3}{|}$$

97%	2%
elimination products	rearrangement product

Wittig originally suggested that the rearrangements proceed by concerted [1,2] shifts. However, chiral migrating groups undergo appreciable racemization during Wittig rearrangements.[22] Furthermore, alkyl groups from the alkyllithium reagents can be incorporated into the Wittig rearrangement products, particularly if excess amounts of alkyllithium reagents are employed (Eq. 20).[23]

$$\text{Ph}-CH_2OCH_3 \xrightarrow{C_4H_9Li} \text{Ph}-\underset{\displaystyle CH_3}{\underset{|}{CH}}-OLi \;+\; \text{Ph}-\underset{\displaystyle C_4H_9}{\underset{|}{CH}}-OLi \qquad (20)$$

In addition, when mixtures of two different ethers are subjected to Wittig rearrangement conditions, crossover products containing parts of both ethers can be obtained.[24]

To explain these observations, it was first suggested that the migrating group is ejected as a carbanion, which then adds to the carbonyl group formed in the cleavage process. That would account for incorporation of alkyl groups from alkyllithium reagents into the products (Eq. 21).

$$\text{Ph}-\overset{\ominus}{CH}{-}O{-}R \longrightarrow R^{\ominus} + \text{Ph}-\underset{\displaystyle}{\overset{\displaystyle H}{C}}{=}O$$

$$\downarrow R'Li \qquad\qquad (21)$$

$$\text{Ph}-\underset{\displaystyle R'}{\overset{\displaystyle H}{\underset{|}{\overset{|}{C}}}}{-}OLi$$

---

*Unusual reactions occur in the carbanions formed from tetrahydrofuran, which undergoes a fragmentation reaction,[20] and from the dihydropyran i, which undergoes an internal nucleophilic substitution reaction.[21]

$$\longrightarrow \quad CH_2{=}CH_2 \;+\; CH_2{=}CHO^{\ominus}Li^{\oplus} \qquad\qquad \longrightarrow \quad \text{▷}-CH{=}CH-OLi$$

i

Later studies showed, however, that the yields of Wittig rearrangement products increase as the migrating groups change from primary to secondary and then to tertiary alkyl groups.[25] That is the expected result if the migrating groups were free radicals but is the opposite of what would be expected if the migrating groups were carbanions. That the migrating groups are indeed free radicals is demonstrated by the fact that migration of a 5-hexenyl group in **7** yields the cyclopentylmethyl derivative **8** as one of the products.[26] 5-Hexenyl radicals are known to cyclize rapidly to cyclopentylmethyl radicals, but no such cyclization occurs with 5-hexenyl carbanions.

$$
\begin{array}{c}
\overset{\text{Li}}{\underset{|}{}} \\
(C_6H_5)_2\overset{|}{C}-O-(CH_2)_4CH{=}CH_2
\end{array}
\quad\longrightarrow\quad
\begin{array}{c}
(CH_2)_4CH{=}CH_2 \\
\underset{|}{} \\
(C_6H_5)_2\overset{|}{C}-OLi
\end{array}
$$

7

$$+$$

$$
\begin{array}{c}
CH_2\!-\!\square \\
\underset{|}{} \\
(C_6H_5)_2\overset{|}{C}-OLi
\end{array}
$$

8

(22)

Therefore, it is now generally accepted that Wittig rearrangements proceed, at least in part, by homolytic dissociation of the carbon oxygen bonds, followed by recombination of the resulting alkyl radicals and ketyl radical anions (Eq. 23).

(23)

The incorporation of alkyl groups from alkyllithium reagents can be explained by electron exchange reactions of the lithium reagents with alkyl radicals from the ethers, as shown in Eq. 24.*

$$CH_3\cdot\ +\ CH_3CH_2CH_2CH{:}^{\ominus}Li^{\oplus}\ \rightleftharpoons\ CH_3{:}^{\ominus}Li^{\oplus}\ +\ CH_3CH_2CH_2CH_2\cdot \qquad (24)$$

---

*An interesting argument against the possibility that Wittig rearrangements proceed in part by stereospecific concerted mechanisms and in part by free radical processes leading to racemization is based on the effects of pressure on the reactions. Increasing the pressure from 0.1 to 1000 atmospheres decreases the rate of rearrangement, as would be expected if a major part of the reaction proceeded by a dissociation process. However, the percent of racemization during the reaction is unchanged, while it should decrease if concerted rearrangements, which should be relatively unaffected by the change in pressure, were accompanying the dissociation processes.[27]

In Wittig rearrangements of allylic carbanions, 1,4-shifts can accompany 1,2-migrations (Eqs. 25 and 26).[28–30]

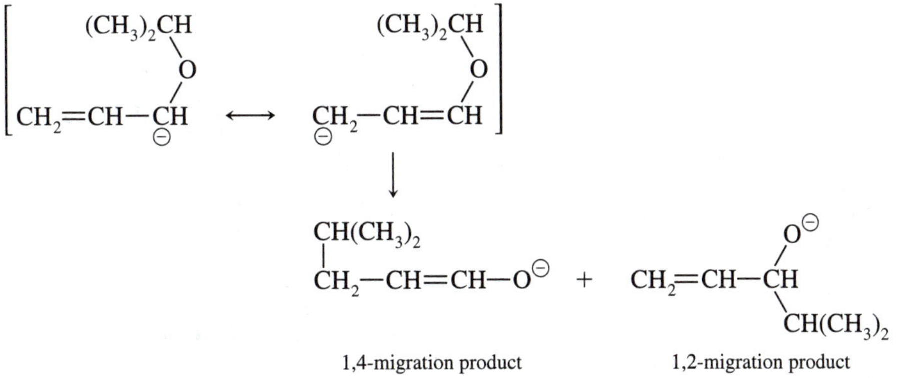

$$
\left[
\begin{array}{cc}
(CH_3)_2CH & (CH_3)_2CH \\
\diagdown & \diagdown \\
O & O \\
\diagup & \diagup \\
CH_2{=}CH{-}\overset{\ominus}{C}H \quad\longleftrightarrow\quad & \overset{\ominus}{C}H_2{-}CH{=}CH
\end{array}
\right]
$$

$$\downarrow \tag{25}$$

$$
\begin{array}{cc}
CH(CH_3)_2 & O^{\ominus} \\
| & \diagup \\
\overset{\ominus}{C}H_2{-}CH{=}CH{-}O^{\ominus} \;+\; & CH_2{=}CH{-}CH \\
& \diagdown \\
& CH(CH_3)_2
\end{array}
$$

1,4-migration product   1,2-migration product

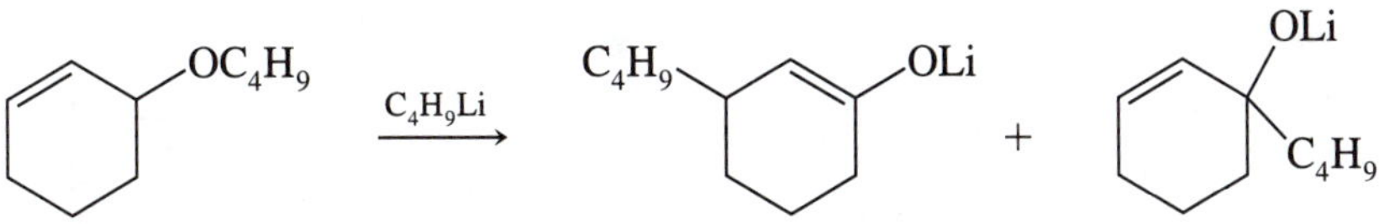

$$
\overset{}{\underset{}{\text{(26)}}}
$$

Suprafacial 1,4-shifts in carbanions are theoretically allowed processes. However, the ratios of 1,4- to 1,2-migration products have been found to increase as the free-radical stabilities of the migrating groups increase,[29] and the percent racemization of a chiral migrating group was found to be the same for 1,4- and 1,2-shifts.[31] Thus, it appears that 1,4-Wittig migrations, like 1,2-migrations, proceed by free-radical dissociation–recombination mechanisms.

[2,3] Wittig rearrangements.   Migrations of allylic groups in Wittig rearrangements usually proceed principally via [2,3] paths and result in inversion of the allylic structures (Eq. 27). 1,2-Migrations may also occur, particularly at relatively high temperatures which would favor homolytic dissociation steps.[32,33]

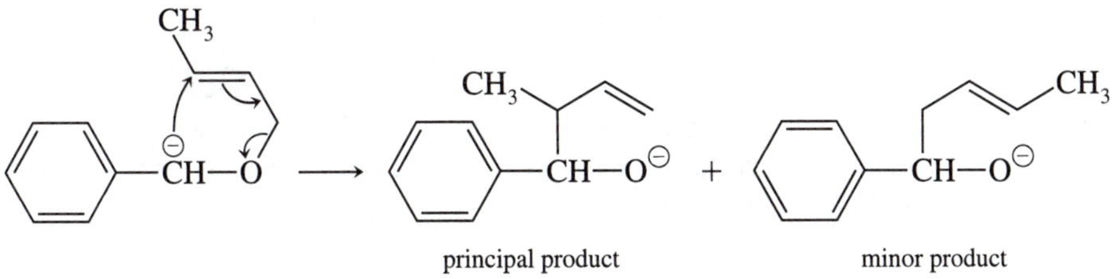

$$\text{(27)}$$

principal product   minor product

Since allylic protons are more acidic than protons on alkyl groups, it is often difficult to form the precursors for [2,3] Wittig rearrangements by abstraction of protons from allyl alkyl ethers. However, the appropriate lithium reagents may be formed by electron-transfer reductions of thioethers (Eq. 28)[34] or by reactions of organotin derivatives with organolithium reagents.[35]

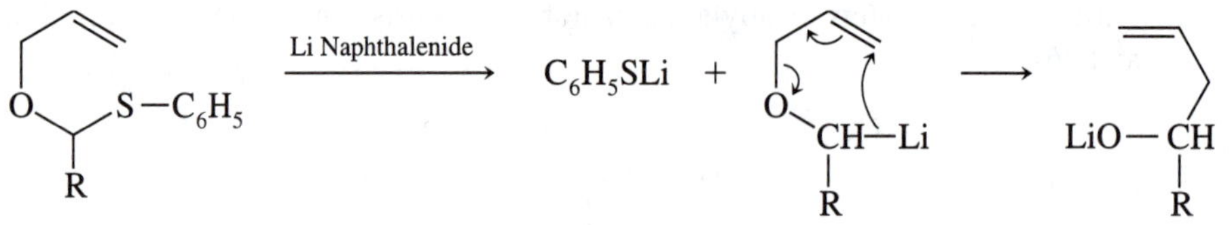

$$\text{(28)}$$

The latter reaction, shown here as Eq. 29, is sometimes known as the *Wittig-Still* rearrangement.

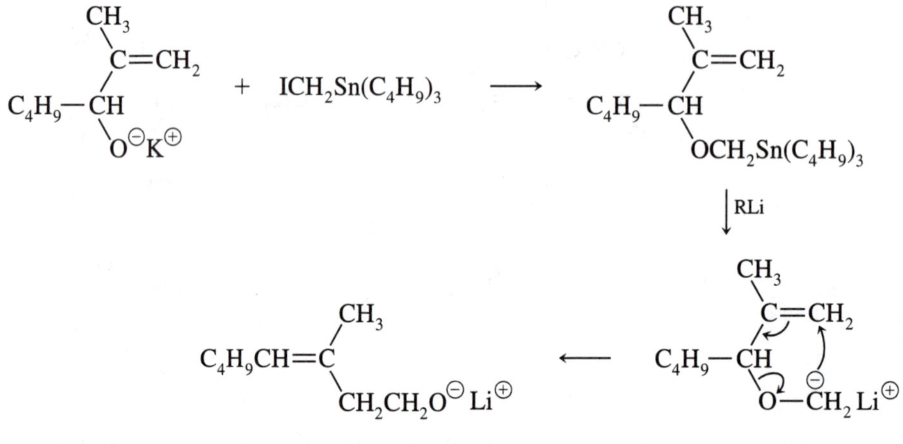

$$\text{(29)}$$

Use in synthesis.   Many studies have demonstrated that if a carbon–oxygen bond to a chiral center is broken in a [2,3] Wittig rearrangement, the new carbon–carbon bond will be formed with nearly complete transfer of chirality.[32,33] In the reaction in Eq. 30, for instance, the allylic chiral center in the product is nearly 100% enantiomerically pure.

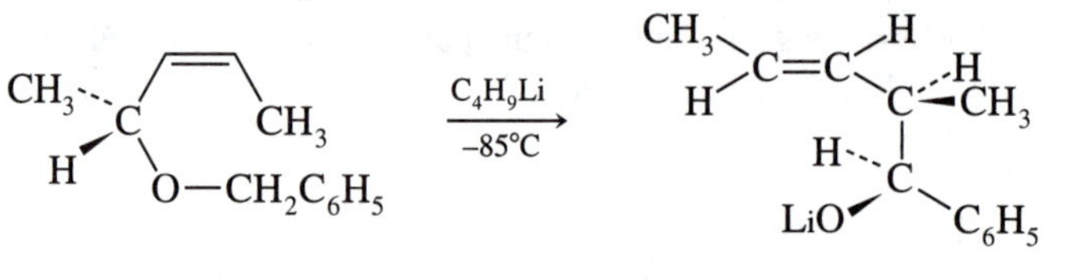

$$\text{(30)}$$

[In addition, the carbinol carbon is formed with greater than 80% enantiomeric purity. If the lithium reagent is formed from a chiral organotin reagent, the rearrangement proceeds with complete inversion at the "anionic" carbon (Eq. 31).[36]]

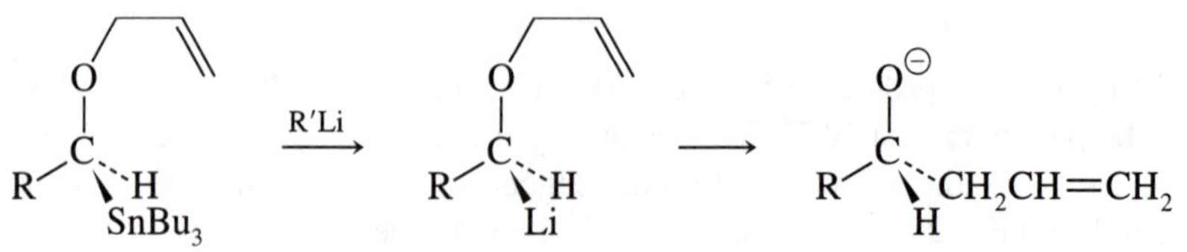

$$\text{(31)}$$

Since [2,3] Wittig rearrangements can be employed to form new chiral centers with high stereoselectivities, they are now widely used synthetic procedures. Such stereospecific processes would be highly unlikely if the rearrangements began by dissociation of the carbanions into free radicals. It is generally accepted, therefore, that [2,3] Wittig rearrangements proceed by concerted, cyclic mechanisms. The use of brackets in describing the order of the rearrangements is quite proper.

Rearrangements of Ylides

Stevens rearrangements.   Molecules in which two oppositely charged atoms are bonded to each other are known as *ylides*. Organic ylides are carbanions that are stabilized by positive charges on neighboring hetero atoms, and possibly, in some cases, by interactions with *d* orbitals on those atoms.

Organic ylides are most commonly formed by reacting quaternary ammonium or phosphonium salts or trialkylsulfonium salts with bases. Ylides in which the carbanions are stabilized by carbonyl groups, in addition to the positively charged hetero atoms, can be formed using common bases such as sodium hydroxide or alkoxides, as shown in Eq. 32b. The formation of less-stabilized ylides usually requires much stronger bases, such as sodium hydride, sodium amide in liquid ammonia, or organolithium reagents (Eqs. 32a and 32c).

$$(CH_3)_4\overset{\oplus}{N}\ \overset{\ominus}{Br} \quad \xrightarrow{C_4H_9Li} \quad (CH_3)_3\overset{\oplus}{N}{-}\overset{\ominus}{CH_2} \quad (+\ C_4H_{10} + LiBr) \qquad (32a)$$

$$\overset{\displaystyle O}{\underset{\displaystyle }{C_6H_5\overset{\|}{C}CH_2\overset{\oplus}{N}(C_2H_5)_3}} \quad \xrightarrow{NaOH} \quad \overset{\displaystyle O}{\underset{\displaystyle }{C_6H_5\overset{\|}{C}{-}\overset{\ominus}{C}H{-}\overset{\oplus}{N}(C_2H_5)_3}} \qquad (32b)$$

$$(CH_3)_3\overset{\oplus}{S}\ \overset{\ominus}{I} \quad \xrightarrow{NaH} \quad (CH_3)_2\overset{\oplus}{S}{-}\overset{\ominus}{CH_2} \qquad (32c)$$

Less commonly, ylides have been obtained from reactions of organic sulfides or of tertiary amines with carbenes or benzyne (Eqs. 33 and 34).[38]

$$(33)$$

$$(34)$$

Alkyl and benzyl groups in ammonium[39] and sulfonium[40] ylides can migrate from the positively charged nitrogen or sulfur atoms to the adjacent negatively charged carbon atoms, as shown in Eqs. 35 and 36.*

$$C_6H_5\overset{\overset{\text{O}}{\|}}{C}-\overset{\ominus}{\underset{\underset{\text{CH}_2\text{C}_6\text{H}_5}{|}}{C}}H-\overset{\oplus}{N}(CH_3)_2 \quad \longrightarrow \quad C_6H_5\overset{\overset{\text{O}}{\|}}{C}-\underset{\underset{\text{CH}_2\text{C}_6\text{H}_5}{|}}{C}H-N(CH_3)_2 \qquad (35)$$

$$\underset{H_3C}{\overset{H_3C}{>}}\overset{\oplus}{S}-\overset{\ominus}{C}H_2 \quad \longrightarrow \quad H_3C-S-CH_2CH_3 \qquad (36)$$

These reactions, which are called *Stevens rearrangements,* are often so rapid that the ylides cannot even be detected before they rearrange. Ylides in which the negative charges are stabilized by carbonyl groups usually have longer lifetimes, and may often be isolated as crystalline salts.[41,42]

Stevens rearrangements have proven to be very useful methods for the formation of new carbon–carbon bonds. Sulfur ylides are most commonly used for this purpose, since elimination of sulfides or mercaptans from the products, as shown in Eq. 37,[43] is usually easier than elimination of amines.[40] However, nitrogen ylides have also proven useful in some cases.[44]

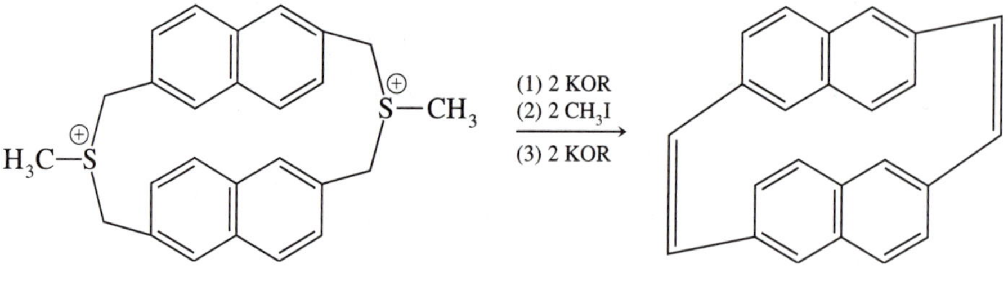

$$\xrightarrow[\text{(3) 2 KOR}]{\substack{\text{(1) 2 KOR} \\ \text{(2) 2 CH}_3\text{I}}} \qquad (37)$$

Ylides containing allylic anions can undergo 1,4- as well as 1,2-Stevens rearrangements (Eq. 38).[45]

$$C_6H_5CH_2-\overset{\oplus}{\underset{\underset{\text{CH}_3}{|}}{\overset{\overset{\text{CH}_3}{|}}{N}}}-\overset{\ominus}{C}H-CH=CH_2$$

$$\downarrow$$

$$(CH_3)_2N-\underset{\underset{\text{CH}_2\text{C}_6\text{H}_5}{|}}{C}H-CH=CH_2 \quad + \quad (CH_3)_2N-CH=CH-CH_2CH_2C_6H_5$$

$$(38)$$

---

*Phosphorus ylides (Wittig reagents) do not undergo Stevens rearrangements.

1,2-Stevens rearrangements of ammonium ylides often proceed with nearly complete retention of the configurations of chiral migrating groups.[45,46] This initially led to the conclusion that the reactions were intramolecular and concerted. However, it was later shown that when Stevens rearrangements were carried out in less viscous solvents, where free radicals could more rapidly move out of the "solvent cages" in which they were formed, the products were partially racemized.[47] 1,4-Migrations resulted in the partial racemization of chiral migrating groups even in relatively viscous solvents.

Further evidence that 1,2-Stevens rearrangements of ammonium ylides proceed by dissociation–recombination mechanisms came from the demonstration that simultaneous rearrangements of two ylides led to the formation of products incorporating parts of both ylides. The percentages of such cross-products increased as the viscosities of the solvents decreased.[47]

Stevens rearrangements of sulfonium ylides also appear to proceed by free-radical dissociation–recombination processes. They result in significant racemization of chiral migrating groups, yield intermolecular cross-products from rearrangements of mixtures of ylides, and yield byproducts that would result from recombination of free radicals, as shown in Eq. 39.[48,49] They also have large positive entropies of activation, as would be expected if the initial steps are homolytic dissociations.[50]

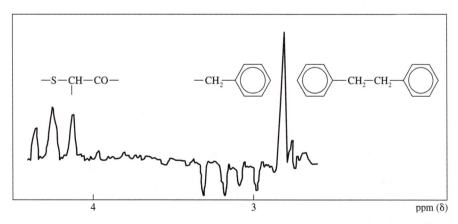

Perhaps the most convincing evidence that Stevens rearrangements proceed by way of free-radical intermediates is the fact that the products from both ammonium and sulfonium ylides show CIDNP (*chemically induced dynamic nuclear polarization*) peaks in their ^{1}H NMR spectra.[48,50,51] CIDNP peaks are exceptionally strong signals that result from the emission of energy rather than its absorption, and thus appear as negative peaks in NMR spectra. For instance, negative peaks are evident in the ^{1}H NMR spectrum for the reaction in Eq. 39 (Figure 8-1).

FIGURE 8.1 ^{1}H NMR spectrum (showing CIDNP peaks) taken in the course of the reaction in Eq. 39.[48] (Reprinted with permission from Elsevier Science, Ltd.)

The presence of CIDNP signals in NMR spectra is incontrovertible evidence that a reaction product has been formed, at least in part, by the combination or interaction of two free radicals.[52]*

Sommelet–Hauser rearrangements.   [2,3] Migrations of allylic groups in ylides are known as *Sommelet–Hauser rearrangements*. As indicated by the reactions in Eqs. 40 and 41, ylides will normally undergo concerted Sommelet–Hauser rearrangements, if possible, in preference to Stevens rearrangements.

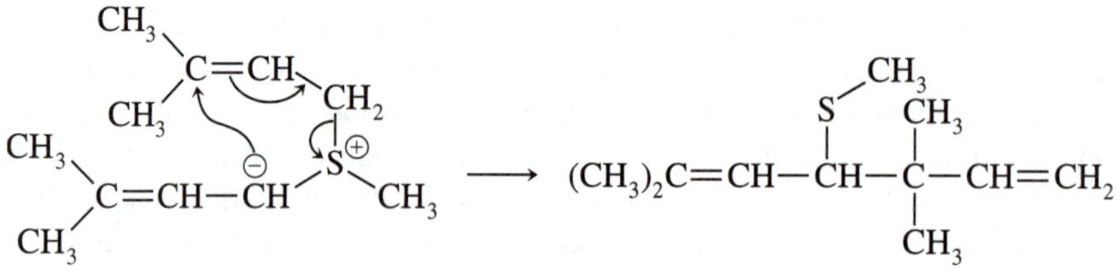

$$(40)$$

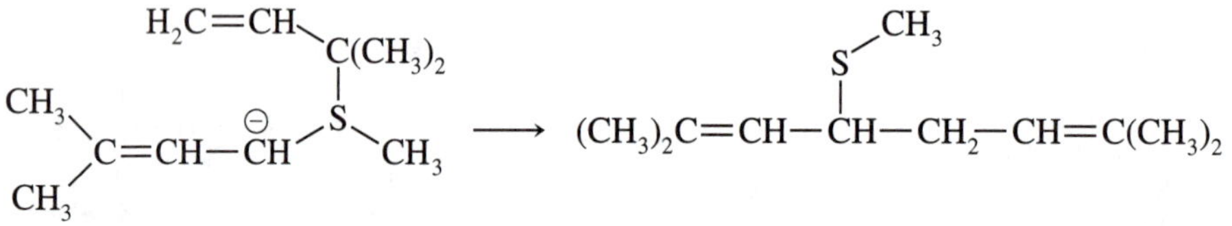

$$(41)$$

Sommelet–Hauser rearrangements usually take precedence over Stevens rearrangements even when the double bond participating in the reaction is part of an aromatic ring, as in the rearrangement of ylide **9**.[53,54]

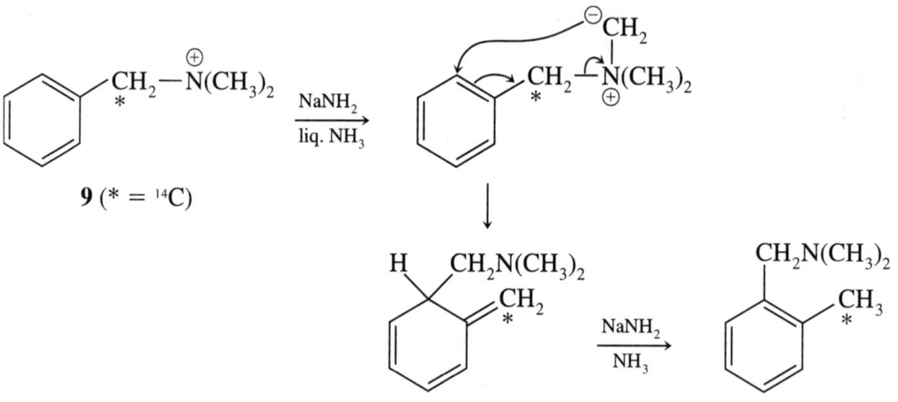

9 (* = ^{14}C)

$$(42)$$

10

---

*CIDNP signals are not seen in the spectra of products from free-radical chain reactions, since those products are principally formed by reactions of free radicals with fully bonded molecules, rather than by the combination of free radicals. Even when the products of a reaction are formed by the combination of free radicals, however, CIDNP peaks cannot always be observed in their NMR spectra. CIDNP peaks usually have very short lifetimes, so that CIDNP spectra are typically taken during the course of a reaction rather than after its completion. Even so, the peaks may exist for too brief a time to be detected.

That the Sommelet–Hauser rearrangement of **9** does indeed proceed by a [2,3] migration, rather than by a [1,4] shift in the benzylic ylide as in Eq. 43, has been demonstrated by labeling the benzylic carbon with ^{14}C. As predicted by the [2,3] migration mechanism, the radioactive carbon in the product appeared in the methyl group on the aromatic ring.[53,54]

$$
\begin{array}{ccc}
\text{H}_3\text{C} - \overset{\oplus}{\text{N}}(\text{CH}_3)_2 & & \text{CH}_3 \\
\ominus \text{CH} & \longrightarrow\!\!\!/\!\!\!\longrightarrow & \text{CH}_2\text{N}(\text{CH}_3)_2 \\
* & & *
\end{array}
\tag{43}
$$

(Does not occur)

In some instances, Stevens rearrangements accompany Sommelet–Hauser rearrangements, but the products from Sommelet–Hauser rearrangements usually predominate. The pronounced preference for Sommelet–Hauser rearrangements is particularly striking since the [2,3] migration mechanism may require forming an ylide by abstracting a proton from a methyl group, even though a benzylic ylide should be far more stable. In fact, it has been shown that the benzylic ylides are formed preferentially and can be trapped by reaction with benzophenone (Eq. 44).[55] However, trace amounts of the methyl ylides in equilibrium with the benzylic ylides are sufficient to yield rapid Sommelet–Hauser rearrangements.

$$
\overset{\ominus}{\text{Br}} \quad \text{CH}_2 - \overset{\oplus}{\text{N}}(\text{CH}_3)_3 \;+\; \text{C}_6\text{H}_5\overset{\text{O}}{\overset{\|}{\text{C}}}\text{C}_6\text{H}_5 \xrightarrow{\text{NaNH}_2}
\begin{array}{c}
\overset{\oplus}{\text{N}}(\text{CH}_3)_3 \\
\text{CH} \\
\text{C}(\text{C}_6\text{H}_5)_2 \\
\overset{\ominus}{\text{O}}
\end{array}
\tag{44}
$$

If the *ortho* positions of the aromatic ring in the ylide are substituted with alkyl groups, the final tautomerism steps cannot take place, and methylenecyclohexadienes are formed instead of aromatic products (Eq. 45).[56]

$$
\begin{array}{c}
\text{CH}_2 - \overset{\oplus}{\text{N}}(\text{CH}_3)_3 \\
\text{H}_3\text{C} \qquad \text{CH}_3
\end{array}
\xrightarrow[\text{liq. NH}_3]{\text{NaNH}_2}
\begin{array}{c}
\text{CH}_2 \\
\text{H}_3\text{C} \qquad \text{CH}_2\text{N}(\text{CH}_3)_2 \\
\text{CH}_3
\end{array}
\tag{45}
$$

Even methylenecyclohexadiene **10** (page 228), which is very easily converted to its aromatic tautomer, can be obtained if the rearrangement of **9** is carried out employing a minimal amount of base.[57]

Rearrangements Resulting from Intramolecular Substitution Reactions

Favorskii rearrangements. The reactions of α-haloketones[58] or α,β-epoxyketones[59] with hydroxide, alkoxide, or amide anions often yield carboxylic acid derivatives resulting from the migration of an alkyl group from the carbonyl group to the α-carbon, as in Eqs. 46–48.

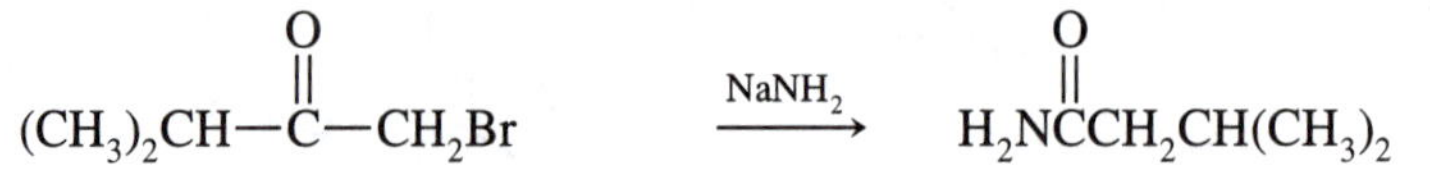

O
(CH3)2CH—C—CH2Br
NaNH2
O
H2NCCH2CH(CH3)2

(46)

H3C
O
O
CH(CH3)2
KOH
H3C OH
CH(CH3)2
CO2

(47)

O
O
Br
Br
KOH
Δ
CO2
O2C

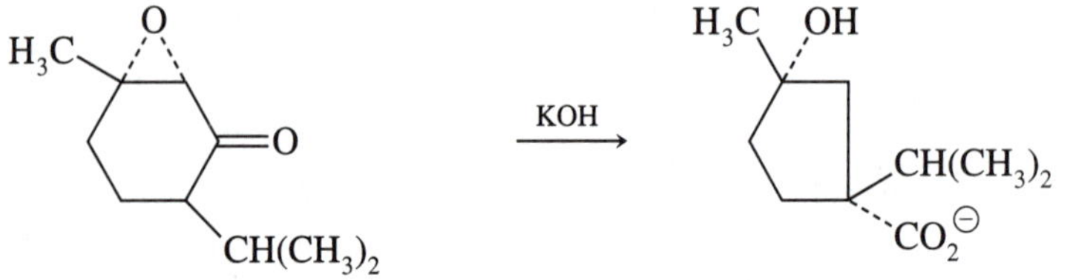

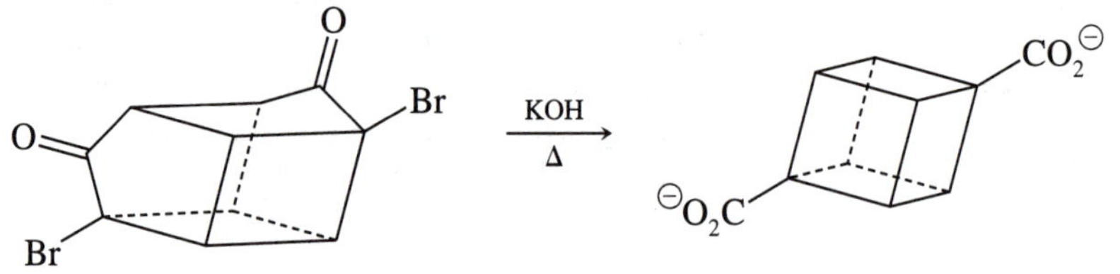

(48)

To distinguish among several possible mechanisms for Favorskii rearrangements, R.B. Loftfield prepared 2-chlorocyclohexanone labeled equally with radioactive ^{14}C at the carbonyl carbon and at C2. The positions of the radioactive carbons were unchanged in samples of 2-chlorocyclohexanone recovered after reaction of the labeled ketone with solutions of sodium alkoxides. However, in cyclopentanecarboxylic acid derivatives resulting from Favorskii rearrangements, almost exactly 50% of the radioactivity was located in the carboxyl group, and 25% was at C1 of the cyclopentane ring. The remaining 25% was found at C2 and C5 of the ring (Eq. 49).[60] (The symmetry of the molecule made it impossible to say whether the radioactivity was divided between the two carbons or located at only one carbon.)

O
Cl
RONa
KOH
25%
O
C—OH
50%
(25%)

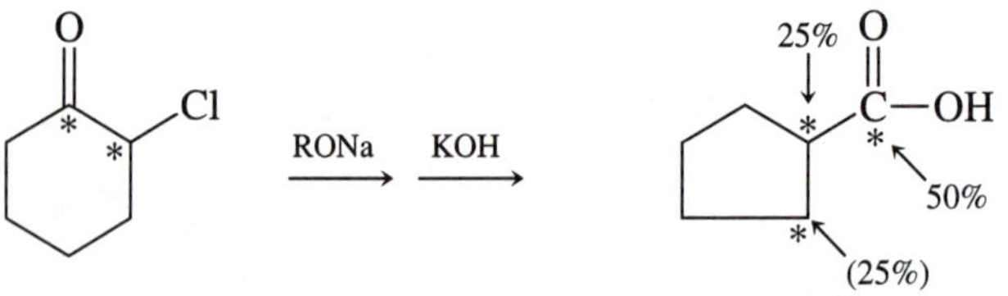

(49)

Loftfield's results clearly ruled out the direct migration of the methylene group from the carbonyl group to the α-carbon, since that would have left 50% of the total radioactivity at C1 of the cyclopentane ring. Instead, the nearly equal distribution of radioactive carbon between C1 and the neighboring ring carbons supported an old suggestion that the products were formed from a cyclopropanone intermediate. Loftfield proposed that the symmetrical cyclopropanone ring opened on reaction with base to distribute the labeled carbons between C1 and C2 of the resulting cyclopentane ring (Eq. 50).[60]

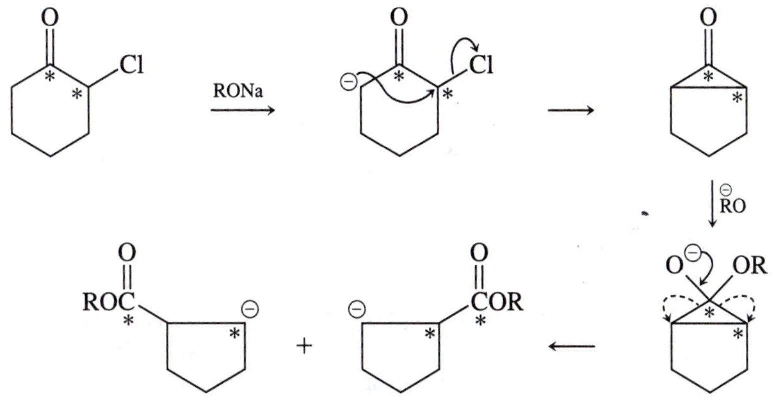

(50)

The formation of cyclopropanones under Favorskii conditions has since been clearly established. For instance, 2,3-di-*tert*-butylcyclopropanone, which is relatively unreactive to base, can be directly isolated from reaction of an α-chloroketone with sodium hydroxide (Eq. 51).[61]

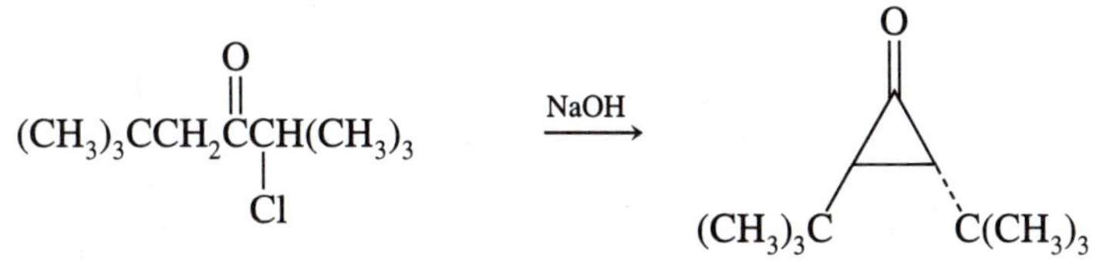

(51)

Cyclopropanones with less bulky substituents cannot be isolated from strong base solutions. At short reaction times, however, the reaction of 2-chloropentan-3-one with sodium methoxide in methanol yielded a hemiacetal, which could be acylated to form a stable derivative of a cyclopropanone (Eq. 52).[62]

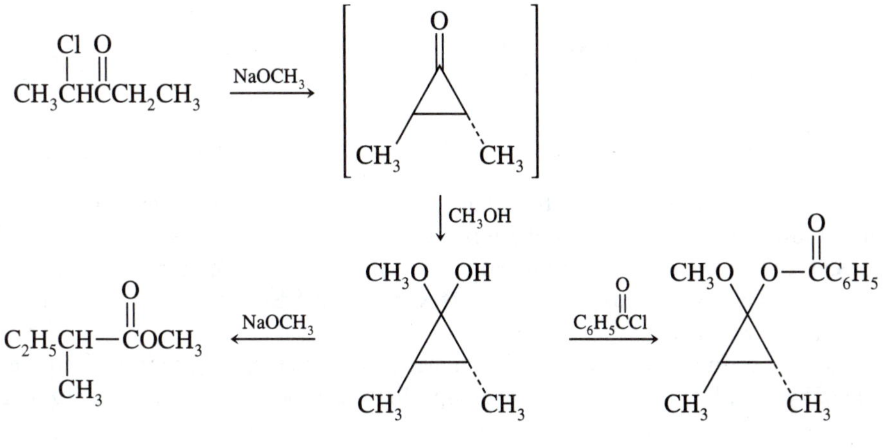

(52)

The reactions of α,α'-dibromoketones with triethylamine yield stable cyclopropenone derivatives (e.g., Eq. 53).

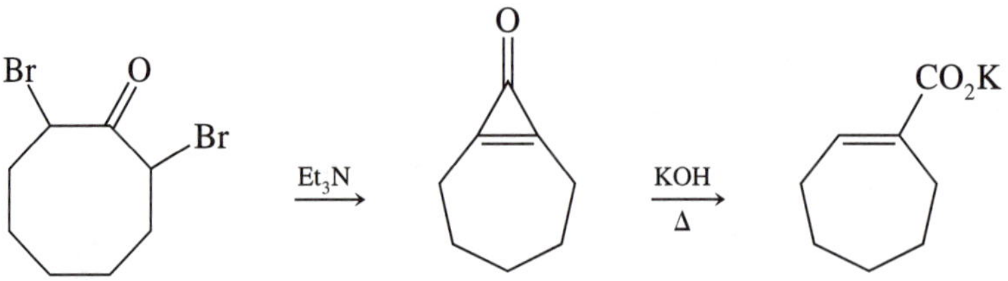

(53)

On heating in potassium hydroxide solutions, the cyclopropenones open to form unsaturated carboxylic acids, which can also be obtained directly from the starting dihaloketones.[63]

If the cyclopropanone intermediates in Favorskii rearrangements are formed by "internal S_N2" displacements of halide ions, the reactions should result in inversions of the configurations of the halogenated carbons. Reaction of ketones **11a** and **11b** with sodium methoxide in ether or dimethoxyethane did indeed yield their inversion products **12a** and **12b**.[64]

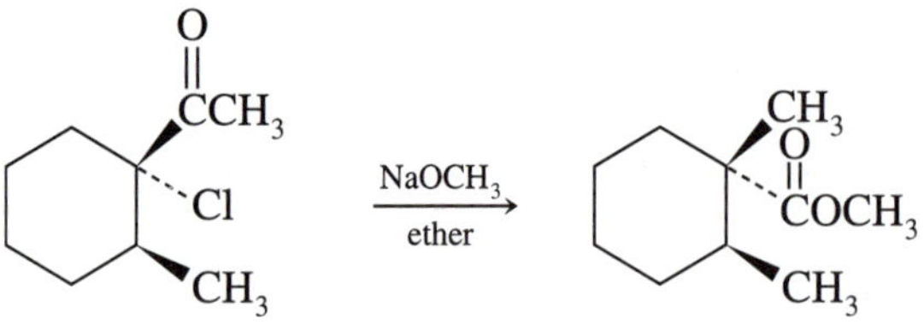

(54)

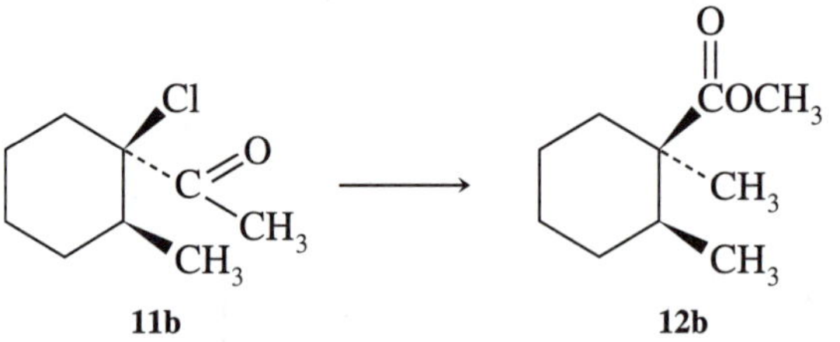

(55)

However, reaction of **11a** with sodium methoxide in the more polar, hydroxylic, solvent methanol resulted in the formation of a mixture of **12a** and **12b**.[65] Thus, it appears that the displacements of halide ions in Favorskii rearrangements proceed by intramolecular analogs of S_N2 reactions in nonpolar solvents but by processes resembling S_N1 reactions of enolate anions in better-ionizing solvents.

The ring-opening steps in the Loftfield mechanism are unusual, since they require the formation of hydrocarbon anions as leaving groups. In the absence of resonance stabilization of one of the two possible anions, the less-substituted alkyl anion is normally formed, as shown in Eq. 56.[59]

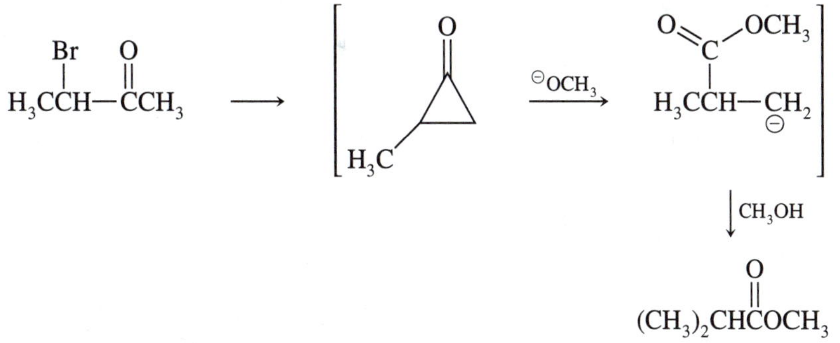

$$\text{Br} \quad \text{O}$$
$$H_3CCH{-}CCH_3 \longrightarrow \left[\begin{array}{c} \text{O} \\ \overset{\displaystyle \triangle}{\underset{H_3C}{}} \end{array} \xrightarrow{\ominus OCH_3} \begin{array}{c} \text{O}{=}\overset{OCH_3}{\underset{|}{C}} \\ H_3CCH{-}\underset{\ominus}{CH_2} \end{array} \right] \tag{56}$$

$$\Big\downarrow CH_3OH$$

$$\text{O}$$
$$(CH_3)_2CHCOCH_3$$

These reactions are similar to ring-opening reactions of cyclopropoxide anions (see page 219 and Eq. 57).[66] In both cases, relief of angle strain in the three-membered ring allows the ejection of hydrocarbon anions as leaving groups.

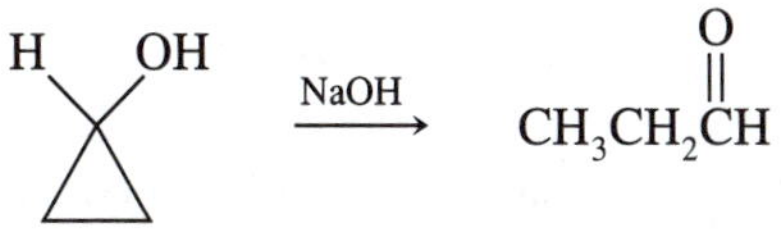

$$\begin{array}{c} H \quad OH \\ \overset{\displaystyle \triangle}{\bigtriangleup} \end{array} \xrightarrow{NaOH} \begin{array}{c} \text{O} \\ CH_3CH_2\overset{\|}{CH} \end{array} \tag{57}$$

Quasi-Favorskii rearrangements. The Loftfield mechanism requires that ketones be converted to their enolate anions as the initial steps in Favorskii rearrangements. However, many ketones lacking hydrogens on their unhalogenated α-carbons undergo reactions that yield products of Favorskii rearrangements (as in Eq. 58), though often at comparatively slow rates.

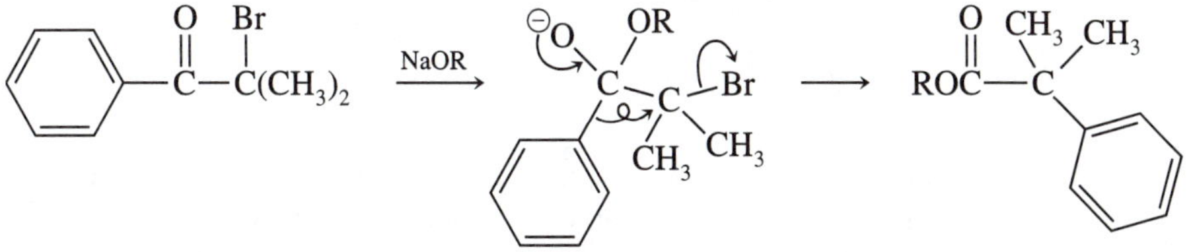

$$\tag{58}$$

These reactions, which are called quasi-Favorskii (or semi-Favorskii) rearrangements, may also occur in bicyclic ketones such as **13**, in which enolate anion formation is inhibited because a double bond at the bridgehead would be twisted away from planarity,[67] and in some enolizable ketones, such as cyclobutanones, in which the cyclopropanones formed via the Loftfield mechanism would be exceptionally strained.[68]

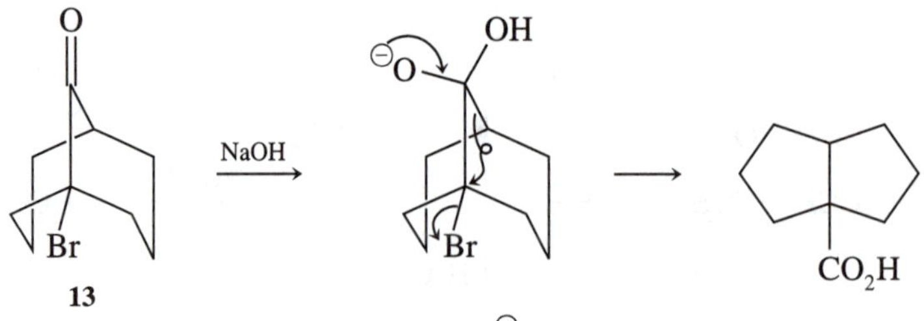

$$(59)$$

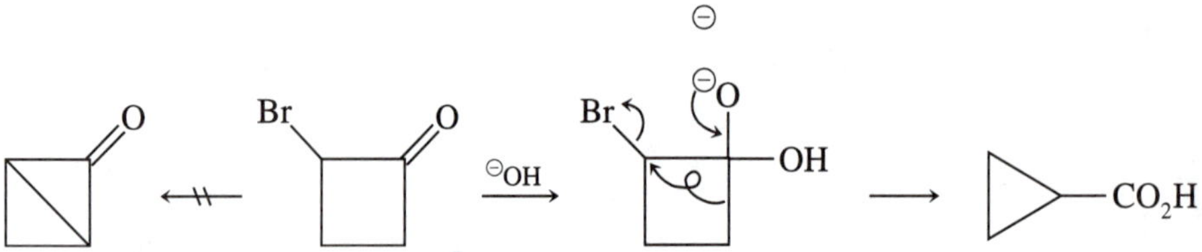

$$(60)$$

Despite their apparent similarity to Favorskii rearrangements, quasi-Favorskii rearrangements actually belong among the migrations to electron-deficient centers discussed in Chapter 6. The rearrangements appear to proceed by addition of bases to the carbonyl groups, followed by [1,2] shifts of the migrating groups accompanying loss of the halide ions. They are closely related to other base-catalyzed [1,2] shifts, such as benzilic acid rearrangements, as well as to semipinacolic rearrangements.

The Ramburg–Bäcklund reaction. In a *Ramburg–Bäcklund reaction* an α-halosulfone reacts with base to yield an alkene formed by joining the two alkyl groups of the sulfone by a double bond.[69] Sulfur dioxide is eliminated during the reaction. Eqs. 61 and 62 illustrate this type of reaction.

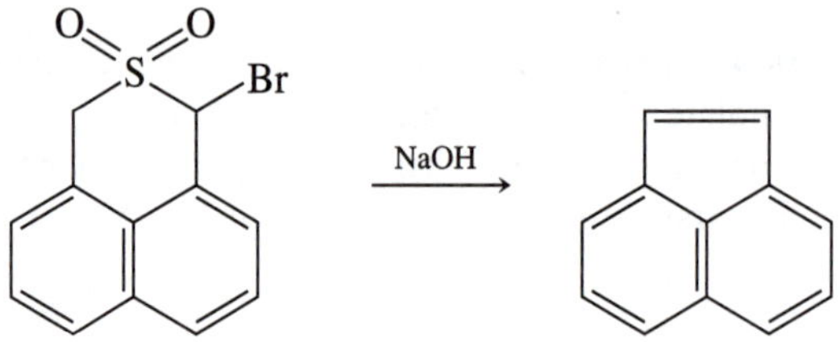

$$(61)$$

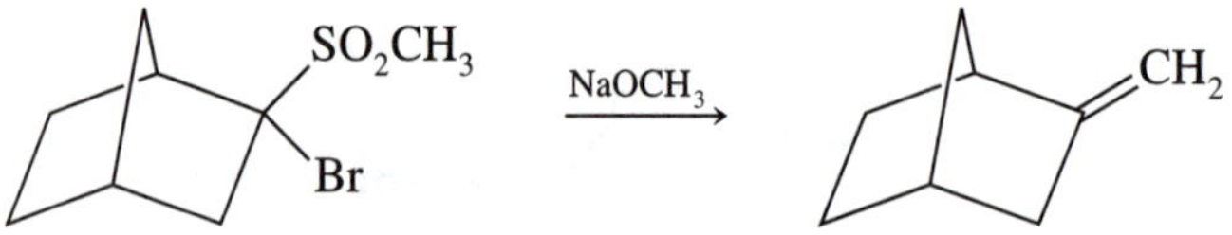

$$(62)$$

Ramburg–Bäcklund reactions appear to be closely related to Favorskii rearrangements. After the initial formation of α-sulfonyl carbanions, internal nucleophilic displacements of halide anions result in the formation of three-membered sulfone rings (Eq. 63).

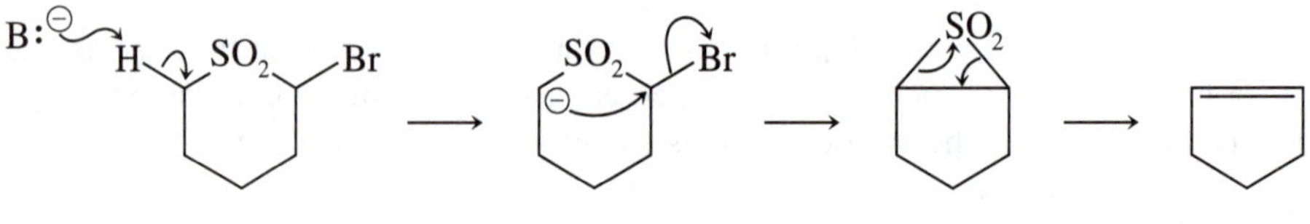

$$(63)$$

(There are no analogs to quasi-Favorskii rearrangements among Ramburg–Bäcklund rearrangements. Ramburg–Bäcklund rearrangements will not take place unless the starting halo-sulfones can initially be converted to carbanions that can displace halide ions from the other α-positions.[69])

In at least one case, a three-membered ring sulfone was isolated when an α-halosulfone was treated with base at −78°C.[70] Usually, however, the cyclic sulfone intermediates in Ramburg–Bäcklund reactions rapidly lose sulfur dioxide to form double bonds. Unlike cyclo-propanones, or the intermediates from reactions with dihalosulfones (to be discussed next), they do not react with bases to form derivatives of acidic products.

Ramburg–Bäcklund reactions of dihalosulfones can result in the formation of alkynes, vinyl halides, and salts of sulfonic acids (Eq. 64).[69]

$$C_6H_5CH_2SO_2CHBr_2 \xrightarrow[\text{HOCH}_3]{\text{NaOCH}_3} \quad \underset{C_6H_5}{\overset{\overset{\displaystyle SO_2}{\diagup \quad \diagdown}}{CH-CHBr}}$$

(64)

$$C_6H_5C{\equiv}CH \; + \; C_6H_5CH{=}CHBr \; + \; C_6H_5CH{=}CHSO_3^{\ominus}Na^{\oplus}$$

If tertiary amines are employed as the bases, however, unsaturated cyclic sulfones (thiirene 1,1-dioxides) can be isolated.[71] The thiirene dioxides will lose sulfur dioxide on heating to form alkynes (Eq. 65).

$$\underset{C_6H_5CH-SO_2-CH-C_6H_5}{\overset{\overset{\displaystyle Br \qquad\quad Br}{|\qquad\qquad\;|}}{}} \xrightarrow{\text{Et}_3N} \quad \underset{C_6H_5 \qquad\quad C_6H_5}{\overset{\overset{\displaystyle SO_2}{\triangle}}{}}$$

(65)

$$\downarrow 120°C$$

$$C_6H_5C{\equiv}CC_6H_5 \; + \; SO_2$$

The Neber rearrangement. Unlike the Favorskii and Ramburg–Bäcklund rearrangements, the *Neber rearrangement* does not result in migrations of alkyl groups. Instead, the reactions of oxime tosylates or of N-chloroimines with strong bases result in migrations of nitrogen atoms to form α-amino ketones (Eq. 66).[72]

(66)

When Neber rearrangements are carried out at low temperatures[73] or in the absence of hy-droxylic solvents,[74] it is possible to isolate azirines (cyclic, three-membered ring imines) from the reactions (Eq. 67). The azirines can be converted to α-aminoketones on hydrolysis, demonstrating that they are intermediates in Neber rearrangements.*

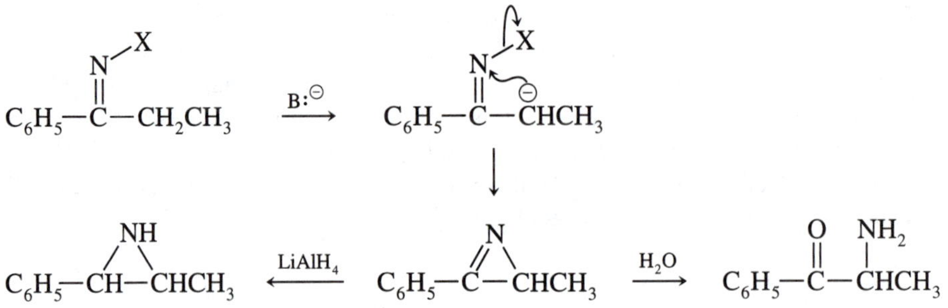

(67)

8.2 REARRANGEMENTS OF FREE RADICALS

The Formation of Free Radicals

Decarbonylation of aldehydes. Organic free radicals are often formed by the reactions of other radicals or atoms with carbon–hydrogen bonds, as shown in Eq. 68.

(68)

However, hydrogen abstraction processes are seldom useful for forming free radicals with specific structures, since there are usually several different positions from which hydrogens can be removed.

The reactions of free radicals with aldehydes, however, usually result in selective attack at the hydrogens bonded to the carbonyl groups. Thus, acyl radicals are formed when aldehydes are heated in the presence of free-radical chain initiators, such as organic peroxides or azoisobutyronitrile (AIBN) (as in Eq. 69). The acyl radicals usually lose carbon monoxide to form alkyl or aryl radicals of known structure whose reactions, including possible rearrangements, can be studied. The alkyl radicals, in turn, abstract hydrogen atoms from aldehyde carbonyls, resulting in efficient chain processes that lead to the decarbonylation of the aldehydes.

---

*The mechanism by which azirine rings are formed in the Neber rearrangement is uncertain. Although the ring is shown in Eq. 67 as being formed by a one-step displacement process, there is no evidence demonstrating that the anion does not lose a halide or tosylate anion (as shown below) to form a nitrene (see Chapter 9), which then cyclizes to form the azirine.

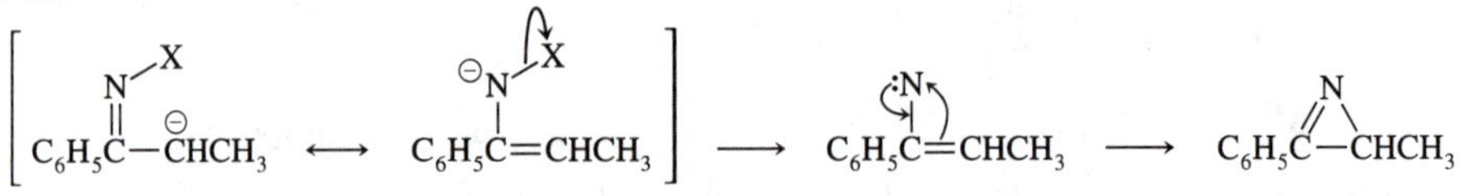

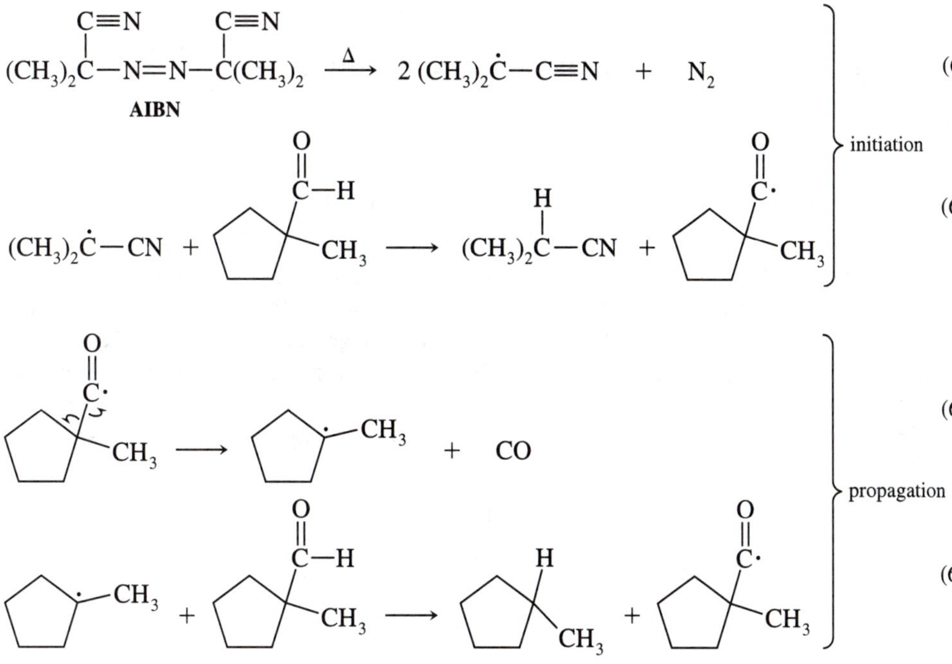

$$(CH_3)_2\overset{\underset{|}{C\equiv N}}{C}-N=N-\overset{\underset{|}{C\equiv N}}{C}(CH_3)_2 \xrightarrow{\Delta} 2\,(CH_3)_2\dot{C}-C\equiv N \ + \ N_2 \qquad (69a)$$

AIBN

$$(CH_3)_2\dot{C}-CN \ + \ \text{[cyclopentane with C(=O)H and CH}_3\text{]} \longrightarrow (CH_3)_2\overset{\underset{|}{H}}{C}-CN \ + \ \text{[cyclopentane with C·(=O) and CH}_3\text{]} \qquad (69b)$$

initiation

$$\text{[cyclopentane with C·(=O) and CH}_3\text{]} \longrightarrow \text{[cyclopentane with ·CH}_3\text{]} \ + \ CO \qquad (69c)$$

propagation

$$\text{[cyclopentane with ·CH}_3\text{]} \ + \ \text{[cyclopentane with C(=O)H and CH}_3\text{]} \longrightarrow \text{[cyclopentane with CH}_3\text{, CH}_3\text{]} \ + \ \text{[cyclopentane with C·(=O) and CH}_3\text{]} \qquad (69d)$$

Reduction by tin hydrides. Organic radicals of known structure are frequently prepared by reacting organic halides with trialkyltin hydrides. Catalytic amounts of free-radical initiators are employed to initiate the free-radical chain reactions (Eq. 70).

$$(C_4H_9)_3Sn-H \ + \ X\cdot \longrightarrow XH \ + \ (C_4H_9)_3Sn\cdot \qquad (70a)$$

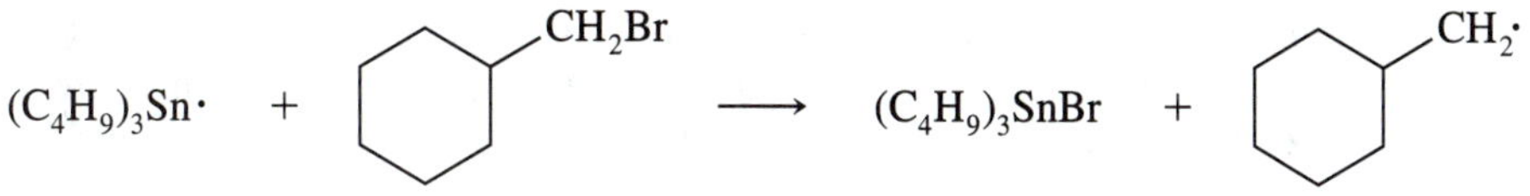

$$(C_4H_9)_3Sn\cdot \ + \ \text{[cyclohexane-CH}_2Br\text{]} \longrightarrow (C_4H_9)_3SnBr \ + \ \text{[cyclohexane-CH}_2\cdot\text{]} \qquad (70b)$$

$$\text{[cyclohexane-CH}_2\cdot\text{]} \ + \ (C_4H_9)_3Sn-H \longrightarrow \text{[cyclohexane-CH}_3\text{]} \ + \ (C_4H_9)_3Sn\cdot \qquad (70c)$$

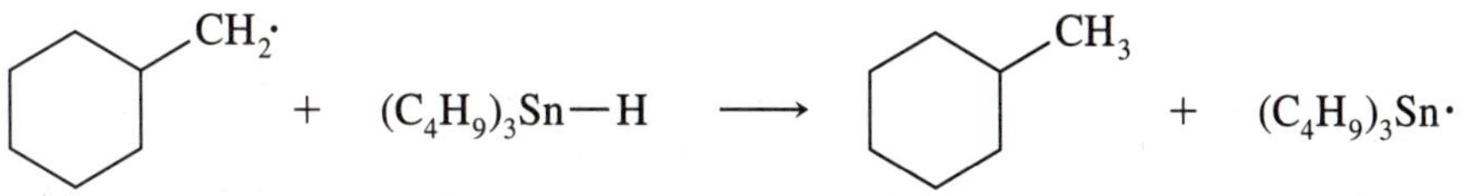

The overall result of the reaction is to reduce carbon–halogen bonds to carbon–hydrogen bonds.[75]

Decomposition of peroxides.   Thermal cleavage of the weak oxygen–oxygen bonds in diacyl peroxides results in the formation of carbalkoxy radicals. The carbalkoxy radicals normally lose carbon dioxide to form hydrocarbon radicals, as shown in Eq. 71b, unless the hydrocarbon radicals are of unusually high energy.[75]

$$(CH_3)_2CH-\overset{\overset{\displaystyle O}{\|}}{C}-O-O-\overset{\overset{\displaystyle O}{\|}}{C}CH(CH_3)_2 \quad \overset{\Delta}{\longrightarrow} \quad 2\ (CH_3)_2CH-\overset{\overset{\displaystyle O}{\|}}{C}-O^{\cdot} \tag{71a}$$

$$2\ (CH_3)_2CH-\overset{\overset{\displaystyle O}{\|}}{C}-O^{\cdot} \quad \longrightarrow \quad 2\ CO_2 \ + \ 2\ (CH_3)_2CH\cdot \tag{71b}$$

Unlike reactions of organic halides with trialkyltin hydrides, formation of free radicals by decomposition of diacyl peroxides does not provide an automatic way to "cap" the resulting radicals. The alkyl radicals may abstract hydrogen or halogen atoms from solvents or from other molecules in solution or may add to double bonds, thereby initiating free-radical chain processes. In the absence of other reagents, the radicals formed by the decomposition of the peroxides may form dimerization or disproportionation products, as shown in Eqs. 72 and 73, respectively.

$$2\ (CH_3)_2CH\cdot \quad \longrightarrow \quad (CH_3)_2CH-CH(CH_3)_2 \tag{72}$$
dimerization product

$$(CH_3)_2CH\overset{\overset{\displaystyle H}{|}}{}\quad CH_2-CH-CH_3 \quad \longrightarrow \quad CH_3CH_2CH_3 \ + \ CH_2=CH-CH_3 \tag{73}$$
disproportionation products

Despite the complexity of the product mixtures which may arise from thermal decompositions of diacyl peroxides, significant information about rearrangements of free radicals has been obtained from radicals formed by these reactions.

Sigmatropic Shifts

[1,n] Rearrangements.   Suprafacial [1,2] shifts in free radicals are normally forbidden, since the HOMOs of organic radicals and of carbanions are the same. In fact, there are no clear examples of [1,2] shifts of hydrogen atoms or alkyl groups in uncharged organic radicals.*

---

*However, the energy barriers to suprafacial [1,2] shifts in radicals should be lower than those in carbanions since only one electron would have to be raised to an antibonding orbital in the transition state for the rearrangement of a free radical. Molecular orbital calculations indicate that the barriers to the migrations would be further lowered in radicals with strongly electron-withdrawing substituents, which would lower the energies of antibonding HOMOs in the transition states for "forbidden" rearrangements.[77] (To put it another way, the more closely a free radical resembles a carbocation, the lower the barrier to a suprafacial [1,2] migration.) 1,2-Shifts of hydrogen atoms do appear to have been observed in the mass spectra of radicals containing protonated carbonyl groups, as in the reaction below.[78]

$$CH_3\overset{\cdot}{C}HCH-\overset{\overset{\displaystyle \overset{\oplus}{O}H}{\|}}{C}-OH \quad \longrightarrow \quad CH_3CH_2\overset{\cdot}{C}-\overset{\overset{\displaystyle \overset{\oplus}{O}H}{\|}}{C}OH$$
| |
CH_3   CH_3

The orbital symmetry barriers to 1,3- ("homo [1,2]") and 1,4- ("bishomo [1,2]") shifts should be lower than those for [1,2] shifts, since the atomic orbitals at the migration termini would have less interaction with the orbitals at the migration origins than in [1,2] migrations. Such rearrangements are very rare, but 1,3- and 1,4-shifts of hydrogen atoms do appear to take place if the rearrangement steps are highly exothermic.

Traces of benzyl chloride, presumably resulting from 1,3-hydrogen migrations in 2-methylphenyl radicals, are obtained from the decomposition of 2-methylbenzoyl peroxide in carbon tetrachloride (Eq. 74). CIDNP spectra of benzyl radicals can be observed in that reaction.[79]

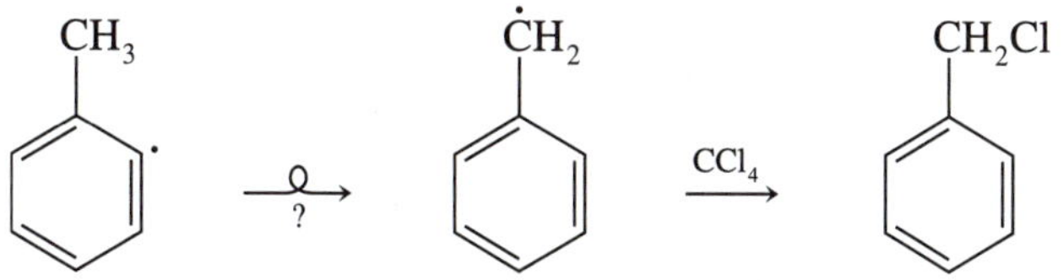

(74)

Phenoxy radicals are similarly obtained from reactions that should form 2-hydroxyphenyl radicals (Eq. 76). However, the rearrangements will take place only at temperatures above 300°C,[80a] because the reaction is less exothermic than that in Eq. 74.

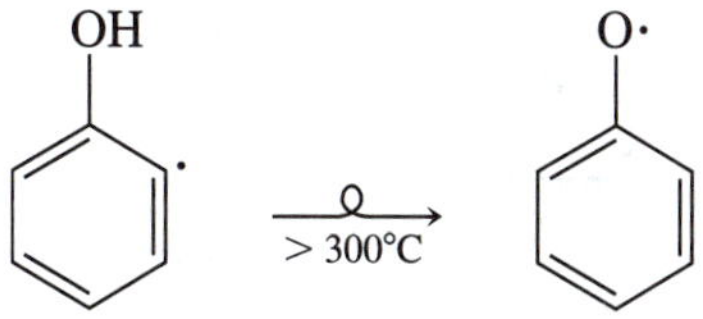

(75)

Eqs. 76 and 77 show examples of 1,4-hydrogen migrations in free radicals.[80]

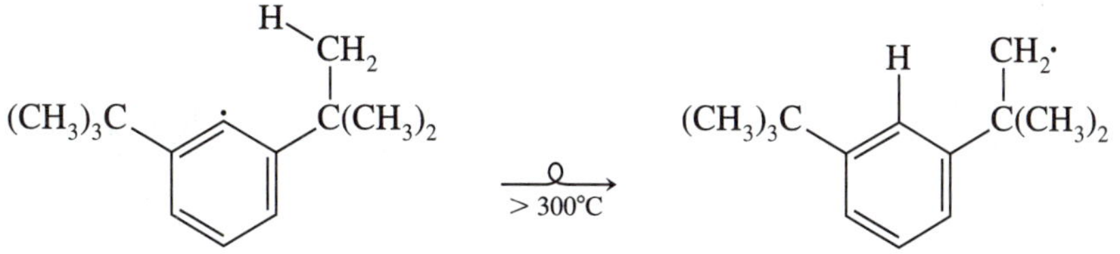

(76)

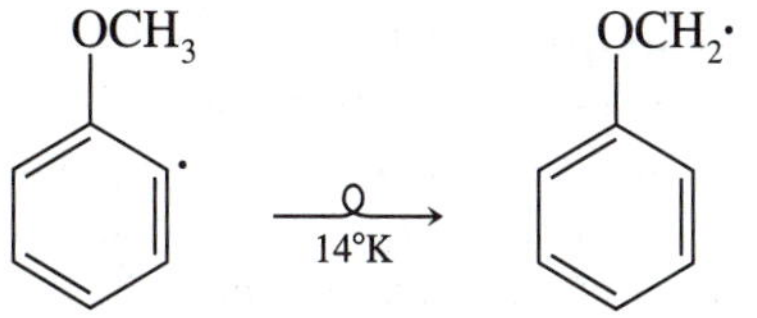

(77)

In contrast to 1,3- and 1,4-migrations, which are rare, 1,5-migrations of hydrogen atoms in free radicals are common reactions[81] (e.g., Eqs. 78–80).

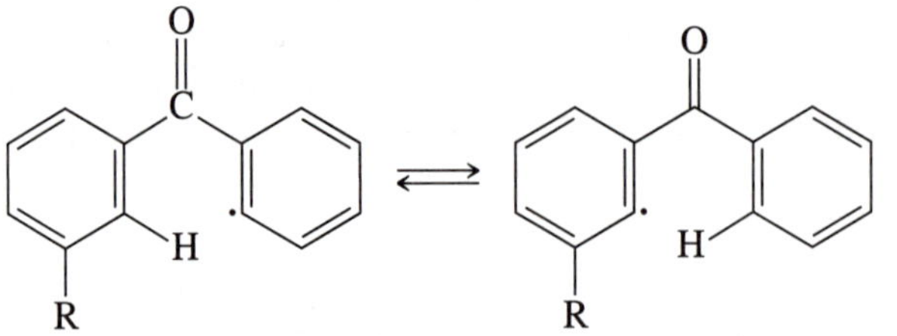

$$(78)$$

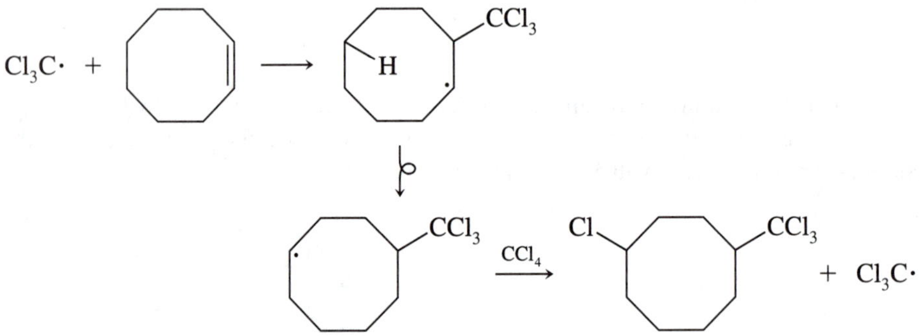

$$(79)$$

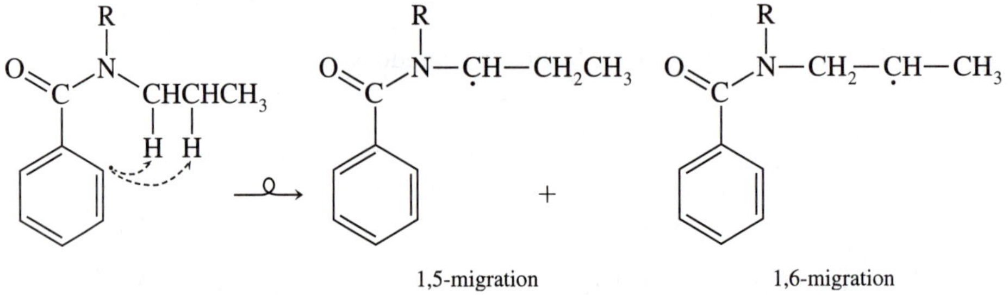

$$(80)$$

1,5-migration   1,6-migration

1,6-Migrations of hydrogen atoms sometimes accompany 1,5-migrations. In fact, the enthalpies of activation appear to be lower for 1,6- than for 1,5-migrations.[82] However, the preferred route for radicals to attack hydrogen–carbon bonds is in a linear arrangement of the three atoms. Thus, enthalpies of activation should be lowest for rearrangements with transition states resembling large rings, while entropies of activation would be most favorable (or least unfavorable) for transition states resembling smaller rings. Rearrangements via transition states resembling six-membered rings (1,5-migrations) usually turn out to be the best compromise.

1,5-Hydrogen and 1,6-hydrogen migrations play important, though often undesired, roles in free-radical polymerization processes, since they can introduce branching into what would otherwise be linear polymers (Eq. 81).

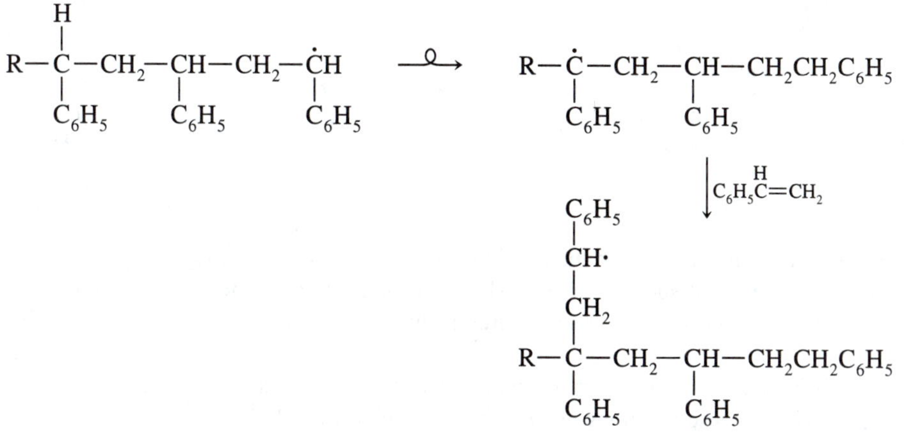

$$
\begin{array}{c}
\text{H} \\
\mid \\
\text{R} - \overset{\mid}{\underset{\mid}{\text{C}}} - \text{CH}_2 - \overset{\mid}{\underset{\mid}{\text{CH}}} - \text{CH}_2 - \overset{\bullet}{\text{CH}} \\
\quad\;\; \text{C}_6\text{H}_5 \quad\; \text{C}_6\text{H}_5 \quad\; \text{C}_6\text{H}_5
\end{array}
\xrightarrow{\quad}
\begin{array}{c}
\text{R} - \overset{\bullet}{\underset{\mid}{\text{C}}} - \text{CH}_2 - \overset{\mid}{\underset{\mid}{\text{CH}}} - \text{CH}_2\text{CH}_2\text{C}_6\text{H}_5 \\
\quad\;\; \text{C}_6\text{H}_5 \qquad \text{C}_6\text{H}_5
\end{array}
$$

$$
\downarrow \; \text{C}_6\text{H}_5\overset{\text{H}}{\text{C}}{=}\text{CH}_2
$$ (81)

$$
\begin{array}{c}
\text{C}_6\text{H}_5 \\
\mid \\
\text{CH}\cdot \\
\mid \\
\text{CH}_2 \\
\mid \\
\text{R} - \overset{\mid}{\underset{\mid}{\text{C}}} - \text{CH}_2 - \overset{\mid}{\underset{\mid}{\text{CH}}} - \text{CH}_2\text{CH}_2\text{C}_6\text{H}_5 \\
\quad\;\; \text{C}_6\text{H}_5 \qquad \text{C}_6\text{H}_5
\end{array}
$$

1,5-Migrations of hydrogen atoms in alkoxy radicals have been utilized to selectively introduce oxygen and nitrogen atoms onto unactivated positions of steroid molecules. In the *Barton reaction*, the alkoxy radicals are formed by the photolysis of nitrite esters (Eq. 82).[83]

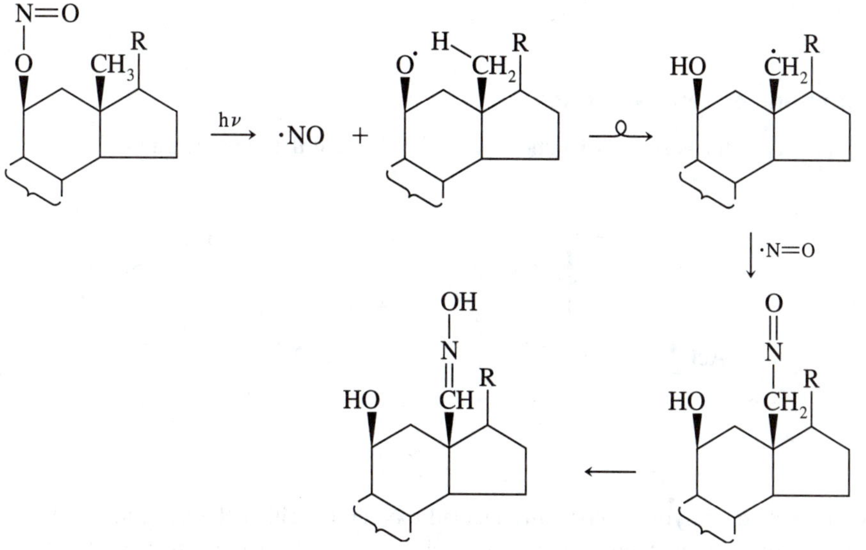

(82)

[2,3] Rearrangements. There appear to be no examples of [2,3] migrations of allylic groups in hydrocarbon radicals. Acyloxy groups, however, rapidly migrate to adjacent radical centers. The bulk of the evidence supports the view that these rearrangements proceed by concerted [2,3] shifts. The positions of the carbonyl and alkoxy oxygen atoms are usually interchanged, as shown in Eq. 83.[84]

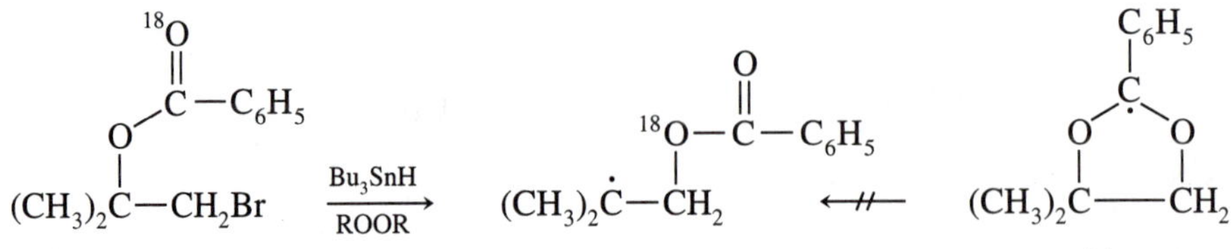

A two-step mechanism, in which the trivalent carbon attacks the carbonyl oxygen of the ester to form a dioxanyl radical such as **14**, was shown to be incorrect, since dioxanyl radicals do not open to acyloxy groups at the temperatures at which the migrations of acyloxy groups take place.[85]

[2,3] Carboxyl migrations are geometrically impossible in small ring lactones. Isotopic labelling shows that these reactions proceed by 1,2-shifts (Eq. 84).[86] There seems to be little evidence about the mechanisms of these reactions.[86]

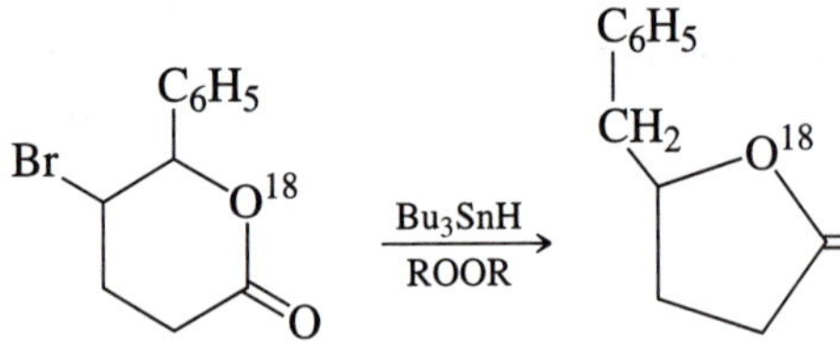

Migrations of Peroxy Groups

Allylperoxy radicals such as **15** undergo allylic shifts of the dioxygen units.[87]

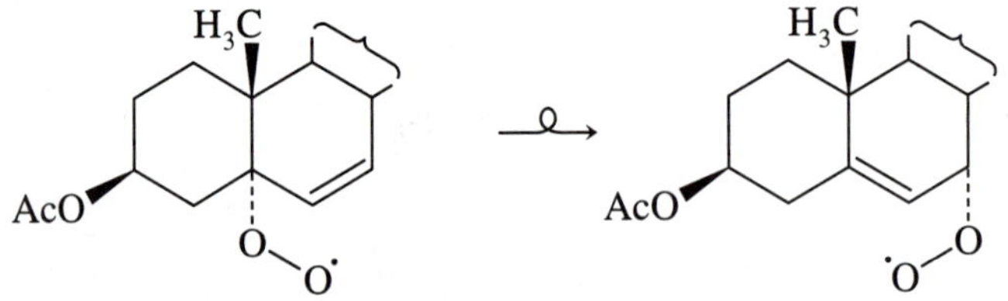

Experiments employing isotopically labeled oxygen (in either the peroxide or the atmosphere) initially indicated that there is almost no exchange of oxygen atoms between the peroxy radicals and atmospheric oxygen.[88] Furthermore, in the rearrangement of **15**, the peroxy function initially remained on the same face of the ring system in the product as in the starting material, although at longer reaction times, some migration to the opposite face occur red.[89] It was also demonstrated that the rearrangement cannot proceed via an intermediate carbon-centered radical such as 16, because 16 would react with atmospheric oxygen more rapidly than it would reopen to form an allylperoxy radical.[90] These results were consistent with a concerted [2,3] migration mechanism for the allylperoxy radical rearrangement.

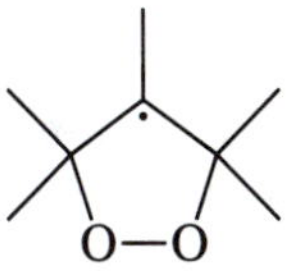

16

It was later shown, however, that oxygen exchange between the peroxides and atmospheric oxygen is indeed possible. The extent of oxygen exchange depends on the viscosity of the solvent, with exchange being most rapid in the least viscous solvents. Loss of facial selectivity in the migration of the peroxy unit is similarly dependent on solvent viscosity.[91] This evidence supports a mechanism in which the allylperoxy radical dissociates to form molecular oxygen and an allylic radical. The two fragments may recombine to form the rearranged allylperoxy radical or, in less-viscous solvents, may diffuse out of the solvent cage, resulting in an intermolecular rearrangement process (Eq. 86).

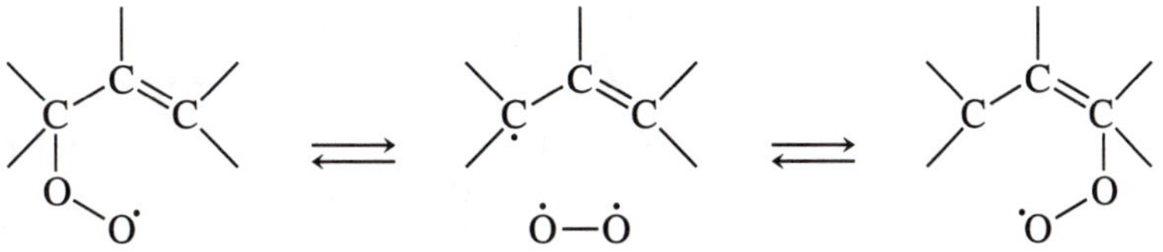

(86)

Rearrangements by Addition–Elimination Mechanisms

Vinyl groups undergo rapid 1,2-migrations in free radicals. The rearrangements proceed by cyclization of the original homoallylic radicals to form cyclopropylmethyl radicals, which may then reopen to form rearranged radicals, as shown in Eq. 87.[92]

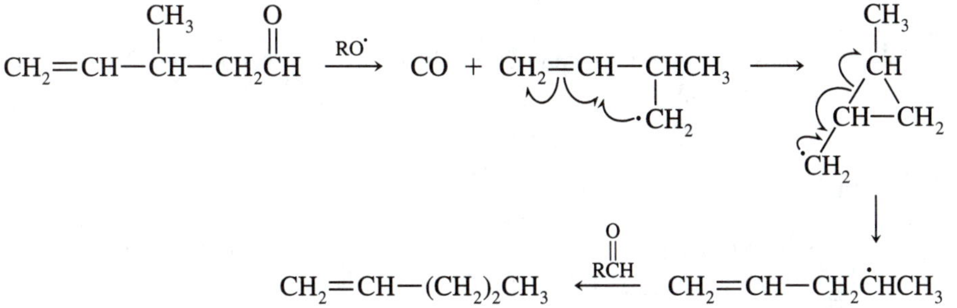

(87)

However, it is possible to trap the cyclized form if that radical is particularly stable, as shown in Eq. 88.[93]

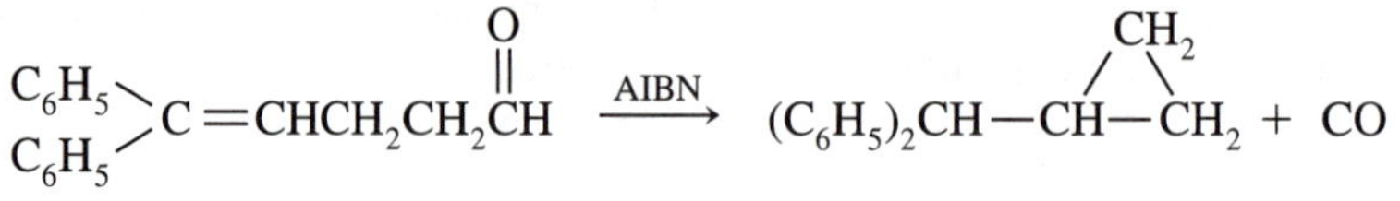

$$\begin{array}{c} C_6H_5 \\ C_6H_5 \end{array}\!\!>\!\!C=\!CHCH_2CH_2\overset{O}{\overset{\|}{C}}H \xrightarrow{\text{AIBN}} (C_6H_5)_2CH\!-\!CH\!\overset{CH_2}{\underset{\diagdown}{\diagup}}CH_2 + CO \qquad (88)$$

1,2-Migrations of aromatic rings occur readily in *neophyl radicals* (Eq. 89) and other radicals with aromatic rings on carbon atoms joined to radical centers, as shown in Eqs. 90 and 91.[94]

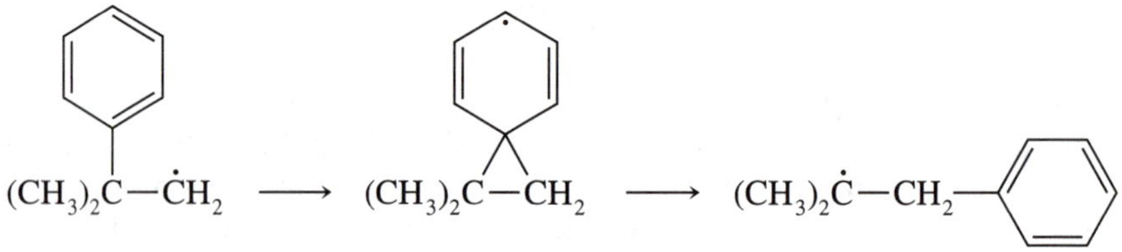

$$(CH_3)_2\overset{\cdot}{C}\!-\!\dot{C}H_2 \longrightarrow (CH_3)_2C\!-\!CH_2 \longrightarrow (CH_3)_2\dot{C}\!-\!CH_2\!-\!\!\bigcirc \qquad (89)$$

$$(C_6H_5)_3C\!-\!O^{\cdot} \longrightarrow (C_6H_5)_2\dot{C}\!-\!OC_6H_5 \qquad (90)$$

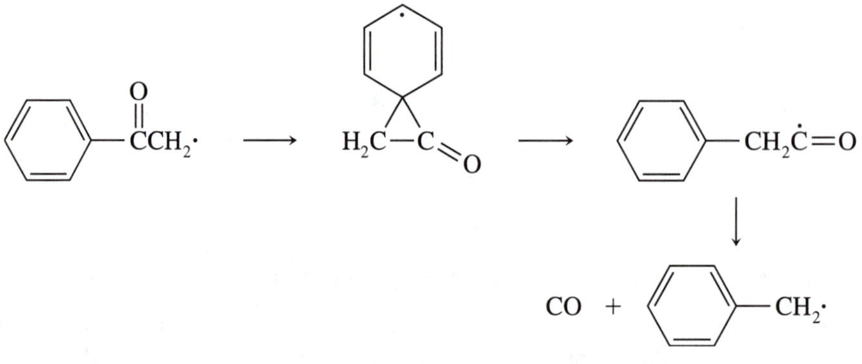

$$(91)$$

$$CO + \bigcirc\!-\!CH_2{\cdot}$$

The rearrangements are favored by the presence of radical stabilizing groups at the migration origins and by substituents on the migrating rings which would help stabilize intermediates resulting from radical attack on the rings. However, it has not been possible to trap the postulated cyclic intermediates.

1,4-Migrations of aromatic rings in free radicals are also well known. Two examples are illustrated in Eqs. 92 and 93.[95]

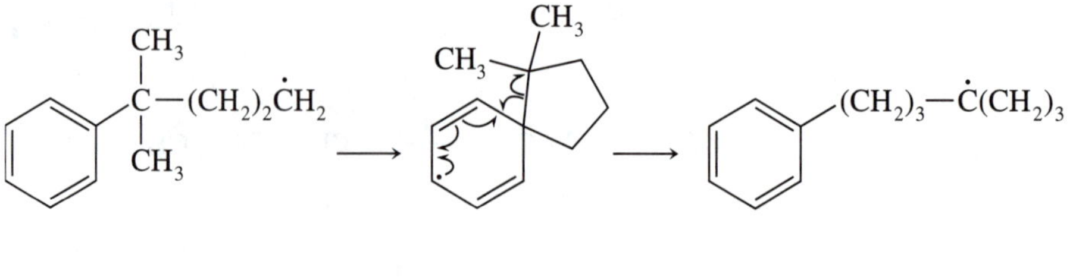

$$(92)$$

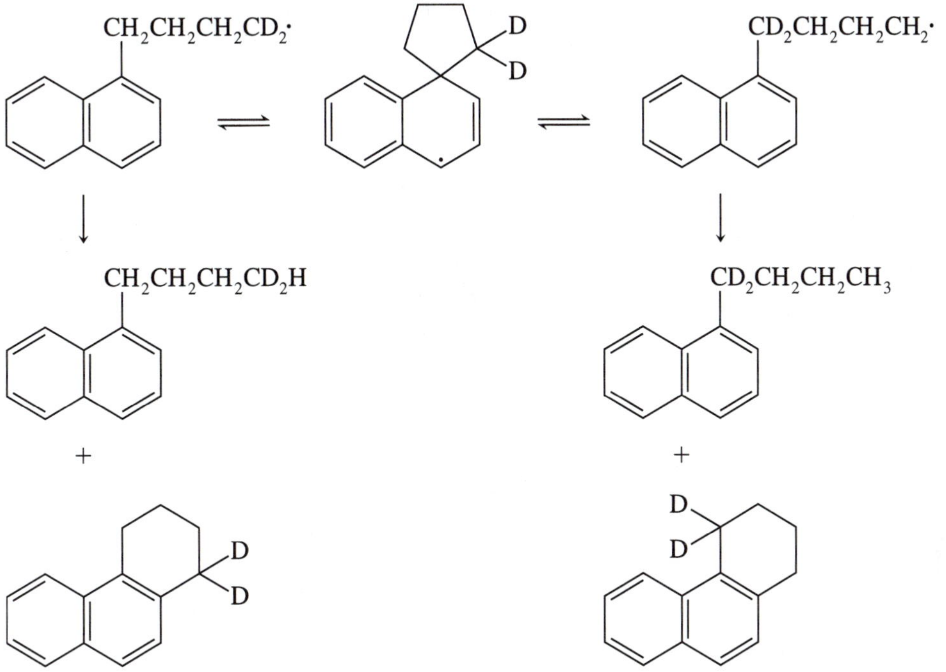

(93)

A transannular 1,5-migration of a phenyl group has also been reported (Eq. 94).[96]

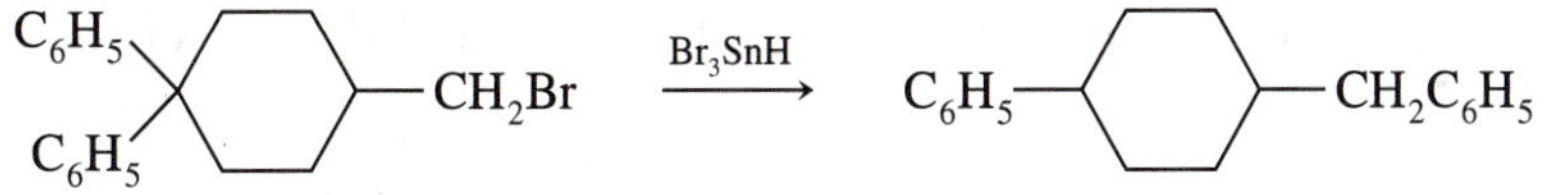

$$C_6H_5\text{-cyclohexyl-}CH_2Br \xrightarrow{Br_3SnH} C_6H_5\text{-cyclohexyl-}CH_2C_6H_5 \quad (94)$$

1,2-Migrations of carbonyl groups, as in Eq. 95, have been observed. However, it is not clear whether the reactions proceed by intramolecular addition–elimination mechanisms or by intermolecular elimination–addition paths.[97]

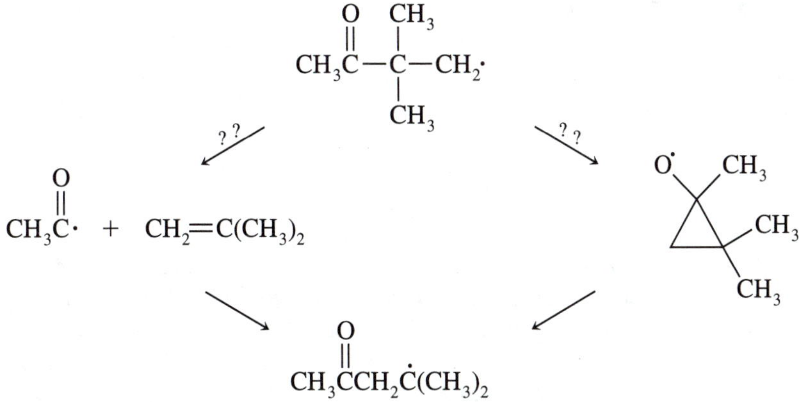

(95)

The degenerate 1,4-migration of an acetyl group shown in Eq. 96 does appear to be an intramolecular process,[98] as do several 1,4-migrations of nitrile groups in free radicals (such as illustrated in Eq. 97).[99]

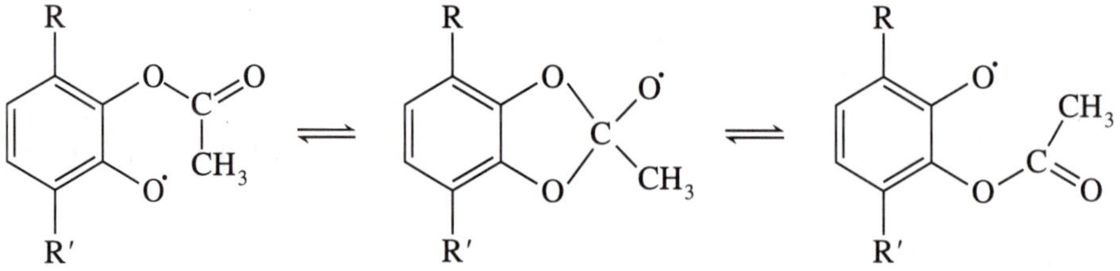

(96)

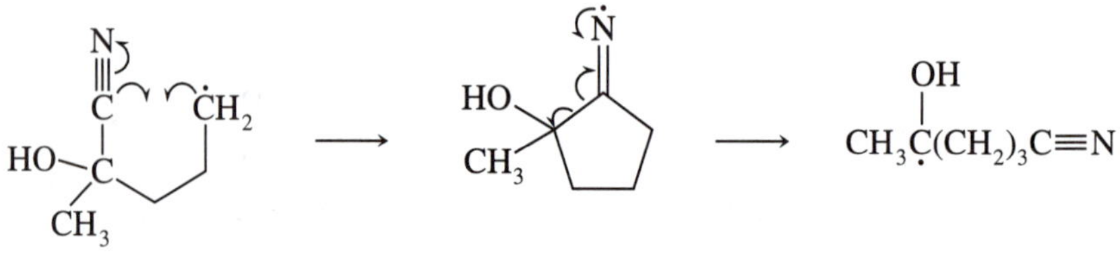

(97)

8.3 THE BERGMAN AND MYERS–SAITO REACTIONS

The Bergman Reaction. In 1972, R.G. Jones and R.G. Bergman showed that when deuterium labeled hexa-3-ene-1,5-diyne (**17**) was heated at 200°C both deuterium atoms shifted from the terminal acetylene positions to vinyl positions at the interior of the chain, as shown in Eq. 98.[100]

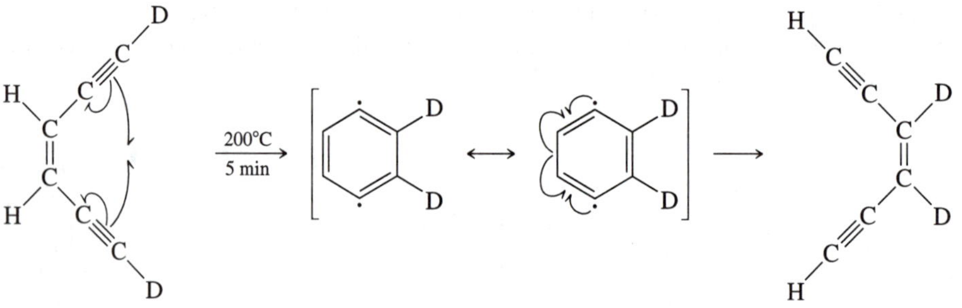

(98)

When the reaction was carried out in the presence of 1,4-dihydrobenzene the starting material was converted to benzene, while in carbon tetrachloride solution *p*-dichlorobenzene was formed. These observations demonstrated that the rearrangement had proceeded by the initial formation of an aromatic diradical ("1,4-dehydrobenzene") that then opened to form the product.[100] Formation of aromatic diradicals in this way has become known as the *Bergman reaction*.

Later work demonstrated that Bergman reactions could proceed at much lower temperatures if the enediyne system was contained in a relatively small ring, as shown in Eq. 99.[101]

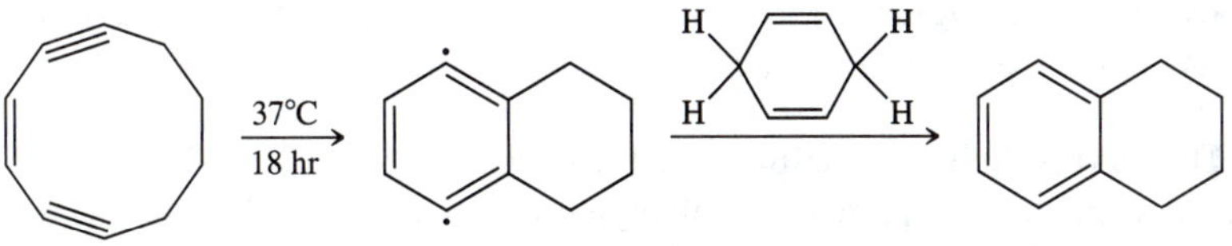

$$\text{(99)}$$

Bergman reactions are also possible if the double bonds of the enediynes comprise parts of aromatic rings, as shown in Eq. 100.[102] Of course, in those cases, the intermediate diradicals, unless "capped" by hydrogen or halogen donors can reopen but only in open in one direction, reforming the starting materials.

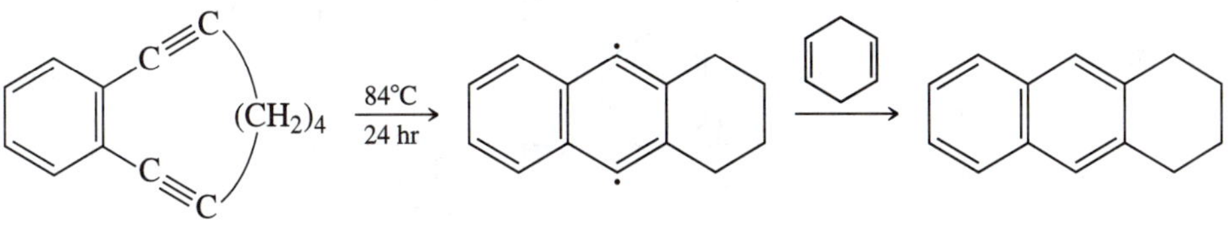

$$\text{(100)}$$

Biological Examples.   Bergman reactions would probably have remained of interest to only a small number of organic chemists had it not been discovered that a series of powerful antibiotics and anticancer agents, including calchicheamins, esperamicins, and dynamicins, owe their biological powers to the formation of diradicals via Bergman reactions.[103]

These biological examples of Bergman reactions are extraordinary in the ingenuity with which they are expressed. The reactions of calchicheamins[104] can serve as examples. Calchicheamins do not, by themselves, readily undergo Bergman reactions at moderate temperatures. In biological systems, however, nucleophiles attack the trisulfide moieties of the calchicheamins, releasing mercaptide anions. The mercaptide anions then add to β carbons of the conjugated carbonyl systems.

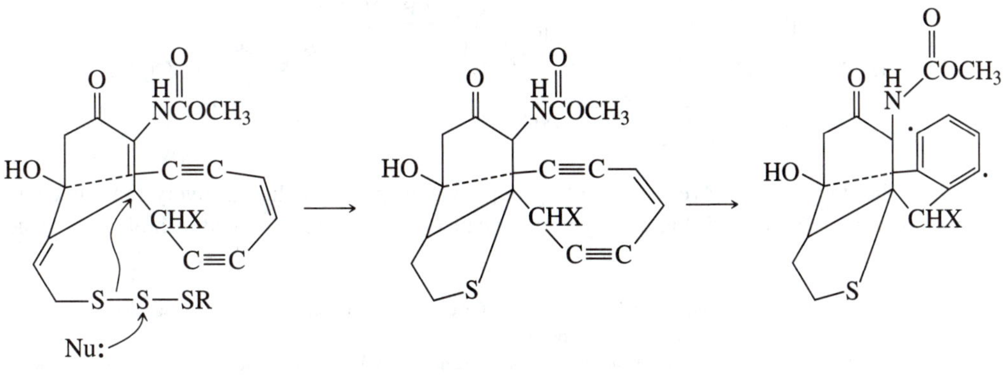

Biological Activation of Calchicheamins

Conversion of a vinyl β carbon to an sp^3 hybridiized form reduces the distance between positions 1 and 6 of the enediyne moiety, thus facilitating bond formation between the two acetylenic groups to form aromatic diradicals.[103] The diradicals can then abstract hydrogens from the deoxyribose rings of DNA (see Section 11.2), thus destroying the DNA. That is a desirable result if the DNA is that of a dangerous microbe or cancer cell.*†

The Myers–Saito Reaction. In 1981, groups led by A.G. Myers in the United States and I. Saito in Japan reported that allenylalkynes (penta-1,2-diene-4-ynes), such as **18**, are converted to diradicals at temperatures barely above those in mammalian bodies.[107]

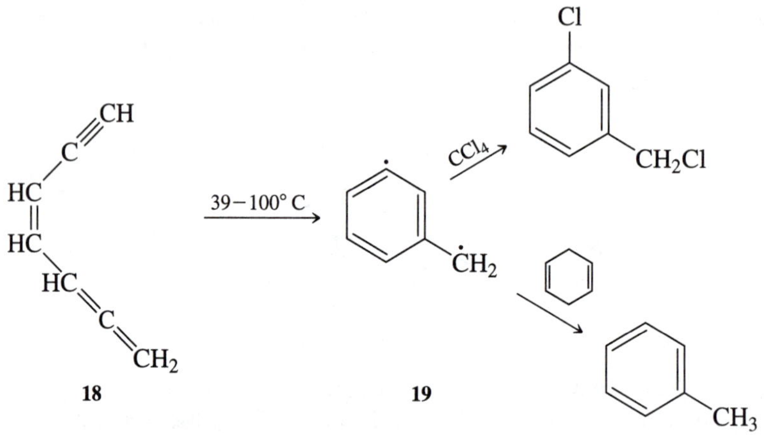

(101)

While Myers–Saito reactions in nonpolar solvents yield products, such as those shown in Eq. 101, that appear to result from diradical intermediates, reaction of **18** in methanol yields benzyl methyl ether (Eq. 102). This suggests that the intermediate "diradical" has dipolar properties[107a], such as those that would result from resonance form **19b**.

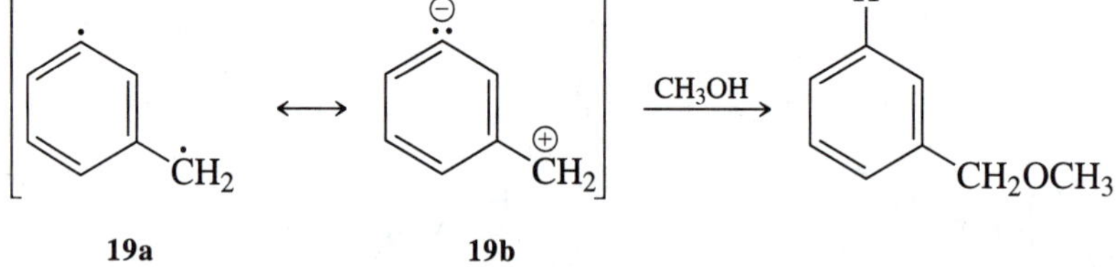

(102)

The work of the Myers and Saito groups was inspired by the discovery that neocarzinostatins, which are powerful antibiotic and anticancer agents, are activated in biological systems

---

*The substituents designated "X" in the formula for calchicheamins are complex oligosaccharide units, which are believed to facilitate intercalation into DNA helices.[102]

†The biological effectiveness of diradical-forming agents is believed to be due to their ability to attack both chains of the DNA double helix, thus preventing reformation of the molecule which might follow destruction of just one strand.[106]

by conversion to allenic structures.[108] As with calchicheamins, the resulting diradicals are effective in destroying DNA molecules.

Biological Activation of Neocarzinostatins

REFERENCES

[1] R.B. Bates, S. Brenner, W.H. Deines, D.A. McCombs, and D.E. Potter, *J. Am. Chem. Soc., 92,* 6345 (1970).

[2] P. Powell, *Adv. Organomet. Chem., 26,* 125 (1986).

[3] S.H. Goel and E. Grovenstein, Jr., *Organomet., 11,* 1565 (1992).

[4] S.W. Staley, G.M. Cramer, and W.G. Kingsley, *J. Am. Chem. Soc., 95,* 5052 (1973).

[5] J.E. Baldwin and F.J. Urban, *Chem. Comm.,* 165 (1970).

[6] E. Grovenstein, Jr., and A.B. Cottingham, *J. Am. Chem. Soc., 99,* 1881 (1977).

[7] M.S. Silver, P.R. Shafer, J.E. Nordlander, C. Ruechardt, and J.D. Roberts, *J. Am. Chem. Soc., 82,* 2646 (1960).

[8] S.E. Wilson, *Tetrahedron Lett.,* 4651 (1975).

[9] E. Grovenstein, Jr., K.-W. Chiu, and B.B. Patil, *J. Am. Chem. Soc., 102,* 5848 (1980).

[10] B.J. Wakefield, *The Chemistry of Organolithium Compounds.* Pergamon Press, New York (1974).

[11] For a review, see E. Grovenstein, Jr., *Adv. Organomet. Chem., 16,* 167 (1977).

[12] H.E. Zimmerman and A. Zweig, *J. Am. Chem. Soc., 83,* 1196 (1961).

[13] J.J. Eisch, C.A. Kovacs, P. Chobe, and M.P. Boleslawski, *J. Org. Chem., 52,* 4427 (1987).

[14] A. Nickon, J.L. Lambert, R.O. Williams, and N.H. Werstiuk, *J. Am. Chem. Soc., 88,* 3354 (1966).

[15] M.B. Rampersad and J.B. Stothers, *Chem. Comm.,* 709 (1976); A. Nickon, J.L. Lambert, J.F. Oliver, D.F. Covey, and J. Morgan, *J. Am. Chem. Soc., 98,* 2593 (1976).

[16] J.P. Freeman and J.H. Plonka, *J. Am. Chem. Soc., 88,* 3662 (1966).

[17] A.K. Cheng, A.K. Ghosh, and J.B. Stothers, *Can. J. Chmn., 62,* 1385 (1984).

[18] G. Wittig and L. Loehman, *Liebigs Ann., 550,* 260 (1942).

[19] R.B. Bates, L.M. Kroposki, and D.E. Potter, *J. Org. Chem., 37,* 560 (1972).

[20] A. Maercher and W. Demuth, *Liebigs Ann.,* 1909 (1977); A. Maercher, *Angew. Chem. Intl. Ed. Engl., 26,* 972 (1987).

[21] V. Rautenstrauch, *Helv. Chim. Acta, 55,* 594 (1972).

[22] For a review, see U. Schoellkopf, *Angew. Chem. Intl. Ed. Engl., 9,* 10 (1970).

[23] P.T. Lansbury and V.A. Pattison, *J. Am. Chem. Soc., 84,* 4295 (1962), and *J. Org. Chem., 27,* 1933 (1962).

[24] U. Schoellkopf and D. Walter, *Liebigs Ann., 654,* 27 (1962).

[25] P.T. Lansbury, V.A. Pattison, J.D. Sidler, and J.B. Bieber, *J. Am. Chem. Soc., 88,* 18 (1966).

[26] J.F. Garst and C.D. Smith, *J. Am. Chem. Soc., 98,* 1526 (1976).

[27] E. Hebert, Z. Welvart, M. Ghelfenstein, and H. Swarc, *Tetrahedron Lett.,* 1381 (1983).

[28] H. Felkin and A. Tambute, *Tetrahedron Lett.,* 821 (1969).

[29] M. Schlosser and S. Strunk, *Tetrahedron, 45,* 2649 (1989).

[30] G. Zhong and M. Schlosser, *Tetrahedron Lett., 34,* 6265 (1993).

[31] H. Felkin and C. Frajerman, *Tetrahedron Lett.,* 3485 (1977).

[32] For reviews see T. Nakai and K. Mikami, *Chem. Revs., 86,* 885 (1986).

[33] K. Mikami and T. Nakai, *Synthesis,* 595 (1991).

[34] C.A. Broka and T.J. Shen, *J. Am. Chem. Soc., 111,* 2981 (1989).

[35] W.C. Still and A.J. Miltra, *J. Am. Chem. Soc., 100,* 1927 (1978).

[36] R. Hoffmann and R. Brueckner, *Angew. Chem. Intl. Ed. Engl., 31,* 647 (1992).

[37] A. William Johnson, *Ylid Chemistry.* Academic Press, New York (1966).

[38] V. Franzen, H.I. Joschek, and C. Mertz, *Liebigs Ann., 643,* 24 (1961).

[39] S.H. Pines, *Organic Reactions, 18,* 403 (1970).

[40] E. Block, *Reactions of Organosulfur Compounds.* Academic Press, New York (1978), pp. 118–120, 198–201.

[41] R.W. Jemison et al., *Chem. Comm.,* 1201 (1970).

[42] B.M. Trost and L.S. Melvin, Jr., *Sulfur Ylides.* Academic Press, New York (1975), pp. 23–24.

[43] M. Haenel and H.A. Staab, *Ber., 106,* 2203 (1973).

[44] For example, I.G. Stara, I. Stary, J. Zavada, and V. Hanus, *J. Am. Chem. Soc., 116,* 5084 (1994).

[45] E.F. Jenny and J. Druey, *Angew. Chem. Intl. Ed. Engl. 1,* 155 (1962); see also K. Chantrapromma, W.D. Ollis, and I.O. Sutherland, *Chem. Comm.,* 670 (1978).

[46] Review: I.E. Marko, in *Comprehensive Organic Synthesis,* vol. 3, B.M. Trost and I. Fleming, eds. Pergamon Press, New York (1991), pp. 927–931.

[47] W.D. Ollis, M. Rey, I.O. Sutherland, and G.L. Closs, *Chem. Comm.,* 543 (1975).

[48] U. Schoellkopf, G. Ostermann, and J. Schossig, *Tetrahedron Lett.,* 2619 (1969).

[49] U. Schoellkopf, J. Schossig, and G. Ostermann, *Liebigs. Ann., 737,* 158 (1970).

[50] J.E. Baldwin, W.F. Erickson, R.E. Hackler, and R.M. Scott, *Chem. Comm.,* 576 (1970).

[51] U. Schoellkopf, U. Ludwig, G. Ostermann, and M. Patsch, *Tetrahedron Lett.,* 3415 (1969).

[52] A.R. Lepley, in *Chemically Induced Magnetic Polarization,* A.R. Lepley and G.L. Closs, Eds., Wiley, New York (1973).

[53] S.W. Kantor and C.R. Hauser, *J. Am. Chem. Soc., 73,* 4122 (1951).

[54] N. Jones and C.R. Hauser, *J. Org. Chem., 26,* 2979 (1961).

[55] W.H. Puterbaugh and C.R. Hauser, *J. Am. Chem. Soc., 86,* 1105 (1964).

[56] C.R. Hauser and D.N. Van Eenam, *J. Am. Chem. Soc., 79,* 5512 (1957). See also D.N. Van Eenam and C.R. Hauser, *J. Am. Chem. Soc., 79,* 5520 (1957).

[57] S.H. Pine and B.L. Sanchez, *Tetrahedron Lett.,* 1319 (1969).

[58] See review by J. Mann in *Comprehensive Organic Synthesis,* vol. 3, B.M. Trost and I. Fleming, Eds. Pergamon Press, New York (1991), pp. 839–858.

[59] H.O. House and W.F. Gilmore, *J. Am. Chem. Soc., 83,* 3972 (1961).

[60] R.B. Loftfield, *J. Am. Chem. Soc., 73,* 4707 (1951).

[61] J.F. Pazos, J.G. Pacifici, G.O. Pierson, D.B. Sclore, and F.G. Greene, *J. Org. Chem., 39,* 1990 (1974).

[62] B. Foehlisch et al., *J. Chem. Res.,* 134 (1991).

[63] R. Breslow, J. Posner, and A. Krebs, *J. Am. Chem. Soc., 85,* 234 (1963).

[64] G. Stork and I.J. Borowitz, *J. Am. Chem. Soc., 82,* 4307 (1960).

[65] H.O. House and W.F. Gilmore, *J. Am. Chem. Soc., 83,* 3972 (1961).

[66] C.H. De Puy, *Accounts Chem. Res., 1,* 33 (1968).

[67] A. Gambacorta, S. Turchetta, P. Bovicelli and M. Botta, *Tetrahedron, 47,* 9097 (1991).

[68] J.M. Conia and J.R. Salaun, *Accounts Chem. Res., 5,* 33 (1972).

[69] For reviews, see L.A. Paquette, *Org. Reac., 25,* 1 (1977); S. Braverman, in *The Chemistry of Sulphoxides and Sulphones,* S. Patai, Z. Rappaport, and C.J.M. Stirling, Eds. Wiley, New York (1978).

[70] A.G. Sutherland and R.J.K. Taylor, *Tetrahedron Lett., 30,* 3267 (1989).

[71] L.A. Carpino, L.V. McAdams, R.H. Rynbrandt, and J.W. Spiewak, *J. Am. Chem. Soc., 93,* 476 (1971); J.C. Philips, J.V. Swisher, H. Haidukewych, and O. Morales, *Chem. Comm.,* 22 (1971).

[72] C. O'Brien, *Chem. Revs., 64,* 81 (1964); A. Delgado et al., *Can. J. Chem., 66,* 517 (1988).

[73] M.S. Hatch and D.J. Cram, *J. Am. Chem. Soc., 75,* 38 (1953).

[74] P.W. Neber and A. Burgard, *Liebigs Ann., 493,* 281 (1932); P.W. Neber and G. Huh, *ibid., 515,* 283 (1935).

[75] For a review, see A.L.J. Beckwith and K.U. Ingold, in *Rearrangements in Ground and Excited States,* vol. 1, P. de Mayo, Ed. Academic Press, New York (1980).

[76] On the basis of kinetic data, it was suggested that [1,2]-hydrogen shifts take place in radicals formed by pyrolysis of alkanes at ca. 600°C. See A.S. Gordon, D.C. Tardy, and R. Ireton, *J. Phys. Chem., 80,* 1400 (1976) and references therein.

[77] F.D. Greene, G.R. Van Norman, J.C. Cantrill, and R.D. Gillian, *J. Org. Chem., 25,* 1970 (1960). C.E. Hudson and D.J. McAdoo, *Tetrahedron, 46,* 331 (1990).

[78] T. Weiske and H. Schwarz, *Tetrahedron, 42,* 6245 (1986).

[79] B.C. Childress, A.C. Rice, and P.B. Shevlin, *J. Org. Chem., 39,* 3056 (1974).

[80] (a) P.H. Kasai and D. McLeod, Jr., *J. Am. Chem. Soc., 96,* 2338 (1972); (b) J. Frey, D.A. Nugiel, and Z. Rappaport, *J. Org. Chem., 56,* 466 (1991) and references therein.

[81] T.L. Sordo and J.J. Dannenberg, *J. Org. Chem., 64,* 1922 (1999).

[82] A.E. Dorigo and K.N. Houk, *J. Am. Chem. Soc., 109,* 2195 (1987).

[83] D.H.R. Barton, J.M. Beaton, L.E. Geller, and M.M. Pechet, *J. Am. Chem. Soc., 82,* 2640 (1960) and *83,* 4076 and (1961); D.H.R. Barton, R.H. Hesse, M.P. Pechet, and L.C. Smith, *J. Chem. Soc., Perkin Trans. 1,* 1159 (1979).

[84] A.L.J. Beckwith and C.B. Thomas, *J. Chem. Soc., Perkin Trans. 2,* 861 (1973); A.L.J. Beckwith and P.J. Duggan, *Chem. Comm.,* 1000 (1988).

[85] L.R. Barclay, D. Griller, and K.U. Ingold, *J. Am. Chem. Soc., 104,* 4399 (1982); P. Kucovsky, I. Stary, and F. Turecek, *Tetrahedron Lett., 27,* 1513 (1986).

[86] D. Crich, H. Huang, A.L.J. Beckwith, *J. Org. Chem., 64,* 1762 (1999).

[87] For a review of early work, see A.A. Frimer, *Chem. Revs., 79,* 365 (1979).

[88] N.A. Porter and J.S. Wujek, *J. Org. Chem., 52,* 5085 (1987).

[89] A.L.J. Beckwith et al., *Chem. Comm.,* 475 (1988).

[90] W.F. Brill, *J. Chem. Soc., Perkin Trans. II,* 621 (1984); N.A. Porter and P. Zuraw, *Chem. Comm.,* 1472 (1985).

[91] N.A. Porter, K.A. Mills, S.E. Caldwell, and G.R. Dubay, *J. Am. Chem. Soc., 116,* 6697 (1994).

[92] L.K. Montgomery and J.W. Watts, *J. Am. Chem. Soc., 89,* 6556 (1967).

[93] T.A. Halgren, M.E.H. Howden, M.E. Medof, and J.D. Roberts, *J. Am. Chem. Soc., 89,* 3051 (1967).

[94] See K.U. Ingold and J. Warkentin, *Can. J. Chem., 58,* 348 (1980) and references therein.

[95] S. Winstein, R. Heck, S. Lapporte, and R. Baird, *Experientia, 12,* 138 (1956); M. Julia, *Pure Appl. Chem., 40,* 553 (1974).

[96] J.W. Wilt, R.A. Dabek, and K.C. Weizel, *J. Org. Chem., 37,* 425 (1972).

[97] C.L. Karl, E.J. Maas, and W. Reusch, *J. Org. Chem., 37,* 2834 (1972); F. Bertini, T. Caronna, L. Grossi, and F. Minisci, *Gazz. Chim. Ital., 104,* 471 (1974).

[98] A.I. Prokof'ev et al., *Dokl. Akad. Nauk SSSR, 229,* 1976 (1976).

[99] D.S. Watt, *J. Am. Chem. Soc., 98,* 271 (1976); A. Bury et al., *J. Chem. Soc., Perkin Trans. II,* 1367 (1982).

[100] R.G. Jones and R.G. Bergman, *J. Am. Chem. Soc., 94,* 660 (1972); R.G. Bergman, *Acc. Chem. Res., 6,* 25 (1973).

[101] K.C. Nicolaou, *J. Am. Chem. Soc., 114,* 7360 (1992); I.V. Alabugin, M. Mancharan, and S.V. Kovalenko, *Org. Lett, 4,* 119 (2002).

[102] M.F. Semmelhack, T. Neu, and F. Foubelo, *J. Org. Chem., 59,* 5038 (1994); K.K. Thoes, J.C. Thoen and F.M. Uckun, *Tetrahedron Lett., 41,* 4019 (2000).

[103] See K.C. Nicolaou and W.-M. Dai, *Angew. Chem. Int. Ed. Eng., 39,* 1387 (1991).

[104] M.D. Lee et al., *J. Am. Chem. Soc., 109,* 3464, 3466 (1987); G.A. Ellestad et al., *Tetrahedron Lett., 30,* 3053 (1989).

[105] K.C. Nicolaou et al., *J. Am. Chem. Soc., 110,* 4866 (1988).

[106] G.N. Zein, A.M. Sinha, W.J. McGahren, and G.S. Ellestad, *Science, 240,* 1198 (1989).

[107] (a) A.G. Myers, E.Y. Kuo, and N.S. Finney, *J. Am. Chem. Soc., 111,* 8057 (1989); (b) R. Nagota, H. Yamanaka, E. Okazaki, and I. Saito, *Tetrahedron Lett., 30,* 4995 (1989).

[108] K. Edo et al., *Tetrahedron Lett., 26,* 331 (1985); See also, P.W. Musch and B. Engels, *Angew. Chem. Int. Ed. Engl., 40,* 3833 (2001).

PROBLEMS

8.1 Predict structures for the principal organic products formed in the following reactions.

(a) 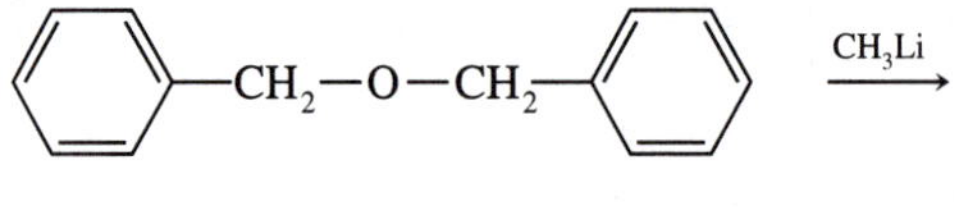
$$\text{CH}_3\text{Li} \longrightarrow$$

(b) 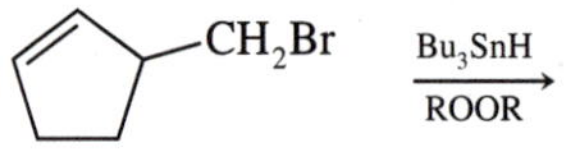 $\xrightarrow[\text{ROOR}]{\text{Bu}_3\text{SnH}}$ a hydrocarbon, whose NMR spectrum shows no peaks for methyl groups

(c) 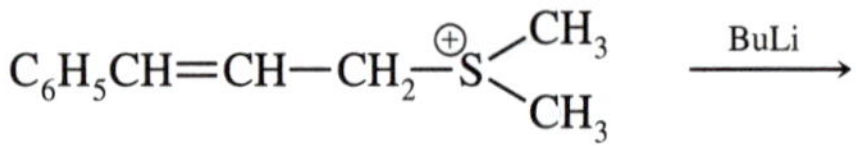 $C_6H_5CH{=}CH{-}CH_2{-}\overset{\oplus}{S}{\overset{\text{CH}_3}{\diagdown}_{\text{CH}_3}}$ $\xrightarrow{\text{BuLi}}$

(d) 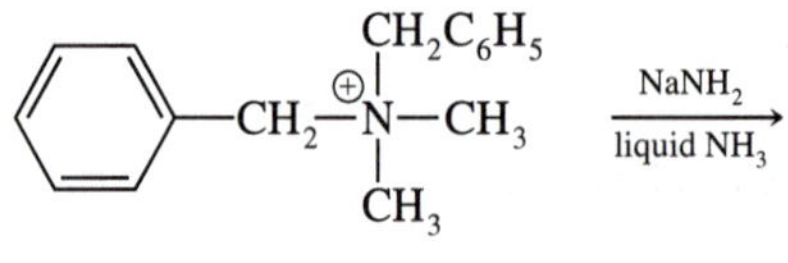 $\xrightarrow[\text{liquid NH}_3]{\text{NaNH}_2}$

(e)

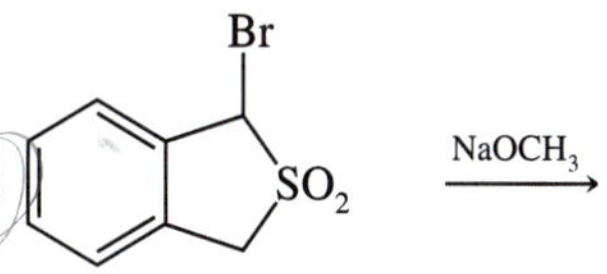

(f) 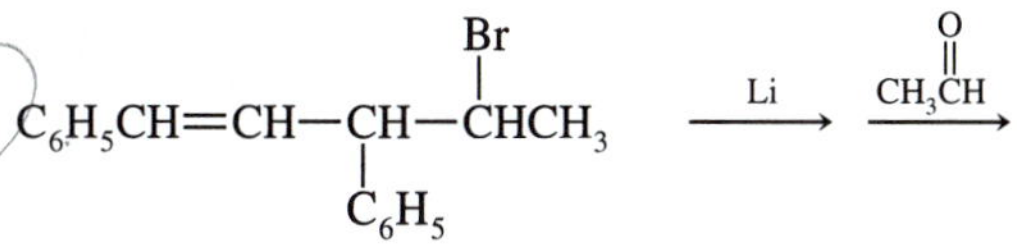 $C_6H_5CH=CH-CH-CHCH_3$ with Br and C_6H_5 substituents → Li, CH_3CH (O)

(g)

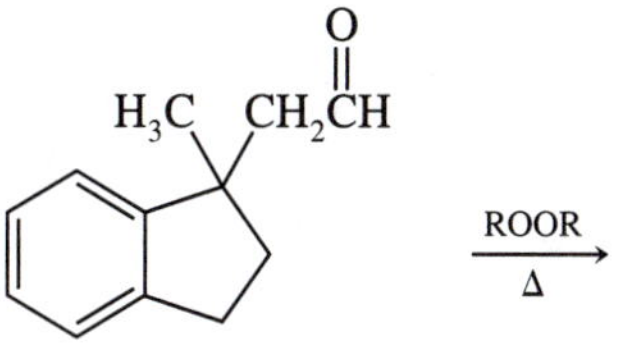

$\xrightarrow[\Delta]{ROOR}$ a hydrocarbon, whose NMR spectrum shows the presence of one methyl group

(h)

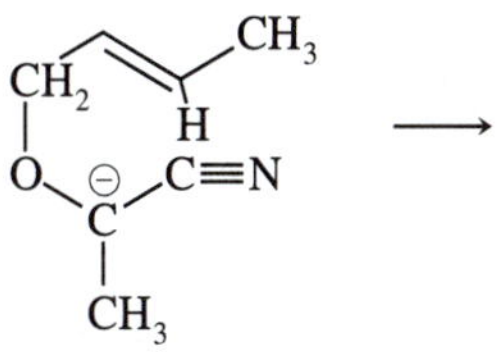

$\longrightarrow$ a ketone

(B. Cazes and S. Julia, *Bull. Soc. Chem. France*, 925 [(1977])

8.2 Suggest reasonable mechanisms for the following reactions.

(a)

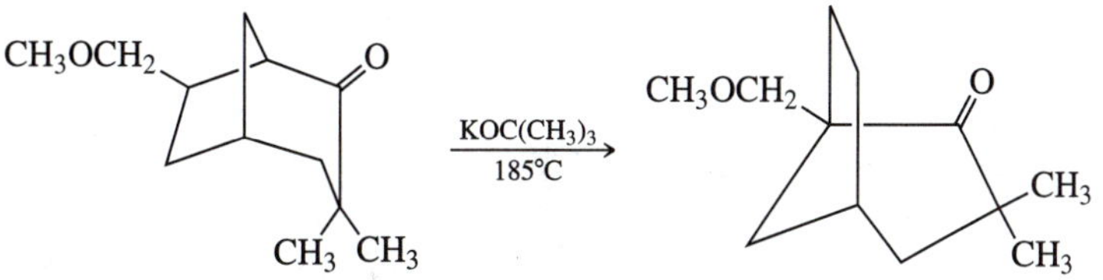

$\xrightarrow[185°C]{KOC(CH_3)_3}$

What does this reaction suggest about the charge distributions in any reaction intermediates?

(Adapted from H.A. Patel, J.B. Stothers, and S.E. Thenas, *Can J. Chem.*, 72, 56 [1994])

(b)

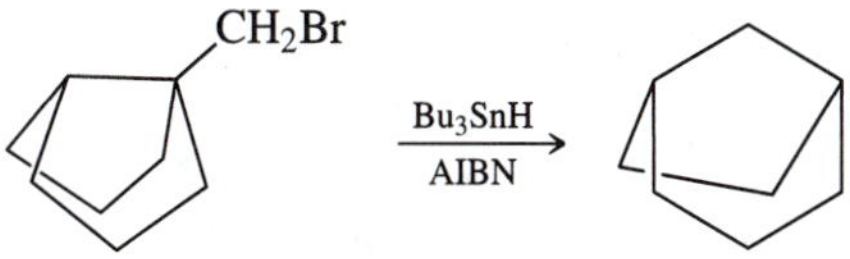

$\xrightarrow[AIBN]{Bu_3SnH}$

(A.M. Mueller and P. Chen, *J. Org. Chem.*, 63, 4581 [1998])

(c)

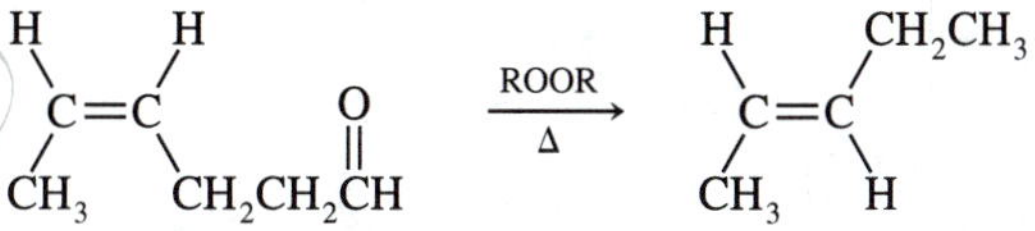

$\xrightarrow[\Delta]{ROOR}$

(d)

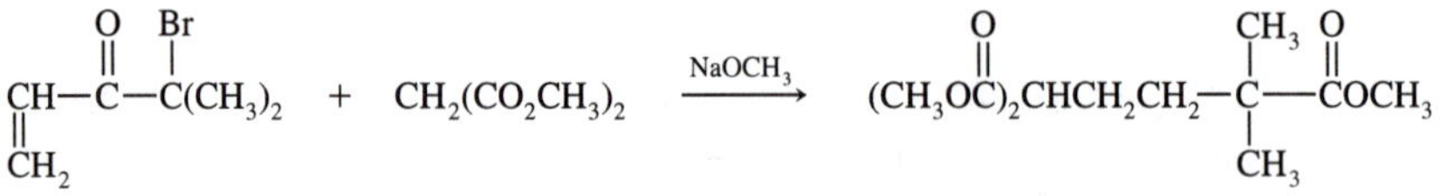

$$\text{CH}-\overset{\text{O}}{\underset{\text{CH}_2}{\overset{\|}{\text{C}}}}-\overset{\text{Br}}{\underset{}{\overset{|}{\text{C}}}}(\text{CH}_3)_2 \quad + \quad \text{CH}_2(\text{CO}_2\text{CH}_3)_2 \quad \xrightarrow{\text{NaOCH}_3} \quad (\text{CH}_3\text{OC})_2\text{CHCH}_2\text{CH}_2-\overset{\text{CH}_3}{\underset{\text{CH}_3}{\overset{|}{\text{C}}}}-\overset{\text{O}}{\overset{\|}{\text{C}}}\text{OCH}_3$$

(e)

$$\xrightarrow{\text{Bu}_3\text{S}-\text{SnBn}}$$

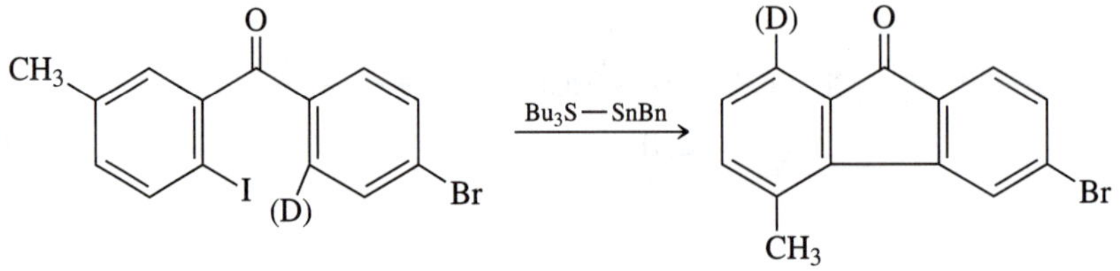

(Adapted from J.M. Cummins et al., *Tetrahedron Lett.*, 40, 6153 [1999])

(f)

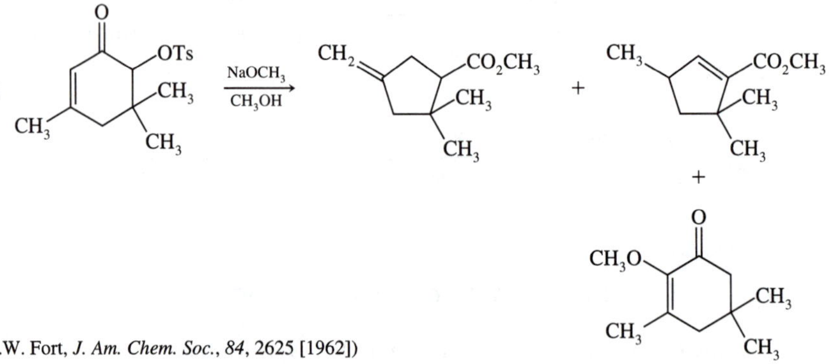

$$\xrightarrow[\text{CH}_3\text{OH}]{\text{NaOCH}_3}$$

(A.W. Fort, *J. Am. Chem. Soc.*, 84, 2625 [1962])

(g)

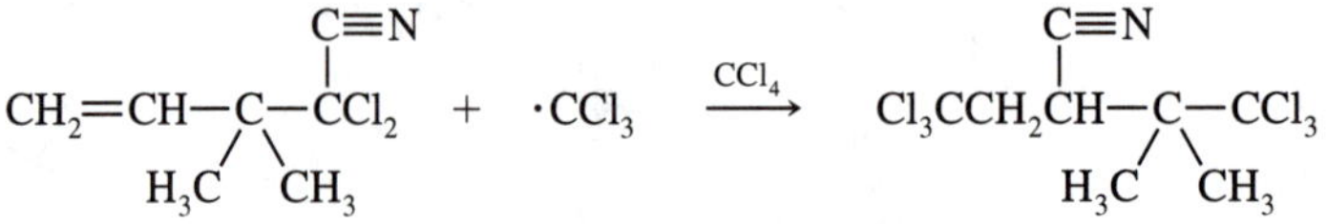

$$\text{CH}_2=\text{CH}-\overset{\text{C}\equiv\text{N}}{\underset{\text{H}_3\text{C}\quad\text{CH}_3}{\overset{|}{\text{C}}}}-\text{CCl}_2 \quad + \quad \cdot\text{CCl}_3 \quad \xrightarrow{\text{CCl}_4} \quad \text{Cl}_3\text{CCH}_2\overset{\text{C}\equiv\text{N}}{\underset{\text{H}_3\text{C}\quad\text{CH}_3}{\overset{|}{\text{CH}-\text{C}}}}-\text{CCl}_3$$

(h)

$$\overset{\ominus}{\text{CH}_2}-\text{CH}=\text{CH}-\text{CH}=\text{CH}-(\text{CH}_2)_3-\text{CH}=\text{CH}_2 \quad \longrightarrow$$

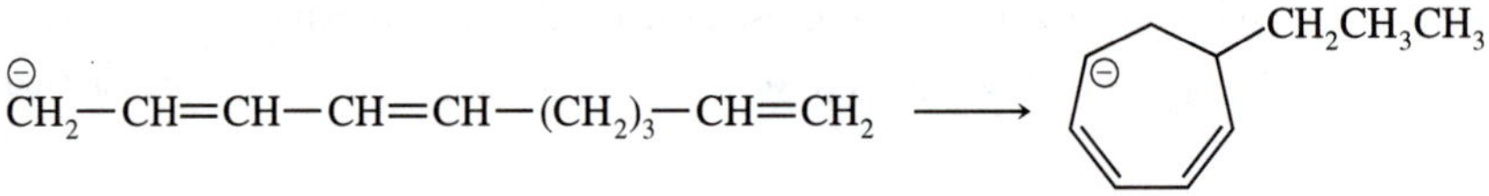

(R.B. Bates et al., *J. Am. Chem. Soc.*, 92, 6345 [1970] and 94, 2130 [1974])

(i)

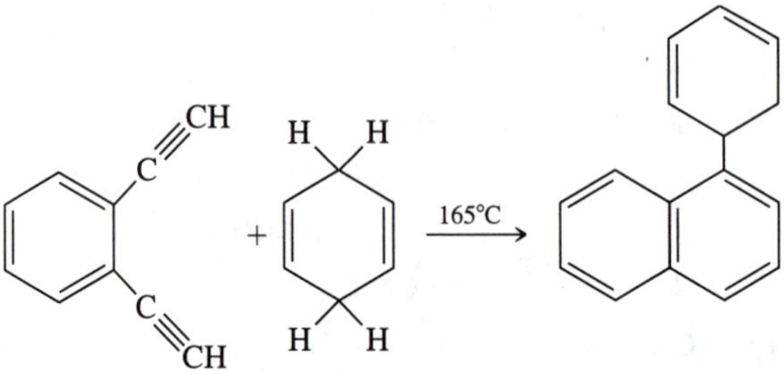

$$\xrightarrow{165°\text{C}}$$

(K.K. Thoes, J.C. Thoen, and F.M. Uckun, *Tetrahedron Lett.*, 41, 4019 [2000])

(j)

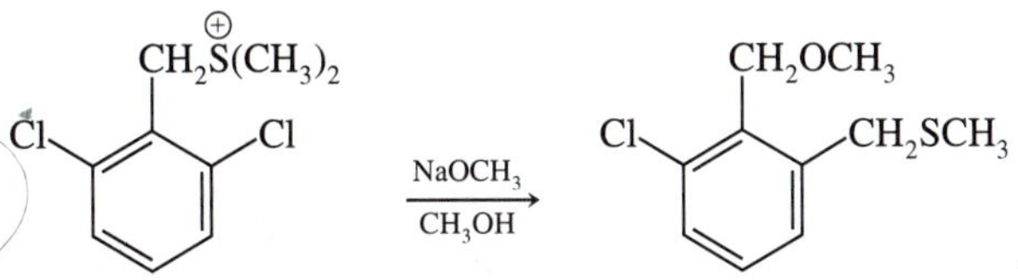

$$\xrightarrow[\text{CH}_3\text{OH}]{\text{NaOCH}_3}$$

(k)

$$\xrightarrow{\text{NaOCH}_3}$$

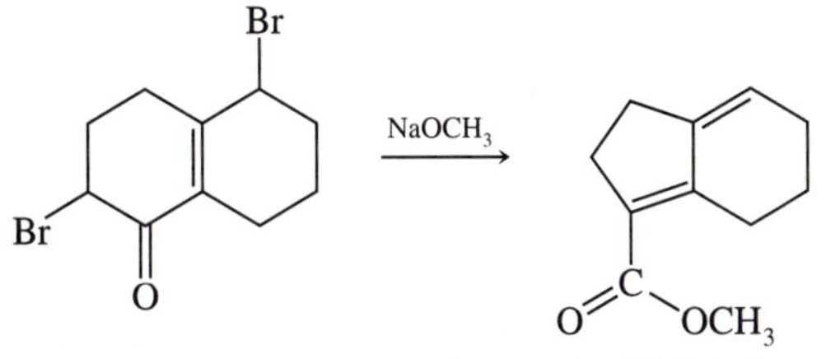

(Suggest two possible mechanisms)

(F. Kurzer and J.N. Patel, *Monats. Chem.*, *118*, 1363 [1987])

(l) $\text{C}_6\text{H}_5\overset{\text{Cl}}{\underset{|}{\text{CH}}}-\overset{\text{O}}{\underset{||}{\text{C}}}\text{CH}_2\text{C}_6\text{H}_5$ + NaOCH$_3$ +

$$\xrightarrow{\text{CH}_3\text{OH}}$$

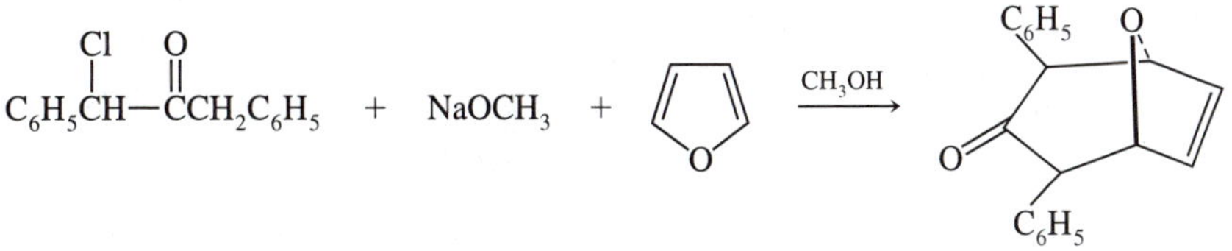

(A.W. Fort, *J. Am. Chem. Soc.*, *84*, 1979 [1962])

(m)

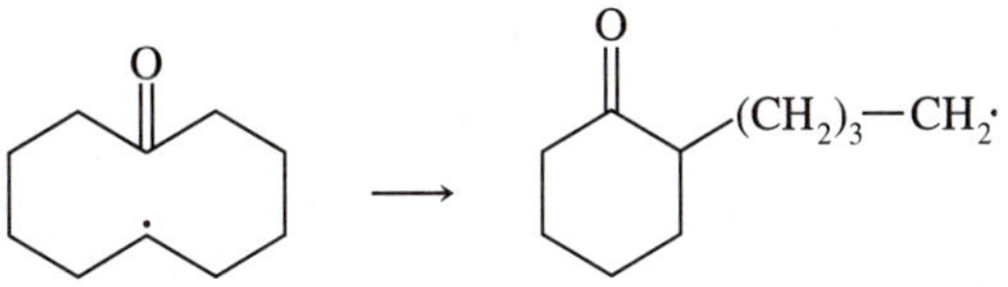

$$\longrightarrow$$

(n)

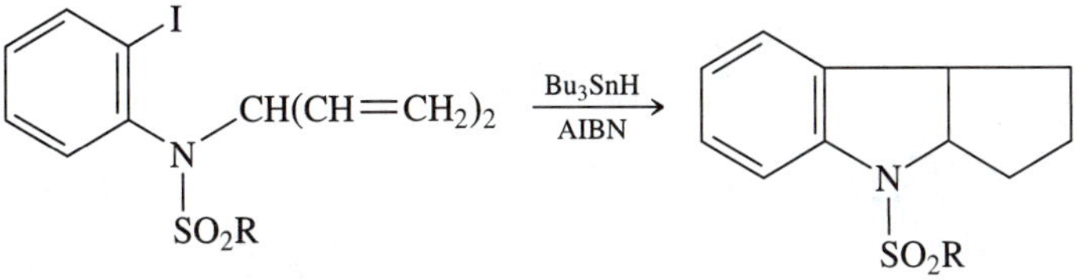

$$\xrightarrow[\text{AIBN}]{\text{Bu}_3\text{SnH}}$$

(S. Bommezjin, C.G. Martin, A.R. Kennedy, D. Lizs, and J.A. Murphy, *Org. Lett.*, *3*, [2001])

(o) $\text{R}-\text{C}\equiv\text{C}-(\text{CH}_2)_2$

$-(\text{CH}_2)_3\text{Br}$

$$\xrightarrow[\text{AIBN}]{\text{Bu}_3\text{SnH}}$$

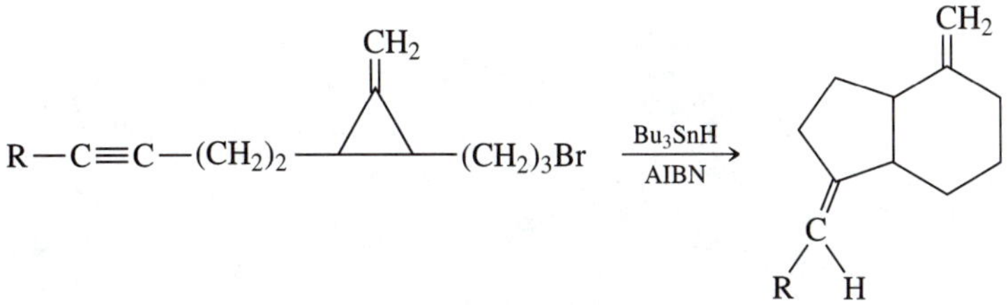

(K.G. Pike, M. Anson, and J.D. Kilburn, *Tetrahedron Lett.*, *39*, 5877 [1998])

Carbenes, Carbenoids, and Nitrenes

9

9.1 INTRODUCTION

Carbenes are molecules containing divalent carbon atoms.[1] Each divalent carbon has two un-shared electrons, which are often shown when writing the structures of carbenes. However, carbenes are neutral molecules, not carbanions.

$$CH_2 \ (:CH_2) \qquad :CH\overset{\overset{\displaystyle O}{\|}}{C}CH_3 \qquad :CCl_2 \qquad :C{=}CHCH_3$$

Some typical carbenes

It is possible to have several divalent carbons in a single molecule. An extreme example is the "hexacarbene," shown below.[2]

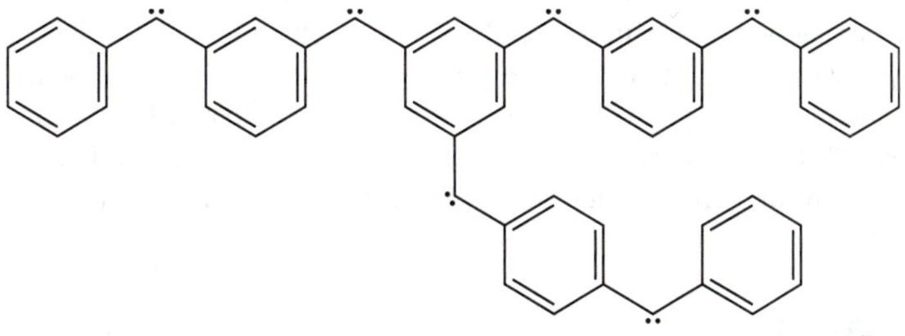

A "hexacarbene"

The rather vague term *carbenoids* is used to refer to molecules in which all the carbons are tetravalent, but which have properties resembling those of carbenes. (Those properties often include the ability to transfer divalent carbons and their substituents to other molecules.) Typically, carbenoids have carbon atoms that are simultaneously bonded both to metal atoms and to halogen atoms. It is often difficult to be certain whether a "carbene" reaction in solution is actually the re-action of a free carbene or the reaction of a carbenoid.

Nitrenes[3] are compounds, such as $:\ddot{N}C_6H_5$, that contain monovalent nitrogen atoms. As with carbenes, it is often difficult to be certain whether a particular reaction actually proceeds via a free nitrene or via a "nitrenoid" containing a nitrogen atom linked to both an electropositive atom and an electronegative atom.

9.2 FORMATION OF CARBENES AND NITRENES

Carbenes can be formed from a wide variety of reactions.[4] However, in most studies of carbenes, the carbenes are obtained either from organodiazo compounds or by "α-elimination" reactions of organic halides.

The earliest evidence for the formation of a ketene was obtained from the thermal decomposition of diazomethane in the presence of carbon monoxide, which resulted in the formation of small amounts of ketene (Eq. 1).[5]

$$H_2C{=}\overset{\oplus}{N}{=}\overset{\ominus}{\underset{\cdot\cdot}{N}}: \quad \xrightarrow[\Delta]{-N_2} \quad H_2C: \quad \xrightarrow{CO} \quad H_2C{=}C{=}O \tag{1}$$

Elimination of nitrogen from organodiazo compounds, as a result either of thermal or of photolytic decomposition of the diazo compounds, remains one of the most important methods for the formation of carbenes (Eq. 2).[6]

$$N_2CH{-}\overset{\displaystyle O}{\overset{\|}{C}}OC_2H_5 \quad \xrightarrow[\text{or } h\nu]{\Delta} \quad N_2 \;+\; :CH\overset{\displaystyle O}{\overset{\|}{C}}OC_2H_5 \tag{2}$$

The decompositions of organodiazo compounds are frequently catalyzed by salts of heavy metals, such as copper, palladium, and rhodium. The products of the catalyzed reactions often resemble those from uncatalyzed reactions. However, the intermediates in the catalyzed reactions should presumably be classified as carbenoids, in which the "carbenic" carbons are bonded to the heavy metal salts in addition to the two substituents of the free carbenes.

Organodiazo compounds may be conveniently prepared by the Bamford–Stevens reaction,[7] in which the *p*-toluenesulfonylhydrazone of a carbonyl compound is converted to its sodium or lithium salt, which then loses a *p*-toluenesulfinate anion (Eq. 3).

$$(CH_3)_2C{=}N{-}\overset{H}{\underset{}{N}}{-}SO_2C_7H_7 \quad \xrightarrow{CH_3Li} \quad (CH_3)_2C{=}N{-}\overset{\overset{\displaystyle Li^{\oplus}}{}}{\underset{}{\overset{\ominus}{N}}}{-}SO_2C_7H_7 \tag{3}$$

$$\downarrow \Delta$$

$$(CH_3)_2C{=}\overset{\oplus}{N}{=}\overset{\ominus}{\underset{\cdot\cdot}{N}}: \quad + \quad LiSO_2C_7H_7$$

Diazo compounds may also be prepared by base-catalyzed decomposition of *N*-nitrosoamides, as shown in Eq. 4,[8]

$$\underset{\substack{\| \\ O}}{\overset{\substack{N=O \\ |}}{CH_3-N-C-OCH_3}} \xrightarrow{\ominus OH} CH_2=N_2 \ + \ CO_2 \ + \ HOCH_3 \ + \ H_2O \quad (4)$$

by oxidation of hydrazones of carbonyl compounds (Eq. 5),[9]

$$(5)$$

or by thermal rearrangements of diazirines (Eq. 6).[10]

$$\xrightarrow{\Delta} \ (CH_3)_2C=\overset{\oplus}{N}=\overset{\ominus}{\underset{..}{N}}: \quad (6)$$

Halocarbenes are commonly prepared by reactions of strong bases with organic polyhalides that lack hydrogens on β-carbons, and therefore cannot undergo the usual β-elimination reactions. Instead, the bases abstract protons from the polyhalogenated carbons (Eq. 7). The resulting carbanions then lose halide ions to form carbenes, as shown in Eq. 8.[11]

$$Cl_3C-H \ + \ K^{\oplus} \ {}^{\ominus}OC(CH_3)_3 \ \longrightarrow \ Cl_3C:^{\ominus} \ K^{\oplus} \ + \ HOC(CH_3)_3 \quad (7)$$

$$Cl_2\overset{\ominus}{\underset{..}{C}}\overset{\frown}{-}Cl \ K^{\oplus} \ \longrightarrow \ Cl_2C: \ + \ Cl \ K^{\oplus} \quad (8)$$

The polyhalides most commonly employed are trihalomethanes (but not trifluoromethane). However, dichloromethanes (Eq. 9),[12] and even benzylic chlorides (Eq. 10),[13] can be converted to carbenes (or, more likely, carbenoids) by reaction with very strong bases, such as organolithium reagents.

$$CH_2Cl_2 \ + \ CH_3Li \ \longrightarrow \ :CHCl \ + \ CH_4 \ + \ LiCl \quad (9)$$

$$-CH_2Cl \ + \ C_4H_9Li \ \longrightarrow \ -\overset{..}{C}H \ + \ LiCl \ + \ C_4H_{10} \quad (10)$$

The occurrence of this type of α-elimination reaction was first demonstrated in 1950 by Jack Hine, who was investigating the unusual reaction of chloroform with aqueous alkali. That reaction results in the formation of carbon monoxide as well as formate anions (Eq. 11).

$$CHCl_3 \ \xrightarrow[H_2O]{NaOH} \ CO \ \text{and} \ \overset{\substack{O \\ \|}}{HCO}{}^{\ominus} \quad (11)$$

Furthermore, chloroform reacts far more rapidly with bases than do either dichloromethane or carbon tetrachloride. These facts, as well as the kinetics of the reaction, can be accounted for by the mechanism shown in Eq. 12.[14]

$$CHCl_3 \ + \ ^{\ominus}OH \ \longrightarrow \ H_2O \ + \ ^{\ominus}:CCl_3$$

$$^{\ominus}:CCl_3 \ \xrightarrow{slow} \ :CCl_2 \ + \ Cl^{\ominus} \qquad (12)$$

$$:CCl_2 \ + \ ^{\ominus}OH \ \xrightarrow{fast} \ CO \ \text{or} \ H\overset{\overset{\displaystyle O}{\|}}{C}O^{\ominus}$$

Polyhalides with bromine or iodine atoms can react with organolithium reagents to form α-halolithium reagents, which are frequently stable at dry ice temperatures (Eqs. 13a and 13b).

$$CH_2BrCl \ + \ C_4H_9Li \ \xrightarrow{-100°C} \ LiCH_2Cl \ + \ C_4H_9Br \qquad (13a)$$

$$CH_2Br_2 \ + \ CH_3Li \ \xrightarrow{-80°C} \ LiCH_2Br \ + \ CH_3Br \qquad (13b)$$

At higher temperatures, they react to yield products similar to those obtained from carbenes formed by other methods. However, the ratios of products can vary depending on the type of halogen,[12,15] suggesting that the α-halolithium compounds act as carbenoids rather than dissociating to form free carbenes.

Carbenoids formed from reactions of dihalides with heavy metals are relatively stable at room temperature, and are more selective in their reactions than most carbenes. The *Simmons–Smith reagent*, which is formed by reaction of diiodomethane with a zinc–copper couple,[16a] is an important reagent of this type. The structure of the reagent is sometimes written in the form shown in Eq. 14, although it probably exists in polymeric clusters. However, theoretical calculations indicate that the reactive species is the monomeric molecule, $IZn\text{-}CH_2\text{-}I$.[16b]

$$CH_2I_2 \ + \ Zn(Cu) \ \longrightarrow \ [CH_2:]ZnI_2(Cu) \qquad (14)$$

Nitrenes are most frequently formed by thermolysis or pyrolysis of organic azides (Eq. 15).[3,17]

$$(15)$$

Like carbenes, however, nitrenes (or "nitrenoids") may also be obtained from α-elimination reactions (Eq. 16).[18]

$$(16)$$

9.3   SINGLET AND TRIPLET CARBENES

Every carbene can exist in two possible electronic structures—as a *singlet* or as a *triplet*. In a singlet structure, the two unshared electrons have paired spins and are located in a single orbital, leaving a second orbital vacant. (In effect, a singlet carbene resembles a carbocation and a carbanion united on the same carbon atom.) In a triplet structure, the two electrons have unpaired spins, and one electron is located in each nonbonded orbital. Triplet carbenes thus resemble diradicals.

According to Hund's rule, triplet forms of carbenes should be more stable than singlet forms. (The energy difference between the singlet and triplet structures for a carbene is called the *singlet–triplet gap*.) Indeed, the triplet form of methylene (CH_2) is about 9 kcal/mol lower in energy than the singlet form.[19] However, most substituents—particularly substituents with unshared electron pairs—stabilize singlet structures more strongly than triplet structures.[20-22] Thus, while most hydrocarbon carbenes are more stable in triplet forms, the singlet forms of carbenes with electron-donating or electron-withdrawing substituents are usually significantly more stable than the triplet forms.

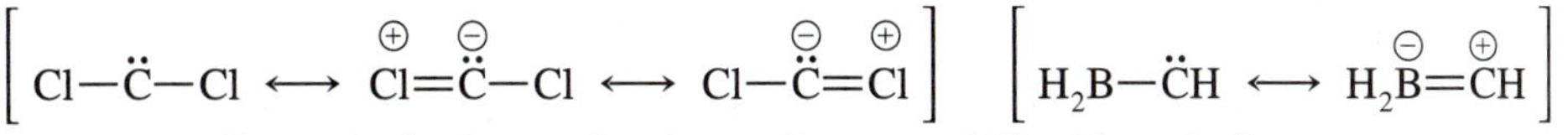

$$\left[Cl-\ddot{C}-Cl \longleftrightarrow \overset{\oplus}{Cl}=\ddot{C}-\overset{\ominus}{Cl} \longleftrightarrow \overset{\ominus}{Cl}-\ddot{C}=\overset{\oplus}{Cl} \right] \quad \left[H_2B-\ddot{C}H \longleftrightarrow \overset{\ominus}{H_2B}=\overset{\oplus}{C}H \right]$$

Some singlet forms of carbenes that are stabilized by substituents

Several heterocyclic carbenes, such as **1a** and **1b**, are indefinitely stable at room temperature and can even be melted at high temperatures and resolidified without decomposition.[23] (In contrast, most carbenes are stable only when frozen into matrices of solid argon at 77°K.)

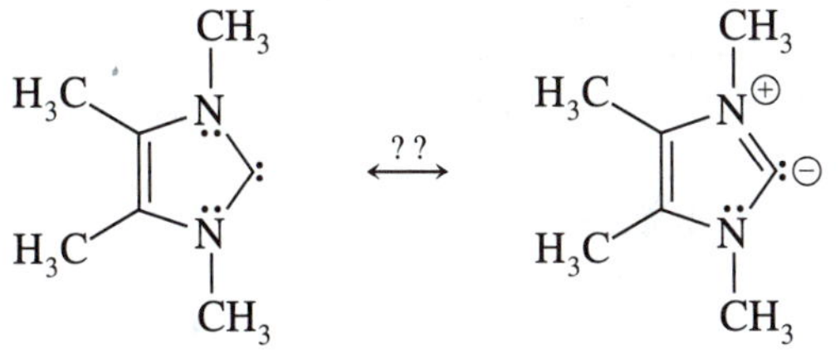

1a 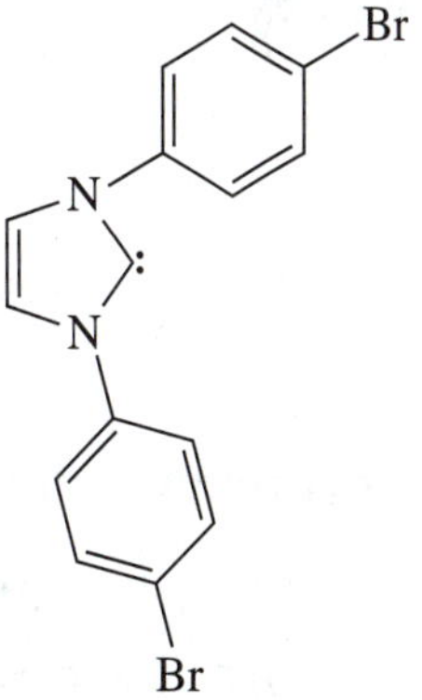   **1b**

Bond lengths in these "carbenes" appear to indicate the presence of isolated single and double bonds[24] rather than the more equal bond lengths expected if they actually have aromatic structures, such as that shown above as a resonance form for 1a. However, theoretical calculations indicate that these molecules do possess a good deal of aromatic stabilization.[25]

Early in the study of carbene reactions, it was suggested that the reactions characteristic of triplet carbenes (which would resemble reactions of diradicals) would differ from those characteristic of singlet carbenes.[26,27] This principle, sometimes known as the *Skell–Woodworth rule*, has largely held up with time. However, in carbenes with relatively small singlet–triplet gaps, intersystem crossing between the singlet and triplet forms is often faster than other reactions of the triplets. In these cases, reactions may proceed via the singlet forms, even though only small amounts of those forms are in equilibrium with the triplet forms.[28] The ratio of reactions attributable to singlet vs. triplet forms may be increased in more polar solvents.[29]

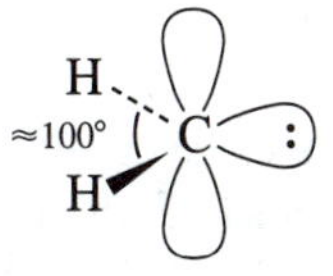

Structure of singlet CH_2[30] 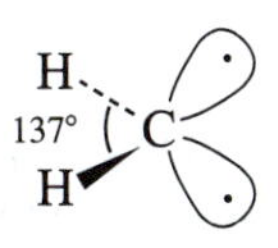   Structure of triplet CH_2[31]

Singlet and triplet forms of carbenes differ in their geometries. The structures of singlet carbenes are bent close to right angles,[30] so that the unshared electron pairs can be located in orbitals with strong s-character. In contrast, as illustrated on page 261, the structures of triplet carbenes are much closer to linear arrangements.[31]

It follows that carbenes that can have relatively linear structures are more likely than bent carbenes to have triplet structures, while otherwise similar but nonlinear carbenes are more likely to be singlets — particularly if a bond angle is relatively small.[32] The singlet–triplet gaps calculated for several carbenes,[33a] are shown below.

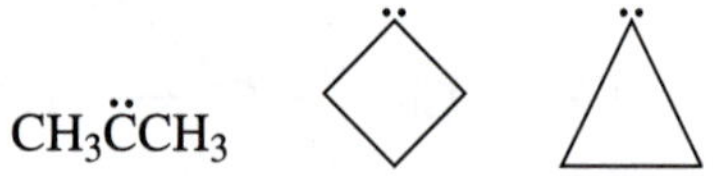

$$CH_3\ddot{C}CH_3$$

Singlet-triplet energy gaps   **1.5**   **5.9**   **13.8**

A similar effect was found in studies of the paracyclophane carbenes **2**. When the number (n) of methylene groups linking the *para* positions of the rings was 9 or 10, the reactions of the carbenes were those expected of singlets. When n was 11 or 12, the reactions observed were those expected of triplet carbenes.[33b]

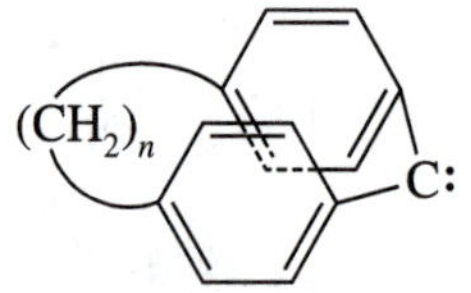

2

9.4   ADDITIONS TO DOUBLE BONDS

Cyclopropane Formation

The current interest in carbene chemistry stems in large part from the demonstration by W. von E. Doering and A.K. Hoffmann, in 1954, that dihalocarbenes can add to alkenes to form cyclopropane derivatives in high yields (Eq. 17).[34]

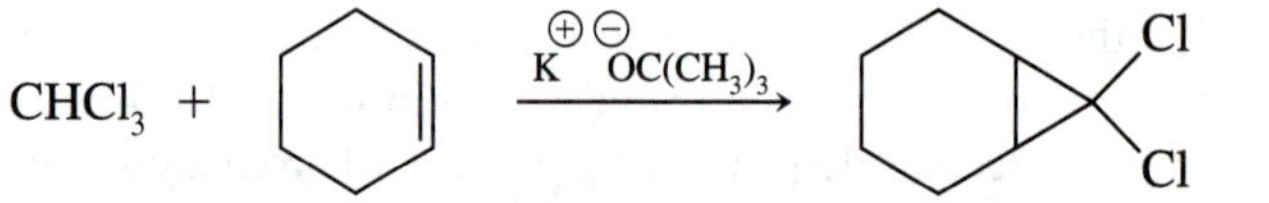

$$(17)$$

Additions to triple bonds can yield cyclopropenes (Eq. 18).[35] Other carbenes can similarly add to π bonds to form three-membered rings, although very reactive carbenes, such as $:CH_2$, may exhibit many side reactions.[1]

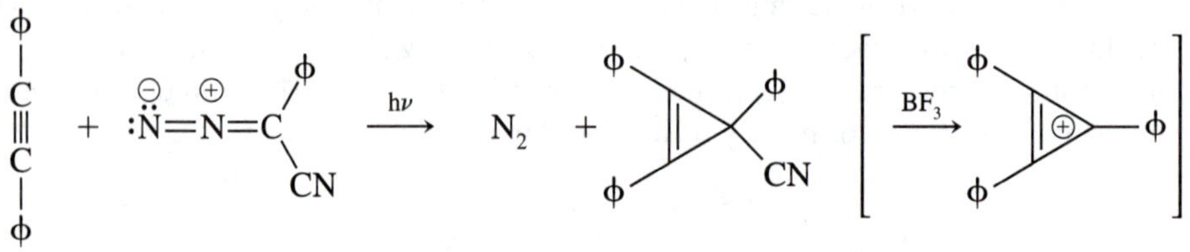

$$(18)$$

Intramolecular addition of carbenes to double bonds has proved to be an extremely useful procedure for the formation of polycyclic molecules, such as those shown in Eq. 19,[24] whose synthesis in other ways would be quite difficult.

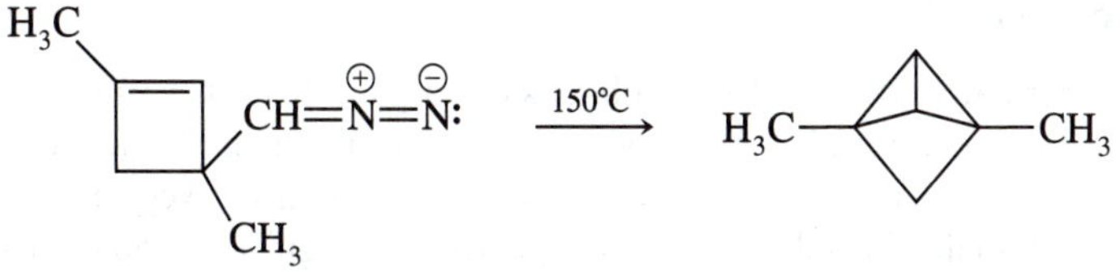

(19a)

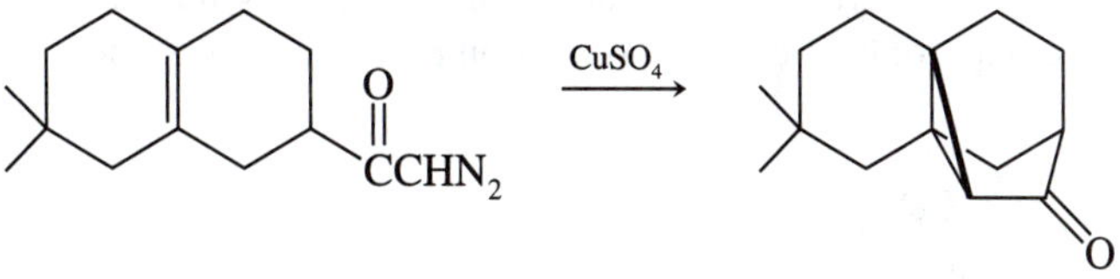

(19b)

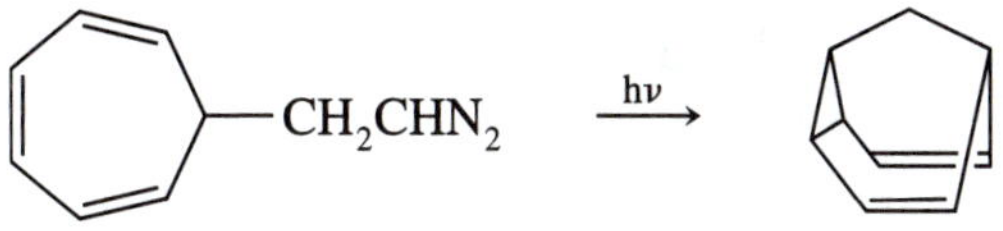

(19c)

It was observed as early as the nineteenth century that reactive carbenes, such as methylene or carbethoxycarbene, even react with double bonds in aromatic rings.[37] The initial bicyclic products from reactions of carbenes with benzene undergo rapid electrocyclic reactions to form cycloheptatriene derivatives, as shown in Eq. 20a.[38]

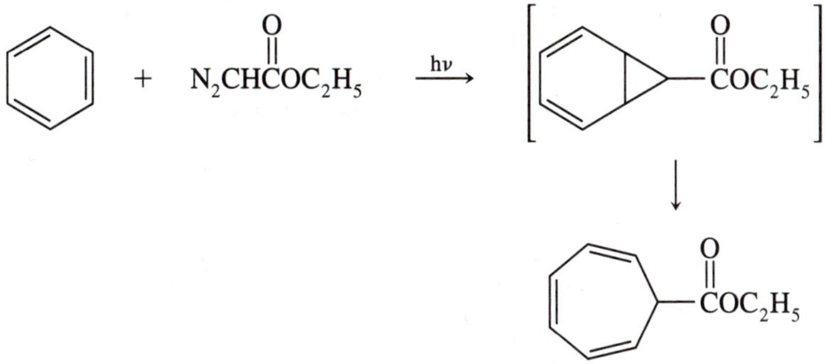

(20a)

However, reactions of carbenes with polycyclic aromatic molecules do yield stable cyclopropane structures (Eq. 20b).[39]

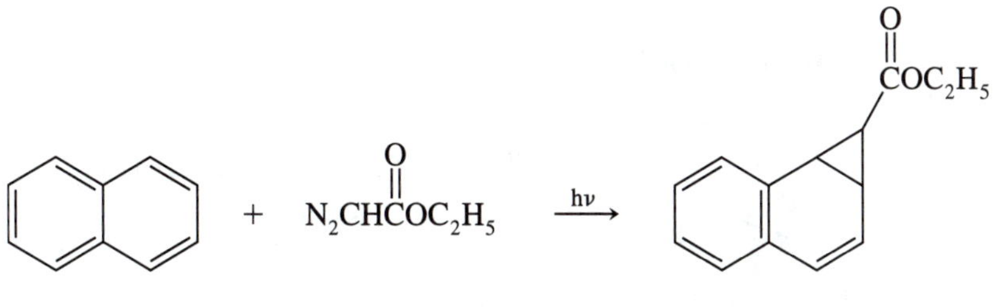

(20b)

1,4- and 1,6-Addition Reactions

Carbenes normally add to individual bonds of conjugated dienes to form cyclopropanes, in 1,2-cycloaddition processes. 1,4-Additions to form five-membered rings are theoretically allowed, but they have been observed only when the diene system is fixed in the *s-cis* coformation, and even then normally in only quite small yields.[40] A significant amount of 1,4-addition *is* observed, accompanying the predominant 1,2-addition, in the reaction of dichlorocarbene with the *s-cis* diene **3**, in which cyclopropane formation is sterically hindered.[41a] In the absence of the bulky methyl groups, only traces of 1,4-addition products are obtained.[40]

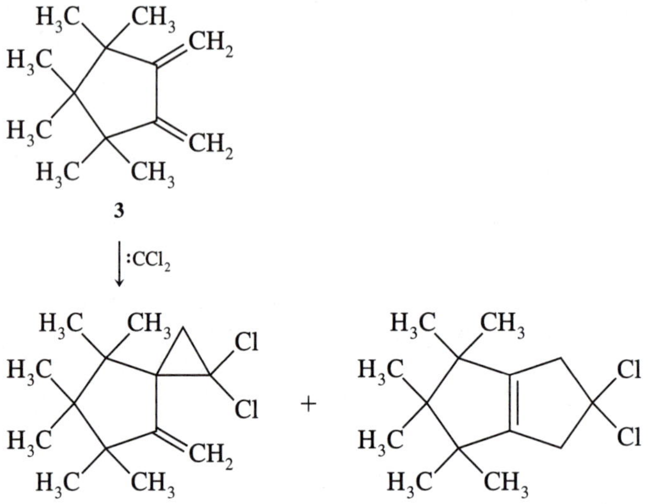

$$\text{(21a)}$$

Interestingly, only 1,4-addition is observed in the reactions of carbene **4**.[41b]

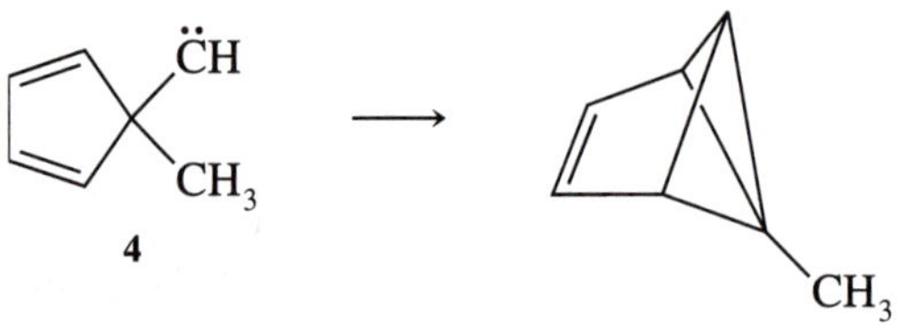

$$\text{(21b)}$$

Retro 1,4-addition reactions of carbenes with dienes can proceed quite easily, provided that the reactions result in the formation of aromatic rings and stable carbenes, as shown in Eq. 22.[42]

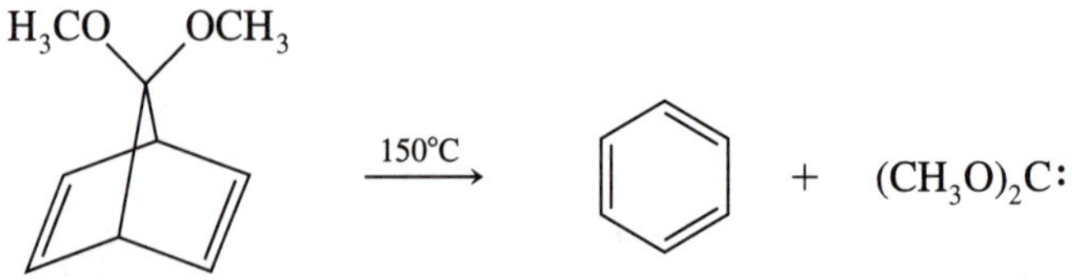

$$\xrightarrow{150°C} \quad + \quad (CH_3O)_2C\colon \qquad \text{(22)}$$

Several reactions that result in overall 1,6-addition have been reported.[43] However, their mechanisms are unclear. The reaction shown in Eq. 23, for instance, has been shown to proceed via a triplet state, implying that the addition is not concerted.[43b]

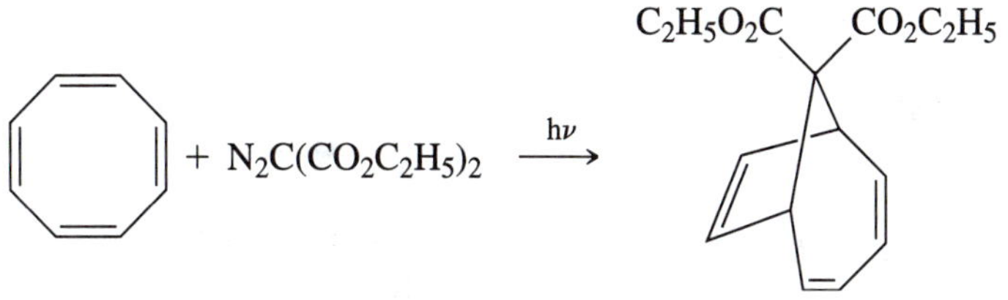

$$\text{(23)}$$

(with structures) $C_2H_5O_2C$ $CO_2C_2H_5$

$+ \ N_2C(CO_2C_2H_5)_2 \xrightarrow{h\nu}$

Stereochemistry of Additions to Double Bonds

Triplet carbenes initially add to double bonds to form triplet diradicals (Eq. 24).

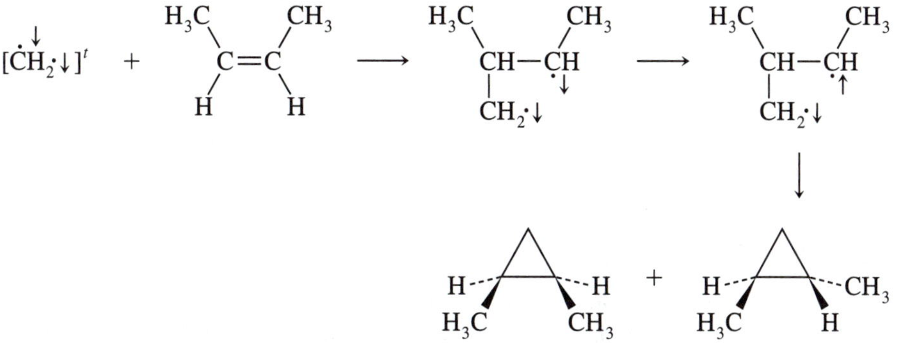

$$[\overset{\cdot}{C}H_2\!\downarrow]^t \ + \ \begin{array}{c} H_3C \quad CH_3 \\ C=C \\ H \qquad H \end{array} \longrightarrow \begin{array}{c} H_3C \quad CH_3 \\ CH-CH \\ CH_2\cdot\downarrow \end{array} \longrightarrow \begin{array}{c} H_3C \quad CH_3 \\ CH-CH \\ CH_2\cdot\downarrow \end{array}$$

$$\text{(24)}$$

$$H\cdots\!\!\triangle\!\!\cdots H \ + \ H\cdots\!\!\triangle\!\!\cdots CH_3$$
$$H_3C \quad CH_3 \qquad\qquad H_3C \quad H$$

The diradicals must undergo intersystem crossing to singlets, in which the spins of the two odd electrons are paired, before they can close to form cyclopropanes. Since rotations around single bonds in the triplet diradicals are usually faster than intersystem crossing, additions of triplet radicals to alkenes are not stereospecific.[16,27,44]

In contrast to triplet carbenes, singlet carbenes usually add stereospecifically to alkenes, forming cyclopropanes that retain the geometries of the alkenes (Eq. 25).[45]*

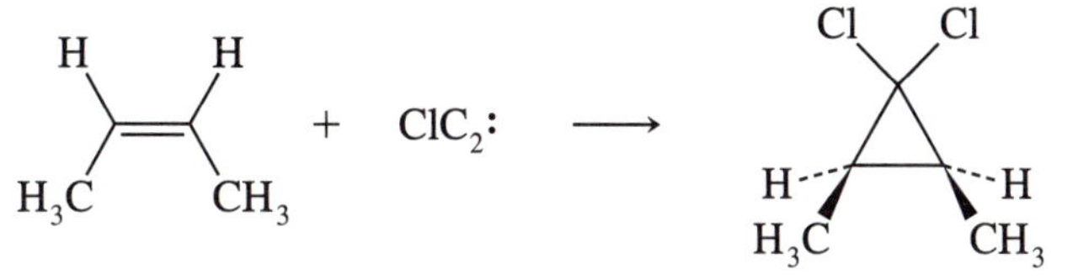

$$\begin{array}{c} H \qquad H \\ C=C \\ H_3C \quad CH_3 \end{array} + \ ClC_2\!: \ \longrightarrow$$

$$\text{(25a)}$$

$$\begin{array}{c} H \qquad CH_3 \\ C=C \\ H_3C \quad H \end{array} + \ ClC_2\!: \ \longrightarrow$$

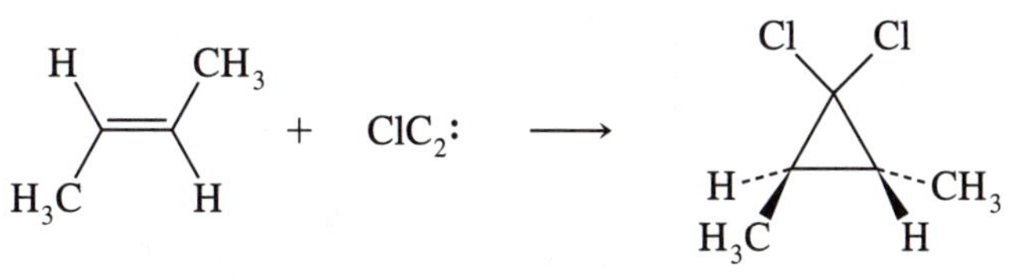

$$\text{(25b)}$$

---

*Singlet carbenes formed photochemically in the gas phase may have such high energies that the resulting cyclopropanes can undergo isomerization, resulting in nonstereospecific cycloadditions.[26] Other cases in which singlet carbenes can add nonstereospecifically to double bonds are discussed in the next section.

Carbenoids, such as the Simmons–Smith reagent invariably react stereospecifically, with retention of the alkene geometry (Eq. 26).

$$
\begin{array}{c}
\text{H} \qquad \text{H} \\
\diagdown\diagup \\
\text{C}=\text{C} \\
\diagup\diagdown \\
\text{H}_3\text{C(CH}_2)_7 \qquad (\text{CH}_2)_7\text{CO}_2\text{CH}_3
\end{array}
\;+\; \text{CH}_2\text{I}_2 \;\xrightarrow{\;\text{Zn(Cu)}\;}\;
\begin{array}{c}
\triangle \\
\text{H}\cdots\cdots\text{CH}_3 \\
\text{H}_3\text{C(CH}_2)_7 \qquad (\text{CH}_2)_7\text{CO}_2\text{CH}_3
\end{array}
\qquad (26)
$$

Nitrenes appear to react in a manner similar to carbenes. Singlet nitrenes add stereospecifically to double bonds, as shown in Eq. 27.[46] Triplet nitrenes, as expected, do not add stereospecifically to double bonds.

$$
\begin{array}{c}
\text{O} \\
\|\\
\text{CH}_3\text{OC}-\ddot{\text{N}}\!:
\end{array}
\;+\;
\begin{array}{c}
\text{H} \qquad \text{H} \\
\diagdown\diagup \\
\text{C}=\text{C} \\
\diagup\diagdown \\
(\text{CH}_3)_2\text{CH} \qquad \text{CH}_3
\end{array}
\;\longrightarrow\;
\begin{array}{c}
\text{O} \\
\|\\
\text{COCH}_3 \\
|\\
\text{N} \\
\triangle \\
\text{H}\cdots\cdots\text{H} \\
(\text{CH}_3)_2\text{CH} \qquad \text{CH}_3
\end{array}
\qquad (27)
$$

Reactivities in Addition Reactions

Most singlet carbenes add to alkenes substituted with electron-donating groups in preference to those substituted with electron-withdrawing groups. Carbenes substituted with halogens or other electron-withdrawing groups show the greatest preference for reaction with electron-rich double bonds (Eq. 28). However, even methylene itself reacts more rapidly with electron-rich than electron-poor double bonds.[47]

$$
\text{ClC}_2\!:
\;+\;
\begin{array}{c}
\text{H}_3\text{C} \qquad \text{CH}_3 \\
\diagdown\diagup \\
\text{C}=\text{C} \\
\diagup\diagdown \\
\text{H}_3\text{C} \qquad \text{C}_2\text{H}_5
\end{array}
\;+\;
\bigcirc
$$

$$\downarrow$$

$$
\begin{array}{c}
\text{Cl} \quad \text{Cl} \\
\diagdown\diagup \\
\triangle \\
\text{H}_3\text{C} \text{CH}_3 \\
\text{H}_3\text{C} \text{C}_2\text{H}_5
\end{array}
\;+\;
\begin{array}{c}
\text{Cl} \\
\text{Cl}
\end{array}
\qquad (28)
$$

product ratio: 23   :   1

The preference of most carbenes for addition to electron-rich double bonds is consistent with theoretical studies indicating that in the formation of a cyclopropane the π bond of an alkene initially interacts with the empty p orbital of a carbene.[48] After the transition state is passed, the unshared electron pair on the original carbenic carbon can participate in forming the cyclopropane ring. The addition of a carbene to a double bond has thus been described as proceeding by a "two-stage" (or "two-phase") approach — an initial "electrophilic stage" followed by a "nucleophilic stage."[49]

However, these two-*stage* reactions appear to be single-*step* reactions. There is no indication that discrete intermediates are formed in the addition of "electrophilic carbenes" to double bonds.

While carbenes will usually add most rapidly to electron-rich double bonds, carbenes substituted with strongly electron-donating groups, such as methoxy or dialkylamino groups, have been found to be nucleophilic reagents.[47,50,51] Like most nucleophiles, they react slowly, if at all, with simple alkenes but will add rapidly to electron-poor double bonds (Eq. 29).[52,53]

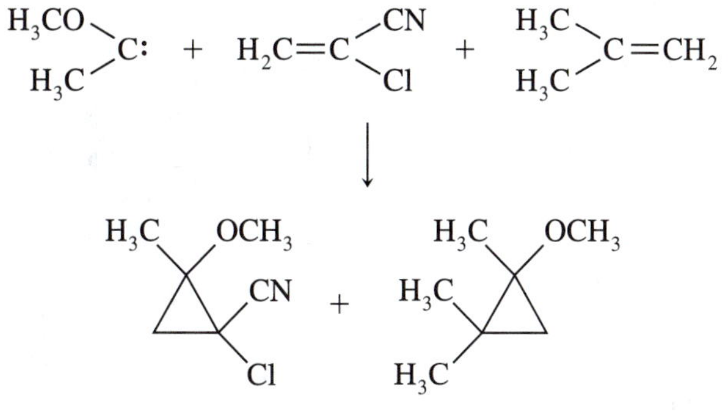

$$
\begin{array}{ccc}
\underset{H_3C}{\overset{H_3CO}{\diagdown}}C: & + & H_2C=C\underset{Cl}{\overset{CN}{\diagup}} & + & \underset{H_3C}{\overset{H_3C}{\diagdown}}C=CH_2 & \qquad (29)
\end{array}
$$

$$\downarrow$$

product ratio: 22,300   :   1

Other carbenes, such as *p*-tolylchlorocarbene, add to both electron-rich and electron-poor double bonds in preference to typical alkene double bonds, as shown in the table below.[54] (Carbenes that react rapidly with both strongly electron-rich and electron-poor double bonds are called *ambiphilic carbenes*.)

RELATIVE RATES OF REACTION WITH 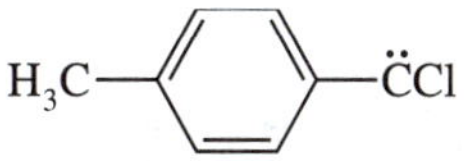 H_3C—⟨ ⟩—$\ddot{C}Cl$	
n-butyl vinyl ether	2.41
1-hexene	1.00
Diethyl fumarate	6.20

There is comparatively little information about the stereochemistry of cyclopropane formation from "nucleophilic" carbenes. However, it was found that the addition of dimethoxycarbene to diethyl maleate yielded the *trans* isomer of the resulting cyclopropane derivative.[50]

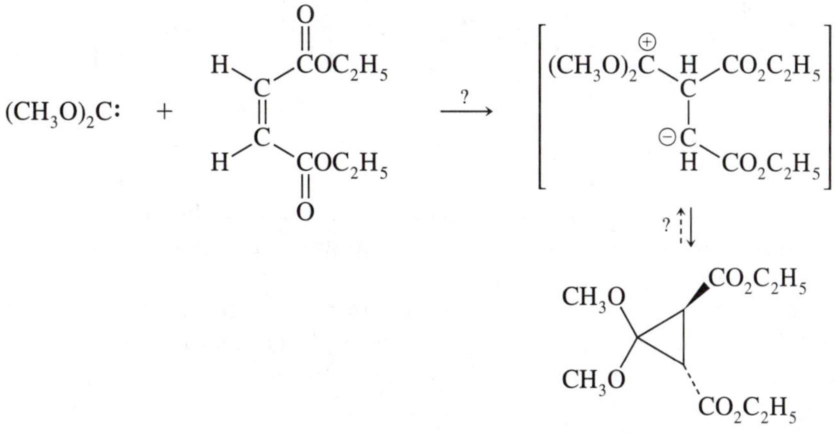

$$
(CH_3O)_2C: \; + \; \cdots \; \xrightarrow{?} \; \left[\cdots \right] \qquad (30)
$$

While this might indicate that the reaction involves the formation of a zwitterionic intermediate, as shown in Eq. 30, it is also possible that the initial addition proceeds stereospecifically to yield the *cis*-substituted cyclopropane, which then opens to form the zwitterion.

Triplet forms of carbenes are less affected by electron-donating or electron-withdrawing substituents on double bonds than are singlet forms. However, triplet carbenes react very rapidly with conjugated double bonds. These reactions can proceed via resonance stabilized diradical intermediates (Eq. 31).[26,27]

$$\phi_2\dot{C}\cdot \;+\; CH_2{=}CH{-}CH{=}CH_2 \;+\; \text{[image]} \;\longrightarrow\; \text{[image]} \;+\; \text{[image]} \tag{31}$$

triplet   product ratio: 25   :   1

9.5   INSERTION REACTIONS

Singlet Carbenes

Singlet methylene, formed by photolysis of diazomethane, can react with alkanes and cycloalkanes by insertion into carbon–hydrogen bonds. Remarkably, those reactions show almost no selectivity between different types of carbon–hydrogen bonds. The reaction of methylene with 2,3-dimethylbutane, for instance, yields nearly the statistically predicted 6 to 1 ratio of products resulting from insertion into the primary and tertiary carbon–hydrogen bonds (Eq. 32).[55]

$$(CH_3)_2\overset{\underset{\displaystyle |}{H}}{C}{-}\overset{\underset{\displaystyle |}{H}}{C}(CH_3)_2 \;+\; :CH_2 \;\longrightarrow\; (CH_3)_2\overset{\underset{\displaystyle |}{H_3C}}{C}{-}\overset{\underset{\displaystyle |}{H}}{C}(CH_3)_2 \;+\; (CH_3)_2\overset{\underset{\displaystyle |}{H}}{C}{-}\overset{\underset{\displaystyle |}{H}}{C}{\Big\langle}^{CH_2CH_3}_{CH_3} \tag{32}$$

product ratio: 17   :   83

Even the reaction of methylene with cyclohexene yields nearly the statistically expected ratio of products from insertion into allylic and vinylic carbon–hydrogen bonds, although the two types of bonds differ markedly in bond strength (Eq. 33).

$$H_2C: \;+\; \text{[image]} \;\longrightarrow\; \text{[image]} \;+\; \text{[image]} \;+\; \text{[image]} \;+\; \text{[image]} \tag{33}$$

product ratios: 11   :   25   :   25   :   40

As a result of these experiments, it was claimed in 1956 that "methylene must be classified as the most indiscrimate reagent known in organic chemistry."[55] Nothing observed since that time requires significant change in that view.

Other singlet carbenes, as well as most carbenoids,[56] are far more selective than methylene. Dichlorocarbene, for instance, can insert into allylic or benzylic carbon–hydrogen bonds but does not react with nonallylic primary or secondary bonds.[57a] It will insert into some, but not all, tertiary carbon–hydrogen bonds. For instance, it will react with *cis*-decalin (Eq. 34) but not with *trans*-decalin.[57b] (*Cis*-decalin is slightly higher in energy than *trans*-decalin, because one C–C bond in *cis*-decalin must be in an axial position.)

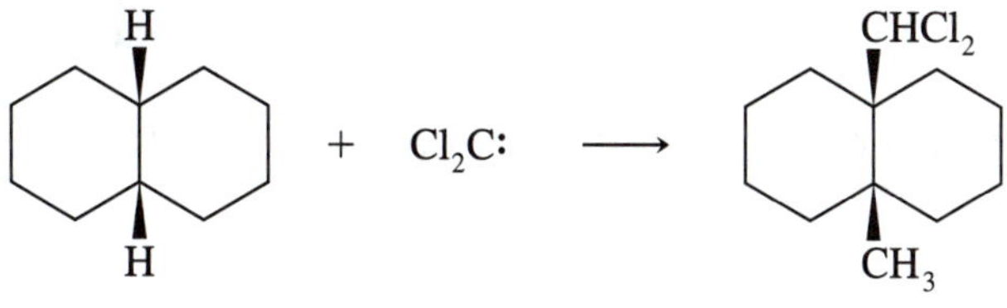

$$(34)$$

Insertions of singlet carbenes into carbon–hydrogen bonds proceed with retention of configuration, as shown in Eqs. 35 and 36.[58,59]

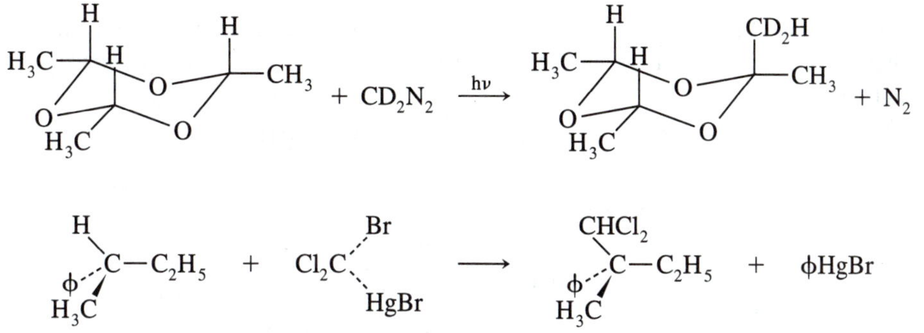

$$(35)$$

$$(36)$$

Carbenes do not insert into unstrained carbon–carbon bonds. However, insertions into very strained carbon–carbon bonds, such as that shown in Eq. 37, do take place.[60]

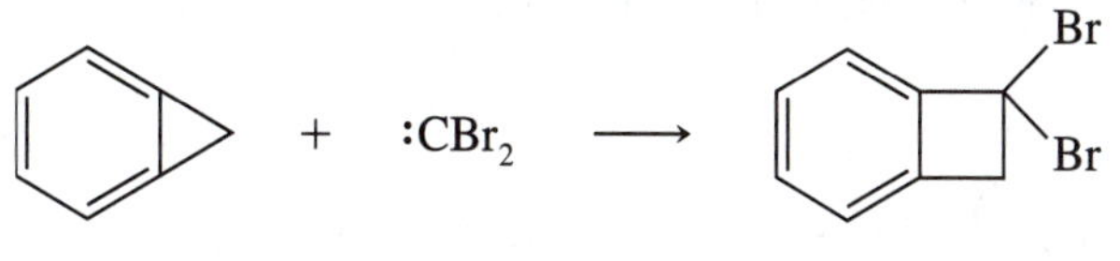

$$(37)$$

Remarkably, even addition to *two* single bonds has been observed (Eq. 38).[61]

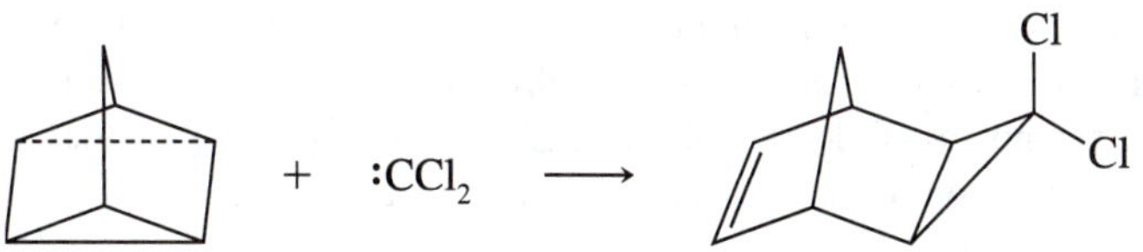

$$(38)$$

Triplet Carbenes

Triplet carbenes can also insert into carbon–hydrogen bonds. Unlike insertions by singlet carbenes, these reactions appear to proceed by multistep processes. Initially, triplet carbenes abstract hydrogen atoms from carbon–hydrogen bonds to form radical pairs, which combine to form the insertion products. If recombination occurs within the solvent cage, the products may show CIDNP peaks, as does the triphenylethane formed in Eq. 39.[62]

$$(C_6H_5)_2C{=}N_2 \xrightarrow{\ h\nu\ } [(C_6H_5)_2\overset{\cdot}{\underset{\cdot}{C}}\cdot] \xrightarrow{\ C_6H_5CH_3\ } [(C_6H_5)_2\overset{\cdot}{C}H + C_6H_5\overset{\cdot}{C}H_2]$$

(triplet)

(39)

$$\downarrow$$

$$(C_6H_5)_2CHCH_2C_6H_5 + (C_6H_5)_2CHCH(C_6H_5)_2 + C_6H_5CH_2CH_2C_6H_5$$

(shows CIDNP peaks)

9.6   REARRANGEMENTS

Hydrogen and Alkyl Group Migrations

1,2-Migrations of hydrogens, such as those in Eqs. 40 and 41, occur extremely easily in singlet carbenes.* (The activation energies for hydrogen migrations in alkyl carbenes appear to be only about 1 kcal/mol.[63])

$$H_3C{-}\overset{\cdot\cdot}{C}H \longrightarrow H_2C{=}CH_2$$

(40)

(41)

Hydrogen migrations in carbenes are usually so rapid that intermolecular addition or insertion reactions of the carbenes cannot compete with their rearrangements to alkenes. However, hydrogen migrations in three-membered ring carbenes (cyclopropylidenes) would yield highly strained double bonds. Therefore, cyclopropylidenes typically undergo ring-opening reactions to form allenes. They can also add to alkenes (Eq. 42).[64]

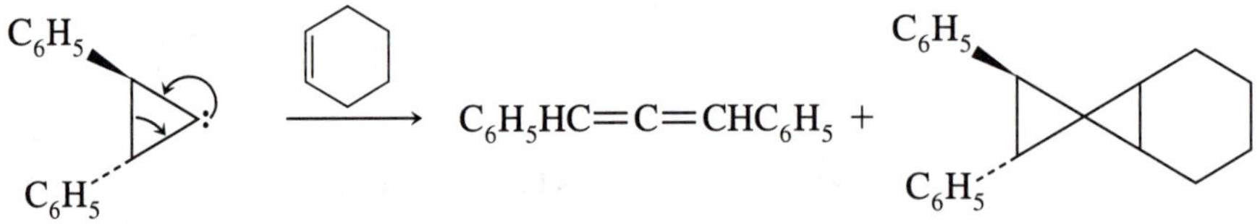

(42)

Alkyl groups in carbenes tend to migrate much more slowly than hydrogens, and their migrations are usually barely detectable if hydrogen migrations can occur instead. When hydrogen migrations are not possible, 1,2-alkyl migrations can occur, but they must compete with intramolecular insertion reactions, as shown in Eqs. 43a and b.

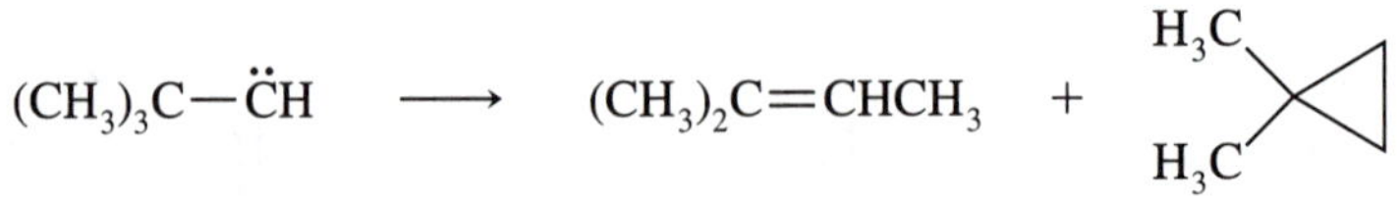

$$(CH_3)_3C{-}\overset{\cdot\cdot}{C}H \longrightarrow (CH_3)_2C{=}CHCH_3 + $$

(43a)

---

*It is often difficult to determine whether a rearrangement takes place in a "free" carbene or whether hydrogen or alkyl migrations are concerted with loss of nitrogen from diazo compounds. For the most part, we will not be concerned with this question but will consider either type of process a "carbene" rearrangement.

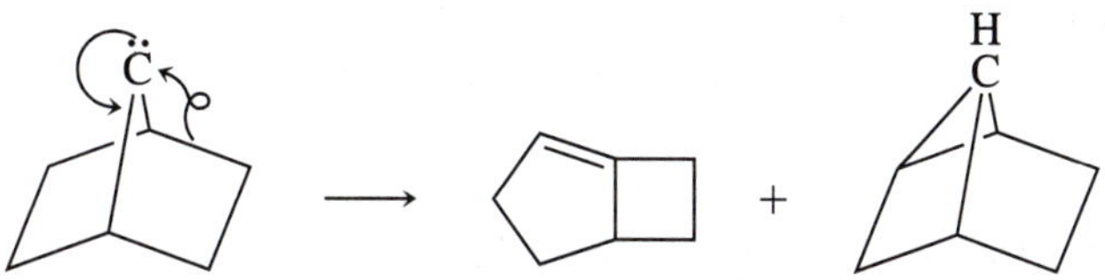

(43b)

Alkyl group migrations are common if they result in the expansion of strained rings, as shown in Eq. 44a.[65a]

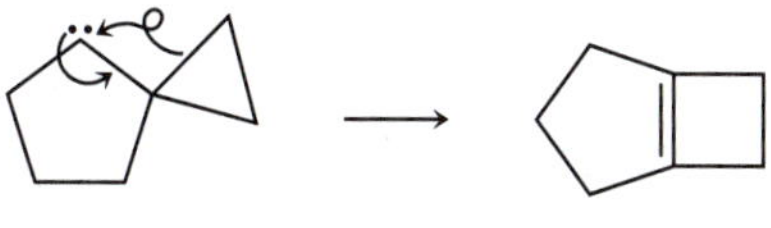

(44a)

Alkyl migration also predominates when a hydroxy substituent can stabilizes formation of an empty orbital at the migration origins, as shown in Eq. 44b.[65b] (That reaction can be regarded as the carbene analog of a pinacol rearrangement.)

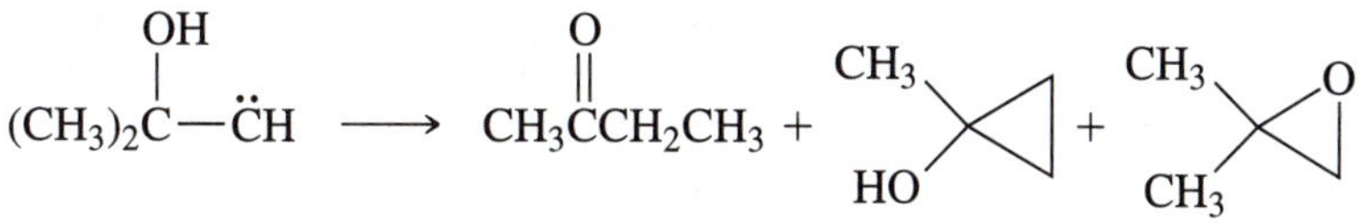

(44b)

69% 28% 3%

Alkyl group migrations in nitrenes are more competitive with hydrogen migrations than in carbenes, but hydrogen migrations still predominate when they are possible, as illustrated in Eq. 45.[66]

$$C_3H_7CH_2N{=}\overset{\oplus}{N}{=}\overset{\ominus}{N}: \xrightarrow{390°C} N_2 + C_3H_7CH_2\ddot{N}: \quad (45)$$

$$\downarrow$$

$$H_2C{=}N{-}C_3H_7 \quad + \quad C_3H_7CH{=}NH$$

15% 85%

The Nature of the Rearrangement Process

1,2-Migrations of hydrogens and alkyl groups in carbenes are often described as insertions of the divalent carbons into neighboring carbon–hydrogen and carbon–carbon bonds. However, it seems more useful to regard these reactions as analogous to carbocation rearrangements.[67]

When the relative migration rates of substituted phenyl groups in carbenes are plotted against Hammett σ^+ values for the substituents, for instance, straight lines with negative slopes are obtained.[68] Migrations of aryl groups in nitrenes show similar substituent effects.[69] (Aryl groups in carbenes and nitrenes usually migrate more rapidly than alkyl groups but less rapidly than hydrogens.) These results are similar to those found in carbocation rearrangements.

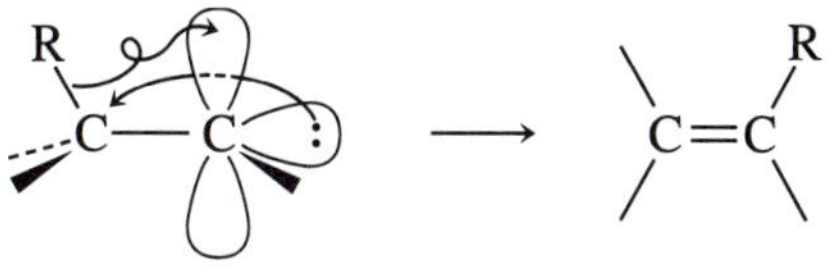

(46)

Migration to an empty orbital of a carbene

1,2-Migrations of hydrogens or alkyl groups in triplet carbenes are essentially unknown. However, there is evidence that aryl groups can migrate in triplet carbenes, as in free-radical rearrangements.[70]

Rearrangements of Carbenoids Formed from Vinyl Halides

Formation of Alkynes.   The reactions of 2,2-disubstituted vinyl bromides with strong bases, such as potassium *t*-butoxide or alkyllithium reagents, can result in the formation of triple bonds, as shown in Eq. 47a.[71a]

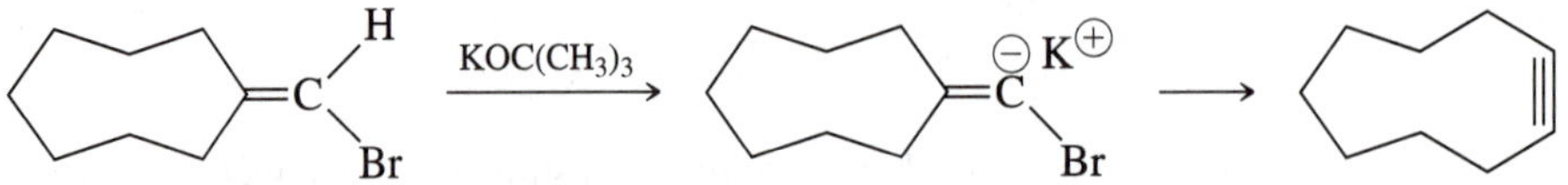

$$(47a)$$

These reactions have sometimes been described as rearrangements of carbanions (and therefore theoretically forbidden), since it has been shown that α-halovinyl carbanions are the first-formed intermediates.[72b] However, they appear to be more accurately described as rearrangements of carbenoids, in which the bonds between the vinyl carbons and the halogen ions are at least partially broken before rearrangement takes place. This can be demonstrated in the reaction shown in Eq. 47b, in which the rearrangement step is slow because it would result in formation of the highly strained molecule cyclohexyne. As a result, the carbenoid intermediate can be intercepted by the addition of alkenes, yielding cyclopropane derivatives (Eq. 47b).[71a]

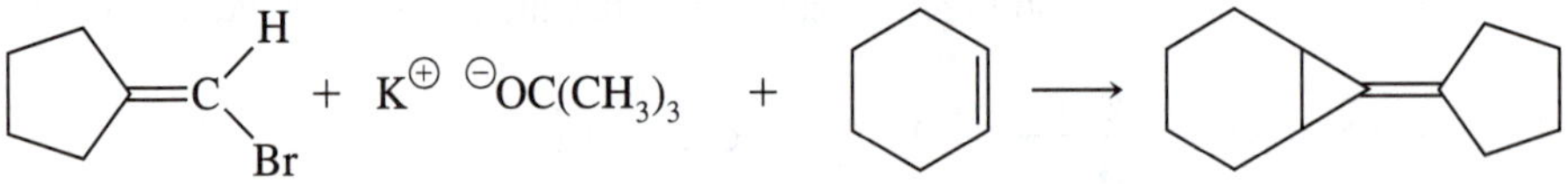

$$(47b)$$

That the intermediates in these reactions are best described as carbenoids, rather than relatively free carbenes, can be demonstrated by the use of isotopically labeled starting materials, as shown in Eq. 48. These studies show that the rearrangements are stereospecific, and that it is invariably the groups *trans* to the halogens that migrate.[73a]

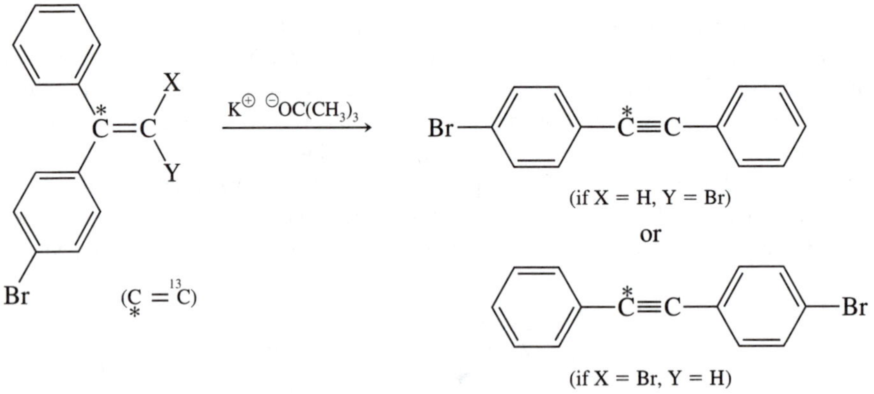

$$(48)$$

It has been demonstrated that migrating groups with electron-donating substituents migrate more rapidly than those bearing electron-withdrawing substituents.[73b] The migrating groups thus appear to be acting as nucleophiles, displacing the halide ions as the rearrangement steps take place.

Formation of 1-Halocyclopentenes.   Cyclopentynes, which might be formed by base-induced rearrangements of halomethylenecyclobutanes, are very strained molecules which are known to react with alkoxide ions to form cyclopentenyl ethers. However, only small amounts of cyclopentenyl ethers are formed from reactions of halomethylenecyclobutanes with alkoxide solutions. Instead, those reactions principally yield 1-halocyclopentenes. The use of isotopically labeled starting materials (Eq. 49) showed that these reactions proceed by two distinct paths: a "single migration" path, in which the halogen atoms remain bonded to the original vinyl carbons, and a (usually predominant) "double migration" path, in which the halogen atom migrates to the ring carbon from which the migrating alkyl group had started. Remarkably, while the alkyl groups *trans* to the halogens migrate, as usual, in the double migration reactions, the groups *cis* to the halogens migrate in the single migration reactions.[75]

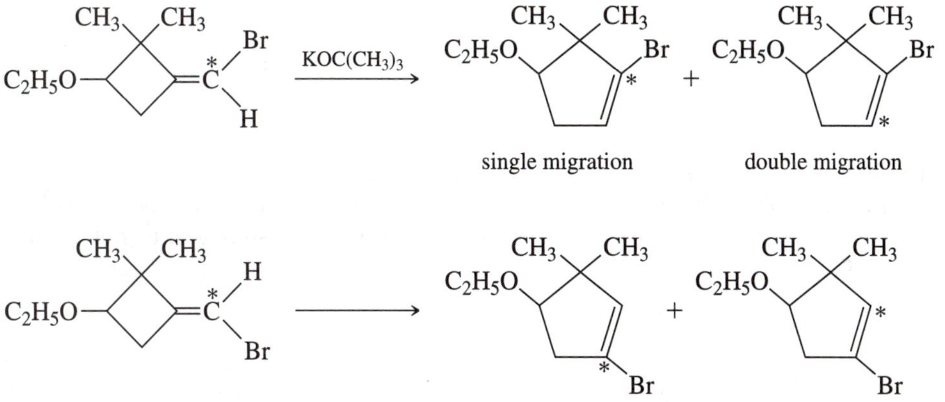

single migration   double migration

(49)

The percentage of double migration is increased when the halide ion is a good leaving group $(I^- > Br^- > Cl^- \gg F^-)$ and when the migrating group is more highly substituted, as well as by higher temperatures.[75]

   "Double migration" rearrangements appear to be closely related to rearrangements leading to acetylenes, except that (due to difficulty in forming cyclopentynes) the departing halide ions attack the sites from which the ring carbons are migrating.

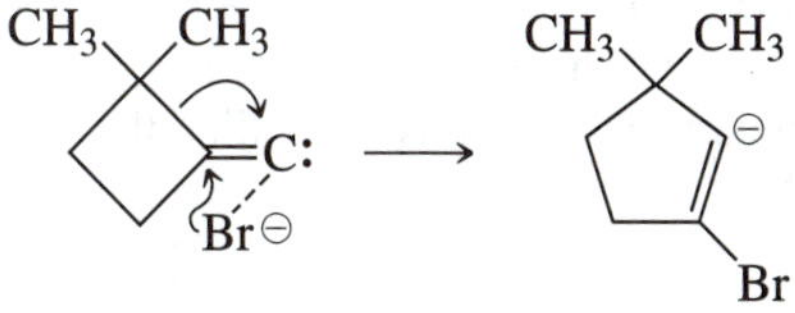

Rearrangement by a "Double Migration" Mechanism

   In contrast, "single migration" reactions appear to be quite different from other reactions of vinyl carbenoids. A unique "carbene anion" mechanism, which will not be discussed further here, has been suggested.[75]

Rearrangements of Acylcarbenes

Wolff rearrangements. When α-diazoketones are photoirradiated, or heated at high temperatures, or reacted with silver oxide or silver salts at room temperature, they lose nitrogen and rearrange to form ketenes (Eq. 50).

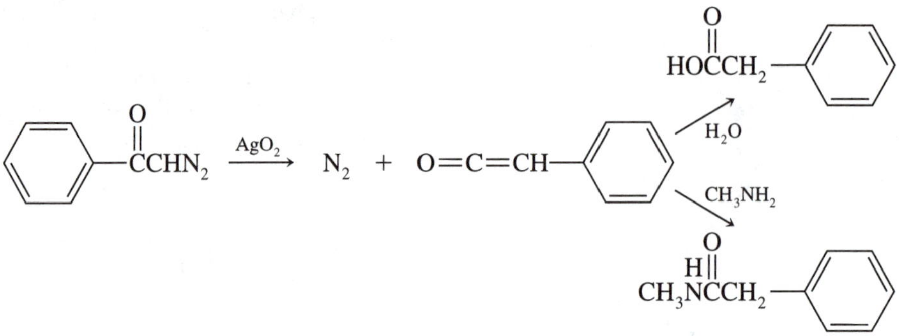

$$\text{(50)}$$

The ketenes react rapidly with water, alcohols, and amines. Therefore, the reactions, called *Wolff rearrangements*, usually result in the formation of carboxylic acids, esters, or amides.[76]

α-Diazoesters can also undergo Wolff rearrangements, resulting in migrations of alkoxy groups (Eq. 51).[77]

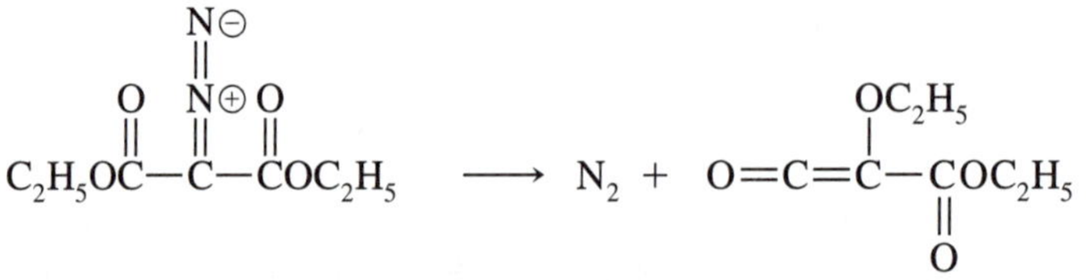

$$\text{(51)}$$

The available evidence suggests that migrations in thermal Wolff rearrangements usually proceed simultaneously with loss of nitrogen and without formation of free carbenes. Some photochemical Wolff rearrangements, however, have been shown to proceed via the free carbenes, which have been identified from their spectra.[78]

Since α-diazoketones can be conveniently prepared by reactions of acid chlorides with diazomethane (Eq. 52), Wolff rearrangements provide a useful method for converting carboxylic acid derivatives to their next-higher homologs. (The entire process, starting with the acid halide and yielding the chain-extended acid derivative, is called the *Arndt–Eistert reaction.*)

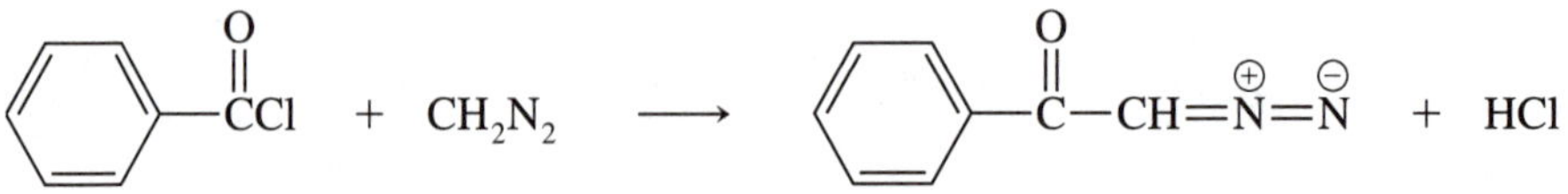

$$\text{(52)}$$

Wolff rearrangements proceed stereospecifically, with retention of the configurations of the migrating groups (Eq. 53).[79]

$$\underset{CH_3CH_2}{\overset{H}{\underset{\diagup}{CH_3\cdots C}}}\overset{O}{\overset{\|}{-CCHN_2}} \;+\; CH_3OH \;\xrightarrow{Ag_2O}\; CH_3O\overset{O}{\overset{\|}{C}}-CH_2-\overset{CH_3}{\underset{CH_2CH_3}{\overset{\diagup}{C}\diagdown H}} \qquad (53)$$

Oxirene formation.   A study of the gas-phase photolysis of α-diazobutanone labeled with isotopic carbon (^{13}C) in the carbonyl group yielded dimethylketene, in which the isotopic label was equally distributed between the carbonyl carbon and the vinyl carbon (Eq. 54).[80]

$$2\;CH_3\overset{O}{\overset{\|}{\underset{*}{C}}}-\overset{}{\underset{\underset{N_2}{\|}}{C}}-CH_3 \;\xrightarrow{h\nu}\; O=\overset{}{\underset{*}{C}}=C\overset{CH_3}{\diagdown CH_3} \;+\; O=C=\overset{}{\underset{*}{C}}\overset{CH_3}{\diagdown CH_3} \;+\; 2\,N_2 \qquad (54)$$

$$(\underset{*}{C} = {}^{13}C)$$

Photolysis of other isotopically labeled α-diazoketones similarly showed the isotopic carbon to be distributed between the carbonyl and vinyl positions, although not necessarily to an equal degree at both positions.[81]

Changes in the apparent locations of carbonyl carbons (more accurately described as changes in the locations of oxygens) in photochemical Wolff rearrangements appear to be due to initial formation of oxirenes that can open to form mixtures of ketocarbenes (Eq. 55).[82]

$$CH_3-\overset{O}{\overset{\|}{\underset{*}{C}}}-\overset{}{\underset{\underset{N_2}{\|}}{C}}CH_3 \;\xrightarrow{h\nu}\; CH_3-\overset{\overset{O}{\diagup\diagdown}}{\underset{*}{C}=C}-CH_3 \qquad (55)$$

dimethyloxirene

$$\downarrow$$

$$CH_3-\overset{O}{\overset{\|}{\underset{*}{C}}}-\ddot{C}CH_3 \;+\; CH_3-\overset{O}{\overset{\|}{\ddot{C}}}-\overset{}{\underset{*}{C}}CH_3$$

In several cases, the formally antiaromatic oxirenes have been identified and can be isolated in solid argon matrices.[83,84]

Oxirene formation appears to occur only when the α-diazoketones maintain conformations in which the carbonyl oxygens are *anti* to the departing nitrogen molecules. Little, if any, oxygen migration occurs in small cyclic diazoketones, in which the oxygen and the diazo group are fixed *syn* to each other, as shown in Eq. 56,[85] or during thermal decompositions of α-diazoketones. Thermal decompositions apparently occur in conformations in which potential migrating groups, rather than carbonyl oxygens, are *anti* to the nitrogens.

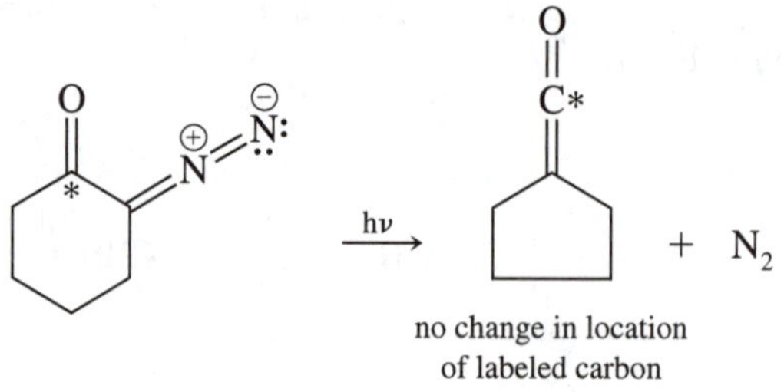

O
N
N:
*
hν
O
C*
+ N₂
no change in location
of labeled carbon

(56)

Retro Carbene Rearrangements

Rearrangements of Strained Rings. Rearrangements of carbenes to form alkenes are typically highly exothermic reactions, and are not readily reversible. However, several compounds containing very highly strained double bonds undergo rapid hydrogen migrations to form carbenes (e.g., Eq. 57).[86]

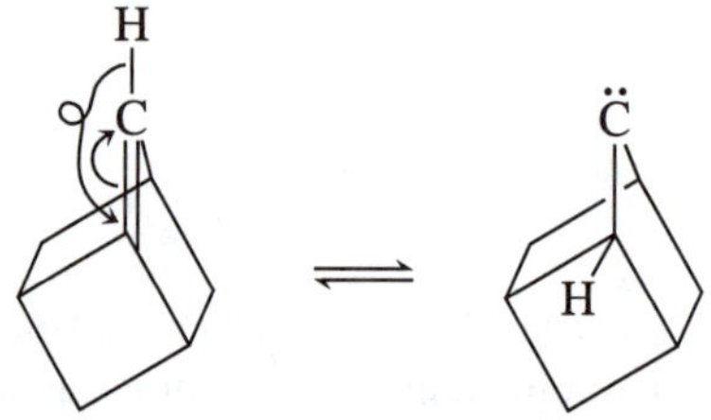

H
C
C
H

(57)

Conversions of π bonds to carbenes by migrations of groups other than hydrogen atoms are even rarer reactions. The migration of an aryl group does occur in the very strained compound **5**.[87]

Br
Br
Br
H

5

(58)

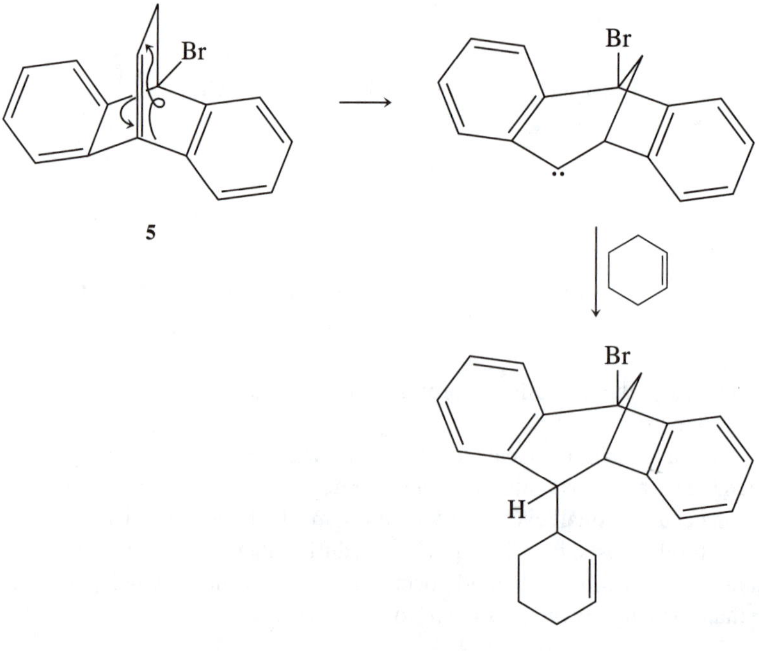

The most common reactions in which strained alkenes are converted to carbenes are in the photochemical or thermal ring openings of cyclopropenes, often resulting in the formation of allylic carbenes.[88] These reactions are reversible, and thus often "invisible." However, in the ring openings of halogen-substituted, cyclopropenes, which can occur at quite low temperatures, the carbenes may be trapped by reactions with alkenes (Eq. 59).[89a]

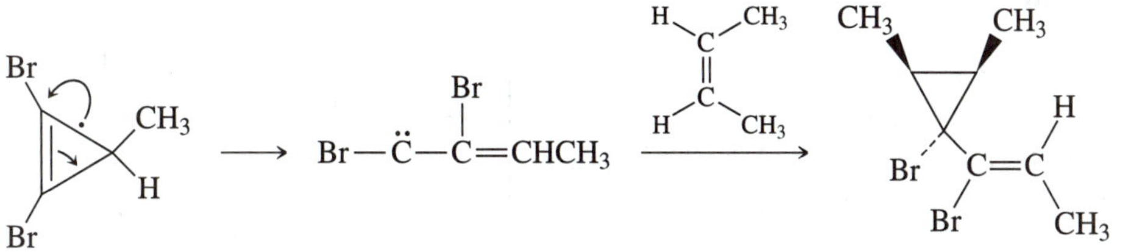

$$
\text{(59)}
$$

Even when the carbenes cannot be trapped, their formation can be inferred from indirect evidence, such as the fact that optically active 1,2-diethylcyclopropene is racemized on heating at *ca.* 175°C (Eq. 60b).[89b]

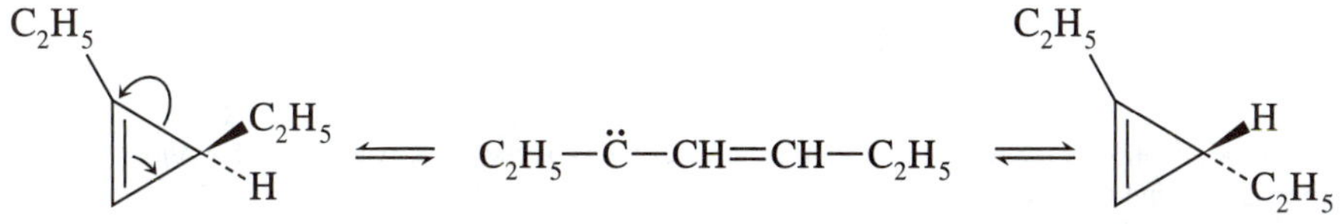

$$
\text{(60)}
$$

When ring opening of a cyclopropene ring results in cleavage of the bond to a vinyl position bearing a hydrogen atom, the ring opening may be accompanied by (or followed by) a hydrogen migration to form a vinyl carbene.[90] This reaction is the reverse of insertion of a carbene into a C–H bond. A second migration can then result in formation of an acetylene, as in Eq. 61.[90]

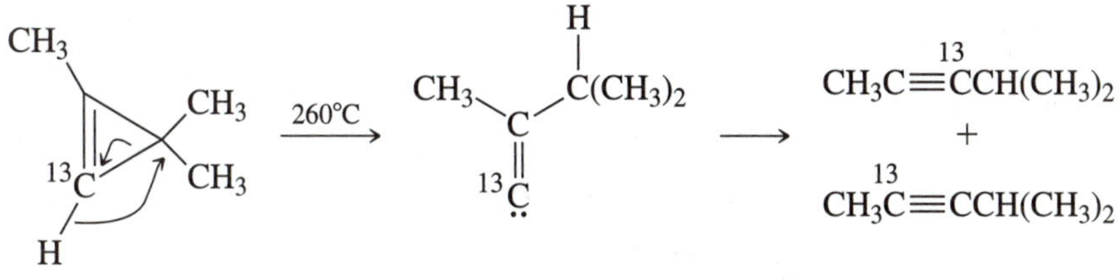

$$
\text{(61)}
$$

Rearrangements of Alkynes.   In principle, at sufficiently high temperatures even molecules with unstrained π bonds might rearrange to form carbenes. In practice, though, such reactions (sometimes called "Roger Brown Rearrangements"[91]) have been observed only in reactions of alkynes.*

Thermolysis of *o*-tolylacetylene at 700°C yields indene as the only significant product. This can best be explained by assuming that an original hydrogen migration forms a vinyl carbene (vinylidene) intermediate (Eq. 62).[92]

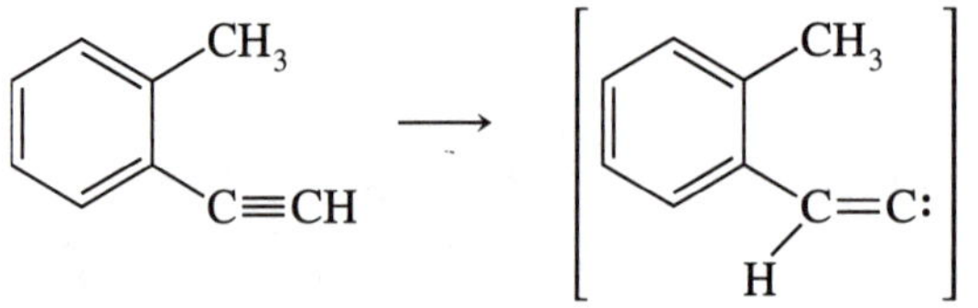

(62)

Even stronger evidence comes from the thermolysis of biphenylacetylene, which yields 1,2-benzoazulene as a major product (Eq. 63).[92] As discussed on page 263, ring expansion is a common result of addition of carbenes to aromatic rings.

---

*A variety of other interesting, low temperature, reactions of alkyne derivatives, such as those shown below, can also give rise to vinyl carbenes.

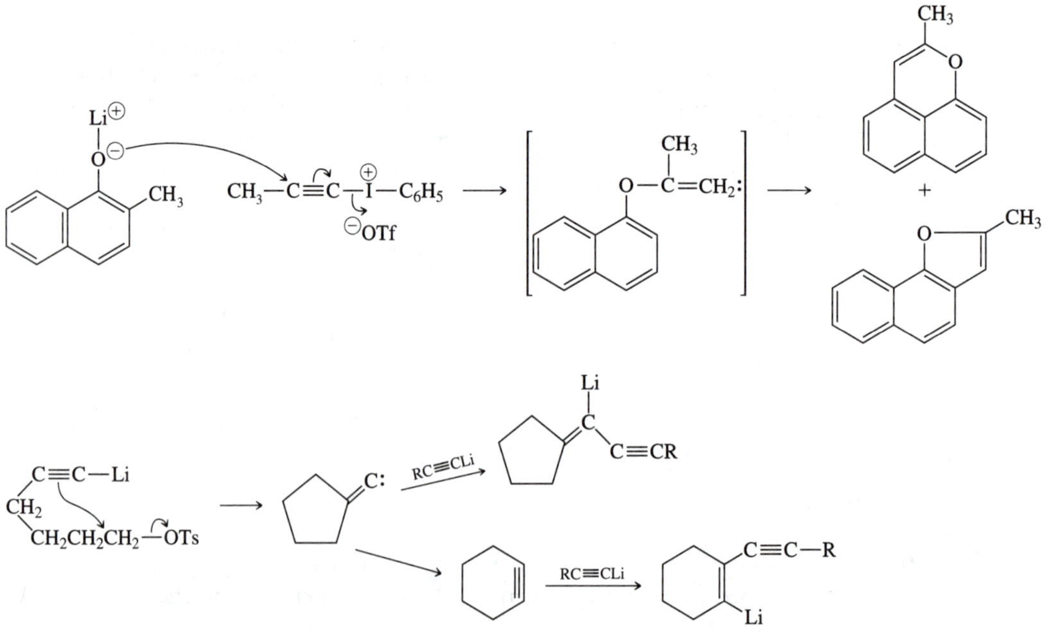

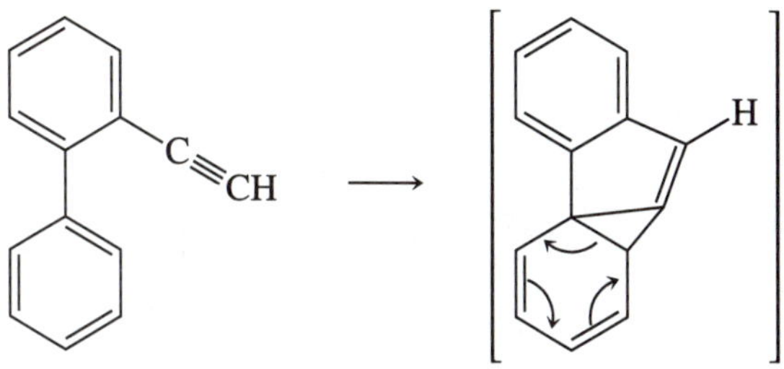

$$(63)$$

The thermolysis of isotopically labeled phenylacetylene results in distribution of the isotopic carbon between the two triply bonded carbons (Eq. 64), presumably as a result of successive migrations of the hydrogen atom and the aromatic ring.[92]

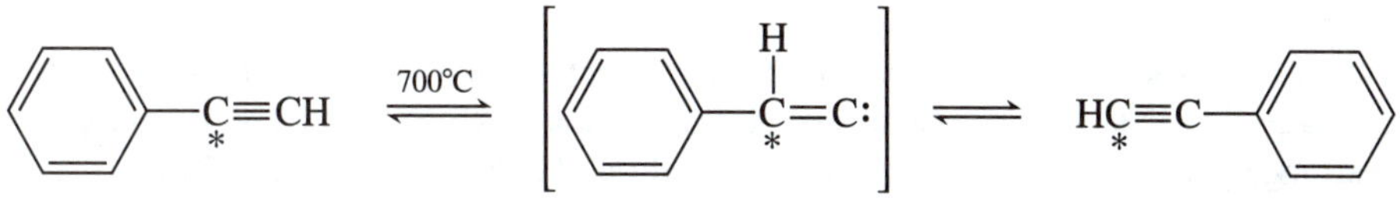

$$(64)$$

As usual in carbene rearrangements, migrations of alkyl groups are much slower than migrations of hydrogens or aryl groups. However, isotopically labeled 1-adamantylacetylene does undergo rearrangement at 780°C (Eq. 65).[93]

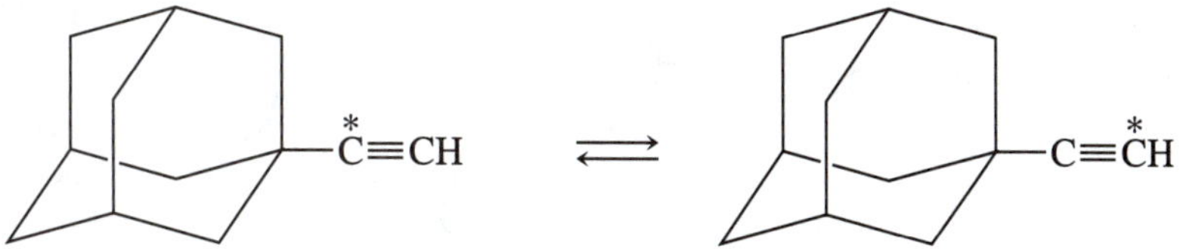

$$(65)$$

Rearrangements of Arylcarbenes and Arylnitrenes

Arylcarbenes can undergo remarkably complicated rearrangements.[94,95] In 1969, it was reported that thermolysis of phenyldiazomethane in the gas phase yields heptafulvalene (**6**), presumably resulting from the rearrangement of phenylcarbene to form a conjugated seven-membered ring carbene, which then dimerizes.[96]

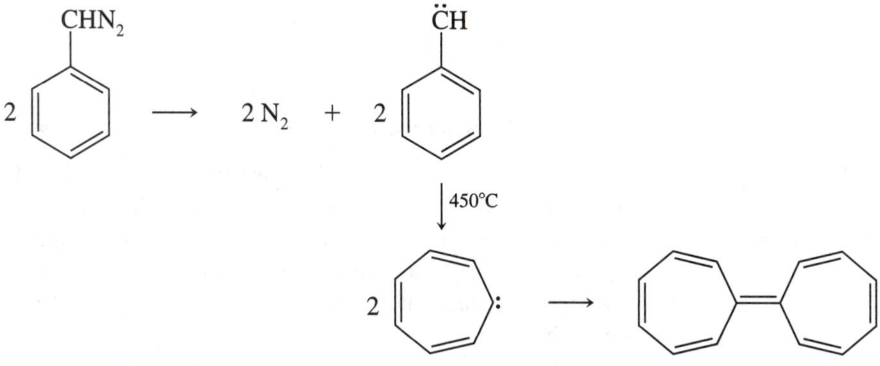

$$(66)$$

6

Shortly thereafter, even more remarkable rearrangements were reported. As Eq. 67 shows, *o*-tolyldiazomethane, *m*-tolyldiazomethane, and *p*-tolyldiazomethane all yielded a mixture of benzocyclobutene and styrene on thermolysis at 420°C.[97]

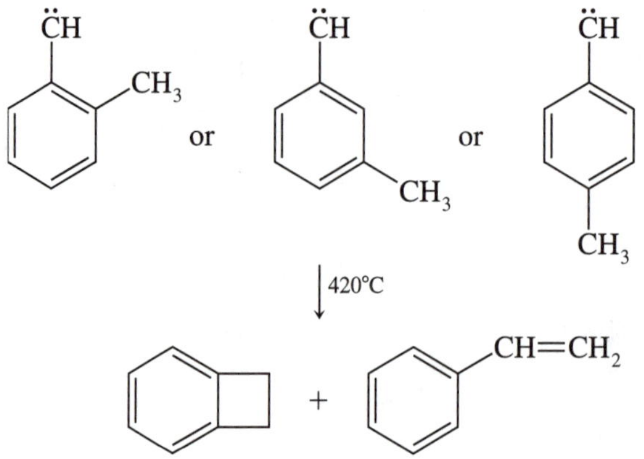

(67)

The yield of benzocyclobutene was much higher from *o*-tolyldiazomethane than from its isomers, suggesting that *o*-tolylcarbene was formed as a discrete intermediate before undergoing the rearrangement steps that resulted in the formation of styrene.

A mixture of benzocyclobutene and styrene was also formed from thermolysis of **7**,[98] again illustrating the ready interconversion of arylcarbenes and molecules with seven-membered rings.

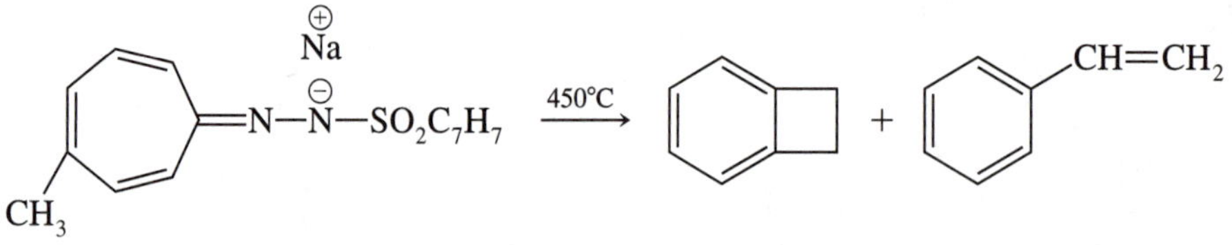

(68)

It seems natural to write the initial product from thermolysis of **7**—or the intermediate leading to formation of heptafulvalene from phenyldiazomethane—as a carbene, as was done in Eq. 66. However, a slight change in bond angles would allow conversion of the carbene to the cyclic allene 1,2,4,6-cycloheptatetraene,

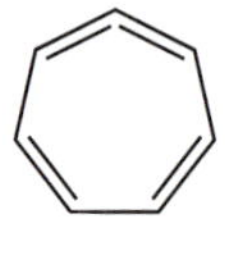

1,2,4,6 - cycloheptatetroene

That allene and its methyl-substituted derivatives were identified from their UV and IR spectra as short-lived products from thermolysis of aryldiazomethanes.[99] Since the allenes are highly strained, they might undergo reactions (e.g., dimerization to form heptafulvalene) that appear to be characteristic of carbenes. Thus, it is not clear whether the seven-membered ring "carbene" forms actually have any independent existence. In fact, if the "carbenes" were slightly deformed from planarity, they would simply be resonance forms of the allenes.

Possible mechanisms for the rearrangements of arylcarbenes would have to account for the fact that *p*-tolylcarbene labeled with ^{13}C at the exocyclic carbon rearranges to form styrene with all the ^{13}C at the *para* position (Eq. 69).[100]

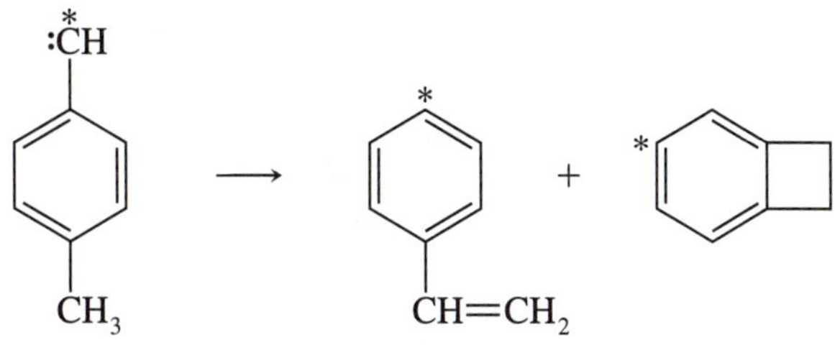

(69)

Several mechanisms, all quite complex, have been proposed. One suggested mechanism for the rearrangement of *p*-tolylcarbene is depicted in Eq. 69a.

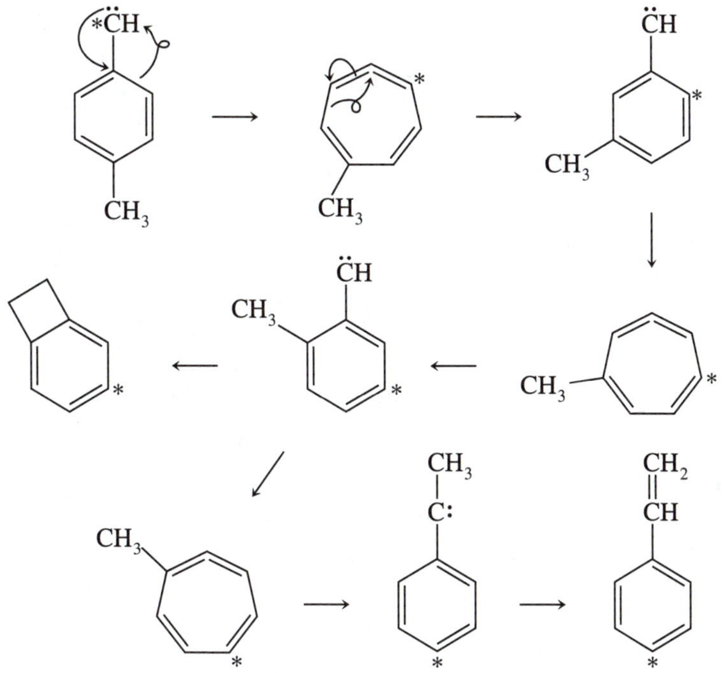

(69a)

Rearrangement of *p*-tolylcarbene labeled with ^{13}C

In the first step, the carbene rearranges to a seven-membered allene (or carbene). The allene could undergo the reverse of this process, reforming the original *p*-tolylcarbene, or could form *m*-tolylcarbene, as shown in Eq. 69a. A second sequence of allene formation and reopening could convert the *m*-tolylcarbene to *o*-tolylcarbene, which could form benzocyclobutene or could continue with a third sequence of allene formation and reopening to form methylphenyl-carbene, which would rearrange to styrene.

In a somewhat more complex mechanism, which was actually the first mechanism proposed, each arylcarbene initially forms a bicyclic cyclopropene, which could then open to form the seven-membered ring allene (or carbene) (Eq. 70).

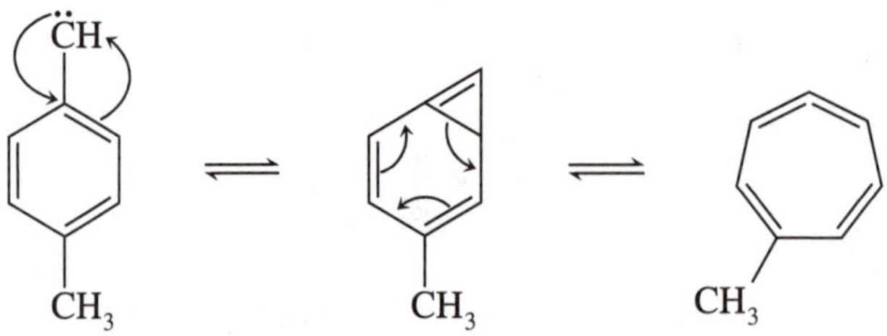

$$(70)$$

In fact, similar bicyclic cyclopropene derivatives can be prepared by quite different methods (Eq. 71) and do undergo rearrangements similar to those of arylcarbenes.[101]

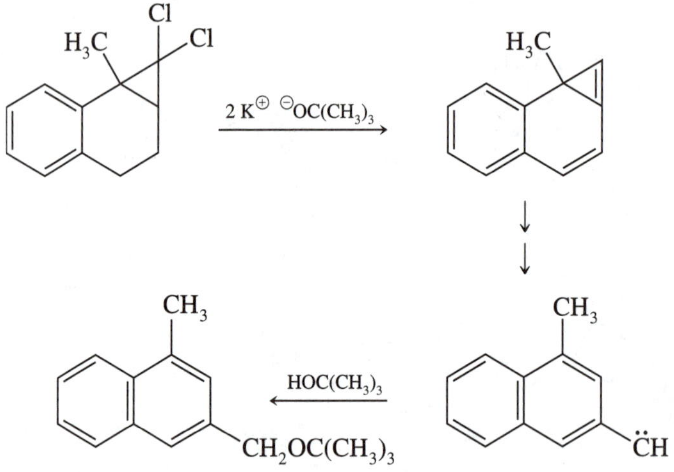

$$(71)$$

In a third possible mechanism, supported by complex arguments,[95] the allenes are in equilibrium with bicyclic cyclopropylidenes, in which the divalent carbons "walk" around the ring (Eq. 72).

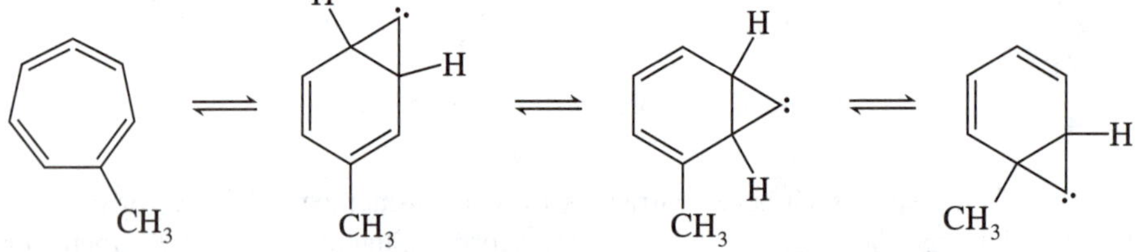

$$(72)$$

Although the precise mechanisms of these remarkable rearrangements have not been established, it has been pointed out that the only intermediates for which direct evidence exists are the arylcarbenes and the cyclic allenes.[99] Thus, *Occam's razor* (a preference for simplicity) favors the first mechanism shown above.

Arylnitrenes undergo rearrangements similar to those of arylcarbenes, forming seven-membered heterocyclic cumulenes. The same intermediates can be obtained from pyridylcarbenes, which are interconvertible with the arylnitrenes (Eq. 73).[102]

$$\ddot{\text{N}}: \qquad \ddot{\text{C}}\text{H}$$

$$\text{(structure 1)} \quad \rightleftharpoons \quad \text{(structure 2)} \quad \rightleftharpoons \quad \text{(structure 3)} \qquad (73)$$

REFERENCES

[1] For a review, see W. Kirmse, *Carbene Chemistry,* 2nd ed. Academic Press, New York (1971).

[2] N. Nakamura et al., *J. Am. Chem. Soc., 114,* 1484 (1992).

[3] See E.F.V. Scriven, Ed., *Azides and Nitrenes.* Academic Press, New York (1984).

[4] For a comprehensive review (in German) of methods for the formation of carbenes, see *Methoden der Organischen Chemie* (Houben Weyl), 4th ed., vol. E19a, M. Regitz, Ed. Georg Thieme Verlag, New York (1989).

[5] H. Staudinger and O. Kupfer, *Ber., 45,* 501 (1912).

[6] P. Helquist, in *Comprehensive Organic Synthesis,* vol. 4, M.L. Semmelhock, Ed. Pergamon Press, New York (1991), pp. 951–997.

[7] R.H. Shapiro, *Org. Reactions, 23,* 405 (1976).

[8] M. Regitz and G. Maas, *Diazo Compounds: Properties and Synthesis.* Academic Press, New York (1986).

[9] L.I. Smith and K.L. Howard, *Org. Syn. Coll., 3,* 351 (1955).

[10] P.S. Engel, *Chem. Rev., 80,* 99 (1980).

[11] V. Nair, in *Comprehensive Organic Synthesis,* vol. 4, B.M. Trost, Ed. Pergamon Press, New York (1991), pp. 999–1006.

[12] W.L. Dilling and F.Y. Edamura, *J. Org. Chem., 32,* 3492 (1967).

[13] G.L. Closs and L.E. Closs, *Tetrahedron Lett.,* 2 (1960).

[14] J. Hine, *J. Am. Chem. Soc., 72,* 2438 (1950).

[15] G. Koebrich and R.H. Fischer, *Tetrahedron, 24,* 4343 (1968).

[16] (a) H.E. Simmons and R.D. Smith, *J. Am. Chem. Soc., 81,* 4256 (1959); (b) W.- H. Fang, D.L. Phillips, D.-Q. Wang, and Y.-L. Li, *J. Org. Chem., 67,* 154 (2002).

[17] E. Leyva, M.S. Platz, G. Persy, and J. Wirz, *J. Am. Chem. Soc., 108,* 3783 (1986); O. Meth-Cohn, *Acc. Chem. Res., 20,* 18 (1987).

[18] K.J. Chapman, L.K. Dyall, and L.K. Frith, *Aust. J. Chem., 37,* 341 (1984).

[19] G.L. Gutsev and T. Ziegler, *J. Phys. Chem., 95,* 7220 (1991).

[20] A.C. Hopkinson and M.H. Lien, *Can. J. Chem., 63,* 3582 (1985).

[21] P. von R. Schleyer, *Pure Applied Chem., 59,* 1647 (1987).

[22] H. Tomioka et al., *J. Am. Chem. Soc., 117,* 6376 (1995).

[23] A.J. Arduengo III, H.V. Rasika, R.L. Harlow, and M. Kline, *J. Am. Chem. Soc., 114,* 5530 (1992).

[24] A.J. Arduengo III, et al., *J. Am. Chem. Soc., 116,* 6812 (1994).

[25] R.R. Sauers, *Tetrahedron Lett., 37,* 149 (1966).

[26] R.M. Ettner, H.S. Skouronek, and P.S. Skell, *J. Am. Chem. Soc., 81,* 1008 (1959).

[27] R.C. Woodworth and P.S. Skell, *J. Am. Chem. Soc., 81,* 3383 (1959).

[28] H. Tomioka, K. Tabayashi, Y. Ozaki, and Y. Izawa, *Tetrahedron, 41,* 1435 (1985); N.T. Turro and Y. Cha, *Tetrahedron Lett., 27,* 6149 (1986).

[29] M.A. Garcia-Gambay, C. Theroff, S.H. Shi, and J. Jernelius, *Tetrahedron Lett., 34,* 8415 (1993).

[30] G. Herzberg and J.W.C. Johns, *Proc. Royal Soc.,* Series A, *295,* 107 (1966); L. Andrews, *J. Chem. Phys., 48,* 979 (1968).

[31] R.A. Bernheim, H.W. Bernard, P.S. Wang, L.S. Wood, and P.S. Skell, *J. Chem. Phys., 5,* 2223 (1971); E. Wasserman, V.J. Kuck, R.S. Hutton, E.D. Anderson, and W.A. Yager, *J. Chem. Phys., 54,* 4120 (1971).

[32] B.C. Gilbert, D. Griller, and A.S. Nozran, *J. Org. Chem., 50,* 4738 (1985).

[33] (a) L.L. Stracener et al., *J. Org. Chem., 65,* 199 (2000); (b) R. Alt, I.R. Gould, H.A. Staab, and N.J. Turro, *J. Am. Chem. Soc., 108,* 6911 (1986).

[34] W. von E. Doering and A.K. Hoffmann, *J. Am. Chem. Soc., 76,* 6162 (1954).

[35] R. Breslow, *J. Am. Chem. Soc., 79,* 5318 (1957); R. Breslow, R. Haynie, and J. Mira, *J. Am. Chem. Soc., 81,* 247 (1959).

[36] G.L. Closs and R.B. Larabee, *Tetrahedron Lett.,* 287 (1965); S.K. Dasgupta, R. Dasgupta, S.R. Ghosh, and U.R. Ghatak, *Chem. Comm.,* 1253 (1969); W. von E. Doering et al., *Tetrahedron, 23,* 3943 (1967).

[37] E. Buchner and T. Curtius, *Ber., 18,* 2337 (1885).

[38] W. von E. Doering et al., *J. Am. Chem. Soc., 78,* 5448 (1956).

[39] R. Huisgen and C. Juppe, *Ber., 94,* 2332 (1961).

[40] L.W. Jenneskens, W.H. de Wolf, and F. Bickelhaupt, *Angew. Chem. Intl. Ed. Eng., 24,* 585 (1985); see also, A.S. Kende and X.-C. Guo, *Tetrahedron Lett., 42,* 1227 (2001).

[41] (a) H. Mayr and U.W. Heigl, *Angew. Chem. Intl. Ed. Eng., 24,* 579 (1985). (b) U. Burger and G. Gardillon, *Tetrahedron Lett., 20,* 4281 (1979).

[42] R.W. Hoffmann, *Acc. Chem. Res., 18,* 248 (1985).

[43] (a) M. Brikhahn, E.V. Dehmlov, and H. Boegge, *Angew. Chem. Int. Ed. Engl., 26,* 72 (1987); (b) W. Pan, M.E. Hendrick, and M. Jones, Jr., *Tetrahedron Lett., 40,* 3085 (1999).

[44] W.J. Baron et al., in *Carbenes,* vol. I, M. Jones, Jr. and R.A. Moss, Eds. John Wiley, New York (1973), p. 1.

[45] P.S. Skell and A.Y. Garner, *J. Am. Chem. Soc., 78,* 3409 (1956); W. von E. Doering and P. La Flamme, *J. Am. Chem. Soc., 78,* 5447 (1956); P.S. Skell and R.P. Woodworth, *J. Am. Chem. Soc., 78,* 4496 (1956).

[46] J.S. McConaghy, Jr. and W. Lwowski, *J. Am. Chem. Soc., 89,* 2357 (1967).

[47] R.A. Moss, in *Carbenes,* vol. I, M. Jones, Jr. and R.A. Moss, Eds. John Wiley, New York (1973), p. 153.

[48] R. Hoffman, *J. Am. Chem. Soc., 90,* 1475 (1968); J.F. Blake, S.G. Wierschke, and W.L. Jorgensen, *J. Am. Chem. Soc., 111,* 1919 (1989).

[49] N.G. Rondan, K.N. Houk, and R.A. Moss, *J. Am. Chem. Soc., 102,* 1770 (1980).

[50] R.W. Hoffman, W. Lilienlum, and B. Dittrich, *Ber., 107,* 3395 (1974).

[51] M. Reiffer and R.W. Hoffman, *Ber., 110,* 37 (1977).

[52] R.A. Moss, *Acc. Chem. Res., 13,* 58 (1980).

[53] R.S. Sheridan et al., *J. Am. Chem. Soc., 110,* 7563 (1988).

[54] N. Soundarajan et al., *J. Am. Chem. Soc., 110,* 7143 (1988).

[55] W. von E. Doering, R.G. Buttery, R.G. Laughlin, and N. Chaudhuri, *J. Am. Chem. Soc., 78,* 3224 (1956).

[56] See M.M. Diaz-Requejo, T.R. Belderain, M.C. Nicasio, S. Trofimenko, and P.J. Perez, *J. Am. Chem. Soc., 104,* 896 (2002) and references therein.

[57] (a) See I.R. Likhotvorik et al., *Tetrahedron Lett., 33,* 911 (1992). (b) E.V. Dehmlow, *Tetrahedron, 27,* 4071 (1971).

[58] W. Kirmse and M. Buschhoff, *Ber., 102,* 1098 (1969).

[59] D. Seyferth and Y.M. Cheng, *J. Am. Chem. Soc., 93,* 4073 (1971).

[60] S. Kagabu and K. Saito, *Tetrahedron Lett., 29,* 675 (1988).

[61] J.E. Jackson, G.B. Mock, M.L. Tetef, G.-X. Zhiang, and M. Jones, Jr., *Tetrahedron, 41,* 1453 (1985).

[62] G.L. Closs and L.E. Closs, *J. Am. Chem. Soc., 91,* 4549 (1969); G.L. Closs and A.D. Trifunac, *J. Am. Chem. Soc., 91,* 4554 (1969).

[63] I.D.R. Stevens, M.T.H. Liu, N. Soundarajan, and N. Paike, *Tetrahedron Lett., 30,* 481 (1989).

[64] W.M. Jones and J.M. Walbrick, *J. Org. Chem., 34,* 2217 (1969).

[65] (a) W. Kirmse, *Angew. Chem. Intl. Ed. Engl., 6,* 594 (1966); (b) R.A. Farlow, D.M. Thamattoor, R.B. Sunoj, and C.M. Haddad, *J. Org. Chem., 67,* 3257 (2002).

[66] W. Pritzkow and D. Timm, *J. Prakt. Chem.* [4], *32,* 178 (1966).

[67] P.B. Sargeant and H. Schechter, *Tetrahedron Lett.,* 3957 (1964).

[68] J.A. Landgrebe and A.G. Kirk, *J. Org. Chem., 32,* 3499 (1967).

[69] W.H. Saunders and J.C. Ware, *J. Am. Chem. Soc., 80,* 3328 (1958).

[70] H. Tomioka, H. Ueda, S. Kondo, and Y. Izawa, *J. Am. Chem. Soc., 102,* 7817 (1980).

[71] K.L. Erickson and J. Wolinsky, *J. Am. Chem. Soc., 87,* 1142 (1965).

[72] See H.E. Zimmerman in *Molecular Rearrangements*, vol. 1, P. de Mayo, Ed. Interscience Publishers, New York (1963), p. 405.

[73] (a) A.A. Bothner-By, *J. Am. Chem. Soc., 77,* 3293 (1955); (b) D.Y. Curtin, E.W. Flynn, and R.F. Nystrom, *J. Am. Chem. Soc., 80,* 4597 (1958).

[74] W.M. Jones and R. Damico, *J. Am. Chem. Soc., 85,* 2273 (1963); H. Rezaei, S. Yamanoi, F. Chemla, and J.F. Normant, *Org. Lett., 2,* 419 (2000).

[75] S.P. Samuel, T.-Q. Nin, and K.L. Erickson, *J. Am. Chem. Soc., 111,* 1429 (1989); Z. Du, M.J. Haglund, L.A. Pratt, and K.L. Erickson, *J. Org. Chem., 63,* 8880 (1998).

[76] See review by G.B. Gill in *Comprehensive Organic Synthesis*, vol. 3, G. Pattendem, Ed. Pergamon Press, New York (1991), pp. 887–910.

[77] J.L. Wang, J.P. Toscano, M.S. Platz, V. Nikolaev, and V. Popik, *J. Am. Chem. Soc., 117,* 5477 (1995).

[78] R.J. McMahon, O.L. Chapman, R.A. Hayes, T.C. Hess, and H.P. Krimmer, *J. Am. Chem. Soc., 107,* 7597 (1985); J.P. Toscano, M.S. Platz, and V. Nikolaev, *J. Am. Chem. Soc., 117,* 4712 (1995).

[79] K. Wiberg and T.W. Hutton, *J. Am. Chem. Soc., 78,* 1640 (1956). See also F.V. Brutcher, Jr. and D.D. Rosenfeld, *J. Org. Chem., 29,* 3154 (1964).

[80] I.G. Csizmadia, J. Font, and O.P. Strausz, *J. Am. Chem. Soc., 90,* 7360 (1968).

[81] K.-P. Zeller, H. Meier, H. Kolshorn, and E. Mueller, *Ber., 105,* 1875 (1972).

[82] For reviews of oxirene intermediates in Wolff rearrangements, see M. Torres, E.M. Lown, H.E. Gunning, and O.P. Strausz, *Pure Applied Chem., 52,* 1623 (1980); and C. Wentrup, *Reactive Molecules.* Wiley-Interscience, New York (1984).

[83] M. Torres, J.L. Bourdelande, A. Clement, and O.P. Strausz, *J. Am. Chem. Soc., 105,* 1698 (1983).

[84] C. Bachmann et al., *J. Am. Chem. Soc., 112,* 7488 (1990).

[85] U. Timm, K.-P. Zeller, and H. Meier, *Tetrahedron, 33,* 453 (1977).

[86] N. Chen, M. Jones, Jr., W.R. White, and M.S. Platz, *J. Am. Chem. Soc., 113,* 4981 (1991).

[87] T.H. Chan and D. Masuda, *J. Am. Chem. Soc., 99,* 936 (1977).

[88] See A. Padwa, *Organic Photochem., 4,* 261 (1979); J.J. Gajewski, *Hydrocarbon Thermal Isomerizations.* Academic Press, New York (1981). See also W.R. Dobier, Jr., G.R. Shelton, M.A. Battiste, J.F. Stanton, and D.R. Price, *Org. Lett., 4,* 233 (2002).

[89] (a) E.J. York, W. Dittmer, J.R. Stevenson, and R.G. Bergman, *J. Am. Chem. Soc., 95,* 5680 (1973). (b) See J.R. Al Dulayymi and M.S. Baird, *Tetrahedron Lett., 36,* 3393 (1995) and references therein.

[90] I.R. Likhotvorik, D.W. Brown, and M. Jones, Jr., *J. Am. Chem. Soc., 116,* 6175 (1994); see also W. Graf von der Schulenberg, H. Hopf, and R. Walsh, *Angew. Chem. Int., Ed. Engl., 38,* 1128 (1999).

[91] (a) K.S. Feldman and A.L. Perkins, *Tetrahedron Lett., 42,* 6031 (2001). (b) T. Harada, K. Iwazaki, T. Otani, and A. Oku, *J. Org. Chem., 63,* 9007 (1998).

[92] R.F.C. Brown, F.W. Eastwood, K.J. Harrington, and G.L. McMullen, *Aust. J. Chem., 27,* 2393 (1974).

[93] R.F.C. Brown, F.W. Eastwood, and G.P. Jackman, *Aust. J. Chem., 30,* 1757 (1977).

[94] Reviews: W.M. Jones, *Acc. Chem. Res., 10,* 353 (1977); W.M. Jones, in *Rearrangements in Ground and Excited States,* P. de Mayo, Ed. Academic Press, New York (1980).

[95] P.P. Gaspar, J.P. Hsu, and S. Chari, *Tetrahedron, 41,* 1479 (1985).

[96] R.C. Joines, A.B. Turner, and W.M. Jones, *J. Am. Chem. Soc., 91,* 7755 (1969); P. Schissel, M.E. Kent, D.J. McAdoo, and E. Hedaya, *J. Am. Chem. Soc., 92,* 2147 (1970).

[97] W.J. Baron, M. Jones, Jr., and P.P. Gaspar, *J. Am. Chem. Soc., 92,* 4739 (1970); G.G. Vander Stouw, A.R. Kraska, and H. Schechter, *J. Am. Chem. Soc., 94,* 1655 (1972).

[98] W.M. Jones, et al., *J. Am. Chem. Soc., 95,* 826 (1973).

[99] P.R. West, O.L. Chapman, and J.-P. LeRoux, *J. Am. Chem. Soc., 104,* 1779 (1982). See also R.J. McMahon et al., *J. Am. Chem. Soc., 109,* 2456 (1987); and O.L. Chapman, J.W. Johnson, R.J. McMahon, and P.R. West, *J. Am. Chem. Soc., 110,* 501 (1988).

[100] E. Hedaya and M.E. Kent, *J. Am. Chem. Soc., 93,* 3283 (1971).

[101] W.E. Billups and L.E. Reed, *Tetrahedron Lett.,* 2239 (1977).

[102] Review: C. Wentrup, *Adv. Hetero Chem., 28,* 232 (1981).

PROBLEMS

9.1 Calculations indicate that the singlet forms of carbenes **a–c** below are more stable than the triplet forms, and that the singlet–triplet energy gaps increase in the order **a** $\ll$ **b** < **c**. [P.K. Freeman and J.K. Pugh, *J. Org. Chem., 65,* 6107 (2000).] Account for these facts.

a   **b**   **c**

9.2 Account for the different results from the following reactions:

$$\underset{\underset{\text{C}_6\text{H}_5\overset{\displaystyle\|}{\text{C}}\text{CH}_2\text{OCH}_3}{}}{\overset{\text{N}_2}{}} \quad \xrightarrow{\Delta} \quad \underset{\approx 90\%}{\text{C}_6\text{H}_5\text{CH}=\text{CHOCH}_3}$$

$$\underset{\text{C}_6\text{H}_5\overset{\displaystyle\|}{\text{C}}\text{CH}_2\text{SCH}_3}{\overset{\text{N}_2}{}} \quad \xrightarrow{\Delta} \quad \underset{\approx 90\%}{\overset{\text{SCH}_3}{\text{C}_6\text{H}_5\overset{\displaystyle|}{\text{C}}=\text{CH}_2}}$$

(J.H. Robson and H. Schechter, *J. Am. Chem. Soc.*, *89*, 7112 [1967])

9.3 Propose reasonable mechanisms for the following reactions:

(a)

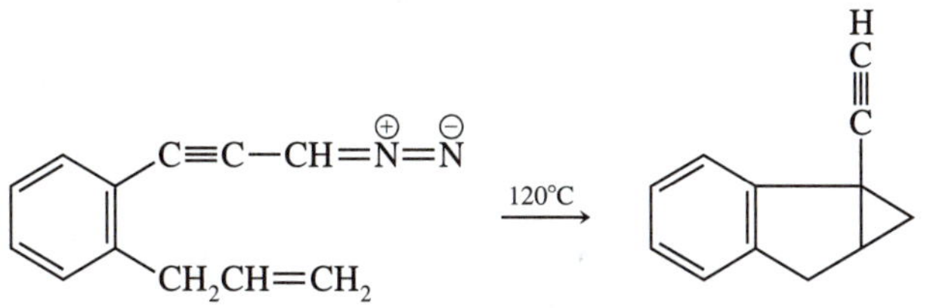

$$\xrightarrow{120°\text{C}}$$

(A. Padwa, Y. Gareau, and S.L. Xu, *Tetrahedron Lett.*, *32*, 983 [1991])

(b)

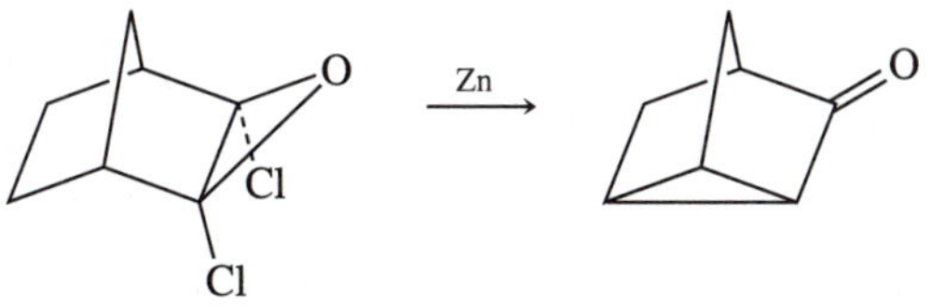

$$\xrightarrow{\text{Zn}}$$

(E. Lewars and S. Siddiqui, *J. Org. Chem.*, *50*, 135 [1985])

(c)

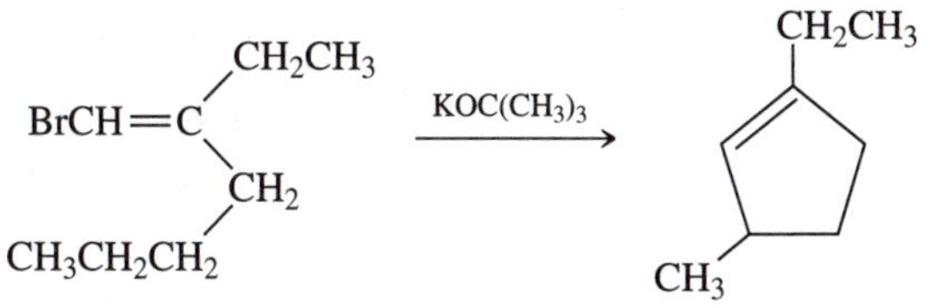

$$\xrightarrow{\text{KOC(CH}_3)_3}$$

(K.L. Erickson and J. Wolinsky, *J. Am. Chem. Soc.*, *87*, 1142 [1965])

(d)

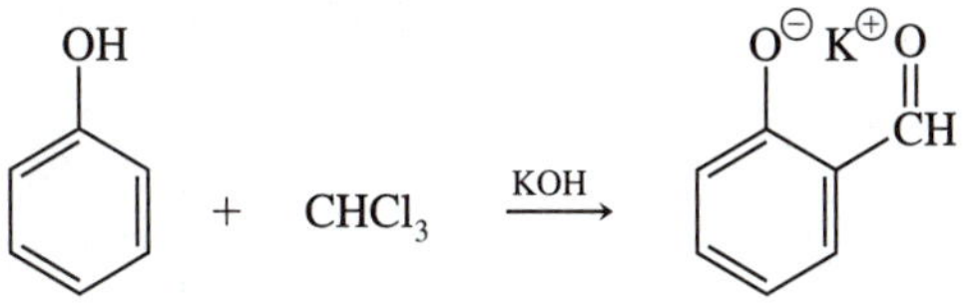

$+$ CHCl$_3$ $\xrightarrow{\text{KOH}}$

(the Reimer-Tiemann reaction)

(e)

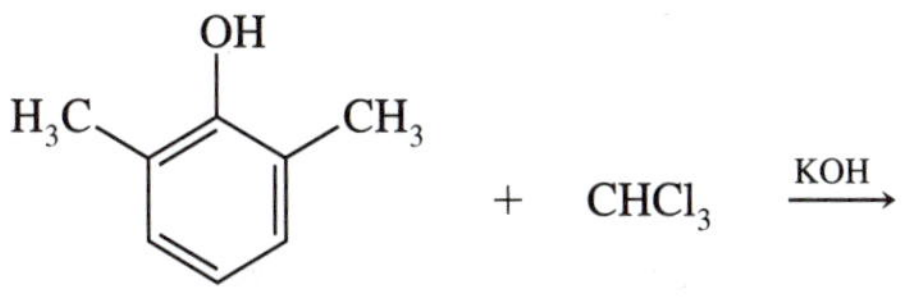

$+$ CHCl$_3$ $\xrightarrow{\text{KOH}}$ a ketone

(f)

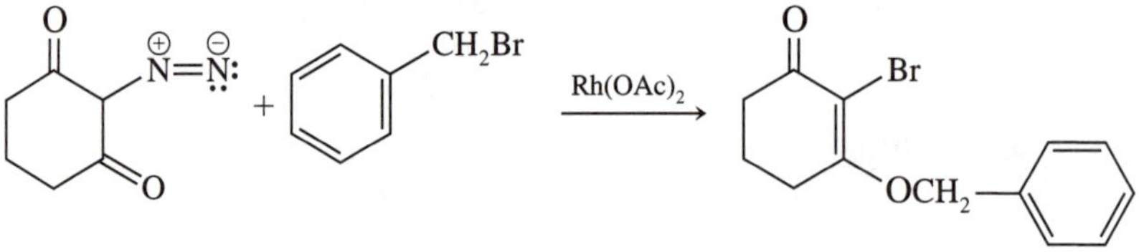

(Y.R. Lee and D.H. Kim, *Tetrahedron Lett.*, *42*, 6561 [2001])

(g)

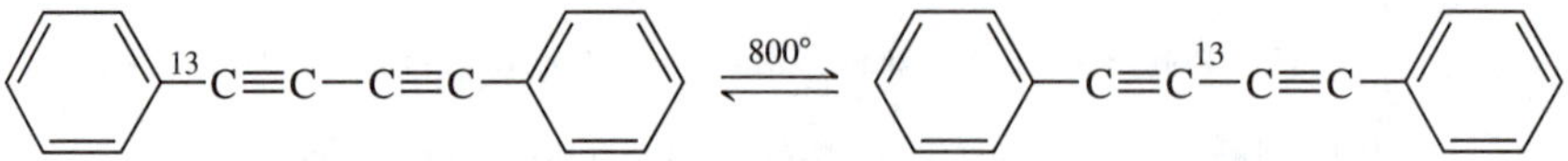

(J. Mabry and R.P. Johnson, *J. Am. Chem. Soc.*, *104*, 6497 [2002])

(h) 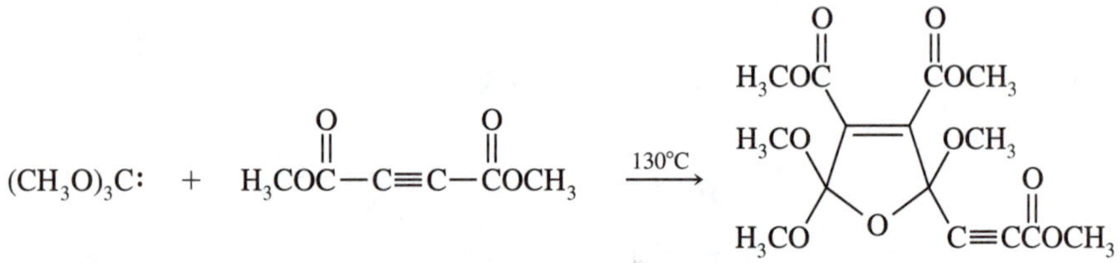 $(CH_3O)_3C\colonsep$ +

(R.W. Hoffmann, W.L. Lilenblum, and B. Dittrich, *Ber.*, *107*, 3395 [1974])

(i)

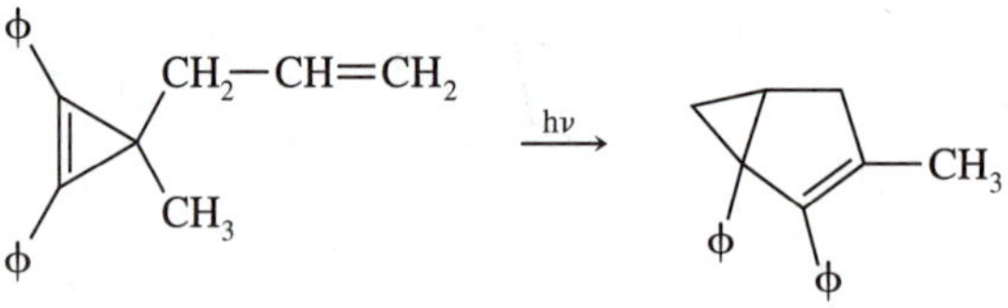

(A. Padwa and T.J. Blacklock, *J. Am. Chem. Soc.*, *99*, 2345 [1977])

(j) 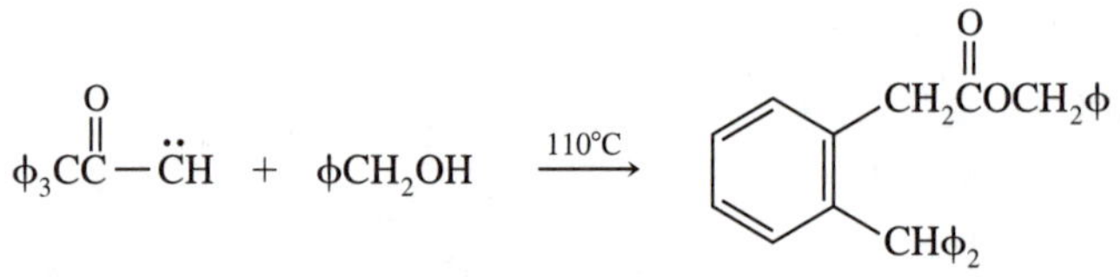 $\phi_3CC-\ddot{C}H$ + ϕCH_2OH $\xrightarrow{110°C}$

(A.L. Wilds et al., *J. Am. Chem. Soc.*, *84*, 1503 [1962])

(k)

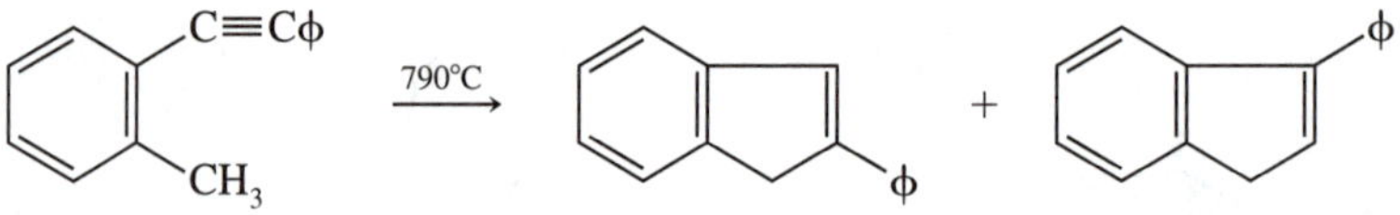

(R.F.C. Brown, F.W. Eastwood, and G.P. Jackman, *Aust. J. Chem.*, *30*, 1757 [1977])

(l)

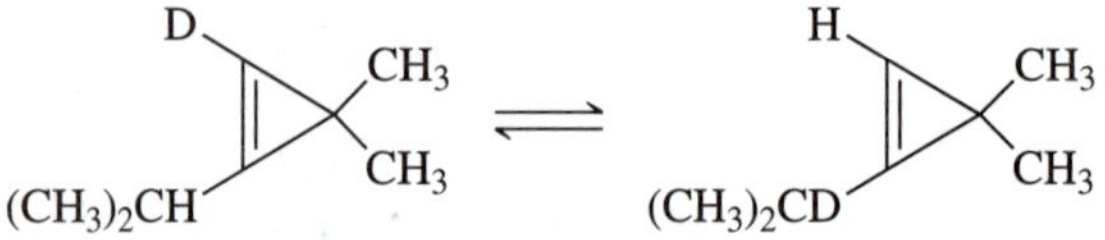

(I.R. Likhotvorik, D.W. Brown, and M. Jones, Jr., *J. Am. Chem. Soc.*, *116*, 6175 [1994])

(m) 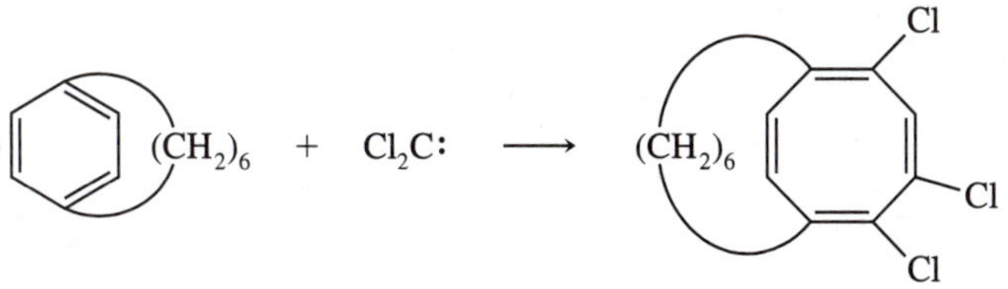 $(CH_2)_6$ + Cl_2C: $\longrightarrow$

(W. Koenigstein and W. Tochtermann, *Tetrahedron Lett.*, *28*, 3483 [1987])

9.4 It was suggested that the reaction below proceeds via a {1,6} addition process. Suggest a reasonable alternative mechanism, employing only concerted steps after loss of nitrogen to form a carbenoid.

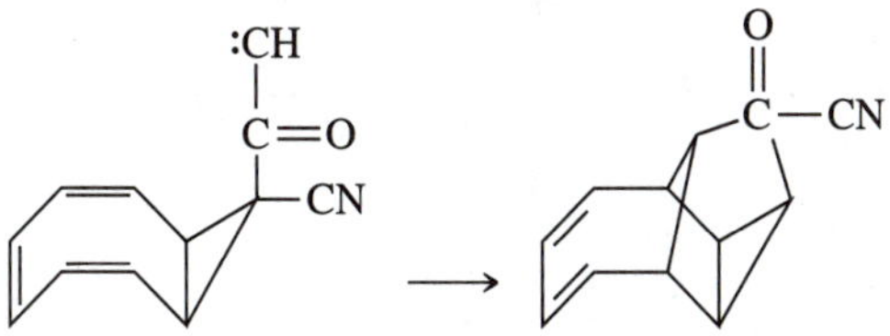

(M. Brixhahn, E. V. Dehmlov, and H. Boegge, *Angew. Chem. Int. Ed. Engl.*, *26*, 72 [1987])

Photochemistry

10.1 INTRODUCTION

Physical Principles

In earlier chapters we looked at pericyclic reactions carried out under photochemical conditions. In this chapter we will discuss several other types of photochemical reactions.

Absorbing Light. Almost all photochemical reactions are governed by the Einstein–Stark law, which, in terms of modern quantum theory, states that chemical reactions caused by light occur as the result of absorption of a single quantum of energy. In most instances, the energy absorbed will correspond to the difference in energy between the ground state and the lowest level excited state (although a few reactions do appear to proceed from higher level excited states). The energy of a quantum of light is related to the frequency, v, or wavelength, λ, of the light by the equation:

$$E = hv = hc/\lambda$$

where h (Planck's constant) $= 6.63 \times 10^{-34} J$
 c is the speed of light
Thus, the longer the wavelength of the light, the lower the amount of energy absorbed.

It should be noted that the "ground state" of any compound actually consists of the ground states of all the individual molecules, each in slightly different conformations and rotational energy levels. Thus the spectrum of any compound covers a *range* of wavelengths, with the maximum absorption corresponding to the most probable energy state of the molecules. The probability of light at any wavelength being absorbed is indicated by the magnitude of the *extinction coefficient*, ϵ, at that wavelength.

Releasing Energy. Energy that is absorbed in photochemical processes must eventually be released. This may occur by nonradiative processes; for example a molecule may transfer much of its energy by collisions with other molecules without exciting them to higher electronic states. This is, indeed, the most common way in which molecules in condensed states release energy they have absorbed.

Another possibility is that energy may be emitted, as it was absorbed, in the form of a quantum of light. If the excited molecule is in a singlet state the process of reemitting light is called *fluorescence*. If it is in a triplet state, the process is called *phosphorescence*. (The general term *luminescence* includes both processes.)

In neither case will the light emitted necessarily have the same wavelength as the light absorbed. According to the *Franck–Condon principle*, the absorption of light is a "vertical process," which is so rapid (in the order of 10^{-15} sec) that the positions of all the atoms remain unchanged on going to the excited state. Since the most stable geometries of molecules in excited states are unlikely to be the same as in their ground states, the excited molecules will usually be able to "decay" to their lowest energy geometries by internal rotations and bond angle changes before emitting light.

Light emitted by a molecule in an excited state will thus typically have longer wavelengths (that is, be of lower energy) than the light absorbed. It is true that light emitted during fluorescence will not usually be greatly different from the light absorbed. However, light emitted during phosphorescence will typically have much longer wavelengths than the light absorbed, since triplet states of molecules are usually lower in energy than excited singlet states.

Sensitization. It is frequently possible for a molecule in an excited state to transfer energy to another molecule, raising the second molecule to an excited state while itself dropping down to the ground state. Transfer of energy in this way is called *sensitization*. If the *sensitizer* is in a singlet state, the newly excited molecule will also be a singlet, while if the initially excited molecule is a triplet, the newly excited molecule will also be a triplet.

Sensitization might occasionally result from light emitted by one molecule being reabsorbed by other molecules which have electronic transitions corresponding to the emitted wavelengths. Most of the emitted light would be likely to escape the solution without being readsorbed, however, so that would be an inefficient way of transmitting electronic energy (except, possibly, in the solid state or in very concentrated solutions).

More efficient energy transfer can occur when a molecule in an excited state forms a complex (an *exciplex*[1]) with another molecule. When the exciplex dissociates, the molecule that was originally in the ground state could be converted to its excited state, while the molecule that was originally in the excited state (the sensitizer) would be released in its ground state.

Photochemical Equilibria. Of most interest to us, as chemists, are those cases in which molecules in excited states undergo chemical changes prior to releasing photochemical energy they had absorbed.

The products of photochemical reactions, like those of every other reaction, will be in equilibrium with the starting materials. However, unlike ground state reactions, the ratio of starting materials to products will not normally be governed by the relative thermodynamic energies of the two compounds. Instead, (at least in reactions not involving sensitizers)* it will principally depend on the relative extinction coefficients of the starting materials and products. The molecule with the higher extinction coefficient will absorb more light and (assuming equal quantum efficiencies for the forward and back reactions) will be converted to the form with the lower extinction coefficient. This will continue until enough molecules with low extinction coefficients have been formed to compensate for the lower probability of their absorbing light.

---

*The equilibrium in reactions involving sensitizers may depend on the efficiency of energy transfer from the sensitizer to the substrate molecule.

10.2 REACTIONS OF CARBON–CARBON DOUBLE BONDS

Radiation in the visible range (ca. 400–800 nm) has energies ranging from 72–36 kcal per Einstein (that is, per "mol" of photons), while radiation in the near UV range (ca. 200–400 nm) has energies of 143–72 kcal per Einstein.

It might thus appear that absorption of near UV light could result in cleaving carbon–hydrogen bonds and carbon–carbon σ bonds, since the dissociation energies of these bonds (ca. 80–85 kcal/mol for typical C–C and ca. 90–98 kcal/mol for typical C–H bonds) are well within the range of energies available from radiation in the near UV region. However, as discussed in Section 10.1, absorption of light occurs far more rapidly than any changes in the positions of atoms. Thus, the two atoms formerly forming a bond cannot immediately move away from each other when the bond is broken. Repulsions between the unpaired electrons would, therefore, raise the energies of the excited states for photoexcitation of saturated carbon–carbon and carbon–hydrogen bonds well above the energy levels of two separated free radicals.* As a result, saturated hydrocarbons, in which all valence electrons are tightly held in σ bonds, do not absorb radiation with wavelengths longer than those in the furthest UV regions.

The Spectra of Alkenes.   Most simple, unconjugated alkenes absorb light mainly in the far UV region, at ca. 175–190 nm. Since it is difficult to obtain solvents that are transparent in that region, photochemical reactions of molecules in the far UV are usually confined to the gas phase, and are rarely synthetically useful processes. (The spectra of most alkenes, particularly those having several alkyl groups as substituents on the double bonds, do have tails which extend into the near UV region. However, their extinction coefficients in that region, and thus the efficiency with which they absorb light, are usually quite low.)

In Chapter 2, we discussed the absorption of light resulting in raising electrons from π orbitals to π^* orbital of double bonds.[†] Such $\pi \rightarrow \pi^*$ transitions are indeed the lowest energy transitions for most *conjugated* double bonds, and the reactions of conjugated double bonds do appear to result from molecules in π^* states.

The spectra of many unconjugated alkenes, however, suggest that transitions to *Rydberg states* may occur at energies similar to, and sometimes lower than, those required for $\pi \rightarrow \pi^*$ transitions.[2]

A Rydberg state is an excited state in which an electron is in an orbital far from the immediate neighborhood of the bond being excited. (Thus, it is in what is effectively a nonbonding orbital, rather than in an antibonding orbital.) Rydberg states are designated by the symbol "R". The lowest-lying Rydberg state from photoirradiation of an alkene is designated π, R(3s), to indicate that one electron from the π bond remains in the π orbital, and the other has been raised to the Rydberg orbital, which resembles the 3s orbital of a helium atom.[2]

A Rydberg state of an alkene can thus be considered to have a partially ionized structure such as that shown on page 294, with a net positive charge on a carbon atom. (Assignment of that type of structure to Rydberg states is supported by the fact that alkyl substituents decrease the energy necessary for $\pi \rightarrow$ R(3s) transitions to a much greater extent than for $\pi \rightarrow \pi^*$ transitions.)[2]

---

*As will be illustrated in Section 10.3, carbon–carbon single bonds *can* be broken by photoirradiation. However, these reactions normally proceed by secondary photochemical processes that occur after absorption of light by some *chromophore* (some light-absorbing group), such as a double bond or carbonyl group, rather than by the single bond itself.

[†]Spectroscopists use the symbols N(*normal*) $\rightarrow$ V to describe $\pi \rightarrow \pi^*$ transitions.

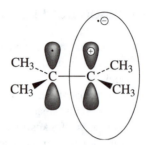

A Possible Representation of a Rydberg State of Tetramethylethylene

The extinction coefficients for transitions to Rydberg states are appreciably lower than for $\pi \rightarrow \pi^*$ transitions. Nonetheless, there is evidence that some photoreactions of alkenes do arise from Rydberg states. (These appear to include ring openings of cyclobutenes,[3] which, as was noted in Section. 2.5, do not appear to follow the Woodward–Hoffmann rules.)

Cis,Trans Isomerizations[4]

Photoirradiation frequently results in the rapid interconversion of *cis* and *trans* isomers of alkenes. These isomerizations are far more rapid than other photoreactions of alkenes, and frequently precede other types of reactions. Even cyclohexenes and cycloheptenes can undergo *cis, trans* isomerization on photoirradiaiton, although their *trans* isomers are too unstable to be isolated.[5a] However, irradiation of either *cis-* or *trans*-cylooctene yields mixtures containing both stereoisomers (Eq. 1).[5b,c]

$$\text{(structure)} \underset{\longleftarrow}{\overset{h\nu}{\longrightarrow}} \text{(structure)} \qquad (1)$$

Because their extinction coefficients are approximately equal, photoisomerizations of *cis*- and *trans*-cyclooctenes yield approximately equal amounts of the two isomers, despite the fact that the very strained *trans* form is far less thermodynamically stable than the *cis* isomer.

However, *cis* and *trans* isomers do not always have similar extinction coefficients. The ratio of stereoisomers obtained on photoirradiation of stilbene (1,2-diphenylethene), for instance, depends on the wavelength employed. A photostationary state containing more than 90% of the thermodynamically less stable *cis* isomer, for instance, can be obtained by irradiation at long wavelengths (ca. 310–320 nm); wavelengths at which the *trans* isomer has the stronger absorption. At shorter wavelengths, more of the *trans* isomer would be obtained.

$$\text{(structure)} \underset{\longleftarrow}{\overset{315 \text{ nm}}{\longrightarrow}} \text{(structure)} \qquad (2)$$

(The fact that *trans*-stilbene absorbs light of longer wavelengths than *cis*-stilbene is believed to be due to steric interference between the *ortho* hydrogens of the two phenyl rings in the *cis* isomer, which forces that isomer to adopt a nonplanar, less highly conjugated, conformation.)

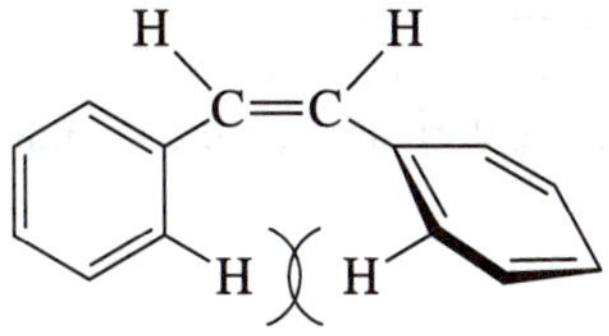

cis-Stilbene must Adopt a Nonplanar Conformation

Cis, trans isomerizations of alkenes can occur from either singlet or triplet states. In either case, after vertical excitation of the alkene to the π^* state (at least for conjugated alkenes*), a 90° rotation around the central bond places the two unshared electrons into p orbitals at right angles to each other. This "orthogonal conformation" of the excited state (also called the p state) is normally the lowest energy conformation, since repulsions between the unshared electrons would be minimized. A molecule in the p state can then undergo bond rotation in either direction to form the two stereoisomers.

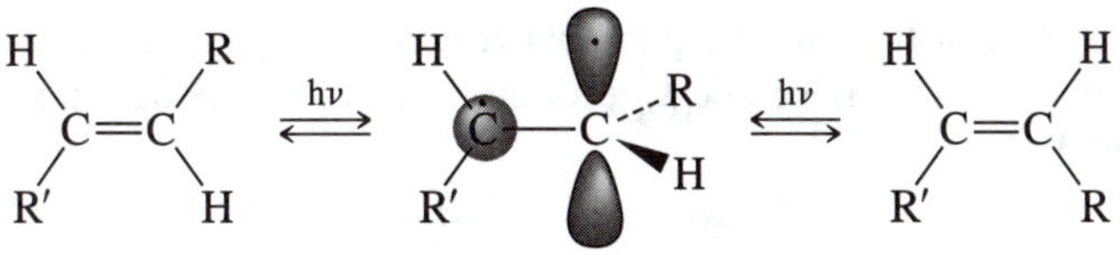

The *cis,trans* Isomerization of an Alkene via an Orthogonal Excited State

Protonation of Initial Photoirradiation Products.   Photoirradiations of cycloalkenes with rings of 6–8 carbons in alcohol solvents (particularly under mildly acidic conditions) yield the products of Markownikoff addition of the alcohols to the double bonds.[6] The addition reactions are frequently accompanied by compounds that appear to result from [1,3] hydrogen migrations.

These reactions apparently begin with *cis,trans* isomerizations, although *trans*-cyclohexene and *trans*-cycloheptene are too unstable to be isolated. Protonation of the *trans*-cycloalkenes then yields the observed products, as shown in Eq. 3.

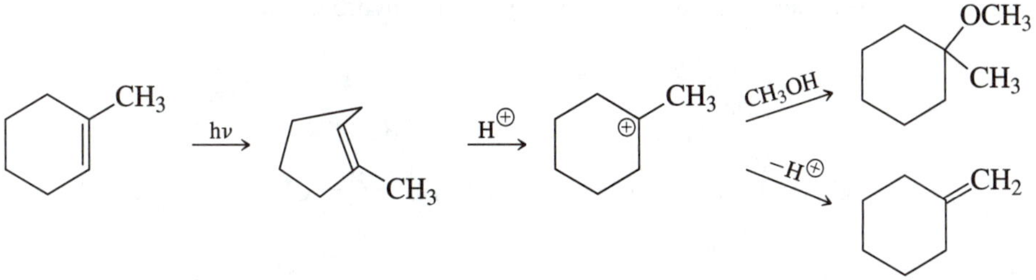

(3)

---

*In contrast to isomerizations of conjugated alkenes such as stilbene, there is evidence that at least some unconjugated alkenes may undergo one-stage *cis-trans* isomerizations, possibly via Rydberg states.[3b]

This mechanism accounts for the observation that if cycloheptene is irradiated in pentane at −78°C, allowed to stand for several minutes, and acidic methanol is then added, cyclohexyl methyl ether can be isolated from the reaction. (Precise yields depend on the temperature of the solution and the time elapsed before the acidic methanol solution is added).[6a]*

Formation of an intermediate *trans*-cyclohexene derivative, followed by protonation to form a carbocation, appears to account for the interesting fragmentation reaction shown in Eq. 4. The corresponding cyclopentene derivative, which cannot be converted to its *trans* isomer, does not undergo a similar reaction.[6a]

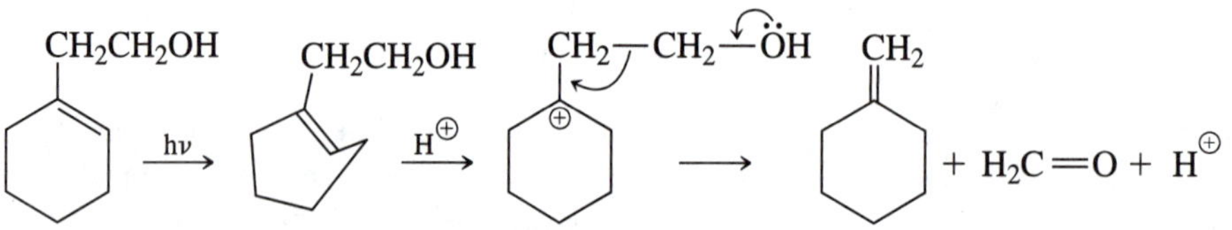

$$+ H_2C=O + H^{\oplus} \quad (4)$$

Formation of Carbenes

While *cis-trans* isomerizations are the principal processes observed on photoirradiation of many alkenes, small amounts of products of other reactions are almost always obtained. On prolonged irradiation, those can become the principal reactions. The products from irradiation of tetra-methylethene, for instance, include **1**, resulting from a [1,3] hydrogen shift (see Section 4.2), as well as **2** and **3**.[6,7]

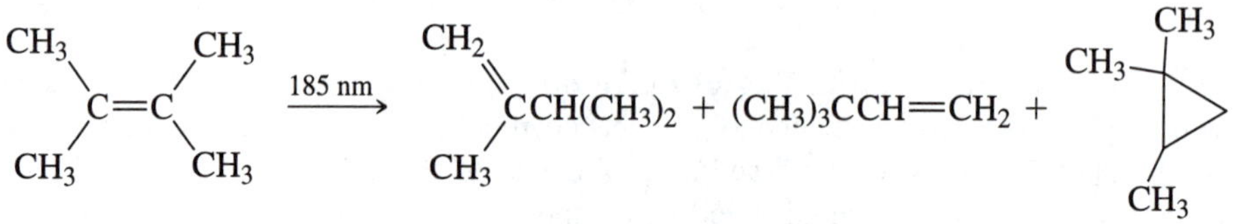

Products **2** and **3** appear to arise via an initial [1,2] methyl migration to form a carbene. Product **2** would then result from a [1,2] hydrogen migration in the carbene, and product **3** from insertion of the divalent carbon into a nearby carbon–hydrogen bond.[6b-8]

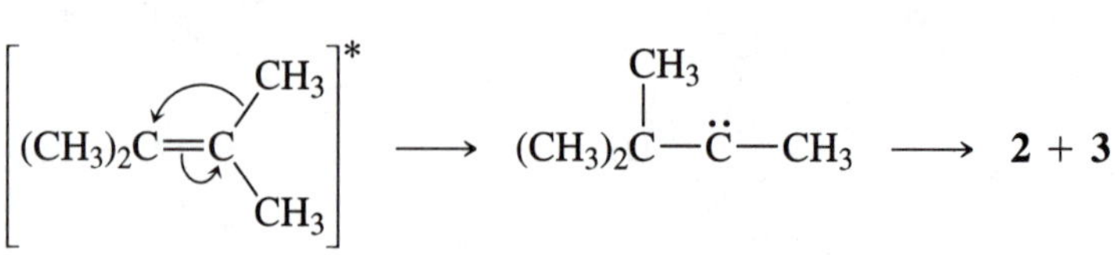

---

*The half-life of *trans*-cycloheptene was determined by this method to be 23 min at −10°C.[6a]

Formation of a carbene on photoirradiation of 1,2-dimethylcyclohexene might result from migration of either a methyl group or a ring methylene group. In fact the predominant reaction is migration of the ring carbon.[6a,7]

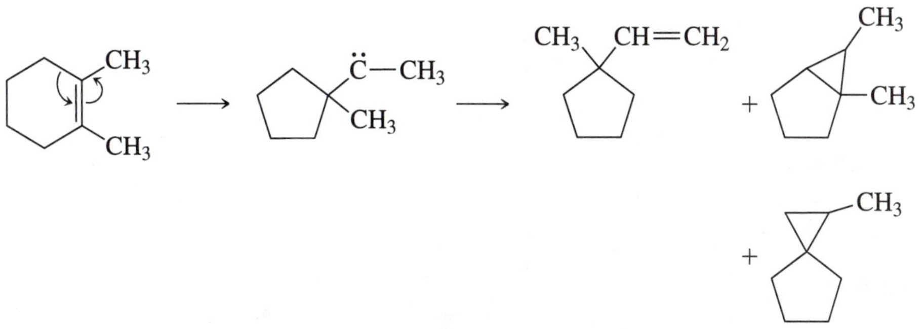

Formation of carbenes by photoirradiation of alkenes appear to proceed via $\pi,R(3s)$ excited states.[6a,7-9] It was previously pointed out that the electron distribution in a Rydberg state resembles that in a carbocation, and the alkyl migrations leading to carbene formation, including preferential migration of the more substituted carbons, are characteristic of migrations in carbocations.

Di-π-Methane Rearrangements[10]

Photoirradiation of a wide variety of molecules in which a single tetravalent carbon is bonded to two double bonds have been found to undergo apparent 1,2-vinyl migrations, giving rise to vinylcyclopropanes, as shown in Eqs. 8–10.[11,12]

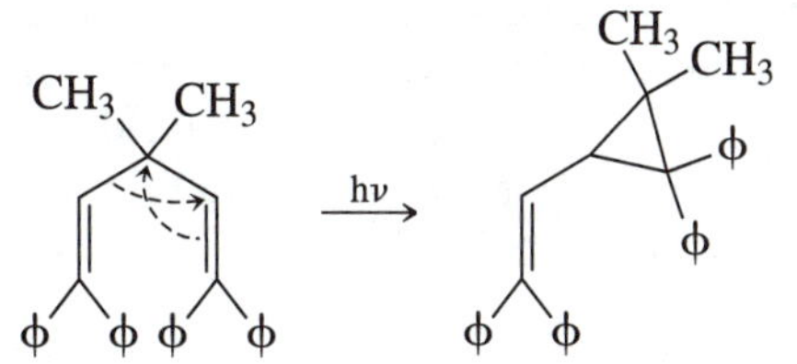

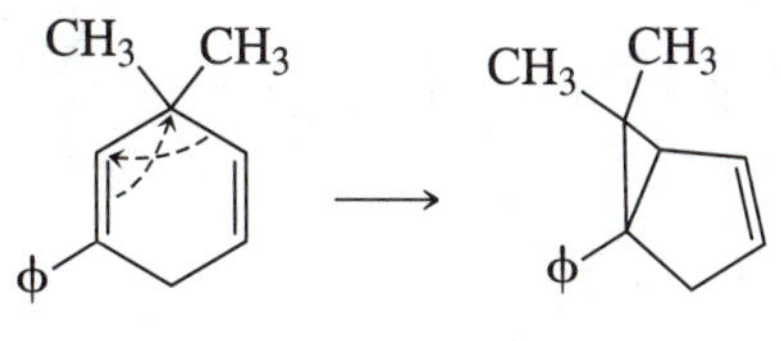

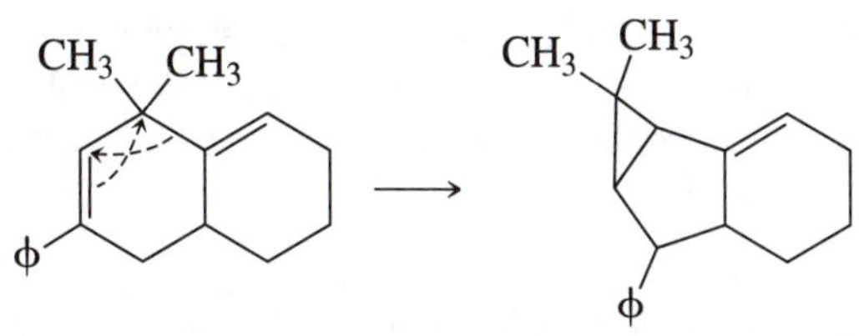

In these reactions, one double bond may be part of an aromatic ring:[13,14]

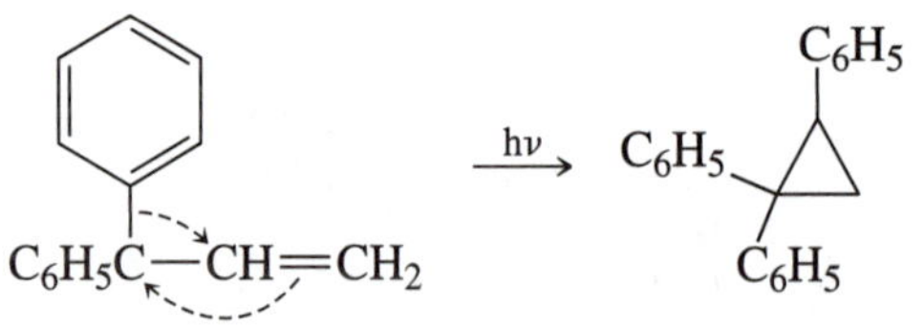

(11)

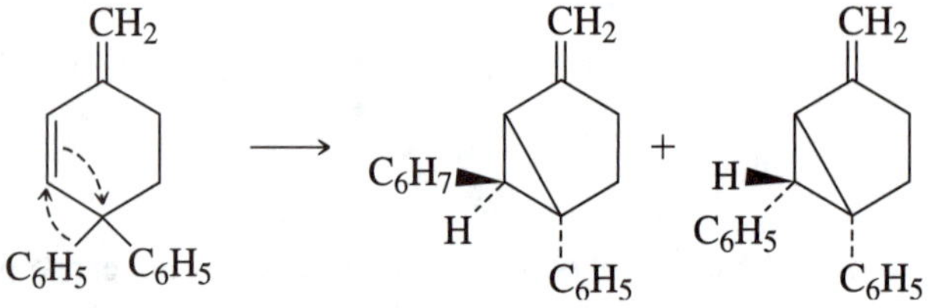

(12)

These reactions are called di-π-methane rearrangements.[10] (They have also been called "Zimmerman rearrangements," after Howard Zimmerman of the University of Wisconsin, whose research group has been a principal source of information about them.)

Mechanisms. Di-π-methane rearrangements may occur from either singlet or triplet states. Reactions via triplets proceed best when the double bonds are contained in small ring structures, in which dissipation of energy via *cis-trans* isomerizations is relatively difficult. Reactions via singlet intermediates, on the other hand, are most likely when the double bonds are *not* parts of rings.

According to the Zimmerman mechanism, di-π-methane rearrangements from triplet states appear to proceed in three "stages", which are illustrated in Eq. 13. In stage A, the two double bonds interact to form a diradical (or "diradical-like" structure) containing a cyclopropane ring. In stage B, the cyclopropylmethyl radical opens to form a double bond and a new diradical, which closes to form the new cyclopropane ring in Stage C.*

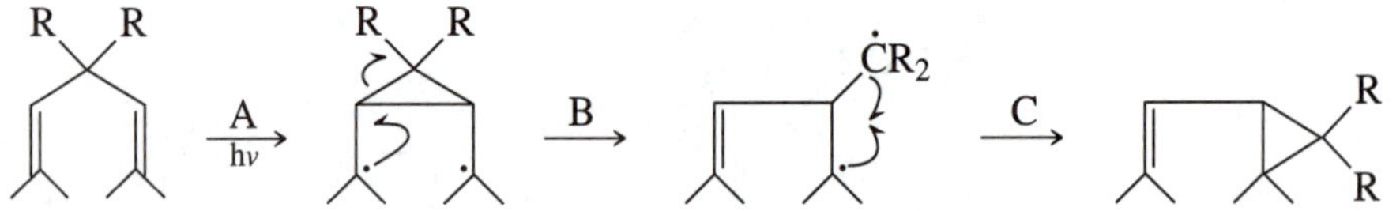

(13)

The existence of free radical intermediates in triplet state rearrangements is consistent with the results of *ab initio* calculations.[15a] In contrast, *ab initio* calculations suggest that in rearrangements proceeding via singlets, stages A and B are replaced by a 1,2 vinyl migration, yielding a very short-lived diradical which collapses to form the cyclopropyl ring without any activation barrier.[15b]

---

*It is not known which of these stages proceed from excited states and which, if any, proceed from ground states.

Regiochemistry and Stereochemistry.   The regiochemistries of di-π-methane rearrangements can be understood by assuming that the various stages are, or at least closely resemble, free-radicals. For instance, di-π-methane rearrangements proceed most rapidly when phenyl groups or other groups that would strongly stabilize free-radical intermediates are present. Furthermore, the "diradical" formed after Stage B (or after the "migration stage" in a singlet reaction) is usually the more stable of the two possible "diradicals." (That means that the vinyl group that would be converted to the *less*-stable radical would usually be the migrating group. An aromatic ring will always migrate in preference to a double bond.)

The stereochemistries of di-π-methane rearrangements are quite complex, since the geometries of both double bonds and of the central tetravalent carbon might all be inverted or retained.

Zimmerman and his coworkers have demonstrated that when the rearrangement proceeds via a singlet state (but not if it proceeds via a triplet),[16] the geometry of the remaining double bond is retained in the product. Furthermore, as shown in Eq. 14, the geometry of the double bond which is incorporated into the cyclopropyl ring is also retained in the product.[17]

$$(14)$$

A quick glance at Eq. 15, showing the results of a di-π-methane rearrangement of a compound with a chiral central carbon,[18] might suggest that the geometry at the central carbon (C3) remains unchanged. However, a closer look will show that the bond between C3 and C4 has been replaced by a bond between C3 and C1. In fact, the reaction has proceeded with inversion at C3.

$$(15)$$

It is interesting to note that the singlet state di-π-methane rearrangement is a six-electron process (involving two σ-bonds and one π-bond) that results in a single inversion of configuration. This is the result that would be predicted by orbital symmetry rules for a photochemical process which proceeds in a single step, or in which orbital symmetry relationships are maintained in any intermediates.[10]

Tri-π-Methane Rearrangements.   No reaction is so complicated that it cannot be made still more complex. The presence of an additional vinyl group on the central carbon, as shown in

Eq. 16, allows formation of a cyclopentene derivative, **5** (the product of a "tri-π-methane rearrangement"), as well as of the normal cyclopropane derivative, **4**.[19]

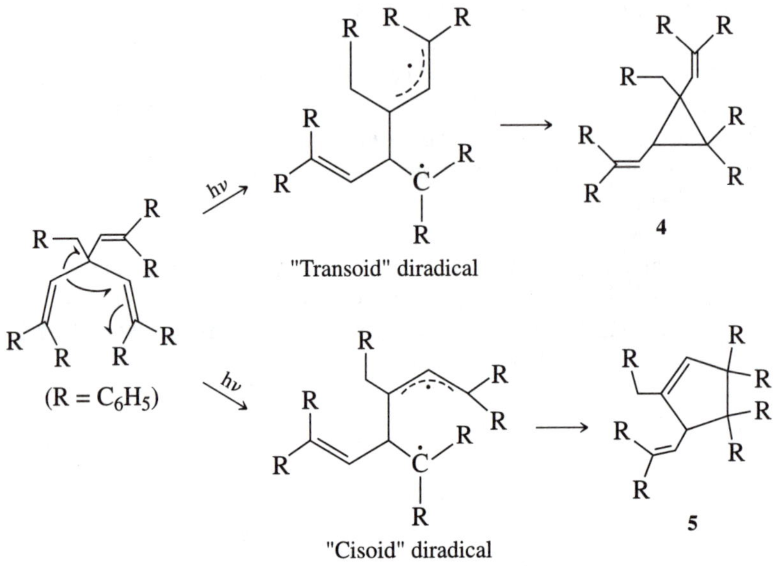

"Transoid" diradical

(16)

(R = C$_6$H$_5$)

"Cisoid" diradical

However, **5** is formed, together with **4**,* only from the singlet reactions. Triplet reactions yield only **4**.

Formation of **5** must proceed via a "cisoid" intermediate diradical. Formation of that diradical as a singlet could immediately be followed by ring closure, while any "cisoid" triplet diradical would have time to rotate to form the more stable "transoid" diradical, which can only yield the cyclopropyl product.

10.3 REACTIONS OF MOLECULES CONTAINING CARBONYL GROUPS

Introduction

π → π* Transitions in carbonyl groups, as in unconjugated carbon–carbon double bonds, require light with wavelengths in the far UV region (at ca. 160–180 nm). They are thus difficult to access, particularly in the liquid phase. However, carbonyl groups also have very weak absorptions (ε ca. 20) in the region around 280 nm. These are attributed to $n → π*$ transitions, in which an electron from a nonbonded (n) pair is raised to the π* level.[†]

Despite the very low value of ε for these processes (at least two powers of ten less than those of the π → π* transitions that would be caused by far UV radiation), most photochemical reactions of carbonyl groups proceed via $n → π*$ transitions resulting from absorption of near UV light.

---

*The relative yield of **5** increased as irradiation continued, since **4** undergoes a photochemical [1,3] shift to yield **5**.
[†]Absorption of light in the far UV region[20] may also result in $n → σ^*$ transitions.

One interesting characteristic of the $n \rightarrow \pi^*$ transition is that it inverts the usual polarity of the carbonyl group, actually placing a negative charge on the carbonyl carbon.

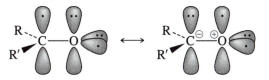

The Electronic Structure of a Carbonyl Group in the n,π^* Excited State

Thus, unlike $\pi \rightarrow \pi^*$ transitions, the energies required for $n \rightarrow \pi^*$ transitions are *increased* by alkyl substituents on the carbonyl carbons, as shown in Table 10.1. (The unusually long wavelength absorption of cyclopropanone is interesting. It is consistent with the stabilities of cyclopropyl anions, due to the high percentage of *s*-orbital character in cyclopropyl bonds.)

TABLE 10.1 WAVELENGTHS OF LIGHT REQUIRED FOR $n \rightarrow \pi^*$ TRANSITIONS IN CARBONYL COMPOUNDS[20]

Compound	Wavelengths (nm)
Formaldehyde	305
Acetaldehyde	290
Acetone	275
Cyclopropanone	330
Cyclopentanone	282
Cyclohexanone	288

Photoexcited carbonyl compounds can exist in n,π^* states either as singlets or triplets. While singlets and triplets often react at very different rates, they usually give the same types of products (although sometimes in quite different ratios.) Therefore, we will not usually attempt to identify the multiplicity of the excited states involved in the reactions discussed below.

Norrish Type I Reactions

Photoexcitation of carbonyl compounds often raises them to electronically excited states with excess vibrational energy. In liquid-phase reactions, the excess energy is usually lost by collisions with other molecules. In the gas phase, however, the excess energy is frequently released by homolytic fission (α-cleavage) of bonds to the carbonyl carbons.

Similar reactions do sometimes occur in the liquid phase—particularly with carbonyl compounds that can form particularly stable free radicals, such as dibenzyl ketone or di-*tert*-butyl ketone. In unsymmetrical ketones, the bond to the alkyl group that can form the more stable free radical is most likely to be broken, as shown in Eq. 17.

$$CH_3\overset{O}{\overset{\|}{C}}CH(CH_3)_2 \xrightarrow{h\nu} CH_3\overset{O}{\overset{\|}{C}}\cdot + \cdot CH(CH_3)_2 \qquad (17)$$

These processes, and the resulting secondary reactions, are collectively known as *Norrish type I reactions*.[21]

Carbon monoxide is frequently released from the acyl radicals resulting from Norrish Type I reactions. The alkyl radicals formed from either the initial α-cleavage or decarbonylation steps can undergo typical dimerization and disproportionation reactions, as shown in Eq. 18.

$$(CH_3)_2CHC \cdot \longrightarrow CO + (CH_3)_2\dot{C}H \longrightarrow \qquad (18)$$

$$(CH_3)_2CHCH(CH_3)_2 + CH_2{=}CHCH_3 + CH_3CH_2CH_3$$

Norrish type I reactions are also frequently observed in reactions of small ring cyclic ketones which, because of geometrical constraints, cannot readily undergo Norrish Type II reactions (described below). In a few instances, particularly those in which loss of carbon monoxide can yield particularly stable free radicals,[22] the initial bond cleavage in a small ring is followed by decarbonylation. However, the fact that scission of the acyl–alkyl bonds in cyclic ketones yields diradicals (rather than two separate free radicals) usually results in interaction of the two radical segments being faster than loss of carbon monoxide.

Aside from simple recombination of the acyl and alkyl radical segments to reform the starting materials (which sometimes results in a change in stereochemistry), the diradicals formed on α-cleavage of a cyclic ketone may undergo two major types of internal disproportionation reactions. Typically, products from both types of reactions are obtained.

1. The acyl radical may abstract a hydrogen atom from a δ carbon to form an unsaturated aldehyde, as shown in Eq. 19.[23] (If no δ hydrogens are available, a hydrogen may be taken from a γ carbon.)

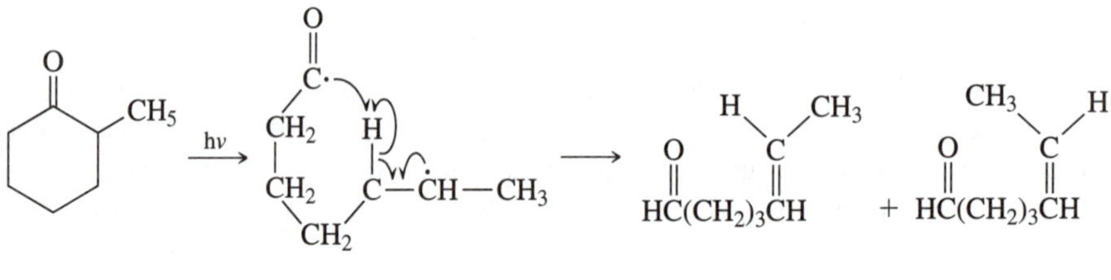

$$(19)$$

2. The alkyl radical may abstract a hydrogen from a position α to the carbonyl group to form a ketene. If the reaction is carried out in an amine solvent, the overall result will be formation of an amide, while reaction in a hydroxylic solvent will yield an ester (or an acid in water, as in Eq. 20).[24]

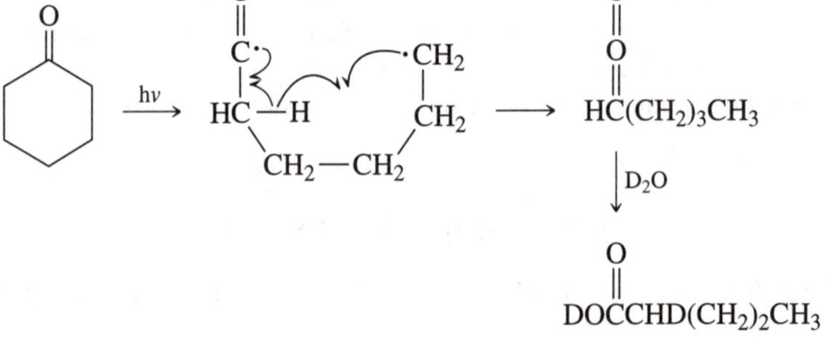

$$(20)$$

In a few cyclic systems with appropriate geometries, the alkyl radical can combine with the carbonyl oxygen to form a carbene, which may, in turn, react with an alcoholic solvent to form an acetal, as shown in Eq. 21.[25]

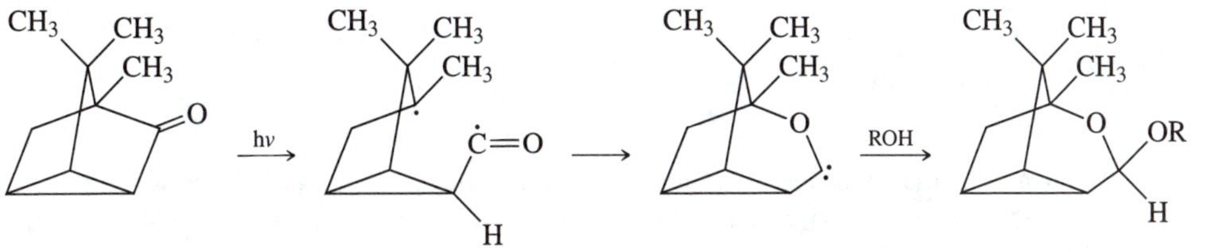

(21)

Hydrogen Abstraction Reactions

A photoexcited carbonyl oxygen will frequently abstract a hydrogen atom from a carbon–hydrogen bond of the solvent. This will result in formation of a pair of free radicals—one from the carbonyl compound and one from the solvent molecule. The two free radicals can then combine or can undergo disproportionation reactions, as shown in Eq. 22a. In either case, the result is reduction of the carbonyl compound to an alcohol.

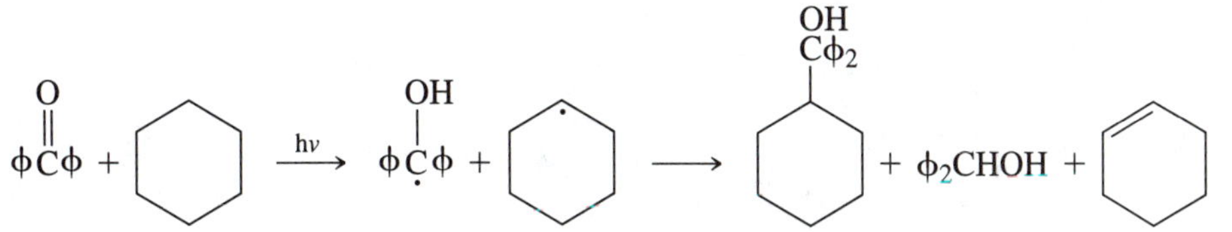

(22a)

It is also possible for two radicals to combine to form dimers. When two radical units derived from the carbonyl compound combine, the result is a photochemical *pinacol reduction*. Photochemical pinacol reductions are particularly efficient when the solvent is the reduced form of the carbonyl group, as shown in Eq. 22b, since only one radical combination product, the pinacol, can be formed. (Hydrogen atoms bonded to oxygenated carbons are usually easily abstracted, since particularly stable free radicals are formed.)

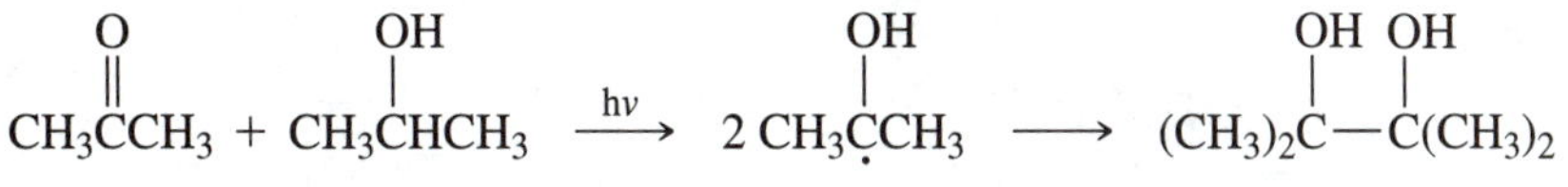

$$CH_3\overset{O}{\overset{\|}{C}}CH_3 + CH_3\overset{OH}{\overset{|}{C}}HCH_3 \xrightarrow{h\nu} 2\ CH_3\overset{OH}{\overset{|}{\underset{\cdot}{C}}}CH_3 \longrightarrow (CH_3)_2\overset{OH}{\overset{|}{C}}-\overset{OH}{\overset{|}{C}}(CH_3)_2 \quad (22b)$$

Norrish Type II Reactions. A photoexcited carbonyl group will often abstract a hydrogen at a γ carbon, forming a 1,4-diradical, as in Eq. 23. Such 1,5-hydrogen transfers are commonly more rapid than intermolecular reactions.

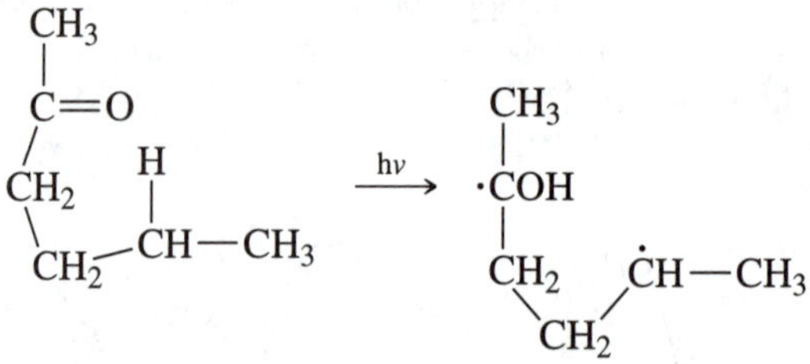

$$
\begin{array}{c}
CH_3 \\
| \\
C=O \\
| \qquad H \\
CH_2 \quad | \\
\diagdown \quad CH-CH_3 \\
CH_2
\end{array}
\xrightarrow{h\nu}
\begin{array}{c}
CH_3 \\
| \\
\cdot COH \\
| \\
CH_2 \quad \dot{C}H-CH_3 \\
\diagdown \quad \diagup \\
CH_2
\end{array}
\qquad (23)
$$

These intramolecular hydrogen transfers, and the reactions resulting from them, are known as *Norrish type II reactions*.[21] (Like other photoreactions of carbonyls, Norrish type II reactions can proceed via singlet or triplet intermediates, but the reactions of triplets appear to be more efficient.)

A diradical formed in a Norrish type II reaction can undergo two principal types of reactions: fragmentation to form an alkene and a new carbonyl compound (initially in the form of its enol), and formation of a cyclobutanol ring.[26]

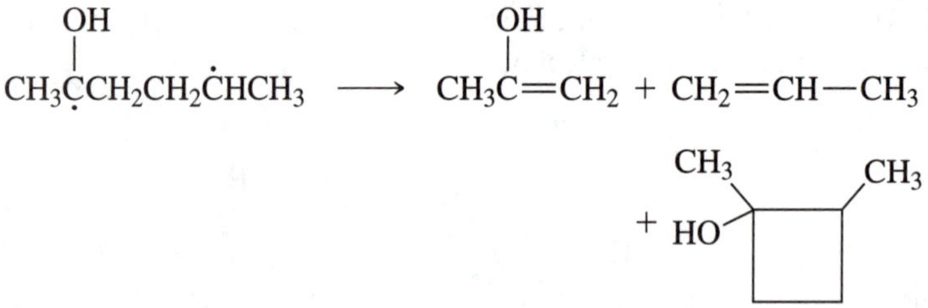

$$
\begin{array}{c}
OH \\
| \\
CH_3\dot{C}CH_2CH_2\dot{C}HCH_3
\end{array}
\longrightarrow
\begin{array}{c}
OH \\
| \\
CH_3C=CH_2
\end{array}
+ CH_2=CH-CH_3
\qquad (24)
$$

$$
+ \; HO
$$

While fragmentation is frequently the major process, high yields of cyclobutanols may be obtained in favorable cases, such as those shown in Eqs. 25 and 26.[27,28]

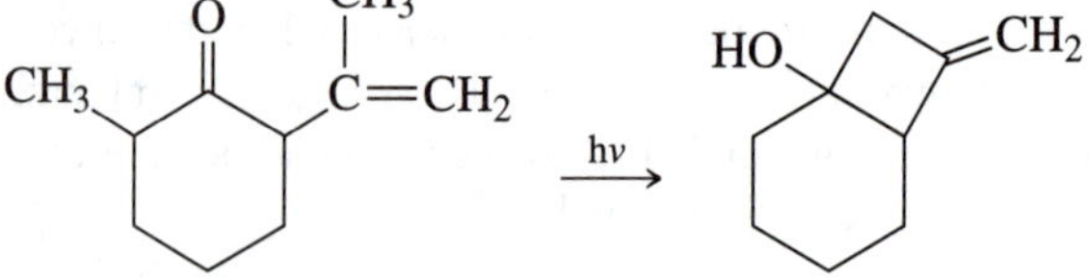

(25)

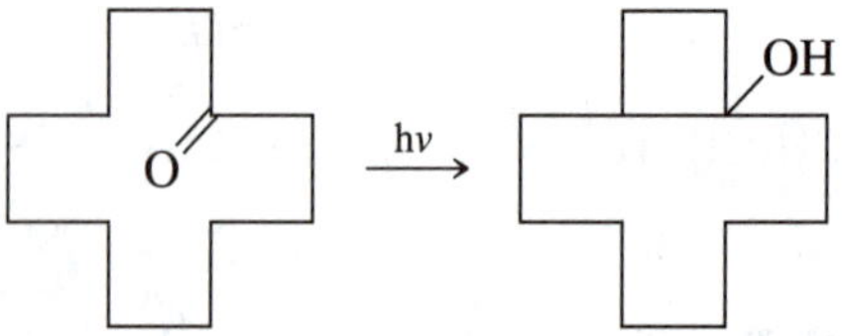

(26)

76%

In a few instances, in which transfers of hydrogens from γ carbons were not feasible, hydrogens from β carbons,[29] or more distant carbons,[30] were attacked.

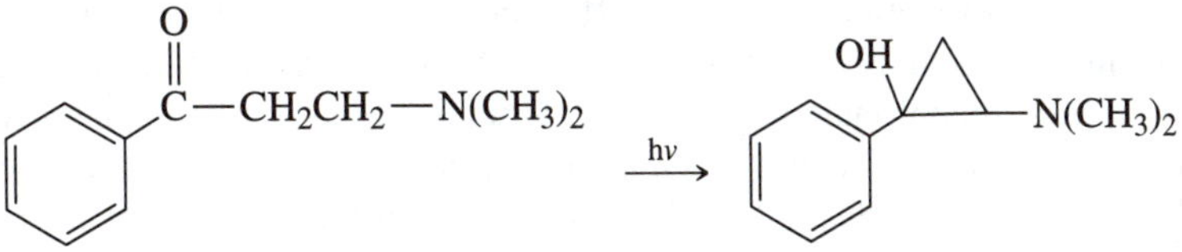

(27)

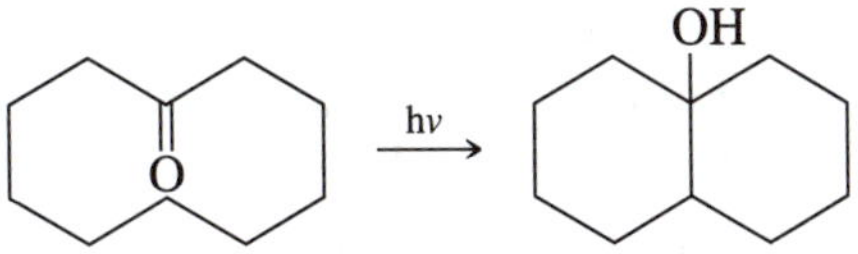

(28)

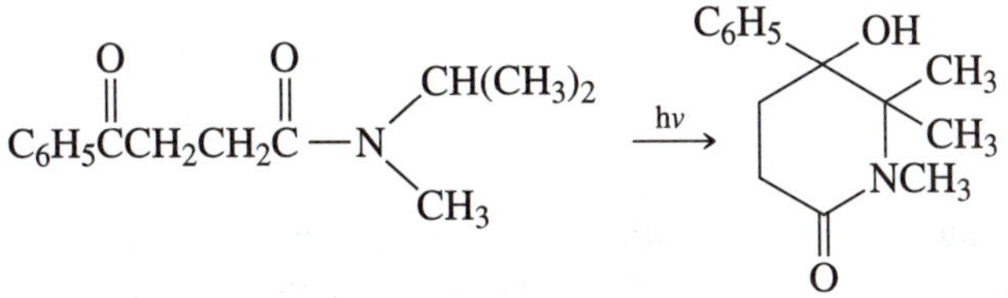

(29)

Reactions of Conjugated Carbonyl Compounds

Conjugated carbonyl compounds are more likely than alkenes and unconjugated carbonyl compounds to react via triplet states, because intersystem crossing is relatively easy for conjugated carbonyls.

Whether reacting via singlet or triplet states, however, conjugated ketones can undergo a remarkable variety of photochemical reactions. They can undergo reactions specific to these systems, in addition to all the reactions of alkenes and unconjugated carbonyl compounds.

Like alkenes, compounds with conjugated carbonyl groups can undergo *cis-trans* isomerizations, including formation of the very reactive *trans* forms of small ring cycloalkenones. They can also undergo [2 + 2] cycloadditions and dimerizations (see Chapter 3). Those reactions are the major processes observed on irradiation of cyclohexenone and cyclopentone. In addition, both intermolecular and intramolecular (Norrish type II) hydrogen transfers are common. Both types of reactions may be complicated, however, by secondary reactions which are not possible with saturated carbonyl compounds. Norrish type II reactions of 6-propylcyclohexenone, for instance, yield small amounts of ketones **6** and **7**, in addition to a mixture of stereoisomeric cyclobutanol derivatives.[31]

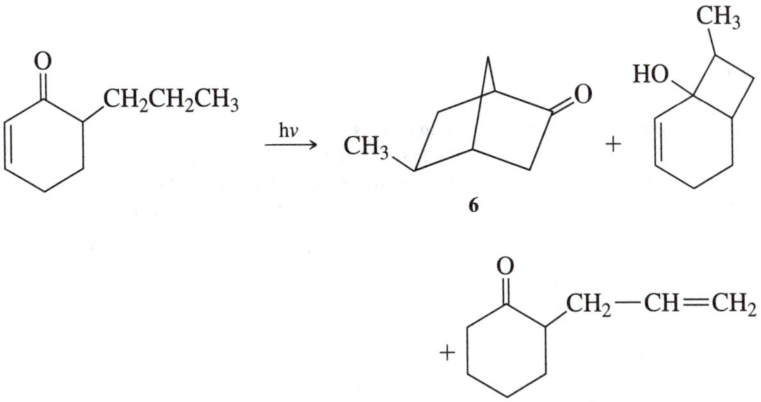

6

(30)

7

In another variant of Norrish type II reactions, conjugated carbonyl groups in open-chain or large ring compounds can abstract protons from δ carbons, converting the carbonyl compounds to their dienol tautomers. Traces of acids or bases may then convert the dienols to β,γ-unsaturated carbonyl compounds, as shown in Eq. 31. (These are very common reactions, though they may be overlooked because of the ease of reconversion of the unconjugated to the conjugated forms.)[32a]*

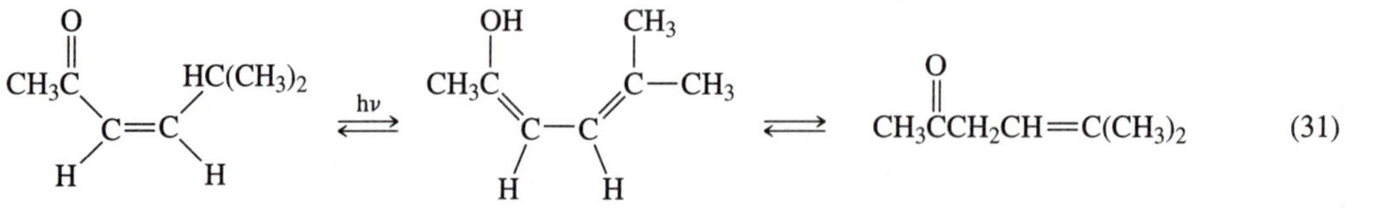

$$CH_3\overset{\overset{O}{\parallel}}{C}\underset{\underset{H}{}}{\overset{HC(CH_3)_2}{\underset{\overset{}{}}{C=C}}}\underset{H}{} \quad \overset{h\nu}{\underset{}{\rightleftharpoons}} \quad CH_3\overset{\overset{OH}{|}}{C}\underset{\underset{H}{}}{C}-\underset{\underset{H}{}}{\overset{\overset{CH_3}{|}}{C}}-CH_3 \quad \rightleftharpoons \quad CH_3\overset{O}{\overset{\parallel}{C}}CH_2CH=C(CH_3)_2 \qquad (31)$$

α,β-Unsaturated ketones with aryl or vinyl groups at C4 can undergo aryl or vinyl migrations, resulting in the formation of three-membered rings, as shown in Eq. 32.[33] These reactions are closely related to di-π-methane rearrangements, except that they are initiated by $n \rightarrow \pi^*$ transitions of the carbonyl groups.

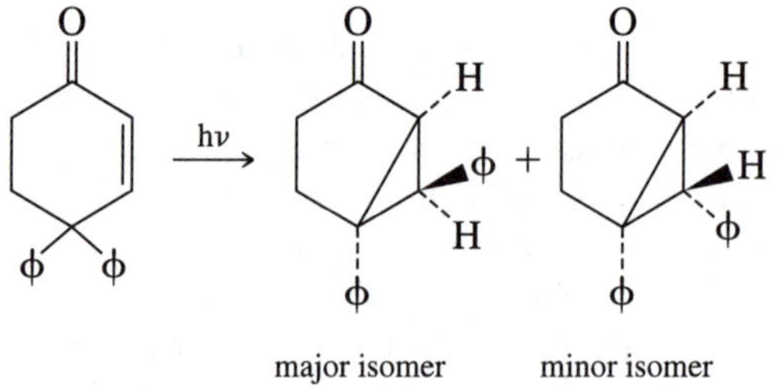

$$\qquad (32)$$

major isomer   minor isomer

Lumiketone Rearrangements.[34]   Cross-conjugated cyclohexadienones, on photoirradiation, characteristically rearrange to form isomers, called *lumiketones*, containing cyclopropyl rings as in Eq. 33.

---

*Δ1,2-Cyclohexenones, which cannot undergo Norrish type II reactions, can nonetheless undergo deconjugation on photoirradiation. These reactions proceed from triplet states, and have been reported to be catalyzed by acids.[32b] It has been suggested[32b] that they proceed by formation of *trans*-cyclohexenones, followed by Markownikoff protonation and loss of proton from γ positions of the resulting carbocations, as shown in the equation below.

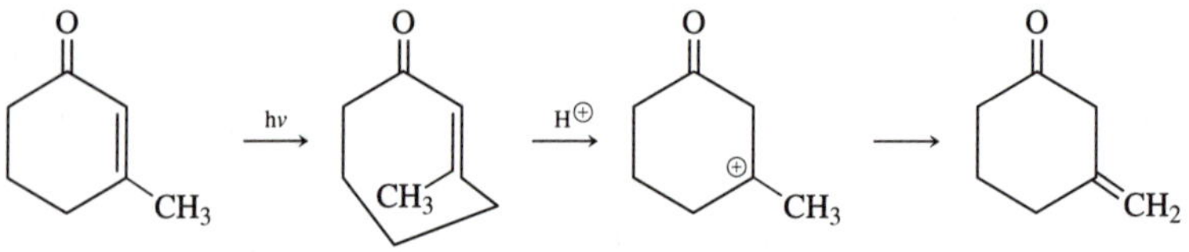

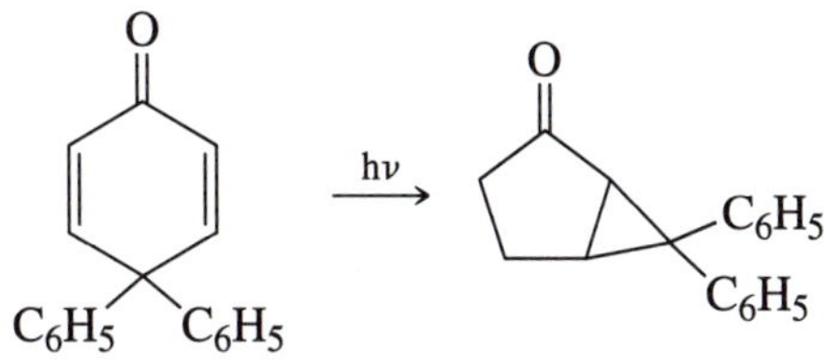

$$(33)$$

Lumiketone rearrangements have been thoroughly investigated in derivatives of natural products. It has been shown that the reactions result in inversion of the geometries at the quaternary carbons, as shown in Eq. 34.[35]

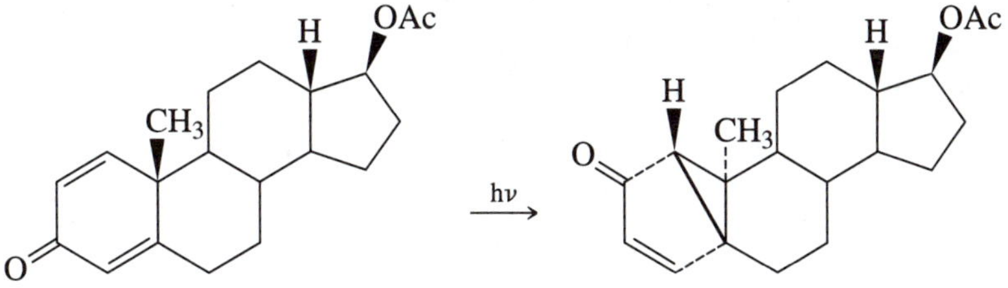

$$(34)$$

Lumiketone rearrangements are extremely efficient reactions, with quantum yields close to 1.0. Like most reactions of conjugated ketones, they have been shown to proceed via triplet states. Zimmerman and his coworkers have again proposed a three-stage mechanism: A) the $n \rightarrow \pi^*$ transition is followed (or possibly accompanied) by interaction of the double bonds to form the three-membered ring; B) an electron returns to the n orbital, yielding a ground state zwitterion; and C) [1,4] bond migration in the zwitterion yields the lumiketone.[36]

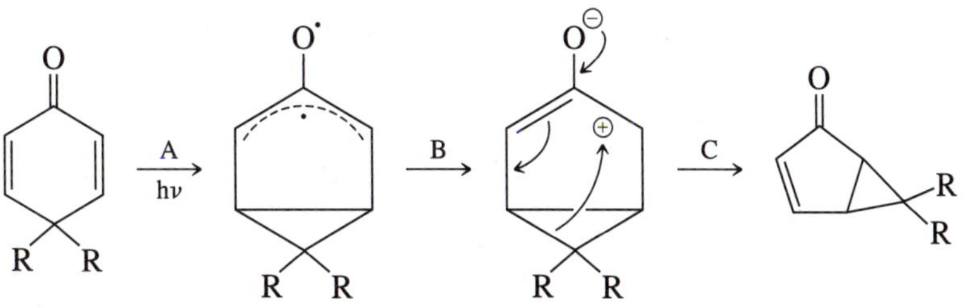

$$(35)$$

This mechanism was supported by the observation that when similar zwitterions were formed in ground state reactions (e.g., under Favorskii rearrangement conditions, as in Eq. 36), they yielded products with lumiketone structures.[37,38] The migration steps proceeded with inversion, as predicted by the Woodward–Hoffmann rules.

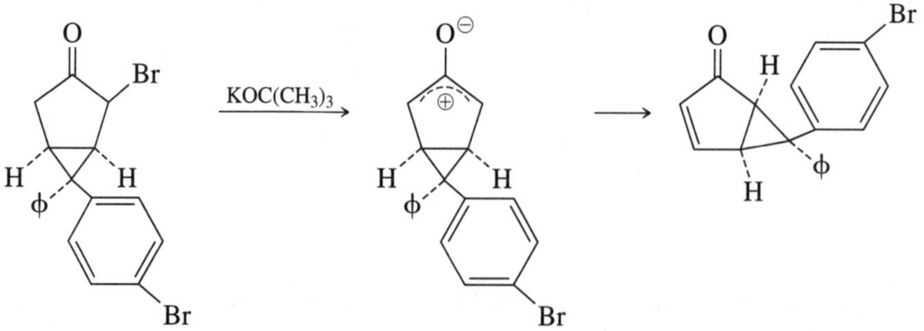

$$(36)$$

Although lumiketones are formed with high efficiencies in nonhydroxylic solvents, the overall yields are frequently reduced by further photorearrangements of the lumiketones. In hydroxylic solvents, particularly in acidic solutions, the intermediate zwitterions may be intercepted prior to formation of the lumiketones, as in Eq. 37.[35]

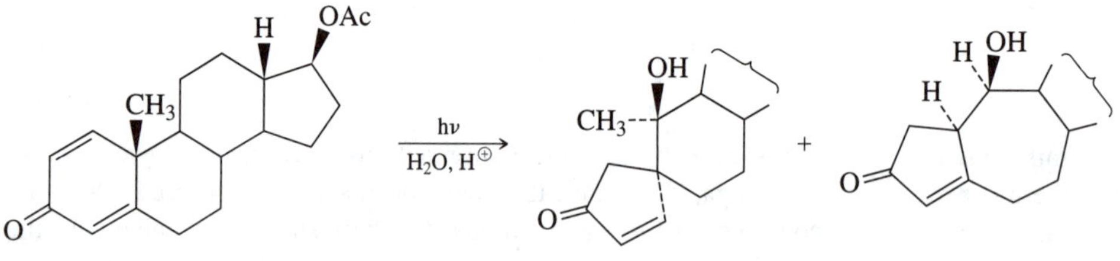

$$(37)$$

Lumiketone-like Rearrangements of Cyclohexenones. Like cyclohexadienones, cyclohexenones can undergo rearrangements to yield products containing cyclopropyl rings provided they have two substituents at C4. The reactions again proceed with inversion at C4, as shown in Eq. 38.[39]

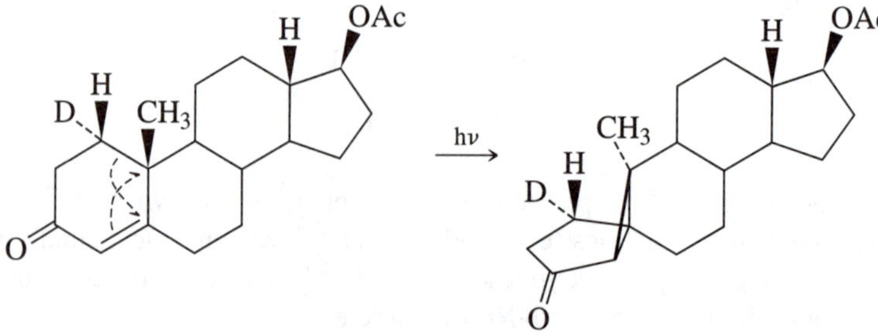

$$(38)$$

While these reactions appear to resemble lumiketone rearrangements, their quantum efficiencies are only about 1% as large as those of cross-conjugated cyclohexadienones.

There appears to be little alternative to these rearrangements proceeding by [1,2] alkyl migrations.

1,2- and 1,3-Acyl Shifts; Oxa-di-π-methane Rearrangements.   β,γ-Unsaturated ketones (particularly small ring ketones that cannot easily dissipate energy by *cis-trans* isomerizations) frequently undergo two different types of migrations of acyl groups: 1,3-migrations (discussed in Section 4.2) and 1,2-migrations. Products of 1,2-acyl migrations are known as oxa-di-π-methane rearrangements.[10c,40] Products resulting from oxa-di-π-methane rearrangements are shown at the left in Eqs. 39 and 40,[41,42] and products of 1,3-acyl shifts are shown at the right.

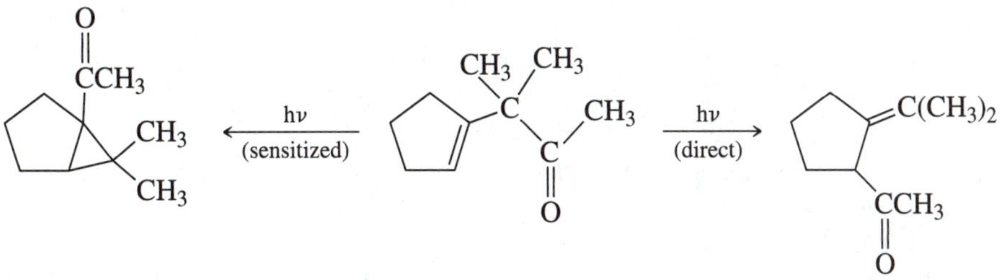

(39)

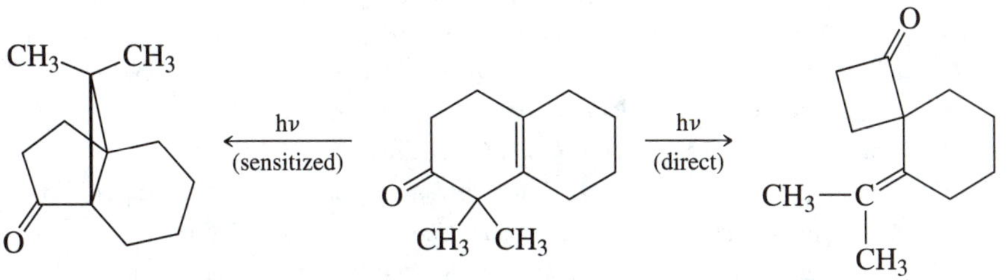

(40)

1,3-Acyl shifts typically arise from singlet states and oxa-di-π-methane rearrangements from triplet states, although there are exceptions to both generalizations. In reactions of simple enones, such as those shown in Eqs. 39 and 40, the initial processes are, as usual for carbonyl compounds, n–π* transitions. However, oxa-di-π-methane rearrangements of compounds containing additional double bonds in conjugation, as in Eq. 41, are particularly efficient. In these reactions, the initial processes are typically π–π* transitions.[43]

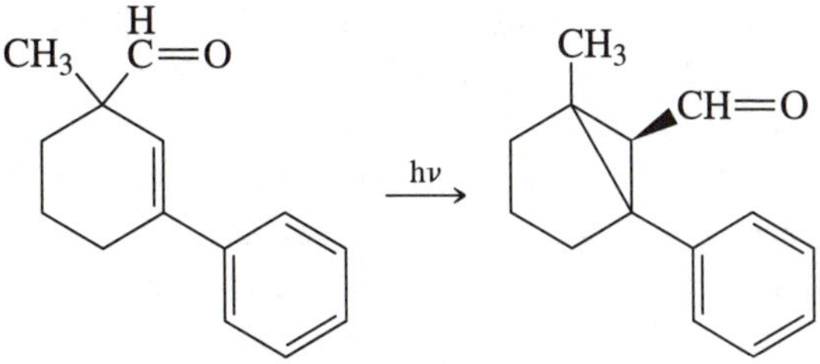

(41)

Although oxa-di-π-methane rearrangements appear to closely resemble di-π-methane rearrangements of hydrocarbons, their stereochemistries appear to be even more complex. Examples of rearrangements proceeding with inversion, with retention, or with racemization of the geometry at the central carbon (C2) have all been observed.

REFERENCES

[1]   *The Exciplex*, M. Gordon and W.R. Ware, Eds., Academic Press, New York (1975).

[2]   For a detailed discussion of Rydberg states, see A.J. Merer and R.S. Mulliken, *Chem. Rev., 69,* 639 (1969).

[3]   W. J. Leigh and B.H.O. Cook, *J. Org. Chem., 64,* 5256 (1999).

[4]   (a) J. Saltiel and J.L. Charlton in *Rearrangements in Ground and Excited States*, vol. 3, P. de Mayo, Ed., Academic Press, New York (1980); (b) H. Gorner and H.J. Kuhn, *Adv. Photochem., 19,* 1 (1995).

[5]   (a) Y. Inoue, S. Takamuku, and H. Sakurai, *Chem. Comm.*, 413 (1976); (b) R. Srinivasan and K.H. Brown, *J. Am. Chem. Soc., 100,* 2589 (1978); (c) A. Squillacote, A. Bergman, and J. DeFelippis, *Tetrahedron Lett., 30,* 6805 (1989).

[6]   (a) P.J. Kropp et al., *J. Am. Chem. Soc., 95,* 7058 (1973); (b) Y. Inoue, S. Takamuku, and H. Sakurai, *J. Chem. Soc., Perkin Trans., 2,* 1635 (1977).

[7]   T.R. Fields and P.J. Kropp, *J. Am. Chem. Soc., 96,* 7559 (1974).

[8]   P.J. Kropp in *Organic Photochemistry*, vol. 4, A. Padwa, Ed. Marcel Dekker, New York (1979).

[9]   R. Srinivasan and K.H. Brown, *J. Am. Chem. Soc., 100,* 4602 (1978).

[10] See (a) S.S. Hixson, P.S. Mariano, and H.E. Zimmerman, *Chem. Revs., 73,* 531 (1973); (b) H.E. Zimmerman in Ref. 3a; (c) H.E. Zimmerman and D. Armeesto, *Chem. Revs., 96,* 3065 (1996).

[11] H.E. Zimmerman and P.S. Mariano, *J. Am. Chem Soc., 91,* 1718 (1969).

[12] H.E. Zimmerman and L.M. Tolbert, *J. Am. Chem. Soc., 97,* 5497 (1975).

[13] G.W. Griffin, A.F. Marcantonio, and H. Kristinsson, *Tetrahedron Lett., 2,* 951 (1965).

[14] H.E. Zimmerman and G.E. Samuelson, *J. Am. Chem. Soc., 91,* 5307 (1969).

[15] (a) M. Reguero et al., *J. Am. Chem. Soc., 115,* 2073 (1993); (b) K. Quenemon, W.T. Borden, E.R. Davidson, and D. Feller, *ibid., 107,* 5054 (1985).

[16] H.E. Zimmerman and A.C. Pratt, *J. Am. Chem. Soc., 92,* 1408, 6259, 6267 (1970).

[17] H.E. Zimmerman, P. Baeckstrom, and D.W. Kurz, *J. Am. Chem. Soc., 96,* 1459 (1974).

[18] (a) H.E. Zimmerman et al., *J. Am. Chem. Soc., 96,* 4630 (1974);(b) H.E. Zimmerman, R.W. Binkley, R.S. Givens, and M.A. Sherwin, *ibid., 89,* 3932 (1967).

[19] H.E. Zimmerman and V. Cirkva, *Org. Lett., 2,* 2365 (2000); *J. Org. Chem., 66,* 1839 (2001).

[20] (a) E.S. Stern and C.J. Timmons, *Electronic Absorption Spectroscopy in Organic Chemistry*, Edward Arnold, Ltd., London (1970); (b) O.L. Chapman and D.S. Weiss in *Organic Photochemistry*, vol. 3, O.L. Chapman, Ed. Marcel Dekker, New York (1973).

[21] (a) See P.J. Wagner in *Rearrangements in Ground and Excited States*, vol. 3, P. de Mayo, Ed., Academic Press, New York (1980); (b) Norrish and Bamford, *Nature, 140,* 195 (1937).

[22] R.C. Cookson, *J. Chem. Soc., C,* 473 (1967).

[23] C.C. Badcock, M.J. Perona, G.O. Pritchard, and B. Rickborn, *J. Am. Chem. Soc., 91,* 543 (1969).

[24] G.O. Schenck and F. Schaller, *Chem. Ber., 98,* 2056 (1965).

[25] P. Yates and R.O. Loutfy, *Acc. Chem. Res., 8,* 209 (1975).

[26] N.C. Yang, S.P. Elliott, and B. Kim, *J. Am. Chem. Soc., 91,* 7551 (1969).

[27] R.T. Matsui, A. Komatsu, and T. Moroe, *Bull. Chem. Soc. Japan, 40,* 2204 (1967).

[28] D. Mori, K. Matsui, and H. Nozaki, *Tetrahedron Lett.*, 1175 (1970).

[29] W. Weigl and P.J. Wagner, *J. Am. Chem. Soc., 118,* 12858 (1996).

[30] (a) M. Barnard and N.C. Yang, *Proc. Chem. Soc.,* 302 (1958); (b) A.T. Haswegawa, CA *127,* 121403 (1997).

[31] A.B. Smith and W.C. Agosta, *J. Org. Chem., 37,* 1259 (1972).

[32] (a) J.-P. Pete, *Adv. Photochem., 21,* 135 (1996); (b) A. Rudolph and A.C. Weedon, *J. Am. Chem. Soc., 111,* 8756 (1989).

[33] H.E. Zimmerman, R. Keese, J. Nasielski, and J.S. Swenton, *J. Am. Chem. Soc., 88,* 4895 (1966); H.E. Zimmerman, *Tetrahedron, 30,* 1617 (1974).

[34] See K. Schaffner and M. Demuth in *Rearrangements in Ground and Excited States,* vol. 3, P. de Mayo, Ed., Academic Press, New York (1980).

[35] F. Frei et al., *Helv. Chim. Acta, 49,* 1049 (1966).

[36] H. Zimmerman, *Adv. Photochem., 1,* 183 (1963).

[37] H.E. Zimmerman, D.S. Crumrine, D. Doepp, and D.S. Huyffer, *J. Am. Chem. Soc., 91,* 434 (1969).

[38] T.M. Brennan and R.K. Hill, *J. Am. Chem. Soc., 90,* 5614 (1968).

[39] M. Bellus, D.R. Kearns, and K. Schaffner, *Helv. Chim. Acta, 52,* 971 (1969).

[40] D. Schuster in *Rearrangements in Ground and Excited States,* vol. 3, P. de Mayo, Ed., Academic Press, New York (1980).

[41] P.S. Engel and M.A. Schexnayder, *J. Am. Chem. Soc., 97,* 145 (1975).

[42] P.S. Engel, M.A. Schexnayder, H. Ziffer, and J.I. Seeman, *J. Am. Chem. Soc., 96,* 924 (1974).

[43] D. Armesti et al., *J. Org. Chem., 61,* 1459 (1996).

PROBLEMS

10.1 Suggest reasonable mechanisms for formation of Products **4** and **5** in Eq. 30.

10.2 Suggest reasonable mechanisms for the following reactions:

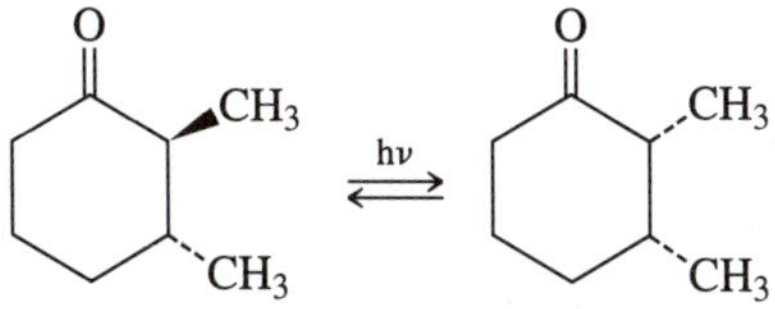

(J.A. Baltrop and J.D. Coyle, *Chem. Comm.,* 1081 [1969])

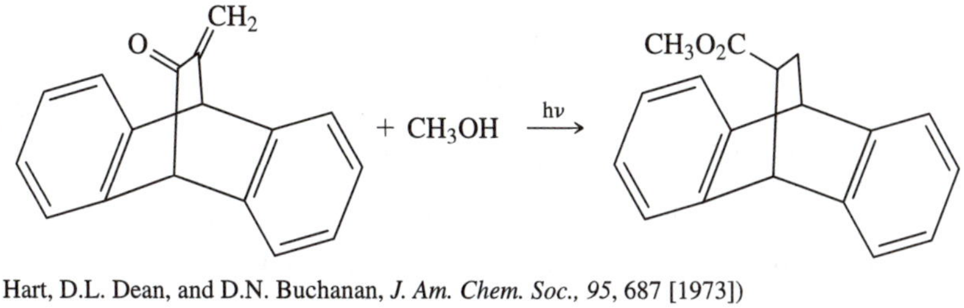

(H. Hart, D.L. Dean, and D.N. Buchanan, *J. Am. Chem. Soc., 95,* 687 [1973])

(c)

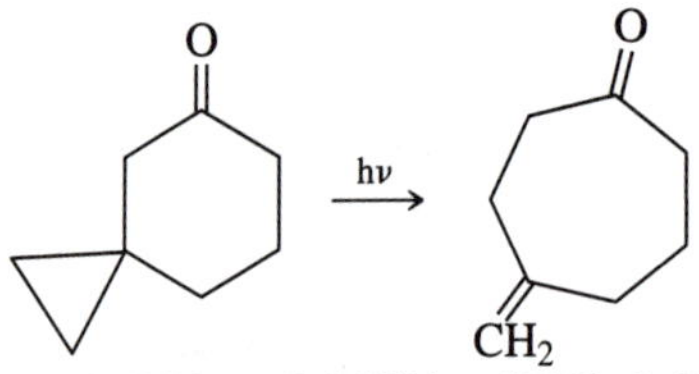

(A. Sonoda, I. Mortoni, J. Miki, and T. Tsuji, *Tetrahedron Lett.*, 3187 [1969])

(d)

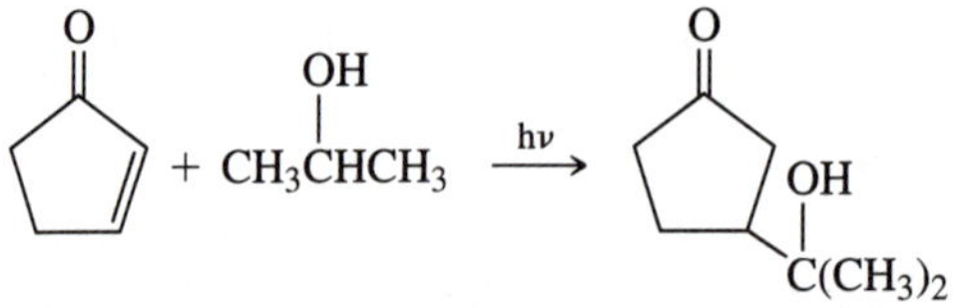

(M. Pfau, R. Dulore, and M. Vilkas, *Compt. Rend.*, 1817 [1962])

(e)

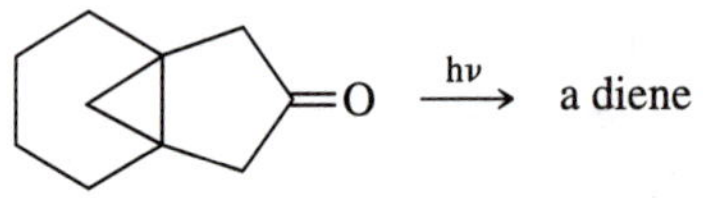

(N. Turro et al., *Tetrahedron Lett.*, 54 [1967])

(f)

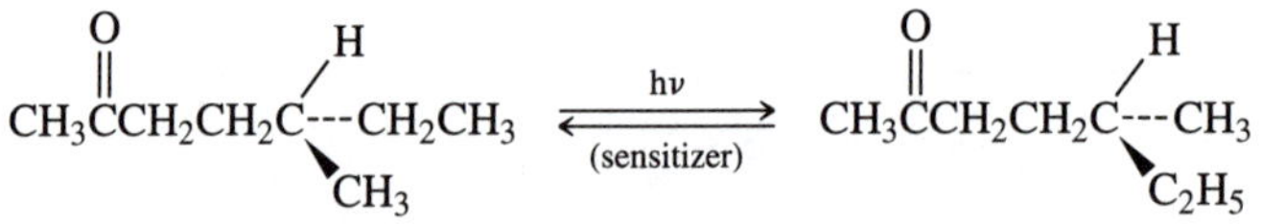

$$CH_3\overset{O}{\overset{\|}{C}}CH_2CH_2\overset{H}{\underset{CH_3}{C---CH_2CH_3}} \underset{(sensitizer)}{\overset{h\nu}{\rightleftharpoons}} CH_3\overset{O}{\overset{\|}{C}}CH_2CH_2\overset{H}{\underset{C_2H_5}{C---CH_3}}$$

(N.C. Yang and S.P. Elliott, *J. Am. Chem. Soc., 91,* 7550 [1969])

(g)

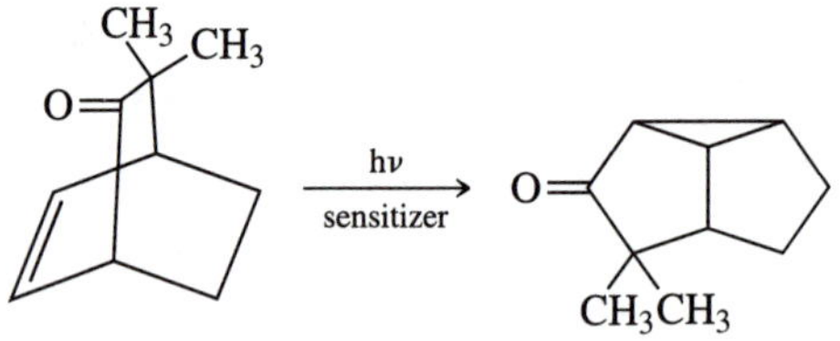

(Adapted from V. Singh and M. Porincho, *Tetrahedron, 52,* 7087 [1996])

(h)

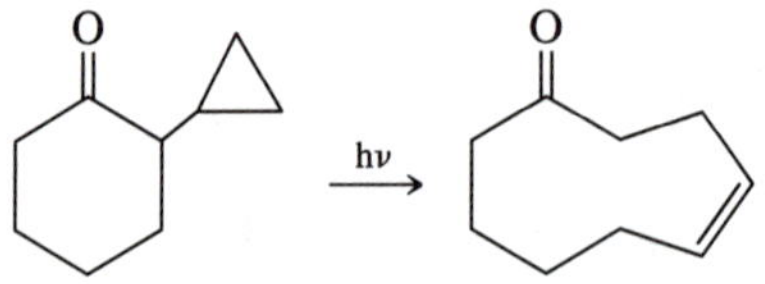

(R.S. Carlson and E.L. Bierswith, *Chem. Comm.,* 1049 [1969])

(i) 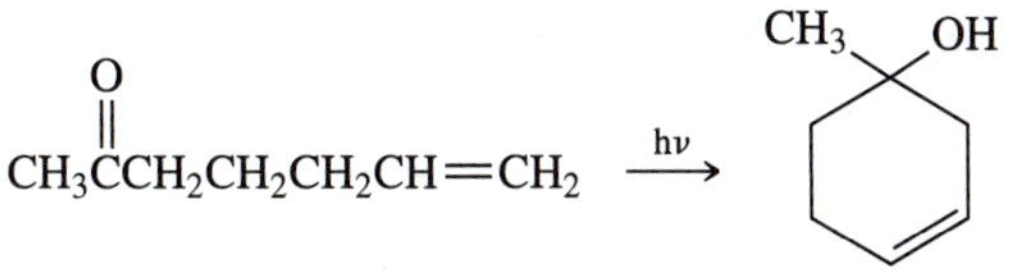 $CH_3\overset{\overset{\displaystyle O}{\|}}{C}CH_2CH_2CH_2CH{=}CH_2$ $\xrightarrow{h\nu}$

(See N.C. Yang, S.P. Elliott, and B. Kim, *J. Am. Chem. Soc., 91,* 7551 [1969])

(j)

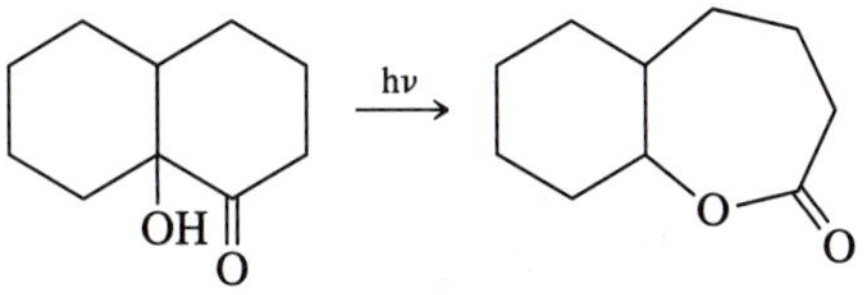

(Adapted from R.C. Cookson, *Pure and Applied Chem., 9,* 575 [1964])

(k)

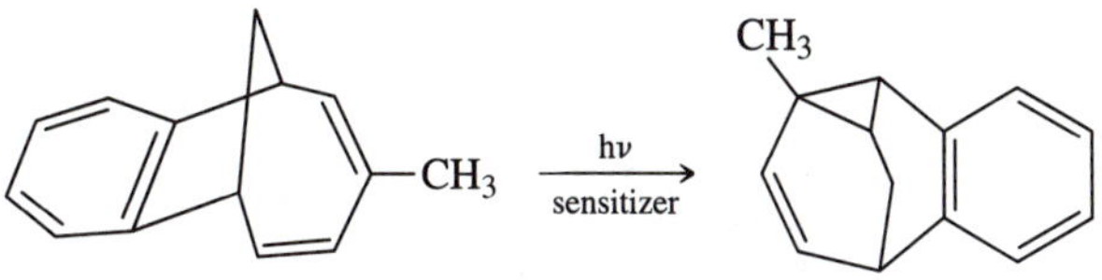

(R.C. Hahn and R.P. Johnson, *J. Am. Chem. Soc., 97,* 212 [1975])

(l) 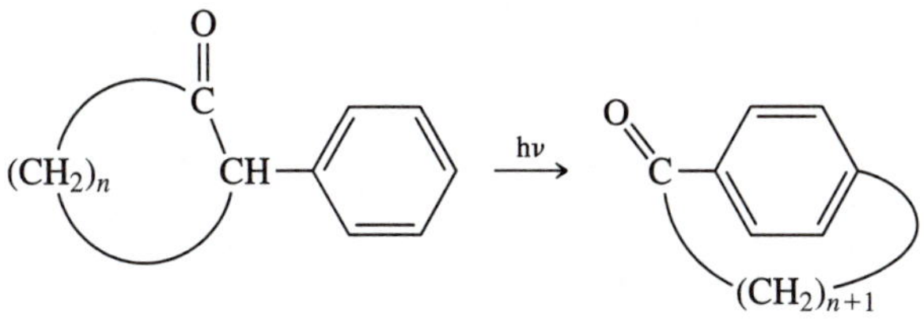 $(CH_2)_n$

(N. Han, X. Lei, and N. Turro, *J. Org. Chem., 56,* 2927 [1991])

(m)

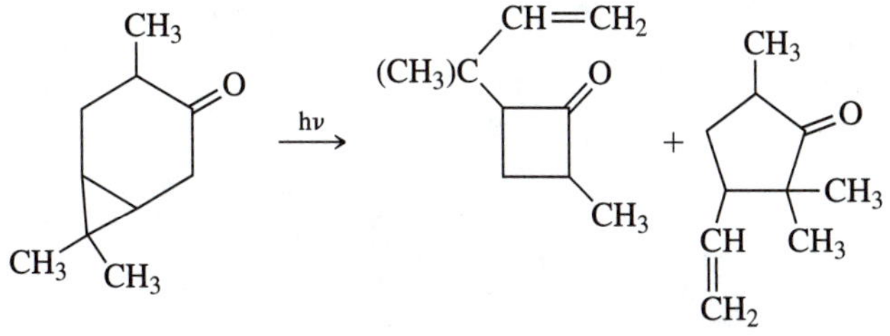

(D.C. Heckert and P.J. Kropp, *J. Am. Chem. Soc., 90,* 4911 [1968])

CHAPTER

Six-Membered Heterocyclic Rings

11

11.1 AROMATIC HETEROCYCLIC MOLECULES

Introduction

Chemical rings composed of two or more different types of atoms are called *heterocyclic rings.* (Rings composed of just one type of atom, such as those in benzene or cyclopropane, are called *homocyclic rings*.) Organic compounds containing heterocyclic rings are so common that the majority of organic molecules contain heterocyclic rings. Most of these molecules closely resemble open-chain molecules containing the same functional groups. The cyclic amine, cyclic ether, cyclic acetal, and cyclic imine shown below, for instance, have very similar properties to those of open-chain molecules in the same classes of compounds.

A cyclic amine   A cyclic ether   A cyclic acetal   A cyclic imine

Heterocyclic molecules are likely to differ markedly from open-chain molecules in only two cases: when they contain small, strained rings such as oxirane (ethylene oxide) or three-membered amine rings; or when the heterocyclic rings have closed, conjugated arrays of orbitals containing $4n + 2\pi$ electrons and can be considered to be aromatic rings.[1] Only heterocyclic rings with aromatic character will be discussed here.

The members of one major class of aromatic heterocycles resemble benzene rings in which one or more carbon atoms have been replaced by hetero atoms. Six-membered analogs of benzene containing phosphorus, arsenic, antimony, and even bismuth atoms, for instance, have been synthesized.[2] However, compounds containing such rings are usually very reactive, air-sensitive substances resembling ylides, since double bonds between carbon and elements in the third and higher rows are quite weak.

In contrast, six-membered aromatic heterocycles containing oxygen atoms are stable and easily prepared. However, in these molecules, which are called *pyrylium salts*,[3] the heterocyclic rings must exist as cations, rather than as neutral rings.

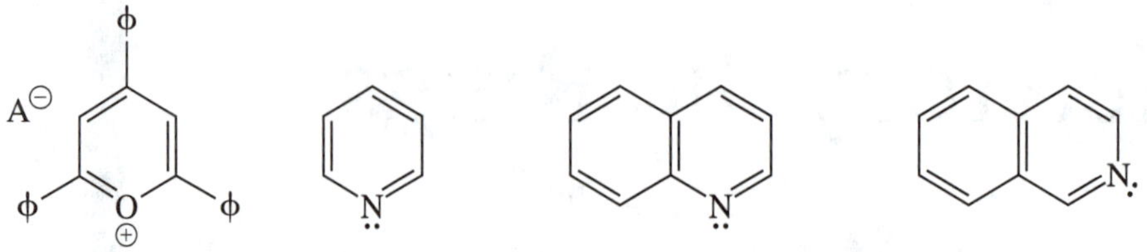

A pyrylium salt   pyridine   quinoline   isoquinoline

The most important six-membered heterocyclic rings, by far, are those containing nitrogen atoms as the hetero atoms. The parent compound of this class of molecules is *pyridine,* which has the molecular formula C_5H_5N. Its derivatives in which the pyridine ring is fused to a second aromatic ring, thus forming heterocyclic analogs of napthalene, are called *quinoline* and *isoquinoline*. (The names quinoline and isoquinoline are derived from quinine, which contains a quinoline ring. Pyridine, quinoline, and isoquinoline rings are also found in a large number of other natural products.[4])

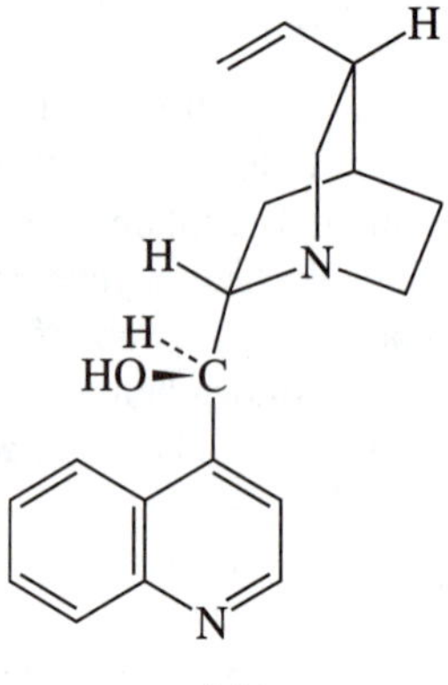

quinine

Pyrazines, which contain two nitrogen atoms *para* to each other, and *pyrimidines*, which have two nitrogen atoms *meta* to each other, are also quite important. Pyrazine and pyrimidine rings are frequently found in naturally occurring molecules. Molecules containing a *pyridazine* ring, with two nitrogen atoms bonded to each other, are less common.

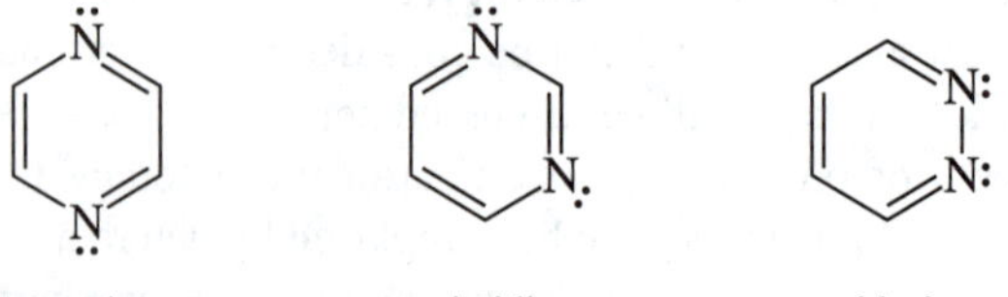

pyrazine   pyrimidine   pyridazine

All of these molecules have conjugated rings containing $4n + 2\pi$ electrons, because the unshared electrons on the nitrogen atoms are in orbitals perpendicular (*orthogonal*) to the orbitals of the double bonds. Thus, the unshared electrons on nitrogens are not counted when deciding whether these rings are aromatic or antiaromatic.

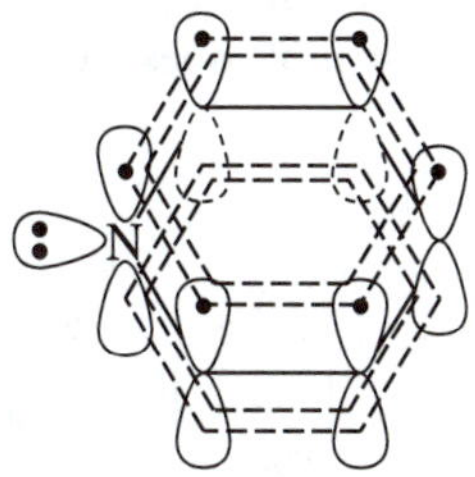

Orbital structure of a pyridine ring

As will be seen, the chemical properties of pyridine and its analogs are consistent with those of molecules with a high degree of aromatic stabilization. The presence of aromatic ring currents can be deduced from the fact that the signals for ring protons (and ring carbons) in the NMR spectra of pyridines lie appreciably downfield from those of analogs lacking closed rings of π electrons.[5]

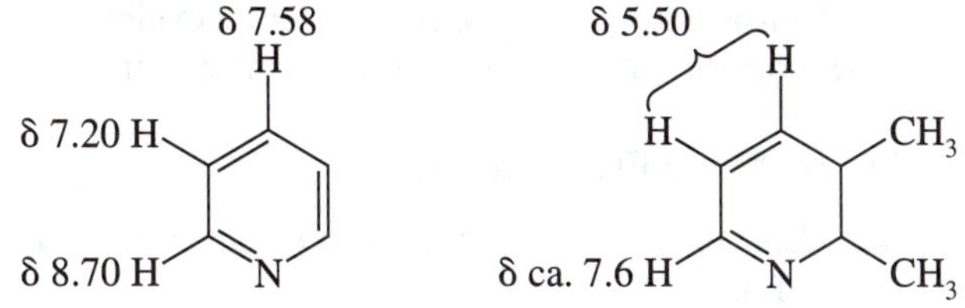

^{1}H NMR shifts in pyridine and a nonaromatic analog

Estimating the aromatic stabilization energies of pyridine and its derivatives is even more difficult than estimating the stabilization energy of benzene. However, a variety of empirical and theoretical methods agree that the stabilization energy of pyridine is close to, although somewhat lower than, that of benzene. The stabilization energies of pyrazine and pyrimidine, in turn, are somewhat lower than that of pyridine but are still sizable.[1]

Naming Molecules Containing Six-Membered Heterocyclic Rings

In each of the heterocyclic ring systems shown above (except for isoquinolines), a nitrogen atom is always designated as atom 1. In naming molecules containing isoquinoline rings, the numbering of ring positions begins at the carbon linking the nitrogen atom to the adjacent ring, so the nitrogen atom is designated as atom 2.

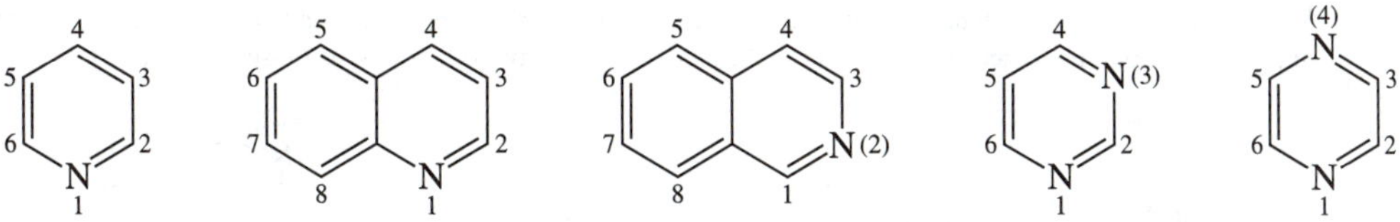

Numbering of ring positions in six-membered heterocyclic rings

The remaining atoms forming the rings are then numbered in sequence. However, in fused aromatic ring systems, such as quinoline and isoquinoline, the atoms common to two rings (which cannot bear substituents whose positions must be specified) are not numbered as part of the normal sequences. In polysubstituted derivatives of pyridine and pyrazine the direction in which the ring position numbers are assigned is chosen to minimize the sum of the numbers. A few examples of how six-membered heterocyclic molecules are named are given in the following diagram.*

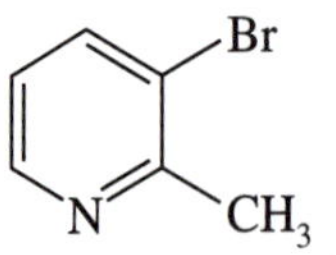

3-bromo-2-methylpyridine 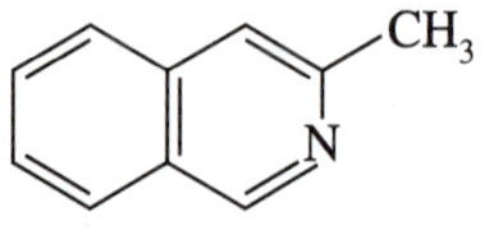   3-methylisoquinoline 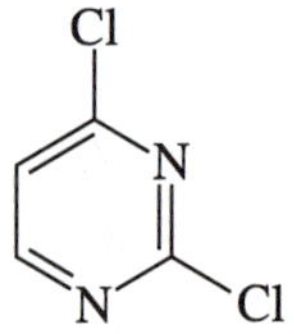   2,4-dichloropyrimidine

11.2 REACTIONS OF PYRIDINE AND ITS ANALOGS

The pyridine ring can be considered to combine two types of functional groups—an aromatic ring and a conjugated, unsaturated imine. To some extent it exhibits the properties of both types of molecules, but the properties of each are affected by the presence of the other.

Reactions at the Unshared Electron Pair

Protonation.   Like most trivalent nitrogen compounds, pyridine is a base and forms salts on reaction with acids (Eq. 1).

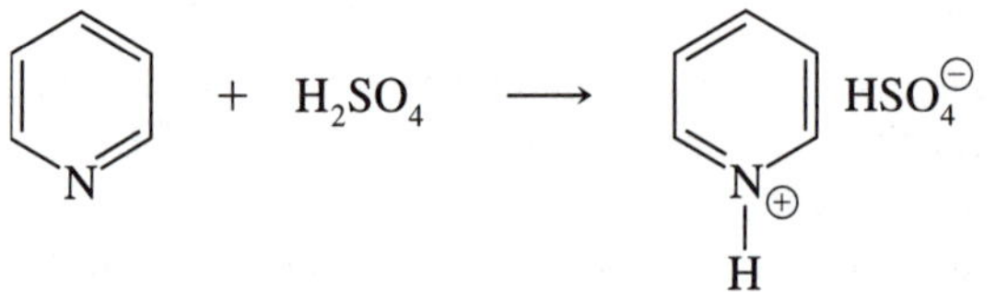

(1)

---

*Atoms common to *two* aromatic rings are identified by the number of the atom that precedes them in the numbering sequence, followed by the letter "a."

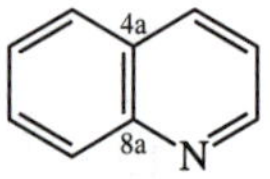

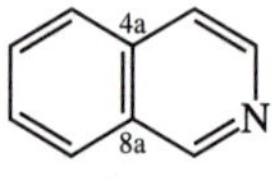

Six-membered heterocyclic molecules containing more than two fused aromatic rings are usually named as derivatives of carbocyclic ring systems in which carbon atoms have been replaced by nitrogen (*aza*) atoms. (An exception is the *acridine* ring system, which is important enough to have its own name.) The number of nitrogen atoms in the rings and their locations are indicated by prefixes to the names of the carbocyclic rings, as illustrated below.[5]

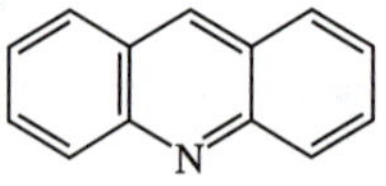

acridine
(9-azaanthracene)   3,9-diazaphenanthrene 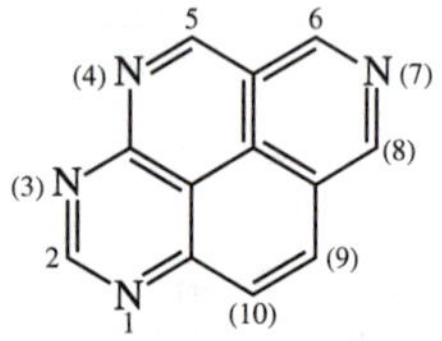   1,3,4,7-tetraazapyrene

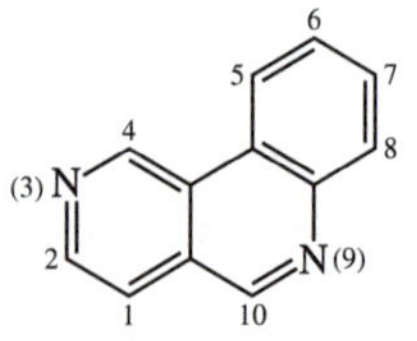

TABLE 11.1   pK_a VALUES OF CONJUGATE ACIDS OF PYRIDINES

Substituent	Position	pK_a of conjugate acid
N(CH$_3$)$_2$	4	9.70
OCH$_3$	4	6.62
OCH$_3$	3	4.88
CH$_3$	4	6.03
CH$_3$	3	5.60
H		5.20
Cl	4	3.83
Cl	3	2.81
Cl	2	0.72
NO$_2$	4	1.61
NO$_2$	3	0.81
Cl, Cl	2 and 6	−2.86

However, as can be seen from Table 11.1, the basicities of pyridine derivatives are strongly affected by substituents. As a result, pyridines with strongly electron-withdrawing substituents (particularly at positions 2 and 6) are not basic by the usual standards, such as solubility in aqueous acid solutions.

Although pyridine and its analogs are usually described as heterocyclic amines, they are actually imines: compounds containing doubly bonded nitrogen atoms. Pyridine is comparable in basicity to other imines and is, therefore, an appreciably weaker base than most aliphatic amines. Its conjugate acid (the *pyridinium ion*) is about 10^5 times as strong an acid as are most alkylammonium ions, which have pK_a values around 10. This difference in acidity can be ascribed to the fact that the unshared electrons of imines are located in sp^2 orbitals, while in aliphatic amines, the unshared electrons are less firmly held in sp^3 orbitals.

Alkylation.   Pyridine and its analogs act as nucleophiles in reactions with alkylating agents, yielding quaternary pyridinium salts via S$_N$2 reactions (Eq. 2).

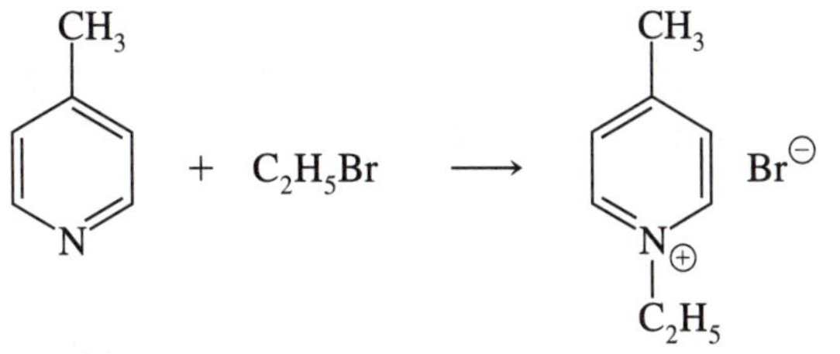

$$\text{(2)}$$

As usual, elimination reactions may accompany substitution. Thus, reactions of pyridines with tertiary halides, in particular, give very low yields of the quaternary salts.

Acylation.   Pyridines react rapidly with acid halides or acid anhydrides to form *N*-acylpyridinium salts (Eq. 3).

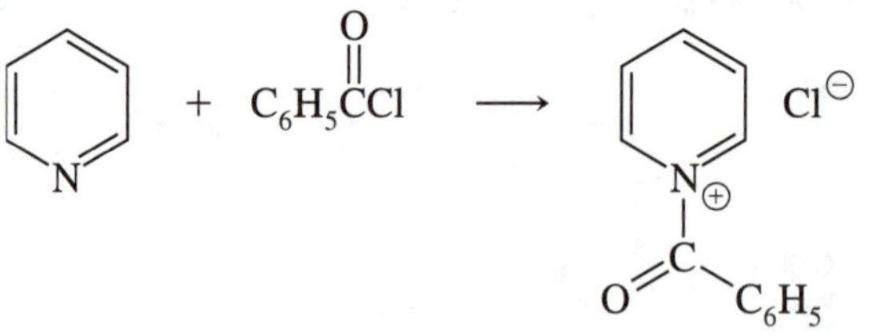

n-benzoylpyridinium chloride

(3)

N-acylpyridinium salts are rarely isolated, since they decompose rapidly on reaction with water, forming salts of the original pyridines. They are very useful acylating agents, which rapidly convert alcohols to esters (Eq. 4).

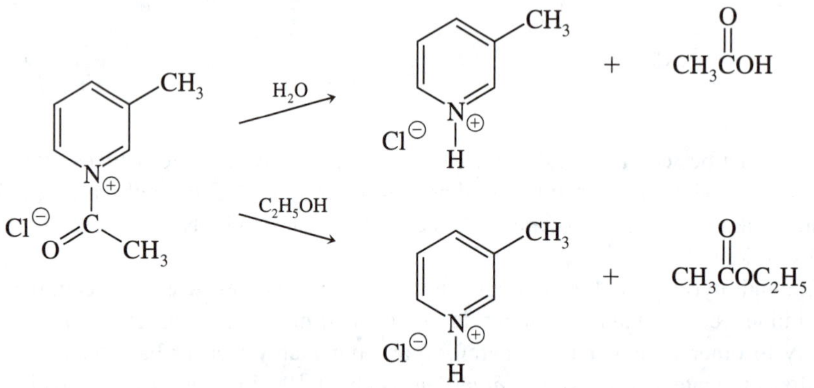

(4)

Derivatives of 4-aminopyridine, such as 4-dimethylaminopyridine (DMAP), are particularly effective catalysts for acylation reactions.[6a,7] Even very acid-sensitive alcohols, which are likely to undergo dehydration under most acylation conditions, can be esterified in high yields using DMAP catalysis (Eq. 5).

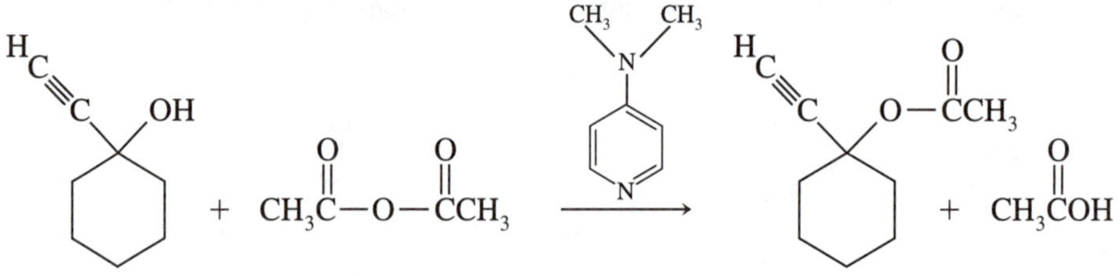

(5)

Mechanism:

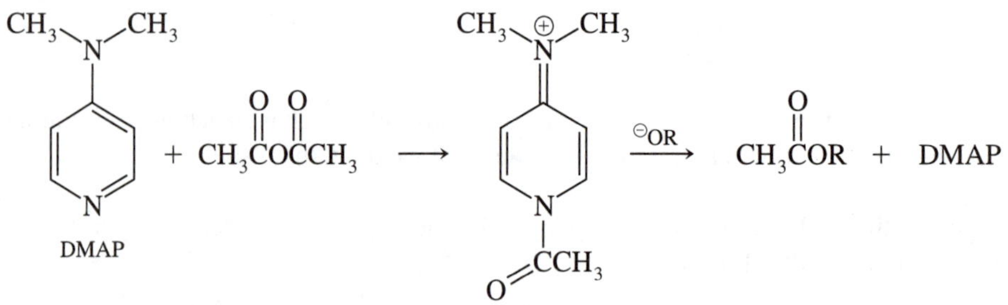

DMAP

(6)

(Reaction times are often dramatically shorter when DMAP is employed as a catalyst than when pyridine is used.)

The effectiveness of DMAP as an acylation catalyst is directly related to its high basicity (see Table 11.1). The use of such a basic catalyst minimizes the possibility of side reactions catalyzed by acidic pyridinium salts and increases the concentration of alkoxide ions that can react with the *N*-acylpyridinium intermediates. (Incidentally, unlike most *N*-acylpyridinium salts, *N*-acyl salts of DMAP are quite stable and can be easily isolated.)

Formation of *N*-oxides.   Pyridine and its analogs react with peracids, or with hydrogen peroxide in the presence of mineral acids, to form *N*-oxides (Eq. 7).

$$\text{(quinoline)} + CH_3\overset{O}{\overset{\|}{C}}-O-OH \longrightarrow \text{(quinoline N-oxide)} + CH_3\overset{O}{\overset{\|}{C}}OH \tag{7}$$

The *N*-oxides have permanent dipolar (ylide) structures and are best represented by formulas that show the charges in the dipoles. For convenience, however, the dipolar structures are often represented by arrows connecting the nitrogen and oxygen atoms.

Electrophilic Substitution Reactions

Deactivation to electrophilic attack.   The reactions of powerful electrophilic reagents with pyridines, as with other aromatic molecules, typically yield substitution, rather than addition, products. In comparison to aromatic hydrocarbons, most pyridine derivatives are highly deactivated toward electrophilic attack. Experimental evidence indicates that pyridine reacts only 10^{-18} to 10^{-20} times as rapidly as benzene in electrophilic substitution reactions.[8]

Since electrophilic substitution reactions are typically carried out in strongly acidic solutions, pyridines are usually nearly completely protonated under the conditions necessary for substitution to take place. Even when protic acids are not originally present, the nitrogen atoms would form strong complexes (e.g., **1** and **2**) with Lewis acid catalysts or with the powerful electrophilic reagents required for aromatic substitution reactions.

1   **2**

Since there are essentially no free pyridine molecules present during most electrophilic substitution reactions of pyridines, it is usually the pyridinium salts that undergo substitution, rather than the pyridine molecules themselves. This conclusion is supported by the fact that substitution reactions of *N*-alkylpyridinium salts are as rapid as those of the corresponding pyridines. For instance, nitration of *N*-methyl-2,4,6-trimethylpyridinium sulfate (Eq. 8) occurs at almost the same rate as nitration of 2,4,6-trimethylpyridine, strongly indicating that substitution is taking place on the cation in each case.[8]

$$HNO_3 + \text{[structure]} \quad HSO_4^{\ominus} \xrightarrow[\text{H}_2\text{SO}_4]{\text{fuming}} \text{[structure]} \quad HSO_4^{\ominus} \quad (8)$$

Since reactions of electrophilic agents with pyridinium cations or cationic complexes result in formation of very high energy, doubly charged intermediates such as **3** and **4**, it is not surprising that electrophilic substitution reactions of pyridines are so slow.

$$2\ HSO_4^{\ominus} \quad \text{[structure]} \qquad AlBr_4^{\ominus} \quad \text{[structure]}$$

3   **4**

However, the presence of the electronegative, $sp2$ hybridized, nitrogen atom in the ring appears by itself to be sufficient to cause significant deactivation of pyridine to attack by electrophilic reagents. Experimental evidence indicates that even if it were not protonated during the reactions, pyridine would be, at most, only 10^{-6} to 10^{-7} times as reactive to electrophilic reagents as is benzene.[9]

If attack of electrophiles were to take place at positions *ortho* or *para* to the nitrogen atoms of pyridines, the intermediate carbocations would each have one exceptionally poor resonance form (e.g., **5b**), in which both positive charges are located on the nitrogen atom.

$$\left[\text{[structure 5a]} \longleftrightarrow \text{[structure 5b]} \longleftrightarrow \text{[structure 5c]} \right] \quad (9)$$

5a   **5b**   **5c**

Therefore, electrophilic substitution reactions of pyridines, when they do occur, normally take place at the *meta* positions of the rings (C3 and C5), via intermediates such as **3** and **4**.

Nitration, sulfonation, and halogenation. Pyridine and its analogs are so unreactive toward electrophilic reagents that they do not undergo Friedel–Crafts alkylation reactions at all. Furthermore, they do not readily undergo nitration, sulfonation, or halogenation, even under the severe conditions at which benzene undergoes these substitution reactions.

Pyridines *can* be nitrated, sulfonated, and halogenated at their *meta* positions under exceptionally strenuous conditions, but the yields are usually poor. For instance, nitration of pyridine requires heating at 300°C with nitric acid in fuming sulfuric acid and gives only a 15% yield of 3-nitropyridine (Eq. 10).[10]

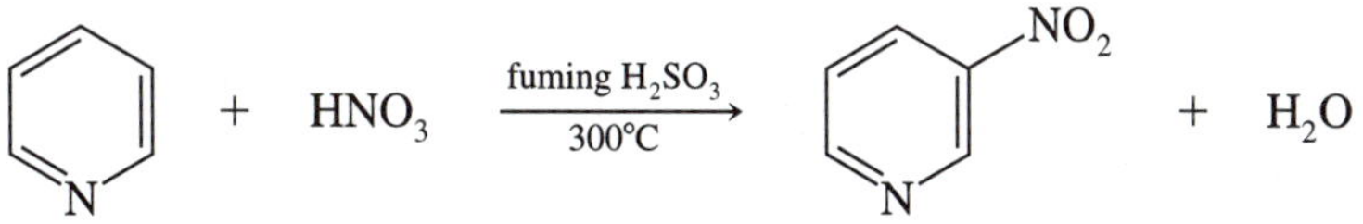

$$\text{pyridine} + HNO_3 \xrightarrow[\text{300°C}]{\text{fuming } H_2SO_3} \text{3-nitropyridine} + H_2O \qquad (10)$$

Remarkably, after more than a century of frustration, it was discovered that pyridine can be rapidly converted to 3-nitropyridine in high yields, at temperatures as low as $-20°C$, by reacting N_2O_5 with aqueous pyridine solutions containing sulfur dioxide or sodium bisulfite.[11a] The reaction appears to proceed by addition of a bisulfite anion to C2 of the N-nitropyridinum salt,* followed by a [1,5] migration of the nitro group to C3 and elimination of sulfurous acid (Eq. 11).[11b]

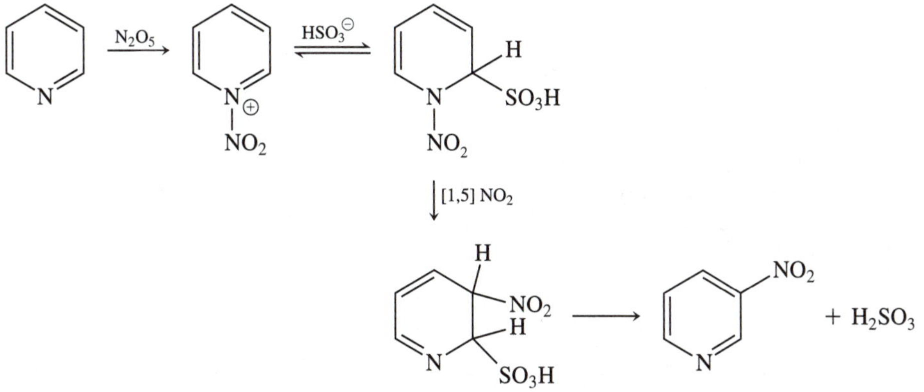

Halogenation reactions of pyridines also require unusually high temperatures and proceed in low yields of one mole or less of a Lewis acid catalyst is used per mole of halogen.[12a] However, bromination and chlorination can take place at moderate temperatures (80–115°C) if 3 to 4 moles of aluminum halides are employed for each mole of halogen (Eq. 12.) (This is called the *swamping catalyst effect*.) Even with such large amounts of catalysts, however, the yields of 3-halogenated pyridines are not outstanding.[12b]

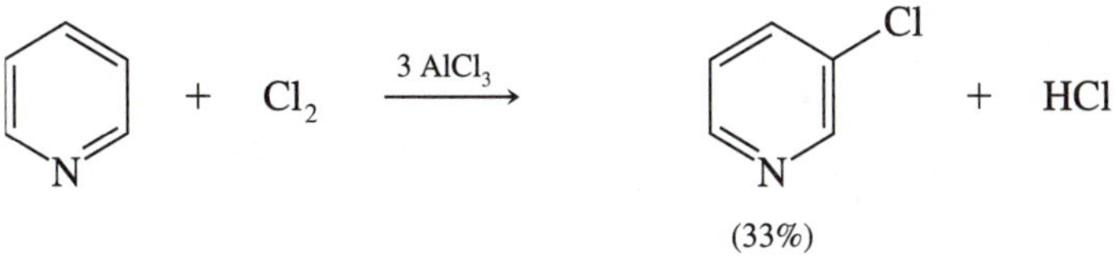

$$\text{pyridine} + Cl_2 \xrightarrow{3 \text{ AlCl}_3} \text{3-chloropyridine} + HCl \qquad (12)$$

(33%)

It has been suggested that excess aluminum halide is required to convert the halogen to a reagent capable of reacting with the aromatic ring because the first mole of aluminum halide forms a complex with the pyridine.

Bromination of pyridines in the gas phase at 300–350°C results in the formation of *meta*-bromopyridines. However, if the temperature is allowed to rise to approximately 500°C, bromine is introduced at *ortho* positions rather than *meta* positions of the pyridine rings (Eq. 13a).[13]

---

*Bisulfite adds to both C2 and C4 of the pyridinum salt. However, since the intermediate formed from additional at C4 cannot undergo the [1,5] NO_2 shift it can only lose bisulfite to reform the salt.[11b]

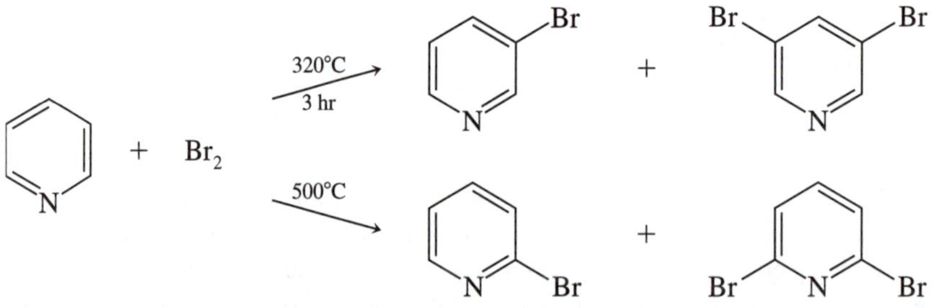

$$(13a)$$

Apparently, at such high temperatures, a free-radical mechanism (Eq. 13b) replaces the ionic mechanism.

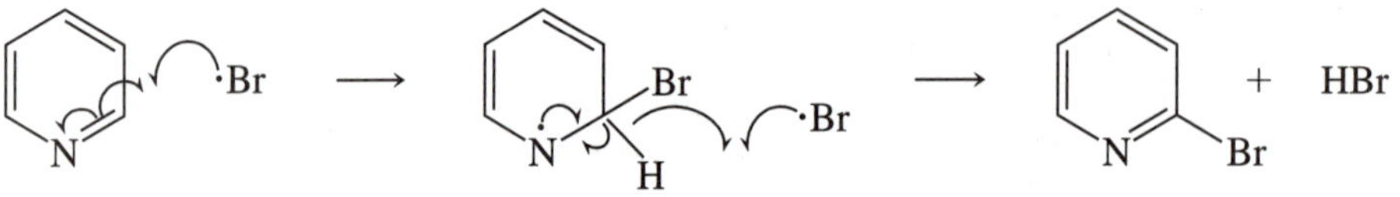

$$(13b)$$

As might be expected, electron-donating substituents on pyridine rings increase the rates of electrophilic substitution as well as the yields of substitution products. 2,4,6-Trimethylpyridine, for instance, can be converted to its 3-nitro derivative in 94% yield at only 100°C (Eq. 14).[14]

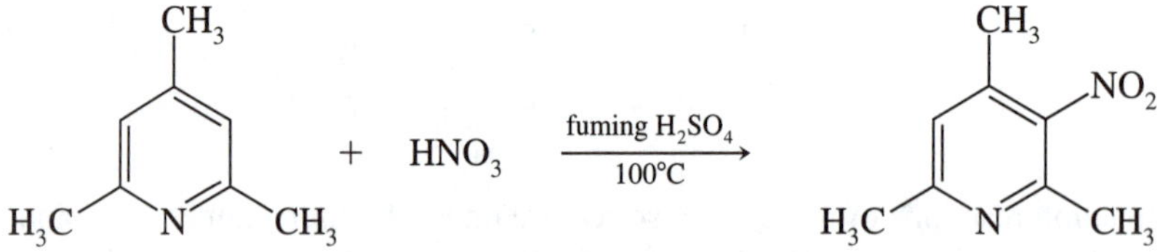

$$(14)$$

Attack on free pyridines.   In a few exceptional cases, it does appear that neutral pyridine molecules, rather than their protonated forms, undergo electrophilic attack. For instance, as shown in Eq. 15, 2,6-di-*tert*-butylpyridine is converted to its 3-sulfonic acid by reaction with sulfur trioxide in sulfur dioxide solution at −10°C.[15]

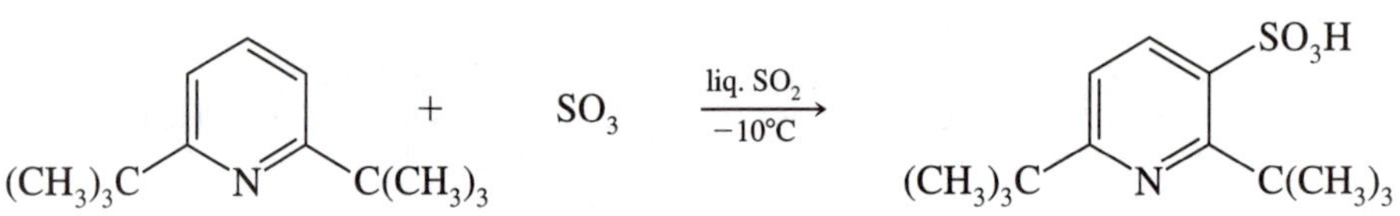

$$(15)$$

Apparently, the bulky *tert*-butyl groups prevent (or minimize) complex formation between the nitrogen atom and electrophiles, so that the pyridine is at least partially present in its relatively reactive neutral form.

Surprisingly, 2,6-dichloropyridine undergoes nitration under much milder conditions than those necessary for nitration of pyridine itself (Eq. 16),[8] even though the electron-withdrawing chlorine atoms would be expected to decrease the reactivity of the pyridine ring.

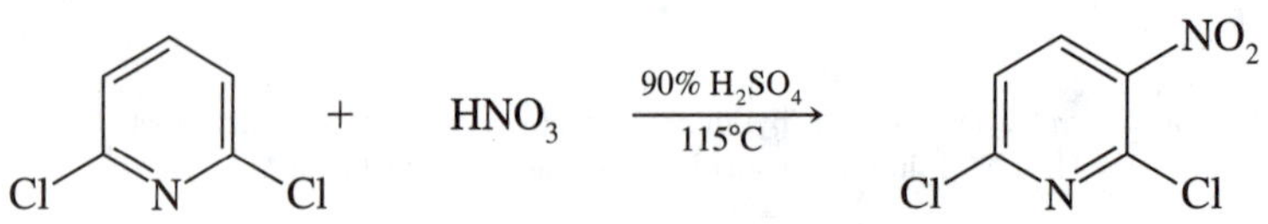

$$(16)$$

However, 2,6-dichloropyridine is a much weaker base than pyridine (see Table 11.1). Apparently, enough unprotonated 2,6-dichloropyridine remains in solution to react relatively rapidly with electrophilic reagents. Other very weakly basic pyridine derivatives with substituents at C2 and C6 also appear to undergo electrophilic substitution in their unprotonated forms.[9]

Reactions of quinolines and isoquinolines. Electrophilic substitution reactions of quinoline take place much more rapidly than similar reactions of pyridine. However, the electrophiles normally attack the nitrogen-free rings of quinolines. As with naphthalene, substitution at an α carbon (C5 or C8), as in Eq. 17, is usually more rapid than at a β carbon.

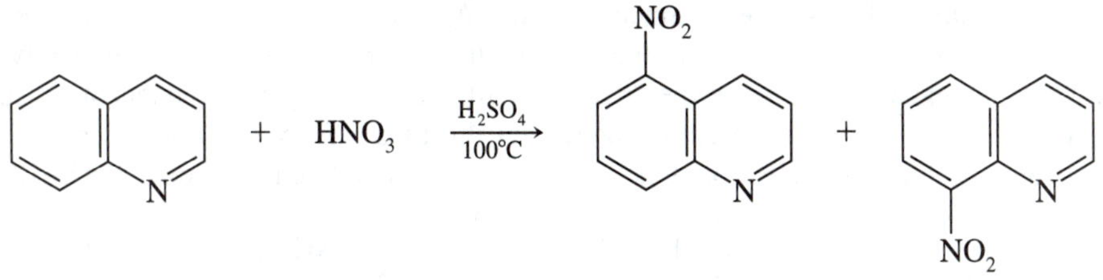

$$\text{(17a)}$$

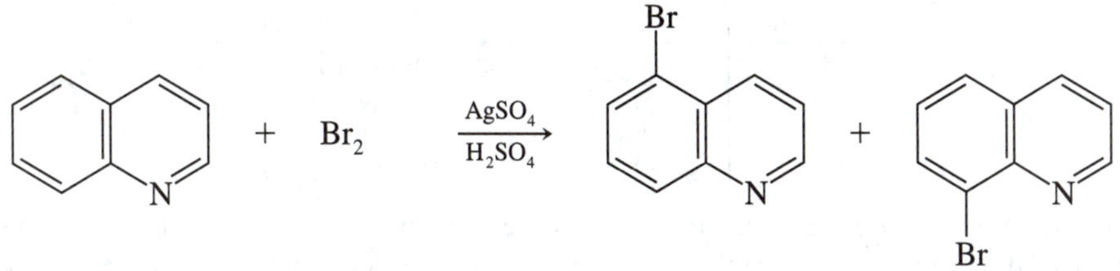

$$\text{(17b)}$$

Electrophilic attack at α carbons yields carbocations (such as **6**) for which two resonance structures can be drawn without disrupting the aromaticity of the pyridine rings. Attack at either β position would yield carbocations (such as **7**) with only one such resonance form.

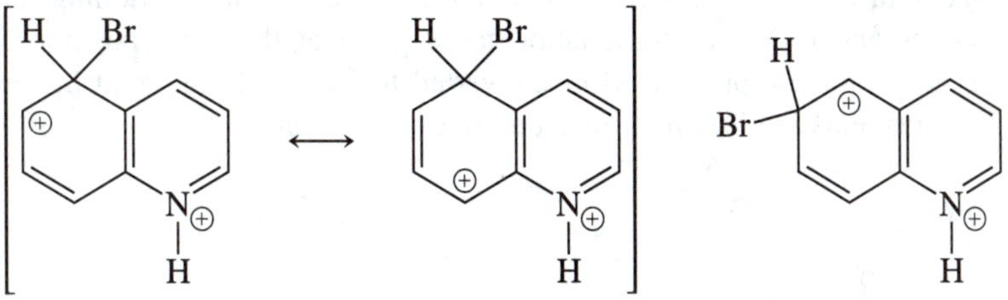

$$\text{(18)}$$

Electrophilic attack on an isoquinoline ring usually yields predominantly substitution at C5, with only small amounts of C8 substitution products (Eq. 19).[16]

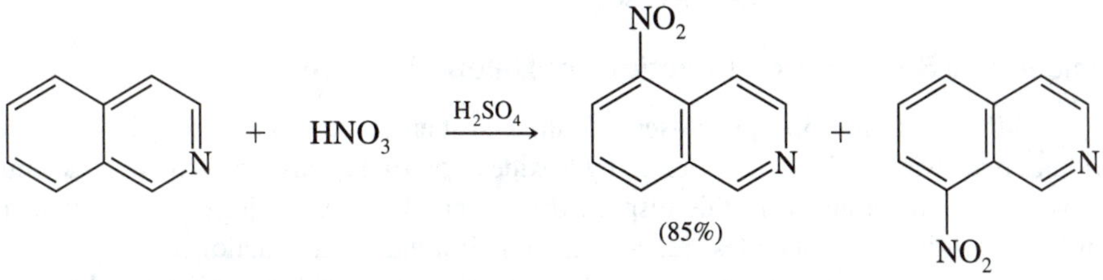

$$\text{(19)}$$

The preference for attack at C5 can be accounted for on the basis that one resonance form (**8**) of the intermediate dication from attack at C8 would have both positive charges located on the nitrogen atom, while all resonance structures from the carbocation formed by attack at C5 would have positive charges on separate atoms.

8

Reactions of N-oxides. Since pyridine N-oxides have positively charged nitrogen atoms, it might be expected that, like pyridinium salts, they would be unreactive toward electrophilic reagents and that, if they reacted at all, they would yield products of substitution at *meta* positions. However, electrophilic attack at C4 (or C2) of a pyridine N-oxide would yield a carbocation stabilized by electron donation from the negatively charged oxygen (Eq. 20).

$$(20)$$

The formation of a nitrogen–oxygen π bond to replace the carbon–carbon π bond broken by the attack of the electrophile more than compensates for the inductive effect of the positive charge on nitrogen. As a result, pyridine N-oxide reacts with nitric acid in sulfuric acid at temperatures as low as 90°C, and yields principally 4-nitropyridine N-oxide.[17] (If desired, the, N-oxide can later be reduced to 4-nitropyridine.)

In contrast to its nitration, sulfonation of pyridine N-oxide by fuming sulfuric acid requires temperatures above 200°C. Sulfonation takes place at the *meta* position.[18] Presumably, the N-oxide is completely protonated or converted to its O-sulfonic acid in fuming sulfuric acid (Eq. 21), thus reducing its ability to act as an electron donor.

$$(21)$$

$(X = H \text{ or } SO_3H)$

Nucleophilic Substitution Reactions of Halopyridine Rings

Addition–elimination processes. Unlike most aromatic halides, many halogen-substituted pyridines react with nucleophiles such as hydroxide or alkoxide ions (or even aliphatic amines[19]) at relatively low temperatures. In this respect, they resemble other conjugated, unsaturated imines. However, pyridines are much less reactive than their nonaromatic analogs.

2-Halo- and 4-halopyridines, as well as 2-halo- and 4-haloquinolines and isoquinolines, undergo nucleophilic attack in strong bases, such as solutions of sodium methoxide, to yield substitution products (Eq. 22).[20]

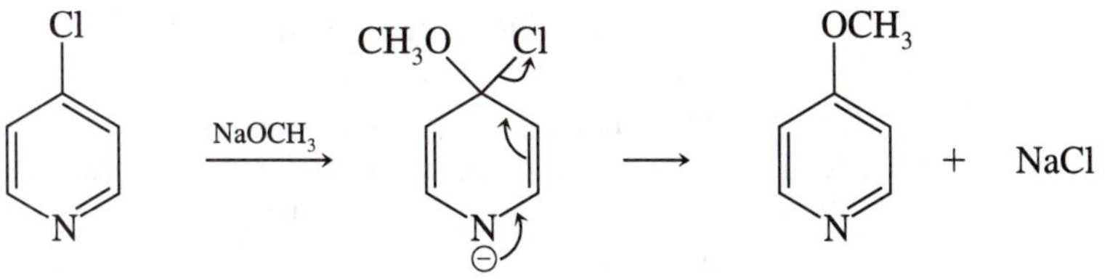

$$+ \quad NaCl \qquad (22a)$$

$$+ \quad NaBr \qquad (22b)$$

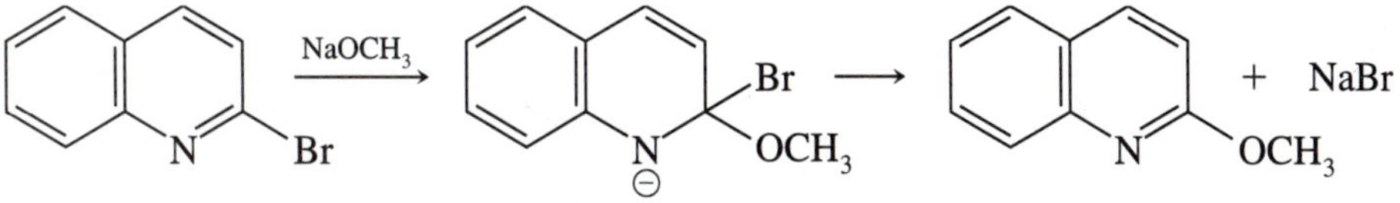

These reactions are considered to proceed by two-step (*addition–elimination*) mechanisms, as shown in Eq. 22, rather than by direct substitution reactions.

Since additions of nucleophiles to pyridine rings disrupt the aromatic systems, the first (addition) steps in the reactions are normally the rate-limiting steps. Not surprisingly, therefore, reactions of nucleophiles with *N*-alkylpyridinium salts of 2-halo- and 4-halopyridines (Eq. 23) occur billions of times as rapidly as reactions of the neutral pyridines.[21]

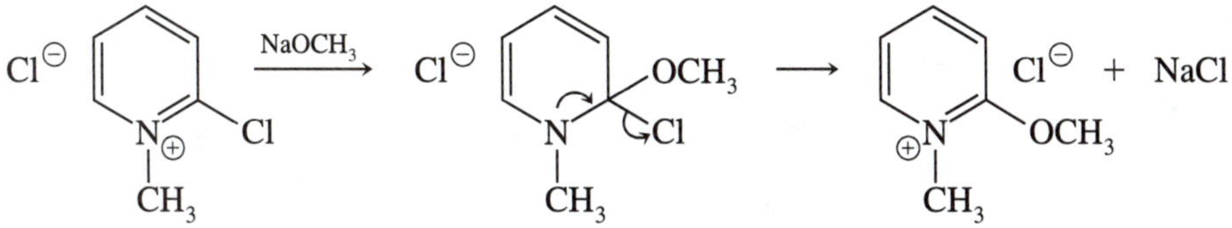

$$+ \quad NaCl \qquad (23)$$

In general, 4-halopyridines are somewhat more reactive toward nucleophilic attack than are 2-halopyridines. 4-Bromopyridine, for instance, is so reactive that it spontaneously dimerizes to form a quaternary salt on standing at 0°C (Eq. 24).

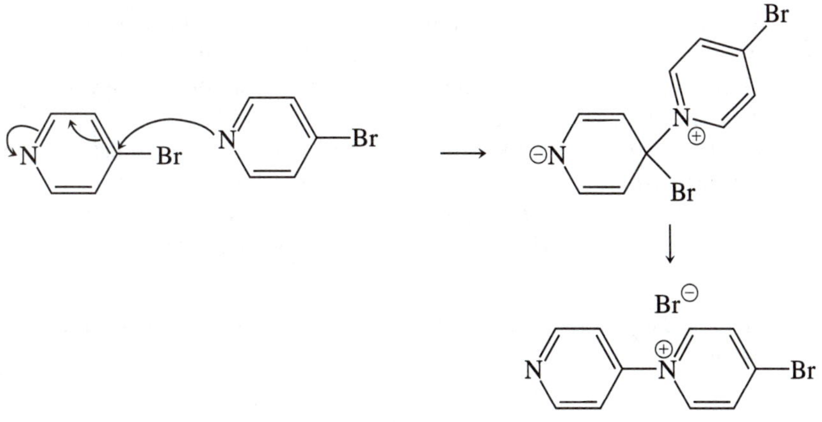

$$(24)$$

In contrast, displacements occur more rapidly at C2 than at C4 of *N*-alkylpyridinium and *N*-alkylquinolinium salts.[22]

Elimination–addition (pyridyne) processes. The addition of a nucleophile to C3 of a pyridine derivative cannot yield an intermediate in which the negative charge is distributed onto the nitrogen atom. As a result, 3-halopyridines are only about 10^{-4} to 10^{-5} times as reactive towards nucleophiles as are 2-halo- and 4-halopyridines.[21]

In general, 3-halopyridines do not react with hydroxide or alkoxide solutions at temperatures below about 150°C. However, they do react with much stronger bases, such as solutions of sodium amide in liquid ammonia. These reactions yield mixtures of 3-aminopyridines and 4-aminopyridines in approximately 1:2 ratios (Eq. 25).[23]

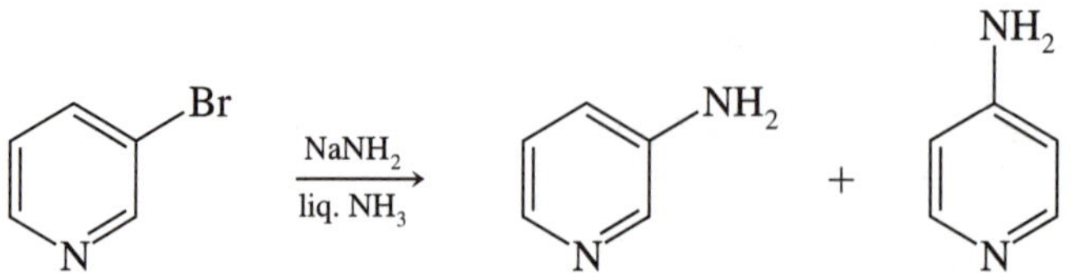

(25)

The formation of 4-aminopyridines cannot be accounted for by either direct substitution or addition–elimination mechanisms. Instead, analogy with the formation of benzyne intermediates from halobenzenes suggests that reactions of 3-halopyridines with very strong bases proceed by initial elimination reactions, forming *3,4-pyridynes*. The pyridynes (also called dehydropyridines) then add ammonia to form the observed products (Eq. 26).

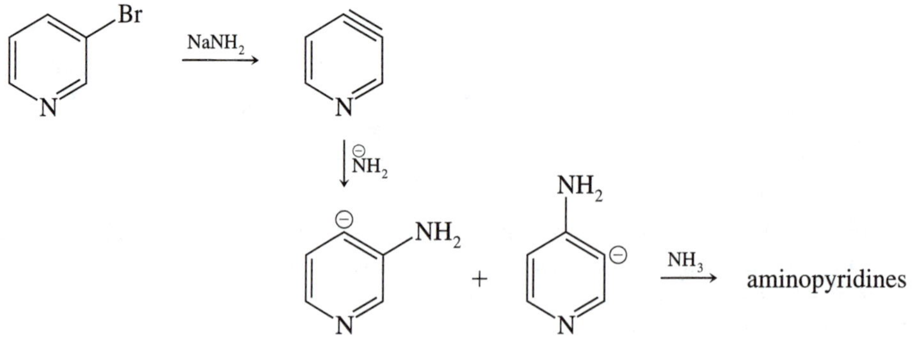

(26)

Although 2,3-pyridyne might, in principle, have been formed as well as 3,4-pyridyne, no 2-aminopyridine, which might have been formed from 2,3-pyridyne, is obtained from the reaction of 3-bromopyridine with sodium amide. Molecular orbital calculations indicate that 2,3-pyridyne should be appreciably less stable than 3,4-pyridyne.[24]

Interestingly, such weak nucleophiles as water and fluoride displace nitrogen very rapidly from 2-pyridinediazonium salts—reactions that presumably proceed via carbocation **9**. The ease of formation of this cation suggests resonance contribution from the 1,2-pyridyne form (Eq. 27)[25a]— a formulation which is supported by theoretical calculations.[25b]*

---

*When formed in the mass spectrometer, the 2-pyridinium cation can apparently undergo Diels–Alder-like reactions with butadienes.[25c]

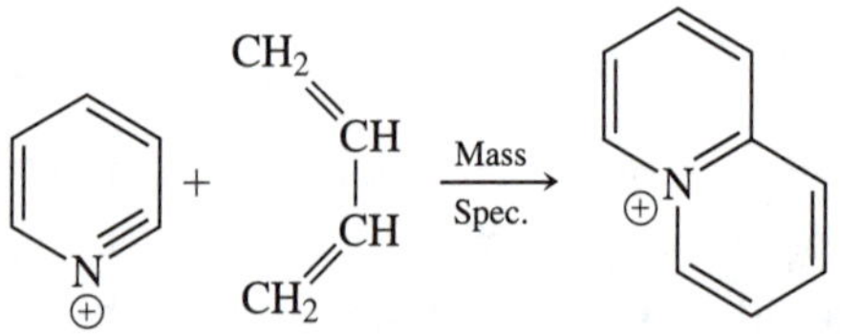

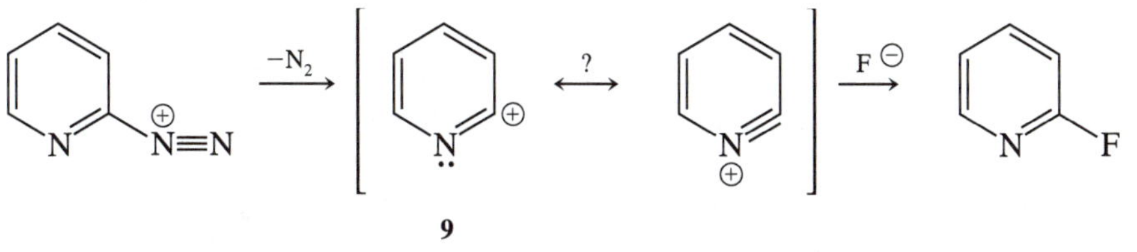

$$(27)$$

9

Hydroxypyridine–Pyridone Tautomerism

Reactions of 2-halopyridines or 4-halopyridines with solutions containing hydroxide ions yield anions of 2- or 4-hydroxypyridines. On protonation, either the hydroxypyridines or their tautomers, 2-pyridone and 4-pyridone, might be formed (Eqs. 28 and 29).

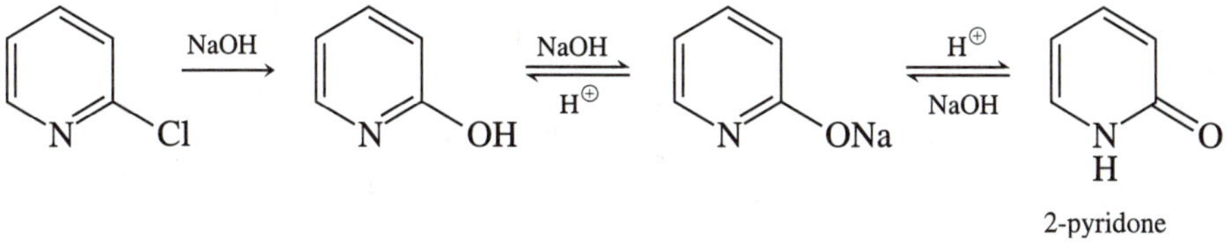

$$(28)$$

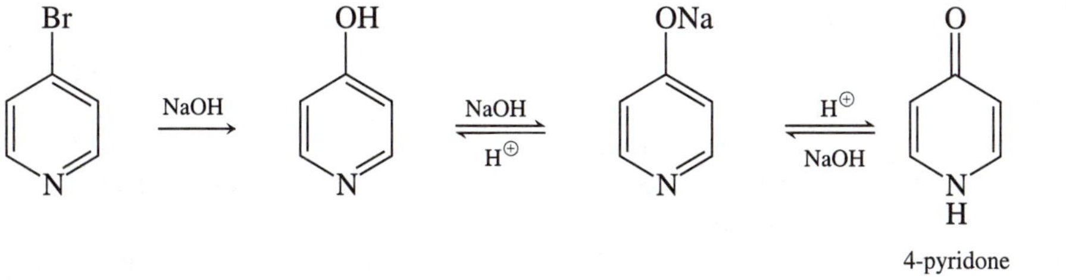

2-pyridone

$$(29)$$

4-pyridone

Since equilibration of the tautomeric forms should be rapid in base, the relative amounts of hydroxypyridines and pyridones obtained would depend on the stabilities of the two types of tautomers.

The position of the equilibrium between hydroxypyridine and pyridone tautomers was debated for many years, with the hydroxypyridine structures supported by the fact that many reactions, such as acetylation with acetic anhydride, result in substitution on oxygen to give derivatives of hydroxypyridines (Eq. 30).

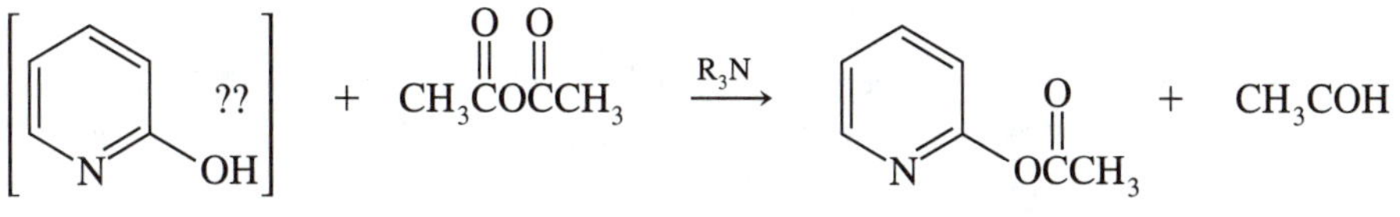

$$(30)$$

However, the UV spectrum of "2-hydroxypyridine" was found to closely resemble that of N-methyl-2-pyridone rather than that of 2-methoxypyridine.

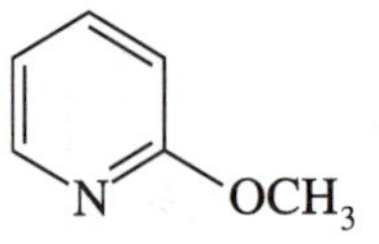

2-methoxypyridine 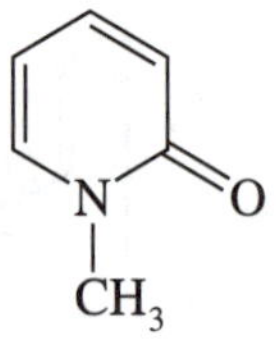   N-methyl-2-pyridone

In fact, in aqueous solutions the 2-pyridone tautomer typically predominates over the 2-hydroxypyridine tautomer by a ratio of about 1000:1. Similarly, "4-hydroxypyridines" in solution usually exist primarily as their 4-pyridone tautomers.[26]

These facts may seem surprising, because hydroxypyridines are aromatic molecules, and pyridones may not, at first glance, appear to be aromatic. However, both 2-pyridone and 4-pyridone have dipolar resonance forms (**10b** and **11b**) that do have aromatic character and contribute greatly to the stability of the pyridone tautomers.

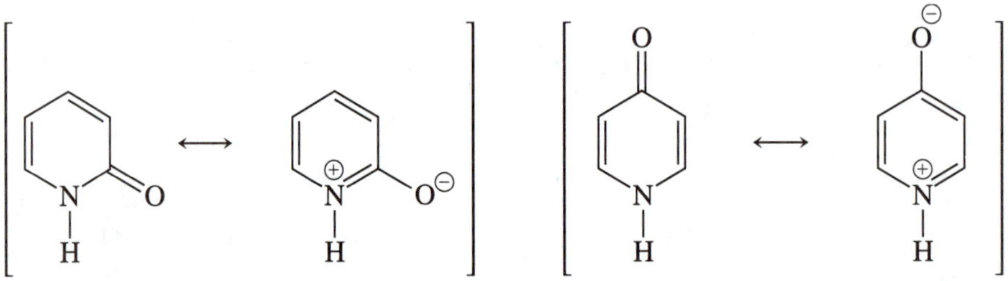

(31)

10a   **10b**   **11a**   **11b**

Dipolar resonance structures should be less important for the hydroxypyridine forms, because the electronegative oxygen atoms would have to bear the positive charges and the nitrogen atoms the negative charges in those resonance forms.

While we have been considering hydroxypyridine and pyridone tautomers as individual molecules, it is important to note that 2-pyridones exist largely as hydrogen-bonded dimers (e.g., **12**) in concentrated solutions, and both 2-pyridones and 4-pyridones can form strong hydrogen bonds to polar solvents.

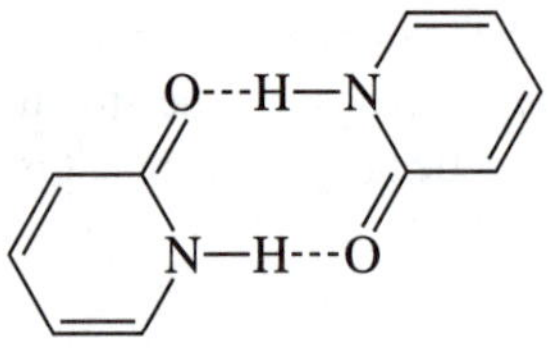

12

Hydrogen-bonding and solvation energies play major roles in determining the relative stabilities of pyridone and hydroxypyridine tautomers in solutions. In very dilute solutions in nonpolar solvents or in the gas phase, in which these compounds exist as individual, weakly solvated (or unsolvated) molecules that are not strongly hydrogen bonded, the hydroxypyridine tautomers are actually more stable than the pyridone forms![27]

In pyridines substituted with sulfur atoms at C2 or C4, the thione forms predominate over the thiol tautomers.

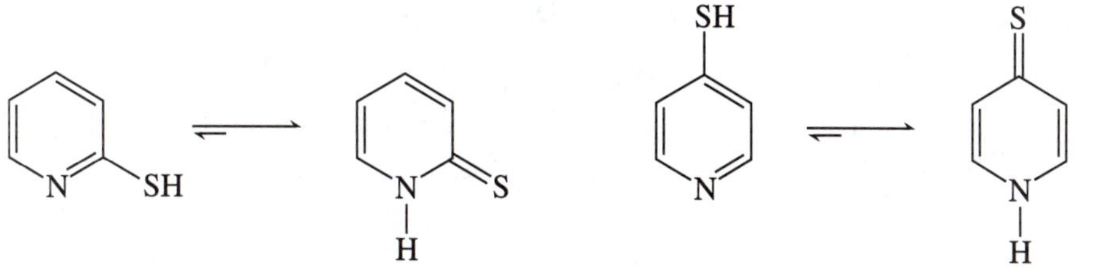

$$(32)$$

principal tautomer   principal tautomer

However, 2-aminopyridine and 4-aminopyridine are more stable than their tautomers, as shown in Eq. 33.

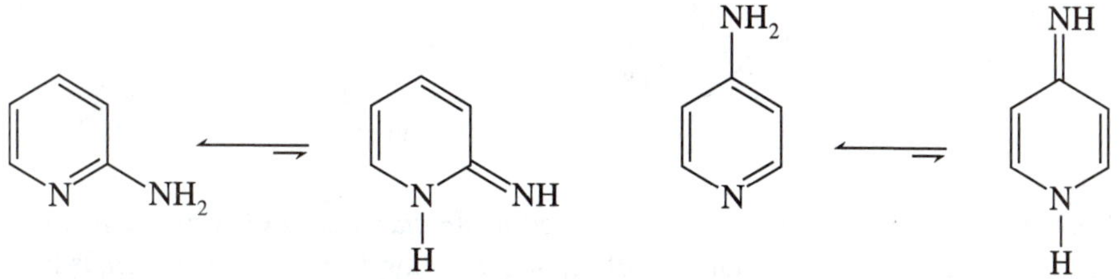

$$(33)$$

principal tautomer   principal tautomer

Dipolar resonance forms for the tautomers with exocyclic carbon–nitrogen double bonds would contribute less to stabilizing those forms than similar resonance forms contribute to 2- and 4-pyridones, because negative charges would be placed on nitrogen atoms rather than oxygen atoms.

"2-Hydroxyquinoline" and "4-hydroxyquinoline" exist predominantly as their amide tautomers, as do the hydroxy derivatives of pyrazine[28] and pyrimidine.[26]

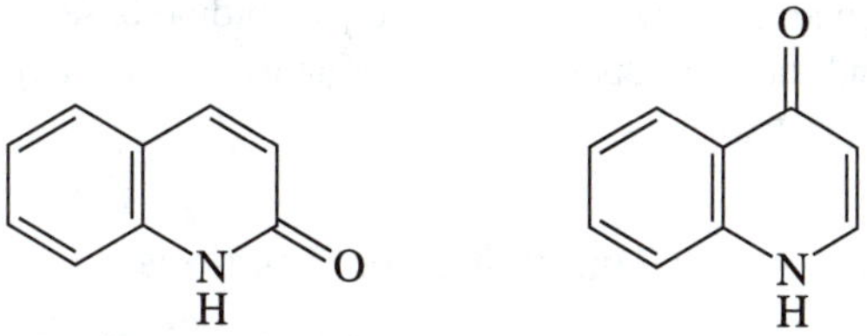

principal tautomers of hydroxyquinolines

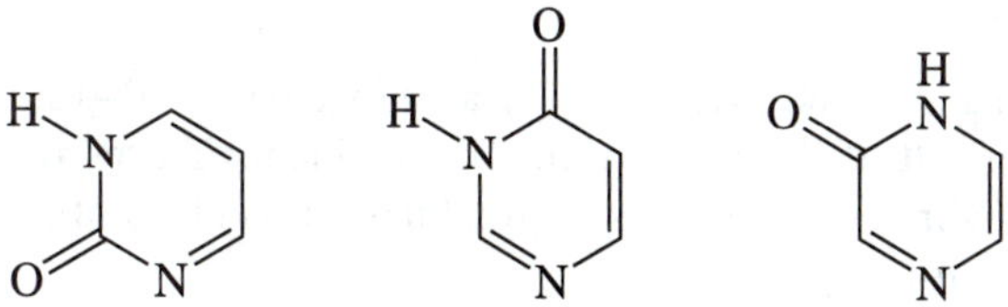

principal tautomers

The latter fact has particular biological significance, because the *pyrimidines* cytosine and thymine, as well as the *purines* guanine and adenine, form critical parts of deoxyribonucleic acid (DNA), the material of which genes are composed. (Uracil replaces thymine in ribonucleic acid, RNA, the "messenger" molecule that carries genetic instructions from DNA to the rest of the cell.)

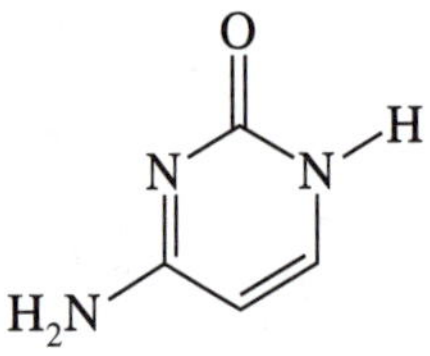

cytosine 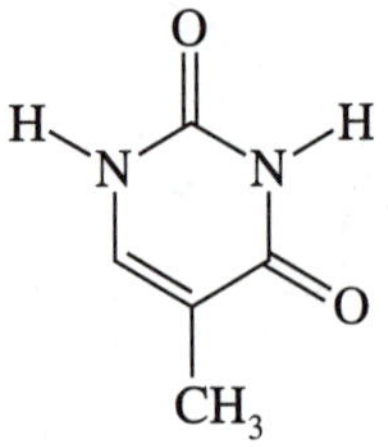   thymine 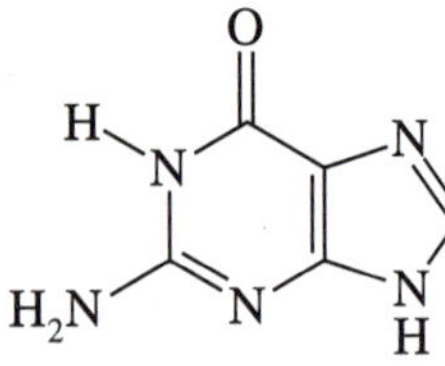   guanine

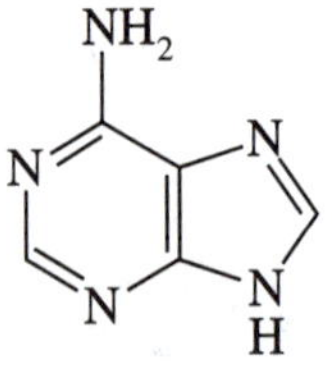

adenine 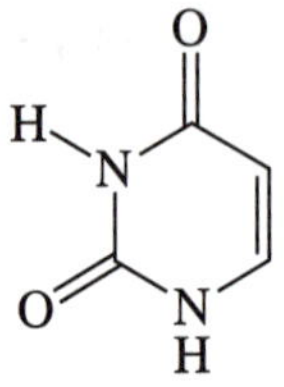   uracil

Both RNA and DNA are polymers of units called *nucleotides,* each of which contains a phosphoric acid molecule linked to a molecule of the sugar deoxyribose, which, in turn, is linked to a purine or pyrimidine.

In his fascinating book *The Double Helix,*[29] biologist James Watson described how he and Francis Crick encountered repeated frustration in attempting to construct a reasonable structure for DNA by putting together models of DNA components. Fortunately, a visiting chemist pointed out that they were using models of the wrong tautomers of the purines and pyrimidines in DNA. Watson and Crick replaced their hydroxypyrimidine models with models of the pyrimidone tautomers, and shortly thereafter they were able to propose a reasonable structure for DNA—a proposal that won them the Nobel Prize.

Hydrogen bonding between purine and pyrimidine bases plays a critical role in the replication of DNA and the reproduction of all organisms. The subject is discussed in detail in all modern biochemistry texts.[30]

Nucleophilic Addition to Unsubstituted Positions

The Chichibabin reaction. Nucleophilic substitution reactions of pyridine derivatives are most common when good leaving groups, such as halide ions, can be displaced. However, even hydride ions can act as leaving groups in reactions of very strong bases with pyridine and its derivatives.[31]

If sodium amide and pyridine are heated together at 70–100°C, hydrogen gas is evolved and the sodium salt of 2-aminopyridine is obtained. (Traces of 4-aminopyridine have also been detected.) Similarly, reactions of pyridines with anions of primary amines yield 2-alkylaminopyridines (Eq. 34).

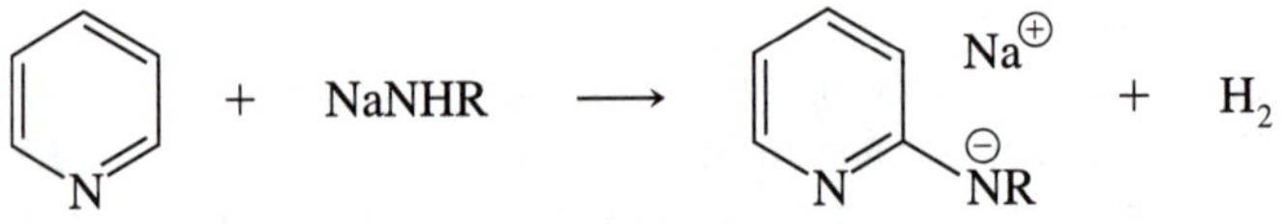

$$\text{pyridine} + \text{NaNHR} \longrightarrow \text{product} + H_2 \qquad (34)$$

(R = H or alkyl)

These reactions, which are called *Chichibabin* (or *Tschitschibabin*) *reactions*,[32] are initiated by the addition of amide anions to the imino function of the pyridine ring to form adducts, such as **13**.

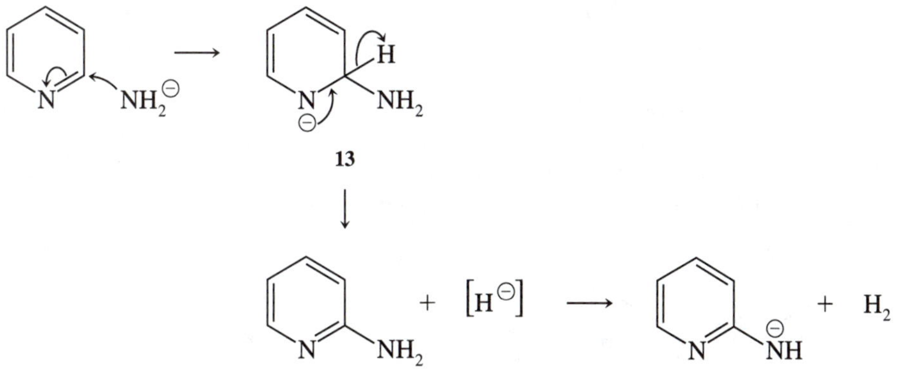

13

(35)

These adducts can be detected spectroscopically. The formation of the adduct is followed by expulsion of a hydride ion (or by reaction of the incipient hydride ion with a proton source in the solution) to form hydrogen gas.

Organolithium reagents also add to pyridines. Since the intermediates formed by the addition of lithium reagents are less basic than those formed by reactions of pyridines with sodium amide (or potassium amide), the ejection of hydride anions is relatively difficult. However, if the reaction mixtures are heated to 100–140°C, lithium hydride precipitates from the solutions, and 2-alkylpyridines or 2-arylpyridines are formed (Eq. 36).[33]

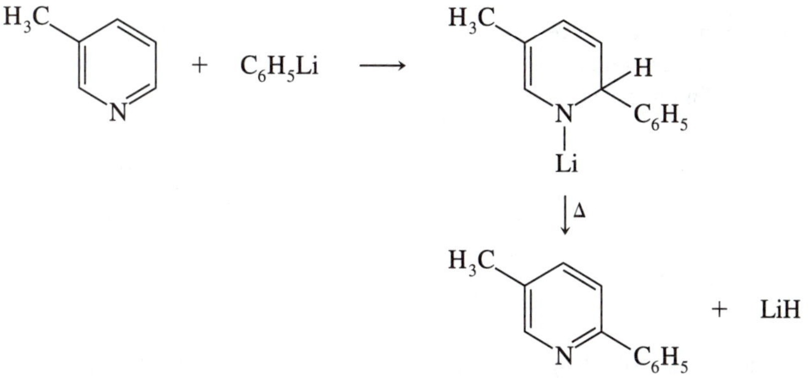

(36)

Reactions requiring oxidation steps.   The reactions of bases with six-membered rings containing two nitrogen atoms are rapid compared to their reactions with pyridines. However, the resulting heterocyclic anions are so stable that hydride ions will not be ejected. It requires the addition of oxidizing agents, such as potassium permanganate, to convert the anions to neutral aromatic molecules, as shown in Eqs. 37 and 38.[34]

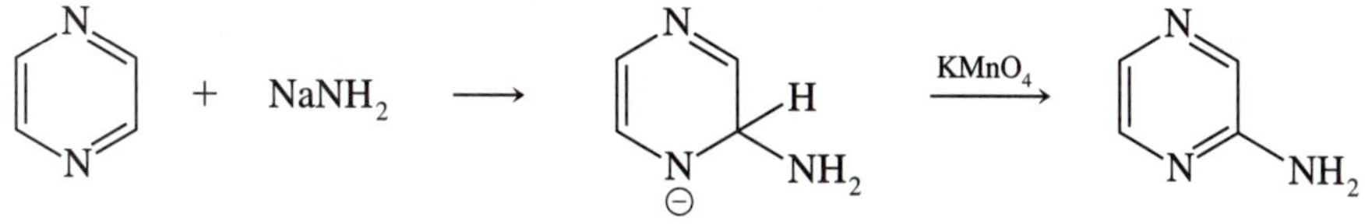

(37)

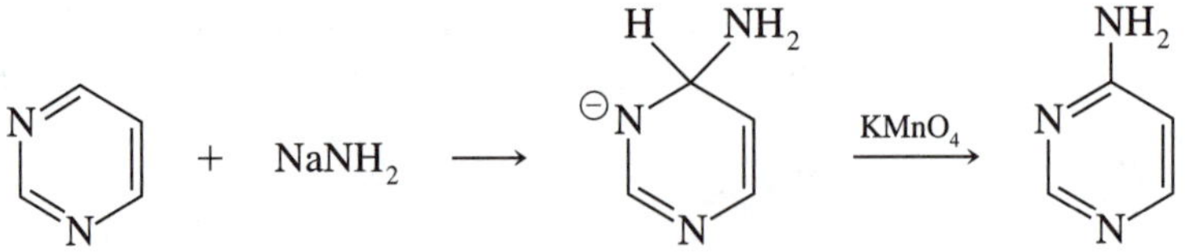

(38)

Amide anions also add to quinoline more rapidly than to pyridine, and the reactions can be carried out at temperatures as low as −65°C. Addition to C2 of quinoline is normally more rapid than to C4. However, the reaction is reversible, and if the temperature is raised to 15–20°C, the anion from addition at C4 becomes the principal product (Eq. 39).

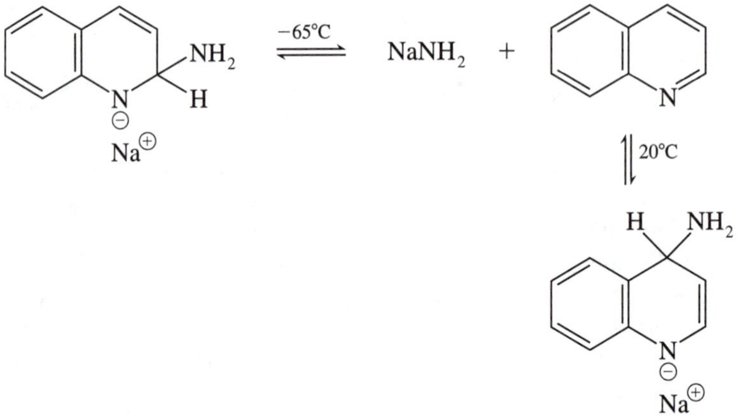

(39)

Organolithium reagents react with *N*-alkylpyridinium salts to yield products from addition at C2 and C4, with addition at C2 predominating.[35] Oxidation will convert the products back to substituted pyridinium salts (Eq. 40).

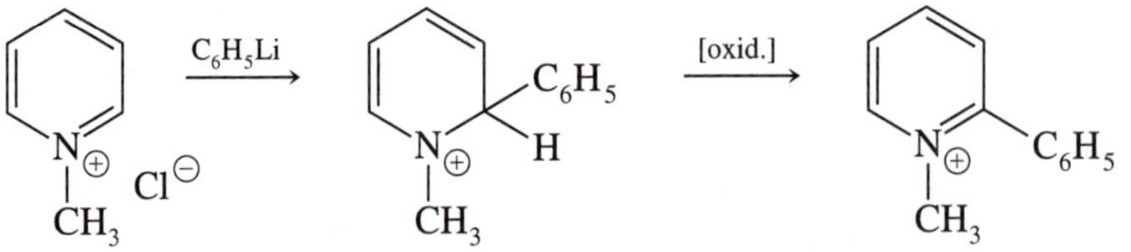

(40)

Formation of dihydropyridines. A variety of reducing agents, including lithium aluminum hydride and sodium borohydride, reduce *N*-alkylpyridinium salts to 1,2- and 1,4-dihydropyridines, as shown in Eqs. 41a and 41b.[35] The dihydropyridines, in turn, are easily oxidized back to pyridinium salts.[36]

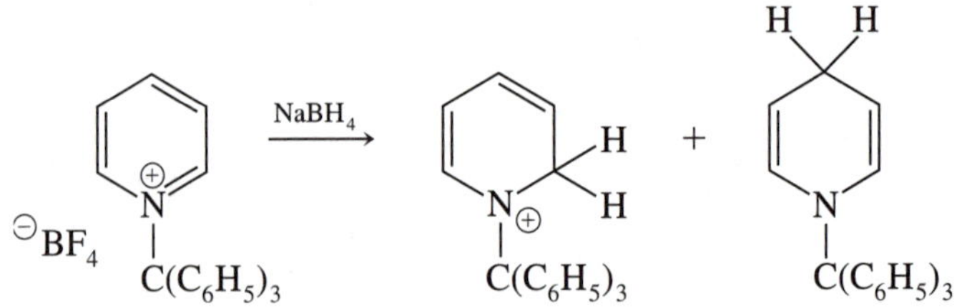

(41a)

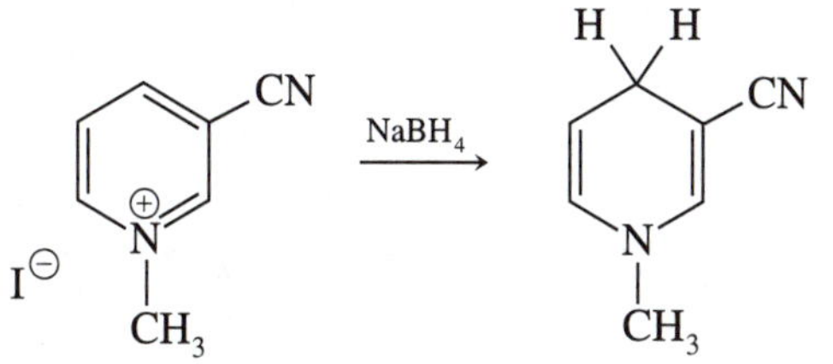

(41)

The reduction reactions of pyridinium salts are of particular interest because of their relation to important biological processes. A dihydropyridine ring in the coenzyme nicotinamide adenine dinucleotide (*NADH)* serves as the source of hydrogen in enzyme-catalyzed biological reductions of carbonyl compounds to alcohols (Eq. 42), while the resulting quaternary pyridinium salt, NAD$^+$, is employed in biological oxidations of alcohols to carbonyl compounds.

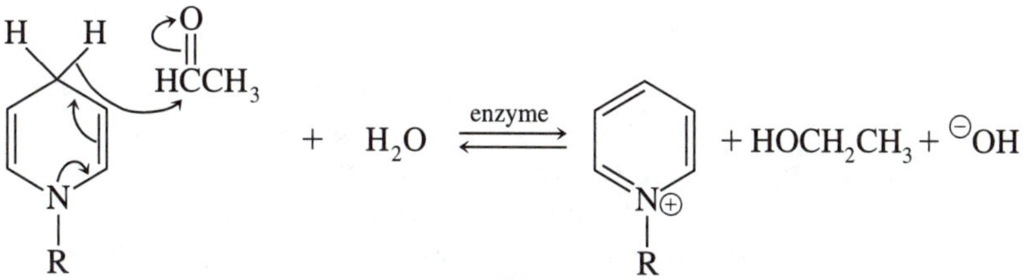

(42)

Substitutions Involving Ring Opening and Reclosure

Substitution by ANRORC mechanisms.   In the 1970s, a Belgian group led by Henk van der Plas found that some apparently straightforward substitution reactions in heterocycles containing two or more nitrogen atoms actually proceed by rather complex mechanisms.[37]

Reaction of 4-phenyl-6-bromopyrimidine with potassium amide in liquid ammonia yielded, as expected, the 6-amino derivative. However, when the starting material was labeled with ^{15}N (the isotopic nitrogen being equally distributed between the two nitrogens in the ring), 50% of the ^{15}N in the product was located in the exocyclic amino group (Eq. 43a)!

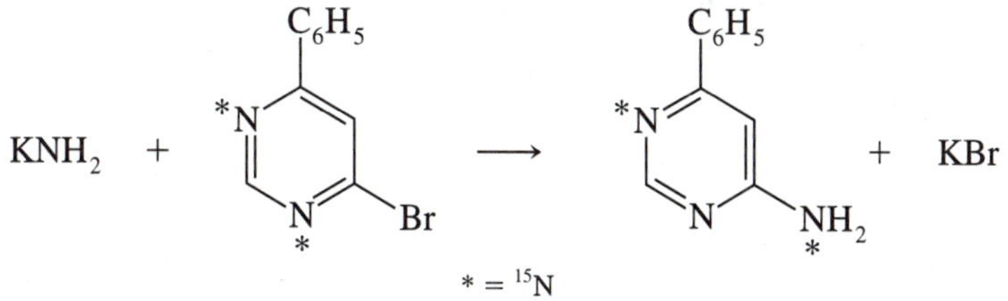

(43a)

$* = ^{15}$N

To explain this result, it was postulated that the amide anion originally added to C2, and that the resulting anion then underwent electrocyclic ring opening, ejection of the bromide anion, and ring closure to yield the 6-aminopyrimidine, as shown in Eq. 43b.

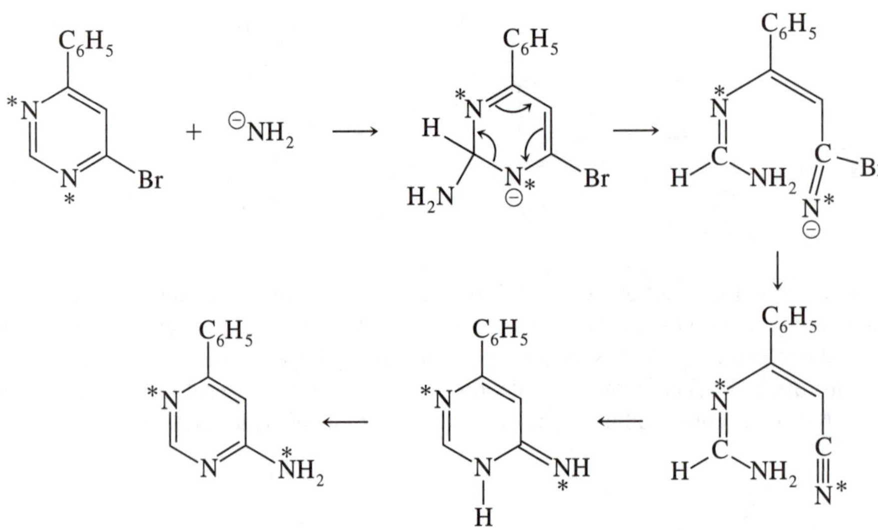

(43b)

(The exact stages at which proton transfers occur is not certain.) The mechanism was designated an ANRORC process: addition of nucleophile, ring opening, ring closure.[37a]

Evidence for the ANRORC mechanism for the reaction in Eq. 43a was provided by the observation that reaction of 4-phenyl-6-bromopyrimidine with the lithium salt of a secondary amine yielded an open-chain compound (Eq. 44), because ring closure would not form a stable product.

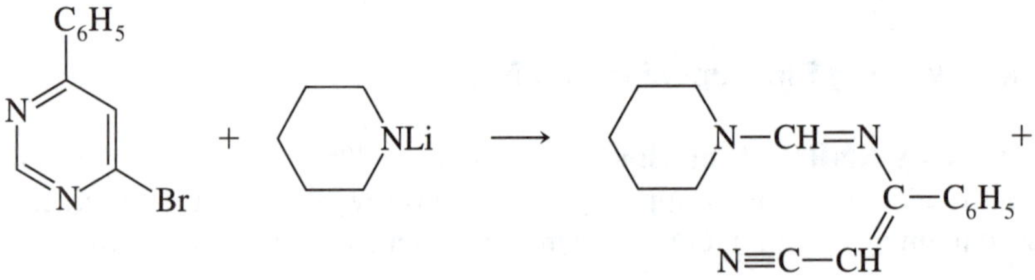

(44)

Even 2-halopyrimidines undergo substitution primarily by ANRORC mechanisms.[37] Again, the initial step is addition of a base to an unsubstituted position (see Eq. 45).

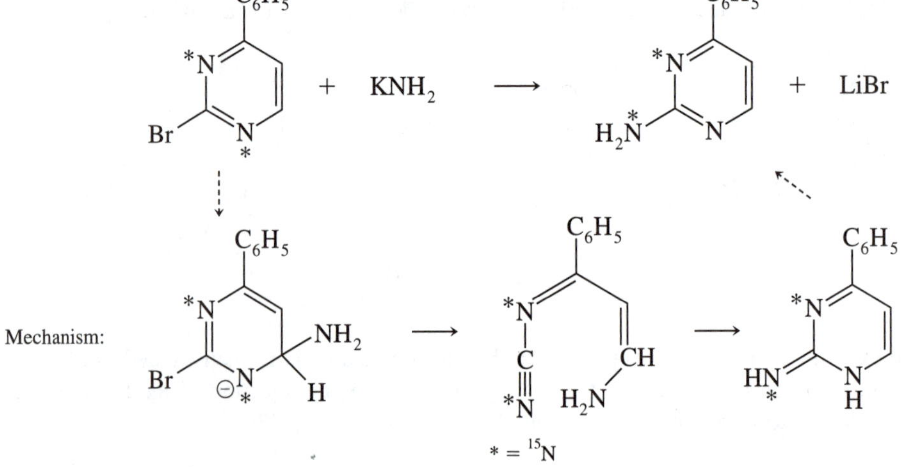

(45)

Mechanism:

$* = {}^{15}N$

Reactions of 5-halopyrimidines introduce a new complication: substitution with rearrangement. The amino group in the product appears at C6 rather than at C5. Isotopic labeling experiments indicate that the products are formed by two different types of reactions that proceed at approximately the same rate. In one type of reaction, addition of the amide anion occurs at C6, and a hydrogen halide is eliminated in a subsequent step (Eq. 46).

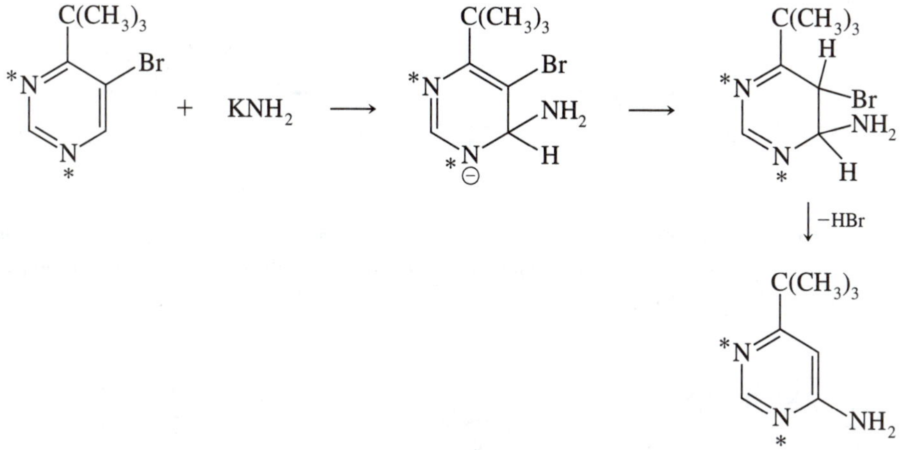

(46)

The second type of reaction proceeds by an ANRORC mechanism (Eq. 47).[37a]

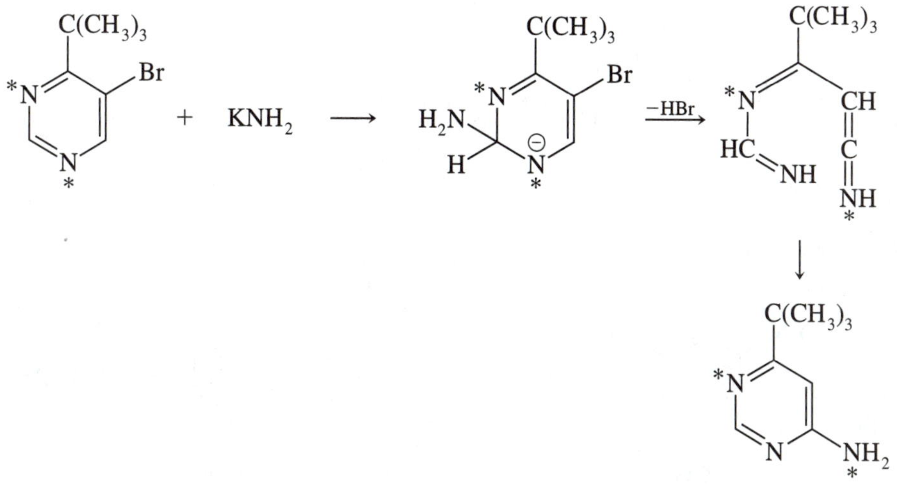

(47)

Similarly, substitution reactions in pyrazines and pyridazines usually proceed by ANRORC mechanisms.[37] Even pyridine derivatives may undergo substitution by ANRORC mechanisms if the rings bear strongly electron-withdrawing groups that can stabilize the ring-opened anions. The open-chain intermediate **14**, for example, was detected spectroscopically in the substitution reaction shown in Eq. 48.[38]

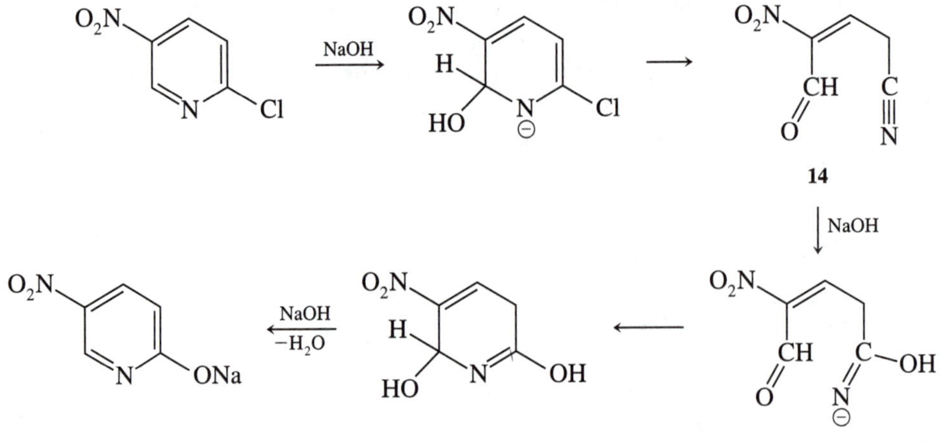

$$\text{(48)}$$

Substitution for heteroatoms in the ring. The reaction of an amine with a pyridinium salt bearing strongly electron-withdrawing substituents on the nitrogen or carbons of the ring can result in the exchange of the ring nitrogen atom, as shown in Eqs. 49 and 50.[38]

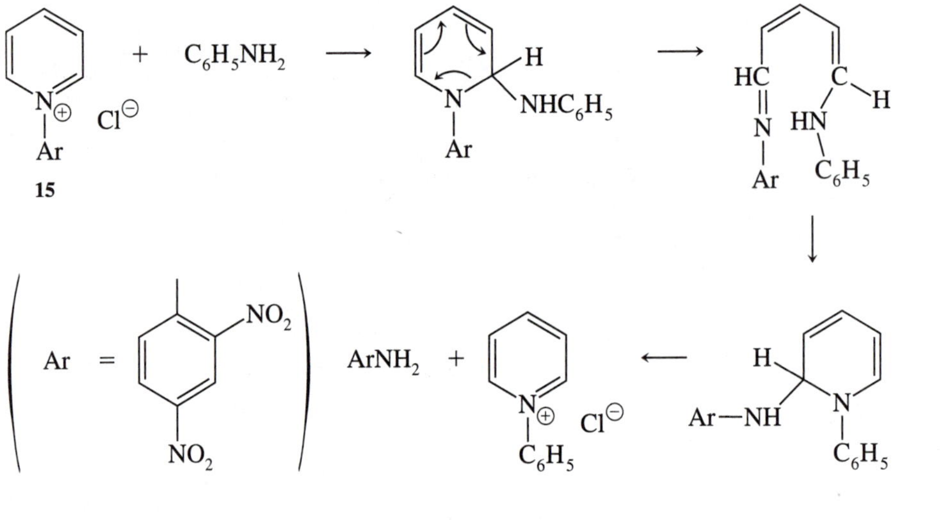

$$\text{(49)}$$

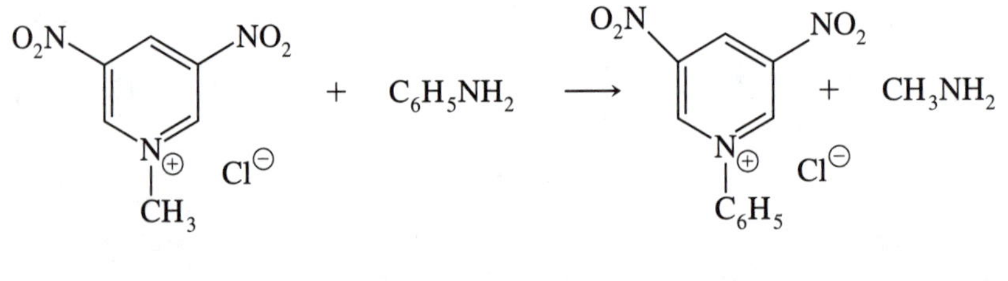

$$\text{(50)}$$

Substitution of aromatic amines for alkylated ring nitrogen atoms is a useful method for synthesizing N-arylpyridinium salts, which, with the exception of salts such as **15** (which can be prepared by reactions of pyridines with 2,4-dinitrohalobenzenes, as in Eq. 51), are not otherwise available.

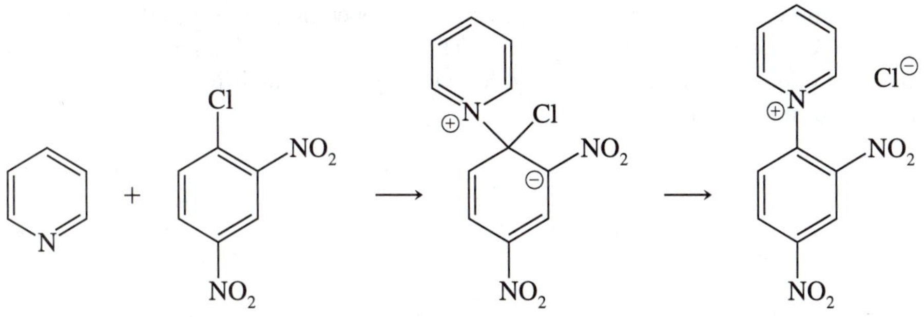

$$(51)$$

15

11.3 SYNTHESIS OF SIX-MEMBERED HETEROCYCLIC RINGS

By Carbonyl Addition Reactions

Formation of pyridine rings. The most common procedures for forming pyridine rings involve reactions of ammonia with dicarbonyl derivatives. For instance, the reaction of ammonium acetate with the unsaturated dialdehyde **16** yields pyridine (Eq. 52).[39]

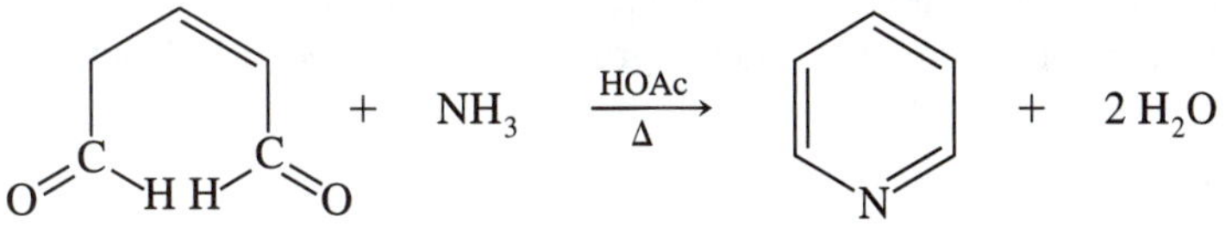

$$(52)$$

16

Pyridines can also be formed by condensing 1,5-dicarbonyl compounds with hydroxylamine (Eq. 53).[40]

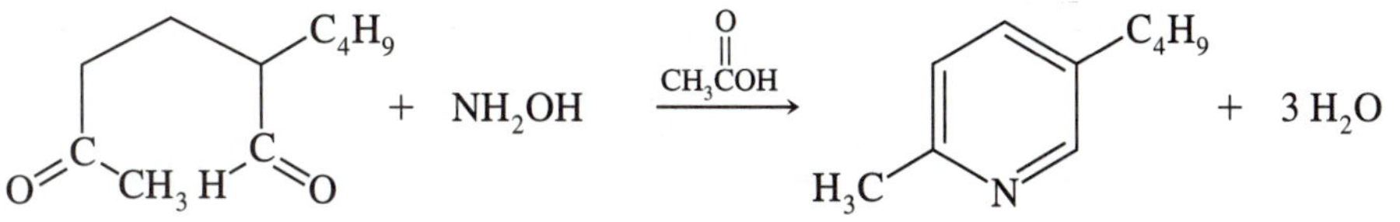

$$(53)$$

Unsaturated 1,5-dicarbonyl compounds with *cis*-configurations are generally difficult to prepare, however, and hydroxylamine is a relatively expensive starting material. Thus, a more common procedure for the formation of pyridine rings is to condense ammonia with 1,5-dicarbonyl compounds (or compounds that are readily converted to 1,5-dicarbonyl compounds) in the presence of mild acid catalysts (Eq. 54).

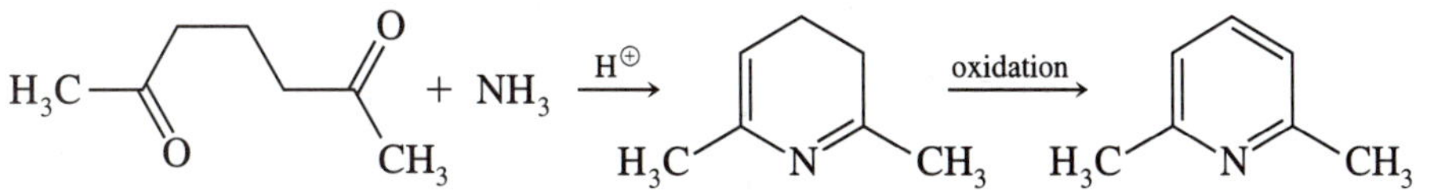

$$(54)$$

The resulting dihydropyridine derivatives are easily oxidized to pyridines by a wide variety of oxidizing agents, such as air, iodine, or chromic acid. Indeed, it is frequently difficult to isolate the dihydropyridines, and pyridines are often obtained directly from the reaction mixtures.

(Reaction of ammonia with cyclohexenes at temperatures above 200°C forms pyridine rings, but unfortunately in rather low yields.[41] This interesting reaction presumably proceeds by conjugate addition of ammonia to the ring, followed by a retroaldol-like ring opening and recyclization to form the pyridine, as shown in Eq. 55.)

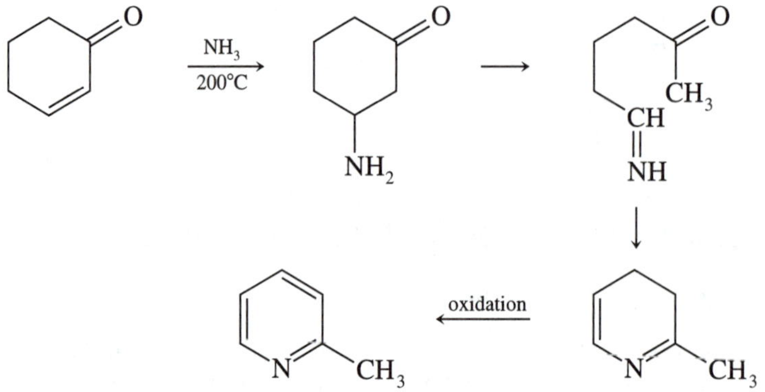

(55)

Among the most frequently used methods for preparing pyridines is the *Hantzsch synthesis*. The condensation of ammonia, an aldehyde, and a β-ketoester or β-diketone at about pH 4 results in the formation of 1,4-dihydropyridines, from which pyridines are obtained on oxidation (Eq. 56).

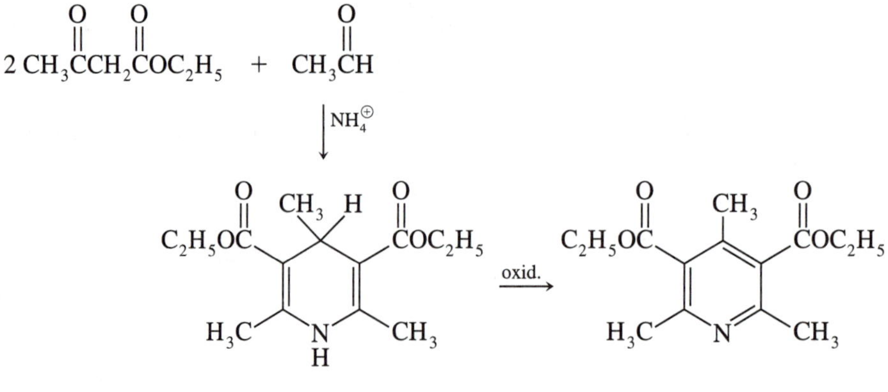

(56)

The mechanism of the Hantzsch synthesis appears to involve the formation of unsaturated ketones and enamines as intermediates. Compounds **17** and **18**, for instance, were isolated from a Hantzsch reaction of benzaldehyde, methyl acetoacetate, and ammonia, and were shown to condense to form a dihydropyridine (Eq. 57).[42]

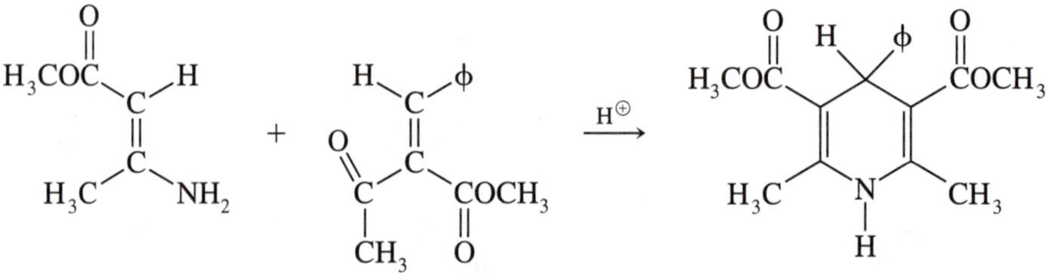

(57)

17   18

Synthesis of rings with two nitrogen atoms.   Pyrazines can be prepared by reactions of 1,2-dicarbonyl compounds with 1,2-diamines, followed by dehydrogenation (Eq. 58).[43]

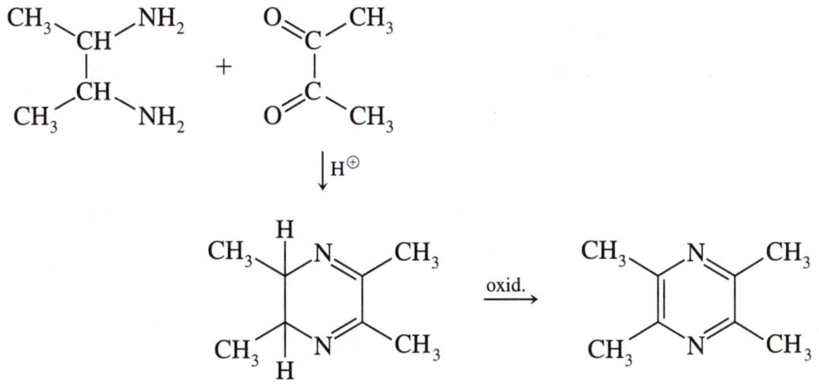

(58)

Pyridazines are similarly obtained by condensing hydrazine with 1,4-dicarbonyl compounds and dehydrogenating the resulting dihydropyridazines (Eq. 59).[44] Pyrimidines are obtained by condensing amidines (**19**, R = H, alkyl, or aryl) or guanidines (**19**, R = NH_2 or NR_2') with 1,3-dicarbonyl derivatives (Eq. 60).

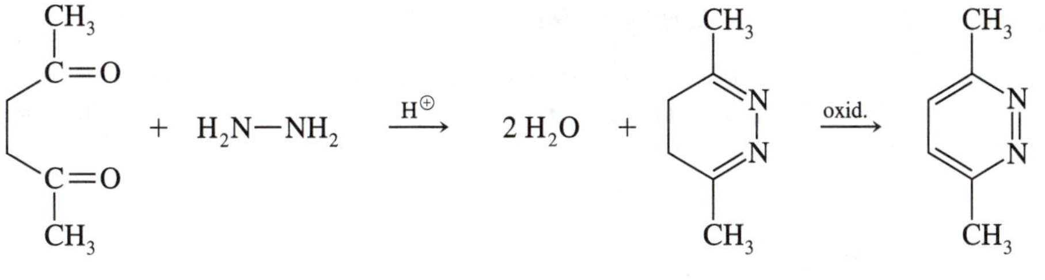

(59)

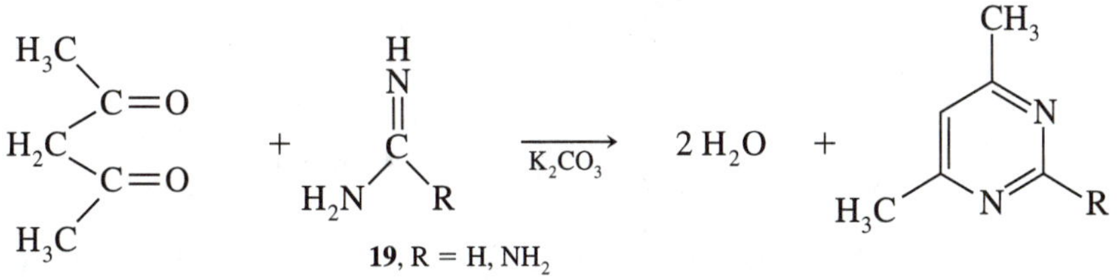

19, R = H, NH_2

(60)

The latter reaction is unusual because it requires basic conditions, while most reactions of ammonia or amines with carbonyl compounds require mild acid catalysts to assist in the loss of water from the initial addition products. Amidines and guanidines are strongly basic because of resonance stabilization of the protonated forms (see Eq. 61). Therefore, they may inhibit the protonation of

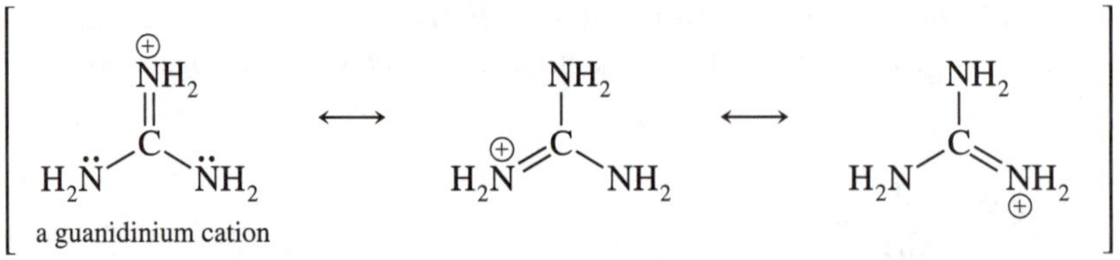

$$ (61) $$

a guanidinium cation

hydroxy groups necessary for acid-catalyzed dehydration of the initial addition products. In contrast, base catalysis of the dehydration steps (Eq. 62) would be assisted by resonance stabilization of *anions* of amidines or guanidines.

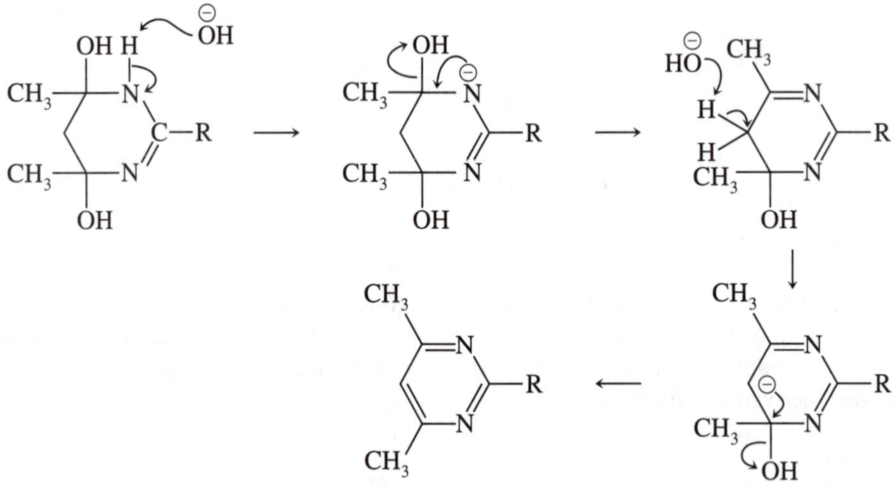

$$ (62) $$

Via Electrocyclic and Cycloaddition Reactions

Formation of pyridine rings by pericyclic reactions is much less common than by carbonyl addition reactions. However, imines of *ortho*-vinylanilines are converted to quinolines in good yields on heating at 155°C (Eq. 63).[45]

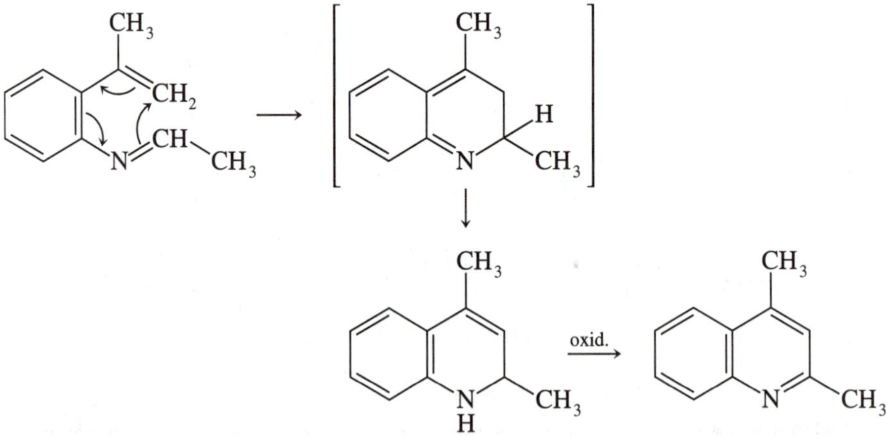

$$ (63) $$

oxid.

Cyano groups bearing strongly electron-withdrawing substituents can act as dienophiles in reactions with 1-alkoxybutadienes. Rapid loss of alcohols from the cycloaddition products forms pyridines in good yields (Eq. 64).[46]

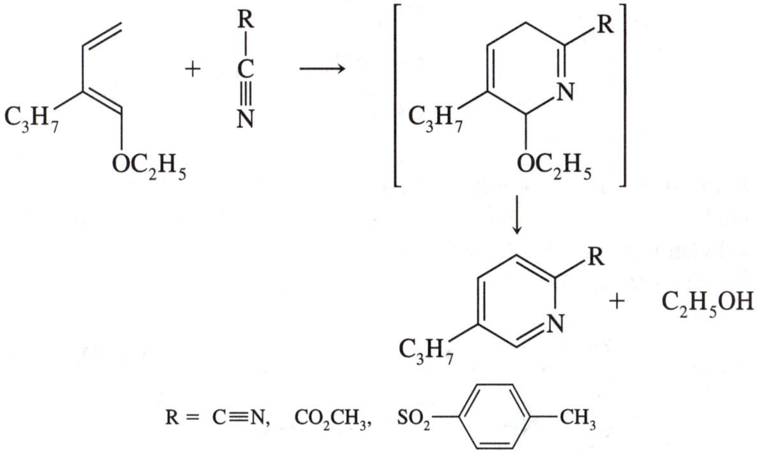

(64)

$$R = C{\equiv}N, \quad CO_2CH_3, \quad SO_2-\!\!\langle\ \rangle\!\!-CH_3$$

Synthesis of Quinolines and Isoquinolines

Preparation of quinolines.   Heterocyclic rings in quinolines and isoquinolines are usually prepared by electrophilic attack upon preexisting benzene rings.

The most common such method of preparing quinolines is by condensing conjugated carbonyl derivatives with anilines in the presence of a strong acid (the *Doebner–von Miller synthesis*).[47] This process most likely proceeds by conjugate addition of the amine to the unsaturated carbonyl compound, followed by electrophilic attack of the protonated carbonyl on the aromatic ring.[48] The resulting 1,2-dihydroquinoline is usually converted to a quinoline during the course of the reaction. Nitrobenzene, which is commonly employed as the solvent, can act as the oxidizing agent. (See Eq. 65.)

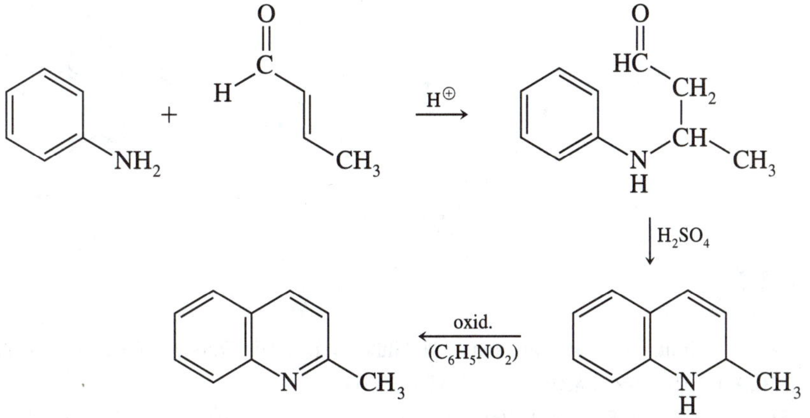

(65)

The Doebner–von Miller synthesis is frequently called the *Skraup reaction*, although, strictly speaking, the Skraup reaction is a version of the process in which glycerol is used in place of the conjugated carbonyl (Eq. 66).[49] Glycerol is converted to acrolein under the strong acid conditions used for the reaction.

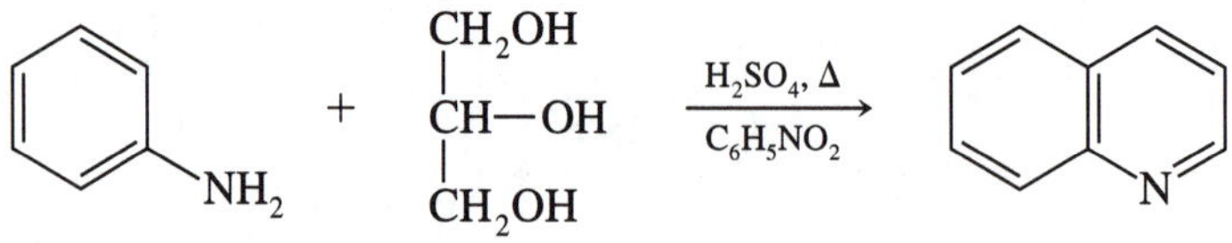

$$\text{(66)}$$

Formation of isoquinoline rings. In the Bischler-Napieralski reaction,[50] powerful dehydrating agents such as phosphorus pentoxide or phosphorus pentachloride convert N-2-phenylethylamides to 2,3-dihydroisoquinolines, which are dehydrogenated by heating with palladium on charcoal (Eq. 67).

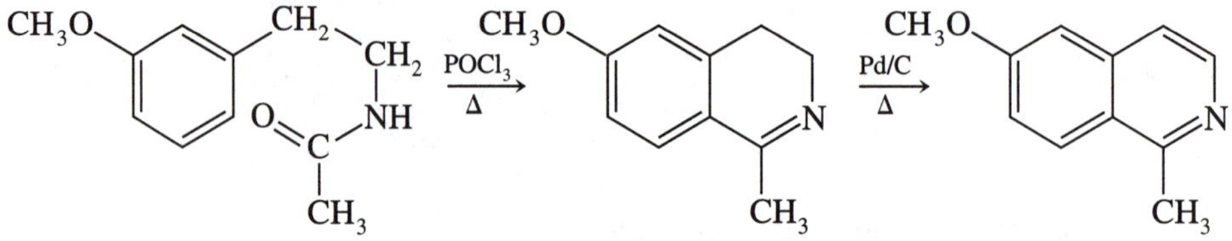

$$\text{(67)}$$

The Pictet–Spengler condensation is a closely related process in which aldehydes are condensed with 2-phenylethylamines in the presence of strong acids. The resulting tetrahydroisoquinolines can then be converted to isoquinolines by heating with palladium on charcoal (Eq. 68).

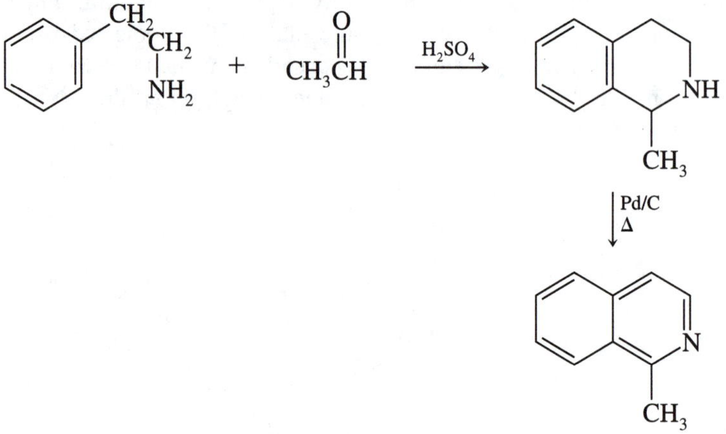

$$\text{(68)}$$

REFERENCES

[1] B.Y. Simkin, V.I. Minkin, and M.N. Glukhovstov, *Adv. Hetero. Chem., 56,* 303 (1993).

[2] A.J. Ashe, *Acc. Chem. Res., 11,* 153 (1978).

[3] See *Pyrylium Salts,* A.T. Balaban et al., Eds. Academic Press, New York (1982).

[4] For example, M. Alverez and M. Salas, *Heterocycles, 32,* 759 (1991).

[5] See A.M. Patterson, L.T. Capell, and D.F. Walker, *The Ring Index,* 2nd ed. American Chemical Society, Washington, D.C. (1960).

[6] Sources of pK_a values are (a) G. Hoefle, W. Steglich, and H. Vorbruggen, *Angew. Chem. Intl. Ed. Engl., 17,* 569 (1978); (b) *Comprehensive Organic Chemistry,* vol. 4, P.G. Sammes, Ed.

Pergamon Press, New York (1979), p. 5; (c) *Comprehensive Heterocyclic Chemistry,* vol. 2, A.J. Boulton and A. McKillop, Eds. Pergamon Press, New York (1984), p. 171.

[7] E.V. Scriven, *Chem. Soc. Rev., 12,* 129 (1983).

[8] C.D. Johnson, A.R. Katritzky, B.J. Ridgewell, and M. Viney, *J. Chem. Soc. (B),* 1204 (1967).

[9] A.R. Katritzky and B.J. Ridgewell, *J. Chem. Soc.,* 3753 (1963); A.R. Katritzky and B.J. Ridgewell, *J. Chem. Soc.,* 1204 (1967).

[10] F. Friedl, *Ber., 45,* 428 (1912).

[11] (a) J.M. Bakke, E. Ranes, J. Riha, and H. Svensen, *Acta. Chem. Scand., 53,* 141 (1999); (b) J.M. Bakke and J. Riha, *Acta. Chem. Scand., 53,* 356 (1999).

[12] (a) S.M. McElvain and M.A. Goese, *J. Am. Chem. Soc., 65,* 2233 (1943); (b) E. Pearson, W.M. Hargrove, J.K.T. Chow, and B.R. Suthers, *J. Org. Chem., 26,* 789 (1961).

[13] J.P. Wibaut and H.J. den Hertog, *Rec. Trav. Chim., 64,* 55 (1945).

[14] E. Plazek, *Ber., 72,* 577 (1939).

[15] H.C. Brown and B. Klanner, *J. Am. Chem. Soc., 75,* 3865 (1953); H.C. van der Plas and H.J. den Hertog, *Rec. Trav. Chim., 81,* 841 (1962).

[16] M.J.S. Dewar and P.M. Maitlis, *J. Chem. Soc.,* 2521 (1957).

[17] H.J. den Hertog and J. Overhoff, *Rec. Trav. Chim., 69,* 468 (1950); E. Ochiai, *J. Org. Chem., 18,* 534 (1953).

[18] H.S. Mosher and F.J. Welch, *J. Am. Chem. Soc., 77,* 2902 (1955).

[19] K. Matsumoto, S. Hashimoto, and S. Otani, *Chem. Comm.,* 306 (1991).

[20] D.L. Comis and S. O'Connor, *Adv. Hetero. Chem., 44,* 199 (1988).

[21] L. Liveris and J. Miller, *J. Chem. Soc.,* 3486 (1963).

[22] For example, J.W. Bunting and N.P. Fitzgerald, *Can. J. Chem., 63,* 1301 (1984).

[23] M.J. Pieterse and H.J. den Hertog, *Rec. Trav. Chim., 80,* 1376 (1961).

[24] W. Adam, A. Grimison, and R. Hoffmann, *J. Am. Chem. Soc., 91,* 2590 (1969).

[25] (a) J.F. Bunnett and P. Singh, *J. Org. Chem., 46,* 4567 (1981); (b) F.C. Gozzo and M.J. Eberlin, *J. Org. Chem., 64,* 2188 (1999); (c) R. Sparrapan et al., *Chem. Eur. J., 6,* 321 (2000).

[26] See J. Elguero, C. Marzin, A.R. Katritzky, and P. Linda, *The Tautomerism of Heterocycles.* Academic Press, New York (1976).

[27] P. Beak, *Acc. Chem. Res., 10,* 186 (1977); P. Beak, J.B. Covington, S.G. Smith, J.M. White, and J.M. Ziegler, *J. Org. Chem., 45,* 1354 (1980).

[28] S. Tobias and H. Guenther, *Tetrahedron Lett., 23,* 4785 (1982).

[29] J.D. Watson, *The Double Helix: A Personal Account of the Discovery of the Structure of DNA.* New American Library, New York (1969).

[30] For an authoritative discussion, see J.D. Watson, N.H. Hopkins, J.W. Roberts, J.A. Steite, and A.M. Weiner, *Molecular Biology of the Gene,* 4th ed. Benjamin Cummings, Menlo Park, CA (1987).

[31] C.K. McGill and J.A. Rappa, *Adv. Hetero. Chem., 44,* 1 (1988).

[32] O.N. Chupakhin, V.N. Charushin, and H.C. van der Plas, *Nucleophilic Aromatic Substitution of Hydrogen.* Academic Press, New York (1994).

[33] K. Ziegler and H. Zeiser, *Ber., 63,* 1847 (1930).

[34] H. Hara and H.C. van der Plas, *J. Hetero. Chem., 19,* 1285 (1982).

[35] See, for example, D.M. Stout and A.I. Meyers, *Chem. Rev., 82,* 223 (1982); J.G. Keay, *Adv. Hetero. Chem., 39,* 1 (1986); J.R. Burke and P.A. Frey, *J. Org. Chem., 61,* 530 (1966).

[36] A. Sausins and G. Duburs, *Heterocycles,* 291 (1988).

[37] For reviews, see (a) H.C. van der Plas, *Acc. Chem. Res., 11,* 462 (1978); (b) A.N. Kost, S.P. Gromov, and R.S. Sagitullin, *Tetrahedron, 37,* 3423 (1981).

[38] J.D. Reinheimer, L.L. Magle, G.G. Dolnikowski, and J.T. Gerig, *J. Org. Chem., 45,* 3097 (1980).

[39] K. Konno, K. Hashimoto, H. Shirahama, and T. Matsumoto, *Tetrahedron Lett., 27,* 3865 (1986).

[40] Y.I. Chumakov and V.P. Sherstyuk, *Tetrahedron Lett.,* 129 (1965).

[41] R.L. Frank and R.W. Meikle, *J. Am. Chem. Soc., 72,* 4184 (1950).

[42] A. Sausins and G. Duburs, *Heterocycles, 27,* 269 (1988).

[43] See the chapter by A.E.A. Porter in *Comprehensive Heterocyclic Chemistry,* vol. 3, A.J. Boulton and A. McKillop, Eds. Pergamon Press, New York (1984).

[44] M. Tisler and B. Stanovnik, *Adv. Hetero. Chem., 49,* 385 (1990).

[45] L.G. Qiang and N.H. Baine, *J. Org. Chem., 53,* 4218 (1988).

[46] B. Potthoff and E. Breitmaier, *Synthesis,* 584 (1986); U. Rueffer and E. Breitmaier, *Synthesis,* 623 (1989).

[47] O. Doebner and W. von Miller, *Ber., 14,* 2812 (1881).

[48] For an alternative, and more complex, mechanism, see J.J. Eisch and T. Dluznieswski, *J. Org. Chem., 54,* 1269 (1989).

[49] R.F. Manske and M. Kulka, *Org. Reactions, 7,* 59 (1953).

[50] W.M. Whaley and T.R. Govindachari, *Org. Reactions, 6,* 74 (1951). See also R.D. Larsen et al., *J. Org. Chem., 56,* 6034 (1991).

PROBLEMS

11.1 Draw structures for all organic products that should be formed in significant yields from the reactions below, or write "no reaction."

(a) 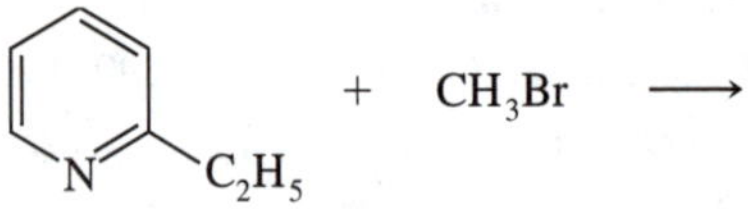 [pyridine with C_2H_5 at 2-position] + CH_3Br $\longrightarrow$

(b) 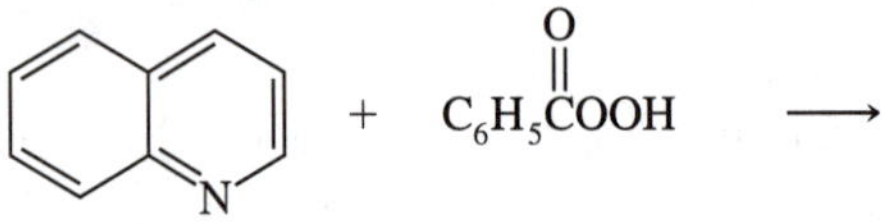 [quinoline] + $C_6H_5\overset{O}{\overset{\|}{C}}COOH$ $\longrightarrow$

(c) [4-chloropyridine] + HNO_3 $\xrightarrow{H_2SO_4}$

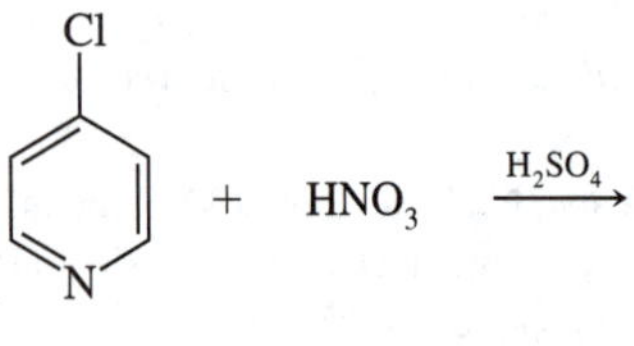

(d) 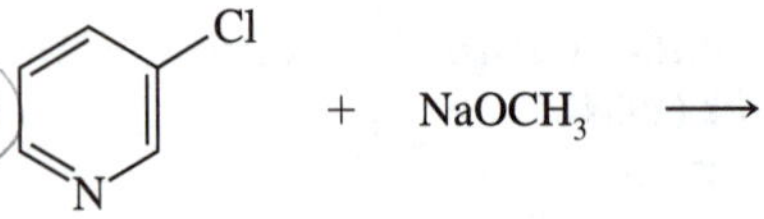 [3-chloropyridine] + $NaOCH_3$ $\longrightarrow$

(e) 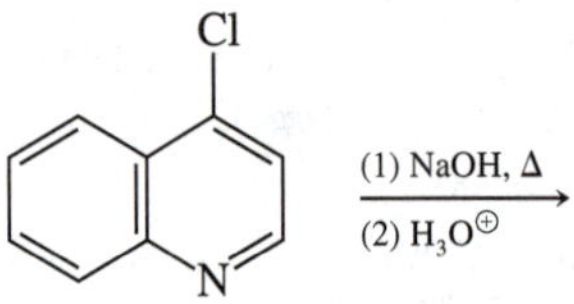 [4-chloroquinoline] $\xrightarrow[\text{(2) } H_3O^{\oplus}]{\text{(1) NaOH, } \Delta}$

(f)

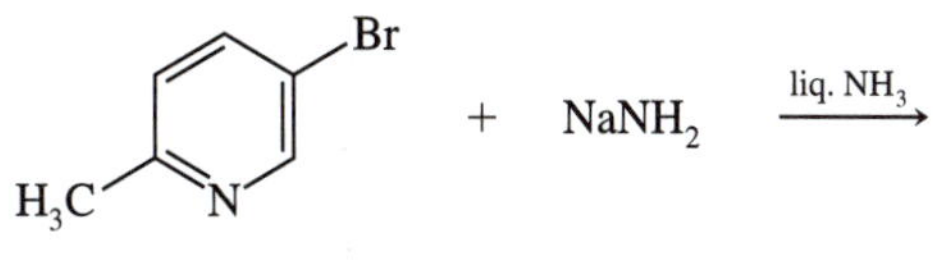

$$+ \quad NaNH_2 \xrightarrow{\text{liq. } NH_3}$$

(g)

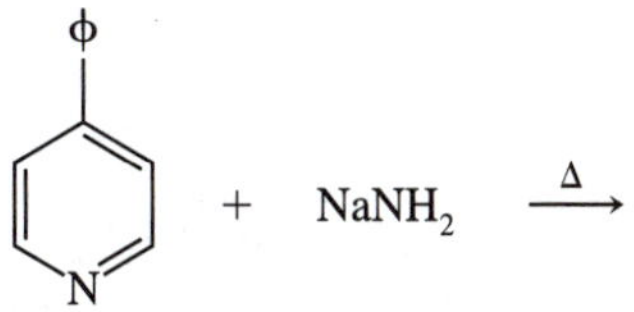

$$+ \quad NaNH_2 \xrightarrow{\Delta}$$

(h)

$$+ \quad C_4H_9Li \longrightarrow$$

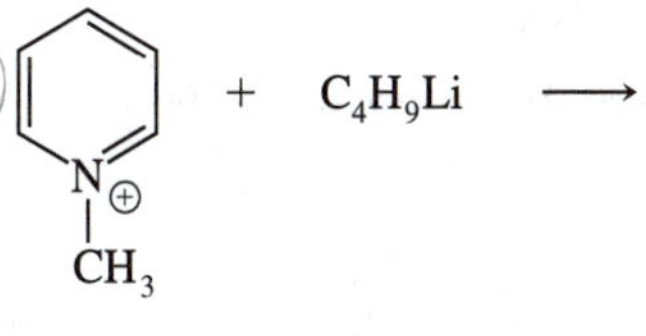

11.2 Write reasonable mechanisms for the following reactions, using curved arrows to show electron motions. Do not combine any two steps into one.

(a)

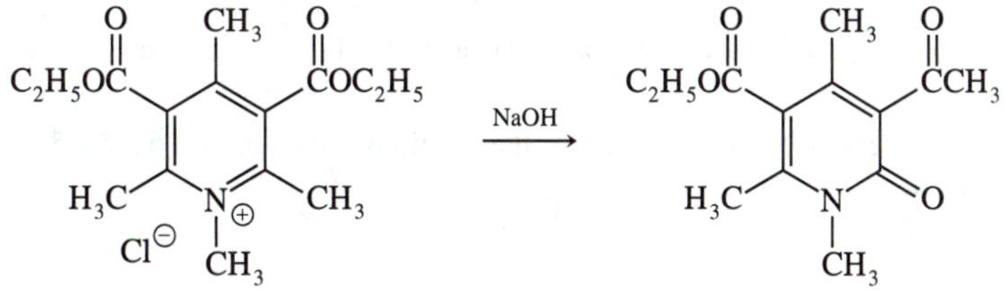

$$\xrightarrow{NaOH}$$

(b)

$$\xrightarrow[\substack{(1)\ NaR_3BH \\ (2)\ C_6H_5CH=O \\ (3)\ H_3O^\oplus}]{}$$

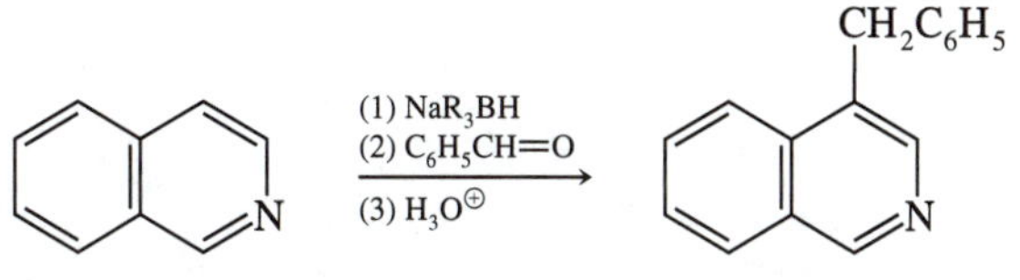

(D.E. Minter and M.A. Ree, *J. Org. Chem.*, *53*, 2653 [1988])

(c)

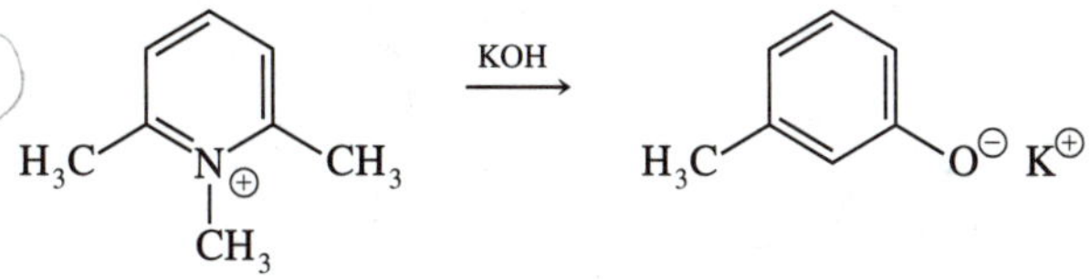

$$\xrightarrow{KOH}$$

(d)

$$+ \quad NH_2{-}NH_2 \longrightarrow$$

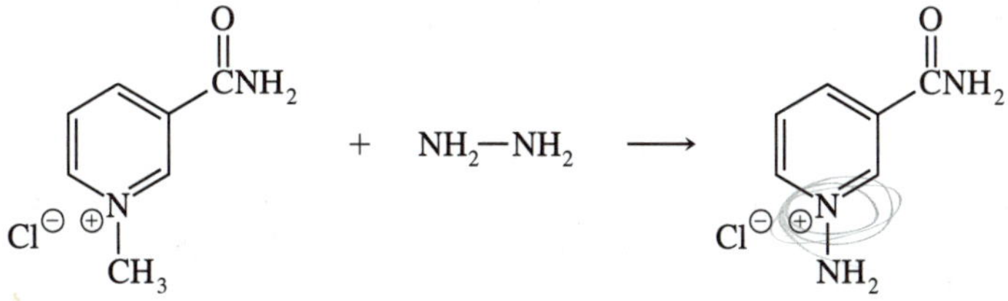

(e)

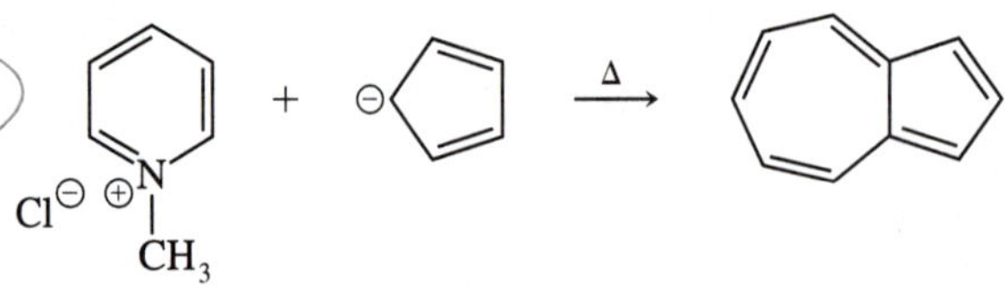

(See K. Hafner, *Angew. Chem., 70,* 419 [1958])

(f)

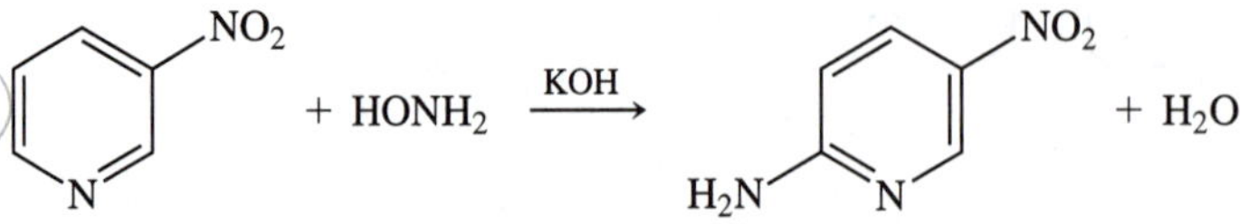

(J.M. Bakke, H. Svensen, and R. Trevison, *J. Chem. Soc., Perkin 1,* 376 [2000])

(g)

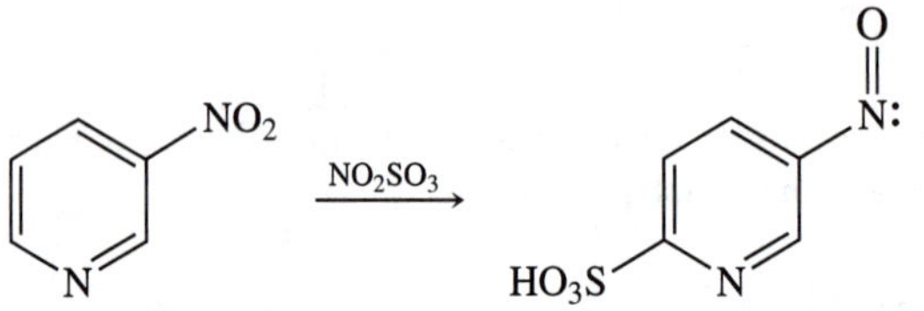

(Adapted from J.M. Bakke, E. Ranes, C. Ramming, and I. Sletvold, *J. Chem. Soc., Perkin 1,* 1241 [2000])

11.3 Suggest reasonable mechanisms for the reactions shown in Eqs. 52, 53, 56, 58, and 65–68 in this chapter.

Five-Membered Heterocyclic Rings

12.1 INTRODUCTION

The principal types of unsaturated, five-membered heterocyclic rings are found in the molecules *furan, pyrrole,* and *thiophene.*

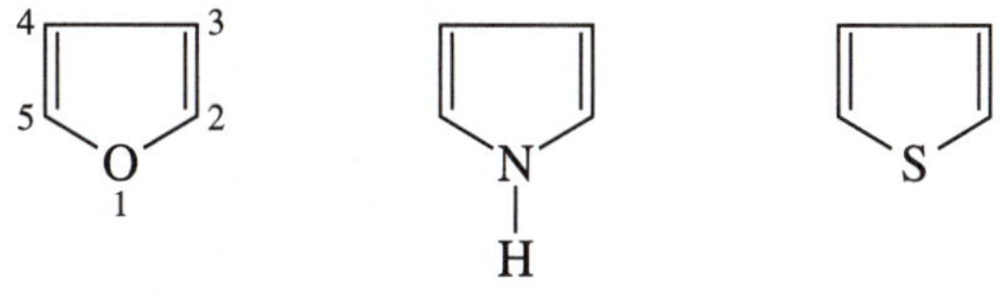

furan   pyrrole   thiophene

The locations of substituents on these rings are specified by numbering the heteroatom as position 1* and the carbon atoms successively as positions 2 through 5. Atoms 2 and 5 of each ring are also called the α-*positions,* and atoms 3 and 4 the β-*positions.*

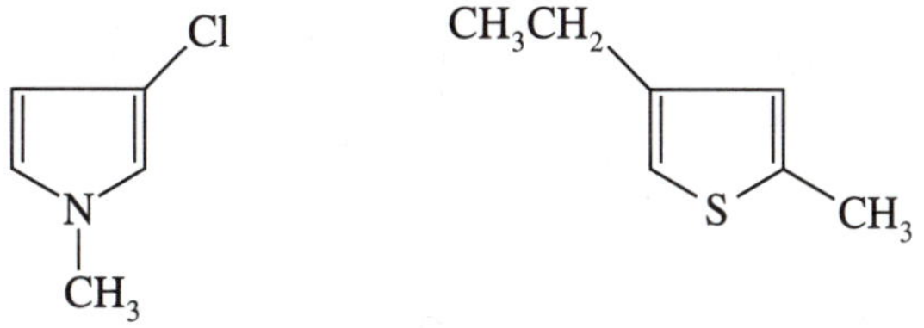

3-chloro-1-methylpyrrole   4-ethyl-2-methylthiophene

*Pyrrole is more precisely named *1H-pyrrole*, to distinguish it from its two unstable tautomers, 2H-pyrrole and 3H-pyrrole.

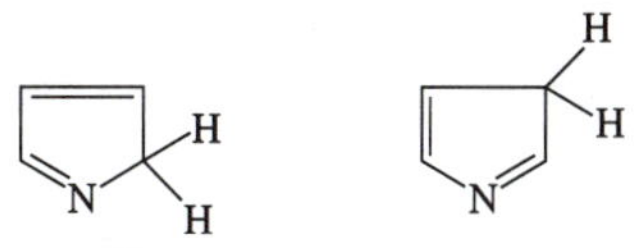

2H-pyrrole   3H-pyrrole

The five-membered heterocyclic rings may be fused to carbocyclic aromatic rings. The fused ring molecules containing oxygen and sulfur are called *benzofuran* and *benzothiophene,* respectively. The nitrogen-containing molecule was known before its structure was established, so that it is called *indole,* rather than benzopyrrole.*

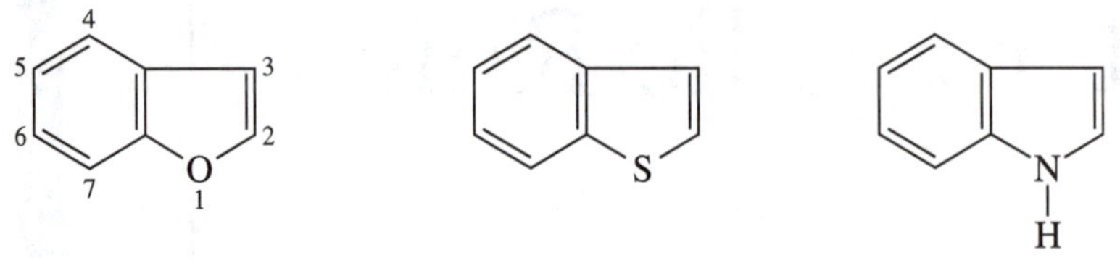

benzofuran   benzothiophene   indole

The positions of atoms in these bicyclic molecules are numbered starting with the heteroatoms and continuing around the heterocyclic and homocyclic rings, but skipping the ring junctures, as was done with quinoline and isoquinoline.

Of course, five-membered rings may contain more than one hetero atom. Undoubtedly, the most important ring system containing two hetero atoms is that of *imidazole,* which is found in the purine units of DNA and RNA, as well as in the amino acid *histidine.*

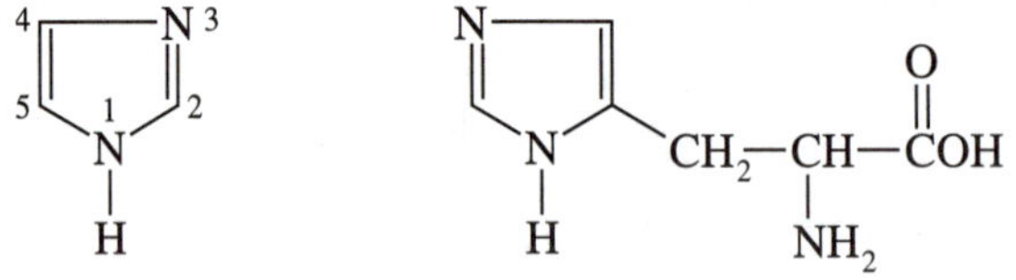

imidazole   histidine

Another important ring system is that of *thiazole,* which is found in the coenzyme thiamine (vitamin B_1), and other naturally occurring molecules.

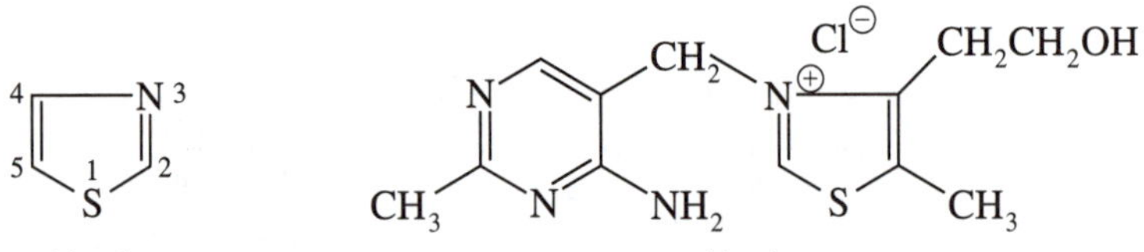

thiazole   thiamine

---

*To be precise, the molecules shown above should be named *benzo[b]furan* and *benzo[b]thiopene* to distinguish them from their far less common isomers, benzo[c]furan and benzo[c]thiophene. (The designations [b] and [c] in the names indicate that the carbocyclic rings are fused to Atoms 2 and 3, and 3 and 4, respectively, of the heterocyclic rings.[1]) The corresponding isomer of indole is named *isoindole.*

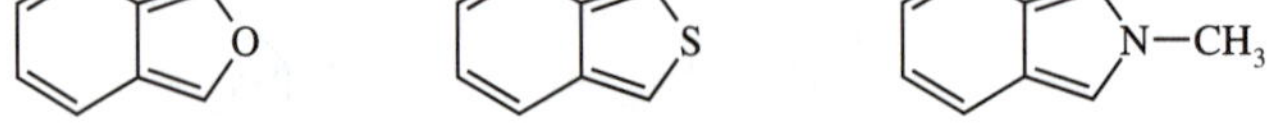

benzo[c]furan   benzo[c]thiophene   N-methylisoindole

Benzo[c]furan, benzo[c]thiopene, and isoindole have only brief lifetimes under ordinary conditions, but some of their derivatives are stable molecules.

12.2 ELECTRON DELOCALIZATION AND AROMATICITY

In a five-membered heterocyclic ring, a pair of electrons on the heteroatom forms part of the aromatic sextet. In order to achieve maximum overlap of this electron pair with the electrons of the π bonds, the heteroatom is presumably rehybridized so that one pair of electrons is located in a p orbital. (The other pair of unshared electrons in furan or thiophene is located in an sp^2 orbital orthogonal to the orbitals of the aromatic sextet.) In an N-substituted pyrrole, the nonring atom bonded to the nitrogen (R in the structure below) is in the same plane as the ring, showing that the nitrogen has sp^2 hybridization.[2]

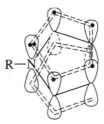

The structure of a pyrrole ring

In contrast to pyridine and other six-membered aromatic heterocyclic rings, the hetero atoms in furan, pyrrole, and thiopene act as electron donors. This can be concluded, for instance, from molecular orbital calculations, which show that each ring carbon in these molecules has a π-electron density somewhat greater than 1.0, while the hetero atoms have π-electron densities less than 2.0.[3]

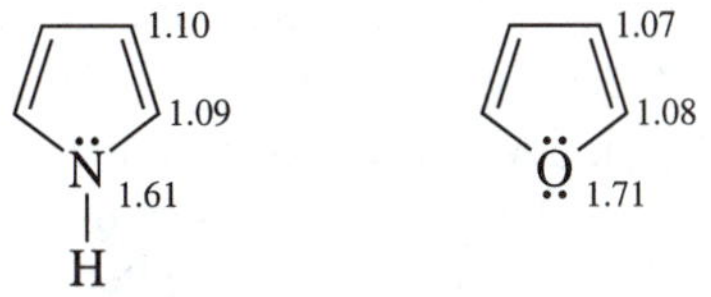

1.10
1.09
N 1.61
H

1.07
1.08
Ö 1.71

Calculated π–electron densities in pyrrole and furan

The qualitative conclusion that hetero atoms in furans, pyrroles, and thiophenes donate electrons to ring carbons can also be derived from an examination of resonance forms. In contrast to benzene and pyridine, each of which has two nonpolar Kekulé resonance forms of equal importance, furan, pyrrole, and thiophene each have only one nonpolar resonance form, but have four dipolar resonance forms. In effect, one pair of electrons on the heteroatom of each of these rings is donated into the ring and shared among the four carbon atoms.

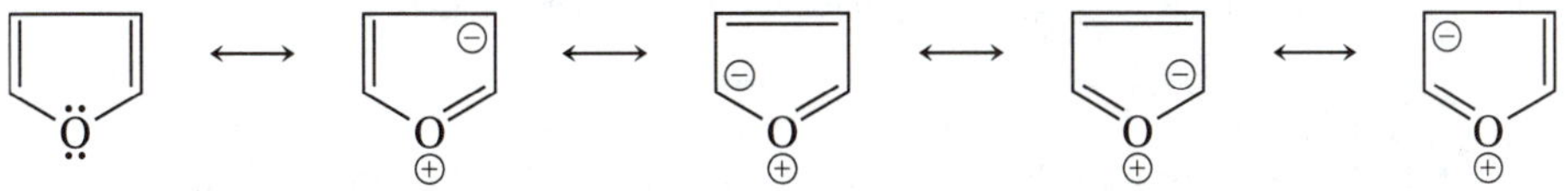

$$(1)$$

Resonance forms of furan

Furan, pyrrole, and thiophene are thus often described as having π-*excessive* rings. In contrast, pyridine and its derivatives are described as having π-*deficient* rings, because the nitrogen atoms, particularly in their protonated forms, draw electrons away from the other atoms of the rings.

Estimates of the "aromatic stabilization energy" or "resonance energy" of benzene have varied greatly depending on the methods used. Estimates of the stabilization energies of five-membered heterocyclic rings, whether based on experimental measurements or theoretical calculations, have varied even more widely. (Estimates of the stabilization energy of furan, for instance, have ranged from about 22 kcal/mol down to as little as 2 kcal/mol.[4]) Almost all investigations agree, however, that pyrrole and thiophene have large stabilization energies (perhaps two-thirds as large as that of benzene), while furan has a much lower stabilization energy.[4]

It is hardly surprising that the resonance energy of furan should be significantly lower than that of pyrrole, since electrons must be drawn away from the very electronegative oxygen atom to be distributed among the carbon atoms of the furan ring. However, the stability of the thiophene ring might seem surprising. Overlap of p orbitals of sulfur with π bonds is poor, so resonance forms involving carbon–sulfur double bonds are not expected to contribute greatly to the stabilization of thiophene rings. It has sometimes been suggested that donation of electrons from the π bonds into d orbitals of sulfur atoms might help stabilize thiophene rings, but the d orbitals of neutral sulfur atoms are so high in energy that they should not interact strongly with carbon–carbon π bonds. At present, there does not seem to be a generally accepted explanation for the stability of thiophene rings.

12.3 REACTIONS OF FIVE-MEMBERED HETEROCYCLIC RINGS

Protonation

The most characteristic property of amines is that they are basic and form salts with even dilute solutions of mineral acids. In contrast, pyrroles are not basic. (The conjugate acid of pyrrole has been estimated to have a pK_a of about -10, and is a very strong acid indeed.[5]) In this respect, pyrroles resemble amides rather than amines. The lack of basicity of amides and pyrroles is partially due to the inductive effects of the carbonyl groups of amides and of the two double bonds in pyrroles. It can also be attributed to the fact that both types of molecules are strongly stabilized by resonance. This stabilization would be almost totally lost if the nitrogens were protonated.

high resonance energy   low resonance energy   high resonance energy   low resonance energy

The N–H bonds of pyrroles are rapidly converted to N–D bonds by reaction of pyrroles with D_2SO_4, showing that it is, in fact, possible to protonate the nitrogen atoms of pyrroles. However, the N-protonated cations, although formed rapidly, are present in such low concentrations that they cannot be directly observed. In contrast, cations formed by the protonation of carbon atoms of pyrroles are stable enough to be observed, and their structures determined, by NMR spectroscopy. If the pyrroles have unsubstituted α positions, only the cations formed by α protonation can be observed (Eq. 2).[6] When both α positions are substituted, as in 2,5-di-methylpyrrole, a mixture of the α-protonated and β-protonated cations is formed (Eq. 3).[7]

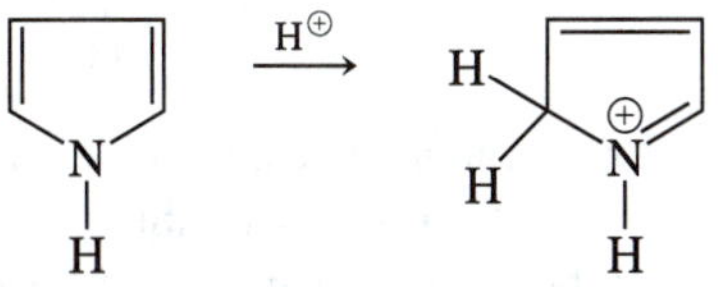

$$ (2) $$

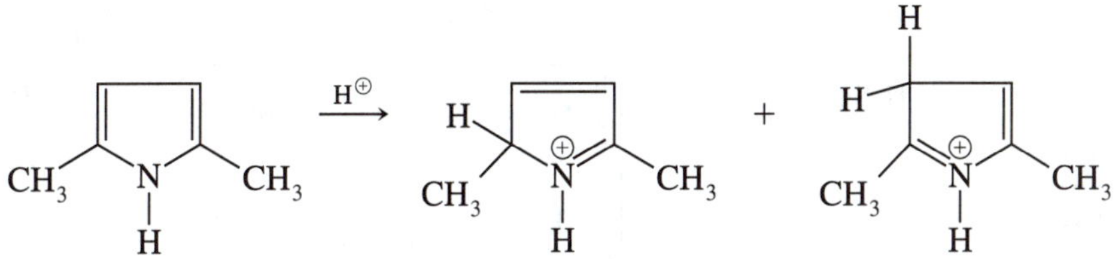

$$ (3) $$

The preference for protonation of pyrroles at α positions rather than β positions is easily understandable, since protonation at an α position yields a "carbocation" (more properly described as an immonium salt) with three resonance forms, while protonation at a β position yields a cation with only two resonance forms. However, both cations are quite easily formed, since a new nitrogen–carbon π bond is formed in each case to replace the carbon–carbon bond broken on protonation.

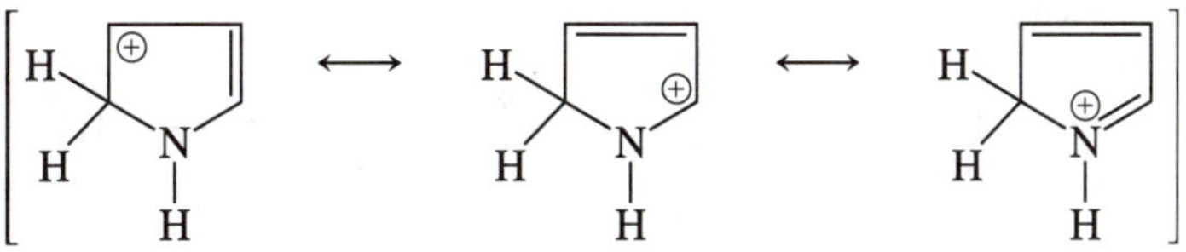

cation from α-protonation

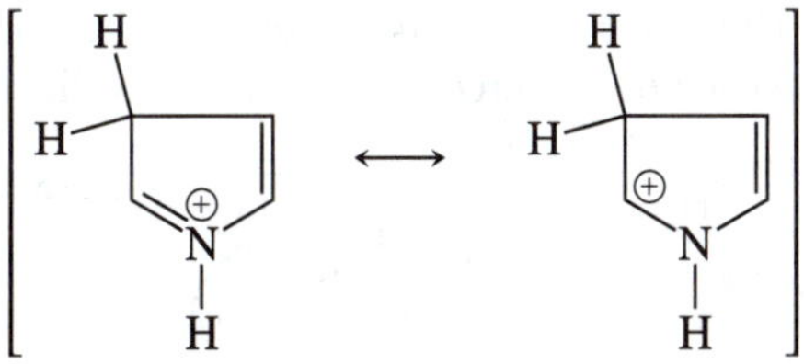

cation from β-protonation

Although monomeric cations can be detected when pyrroles are dissolved in acids for short periods, the pyrroles are converted into oligomers, and then into polymers, after longer reaction times.[8] The structures of the polymers have not been established, but the structure of a trimer, isolated in low yield from the reaction of pyrrole with hydrochloric acid was established. (Eq. 4).

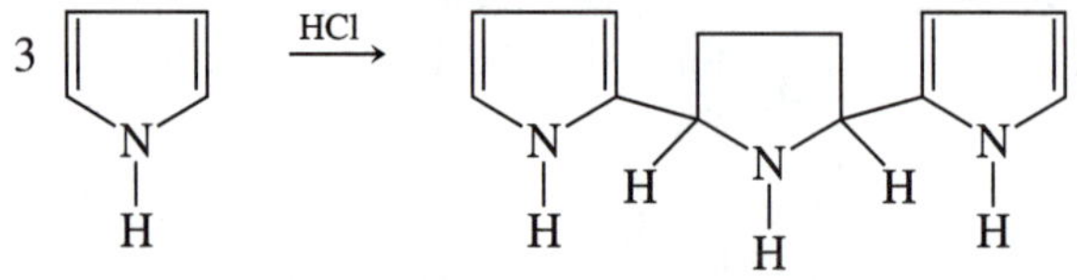

$$3 \quad \text{[pyrrole]} \xrightarrow{\text{HCl}} \text{[trimer structure]} \qquad (4)$$

While pyrroles are not basic, imidazoles, which contain amidine functional groups, form stable salts on reaction with mineral acids. The doubly bonded (imino) nitrogens are protonated, because the resulting cations retain the resonance energies of the amidine groups as well as the stabilities of aromatic rings (Eq. 5).

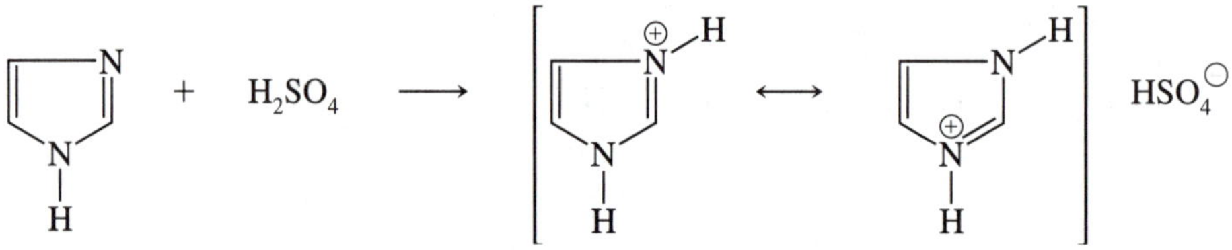

$$\text{[imidazole]} + H_2SO_4 \longrightarrow \left[\text{[resonance structures]} \right] HSO_4^{\ominus} \qquad (5)$$

Furan and thiophene react similarly to pyrrole in acidic solutions. Both will initially undergo proton exchange at C2 and C5 and then undergo dimerization and polymerization on continued exposure to strong acids (Eq. 6).

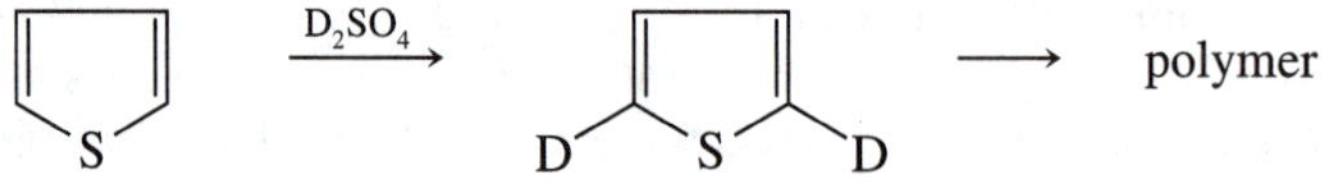

$$\text{[thiophene]} \xrightarrow{D_2SO_4} \text{[2,5-D-thiophene]} \longrightarrow \text{polymer} \qquad (6)$$

In addition, in aqueous solutions, furan derivatives can undergo hydrolysis to form 1,4-dicarbonyl compounds (Eq. 7). (That reaction is the reverse of a process frequently used to synthesize furans, as shown in Eq. 55.)

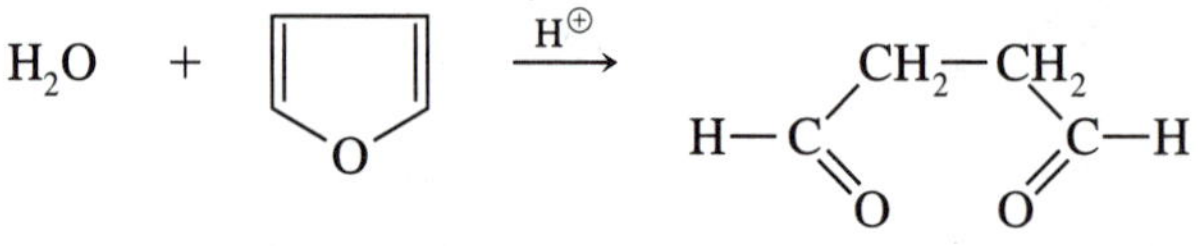

$$H_2O + \text{[furan]} \xrightarrow{H^{\oplus}} \text{[1,4-dicarbonyl compound]} \qquad (7)$$

Pyrrole reacts with acids far more rapidly than furan does, while furan is much more reactive than thiophene. This order of reactivity parallels the relative stability of the cations that would be formed from each of the three types of heterocyclic rings.

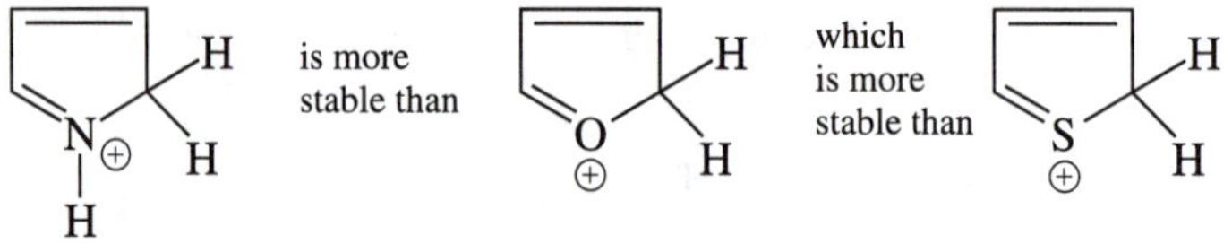

[pyrrole cation] is more stable than [furan cation] which is more stable than [thiophene cation]

Halogenation, Nitration, and Sulfonation

Halogenation.   One of the classic criteria for aromaticity is that aromatic molecules undergo substitution reactions, rather than addition reactions, with electrophilic reagents. This criterion is of course far from definitive, even with hydrocarbons, because many polycyclic aromatic molecules (such as anthracene; see Eq. 8) can undergo addition reactions. However, the general occurrence of electrophilic substitution reactions is still an important characteristic of most aromatic molecules.

$$+ \quad Br_2 \quad \longrightarrow \qquad\qquad\qquad\qquad (8)$$

In contrast to pyridine and its analogs, in which it is difficult to observe electrophilic substitution because the molecules are so resistant to electrophilic attack, it is often difficult to observe clean electrophilic substitution reactions of five-membered aromatic rings because of their great reactivity. Since electrophilic reactions are typically carried out in strong acid solutions, decomposition or polymerization of five-membered heterocyclic rings may compete with substitution reactions. Furthermore, "π-excessive" molecules are easily oxidized, and many electrophilic reagents, such as halogens and nitric acid, are powerful oxidizing agents.

Possibly as a result of these factors, the reactions of pyrrole with chlorine and bromine yield complex mixtures that often contain products of multiple halogenation. 2-Chloropyrrole and 2-bromopyrrole can be prepared by reaction of pyrrole with N-chlorosuccinimide (NCS) and N-bromosuccinimide (NBS), respectively (Eq. 9).[9]

$$(9)$$

NCS, X = Cl
NBS, X = Br

The actual halogenating agents in these reactions appear to be molecular chlorine and bromine, respectively, since NCS and NBS react with the hydrogen chloride and hydrogen bromide formed on halogenation of pyrroles to yield the halogens in low concentrations (Eq. 10).

$$N-X \quad + \quad HX \quad \longrightarrow \quad N-H \quad + \quad X_2 \qquad\qquad (10)$$

X = Cl, Br

Chlorination and bromination of pyrroles with these reagents can thus be carried out in acid-free conditions and under conditions that should minimize multiple halogenations.

(It may be noted that the preference for attack at α positions would not be predicted simply by examining electron densities in pyrrole [see Section 11.2]. The transition states for electrophilic substitution reactions of pyrrole appear to resemble the cationic intermediates more closely than they resemble the starting materials. The greater stabilization of the intermediate from attack at an α position is, therefore, more important than electron distributions in the starting material.)

Halogenation of pyrroles can be directed to the β positions by placing very bulky substituents, such as triisopropylsilyl groups, on the nitrogen atoms. The silyl substituents are easily introduced, and they are easily removed by reaction with fluoride salts (Eq. 11).[10]

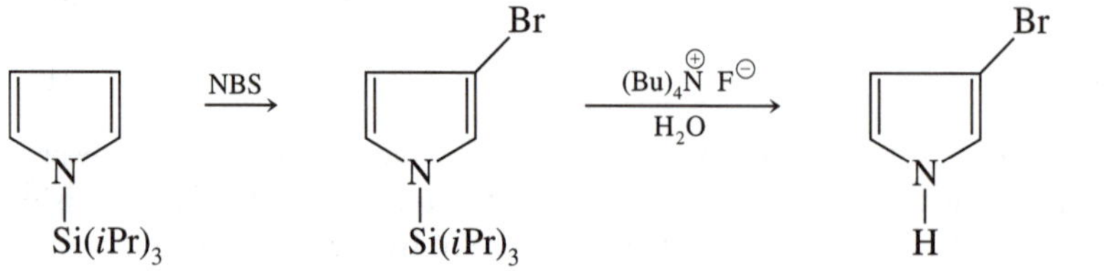

$$\text{(11)}$$

Halogenation of pyrroles can take place even if all the ring carbons bear substituents. However, the products of the reactions have halogens on α-carbons of side chains rather than on the rings. It has been demonstrated that the reactions initially involve electrophilic attack on the ring to form carbocation intermediates, which can be detected by their UV and NMR spectra. (All four possible cations can be observed, although only one is shown in Eq. 12.) The cations then rearrange to form the side-chain halogenated products. The mechanisms of the rearrangements have not yet been established.[11]

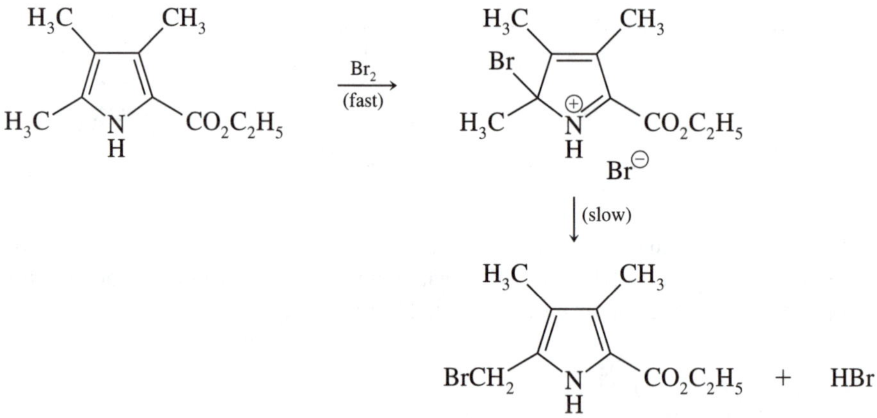

$$\text{(12)}$$

Careful addition of bromine to furan at $-5°C$ in dioxane solution gives 2-bromofuran in good yield. However, a study of the reaction of furan with bromine at $-50°C$ showed that 1,4-addition products are first formed.[12] Loss of HBr on workup yields the substitution product (see Eq. 13).

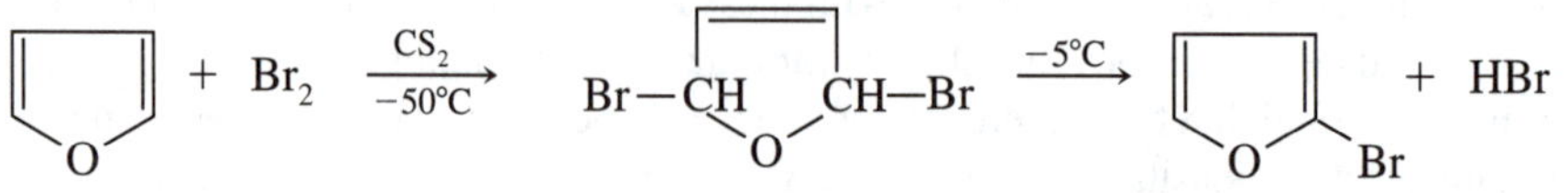

$$\text{(13)}$$

The reaction of furan with bromine in the presence of potassium acetate yields a 2,5-diacetoxy derivative (Eq. 14).[13] This presumably results from displacement of bromide ions from the initial reaction products.

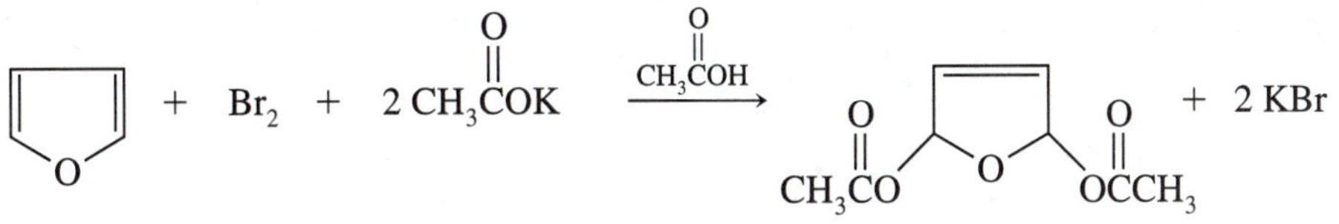

$$\text{[furan]} + Br_2 + 2\ CH_3COK \xrightarrow{CH_3COH} \text{[2,5-diacetoxy derivative]} + 2\ KBr \qquad (14)$$

The reaction of thiophene with chlorine produces complex mixtures of addition and substitution products. Careful bromination of thiophene yields 2,5-dibromothiophene, together with a lower yield of 2-bromothiophene (Eq. 15).[14]

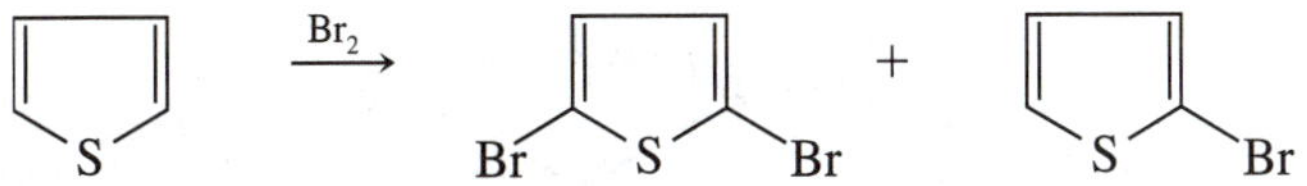

$$\text{[thiophene]} \xrightarrow{Br_2} \text{[2,5-dibromothiophene]} + \text{[2-bromothiophene]} \qquad (15)$$

(The existence of thiophene was discovered when it was found that benzene obtained from coal oils, which contained small amounts of thiophenes, reacted with bromine, whereas benzene prepared by decarboxylation of benzoic acid did not.)

Nitration and sulfonation.   Nitration of thiophene in sulfuric acid solution is possible but difficult to control. Nitration in acetic acid–acetic anhydride solution yields predominantly 2-nitrothiophene, together with some 3-nitrothiophene (Eq. 16).[14]

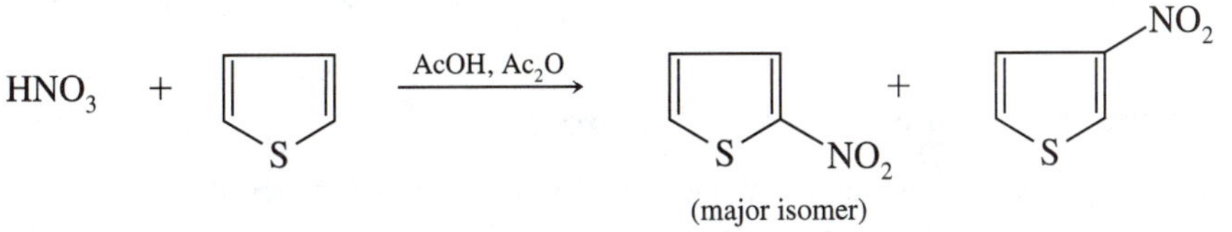

$$HNO_3 + \text{[thiophene]} \xrightarrow{AcOH,\ Ac_2O} \underset{\text{(major isomer)}}{\text{[2-nitrothiophene]}} + \text{[3-nitrothiophene]} \qquad (16)$$

Pyrrole undergoes polymerization in sulfuric acid but may be nitrated in acetic acid or acetic anhydride solution, yielding principally 2-nitropyrrole (Eq. 17). Acetyl nitrate $(CH_3CO_2NO_2)$ appears to be the actual nitrating agent.[10]

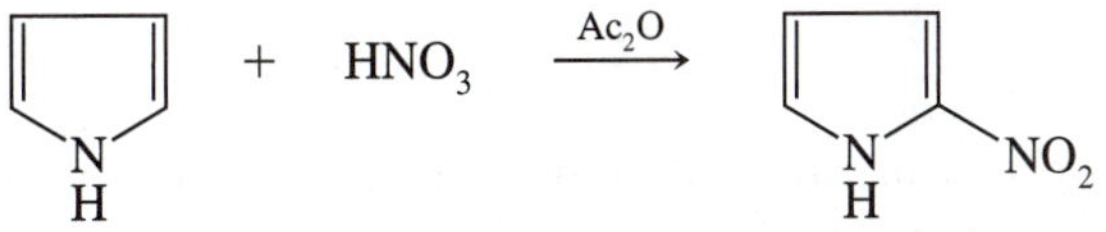

$$\text{[pyrrole]} + HNO_3 \xrightarrow{Ac_2O} \text{[2-nitropyrrole]} \qquad (17)$$

Imidazoles are essentially completely protonated under the conditions for most electrophilic substitution reactions, and therefore, undergo substitution only with great difficulty (Eq. 18).

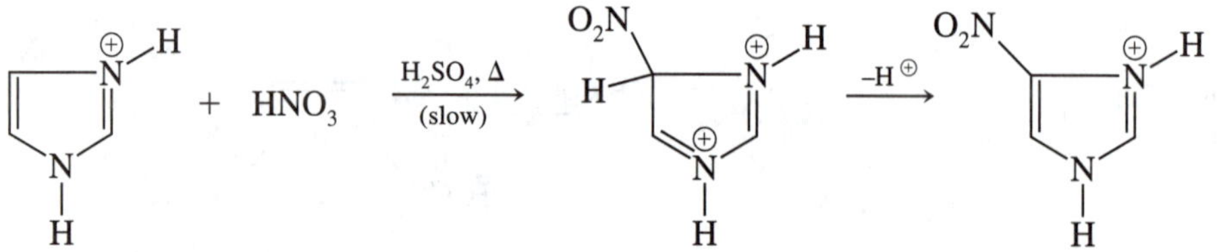

(18)

Nitration of furan, which has a lower stabilization energy than pyrrole or thiophene, initially yields an addition product, which is converted to 2-nitrofuran on warming or on reaction with pyridine (Eq. 19).[15]

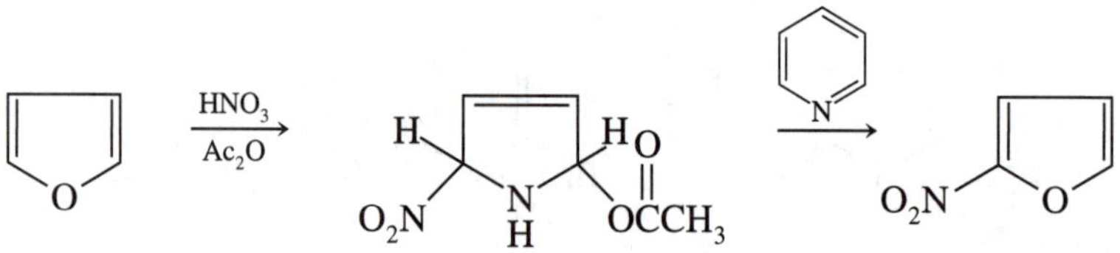

(19)

Furan, pyrrole, and thiophene can be converted to their 2-sulfonic acids by reaction with sulfur trioxide in dioxane or pyridine (Eq. 20). If both α positions of the ring are occupied, sulfonation will take place at a β position.[14,16,17]

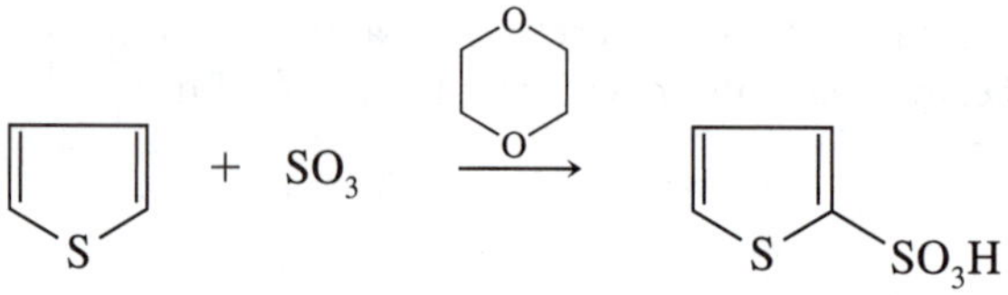

(20)

Reactions with Carbon-Centered Electrophiles

Acylation. Furan and thiophene can be converted to their 2-acyl derivatives (Friedel–Crafts acylation) by reaction with acylating agents in the presence of Lewis acid catalysts (Eq. 21).

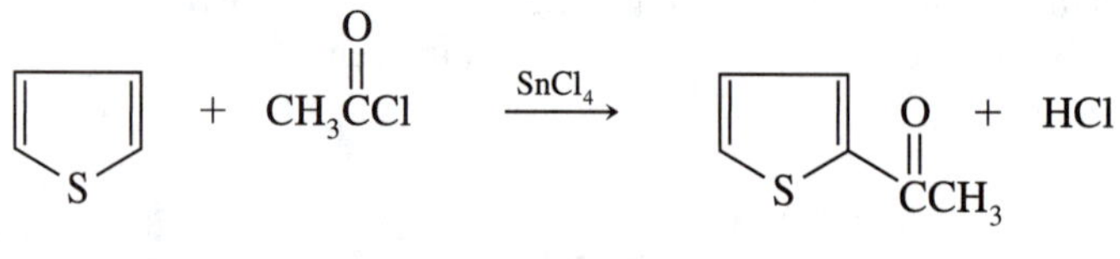

(21)

Pyrroles, which are more reactive, can be converted to 2-acylpyrroles by reaction with acid chlorides or with reactive acid anhydrides, such as trifluoroacetic anhydride (Eq. 22), even in the absence of catalysts.[9]

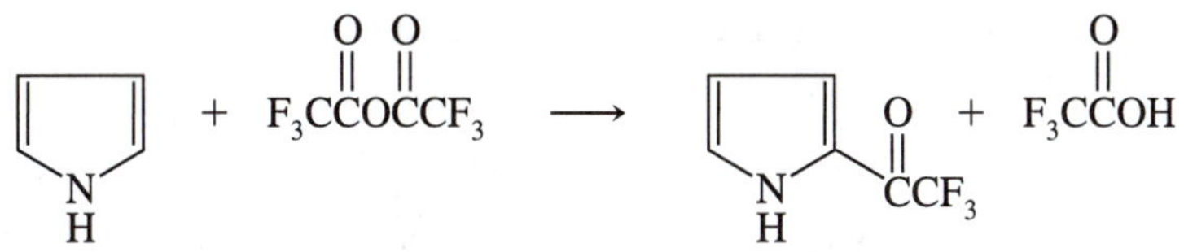

$$\text{(pyrrole-NH)} + \underset{\substack{\| \quad \| \\ O \quad O}}{F_3CCOCCF_3} \longrightarrow \text{(pyrrole)} + \underset{\substack{\| \\ O}}{F_3CCOH} \qquad (22)$$

Pyrroles with unsubstituted nitrogen atoms differ from other heterocyclic rings in that they can be wholly or partially converted to their anions by reaction with bases. (The pK_a of pyrrole in water is about 17.5, compared to about 40 to 45 for most amines.[18]) Alkali metal salts of pyrrole are usually acylated on nitrogen (Eq. 23), but Grignard salts usually react to give 2-acyl derivatives (Eq. 24).[9]

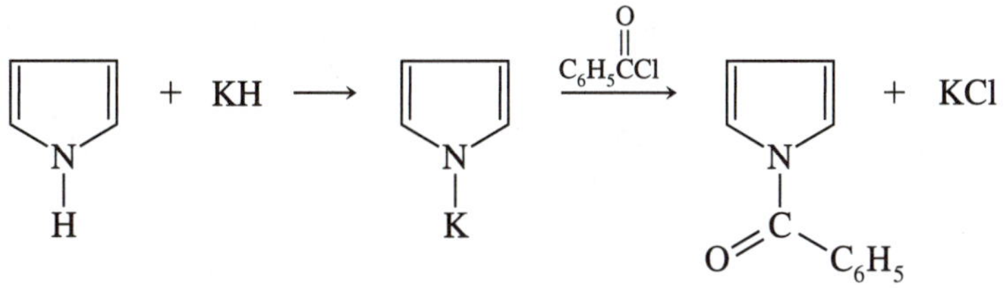

$$\text{(pyrrole-NH)} + KH \longrightarrow \text{(pyrrole-K)} \xrightarrow{C_6H_5CCl} \text{(N-benzoyl pyrrole)} + KCl \qquad (23)$$

$$\text{(pyrrole-NH)} + CH_3MgBr \longrightarrow \text{(pyrrole-MgBr)} \xrightarrow{CH_3CCl} \text{(2-acetyl pyrrole)} + MgBrCl \qquad (24)$$

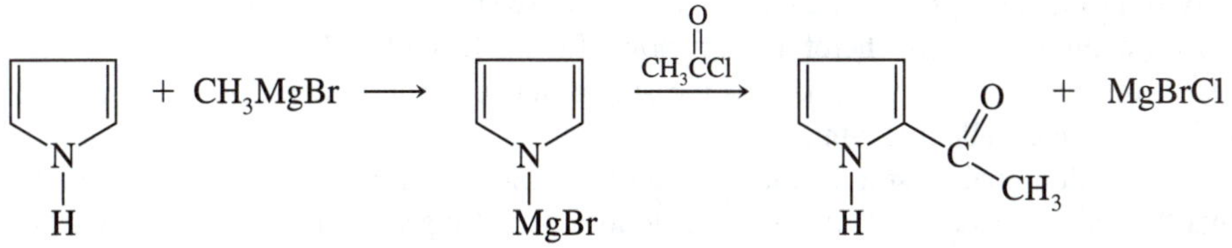

Acylation of imidazoles. Imidazoles, which are very much stronger bases than pyrroles, can be rapidly acylated by acid anhydrides and acid halides (including halides of phosphoric and sulfonic acids). (See Eqs. 25 and 26.)

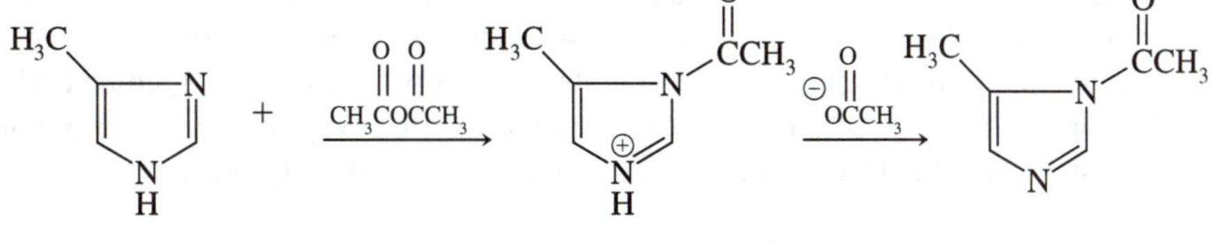

$$\text{(4-methylimidazole)} + \underset{\substack{\| \quad \| \\ O \quad O}}{CH_3COCCH_3} \longrightarrow \text{(N-acetyl intermediate)} \xrightarrow{\overset{\ominus}{O}CCH_3} \text{(N-acetyl-4-methylimidazole)} \qquad (25)$$

$$2\ \text{(imidazole)} + (C_2H_5O)_2PCl \longrightarrow \text{(N-phosphoryl imidazole)} + \text{(imidazolium)}\ Cl^{\ominus} \qquad (26)$$

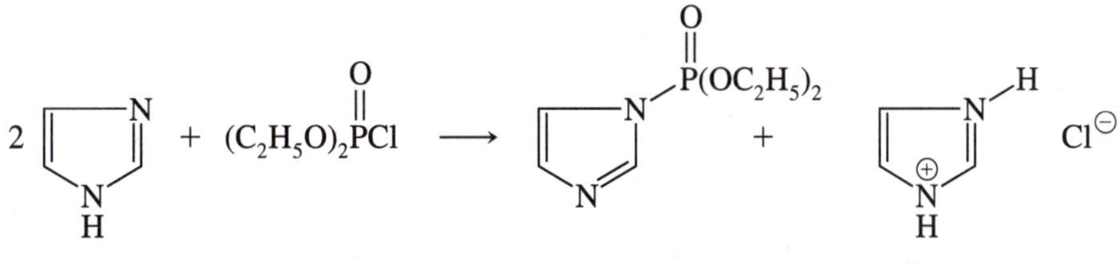

The resulting N-acylimidazoles are themselves excellent acylating agents. They are about as reactive as acid anhydrides,[19] and have found wide use in the synthesis of esters and amides. In contrast to acid halides or anhydrides, no acids are formed as by-products of the reactions (see Eqs. 27 and 28).

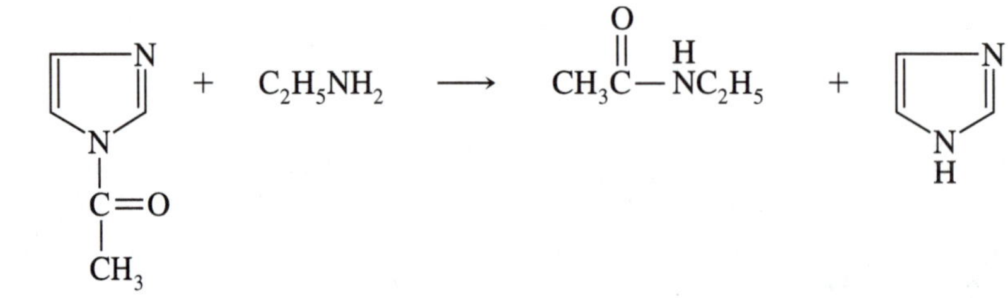

$$\qquad (27)$$

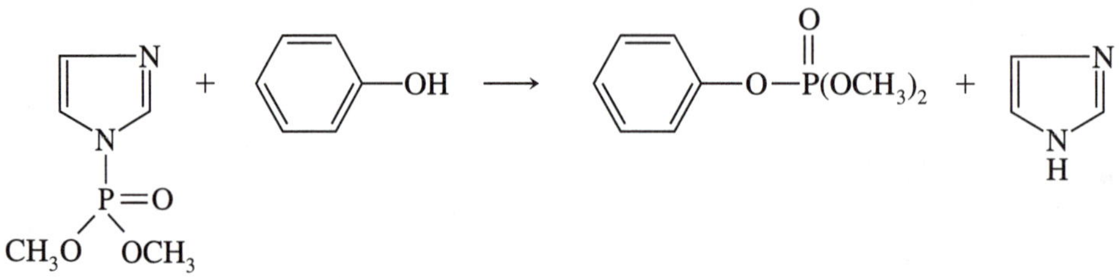

$$\qquad (28)$$

The effectiveness of imidazoles in these reactions is based, in part, on their remarkable duality of properties. As amidine derivatives, they are strong bases compared to most trivalent nitrogen compounds and, therefore, are powerful nucleophiles. However, once acylation takes place, the acyl groups are linked to the single-bonded, nonbasic nitrogens, and the imidazole units become excellent leaving groups.

An additional reason for the high rates at which acylimidazoles react with nucleophiles is that the imidazoles formed during the reactions actually act to catalyze further reaction. In fact, imidazoles are excellent catalysts for the hydrolysis of esters, as well as of more reactive carboxyl derivatives. (Imidazole catalysis of the hydrolysis of esters has been the subject of intense study, because the amino acid histidine, which contains an imidazole unit, appears to be the active site in enzymes such as chymotrypsin, which hydrolyze esters in biological systems.)

Careful investigations have identified several mechanisms by which imidazoles act as catalysts for the hydrolysis of esters.[19] They can simply abstract protons from water, forming hydroxide ions that then attack the esters. However, another duality in the properties of imidazoles is that while they are much stronger bases than pyrroles, they are also stronger acids and are easily converted to their anions. (Imidazole itself has a pK_a around 14.4, compared with a pK_a of 17.5 for pyrrole.) The conjugate anions of imidazoles are very effective nucleophiles in reactions with esters, and the resulting acylimidazoles are rapidly hydrolyzed, as shown in Eq. 29.

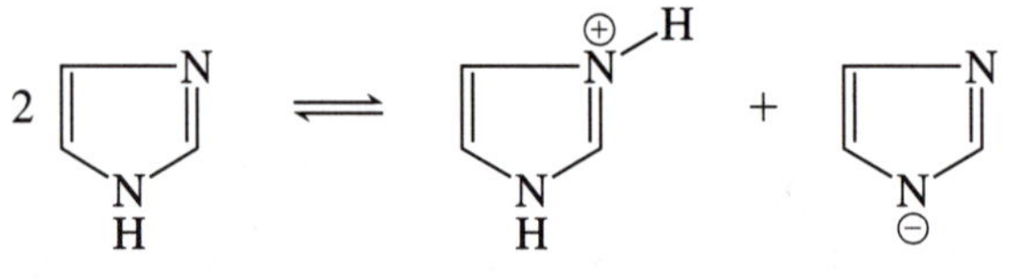

$$\qquad (29a)$$

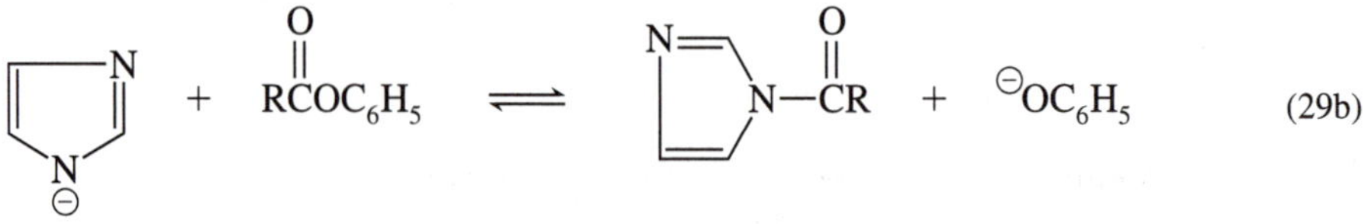

$$\qquad (29b)$$

$$\underset{\underset{\displaystyle N-CR}{\overset{\displaystyle O}{\parallel}}}{\left[\begin{array}{c}N=\\ \\ \end{array}\right]} + H_2O \longrightarrow \underset{\displaystyle NH}{\left[\begin{array}{c}N=\\ \\ \end{array}\right]} + \underset{\displaystyle HOCR}{\overset{\displaystyle O}{\parallel}} \qquad (29c)$$

Neutral imidazole molecules, as well as their anions, can act as nucleophiles in reactions with very reactive esters, such as *para*-nitrophenyl acetate (Eq. 30).

$$\underset{\displaystyle H}{\overset{\displaystyle N}{\left[\begin{array}{c}\\N\\\end{array}\right]}} + \; CH_3\overset{\displaystyle O}{\overset{\parallel}{C}}-O-\!\!\!\bigcirc\!\!\!-NO_2$$

$$\Big\downarrow$$

$$\underset{\displaystyle \overset{\oplus}{N}}{\underset{\displaystyle H}{\left[\begin{array}{c}\overset{\displaystyle CCH_3}{\overset{\parallel}{\overset{\displaystyle O}{}}}\\N\\\end{array}\right]}} + \; {}^{\ominus}O-\!\!\!\bigcirc\!\!\!-NO_2 \qquad (30)$$

$$\Big\downarrow$$

$$\underset{\displaystyle N}{\left[\begin{array}{c}\overset{\displaystyle CCH_3}{\overset{\parallel}{\overset{\displaystyle O}{}}}\\N\\\end{array}\right]} + \; HO-\!\!\!\bigcirc\!\!\!-NO_2$$

Alkylation.   The strongly acidic conditions required for Friedel–Crafts alkylations often destroy five-membered ring heterocycles. Even when initial alkylation at C2 is successful, the products are susceptible to further alkylation. (This contrasts with acylation, in which the acyl group deactivates the ring to further reaction with electrophiles.) Thus, while some 2-alkyl and 2,5-dialkylfurans and pyrroles can be isolated from Friedel–Crafts reactions,[9] the yields are low, and these are generally not useful synthetic procedures.

Friedel–Crafts alkylation of thiophene is somewhat more successful, but yields are still low, as shown in Eq. 32.[12]*

------

*Pyrrole reacts with carbocations in the gas phase to yield the three possible monoalkylation products (Eq. 31).

$$\underset{\underset{\displaystyle H}{\displaystyle N}}{\left[\!\!\left[\begin{array}{c}\\\\\end{array}\right.\!\!\right]} + \; CH_3CH_2^{\oplus} \longrightarrow \underset{\underset{\displaystyle H}{\displaystyle N}}{\overset{\displaystyle CH_2CH_3}{\left[\!\!\left[\begin{array}{c}\\\\\end{array}\right.\!\!\right]}} + \underset{\underset{\displaystyle H}{\displaystyle N}\;CH_2CH_3}{\left[\!\!\left[\begin{array}{c}\\\\\end{array}\right.\!\!\right]} + \underset{\underset{\displaystyle CH_2CH_3}{\displaystyle N}}{\left[\!\!\left[\begin{array}{c}\\\\\end{array}\right.\!\!\right]} \qquad (31)$$

product ratios: 69   :   22   :   9

(The carbocations were formed by γ-ray irradiation of alkanes, so the reactions were acid free.) Unlike most other electrophilic substitution reactions of pyrrole, attack occurs principally at the β positions. It has been suggested that the product composition is determined by the initial electrostatic interaction between the reagents, so that reaction occurs principally at the site of highest electron density in pyrrole.[20] Gas-phase alkylations of furan occur principally at the α-carbons, while alkylation of thiophene occurs essentially equally at the α- and β-carbons.[20]

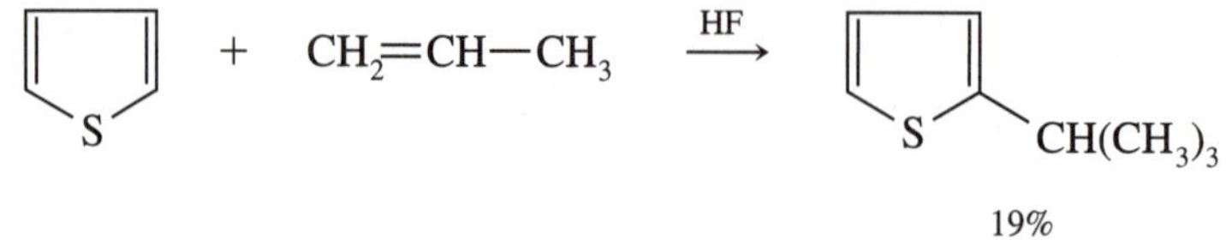

$$\text{(32)}$$

Pyrroles normally do not react with alkyl halides in the absence of catalysts. However, the more basic imidazole ring does react readily with alkylating agents. It is invariably the imino nitrogen atom that is alkylated, so that the resonance energies of the aromatic ring and of the amidine group are retained (Eq. 33).

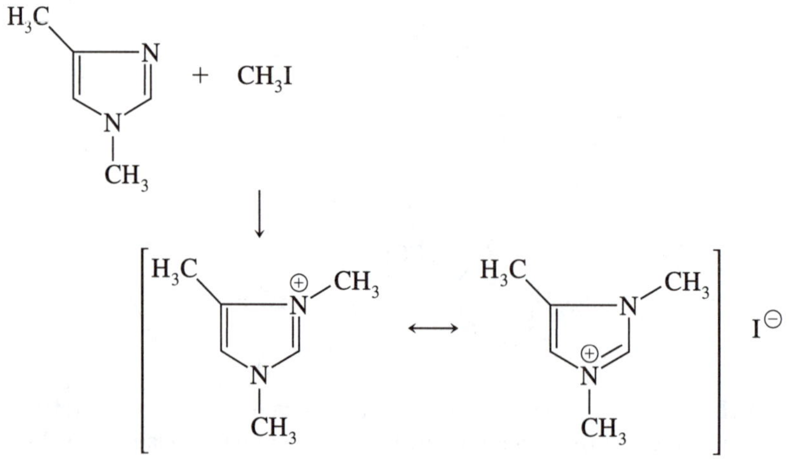

$$\text{(33)}$$

Reactions with carbonyl groups. In contrast to benzene and other aromatic hydrocarbons, π-excessive heterocyclic rings can react with aldehydes and ketones in the presence of acidic catalysts. However, the resulting hydroxyalkyl-substituted molecules (e.g., **1**) are very easily converted to "carbocations" such as **2** (Eq. 34).

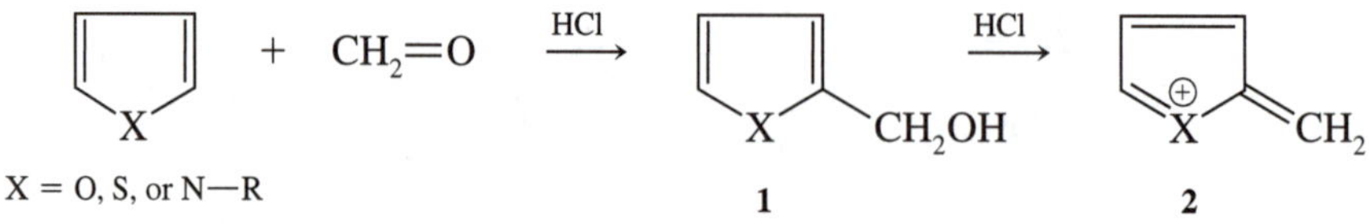

$$\text{(34)}$$

X = O, S, or N—R

These cations can themselves act as electrophiles in reactions with heterocyclic rings. Thus, the reaction of thiophene with formaldehyde and hydrochloric acid or zinc chloride forms compound

3, as well as smaller amounts of the chloromethylation product **4**. (Similar reactions can take place with derivatives of furan and pyrrole, as shown in Eq. 36.[14,16,17])

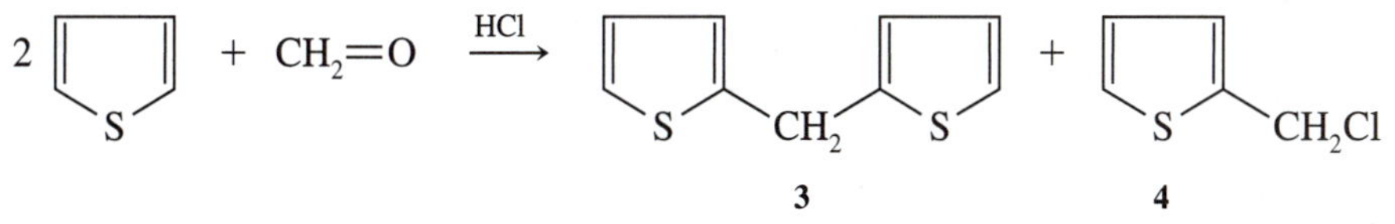

$$
2 \;\bigg[\!\!\bigg]_{S} \;+\; CH_2{=}O \;\xrightarrow{HCl}\; \underset{3}{\bigg[\!\!\bigg]_{S}\!CH_2\!\bigg[\!\!\bigg]_{S}} \;+\; \underset{4}{\bigg[\!\!\bigg]_{S}\!CH_2Cl} \tag{35}
$$

$$
2 \; H_3C\!\bigg[\!\!\bigg]_{\substack{N\\H}} \;+\; CH_2{=}O \;\xrightarrow{HCl}\; H_3C\!\bigg[\!\!\bigg]_{\substack{N\\H}}\!CH_2\!\bigg[\!\!\bigg]_{\substack{N\\H}}\!CH_3 \;+\; H_2O \tag{36}
$$

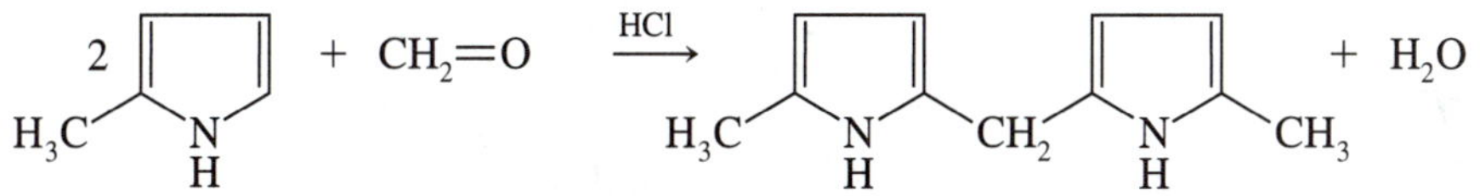

These processes need not stop at reaction of two heterocyclic rings. If the second α position of a pyrrole is not already occupied, for instance, the reaction is likely to continue, yielding tetrameric products (porphyrinogens), as shown in Eq. 37.

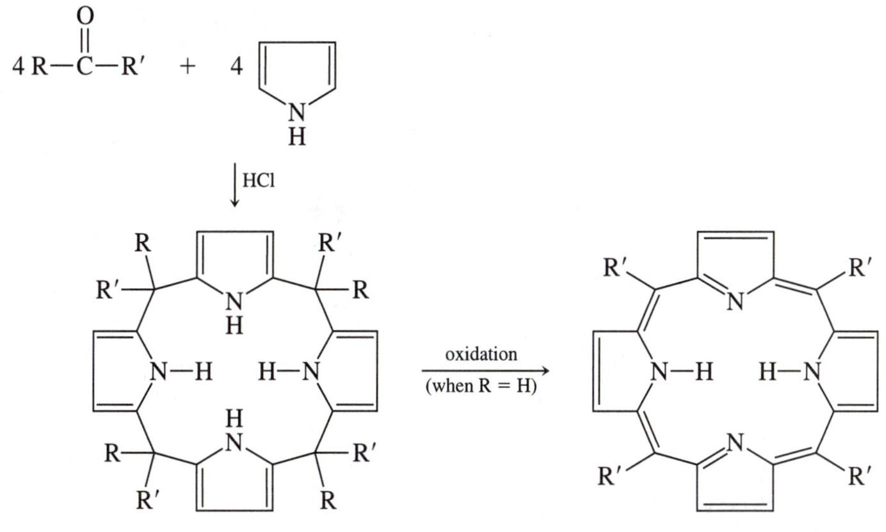

a porphyrin

The tetramers from reactions with aldehydes are usually immediately oxidized with loss of six hydrogen atoms to yield molecules called *porphyrins*.[21] The simplest member of this class (R′ = H) is called *porphin*.

It may appear that only one of the four heterocyclic rings in a porphyrin retains a pyrrole structure, but the molecule as a whole contains an 18 π-electron aromatic ring. The 18-electron aromatic ring system remains intact even if the double bonds between β positions of several of the five-membered rings are reduced, because unshared electrons on nitrogen atoms remain part of the 18-electron system.

The dianions of porphyrins form very strong complexes with metal ions. Several of these complexes are of enormous importance in biological systems. Hemes (iron complexes), the colored portions of hemoglobins, and chlorophylls (magnesium complexes) are perhaps the most obvious examples.

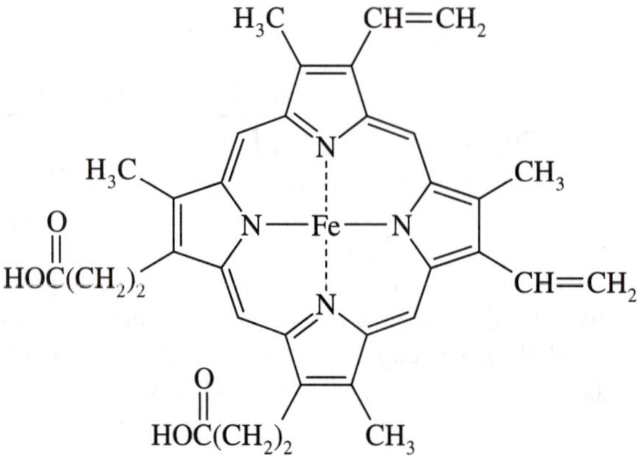

heme

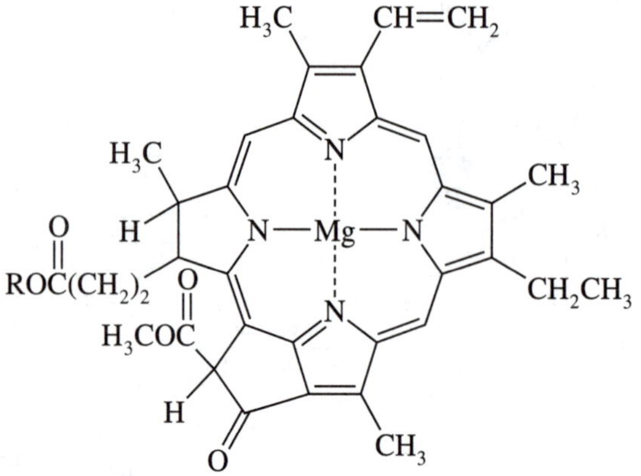

chlorophyll-a

Analogs of Mannich reactions.   Acid-catalyzed reactions of five-membered ring heterocycles with formaldehyde and secondary amines frequently yield 2-substituted derivatives, in a variation of the Mannich reaction of ketones with formaldehyde and amines.[22] The Mannich reaction is usually considered to proceed via the initial formation of an immonium salt, as shown in Eqs. 38 and 39.

$$(CH_3)_2NH \quad + \quad CH_2{=}O \quad \xrightarrow{HCl} \quad H_2C{=}\overset{\oplus}{N}(CH_3)_2 \quad Cl^{\ominus} \quad + \quad H_2O \qquad (38)$$

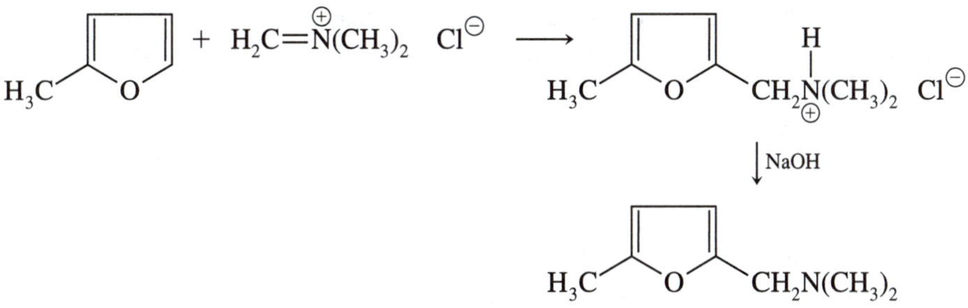

$$
\text{(39)}
$$

Reactions with conjugated carbonyls. Pyrroles undergo conjugate addition reactions with conjugated carbonyl compounds. As usual, reaction normally takes place at α positions, if possible. These reactions can be catalyzed by acids but may also proceed without catalysts, as shown in Eqs. 40a and 40b.

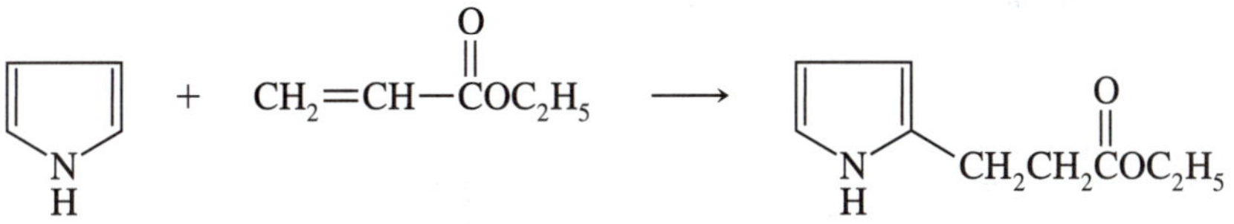

$$
\text{(40a)}
$$

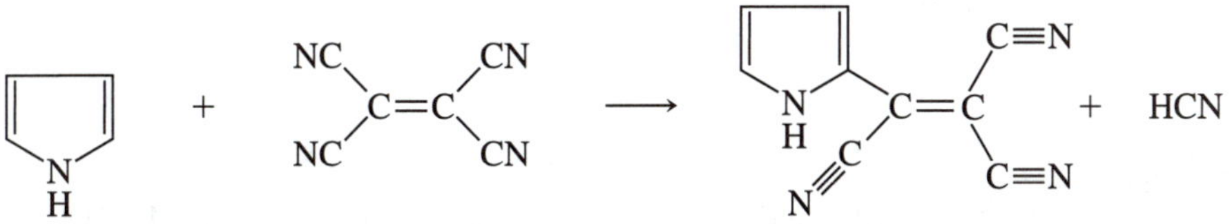

$$
\text{(40b)}
$$

N-substituted pyrroles react with conjugated acetylenes, such as acetylene dicarboxylic ester, to yield several different types of products. When the substituents on nitrogen are electron donating, products resulting from electrophilic substitutions, and retaining the pyrrole ring, are favored. When the substituents are electron-withdrawing, products apparently resulting from [4 + 2] cycloadditions are favored. However, as shown in Eq. 41, both types of products appear to arise from the same dipolar intermediates, the relative yields depending on the rates of ring closure versus hydrogen transfer.[23a]

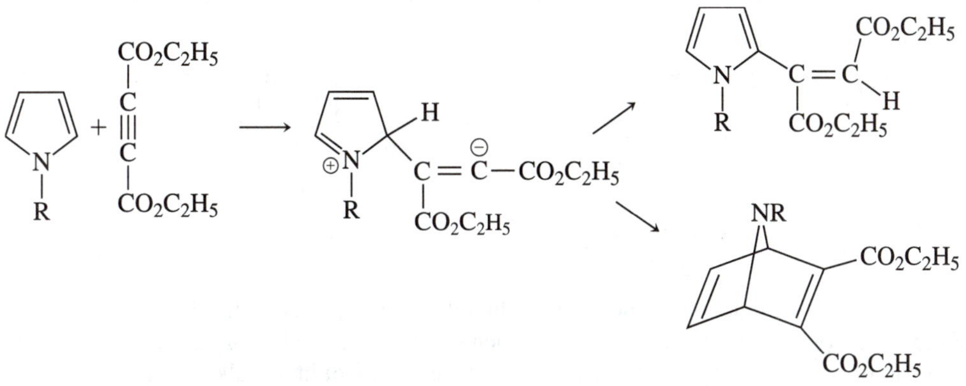

$$
\text{(41)}
$$

Furan, which has much lower aromatic stabilization than pyrrole, behaves like a simple diene in reactions with conjugated carbonyl compounds and other dienophiles, giving Diels–Alder adducts (Eq. 42).[23b]

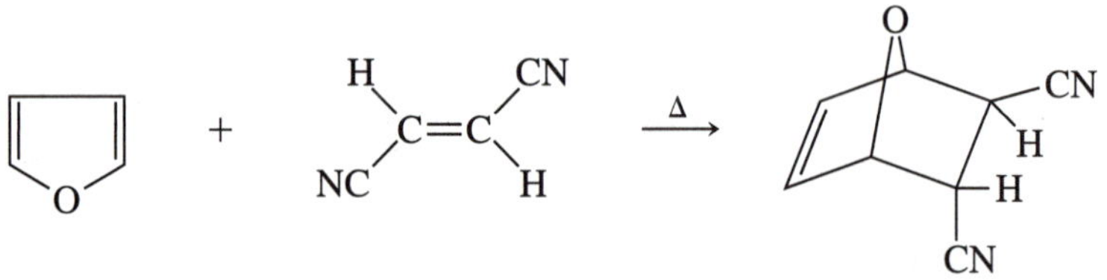

$$(42)$$

Reactions of fused-ring systems. In contrast to reactions of monocyclic five-membered aromatic rings, indole, benzofuran, and benzothiophene undergo electrophilic substitution predominantly at C3 (Eqs. 43a and 43b).

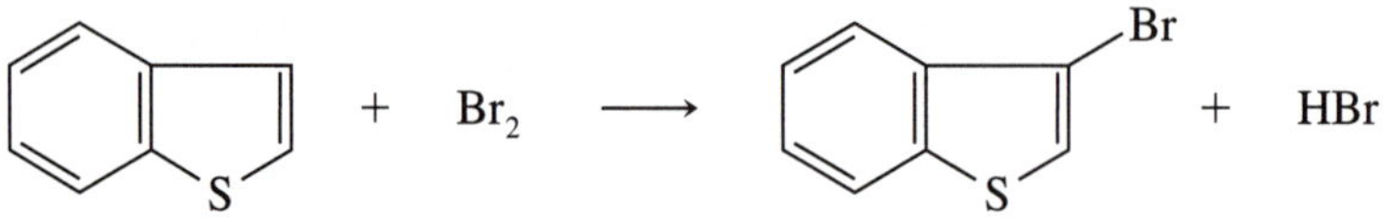

$$+ \quad Br_2 \longrightarrow \quad + \quad HBr \qquad (43a)$$

$$+ \quad C_6H_5\overset{O}{\overset{\|}{C}}ONO_2 \longrightarrow \quad + \quad C_6H_5\overset{O}{\overset{\|}{C}}OH \qquad (43b)$$

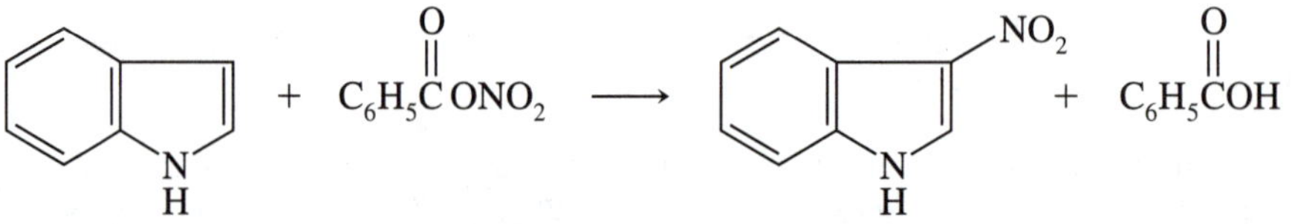

Substitution at C3 in any of the benzo-substituted heterocycles yields a very stable cation in which a new bond to the heteroatom has been formed to replace the carbon–carbon π bond broken by attack of the electrophilic agent. If, instead, C2 were attacked by the electrophile, formation of a π bond to the heteroatom would disrupt the aromatic ring.

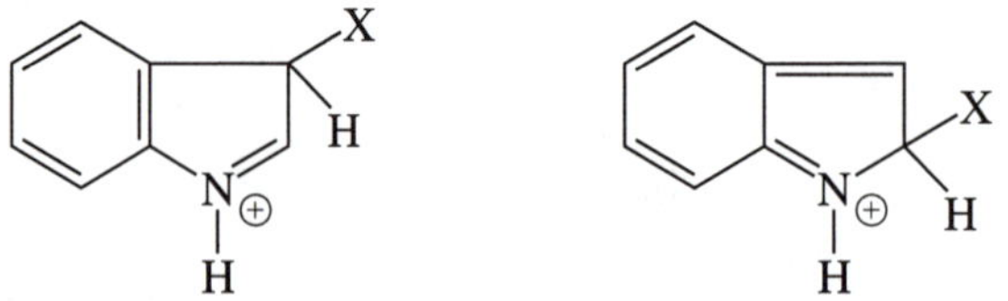

a very stable cation

a less stable cation

The Mannich reaction of indole with dimethylamine yields a salt of the C3-substituted alkaloid *gramine* (one of the flavoring constituents of beer) (Eq. 44). Similar Mannich reactions form part of the biosynthetic routes leading to a wide variety of indole alkaloids.

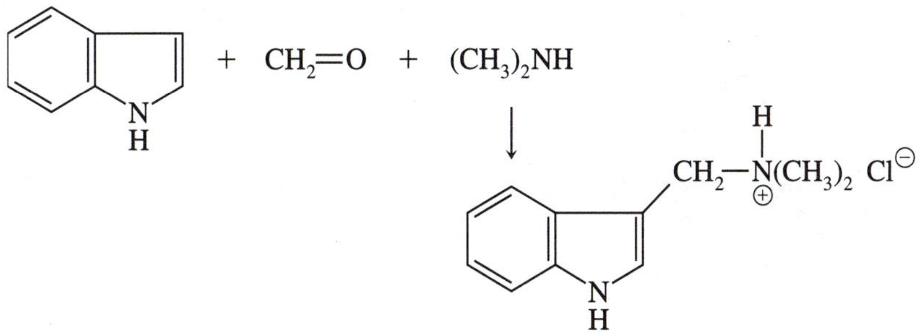

$$+ \quad CH_2{=}O \quad + \quad (CH_3)_2NH$$

$$\downarrow$$

$$CH_2{-}\overset{H}{\underset{\oplus}{N}}(CH_3)_2 \ Cl^{\ominus} \qquad (44)$$

The tendency for indole derivatives to react at C3 is so strong that many products that appear to result from reaction at C2 are actually formed by initial attack at C3, followed by migration of the attacking group to C2. Deuterium labeling demonstrated this to be the case in the reaction shown in Eq. 45.[24]

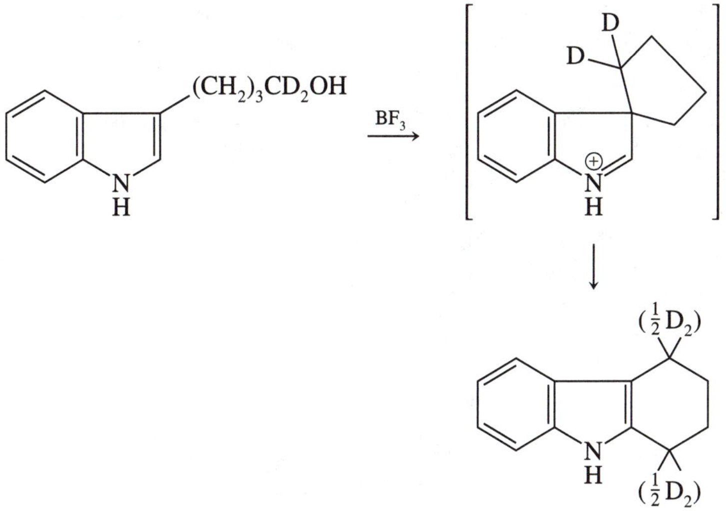

$$(CH_2)_3CD_2OH \quad \xrightarrow{BF_3}$$

$$\downarrow \qquad (45)$$

$$(\tfrac{1}{2}D_2)$$

$$(\tfrac{1}{2}D_2)$$

Reactions with Carbenes and Nitrenes

Heating or photoirradiating ethyl diazoacetate in the presence of furan or thiophene yields cyclopropane derivatives (Eq. 46).[24]

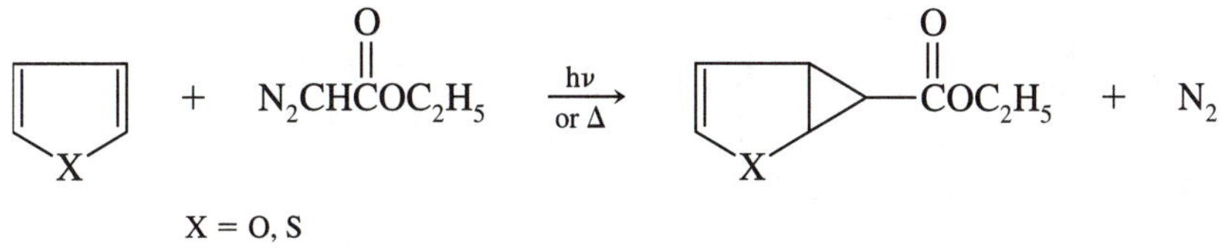

$$\begin{array}{c} \\ \underset{X}{\boxed{}} \end{array} + \ N_2CH\overset{O}{\overset{\|}{C}}OC_2H_5 \ \xrightarrow[\text{or }\Delta]{h\nu} \ \underset{X}{\boxed{}}{-}\overset{O}{\overset{\|}{C}}OC_2H_5 \ + \ N_2 \qquad (46)$$

X = O, S

Pyrrole, however, is converted to 2-carbethoxymethylpyrrole under the same conditions (Eq. 47).[25,26]

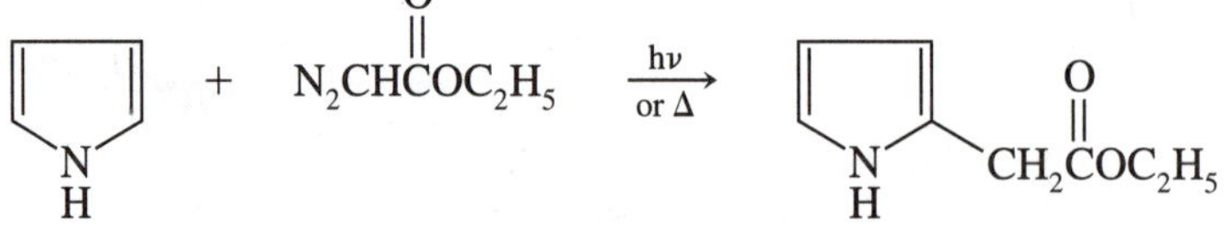

$$(47)$$

Pyrroles react with chloroform or bromoform and bases to form 3-halopyridines (Eq. 48).[14]

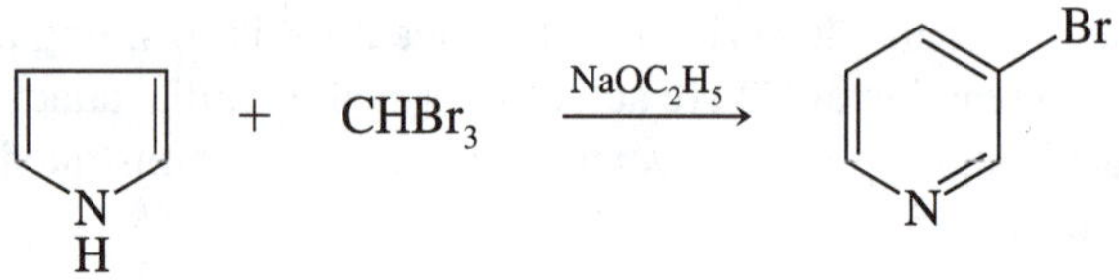

$$(48)$$

This interesting rearrangement appears to proceed by way of a cyclopropane intermediate, formed by the reaction of a dihalocarbene with the anion of the pyrrole. It is not clear whether the cyclopropane is formed in one step or via an intermediate carbanion, as shown in Eq. 50.

$$\text{CHBr}_3 + \text{RO}^{\ominus} \longrightarrow \text{:CBr}_2 + \text{ROH} + \text{Br}^{\ominus} \qquad (49)$$

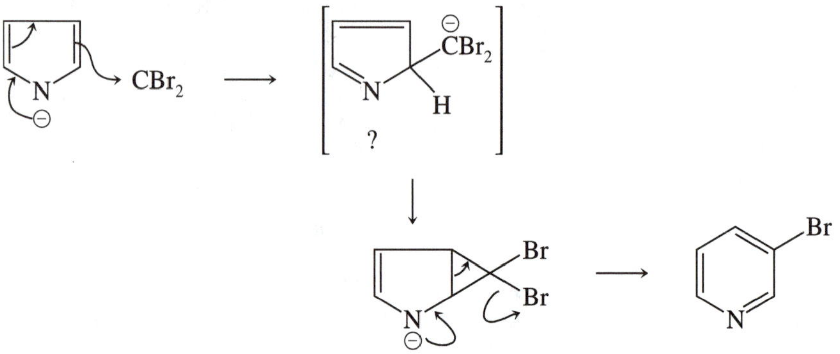

$$(50)$$

Remarkably, reactions of thiophenes, pyrroles, and furans with carbethoxynitrene all result in the formation of products containing pyrrole rings. Sulfur is eliminated in the reaction of thiophene, while the heteroatoms from furan and pyrrole are retained in the products (Eqs. 51 to 53).[27]

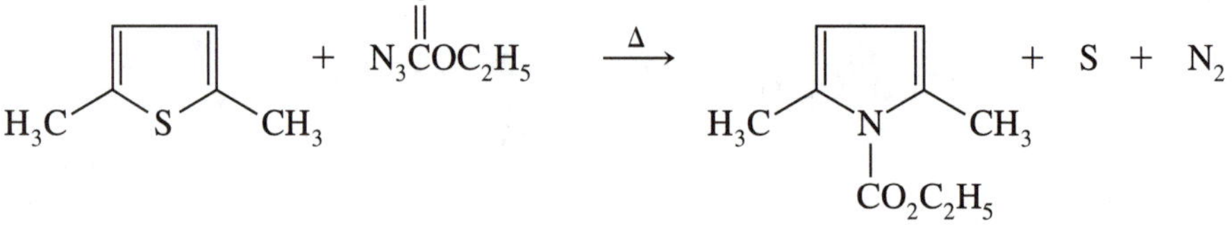

$$(51)$$

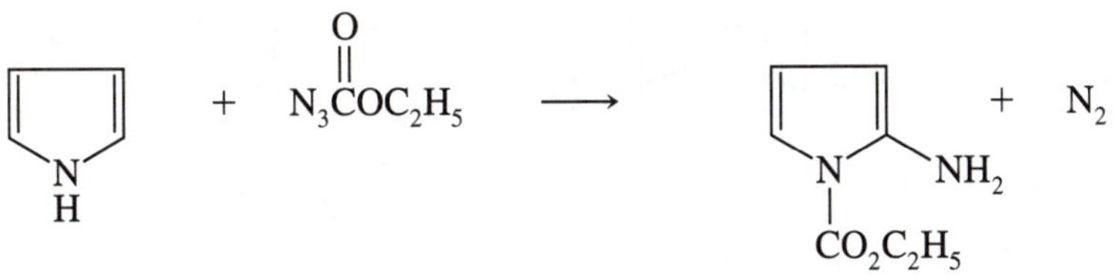

$$
\text{(pyrrole)} \quad + \quad \text{N}_3\overset{\text{O}}{\underset{}{\text{C}}}\text{OC}_2\text{H}_5 \quad \longrightarrow \quad \text{(product with NH}_2\text{, CO}_2\text{C}_2\text{H}_5) \quad + \quad \text{N}_2 \tag{52}
$$

$$
\text{(furan)} \quad + \quad \text{N}_3\overset{\text{O}}{\underset{}{\text{C}}}\text{OC}_2\text{H}_5 \quad \longrightarrow \quad \text{(product with N, O, CO}_2\text{C}_2\text{H}_5) \quad + \quad \text{N}_2 \tag{53}
$$

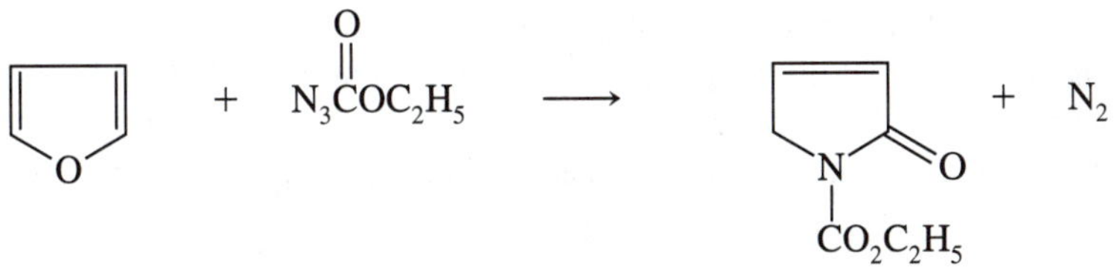

The structures of the reaction products suggest that they are formed by 1,4-additions of the nitrenes to the rings, followed by ring-opening steps. However, 1,4-addition reactions of carbenes are very rare (see Section 9.4). It seems likely, therefore, that initial 1,2-additions are followed by rearrangements to form the 1,4-addition products, as shown in Eq. 54.

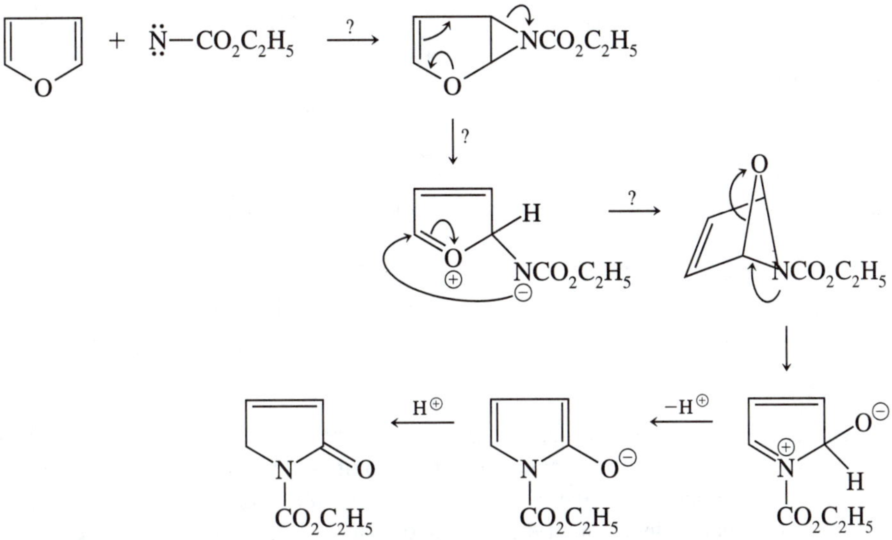

$$
\tag{54}
$$

$$
\tag{55}
$$

12.4 SYNTHESIS OF FIVE-MEMBERED HETEROAROMATIC RINGS

Carbonyl Cyclization Reactions

Furan, pyrrole, and thiophene rings are most commonly prepared from open-chain carbonyl derivatives. Derivatives of furan, for example, are readily prepared by reacting 1,4-dicarbonyl compounds with acids (Eq. 55).

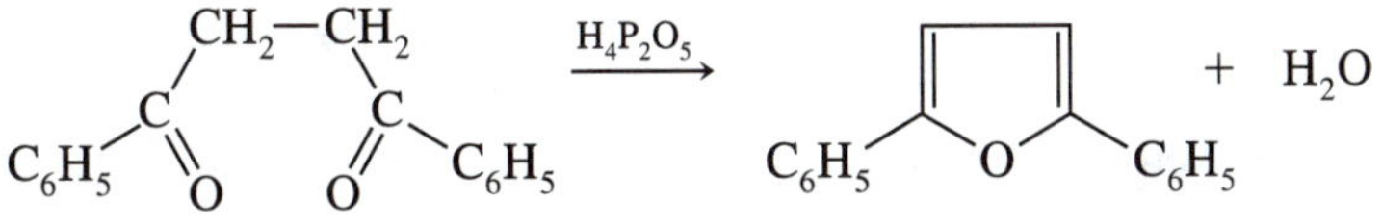

$$
\underset{\text{C}_6\text{H}_5}{\overset{\text{CH}_2-\text{CH}_2}{\underset{\text{O} \quad \text{O}}{\text{C} \qquad \text{C}}}}\text{C}_6\text{H}_5 \quad \xrightarrow{\text{H}_4\text{P}_2\text{O}_5} \quad \underset{\text{C}_6\text{H}_5}{\overset{}{\text{O}}}\text{C}_6\text{H}_5 \quad + \quad \text{H}_2\text{O} \tag{55}
$$

Furan itself is prepared commercially by pyrolysis of furfural (furan-2-carboxaldehyde) in the presence of metal-containing catalysts (Eq. 56).

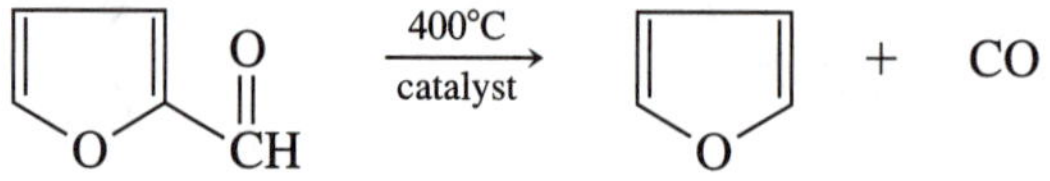

$$\underset{\text{catalyst}}{\overset{400°C}{\longrightarrow}} \qquad + \quad CO \qquad (56)$$

Furfural, in turn, is obtained commercially by treating fibrous vegetable materials, such as oat bran or corncobs, with acid. In the initial stages of this process, carbohydrate polymers are hydrolyzed to pentoses, such as xylose. The pentoses are then dehydrated and cyclized to form furfural (Eq. 57). The details of the mechanism are still not certain, but compound **1** appears to be a key intermediate.[2]

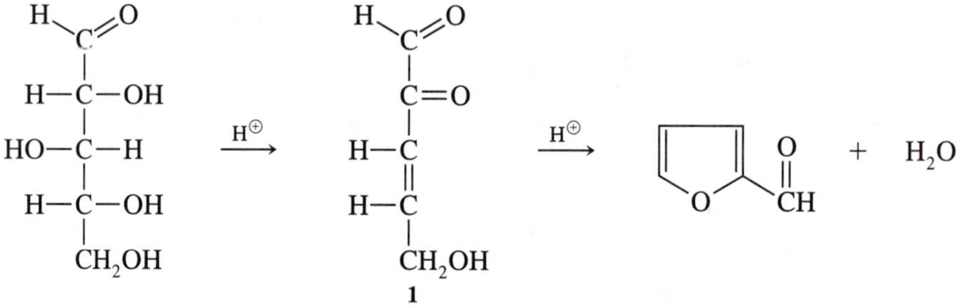

$$\overset{H^{\oplus}}{\longrightarrow} \qquad \overset{H^{\oplus}}{\longrightarrow} \qquad + \quad H_2O \qquad (57)$$

1

Thiophenes may be prepared by reactions of 1,4-dicarbonyl compounds with phosphorus pentasulfide (Eq. 58) or with hydrogen sulfide in the presence of mineral acids.[14]

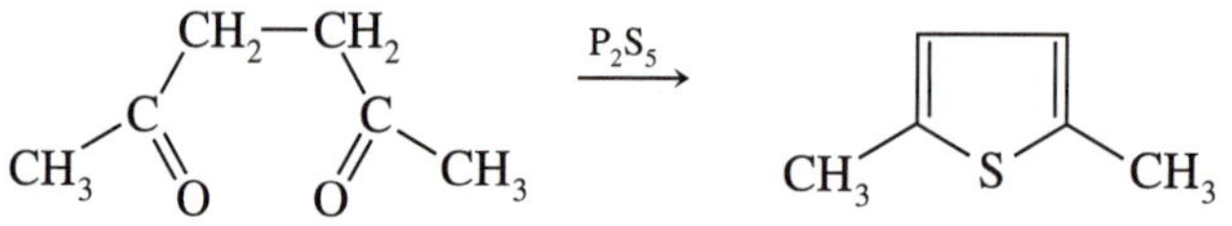

$$\overset{P_2S_5}{\longrightarrow} \qquad (58)$$

Pyrroles are commonly formed by the *Paal–Knorr synthesis,** in which 1,4-dicarbonyl compounds are condensed with ammonia or primary amines.[28] Unlike the similar cyclizations to form furans and thiophene rings, the synthesis of pyrroles takes place under weakly acidic conditions to avoid complete conversion of the amines to their salts. The Paal–Knorr reaction has been applied to the synthesis of "heterocyclophanes," in which C2 and C5 or C2 and C4 of the pyrrole rings are linked to each other so as to form polycyclic ring systems (Eqs. 59a and 59b).[29]

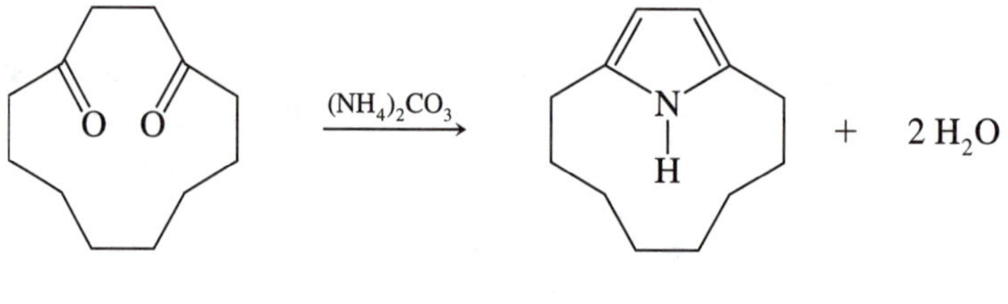

$$\overset{(NH_4)_2CO_3}{\longrightarrow} \qquad + \quad 2\ H_2O \qquad (59a)$$

*The term *Paal–Knorr synthesis* is also applied to the preparation of furans and thiophenes from 1,4-dicarbonyl compounds.

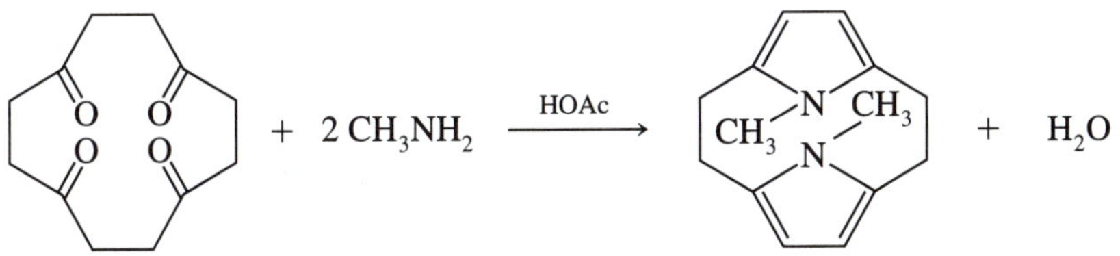

$$\text{(structure)} \quad + \quad 2\ CH_3NH_2 \quad \xrightarrow{\ HOAc\ } \quad \text{(structure)} \quad + \quad H_2O \qquad (59b)$$

In the *Knorr synthesis,* α-aminoaldehydes and α-aminoketones condense with β-dicarbonyl compounds or β-ketoesters at approximately pH 4 to form pyrroles (Eq. 60).

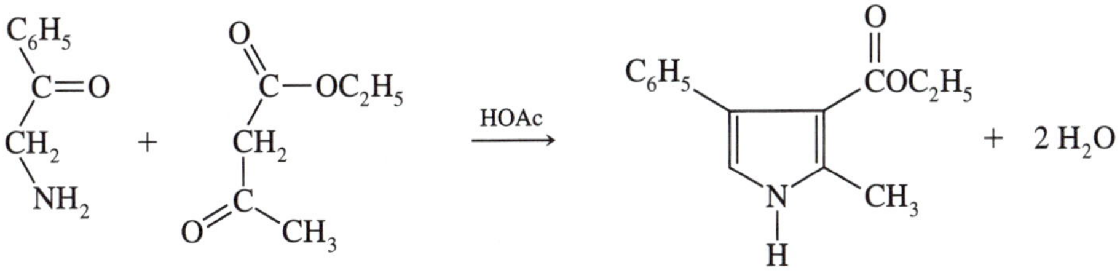

$$\text{(structure)} \quad + \quad \text{(structure)} \quad \xrightarrow{\ HOAc\ } \quad \text{(structure)} \quad + \quad 2\ H_2O \quad (60)$$

One difficulty with this process is that α-aminocarbonyl compounds easily undergo self-condensation on standing, so they must usually be prepared immediately before use. Often, the aminoketones are actually prepared in the presence of the dicarbonyl compounds to minimize the likelihood of self-condensation.[30] One convenient way to do this is by reduction of an α-ketooxime, which may be formed by reaction of a carbonyl compound with an alkyl nitrite in the presence of base, as shown in Eq. 61.

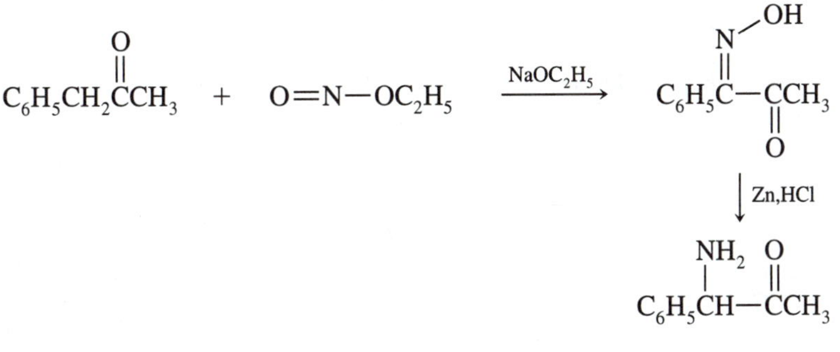

$$C_6H_5CH_2\overset{O}{\overset{\|}{C}}CH_3 \quad + \quad O{=}N{-}OC_2H_5 \quad \xrightarrow{\ NaOC_2H_5\ } \quad C_6H_5\overset{N\!-\!OH}{\overset{\|}{C}}{-}\overset{O}{\overset{\|}{C}}CH_3$$

$$\Big\downarrow Zn, HCl$$

$$C_6H_5\overset{NH_2}{\overset{|}{C}}H{-}\overset{O}{\overset{\|}{C}}CH_3 \qquad\qquad (61)$$

(Reaction of a carbonyl compound with an alkyl nitrite closely resembles a Claisen condensation, except that the nitrite ester replaces a carboxylate ester as the electrophile that reacts with an enolate anion.)

In the *Hantzsch synthesis,* pyrroles are formed by the reaction of ammonia or a primary amine with an α-haloketone and a β-ketoester (Eq. 62).[30] The orientation of substituents in the products indicates that α-aminocarbonyl compounds are not intermediates in the Hantzsch synthesis. The exact sequence of reactions in its mechanism is still not certain, but it presumably starts by addition of the ammonia or amine to the carbonyl group of the ketoester.

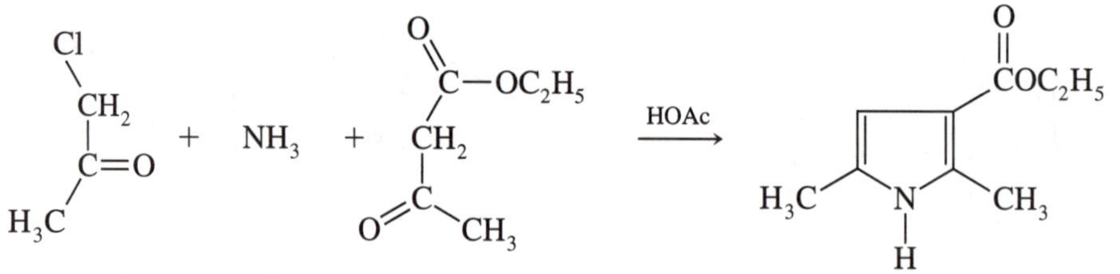

$$\text{(62)}$$

Pyrroles may also be formed by conjugate addition of α-aminocarbonyl compounds to acetylenic ketones or esters[30] (Eq. 63).

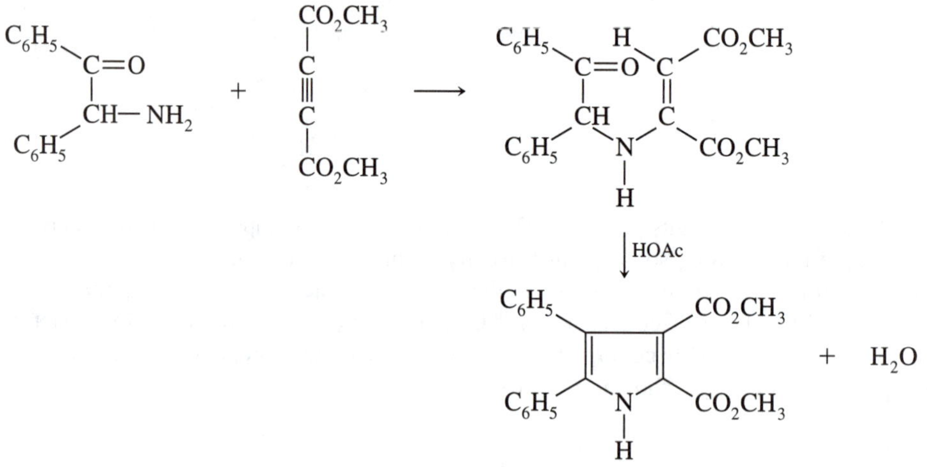

$$\text{(63)}$$

Reductive cyclization. Thiophenes can be prepared by pinacol reductions of α-thiacarbonyl compounds followed by dehydration, as shown in Eq. 64.[31]

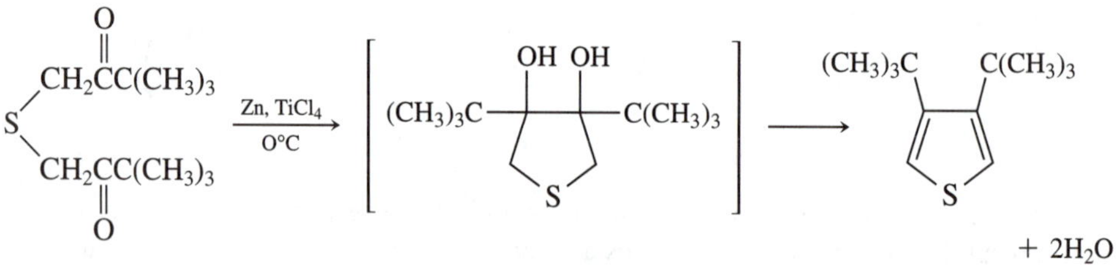

$$\text{(64)}$$

The Fischer Indole Synthesis

In 1883, Emil Fischer discovered that phenylhydrazones of aldehydes and ketones are converted to indoles on reaction with strong acids (Eq. 65). This reaction has since proved to be one of the most useful methods for preparing indoles.[32]

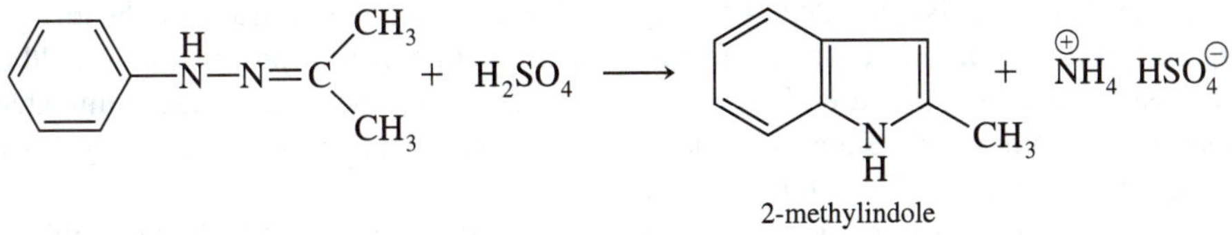

$$\text{(65)}$$

2-methylindole

A key step in the *Fischer indole synthesis* is the conversion of the phenylhydrazone to its enamine tautomer, which can then undergo a [3,3] sigmatropic shift to form a new carbon–carbon bond. Tautomerism of the resulting cyclohexadienimine to an aniline and addition of the amino group of the aniline to the imino group of the side chain yields the indole (Eq. 66).

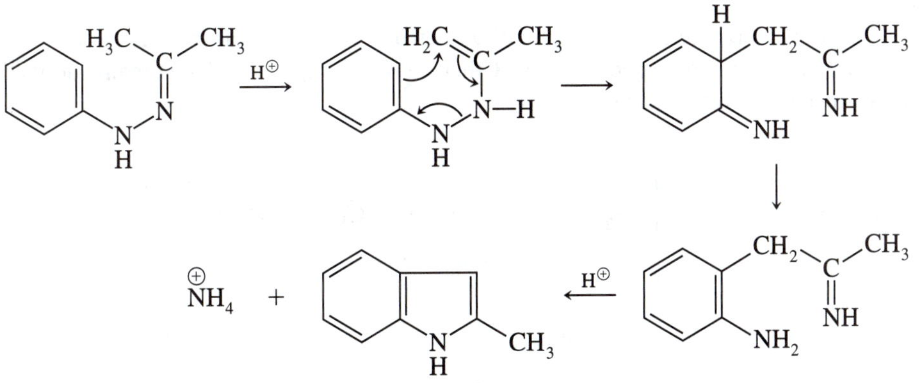

$$\text{(66)}$$

It is worth noting that the sigmatropic shift forms the diimine of a 1,4-dicarbonyl compound, so that the early steps in the Fischer indole synthesis can be regarded simply as an unusual method of preparing 1,4-dicarbonyl derivatives for cyclization to pyrroles.

Dipolar Cycloaddition Reactions

The Hantzsch and Knorr reactions and the Fischer indole synthesis have been used to prepare heterocyclic compounds since the nineteenth century. In contrast, although individual examples of dipolar cycloaddition reactions have been known for many decades, the concept of these reactions as a general class that can be used to prepare a wide variety of heterocyclic rings was first proposed by Rolf Huisgen of the University of Münich in 1960.[33]

In dipolar cycloaddition reactions, a "1,3-dipole" adds to a π bond to form a five-membered heterocyclic ring.[34-36] By analogy with the Diels–Alder reaction, the molecule containing the reacting π bond is called a *dipolarophile* (Eq. 67).

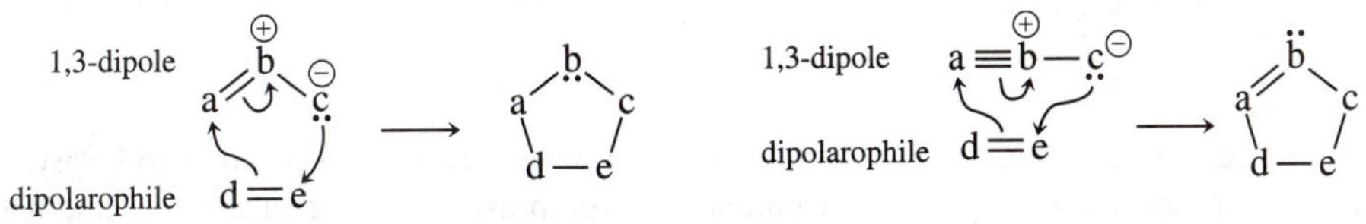

$$\text{(67)}$$

A 1,3-dipole can be regarded as an allylic anion in which the central atom bears a positive charge (at least in its major resonance forms). Thus, 1,3-dipoles bear no net charges, although it is impossible to draw structures for the molecules in which the charges are eliminated. 1,3-Dipoles may also have triple bonds in place of the double bonds, in which case, they resemble propargylic anions rather than allylic anions.

Some 1,3-dipoles exist as reasonably stable molecules. Ozone and hydrazoic acid are well-known inorganic dipoles.

$$\left[\overset{..}{\underset{..}{O}} = \overset{\oplus}{\underset{..}{O}} - \overset{\ominus}{\underset{..}{O}} : \quad \longleftrightarrow \quad : \overset{\ominus}{\underset{..}{O}} - \overset{\oplus}{\underset{..}{O}} = \overset{..}{\underset{..}{O}} : \right]$$

ozone

$$\left[H - \overset{..}{\underset{..}{N}} = \overset{\oplus}{N} = \overset{\ominus}{\underset{..}{N}} : \quad \longleftrightarrow \quad HN \equiv \overset{\oplus}{N} - \overset{\ominus}{\underset{..}{N}} : \right]$$

hydrazoic acid

The best-known organic 1,3-dipoles are diazomethane and its derivatives, such as ethyl diazoacetate (which was first prepared in 1883). Other stable 1,3-dipoles include organic derivatives of hydrazoic acid, such as phenyl azide.

$$\left[CH_2 = \overset{\oplus}{N} = \overset{\ominus}{\underset{..}{N}} : \quad \longleftrightarrow \quad \overset{..}{C}H_2 - \overset{\oplus}{N} \equiv N : \right]$$

diazomethane

$$\left[C_2H_5O\overset{O}{\overset{||}{C}} - CH = \overset{\oplus}{N} = \overset{\ominus}{\underset{..}{N}} : \quad \longleftrightarrow \quad C_2H_5O\overset{O}{\overset{||}{C}} - \overset{\ominus}{\underset{..}{C}}H - \overset{\oplus}{N} \equiv N : \right]$$

ethyl diazoacetate

$$\left[C_6H_5 - \overset{..}{\underset{..}{N}} = \overset{\oplus}{N} = \overset{\ominus}{\underset{..}{N}} : \quad \longleftrightarrow \quad C_6H_5 - \overset{\ominus}{\underset{..}{N}} - \overset{\oplus}{N} \equiv N : \right]$$

phenyl azide

Formation of heterocyclic rings by cycloaddition reactions of diazoalkane derivatives with alkenes was first observed in the nineteenth century, although the nature of the processes involved were somewhat obscured by proton shifts in the initial products (Eq. 68).[37]

(68)

Unlike diazoalkanes and aryl azides, many 1,3-dipoles are too reactive to be isolated at room temperatures, and they must be prepared in the presence of the alkenes or alkynes with which they will react. Among the most useful 1,3-dipoles are "nitrile ylides," which may be prepared by the dehydrochlorination of imidoyl chlorides (Eq. 69) or by the photolysis of 2H-azirines (Eq. 70).[34]

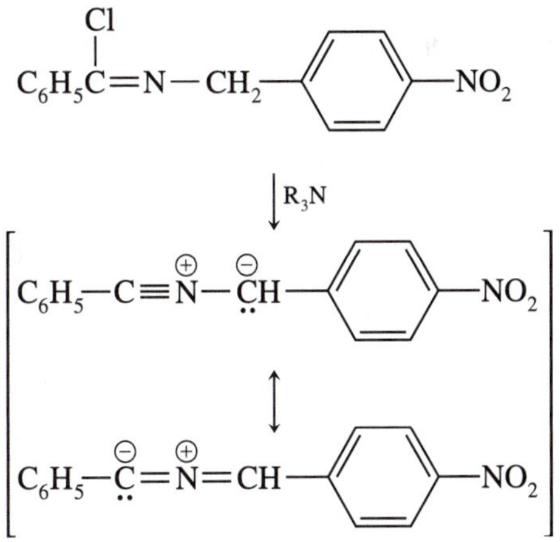

$$
\begin{array}{c}
\underset{\text{Cl}}{\overset{|}{C_6H_5C}}=N-CH_2-\!\!\!\left\langle\!\!\!\bigcirc\!\!\!\right\rangle\!\!\!-NO_2
\end{array}
$$

$$\downarrow R_3N$$

$$
\left[
\begin{array}{c}
C_6H_5-C\overset{\oplus}{\equiv}\overset{\ominus}{N}-\overset{\ominus}{\underset{\cdot\cdot}{C}H}-\!\!\!\left\langle\!\!\!\bigcirc\!\!\!\right\rangle\!\!\!-NO_2 \\
\\
\updownarrow \\
\\
C_6H_5-\overset{\ominus}{\underset{\cdot\cdot}{C}}=\overset{\oplus}{N}=CH-\!\!\!\left\langle\!\!\!\bigcirc\!\!\!\right\rangle\!\!\!-NO_2
\end{array}
\right] \qquad (69)
$$

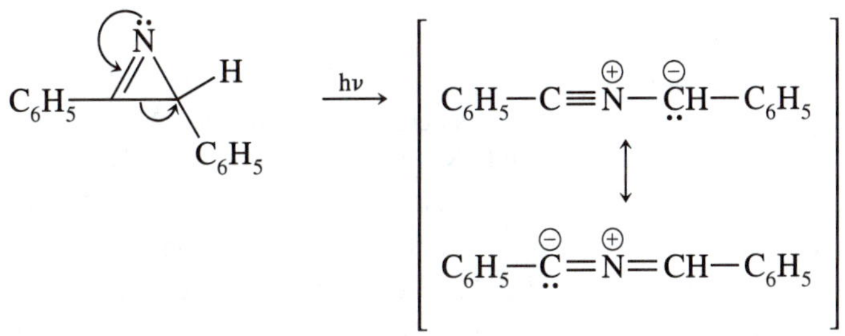

$$
\underset{C_6H_5}{\overset{\ddot{N}}{\diagdown}}\!\!\!\!\!\!\!\!\!\!\bigtriangleup\!\!\!\overset{H}{\underset{C_6H_5}{}}
\quad\overset{h\nu}{\longrightarrow}\quad
\left[
\begin{array}{c}
C_6H_5-C\overset{\oplus}{\equiv}\overset{\ominus}{N}-\overset{}{\underset{\cdot\cdot}{C}H}-C_6H_5 \\
\\
\updownarrow \\
\\
C_6H_5-\overset{\ominus}{\underset{\cdot\cdot}{C}}=\overset{\oplus}{N}=CH-C_6H_5
\end{array}
\right] \qquad (70)
$$

Nitrile ylides cannot be isolated, but they can be trapped by reaction with molecules containing π bonds to form heterocyclic rings, as shown in Eq. 71.

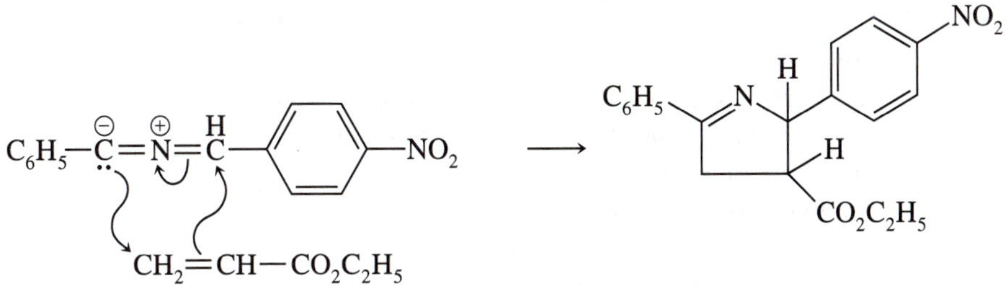

$$
\begin{array}{c}
C_6H_5-\overset{\ominus}{\underset{\cdot\cdot}{C}}=\overset{\oplus}{N}=\overset{H}{\underset{}{C}}-\!\!\!\left\langle\!\!\!\bigcirc\!\!\!\right\rangle\!\!\!-NO_2 \\
\\
CH_2{=}CH-CO_2C_2H_5
\end{array}
\quad\longrightarrow\quad
\begin{array}{c}
\end{array} \qquad (71)
$$

The initial cycloaddition products may be converted to pyrroles by oxidation or by the elimination of leaving groups, if any are present on the rings. Reaction of nitrile ylides with compounds containing triple bonds yields 2H-pyrroles that rapidly tautomerize to 1H-pyrroles (Eq. 72).

(72)

REFERENCES

[1] For further information about the nomenclature of polycyclic ring systems, see A.M. Patterson, L.T. Capell, and D.F. Walker, *The Ring Index*, 2nd ed. American Chemical Society, Washington, D.C. (1960).

[2] N. Suryaprakash, A.C. Kunwar, and C.L. Khetrapal, *Chem. Phys. Lett., 107,* 333 (1984).

[3] Pyrrole: J.M. Andre, D.P. Vercauteren, G.B. Street, and J.L. Bredas, *J. Chem. Phys., 80,* 5643 (1984). Furan: I.G. John and L. Radom, *J. Am. Chem. Soc., 100,* 3981 (1978).

[4] B.Y. Simkin, B.I. Minkin, and M.N. Glukhovtsev, *Adv. Hetero. Chem., 56,* 304 (1993).

[5] A.F. Pozharskii, *Chem. Heterocycl. Comp., 21,* 717 (1985).

[6] Y. Chiang and E.B. Whipple, *J. Am. Chem. Soc., 85,* 2763 (1963).

[7] E.B. Whipple, Y. Chiang, and R.L. Hinman, *J. Am. Chem. Soc., 85,* 26 (1963).

[8] G.F. Smith, *Adv. Hetero. Chem., 2,* 287 (1963).

[9] A.H. Jackson et al., in *The Chemistry of Heterocyclic Compounds: Pyrroles,* vol. 48, part 1, R.A. Jones, Ed. John Wiley & Sons, New York (1990), pp. 295–391.

[10] B.L. Bray et al., *J. Org. Chem., 55,* 6317 (1990).

[11] G. Angelini, C. Giancaspro, G. Illuminati, and G. Sleiter, *J. Org. Chem., 45,* 1786 (1980).

[12] E. Baciocchi, S. Clementi, and G.V. Sebastiani, *Chem. Comm.,* 875 (1975).

[13] N. Clauson-Kaas, *Acta Chem. Scand., 1,* 379 (1947).

[14] R. Taylor, in *The Chemistry of Heterocyclic Compounds: Thiophene and Its Derivatives*, vol. 44, part 1, S. Gronowitz, Ed. John Wiley & Sons, New York (1985), pp. 33–88.

[15] N. Clauson-Kaas and J. Fakstorp, *Acta Chem. Scand., 1,* 210 (1947).

[16] P. Bosshard and C.H. Eugster, *Adv. Hetero. Chem., 7,* 377 (1966).

[17] F.M. Dean, *Adv. Hetero. Chem., 30,* 168 (1982).

[18] G. Yagil, *Tetrahedron, 23,* 2855 (1967).

[19] K. Schofield, M.R. Grimmett, and B.R.T. Keene, *Heteroaromatic Nitrogen Compounds. The Azoles.* Cambridge University Press, New York (1976).

[20] G. Laguzzi and M. Speranza, *Perkin Trans. II,* 857 (1987); G. Laguzzi, R. Bucci, F. Grandinetti, and M. Speranza, *J. Am. Chem. Soc., 112,* 3064 (1990).

[21] J.S. Lindsey et al., *J. Org. Chem., 52,* 827 (1987).

[22] R.F. Holdren and R.M. Hixon, *J. Am. Chem. Soc., 68,* 1198 (1946).

[23] (a) L.R. Domingo, M.T. Picher, and R.J. Zaragoza, *J. Org. Chem., 63,* 9183 (1998); (b) H. Kuehn and O. Stein, *Ber., 70,* 567 (1937).

[24] D.W. Clack, H.L. Jackson, N. Prasitpan, and D.V.R. Shannon, *Perkin Trans. II,* 909 (1982).

[25] W. Seinkop and H. Augestad-Jensen, *Liebigs Ann., 428,* 154 (1922); G.O. Schenk and R. Steinmetz, *Liebigs Ann., 668,* 19 (1963).

[26] C.D. Nenitzescu and E. Solomonica, *Chem. Ber., 64,* 1924 (1931).

[27] K. Haffner and W. Kaiser, *Tetrahedron Lett.,* 2185 (1964).

[28] M.V. Sargent and T.M. Crespi, in *Comprehensive Organic Chemistry,* D. Barton and W.D. Ollis, Eds. Pergamon Press, New York (1979), pp. 693–720.

[29] (a) H. Nozaki, T. Kuyama, and T. Mori, *Tetrahedron, 25,* 5357 (1969); (b) J.F. Haley, S.M. Rosenfeld, and P.M. Keehn, *J. Org. Chem., 42,* 1379 (1977).

[30] G.P. Bean in *The Chemistry of Heterocyclic Compounds: Pyrroles,* vol. 48, part 1, R.A. Jones, Ed. John Wiley & Sons, New York (1990), pp. 105–294.

[31] J. Nakayama et al, *J. Org. Chem., 64,* 6499 (1999).

[32] B. Robinson, *The Fischer Indole Synthesis.* John Wiley & Sons, New York (1982).

[33] R. Huisgen, *Naturwiss. Rundschau, 14,* 63 (1961) and *Proc. Chem. Soc.,* 357 (1961).

[34] A. Padwa, Ed., *1,3-Dipolar Cycloaddition Chemistry,* vol. 1. John Wiley & Sons, New York (1984).

[35] P.N. Confalone and E.M. Huie, *Organic Reactions, 36,* 1 (1988).

[36] A. Padwa, *Acc. Chem. Res., 24,* 22 (1991).

[37] E. Buechner, *Ber., 21,* 2637 (1888); E. Buechner, M. Fritsch, A. Papendieck, and H. Witter, *Liebigs Ann., 273,* 214 (1893).

PROBLEMS

12.1 Draw structures for the principal organic products from the reactions below, or write "no reaction."

(a)

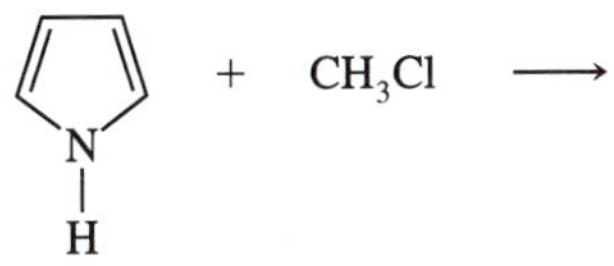

$+$ CH_3Cl $\longrightarrow$

(b)

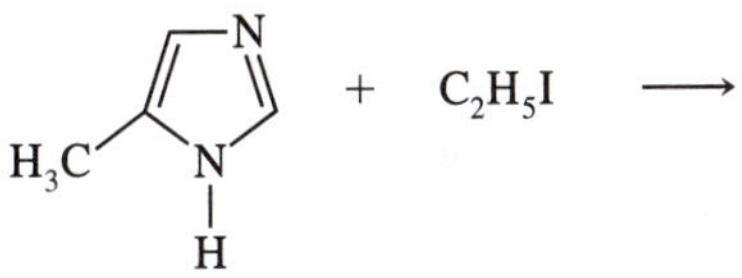

$+$ C_2H_5I $\longrightarrow$

(c)

$+$ HNO_3 $\xrightarrow{Ac_2O}$

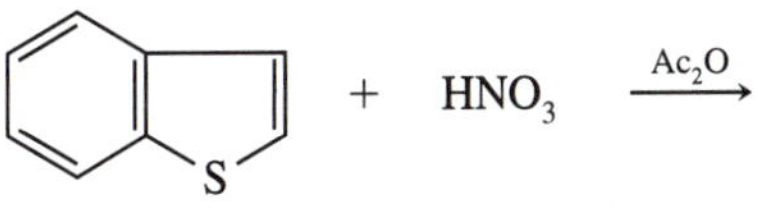

(d) 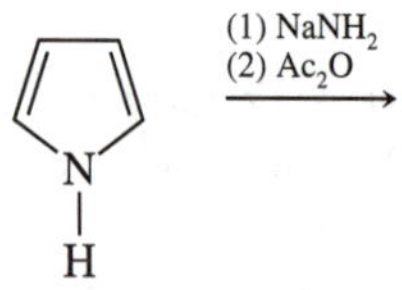 [pyrrole structure with N–H] $\xrightarrow{\text{(1) NaNH}_2 \\ \text{(2) Ac}_2\text{O}}$

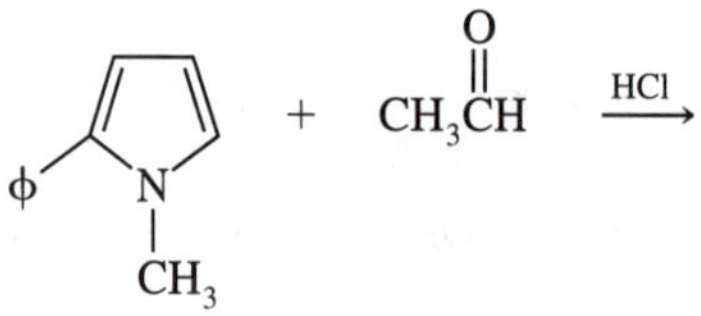

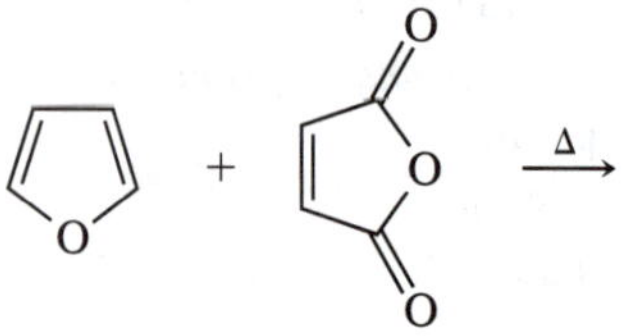

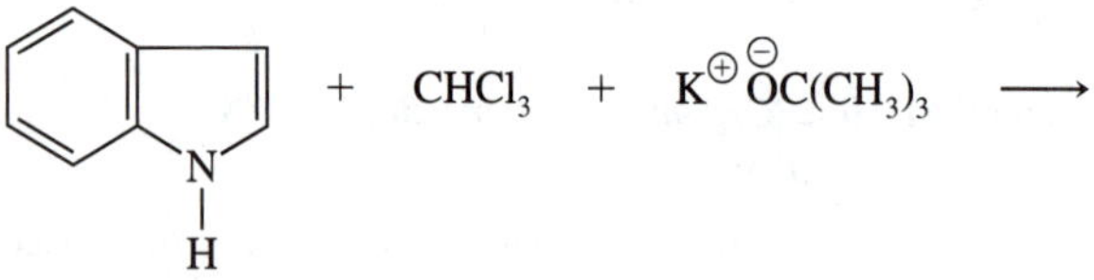

(e) [2-phenyl-1-methylpyrrole structure] $+$ $CH_3\overset{\displaystyle O}{\overset{\|}{CH}}$ $\xrightarrow{\text{HCl}}$

(f) [furan structure] $+$ [maleic anhydride structure] $\xrightarrow{\Delta}$

(g) [indole structure with N–H] $+$ $CHCl_3$ $+$ $K^{\oplus}\overset{\ominus}{O}C(CH_3)_3$ $\longrightarrow$

(h) $\phi-\overset{\displaystyle O}{\overset{\|}{C}}-CH_2NH_2$ $+$ $H_3CO\overset{\displaystyle O}{\overset{\|}{C}}CH_2\overset{\displaystyle O}{\overset{\|}{C}}\phi$ $\xrightarrow{\text{HOAc}}$

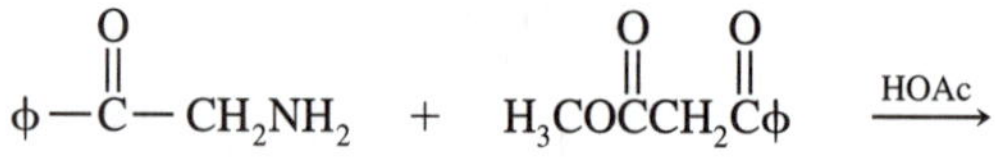

(i) 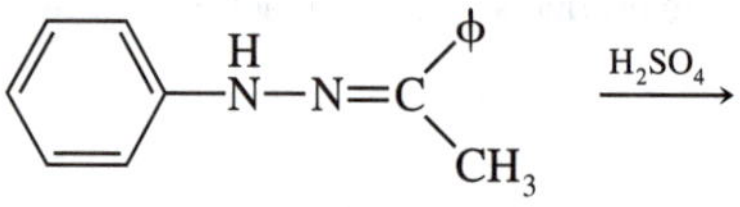 [phenyl]$-\overset{H}{N}-N=C\overset{\phi}{\underset{CH_3}{\diagup}}$ $\xrightarrow{\text{H}_2\text{SO}_4}$

(j) 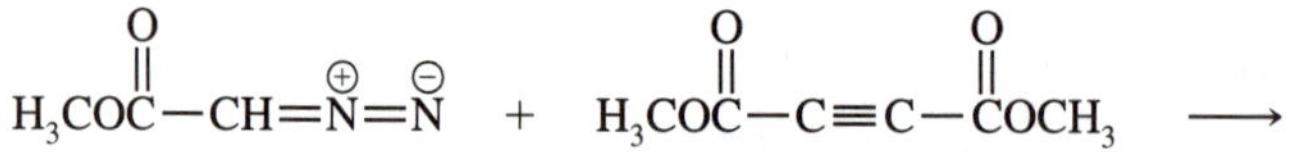 $H_3CO\overset{\displaystyle O}{\overset{\|}{C}}-CH=\overset{\oplus}{N}=\overset{\ominus}{N}$ $+$ $H_3CO\overset{\displaystyle O}{\overset{\|}{C}}-C\equiv C-\overset{\displaystyle O}{\overset{\|}{C}}OCH_3$ $\longrightarrow$

12.2 Would you expect pyrazole to be a stronger base or a weaker base than pyrrole? Should it be a stronger or weaker base than imidazole? Explain the reasons for your answers.

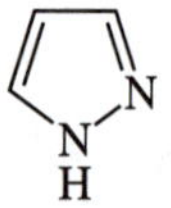

[pyrazole structure with N–N–H]

pyrazole

12.3 Suggest reasonable mechanisms for the reactions in Eqs. 55, 57, 59, 61, 62, and 63 in this chapter.

12.4 Suggest reasonable mechanisms for the following reactions, using curved arrows to represent electron motions. Do not combine any two steps into one.

(a)

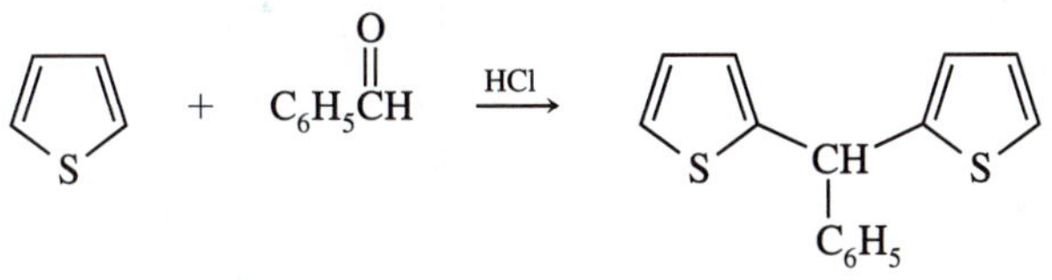

(b)

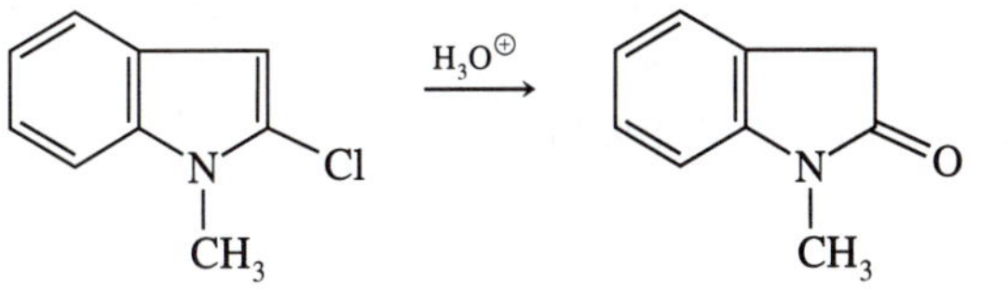

(c)

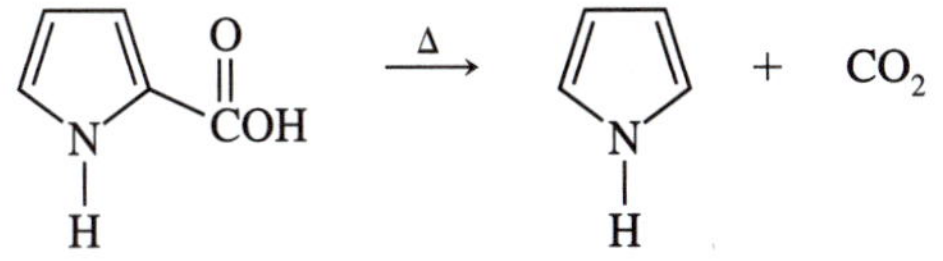

(d)

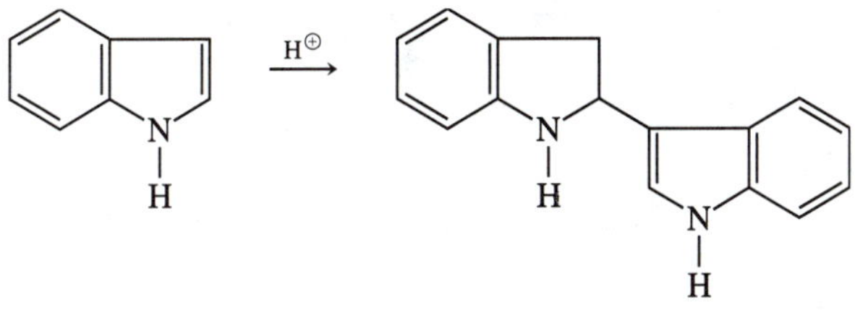

(e)

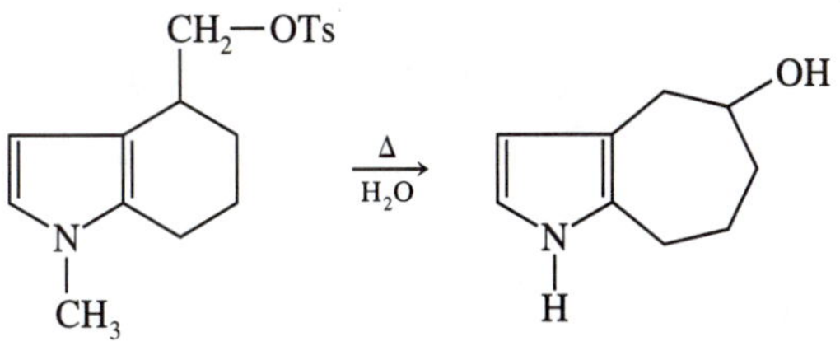

(f)

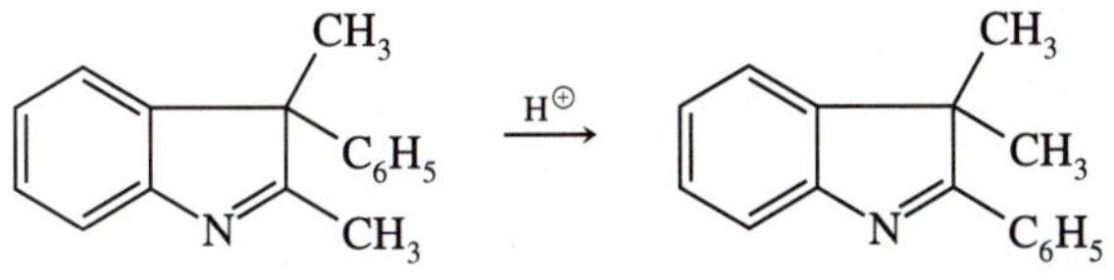

(g)

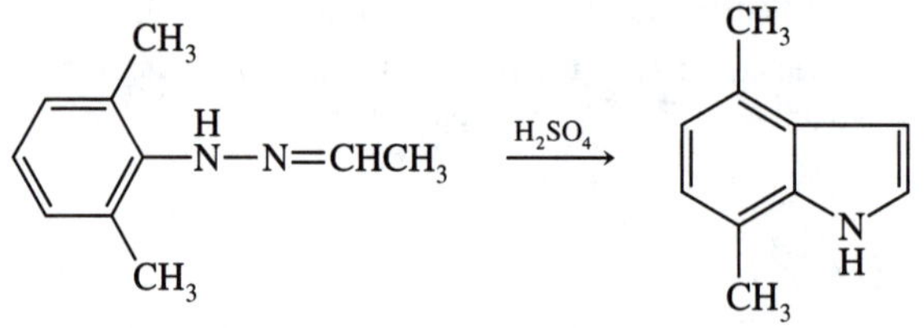

H₂SO₄

(h)

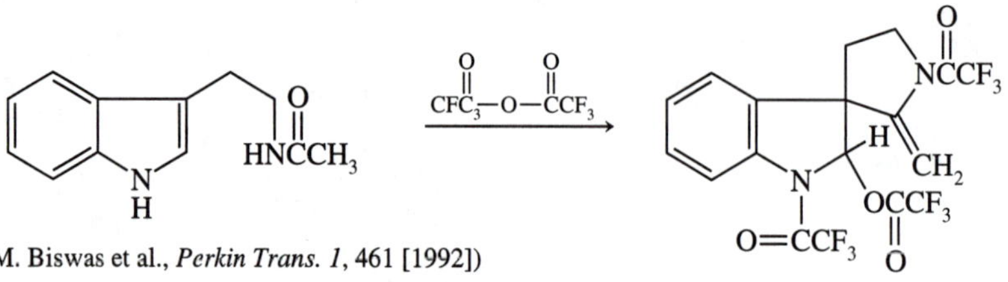

(K.M. Biswas et al., *Perkin Trans. 1*, 461 [1992])

Organophosphorus and Organosulfur Chemistry

13.1 PROPERTIES OF DIVALENT SULFUR AND TRIVALENT PHOSPHORUS DERIVATIVES

Acidities

Organic derivatives of divalent sulfur which contain S–H bonds are called *thiols* or *mercaptans*. They are weakly acidic, with pK_a values of approximately 7 (compared to *ca.* 16 for water and alcohols).[1] Thus, they are easily converted to their anions in aqueous base.

Trivalent phosphorus derivatives (*phosphines*) with P–H bonds, in contrast, are very weak acids. The pK_a of phosphine (PH_3) itself been estimated at *ca.* 27 (compared to *ca.* 40 for ammonia and alkylamines).[1] It thus cannot be converted to its anion in aqueous solution. However, it will react with such strong bases as lithium reagents or alkali amides to form phosphine anions.

$$(C_6H_5)_2PH + NaNH_2 \longrightarrow (C_6H_5)_2\overset{\ominus}{P}\,\overset{\oplus}{Na} + NH_3 \qquad (1)$$

Basicities

Unlike amines, which are basic enough for their basicities to be measured in aqueous solutions, most phosphines are such weak bases that their basicities can only be studied in nonaqueous systems.[2] Although the differences in solvents makes comparison more complicated, a survey of the effects of changing substituents on the basicities of several amines and phosphines (Table 13.1) shows rather remarkable differences.

The basicities of amines usually increase as alkyl substituents are substituted for hydrogens. (Adding a third alkyl group usually results in a small decrease, apparently because it is more difficult to solvate the protonated forms of tertiary amines.[3]) The effects of alkyl substitution are small, however, resulting in less than 10^2 difference in basicities between ammonia and dialkylamines.

In contrast, the effects of increasing alkyl substitution on phosphorus are truly enormous. Replacing a single hydrogen atom on phosphine by an alkyl group increases its basicity by more

TABLE 13.1 pK_a's OF CONJUGATE ACIDS OF AMINES (IN AQUEOUS SOLUTIONS)[3,4] AND PHOSPHINES (IN NITROMETHANE)[2]

Amine	pK_a	Phosphine	pK_a
NH_3	9.25	PH_3	-14
$n\text{-}C_4H_9NH_2$	10.59	$n\text{-}C_8H_{17}PH_2$	0.43
$(n\text{-}C_4H_9)_2NH$	11.25	$(n\text{-}C_8H_9)_2PH$	4.41
$(n\text{-}C_4H_9)_3N$	10.89	$(n\text{-}C_5H_{11})_3P$	8.33
$C_6H_5N(CH_3)_2$	5.06	$C_6H_5P(CH_3)_2$	6.25
$(C_6H_5)_2NH$	0.78	$(C_6H_5)_2PH$	0.03
$(C_6H_5)_3N$	(too weak to measure)	$(C_6H_5)_3P$	2.73

than a factor of 10^{14}, while each further alkyl substituent increases the basicities of phosphines by factors of ca. 10^4.

The dramatic differences in the effects of alkyl substitution on basicities of amines and phosphines have been attributed to differences in the geometries of the two types of molecules.[5] While all aliphatic amines have nearly tetrahedral structures, the bond angles in PH_3 are only 93.5° (showing nearly pure *p* orbital hybridization of phosphorus). The angles increase with alkyl substitution, reaching 100° (indicating more *s* orbital contribution) in trimethylphosphine.[6] Protonation converts phosphines to nearly tetrahedral sp^3 hybridized salts with bond angles of *ca.* 109.5°. The energy gained on protonation is diminished (or the energy required to effect protonation is increased) by the energy needed to rehybridize the phosphines.[5] Since the smaller the number of alkyl groups on phosphorus atoms in phosphines the larger the degree of *p* orbital hybridization, protonation of less-substituted phosphines is more difficult.

Another major difference between amines and phosphines is found in the effects of aromatic substituents. Substituting phenyl groups for alkyl groups (or hydrogen atoms) on nitrogen makes protonation of amines more difficult, since resonance between the nitrogen atoms and the aromatic rings is eliminated on protonation. Thus, converting an aliphatic amine to a similarly substituted aniline derivative decreases the basicity of the amine by a factor of nearly 10^6.

In contrast, substituting phenyl groups for hydrogen atoms markedly *increases* the basicity of phosphines. Substituting phenyl groups for alkyl groups does decrease the basicities of phosphines, but by little more than a factor of 10^2. This relatively small effect can be accounted for entirely by the fact that substituting an sp^2 bond to an aromatic ring for an sp^3 bond to an alkyl group has an appreciable electron-withdrawing inductive effect. Thus, resonance interactions between phosphorus atoms and aromatic rings have little effect on the basicities of arylphosphines. (Of course, resonance interactions involving formation of double bonds between carbons and third-row elements are always small.)

Thiols and dialkyl sulfides, like comparable oxygen-containing molecules, are not basic by normal standards. In fact, divalent sulfur atoms appear to be even less basic than oxygen atoms in comparable environments. Because of the difficulty in measuring basicities of organic molecules containing oxygen and sulfur atoms, there is little evidence as to substituent effects on their relative basicities.

Nucleophilic Reactivities

Although phosphines are much less basic than comparably substituted amines, they are typically far *more effective nucleophiles* in reactions with alkyl halides or sulfonates. Despite its much lower basicity, for example, triethylphosphine reacts with ethyl iodide in acetone (forming a quaternary phosphonium salt) about 20 times as rapidly as does triethylamine.[7] Even triphenylphosphine reacts almost as rapidly with ethyl iodide as does triethylamine,[7] while triphenylamine is quite unreactive under the same conditions.

$$(C_6H_5)_3P\!: \; + \; CH_3CH_2\,I \; \longrightarrow \; (C_6H_5)_3\overset{\oplus}{P}CH_2CH_3 \quad \overset{I^{\ominus}}{} \tag{2}$$

Derivatives of divalent sulfur also show remarkable nucleophilicities toward alkylating agents when compared to similar oxygen compounds. Neutral divalent sulfides, for instance, easily form sulfonium salts on reaction with alkylating agents. In contrast, ethers, alcohols, and water are essentially unreactive under the same conditions, as shown in Eq. 3.

$$CH_3O(CH_2)_3SCH_3 \; + \; CH_3I \; \longrightarrow \; CH_3O(CH_2)_3\overset{\oplus}{S}(CH_3)_2 \quad \overset{I^{\ominus}}{} \tag{3}$$

In fact, dimethyl sulfide (in methanol solution) is more than 10^5 times as reactive a nucleophile toward methyl iodide as is methanol.[8] Mercaptide anions, in water or alcohols, are thousands of times as nucleophilic toward alkylating agents as are similarly substituted oxygen anions.[8] Even *neutral* alkyl and aryl sulfides are much more nucleophilic toward alkylating agents than many anionic oxygen-centered nucleophiles.

The marked superiority of sulfur nucleophiles compared to oxygen nucleophiles is somewhat reduced in reactions with *aryl* halides, although it is still necessary to compare methoxide anions to much less basic thiophenoxide anions to obtain roughly comparable reaction rates.[9] Even smaller differences between sulfur and oxygen nucleophiles are observed in reactions with acylating agents (such as *p*-nitrophenyl acetate)[10] and with heterocyclic halides (such as 2-chloroquinoline).[11]

The relative reactivities of thiophenoxide and methoxide anions in these reactions can be strongly dependent on the nature of the leaving groups. Methoxide anions are frequently more reactive than thiophenoxide anions when fluoride anions are displaced, but less so with other halide anions. This effect is illustrated in the displacement reactions of 2-halothiazoles (Eq. 4).[12]

$$A^{\ominus} \; + \; \underset{X}{\overset{N}{\diagup\!\!\diagdown}}{}_S \; \longrightarrow \; \underset{A}{\overset{N}{\diagup\!\!\diagdown}}{}_S \; + \; X^{\ominus} \qquad (A^{\ominus} = C_6H_5S^{\ominus} \; or \; CH_3O^{\ominus}) \tag{4}$$

$$
\begin{array}{cccc}
X = & F, & Cl, & Br, \quad I \\
kC_6H_5S^{\ominus}/kCH_3O^{\ominus} = 1.1 \times 10^{-2}, & 5.5 \times 10^{-1}, & 1.7, & 29.0
\end{array}
$$

While sulfur-centered nucleophiles are far more reactive than oxygen-centered nucleophiles in reactions with alkylating agents, and (to a lesser extent) in reactions with other carbon-centered electrophiles, this is not true for reactions with all electrophiles. There appears to be no difference, for instance, between sulfur-centered and oxygen-centered nucleophiles in rates of reactions with phosphoryl (and thiophosphoryl) halides. Instead, sulfur and oxygen anions of equal basicities react at essentially the same rates in the thiophosphorylation reaction shown in Eq. 5.[13]

$$A^{\ominus} + (C_6H_5O)_2\overset{\overset{\textstyle S}{\|}}{P}Cl \longrightarrow (C_6H_5O)_2\overset{\overset{\textstyle S}{\|}}{P}A + Cl^{\ominus} \qquad (5)$$

13.2 HARD AND SOFT ACIDS AND BASES

Hardness and Softness

The high reactivities of sulfur-centered nucleophiles in alkylation reactions, their smaller advantage over oxygen-centered nucleophiles in aromatic substitution and acylation reactions, and the equivalent reactivities of the two types of nucleophiles in phosphorylation reactions, can be accounted for by Ralph G. Pearson's principle of *hard and soft acids and bases* (HSAB).[14]

This principle states that "hard acids bind strongly to hard bases, and soft acids bind strongly to soft bases."[14] (It should be noted that the term "acids" in the HSAB principle includes molecules and ions which organic chemists would term electrophiles, and the term "bases" includes nucleophiles. When considering the HSAB principle in relation to reactions of organic molecules, it is often a good idea to mentally substitute the words "electrophiles" and "nucleophiles" for "acids" and "bases.")

Hard acids, in Pearson's classification, are acids in which the reacting atoms are small and bear strong positive charges (or appreciable partial positive charges), and are therefore not easily polarized. Acids in which the reacting atoms are located in lower rows of the periodic table would normally be harder than comparable acids with the central atoms in higher rows. (Protons, located in the first row of the periodic table, are the archetypal hard acids.)

The net charge on the reacting atom, and thus the hardness, of an acid will be increased by the presence of electronegative substituents. Such substituents often increase the oxidation state of the reacting atom. Therefore it is generally true that the higher the oxidation state of the reacting atom, the harder the acid or electrophile. This is illustrated below for sulfur-centered and phosphorus-centered electrophiles.

Increasing Hardness →

$$RSCl, \; RSOCl. \; RSO_2Cl, \; ROSO_2Cl$$
$$R_2PCl, \; R(RO)PCl, \; (RO)_2PCl, \; (RO)_2\,POCl$$

The *softness* of a base (or nucleophile) is inversely affected by the factors which contribute to hardness. When comparing two bases, the reacting atoms of the softer base would typically be lower in electronegativity and be substituted with less strongly electronegative groups. Changing a reacting atom to one in a higher row of the periodic table would result in a large increase in softness of the base. Thus, nucleophiles with sulfur as the "central" (reacting) atom would be much softer than oxygen-centered nucleophiles, and phosphorus-centered nucleophiles would be softer than nitrogen-centered nucleophiles. Among the halide anions, softness would increase in the order F^-, Cl^-, Br^-, I^-.

It may be noted that in the discussion above, hardness and softness were defined entirely in relative terms. No attempt was made to assign numerical values to the hardness or softness of different acids and bases. While many efforts have been made to quantify the concepts of hardness and softness,[15] the HSAB principle has remained most useful as a *qualitative* method of interpreting reactivities in a vast number of reactions.

Applying the HSAB Principle

The relative reactivities of sulfur-centered and oxygen-centered nucleophiles in reactions with different electrophilic reagents can be nicely accounted for by the HSAB principle.*

Carbon atoms in alkyl groups are in low oxidation states and are not strongly electronegative. Most alkylating agents are, therefore, soft electrophiles. They react more rapidly with soft, sulfur-centered nucleophiles than with hard, oxygen-centered reagents. In contrast, acyl carbons are in higher oxidation states and are harder electrophiles. The advantage of sulfur-centered over oxygen-centered nucleophiles is, therefore, lower in acylation than in alkylation reactions.

Aromatic ring carbons, which are sp^2 hybridized, should be harder than sp^3 hybridized carbons. In addition, the hardness of aromatic ring carbons should be markedly increased by nitro groups in *ortho* and *para* positions. Thus, the relative effectiveness of sulfur-centered to oxygen-centered nucleophiles is smaller in reactions with dinitroaryl halides than with alkyl halides.

The thiophosphoryl group, with several electronegative substituents as well as a formal positive charge on phosphorus, is a very hard electrophile. Thus, it is not surprising that the reactivities of both soft, sulfur-centered nucleophiles and hard, oxygen-centered nucleophiles correlate with their basicities—a measure of reactivity toward protons, which are very hard acids.

The HSAB principle also accounts for the effects of leaving groups on the relative nucleophilicities of thiophenoxide and methoxide anions. The more electronegative the potential leaving group, the harder the carbon atom to which it is bonded, and the greater the relative effectiveness of hard, oxygen-centered nucleophiles compared to soft, sulfur-centered nucleophiles.[†]

Solvation effects.   The preceding discussions did not take into account the effects of solvation. The rates of nucleophilic substitution reactions are usually measured in hydroxylic solvents, in which hydrogen bonding plays a major role. Appreciable differences are observed when the reactions are carried out in dipolar, nonhydroxylic solvents, such as dimethyl sulfoxide (DMSO).[16‡] Rates of reactions with anionic nucleophiles are markedly increased in these solvents. The differences in rates among the various nucleophiles are significantly reduced, and the relative reactivity orders are often changed. (For instance, the usual order of nucleophilicities toward alkyl halides, $I^- > Br^- > Cl^-$, is reversed in DMSO.)[16]

These results are consistent with the HSAB principle, since hydrogen bonding involves interactions with hard acids. Thus, the overall differences in reactivity between hard and soft nucleophiles appear to be partially due to solvation effects (including hydrogen bonding) and partially due to intrinsic differences in interactions between nucleophiles and electrophiles.

---

*When comparing nucleophilicities and basicities, we made the tacit assumption that the two properties would be closely related, even though basicities are measures of equilibria, while nucleophilicities are measures of reaction rates. The HSAB accounts for the fact that this assumption often fails. Basicities are measures of equilibria in reactions with protons, which are very hard acids. The basicities of anions should, therefore, be good predictors of their rates of reaction with hard electrophiles, but not with soft electrophiles.

†The equilibrium constants for the first steps in reactions such as that shown in Eq. 4, particularly when fluoride anions are the leaving groups, may also play a role in determining the apparent reactivities of sulfur- and oxygen-centered nucleophiles.

‡Crown ethers may be added to complex with, and minimize the effects of, metallic counterions. Reactions under these conditions have been described as involving "bare" anions.[17] However, bare anions or not, solvation always plays a major (and very complicated) role in reactions in solution.

Competing reactions. In the processes we have considered so far, only one type of reaction has been feasible. Perhaps the most useful application of the HSAB principle for organic chemists is in accounting for choices among several possible competing reaction paths. This will be illustrated in the following sections.

13.3 COMPOUNDS CONTAINING PHOSPHORUS–OXYGEN BONDS

The Phosphoryl Group

Compounds containing *phosphoryl* groups (which are usually represented as P=O groups, as shown in Eqs. 6 and 7) form the largest and most stable groups of organophosphorus compounds. A wide variety of oxidizing agents easily convert tertiary phosphines, as well as other trivalent phosphorus compounds, to compounds containing phosphoryl groups.

$$(CH_3)_3P + Cl_2 + H_2O \longrightarrow (CH_3)_3P{=}O + 2\,HCl \tag{6}$$

$$(C_6H_5O)_3P \xrightarrow{O_3} (C_6H_5O)_3P{=}O \tag{7}$$

Phosphoryl bonds are among the strongest chemical bonds. Their bond dissociation energies (BDEs) in phosphine oxides average about 138 kcal/mole[18]—approximately 50% stronger than typical C–O single bonds. Other phosphoryl bonds can be even stronger. Phosphoryl groups in molecules such as $(C_2H_5N)_3P{=}O$, for instance, have bond energies in the neighborhood of 156 kcal/mole.[18]

As a result, trivalent phosphorus derivatives are excellent reducing agents, abstracting oxygen atoms from a wide variety of oxidizing agents. Amine oxides and sulfoxides, which have structures that, on paper, resemble phosphoryl groups, are reduced by trivalent phosphorus derivatives. At sufficiently high temperatures even pyridine oxides, which are usually quite difficult to reduce, are deoxygenated by triphenylphosphine (Eq. 8).

$$(C_6H_5)_3P + \underset{\underset{O^\ominus}{\overset{|}{\underset{}{N_\oplus}}}}{\bigcirc} \longrightarrow (C_6H_5)_3P{=}O + \underset{N}{\bigcirc} \tag{8}$$

The structures of phosphoryl groups have been debated for decades. In many ways, phosphoryl bonds do appear to resemble double bonds. They are, for instance, appreciably shorter, as well as much stronger, than P–OR single bonds. However, exactly which orbitals are involved in forming the second bond (assuming it exists) is much less clear.

Obviously, phosphoryl bonds in pentavalent phosphorus compounds cannot include $p\pi{-}p\pi$ bonds because no unused p orbitals are available.* It had long been postulated that phosphoryl groups may be represented as resonance hybrids of dipolar structures and of double bonds formed by interaction of p electrons on oxygen with the vacant d orbitals on phosphorus—that is, as $d\pi{-}p\pi$ double bonds.

---

*A few (rather high-energy) *trivalent* phosphorus compounds (e.g., Cl–P=O) with genuine $p\pi{-}p\pi$ bonds between phosphorus and oxygen have been prepared.[20a]

$$\overset{\oplus}{R_3P}{-}\overset{\ominus}{O} \longleftrightarrow R_3P{=}O$$

Are Phosphoryl Groups Formed by $d\pi-p\pi$ Bonding?

However, all recent studies agree that participation of d orbitals of phosphorus plays little, if any, role in phosphoryl groups.[20a] It is not clear, though, exactly what type of bonding *is* involved, and even whether the groups should be represented by doubly bonded structures at all. Several alternative structures have been proposed. One suggestion is that electron pairs on oxygen interact with antibonding σ^* orbitals of phosphorus, rather than with d orbitals.[20b]

In this chapter, we will handle the matter by writing doubly bonded structures for phosphoryl groups, and avoiding further discussion of the subject.

Molecules with hydrogens bonded to phosphoryl groups. Even reagents that do not act as oxidizing agents frequently give reactions that result in formation of phosphoryl groups. For instance, reactions that might have been expected to form trivalent phosphorus derivatives with P–OH groups instead form their tautomers containing phosphoryl groups (Eq. 9).

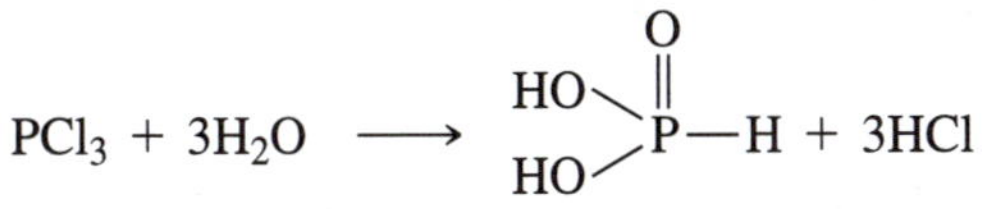

$$PCl_3 + 3H_2O \longrightarrow \overset{\overset{\displaystyle O}{\parallel}}{\underset{HO}{\overset{HO}{>}}P{-}H} + 3HCl \qquad (9)$$

In most instances, no detectable amounts of the hydroxy tautomers can be detected. (Rare exceptions do exist when the phosphorus atom bears very strongly electronegative groups, such as trifluoromethyl groups, as in Eq. 10.[21])

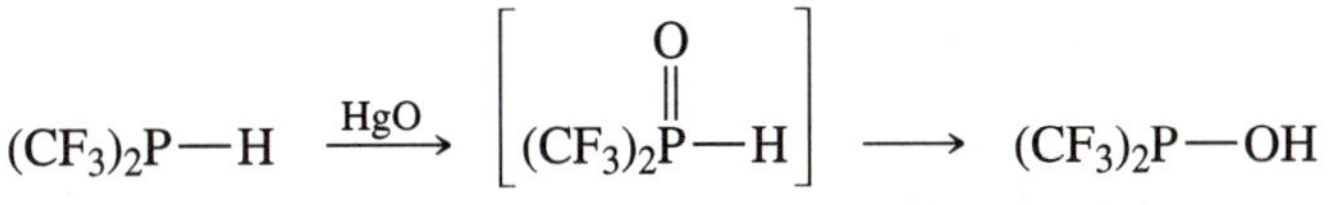

$$(CF_3)_2P{-}H \xrightarrow{\text{HgO}} \left[(CF_3)_2\overset{\overset{\displaystyle O}{\parallel}}{P}{-}H \right] \longrightarrow (CF_3)_2P{-}OH \qquad (10)$$

Phosphorous acid, $(HO)_3P$, is thus unknown, because only the tautomeric form exists. However, its esters, such as trimethyl phosphite, $(CH_3O)_3P$, are common compounds.*

---

*The nomenclature of some derivatives of acids containing phosphorus is briefly illustrated below:

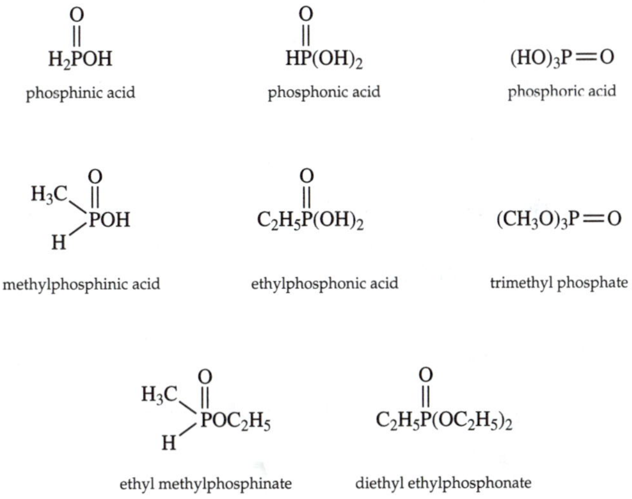

$\overset{\overset{\displaystyle O}{\parallel}}{H_2POH}$	$\overset{\overset{\displaystyle O}{\parallel}}{HP(OH)_2}$	$(HO)_3P{=}O$
phosphinic acid	phosphonic acid	phosphoric acid

$\overset{\overset{\displaystyle O}{\parallel}}{\underset{H}{\overset{H_3C}{>}}POH}$	$\overset{\overset{\displaystyle O}{\parallel}}{C_2H_5P(OH)_2}$	$(CH_3O)_3P{=}O$
methylphosphinic acid	ethylphosphonic acid	trimethyl phosphate

$\overset{\overset{\displaystyle O}{\parallel}}{\underset{H}{\overset{H_3C}{>}}POC_2H_5}$	$\overset{\overset{\displaystyle O}{\parallel}}{C_2H_5P(OC_2H_5)_2}$
ethyl methylphosphinate	diethyl ethylphosphonate

Arbusov reactions.   While phosphines form phosphonium salts on reactions with alkyl halides, the phosphonium salts formed from similar reactions of trivalent phosphorus derivatives containing phosphorus-alkoxy bonds are unstable and undergo dealkylation reactions, yielding compounds containing phosphoryl groups, as in Eq. 11.[22]

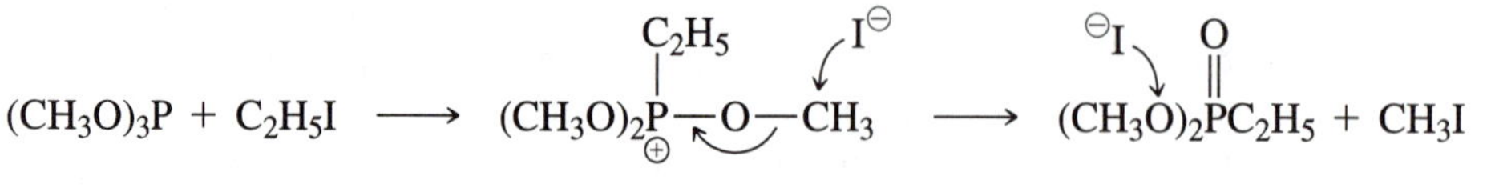

$$(CH_3O)_3P + C_2H_5I \longrightarrow (CH_3O)_2\overset{C_2H_5}{\underset{\oplus}{P}}{-}O{-}CH_3 \longrightarrow (CH_3O)_2\overset{O}{P}C_2H_5 + CH_3I \quad (11)$$

These processes are known as *Arbuzov* (or *Michaelis–Arbuzov*) *reactions*. (They are sometimes called Arbuzov *rearrangements*, particularly when the alkyl group in the alkyl halide is the same as that in the alkoxy group.)

Similar reactions occur when compounds containing P–OR groups react with halogens or hydrogen halides, as shown in Eqs. 12 and 13.[22]

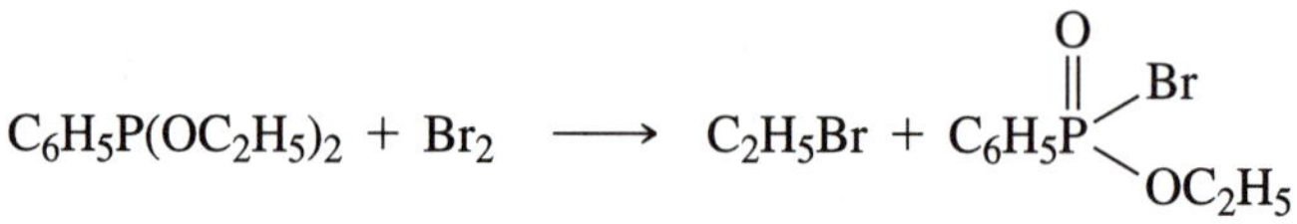

$$C_6H_5P(OC_2H_5)_2 + Br_2 \longrightarrow C_2H_5Br + C_6H_5\overset{O}{P}\overset{Br}{\underset{OC_2H_5}{}} \quad (12)$$

$$(CH_3O)_3P + HCl \longrightarrow (CH_3O)_2\overset{O}{P}H + CH_3Cl \quad (13)$$

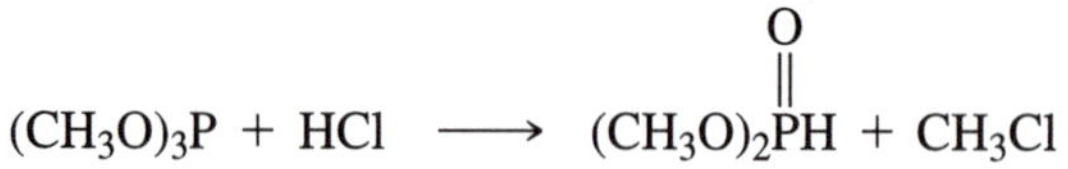

The alkoxyphosphonium salts formed as intermediates in Arbuzov reactions normally decompose too rapidly to be detected. However, the intermediates formed from alkylations of triaryl phosphites are stable and can be isolated.[23] They decompose to phosphonate esters on addition of alcohols or water.

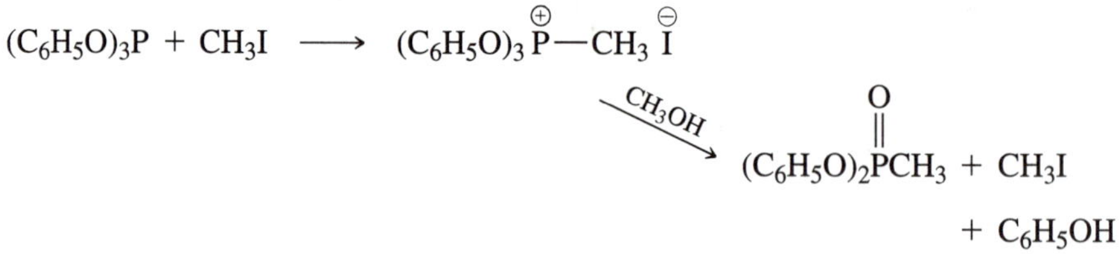

$$(C_6H_5O)_3P + CH_3I \longrightarrow (C_6H_5O)_3\overset{\oplus}{P}{-}CH_3\ \overset{\ominus}{I}$$

$$\xrightarrow{CH_3OH} (C_6H_5O)_2\overset{O}{P}CH_3 + CH_3I \quad (14)$$
$$+ C_6H_5OH$$

Perkow reactions.   Reactions of trialkyl phosphites and related compounds with α-halocarbonyl compounds often yield enol phosphates (*Perknow reaction* products) in addition to, or in place of, phosphonate esters (Eq. 15).[24]

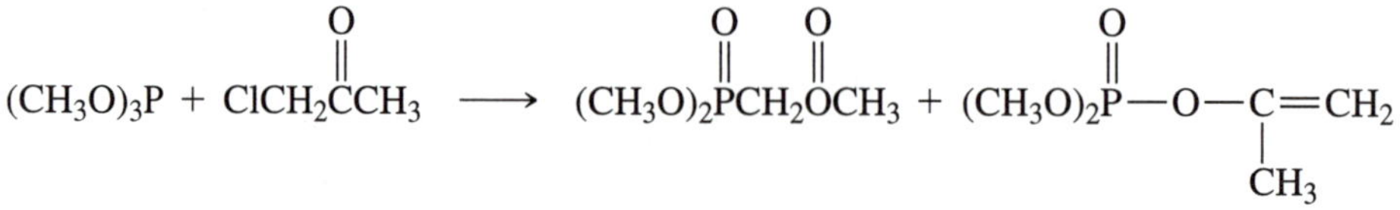

$$(CH_3O)_3P + ClCH_2\overset{O}{C}CH_3 \longrightarrow (CH_3O)_2\overset{O}{P}CH_2\overset{O}{O}CH_3 + (CH_3O)_2\overset{O}{P}{-}O{-}\underset{CH_3}{C}{=}CH_2 \quad (15)$$

The proportion of enol phophates formed decreases when bulky groups inhibit attack at the carbonyl groups. It is therefore greater from reactions with α-haloaldehydes than from reactions with α-haloketones. Furthermore, the carbonyl addition product **1**, in addition to the enol phosphate, has been isolated from the reaction of triethyl phosphite with α-bromoacetophenone in the presence of ethanol.[25]

$$(CH_3O)_3P + BrCH_2\overset{\overset{\displaystyle O}{\|}}{C}\!-\!\!\bigcirc \xrightarrow{CH_3OH} (CH_3O)_2\overset{\overset{\displaystyle O}{\|}}{P}\!-\!\underset{\underset{\displaystyle CH_2Br}{|}}{\overset{\overset{\displaystyle OH}{|}}{C}}\!-\!\!\bigcirc \tag{16}$$

1

$$+ \;(CH_3O)_2\overset{\overset{\displaystyle O}{\|}}{P}\!-\!O\!-\!\underset{}{\overset{\overset{\displaystyle CH_2}{\|}}{C}}\!-\!\!\bigcirc$$

These facts are consistent with a mechanism in which the phosphite initially adds to the carbonyl group and then migrates to oxygen (Eq. 17).

$$(RO)_3P + ClCH_2\overset{\overset{\displaystyle O}{\|}}{C}CH_3 \longrightarrow ClCH_2\!-\!\underset{\underset{\displaystyle (RO)_3P\oplus}{|}}{\overset{\overset{\displaystyle CH_3}{|}}{C}}\!-\!O^{\ominus} \longrightarrow Cl\!-\!CH_2\!\overset{\frown}{\rightarrow}\!\underset{\underset{\displaystyle (RO)_3P}{|}}{\overset{\overset{\displaystyle CH_3}{|}}{C}}\!\begin{smallmatrix}\\\searrow\\O\end{smallmatrix}$$

$$\tag{17}$$

$$RCl + \underset{\underset{\displaystyle O-P(OR)_2}{\|}}{CH_2\!\!=\!\!\overset{\overset{\displaystyle CH_3}{|}}{C}}\!-\!O\!-\!\overset{\overset{\displaystyle O}{\|}}{P}(OR)_2 \longleftarrow CH_2\!\!=\!\!\overset{\overset{\displaystyle CH_3}{|}}{C}\!-\!O\!-\!\overset{\oplus}{P}(OR)_3 \; Cl^{\ominus}$$

The carbonyl addition mechanism appears to be the most common process by which Perkow reactions of monohalocarbonyl compounds occur. However, an alternative mechanism involving attack at halogen to form enolate anions appears to play a significant role in reactions of some polyhaloketones[26] (as in Eq. 18) and even of trihaloesters, as shown in Eq. 19.[27]

$$\begin{array}{c}\overset{\overset{\displaystyle O}{\|}}{Br_2CH_2\overset{}{C}C_6H_5}\\+\\(CH_3O)_3P\end{array} \longrightarrow \begin{array}{c}\overset{\ominus}{\overset{\displaystyle O}{}}\\BrCH\!\!=\!\!CC_6H_5\\+\\(CH_3O)_3\overset{\oplus}{P}Br\end{array} \begin{array}{c}\longrightarrow (CH_3O)_2\overset{\overset{\displaystyle O}{\|}}{P}\!-\!O\!-\!\overset{\overset{\displaystyle C_6H_5}{|}}{C}\!\!=\!\!CHBr\\+ CH_3Br\\[2ex]\xrightarrow{HOR}\; (CH_3O)_2\overset{\overset{\displaystyle O}{\|}}{P}OR + BrCH_2\overset{\overset{\displaystyle O}{\|}}{C}C_6H_5\end{array} \tag{18}$$

$$(CH_3O)_3P + Cl_3CCOC_2H_5 \longrightarrow (CH_3O)_2\overset{O}{\underset{\|}{P}}-O-\overset{OC_2H_5}{\underset{\|}{C}}=CCl_2 + CH_3Cl \qquad (19)$$

This mechanism is supported by the fact that the Perkow reaction of α,α-dibromoace-tophenone proceeds much more rapidly than the reaction of α,α-dichloroacetophenone, even though monochloro ketones usually react more rapidly than monobromo ketones. Furthermore, as shown in Eq. 18, the reduction product, α-bromoacetophenone, is formed when the reaction is carried out in ethanol or acetic acid. In contrast, the dichloro compound gave the enol phosphate even in hydroxylic solvents.[26]

Competition between Perkow, reduction, and alkylation reactions provides an illustration of the usefulness of the HSAB principle in interpreting the choice among competing reactions. Attack by trialkyl phosphites at bromine atoms in α-haloketones, resulting in reductions in protic solvents, is less common than Perkow reactions resulting from attack at carbonyl carbons. In contrast, phosphines, which are softer bases, prefer to attack soft bromine atoms rather than harder carbonyl carbons.[28a] When the halogen is the harder chlorine atom, phosphines tend to attack alkyl carbons (which are harder than bromine atoms but softer than carbonyl carbons), as shown in Eq. 20.[28]

$$(C_6H_5)_3P \quad \begin{array}{c} \xrightarrow{BrCH_2\overset{O}{\underset{\|}{C}}C_6H_5} \\ \\ \xrightarrow{ClCH_2\overset{O}{\underset{\|}{C}}C_6H_5} \end{array} \quad \begin{array}{c} \overset{Br^{\ominus}}{\underset{}{}} \quad \overset{C_6H_5}{\underset{|}{}} \\ (C_6H_5)_3\overset{\oplus}{P}-O-C=CH_2 \\ \\ \overset{Cl^{\ominus}}{\underset{}{}} \quad \overset{O}{\underset{\|}{}} \\ (C_6H_5)_3\overset{\oplus}{P}-CH_2CC_6H_5 \end{array} \qquad (20)$$

13.4 COMPOUNDS CONTAINING SULFUR–OXYGEN BONDS

Sulfoxides and Sulfones

The sulfur–oxygen (sulfinyl) bonds in sulfoxides (R[SO]R′) have bond dissociation energies of ca. 86–89 kcal/mol. The first dissociation of a sulfur–oxygen (sulfonyl) bond in a sulfone (R[SO$_2$]R′) requires about 113 kcal/mol.[29] The BDEs of both types of bonds, of course, are much lower than those of phosphoryl bonds.

The relatively low bond energies and extremely high polarities of sulfonyl and sulfinyl bonds suggest that these bonds have much less *double bond character* than phosphoryl bonds. Therefore, structures of sulfoxides are frequently shown with polarized (S^+-O^-) structures. However, for purposes of simplicity in presentation, equations in this book will represent sulfoxides and sulfones as having double-bond $(S=O)$ structures.

Phosphines and other trivalent phosphorus derivatives are easily oxidized to phosphoryl derivatives, while phosphoryl groups are very difficult to reduce. In contrast, the relatively weak sulfur–oxygen bonds in sulfoxides are easily reduced, yielding divalent sulfur derivatives, as in Eq. 21.

Pummerer rearrangements. While phosphoryl groups are quite stable in acids, the oxygen atoms in sulfoxides are readily replaced by halide ions on reaction with hydrogen halides (Eq. 22).

$$CH_3-\overset{\overset{\displaystyle O}{\|}}{S}-C_6H_5 \quad \overset{(C_6H_5)_3P}{\underset{C_6H_5SH}{\diagup\!\!\!\diagup}}$$

$$(C_6H_5)_3P{=}O + CH_3SC_6H_5$$

$$C_6H_5S-SC_6H_5 + CH_3SC_6H_5 + H_2O \tag{21}$$

$$(C_6H_5)_2S{=}O + HCl \longrightarrow (C_6H_5)_2\overset{\oplus}{S}Cl \ \overset{\ominus}{Cl} + H_2O \tag{22}$$

When methyl or methylene groups are bonded to the sulfur atoms α-halo sulfides are formed, with the halogens bonded to the more substituted α-carbons (Eq. 23).[30]

$$CH_3\overset{\overset{\displaystyle O}{\|}}{S}C_2H_5 + HCl \longrightarrow \left[CH_3\overset{\oplus}{S}{=}CHCH_3\right] \overset{\ominus}{Cl} + H_2O \longrightarrow CH_3S\overset{\overset{\displaystyle Cl}{|}}{C}HCH_3 \tag{23}$$

Similar migrations (known as *Pummerer rearrangements*) take place when sulfoxides are heated with acid anhydrides. The rearrangement processes are often accompanied by formation of vinyl sulfides (Eq. 24).[31]

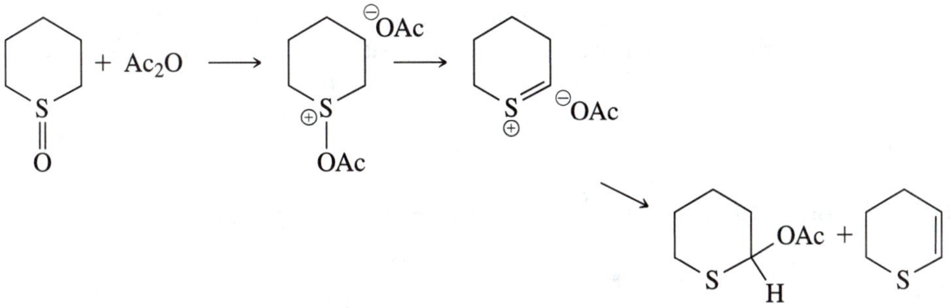

$$\tag{24}$$

Kinetic isotope studies demonstrate that the loss of a proton from the α-carbon, effectively forming a C=S double bond, is concerted with loss of the leaving group from sulfur.[32] The fact that carbon-skeleton rearrangements can occur in favorable cases, such as that shown in Eq. 25, shows that the intermediate has appreciable carbocation-like character.[31]

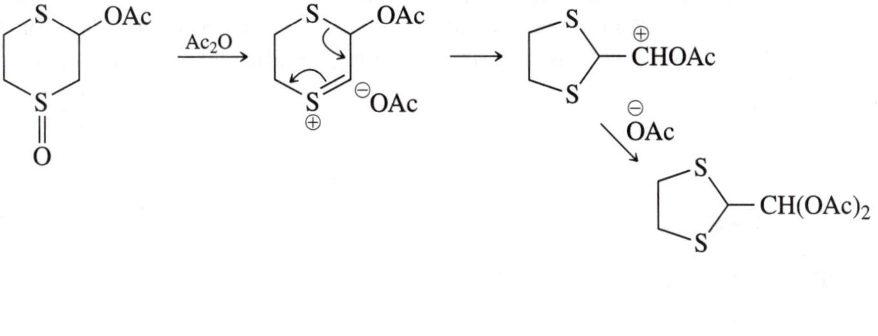

$$\tag{25}$$

Remarkably, the anion in a Pummerer rearrangement may even migrate to a β carbon.[33] It does not appear to be known whether a cyclic intermediate, such as that shown in Eq. 26, is formed in the reaction.

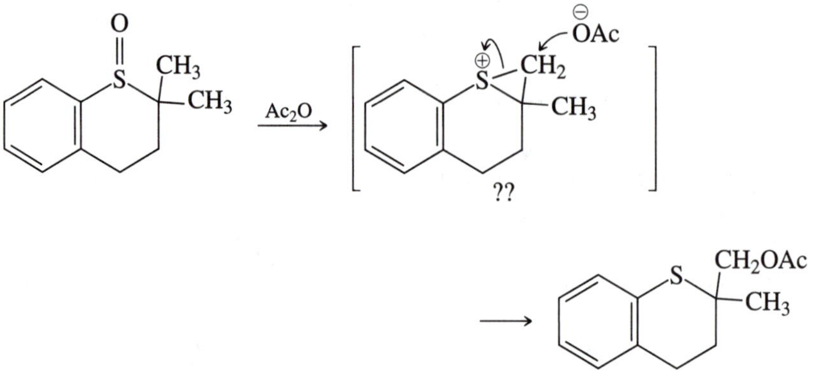

$$\text{(26)}$$

Sulfoxides as oxidizing agents.   Dimethyl sulfoxide oxidizes many alkylating agents to aldehydes or ketones.[34] It is often possible to isolate the intermediate O-alkylsulfoxonium salts, which then eliminate dimethyl sulfide on addition of bases.

The use of DMSO-d_6 has demonstrated that the elimination steps can proceed by two different mechanisms. In reactions of α-halocarbonyl compounds, the base abstracts an acidic proton from the α-carbon, after which a standard β-elimination yields dimethyl sulfide-d_6, as shown in Eq. 27.[35a]

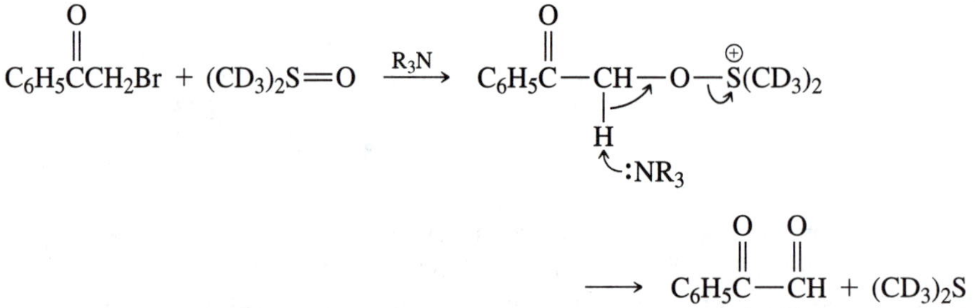

$$\text{(27)}$$

In contrast, the reaction of DMSO-d_6 with a primary alkyl halide resulted in formation of dimethyl sulfide-d_5. It has been proposed that this reaction proceeds by initial abstraction of a proton from a carbon bonded to sulfur, followed by an intramolecular elimination step. (Eq. 28).[35a] (Both mechanisms occurred in the reaction of DMSO-d_6 with a secondary alkyl halide.)[35b]

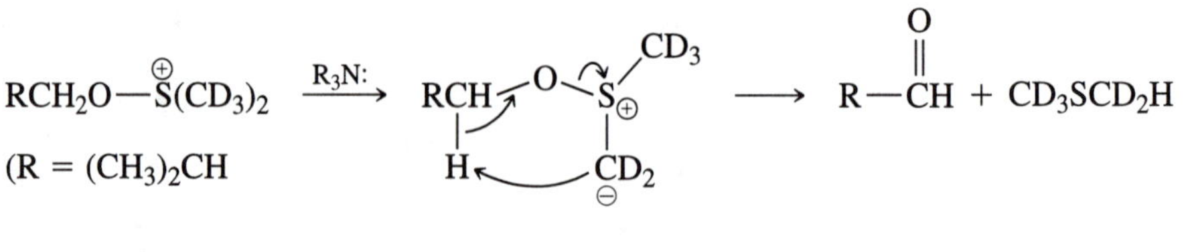

$$\text{(28)}$$

In the presence of acid catalysts, epoxides undergo oxidation reactions similar to those of alkyl halides.[36]

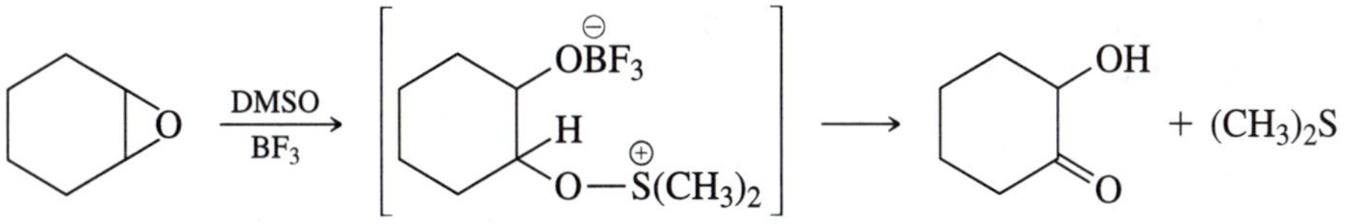

$$(29)$$

In the presence of a variety of electrophilic reagents, including protic acids, acyl halides, and sulfur trioxide, DMSO oxidizes primary and secondary alcohols to carbonyl compounds. The reactions proceed at low temperatures and usually give high (often essentially quantitative) yields.[37] The *Swern oxidation*, using DMSO with oxalyl chloride and tertiary amines, is the most widely used procedure.[37,38] A large number of alcohols have been successfully converted to carbonyls without affecting other easily oxidized functional groups (e.g., Eqs. 30, 31).[38,39]

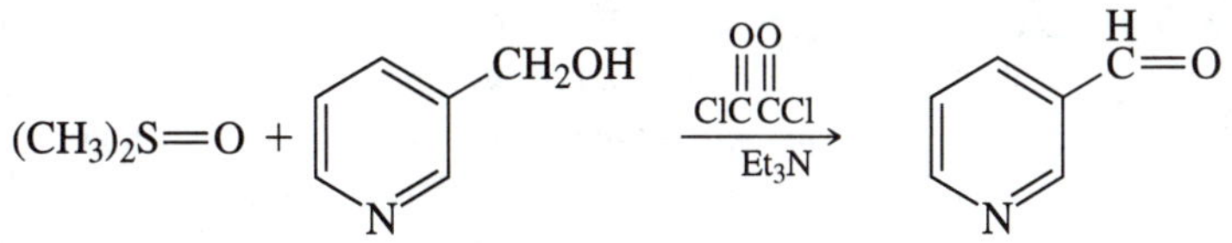

$$(30)$$

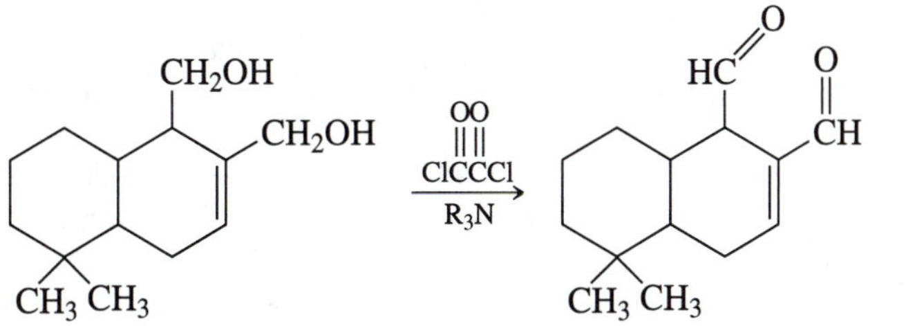

$$(31)$$

In the Swern oxidation, DMSO is "activated" by conversion to the S-chlorosulfonium salt prior to reaction with the alcohol (Eq. 32).[38,40]

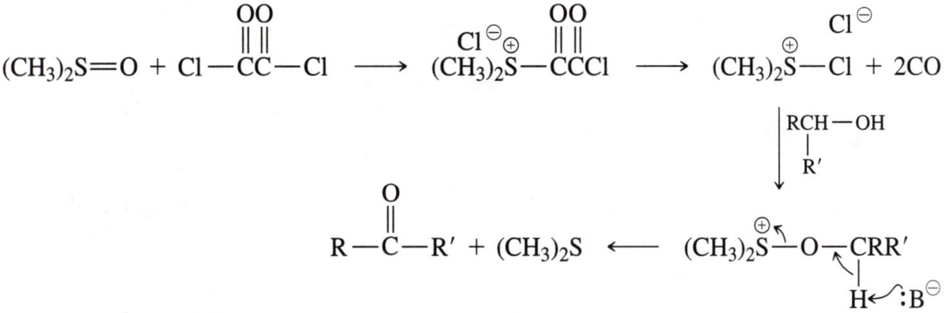

$$(32)$$

13.5 PHOSPHORUS YLIDES

Alkylphosphonium salts are approximately midway in acidity between simple ketones and α-phenyl ketones.[41] Thus, their α hydrogens are easily exchanged in the presence of common bases. However, strong bases, such as organolithium reagents (as shown in Eq. 33), amide anions, or sodium hydride, are necessary to *completely* convert phosphonium salts to their anions. Phosphonium salts with electron-accepting substituents, such as carbonyl groups, on their α-carbons are more acidic, and can be completely converted to their anions by common bases, such as alkoxide anions (Eq. 34).

$$(C_6H_5)_3\overset{\oplus}{P}CH_2CH_3 + n\text{-BuLi} \longrightarrow \left[(C_6H_5)_3P{=}CHCH_3 \longleftrightarrow (C_6H_5)_3\overset{\oplus}{P}{-}\overset{\ominus}{C}HCH_3 \right] \quad (33)$$

$$(C_6H_5)_2\overset{\oplus}{P}{-}CH_2\overset{O}{\overset{\|}{C}}CH_3 + NaOCH_3 \longrightarrow (C_6H_5)_2P{=}CH\overset{O}{\overset{\|}{C}}CH_3 \quad (34)$$

(with CH_3 substituents on P)

The products of these reactions (*Wittig reagents*) are *ylides*—compounds with positive and negative charges on adjacent atoms. Phosphorus ylides that lack strong electron-withdrawing groups on their anionic carbons are called *unstabilized* Wittig reagents, while those with electron-withdrawing substituents are called *stabilized* Wittig reagents. Ylides with negative charges stabilized by phenyl or vinyl groups are called *semistabilized* Wittig reagents.

The nature of bonding in phosphorus ylides has been the subject of the same debates as for phosphoryl and sulfoxide groups. Once again, we will avoid further discussion of the subject, but simply represent their structures as having double bonds between carbon and phosphorus atoms.

Wittig reactions.[42]   In 1953, Georg Wittig and his students in Heidelberg made the remarkable observation that the reactions of phosphorus ylides with aldehydes, and with many ketones, resulted in replacing the carbon–oxygen double bonds by carbon–carbon double bonds.* The ylides, in turn, were converted to phosphine oxides (Eq. 35).[43]

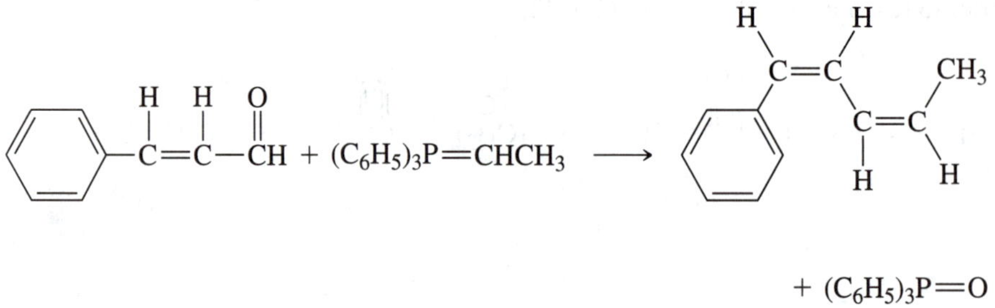

$$+ (C_6H_5)_3P{=}O$$

(35)

---

*Organolithium exchange reactions, such as that shown below, can also result from reactions of phosphonium salts with lithium reagents.

$$(C_6H_5)_3\overset{\oplus}{P}CH_3 \;\; \overset{Br^{\ominus}}{} \;\; + \;\; C_2H_5Li \;\; \longrightarrow \;\; (C_6H_5)_2\overset{\oplus}{P}\overset{CH_3}{\underset{C_2H_5}{<}} \;\; \overset{Br^{\ominus}}{} \;\; + \;\; C_6H_5Li$$

Wittig reactions avoid many of the problems frequently encountered in forming double bonds. Unlike many elimination reactions, which may yield mixtures of alkenes (usually consisting principally of alkenes with the most substituted double bonds), Wittig reactions are regiospecific, yielding products with $C=C$ double bonds specifically replacing $C=O$ groups. Unlike acid-catalyzed elimination processes, Wittig reactions do not result in molecular rearrangements. Wittig reagents, in contrast to many organometallic reagents, seldom undergo conjugate addition reactions with conjugated carbonyl groups (though they may give conjugate addition with conjugated *carboxyl* derivatives.) Wittig reactions can be carried out at temperatures as low as $-78°C$, and can often be carried out successfully in the presence of a wide variety of other functional groups. Eqs. 36 and 37 demonstrate two examples of such reactions in complex systems.[44,45]

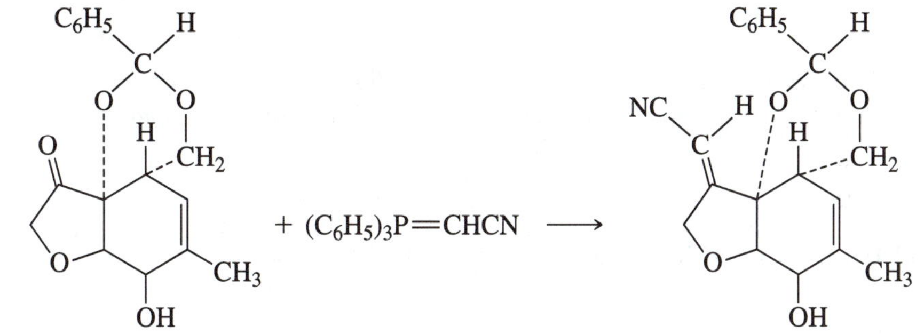

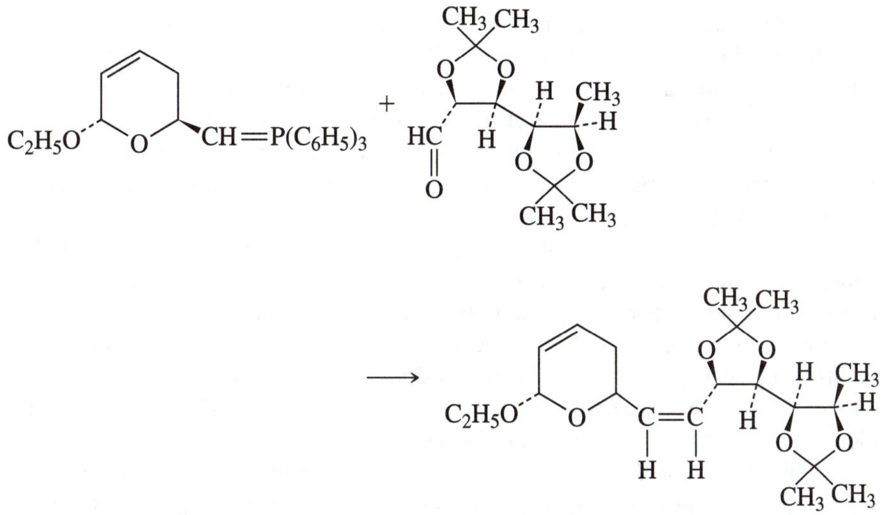

One useful feature of the Wittig reaction is that *unstabilized* Wittig reagents usually yield products in which *Z* (*cis*) configurations, which are usually more difficult to form than *E* (*trans*) configurations, predominate, as shown in Eqs. 35 and 37. The percentage of the *Z* isomer typically increases with increased size of substituents on phosphorus, and can range up to 95%. Stabilized Wittig reagents, on the other hand, usually yield principally double bonds with *E* configurations, while semistabilized Wittig reagents tend to yield products containing significant amounts of both *E* and *Z* stereoisomers.

The mechanism of Wittig reactions. Wittig initially suggested that phosphorus ylides add to carbonyl groups to form four-membered rings (oxaphosphetanes), but later suggested that formation of betaines (molecules with positive and negative charges not located on adjacent atoms) preceded oxaphosphetane formation, as shown in Eq. 38.

$$R_3P{=}CHR' + O{=}CX_2 \xrightarrow{\text{??}} R_3\overset{\oplus}{P}{-}\underset{\underset{R'}{|}}{CH}{-}\overset{\overset{O^{\ominus}}{|}}{CX_2} \longrightarrow \begin{matrix} R_3P{-}O \\ | \quad | \\ CH{-}CX_2 \\ R' \end{matrix} \tag{38}$$

$$R_3P{=}O + R'CH{=}CX_2$$

The proposal that betaines were the initial addition products was supported by the observation that addition of acids to some Wittig reactions yielded β-hydroxyphosphonium salts (Eq. 39).[44]

$$(C_6H_5)_3P{=}CR_2 + X_2C{=}O \xrightarrow{\text{??}} (C_6H_5)_3\overset{\oplus}{P}CR_2\overset{\overset{O^{\ominus}}{|}}{CX_2} \xrightarrow{H^{\oplus}} (C_6H_5)_3\overset{\oplus}{P}CR_2\overset{\overset{OH}{|}}{CX_2} \tag{39}$$

Furthermore, precipitates containing betaines, together with lithium halides, have been obtained from several reactions of Wittig reagents with carbonyl compounds, provided the Wittig reagents were formed from organolithium reagents.[46] However, it has been demonstrated that addition of lithium halides to solutions containing oxaphosphetanes yields identical precipitates.[47] No such precipitates were obtained from "salt-free" (actually lithium-ion-free) Wittig reagents.

However, ^{31}P NMR studies have almost never indicated the formation of betaines during Wittig reactions. In contrast, oxaphosphetanes can be detected at low temperatures in many reactions of unstabilized Wittig reagents and have been isolated in many cases.[42,47]

Theoretical studies show that betaines, particularly in conformations with the oxygen and phosphorus atoms *anti* to each other, are very high-energy structures. Concerted reactions directly forming oxaphosphetanes are found to be preferred to betaine formation.[48,49] However, the calculations also indicate that Wittig reactions are very "asynchronous," with the carbon–carbon bonds being appreciably formed in the transition states while little P–O bond formation has occurred. (The mechanism has been variously described as proceeding by a "borderline two-step" mechanism[49] and via a "quasi-betaine" transition state.[48])

Thus, the available evidence indicates that oxaphosphetanes are directly formed via "four-center transition states" but that the transition states strongly resemble betaines (with the oxygen and phosphorus atoms *syn* to each other). The oxaphosphetanes usually decompose rapidly, at temperatures below 0°C, forming alkenes and phosphine oxides. However, formation of oxaphosphetanes from some Wittig reagents is reversible.

Stereochemistry. Perhaps the aspect of the Wittig reaction mechanism that has produced the most debate is the fact that unstabilized Wittig reagents typically yield alkenes with the thermodynamically less-favored *Z* configurations. Early attempts to explain the stereochemistry of Wittig reactions were based on the assumption that betaines (in *anti* configurations) were intermediates in the reactions. These explanations no longer seem tenable.

E. Vedejs and his coworkers suggested that reactions of reactive, unstabilized Wittig reagents proceed via *early transition states*, in which the positions of the atoms resemble those of the reagents as they approach each other. They further suggested that the $C=O$ and $P=C$ bonds in a Wittig reaction approach each other in a roughly orthogonal manner, rather than in a parallel arrangement.* That would result in a highly puckered transition state with strong nonbonded repulsions between substituents on phosphorus and on carbon, as shown below. Such transition states should lead to oxaphosphetanes with substituents on ring carbons *cis* to each other.[50]

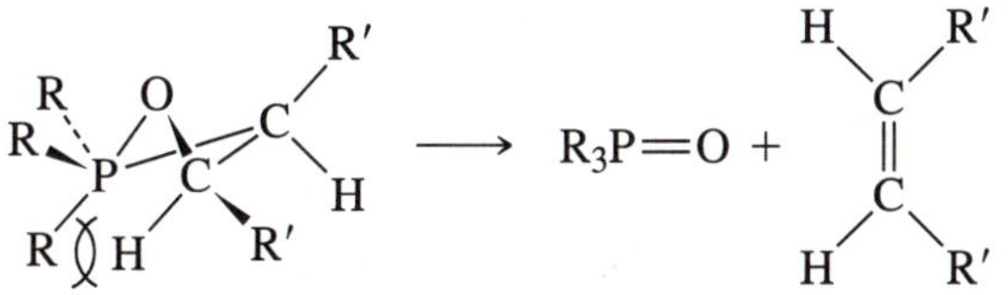

A Proposed Early Transition State Leading to Z Alkenes

In contrast, they suggested, reactions of carbonyl compounds with less reactive, stabilized Wittig reagents proceed by late transition states, with structures that more closely resemble the oxaphosphetane products than the starting materials. It has been observed that oxaphosphetanes with *trans* substituents are more stable than those with *cis* substituents.[†]

Anions of Wittig reagents.   One limitation on the use of Wittig reagents—even unstabilized Wittig reagents—has been the fact that they do not react with highly hindered ketones. However, *tert*-butyl lithium (which is appreciably more basic than primary organolithium reagents) can convert Wittig reagents to their anions. These powerful reagents (which effectively have two negative charges on the same carbon atom) can add even to very hindered ketones, as shown in Eq. 40.[52] Very mild protonation can then convert the addition products to normal Witting reaction intermediates.

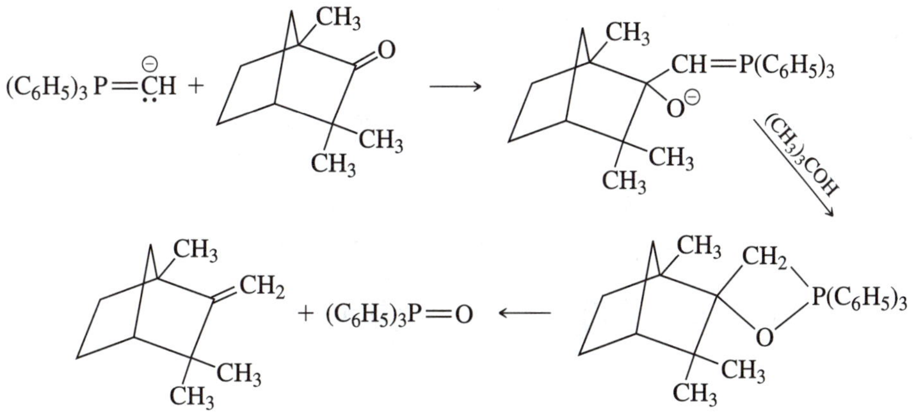

(40)

---

*It was originally suggested that the orthogonal approach of the reagents in a Wittig reaction was necessary to allow a $[_\pi 2_s + _\pi 2_a]$ cycloaddition.[51] Later, steric factors were considered to be responsible.[50]

†Predominant formation of products with *E* configurations from some reactions of unstabilized Wittig reagents was demonstrated to result from reversal of oxaphosphetane formation, resulting in thermodynamic control of stereochemistry.

A second type of anion from Wittig reagents can be prepared by reacting *di*alkylphosphonium salts with two equivalents of lithium reagents. Again, the resulting anions can react even with very hindered ketones.[53]

$$
\underset{\substack{\oplus \\ + \; 2n\text{-BuLi}}}{(C_6H_5)_2\overset{\oplus}{P}(CH_3)_2 \;\; \overset{\ominus}{Br}} \longrightarrow \underset{\substack{\oplus \\ Li \; \overset{\ominus}{CH_2}}}{(C_6H_5)_2P\overset{CH_2}{\diagdown}} \xrightarrow[\text{(2) } H_2O]{\text{(1) } t\text{-Bu}\overset{O}{\overset{\|}{C}}t\text{-Bu}} \quad
\begin{array}{c}
(CH_3)_3C\overset{\overset{\displaystyle CH_2}{\|}}{-}\overset{}{C}-C(CH_3)_3 \\
+ \\
(C_6H_5)_2\overset{O}{\overset{\|}{P}}CH_3
\end{array}
\qquad (41)
$$

The Emmons–Wadsworth reaction.[42b]   While unstabilized Wittig reagents are capable of adding to all but highly hindered ketones, stabilized Wittig reagents react with aldehyde but not with most ketones. More basic analogs of stabilized Wittig reagents can be formed by reacting alkylphosphonate esters with strong bases. The resulting anions (*Emmons–Wadsworth reagents*) will react with most ketones to form carbon–carbon double bonds, as in Eq. 42.

$$
(C_2H_5O)_2\overset{O}{\overset{\|}{P}}CH_2CN \xrightarrow{\text{NaH}} (C_2H_5O)_2\overset{O}{\overset{\|}{P}}\overset{\ominus}{CH}CN \;\; \overset{\oplus}{Na} \qquad (42)
$$

$$
(C_2H_5O)_2\overset{O}{\overset{\|}{P}}\underset{\overset{\oplus}{Na}}{\overset{\ominus}{CH}CN} \;+\; \bigcirc\!\!-\!\!\overset{O}{\overset{\|}{C}}CH_3 \longrightarrow \bigcirc\!\!-\!\!\overset{\overset{\displaystyle CH_3}{|}}{C}\!\!=\!\!CH\!-\!CN
$$
$$
+\;(C_2H_5O)_2\overset{O}{\overset{\|}{P}}ONa
$$

13.6 REACTIONS OF SULFUR YLIDES

Although DMSO yields products resulting from initial O-alkylation on reactions with many alkylating agents (see Section 13.4), it reacts with methyl iodide to form the *S-alkylsulfoxonium* salt.[54]

$$
(CH_3)_2S\!=\!O + CH_3I \;\xrightarrow{\Delta}\; (CH_3)_3\overset{\oplus}{S}\!=\!O \;\; \overset{\ominus}{I} \qquad (43)
$$

Both these salts and *S-alkylsulfonium* salts can be converted to ylides on reaction with strong bases.[55]*

---

*While ylides from S-alkylsulfoxonium salts are reasonably stable, ylides from alkylsulfonium salts are very sensitive and undergo easy elimination reactions even at temperatures below 0°C.[55]

$$
(CH_3)_3\overset{\oplus}{S} \;\overset{\ominus}{I} \;\xrightarrow{\overset{\ominus}{B}}\; (CH_3)_2\overset{\oplus}{S}\!-\!\overset{\ominus}{CH_2} \;\xrightarrow{0°C}\; CH_2\!=\!CH_2 + HSCH_3
$$

$$(CH_3)_3\overset{\oplus}{S}{=}O \quad \xrightarrow[\text{DMSO}]{\text{NaH}} \quad (CH_3)_2\overset{\oplus}{S}{-}\overset{\ominus}{CH_2} \qquad (44)$$

$$C_6H_5{-}\overset{\oplus}{S}(CH_3)_2 \quad \xrightarrow[\text{DMSO}]{\text{NaH}} \quad C_6H_5{-}\overset{\oplus}{S}{-}CH_3 \qquad (45)$$

In contrast to phosphorus ylides, which react with carbonyl compounds to form $P{=}O$ bonds, both types of sulfur ylides form epoxides on reactions with unconjugated carbonyl compounds (Eq. 46).

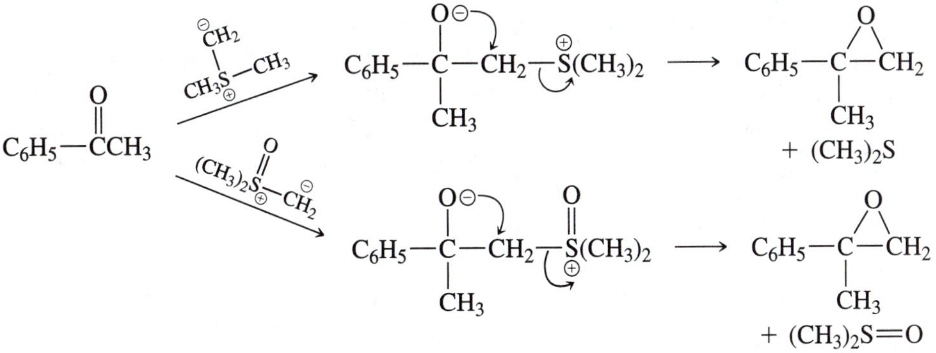

$$(46)$$

The two types of sulfur ylides differ, however, in reactions with conjugated carbonyl compounds. While sulfonium ylides still tend to form epoxides as their major products, sulfoxonium ylides usually undergo conjugate addition to form cyclopropanes, as shown in Eq. 47.[55,56]

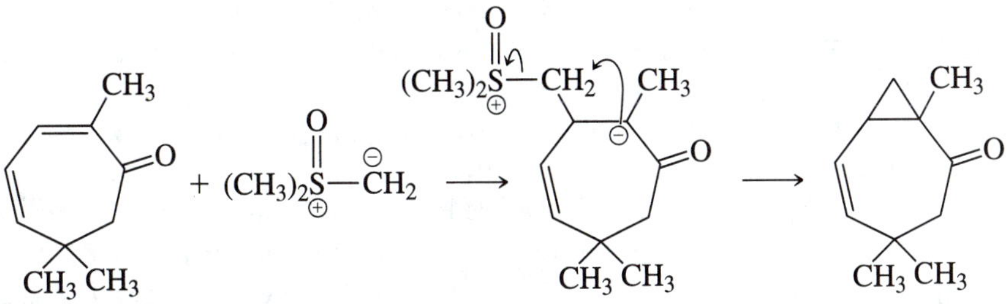

$$(47)$$

Both types of reagents usually add to carbonyl carbons more rapidly than to β carbons of conjugated carbonyl systems. Addition of sulfonium ylides to the carbonyl carbons is essentially irreversible, thus yielding epoxide products. In contrast, addition of the more stable (less basic) sulfoxonium ylides to carbonyl groups is reversible, resulting principally in formation of the thermodynamically favored conjugate addition products.

REFERENCES

[1] D.D. Perrin, Ionisation Constants of Inorganic Acids in Aqueous Solution, 2nd Ed, Pergamon Press, New York.

[2] W.A. Henderson, Jr., and C.A. Streuli, *J. Am. Chem. Soc., 82,* 5791 (1960).

[3] H.C. Brown, H. Bartholomay, Jr., and M.D. Taylor, *J. Am. Chem. Soc., 66,* 435 (1944); H.K. Hall, Jr., *J. Am. Chem. Soc., 79,* 5441 (1957).

[4] (a) H.C. Brown, D.H. McDaniel, and O. Häfliger, in *Determination of Organic Structures by Physical Methods*, vol. I, E.A. Braude and F.C. Nachod, Eds., Academic Press, New York (1955); (b) E.M. Arnett in *Progress in Physical Organic Chemistry, vol. 1*, S.G. Cohen, A. Streitwieser, Jr., and R.W. Taft, Eds., Interscience, New York (1963).

[5] J.H. Gibbs, *J. Chem. Phys., 22,* 1460 (1954).

[6] R.P. Larson and C.A. Kraus, *Proc. Nat. Acad. Sci., 40,* 70 (1954).

[7] Compare W.A. Henderson, Jr., and S.A. Buckler, *J. Am. Chem. Soc., 82,* 5794 (1960) and R.P. Larson and C.A. Kraus, *Proc. Nat. Acad. Sciences, 40,* 70 (1954).

[8] R.G. Pearson, H. Sobel, and J. Songstad, *J. Am. Chem. Soc., 90,* 378 (1968).

[9] See J. Miller, *Aromatic Nucleophilic Substitution*, Elsevier Publishing Co., New York (1968).

[10] C.W.L. Bevan and J. Hirst, *J. Chem. Soc.,* 254 (1956).

[11] G. Illuminati and G. Marino, *Tetrahedron Lett.,* 1055 (1963).

[12] G. Bartoli, L. Di Nunno, L. Forlani, and P.E. Todesco, *Int. J. Sulfur Chem., Part C, 6,* 77 (1971).

[13] B. Miller, *J. Am. Chem. Soc., 84,* 403 (1962).

[14] R.G. Pearson, *J. Am. Chem. Soc., 85,* 3533 (1963).

[15] See *Hard and Soft Acids and Bases*, R.G. Pearson, Ed., Dowden, Hutchison & Ross, Inc., Stroudsburg, PA (1973); R.G. Pearson, *Chemical Hardness*, Wiley VCH, New York (1997).

[16] E.G., R.L. Fuchs and L.L. Cole, *J. Am. Chem. Soc., 95,* 3194 (1973); D. Landini, A. Maia, and F. Montanari, *J. Am. Chem. Soc., 100,* 2796 (1978).

[17] C.L. Liotta and E.E. Grisdale, *Tetrahedron Lett.,* 4205 (1975).

[18] A.P. Clayton and C.L. Mortimer, *J. Chem. Soc.,* 3212 (1962).

[19] E. Howard, Jr., and W.F. Olszewski, *J. Am. Chem. Soc., 81,* 1483 (1959).

[20] For references, see (a) L.D. Quin, *A Guide to Organophosphorus Chemistry*, Wiley, New York (2000), Sec. 10C; (b) D.G. Gilheany, *Chem. Rev., 94,* 1339 (1994).

[21] J.E. Griffiths and A.B. Burg, *J. Am. Chem. Soc., 84,* 3442 (1962).

[22] See R. Engel, *Synthesis of Carbon–Phosphorus Bonds*, CRC Press, Boca Raton, FL (1988).

[23] A. Michaels and T. Becker, *Ber., 30,* 1003 (1897).

[24] See G.B. Borowitz and I.J. Borowitz in *Handbook of Organophosphorus Chemistry*, R. Engel, Ed., Marcel Dekker, Inc., New York (1992).

[25] G. Keglevich, I. Petnehazy, L. Toke, and H.R. Hudson, *Phosphorus and Sulfur, 29,* 341 (1987).

[26] I.J. Borowitz, S. Firstenberg, E.W.R. Casper, and R. Crouch, *Phosphorus, I,* 301 (1972).

[27] (a) I.J. Borowitz, K.C. Kirby, Jr., P.E. Rosek, and E.W.R. Casper, *J. Org. Chem., 36,* 88 (1971); (b) I.J. Borowitz, H. Parnes, E. Lord, and K.C. Yee, *J. Am. Chem. Soc., 94,* 6817 (1972).

[28] See F.W. Lichtenthaler, *Chem. Rev., 61,* 607 (1961).

[29] H. Mackle, *Tetrahedron, 19,* 1159 (1963).

[30] R.H. Rybrandt, *Tetrahedron Lett.,* 3553 (1971).

[31] W.E. Parham and M.D. Bhausan, *J. Org. Chem., 28,* 2686 (1963).

[32] G.E. Wilson, Jr., and R. Albert, *J. Org. Chem., 38,* 2156 (1973).

[33] R.B. Morin, D.O. Spry, and R.A. Mueller, *Tetrahedron Lett.*, 849 (1969).

[34] N. Kornblum, W.J. Jones, and G. Anderson, *J. Am. Chem. Soc., 81,* 4113 (1959); N. Kornblum et al., *J. Am. Chem. Soc., 79,* 6562 (1957).

[35] (a) K. Torssell, *Tetrahedron Lett.*, 445 (1966); (b) F.W. Swart and W.W. Epstein, *J. Org. Chem., 32,* 835 (1967).

[36] T. Cohen and T. Tsuji, *J. Org. Chem., 26,* 1681 (1961); T.M. Santosusso and D. Swern, *Tetrahedron Lett.*, 4261 (1968).

[37] For a review, see T.T. Tidwell, *Organic Reactions, 39,* 297 (1990).

[38] K. Omura and D. Swern, *Tetrahedron, 34,* 1651 (1978).

[39] S.C. Howell, S.V. Ley, M. Mahon, and P.A. Worthington, *Chem. Comm.,* 507 (1981).

[40] A.J. Mancuso, D.S. Brownfain, and D. Swern, *J. Org. Chem., 44,* 4148 (1979).

[41] F.G. Bordwell, *Acc. Chem. Res., 21,* 456 (1988); X.-M. Zhang and F.G. Bordwell, *J. Am. Chem. Soc., 116,* 968 (1994).

[42] For reviews, see (a) B.E. Maryanoff and A.B. Reitz, *Chem. Rev., 89,* 863 (1989); (b) A.W. Johnson, *Ylides and Imines of Phosphorus,* John Wiley & Sons, Inc., New York, 1993.

[43] G. Wittig and G. Geissler, *Liebigs Ann., 580,* 44 (1953).

[44] G. Wittig and U. Schöllkopf, *Ber., 87,* 1318 (1954).

[45] M.T. Crimmins, W.G. Hollis, Jr., and J.G. Lever, *Tetrahedron Lett., 28,* 3467 (1987).

[46] F.W. Lichenthaler, K. Lorenz, and W. Ma, *Tetrahedron Lett., 28,* 37 (1987).

[47] M. Schlosser and K.F. Christmann, *Liebigs Ann., 708,* 1 (1967).

[48] H.J. Bestmann and O. Vostrowsky, *Topics in Current Chemistry, 109,* 85 (1983).

[49] F. Mari, P.M. Lahti, and W.E. McEwen, *J. Am. Chem. Soc., 114,* 813 (1992).

[50] (a) E. Vedejs and C.F. Marth, *J. Am. Chem. Soc., 110,* 3948 (1988); (b) E. Vedejs and M.J. Peterson, *Topics in Stereochemistry, 21,* 1 (1994).

[51] E. Vedejs and K.A. Snoble, *J. Am. Chem. Soc., 95,* 5778 (1973).

[52] E.J. Corey and J. Kang, *J. Am. Chem. Soc., 104,* 4724 (1982); E.J. Corey, J. Kang, and K. Kyler, *Tetrahedron Lett., 26,* 555 (1985).

[53] H.J. Cristau, Y. Ribiell, L. Chiche, and F. Plenat, *J. Organometal. Chem., 352,* C47 (1988).

[54] R. Kuhn and H. Tischmann, *Liebigs Ann., 611,* 117 (1958).

[55] E.J. Corey and M. Chaykovsky, *J. Am. Chem. Soc., 87,* 1353 (1965).

[56] Y. Gololobov, A.M. Nesmaanov, V.P. Lysenko, and I.E. Boldeskul, *Tetrahedron, 43,* 2609 (1987).

PROBLEMS

13.1 Draw structures for the principal organic products expected from the following reactions.

(a) $2 \ (CH_3)_2\overset{S}{\underset{}{P}}{}^{\ominus}\text{=O} \quad Na^{\oplus} \ + \ ClCH_2\text{—}\bigcirc\text{—}\overset{O}{\underset{}{C}}Cl \ \longrightarrow$

(b) $2 \ (CH_3O)_3P \ + \ ClCH_2\text{—}\bigcirc\text{—}\overset{O}{\underset{}{C}}CH_2Cl \ \longrightarrow$

(c) 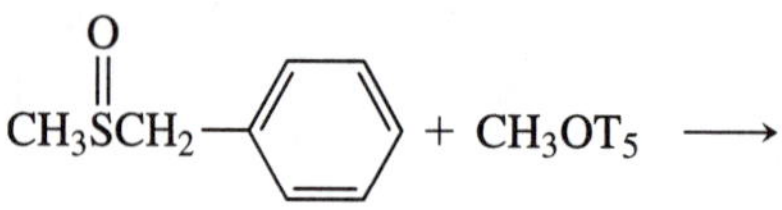 CH₃SCH₂—⟨benzene ring with C=O⟩ + CH₃OT₅ ⟶

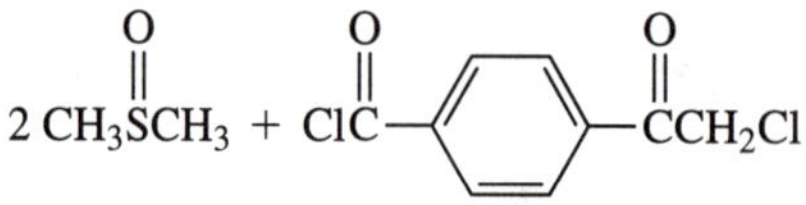

(d) 2 CH₃SCH₃ + ClC⟨benzene ring⟩CCH₂Cl

(e) HCCH₂CH₂CH₂Br + (C₆H₅)₃P ⟶ $\xrightarrow{\text{NaH}}$

(f) φ₃P⁺—CH=CH₂ + CH₃C(=O)—CH⁻—COCH₃ ⟶

13.2 Write reasonable mechanisms for the following reactions.

(a) (C₆H₅)₃P + Br₂ ⟶ $\xrightarrow{\text{C}_2\text{H}_5\text{OH}}$ C₂H₅—Bı

(b) 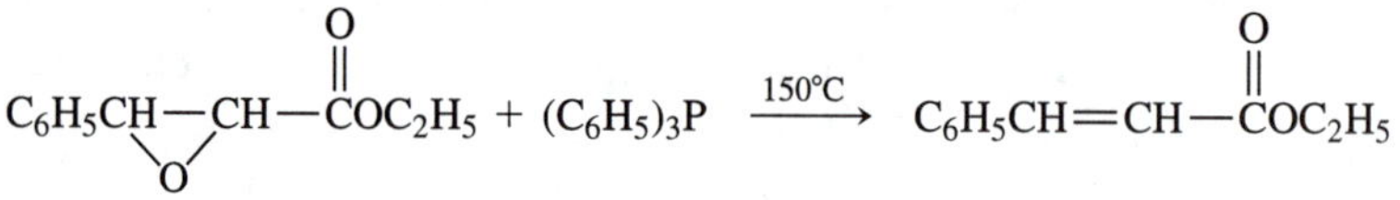 C₆H₅CH—CH—COC₂H₅ + (C₆H₅)₃P $\xrightarrow{150°\text{C}}$ C₆H₅CH=CH—COC₂H₅

(G. Wittig & W. Haag, *Ber.*, *88,* 1654 [1955])

(c) 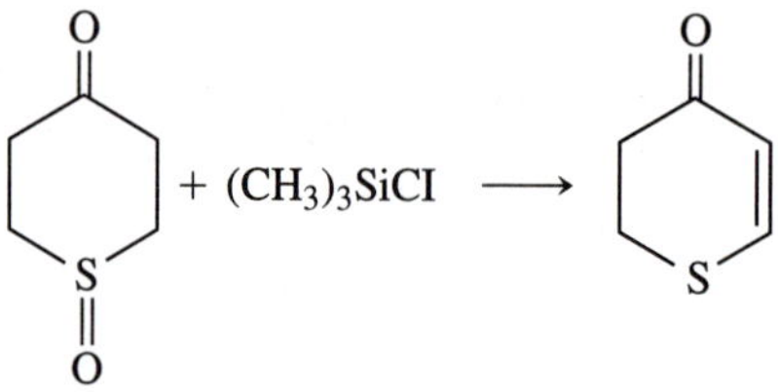 ⟨cyclic ketone with S=O⟩ + (CH₃)₃SiCl ⟶ ⟨cyclic enone with S⟩

(S. Lane, S.J. Quick, & R.J.K. Taylor, *J. Chem. Soc., Perkin Trans. 1*, 2549 [1984])

(d) 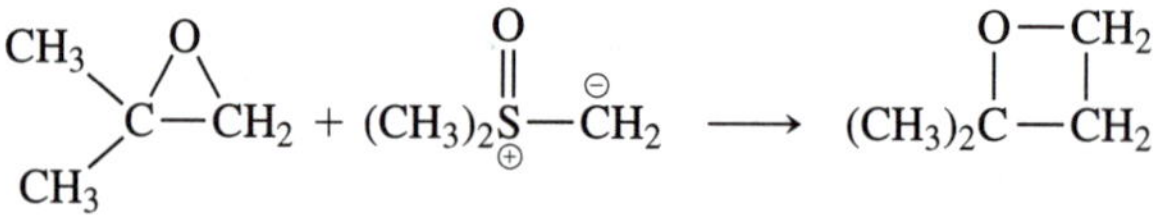 (CH₃)₂C—CH₂ (epoxide) + (CH₃)₂S⁺—CH₂⁻ ⟶ (CH₃)₂C—CH₂ (oxetane)

(e) 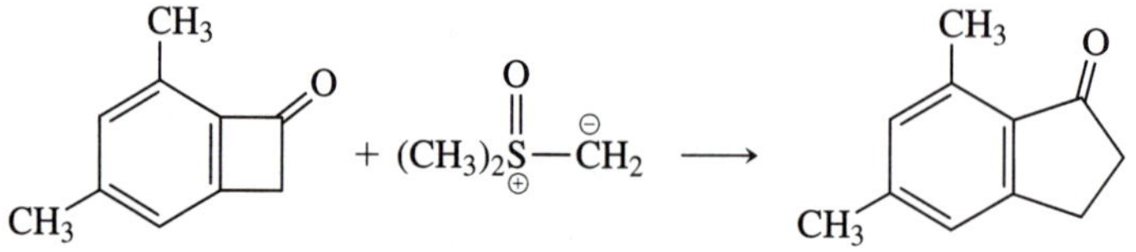 ⟨bicyclic compound with CH₃ groups⟩ + (CH₃)₂S⁺—CH₂⁻ ⟶ ⟨indanone with CH₃ groups⟩

(J. K. Landquist & A. Stanier, *Tetrahedron Lett.*, 1611 [1975])

(f)

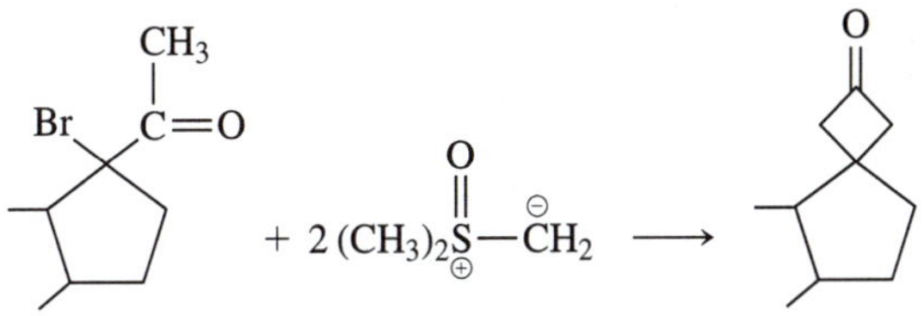

(R. Wiechert, *Angew. Chem., Int. Ed. Engl., 9,* 237 [1970])

(g) 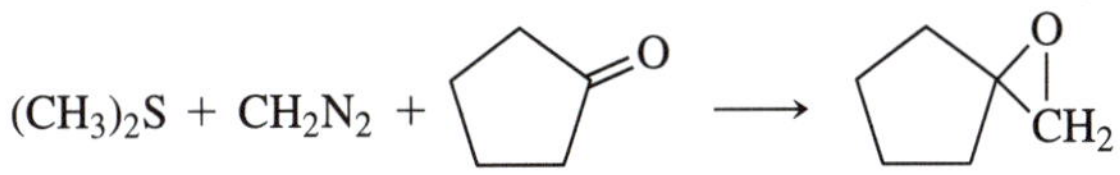 $(CH_3)_2S + CH_2N_2 +$

(h) 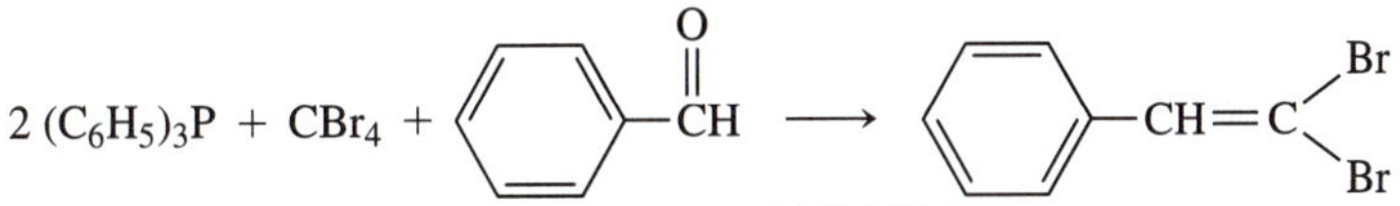 $2 (C_6H_5)_3P + CBr_4 +$

(E.J. Corey & P.L. Fuchs, *Tetrahedron Lett.,* 3769 [1972])

(i)

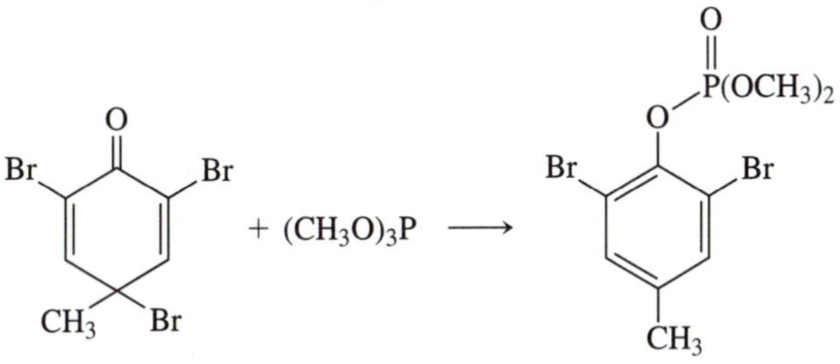

$+ (CH_3O)_3P \longrightarrow$

(B. Miller, *J. Org. Chem., 30,* 1964 [1961])

(j)

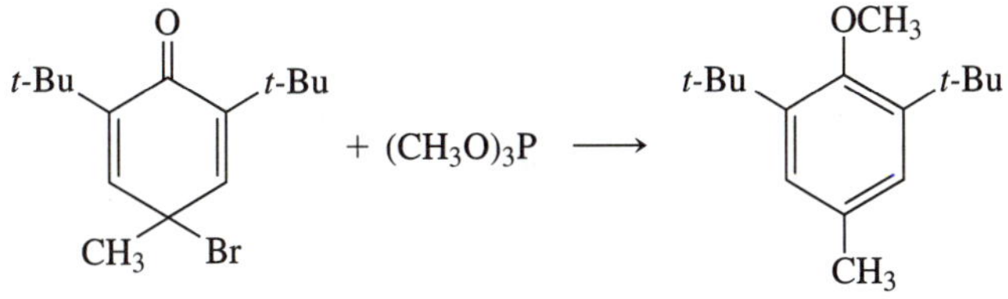

$+ (CH_3O)_3P \longrightarrow$

(B. Miller, *J. Org. Chem., 26,* 4781 [1965])

(k)

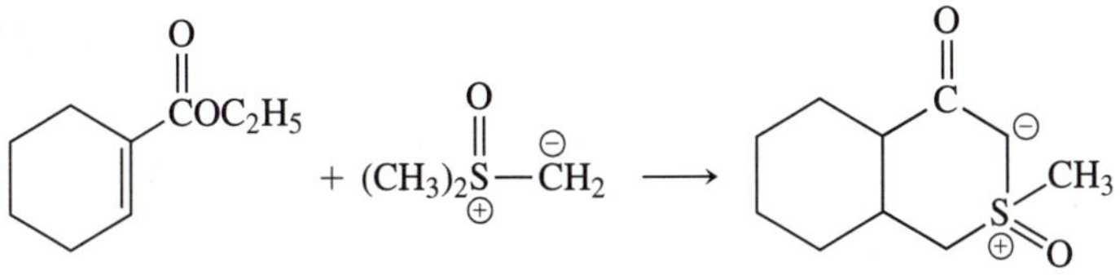

(E.J. Corey & M. Chaykovsky, *J. Am. Chem. Soc., 86,* 1640 [1964])

Index